Springer Climate

Series Editor
John Dodson, Institute of Earth Environment, Chinese Academy of Sciences, Xian, Shaanxi, China

Springer Climate is an interdisciplinary book series dedicated to climate research. This includes climatology, climate change impacts, climate change management, climate change policy, regional climate studies, climate monitoring and modeling, palaeoclimatology etc. The series publishes high quality research for scientists, researchers, students and policy makers. An author/editor questionnaire, instructions for authors and a book proposal form can be obtained from the Publishing Editor.
Now indexed in Scopus® !

Uday Chatterjee • Angela Oyilieze Akanwa
Suresh Kumar • Sudhir Kumar Singh
Abira Dutta Roy
Editors

Ecological Footprints of Climate Change

Adaptive Approaches and Sustainability

Editors
Uday Chatterjee
Department of Geography
Bhatter College, Dantan (Affiliated to
Vidyasagar University)
Paschim Medinipore
West Bengal, India

Angela Oyilieze Akanwa
Department of Environmental Management,
Faculty of Environmental Sciences
Chukwuemeka Odumegwu Ojukwu University
(COOU)
Anambra State, Nigeria

Suresh Kumar
Department of Space, Government of India
Indian Institute of Remote Sensing, Indian
Space Research Organization (ISRO)
Dehradun, Uttarakhand, India

Sudhir Kumar Singh
K. Banerjee Centre of Atmospheric
and Ocean Studies, IIDS, Nehru Science Centre
University of Allahabad
Prayagraj, Uttar Pradesh, India

Abira Dutta Roy
Department of Geography
Bankura Zilla Saradamani Mahila
Mahavidyapith
Bankura, West Bengal, India

ISSN 2352-0698 ISSN 2352-0701 (electronic)
Springer Climate
ISBN 978-3-031-15500-0 ISBN 978-3-031-15501-7 (eBook)
https://doi.org/10.1007/978-3-031-15501-7

© The Editor(s) (if applicable) and The Author(s), under exclusive license to Springer Nature Switzerland AG 2022

This work is subject to copyright. All rights are solely and exclusively licensed by the Publisher, whether the whole or part of the material is concerned, specifically the rights of translation, reprinting, reuse of illustrations, recitation, broadcasting, reproduction on microfilms or in any other physical way, and transmission or information storage and retrieval, electronic adaptation, computer software, or by similar or dissimilar methodology now known or hereafter developed.

The use of general descriptive names, registered names, trademarks, service marks, etc. in this publication does not imply, even in the absence of a specific statement, that such names are exempt from the relevant protective laws and regulations and therefore free for general use.

The publisher, the authors, and the editors are safe to assume that the advice and information in this book are believed to be true and accurate at the date of publication. Neither the publisher nor the authors or the editors give a warranty, expressed or implied, with respect to the material contained herein or for any errors or omissions that may have been made. The publisher remains neutral with regard to jurisdictional claims in published maps and institutional affiliations.

This Springer imprint is published by the registered company Springer Nature Switzerland AG
The registered company address is: Gewerbestrasse 11, 6330 Cham, Switzerland

Dedicated to Young Scholars in the Field of Geography, Environmental Science and Sustainability Science

Foreword

Climate plays a vital role in regulating agriculture productivity and practices, food habits and drinking water needs of our Earth and thus socio-economy of developing nations such as India. Global warming driven-climate change impacts both fauna and flora and thus ecological footprints across the regions. The book entitled 'Ecological footprints of climate change: Adaptive Approaches and Sustainability' edited by Uday Chatterjee, Angela Oyilieze Akanwa, Suresh Kumar, Sudhir Kumar Singh and Abira Dutta Roy to be published by Springer is timely and extremely relevant in the present scenario. The chapters written for this book are outstanding examples of advance works applied to the relevant field.

With fast depleting natural resources due to expanding human requirements and intense economic activities, there is a need to adopt people friendly development models which will help in sustaining human civilization on this mother earth. Overutilization of land resources and rampant urbanization as well as industrialization in developing economies contribute to severe land degradation and contamination of groundwater table worldwide. Such environmental severity impacts both food production and drinking water aquifers.

This volume is a collection and compilation of 30 chapters outlined under six major parts viz.

Part I: Introduction; Part II: Climate change and contemporary issues, challenges and sustainability; Part III: Agriculture and Forestry and Climate Change; Part IV: Food Security and Livelihoods; Part V: Infrastructure and Resilient Cities and Settlements; and Part VI: Global Health and Sustainable and Adaptive Approaches and Sustainability. The topics identified in these parts are diverse and of vital importance for climate resilience and sustainable development.

I would like to congratulate the editors for their noble initiative in bringing out this precious volume of contemporary relevance. This book forms a valuable addition to the existing knowledge and is aimed for university students and researchers in climate change, agriculture, forestry, livelihoods and sustainable development. I hope it will be widely acclaimed by geographers, environmental scientists, climate workers and policy planners as well as decision-makers engaged in dealing with climate change impacts.

J.C. Bose National Fellow, Professor of
Geology, Department of Geology and
Geophysics, Indian Institute of
Technology Kharagpur, Kharagpur
West Bengal, India

Anil K. Gupta FTWAS, FNA, FASc,
FNASc

Preface

The concept of ecological footprint is employed to determine the extent of population pressure on natural resources and the estimated quantity expedient for human satisfaction. It covers the endless demands placed on nature and estimated quantity of renewable resources consumed and the regenerative bio-capacity of the earth. The concept of ecological footprint is globally employed in the process of analysing sustainability assessments. Globally, ecological footprint assessments reflect the vast pressure of human population on the earth in comparison with the earth's renewability potentials. Climate change is a wicked problem sponsored by wanton anthropogenic exploitation of natural resources. Its impacts have been etched deep into the national and global ecosystems leaving intractable ecological footprints. Mankind has been under ecological overshoot since the 1970s, with annual resource requirements surpassing Earth's biocapacity. Assessing the ecological footprint (EF) is basically an expedient to measure and estimate the human demands and impacts on our global environment. According to the 2022 release of the National Footprint and Biocapacity Accounts, humankind would require resources equivalent to 1.75 planets similar to that of the Earth in order to meet their growing demands and absorb human produced garbage.

Earth's average surface temperature has increased by 1.4 °F (0.8 °C) from pre-industrial era. This increase is mainly due to burning of coal and petroleum products in power stations, factories and motor vehicles, which act as major source of carbon dioxide. To regulate the uncontrolled greenhouse gas emission, the Kyoto Protocol was signed in 1997, which is an international agreement intended to cut the greenhouse gas emissions, but many nations failed to limit their emissions. Furthermore, in 2018, the Paris Agreement was signed by over 200 nations with the purpose of preventing global temperatures from reaching a 3.6 °F (2.0 °C) increase and reducing greenhouse gas emissions to a rate where they can be naturally absorbed by the environment between 2050 and 2100. Recently, a total of 190 countries agreed at COP26 to phase out coal power, which is considered as the single largest contributor to human-caused climate change. Its goal was exclusively to continue the endeavors of restricting global warming to 1.5 °C, and the Glasgow Climate Treaty also aided in achieving this goal. Despite these initiatives, outcomes from various climate model simulations proposed that planet's average temperature could be between 2 and 9.7 °F (1.1–5.4 °C) warmer by 2100 than it is today. Hence, efficient framework, sustainable management policies and stringent implementation are very much required for tackling these concerns. Subsequently, delivery of these goals, proper climate resilience through adaptation and mitigation will be possible.

Global implications of climate change adaptation endeavours can trigger the appropriate scientific, adaptive and sustainable approaches. The advances in science and technology have enhanced a nation's ability to plan for the future by investing in adaptive and mitigate measures to monitor present and future changes. Wealth, infrastructure and political stability all contribute to a nation's capacity to anticipate and respond to climate change. The adaptive-based models are essential to reduce the ecological, social and economic costs of environmental management. Adaptive management focuses upon developing alternative approaches or rather unique methods that will identify gaps in knowledge. This will in turn be very useful in setting and updating research and action priorities, applicable for climate change policy, thereby serving as a continual update on knowledge and policy needs in climate change science. At present with the availability of multiple climate footprints, there are immense opportunities to explore all ideas towards evaluating their possibilities in presenting alternative futures through developing sustainable adaptive measures and implementing alternative policies, in order to solve the intractable ecological footprints of climate change. This book attempts to amalgamate all these ideas.

The book includes a broad range of topics covering ecological footprints, climate change, sustainable development, adaptive methodologies and sustainability. Topics on agriculture, forestry, water resources, food insecurity, human settlements, global health and many more have been dealt in the chapters of this book providing adaptation measures for minimizing the footprints of climate change. Climate change and its consequences are being experienced all over the globe, although developing countries are considered as the primary victims, especially in tropical regions where the hydrological cycles are more intense and experience higher exposure to the risk of climate change. India is a developing country located in a tropical region where climate-related disasters (storms, floods, cyclones, extreme precipitation and droughts) are more prominent. Apart from this, Himalayas is considered as the water tower of Asia and plays a vital role in regulating climate as well as downstream water availability. Therefore, it requires special attention to adapt and mitigate the adverse effects of climate change. Thus, we have included book chapters with majority of the case studies from the Indian sub-continent. Consequently, this book will provide the adaptation and mitigation approach which can be implemented over this region.

The chapters in the book have been grouped into six different parts addressing issues of climate change to provide a comprehensive overview. *First (I) part* provides a holistic view of the ecological footprint, climate change and sustainability as well as the linkages between them. Chapter 1 provides an insight on changes of footprint with respect to climate change. Chapter 2 introduces the assessment of global-scale synergy between adaptation, mitigation and sustainable development for projected climate change, whereas Chapter 3 deliberates inclusive concept concerning global warming impacts on environment in the last century. Chapter 4 delivers on the application of the earth system climate model. *Part II* focussed on contemporary issues related to climate change and environment with case studies for in-depth understanding for readers. In this part, Chapter 5 focuses on climate change impact on land degradation in hilly and mountainous landscape and the sustainability issues with adaptation strategies. Chapter 6 introduces impacts of the inherent hazards of climate change on the coastal environment. Chapter 7 sheds light on the assessment of ground water vulnerability to climate change using GIS techniques. Chapter 8 discusses the impact of climate change on water crisis, whereas Chapter 9 reviews the factors affecting governance of disaster management and delivers a comparative study of the Sundarbans. Chapter 10 introduces the application of geospatial technology in understanding seasonal flood hazard events. Chapter 11 is devoted towards application of geospatial techniques in watershed vulnerability to climate change and environmental sustainability.

Third (III) part focuses on climate change induced challenges on vital sectors such as agriculture and forest. Chapter 12 discusses about the application of crop simulation models in determining sustainable agriculture under different climate change scenarios. Chapter 13 reviews the peri-urban farmer's perception of climate change. Chapter 14 introduces the spatiotemporal drivers of agricultural vulnerability to climate change. Chapter 15 deals with forest landscape dynamics and people's livelihood dependency on forest. Chapter 16 uses the forest fire risk modelling and GIS remote sensing for impact assessment. *Fourth (IV) part* briefly explains the food security and livelihood which comprises Chapters 17 and 18 elaborates on the climate smart agricultural interventions for food security and evaluation of carbon neutral project, respectively. *Fifth (V) part* primarily focusses on infrastructure and urban development in the context of prevailing climate change issues. It includes land use/land cover change dynamics and modelling, land surface temperature analysis, urban heat island and climate change using geospatial indicators, site suitability for wasteland utilization by solar power plant installation and tsunami impact assessment, whereas the last chapter in this part discusses waterborne pathogen exposure under climate change and impact of climate change on health (Chapters 19, 20, 21, 22, 23, 24 and 25). *Sixth (VI) part* opens on health issues related to changing climate and adaptive approaches. It introduces health implications, cohort study on ambient air quality, practices of women fisher folk in response to climate change, climate change and health impacts and flood footprints in changing climate and climate related actions (Chapters 26, 27, 28, 29 and 30). Apart from these, the above-mentioned individual parts also focus on sustainability and adaptive approaches in the purview of climate change. Hence, in this way, these chapters can provide a holistic perception of components of ecological footprints.

The environment encompasses the interaction between the living and non-living components where humans exert their influence. It provides various ecosystem services that sustain human existence and civilization. Unfortunately, the unrestrained manipulations have escalated climate change issues. The introduction of application of sophisticated technologies in resource extraction and their use has further increased the ecological footprints. Therefore, there is an urgent need as pursued by this book for proper adaptive management of the ecological resources to bring about balance and human-earth sustainability. We prepared all the chapters in a very transparent and interactive approach. It is hoped that the book as a whole will provide a timely synthesis of a rapidly growing and important field of study but will also bring forward new and stimulating ideas that will shape a coherent and fruitful vision for future work for the community of undergraduates, post-graduates and researchers in the fields of environmental sciences and geography. Research scholars, geographers, environmentalist, climatologists, policymakers, NGOs, corporate sectors, social scientists and government organisations will find this book to be of great value.

Paschim Medinipore, West Bengal, India	Uday Chatterjee
Anambra State, Nigeria	Angela Oyilieze Akanwa
Dehradun, Uttarakhand, India	Suresh Kumar
Allahabad, Uttar Pradesh, India	Sudhir Kumar Singh
Bankura, West Bengal, India	Abira Dutta Roy

Acknowledgments

This book has been inspired by the tremendous hard work that has been put by climate scientists, the grass root level people and the policy makers who are trying their level best to curtail environmental degradation. We express our heartfelt gratitude to the researchers who have made this book a reality through their contributions. We thank the anonymous reviewers for their constructive criticism, which has helped to improve not only the quality of research but also the book. As ever learners, we appreciate the incredible support that our colleagues, students, parents, family members, teachers and collaborators have shown towards easing the efforts we have put day in and day out while editing this book, such that it may create value and contribute positively towards the knowledge of sustainable development. Last but not the least, we would like to acknowledge the continuous assistance provided by our publisher and its publishing editor, Springer.

Disclaimer

The authors of individual chapters are solely responsible for the ideas, views, data, figures, and geographical boundaries presented in the respective chapters of this book, and these have not been endorsed, in any form, by the publisher, the editor, and the authors of forewords, preambles, or other chapters.

Contents

Part I Introduction

1 Ecological Footprints in Changing Climate: An Overview 3
   Suresh Kumar, Uday Chatterjee, and Anu David Raj

2 Assessing Global-Scale Synergy Between Adaptation, Mitigation, and Sustainable Development for Projected Climate Change 31
   Aman Srivastava, Rajib Maity, and Venkappayya R. Desai

3 Global Warming Impacts on the Environment in the Last Century 63
   Sankar Mariappan, Anu David Raj, Suresh Kumar, and Uday Chatterjee

4 Analysis of Low-Flow Indices in the Era of Climate Change: An Application of CanESM2 Model 95
   Mohammadreza Goodarzi and Alireza Faraji

Part II Climate Change and Contemporary Issues, Challenges and Sustainability

5 Climate Change Impact on Land Degradation and Soil Erosion in Hilly and Mountainous Landscape: Sustainability Issues and Adaptation Strategies 119
   Suresh Kumar, Anu David Raj, Justin George Kalambukattu, and Uday Chatterjee

6 Vulnerability Assessment of the Inherent Hazards of Climate Change on the Coastal Environment of the Mahanadi Delta, East Coast of India 157
   Monalisha Mishra, Gopal Krishna Panda, Kishor Dandapat, and Uday Chatterjee

7   Assessment of Groundwater Vulnerability to Climate Change
     of Jalgaon District (M.S.), India, Using GIS Techniques   179
     Kalyani Mawale, Jaspreet Kaur Chhabda, and Arati Siddharth Petkar

8   Impact of Climate Change on Water Crisis in Gujarat (India)   201
     Nairwita Bandyopadhyay

9   Factors Affecting Governance Aspect of Disaster Management:
     Comparative Study of the Sundarbans in India
     and Bangladesh   219
     Srijita Chakrabarty

10   Application of Geospatial Technology in Seasonal
     Flood Hazard Event in Dhemaji District of Assam   247
     Krishna Das, A. Simhachalam, and Ashok Kumar Bora

11   Geospatial Approach in Watershed Vulnerability
     to Climate Change and Environmental Sustainability   271
     Anu David Raj, Justin George Kalambukattu, Suresh Kumar,
     and Uday Chatterjee

Part III Agriculture and Forestry and Climate Change

12   Agro-climatic Variability in Climate Change Scenario:
     Adaptive Approach and Sustainability   313
     Trisha Roy, Justin George Kalambukattu, Siddhartha S. Biswas,
     and Suresh Kumar

13   Peri-urban Farmers' Perception of Climate Change:
     Values and Perspectives – A French Case Study   349
     Marie Asma Ben-Othmen, Juliette Canchel, Lucie Devillers,
     Anthony Hennart, Lucie Rouyer, and Mariia Ostapchuk

14   Determinants and Spatio-Temporal Drivers of Agricultural
     Vulnerability to Climate Change at Block Level, Darjeeling
     Himalayan (Hill) Region, West Bengal, India   373
     Deepalok Banerjee, Jyotibrata Chakraborty, Bimalesh Samanta,
     and Subrata B. Dutta

15   Forest Landscape Dynamic and People's Livelihood
     Dependency on Forest: A Study on Bankura District,
     West Bengal   399
     Abira Dutta Roy and Santanu Mandal

16   Forest Fire Risk Modeling Using GIS and Remote Sensing
     in Major Landscapes of Himachal Pradesh   421
     Shreyasee Dutta, Akanchha Vaishali, Sadaf Khan, and Sandipan Das

Contents

Part IV Food Security and Livelihoods

17 **Climate-Smart Agriculture Interventions for Food
and Nutritional Security**.................................... 445
Manpreet Kaur, D. P. Malik, Gurdeep Singh Malhi,
Muhammad Ishaq Asif Rehmani, and Amandeep Singh Brar

18 **Critical Appraisal and Evaluation of India's First Carbon
Neutral Community Project – A Case of Meenangadi
Panchayat, Kerala, India**.................................... 465
Arunima KT and Mohammed Firoz C

Part V Infrastructure and Resilient Cities and Settlements

19 **Land Use and Land Cover Change Dynamics and Modeling
Future Urban Growth Using Cellular Automata Model
Over Isfahan Metropolitan Area of Iran**...................... 495
Bonin Mahdavi Estalkhsari, Pir Mohammad, and Alireza Karimi

20 **Analysing Spatio-temporal Changes in Land Surface
Temperature of Coastal Goa Using LANDSAT Satellite Data**..... 517
Venkatesh G. Prabhu Gaonkar, F. M. Nadaf, Vikas BalajiraoKapale,
Siddhi Gaonkar, Sumata Shetkar, and Merel D'Silva

21 **Analysing the Relationship Between Rising Urban Heat
Islands and Climate Change of Howrah Sadar Subdivision
in the Past Two Decades Using Geospatial Indicators**......... 543
Parama Bannerji and Radhika Bhanja

22 **Assessment of Site Suitability of Wastelands for Solar
Power Plants Installation in Rangareddy District,
Telangana, India**.. 559
Dhiroj Kumar Behera, Aman Kumari, Rajiv Kumar, Mohit Modi,
and Sudhir Kumar Singh

23 **Integrated Study on Tsunami Impact Assessment
in Cilacap, Indonesia: Method, Approach, and Practice**....... 577
Ranie Dwi Anugrah and Martiwi Diah Setiawati

24 **The Public Health Risks of Waterborne Pathogen
Exposure Under a Climate Change Scenario in Indonesia**....... 607
Martiwi Diah Setiawati, Marcin Pawel Jarzebski, Fuminari Miura,
Binaya Kumar Mishra, and Kensuke Fukushi

25 **Perceived Impact of Climate Change on Health:
Reflections from Kolkata and Its Suburbs**.................... 625
Sudarshana Sinha and Anindya Basu

Part VI Global Health, Sustainable and Adaptive Approaches and Sustainability

26 Health Implications, Leaders Societies, and Climate Change: A Global Review.................................... 653
Ansar Abbas, Dian Ekowati, Fendy Suhariadi, and Rakotoarisoa Maminirina Fenitra

27 A Retrospective Cohort Study on Ambient Air Quality and Respiratory Morbidities............................ 677
Shruti S. Tikhe and Kanchan Khare

28 Coping Practices of Women Fisherfolk in Responses to Climate Change at UNESCO Declared World Heritage Site of Sundarbans.................................... 701
Anisa Mitra and Prabal Barua

29 Climate Change and Health Impacts in the South Pacific: A Systematic Review................................ 731
Mumtaz Alam, Mohammed Feroz Ali, Sakul Kundra, Unaisi Nabobo-Baba, and Mohammad Afsar Alam

30 Changing Climate, Flood Footprints, and Climate-Related Actions: Effects on Ecosocial and Health Risks Along Ugbowo-Benin Road, Edo State, Nigeria................ 749
Angela Oyilieze Akanwa, Ngozi Joe-Ikechebelu, Angela Chinelo Enweruzor, Kenebechukwu Jane Okafor, Fredrick Aideniosa Omoruyi, Chinenye Blessing Oranu, and Uche Marian Umeh

Index.................................... 773

About the Editors

Uday Chatterjee is an assistant professor at the Department of Geography, Bhatter College, Dantan, Paschim Medinipur, West Bengal, India, and Applied Geographer with a Post-Graduate in Applied Geography at Utkal University and Doctoral Degrees in Applied Geography at Ravenshaw University, Cuttack, Odisha, India. He has contributed various research papers published in various reputed national and international journals and edited book volumes. He has authored jointly edited books entitled 'Harmony with Nature: Illusions and Elusions from Geographer's Perspective in the 21st Century' and 'Land Reclamation and Restoration Strategies for Sustainable Development' (November 2021, Edition: 1st, Publisher: Elsevier, Editor: Dr. Gouri Sankar Bhunia, Dr. Uday Chatterjee, Dr. Anil Kashyap, Dr. Pravat Kumar Shit • ISBN: 9780128238950 (https://www.elsevier.com/books/land-reclamation-and-restoration-strategies-for-sustainable-development/bhunia/978-0-12-823895-0). He has also conducted (Convener) one Faculty Development Programme on 'Modern methods of teaching and advanced research methods' sponsored by Indian Council of Social Science Research (ICSSR), Government. of India. His areas of research interest cover urban planning, social and human geography, applied geomorphology, hazards and disasters, environmental issues, land use, and rural development. His research work has been funded by the West Bengal Pollution Control Board (WBPCB), Government of West Bengal, India. He has served as a reviewer for many international journals. Currently, Dr. Uday Chatterjee is the lead editor of Special Issue (S.I) of Urbanism, Smart Cities and Modelling, Geojournal, Springer.

Angela Oyilieze Akanwa is a senior lecturer and presently the Head of Department for the Department of Environmental Management in Chukwuemeka Odumegwu Ojukwu University (COOU) Uli, Anambra State. She is an academia, a consultant and a researcher with focus on the implications of climate change, human hazards and disaster risks perceived from various human activities. These human activities include forest loss, mining/quarrying, solid waste management and water pollution. She has contributed to the development of baseline studies and data gathering as well as being a facilitator to a consultancy firm handling monitoring and evaluation studies, environmental and social impact studies and report for the World Bank/NEWMAP Gully Erosion Projects in Anambra State. She is a member of the Professional body of Environmental Management Association of Nigeria (EMAN). She has taught a variety of courses in environmental management both at the undergraduate and postgraduate levels, and she happens to be the project manager of Nigerian Coalition for Eco-Social Health Research (NCEHR). She has published multiple international journals, book chapters and conference papers in prominent journals such Springer, Elsevier, Intech Open and many more. She has worked in committees and held several leadership positions in professional, volunteer faith-based organisations and community projects both within and outside the university.

Suresh Kumar is Scientist–SG and Group Head, Agriculture and Soils Department at Indian Institute of Remote Sensing (IIRS), Government of India, Indian Space Research Organization (ISRO). He has 27 years of vast experience in applications of Geospatial Technologies in Natural Resource Management with specialisation in soil resource management, land degradation and watershed management. He is graduated in Agriculture and Doctorate in Soil Science from G.B. Pant University of Agriculture and Technology (G.B. P.U. &

T., Pantnagar) in 1993 and since then serving at Indian Institute of Remote Sensing (IIRS) at various scientist positions. He has done commendable research and published 22 research papers in international journal, 30 research papers in national journal and 7 book chapters with CRC/Springer Nature Press. He has attended several international conferences and national seminar/symposium and delivered invited lead talk in the seminar/symposium. He carried out several research projects as PI/ Co-I of the projects such as FAO-AEZ based agricultural land use planning; National soil carbon pool assessment, soil carbon dynamic (SCD) studies, mountain ecosystem processes and services: Studying impact of soil erosion and nutrient loss and its impact on soil quality; Digital soil mapping using environmental covariates for mountainous region, etc. He carried out operational projects of National Land Degradation Mapping, National Wasteland Mapping and Integrated Mission for Sustainable Development (IMSD) projects. He is life members of various professional societies such as Indian Society of Remote Sensing, Dehradun; Indian Society of Soil Survey and Land Use Planning, NBSS&LUP, Nagpur; Indian Association of Soil and Water Conservation, CSWCRTI, Dehradun; Farming Systems Research and Development Association, Modipuram, PDCSR, Modipuram, Meerut; Association of Agrometeorologist, GAU, Anand, India.

Sudhir Kumar Singh PhD, Assistant Professor, offers Remote Sensing and Satellite Meteorology, Environmental Chemistry and Statistical Methods in Climate Research courses to MTech students at K. Banerjee Centre of Atmospheric and Ocean Studies, University of Allahabad, Prayagraj, India. He has worked on the application of geochemical and hydrological model (SWAT) to study the impact of land use/land cover change on water quality and quantity of the drought-affected river basin. He worked as PI and Co-PI funded by University Grants Commission, Department of Science and Technology, New Delhi, India, and Ministry of Earth Sciences, Government of India. He has published many papers in peer-reviewed journals with high impact factors. Also, a part of the BRICS international team to look at the impact of land use/land cover and climate change on river basins. His research focuses on remote sensing and geographical information system applications in the environment with special reference to land use/land cover change and water resource management.

Abira Dutta Roy is currently an assistant professor at Department of Geography, Bankura Zilla Saradamani Mahila Mahavidyapith. She was previously teaching at Adamas' University, Barasat. Her teaching experience is over 6 years and a research experience of over 15 years. She was awarded her Doctoral degree and M.Phil. from the renowned Centre for the Study of Regional Development, Jawaharlal Nehru University. She had completed her Masters from the prestigious Delhi School of Economics, University of Delhi, and her graduation from Banaras Hindu University. Her research area is focused on hydrometeorology, urban landscape, agricultural landscape, glacial dynamics where the application of remote sensing and geoinformatics plays a pivotal role. She is proficient in use of hydrological models and other earth system models. She has worked as research fellows in different Department of Science and Technology, Government of India, and was part of sponsored projects conducted by reputed institutes of the country as well as by various NGO and environmental consultancies. These have provided her knowledge and field experience about the ground reality. She has quite a few publications in reputed books and journals, as well as over a dozen of paper presentation at national and international conferences. Her teaching skills have been well appreciated and hence have been invited to deliver lectures and training sessions in various institutes across the country. She is also a life member of the Indian Association of Hydrologists. She is currently supervising doctoral research fellows from Adamas University and from Bankura University.

Contributors

Ansar Abbas Department of Economics and Business, Jawa Timur, Surabaya, Indonesia

Angela Oyilieze Akanwa Department of Environmental Management, Chukwuemeka Odumegwu Ojukwu University (COOU), Awka, Anambra, Nigeria

Mohammad Afsar Alam Department of Social Sciences, Fiji National University, Suva, Fiji

Mumtaz Alam Department of Social Sciences, Fiji National University, Suva, Fiji

Mohammed Feroz Ali Department of Sports Science and Education, Fiji National University, Suva, Fiji

Ranie Dwi Anugrah Detailed Spatial Planning of Environmental Carrying Capacity Area – Region 1, Ministry of Agrarian Affairs and Spatial Planning, National Land Agency, Jakarta, Indonesia

Vikas BalajiraoKapale Department of Geography, Government College, Khandola, Goa, India

Nairwita Bandyopadhyay Department of Geography, Haringhata Mahavidyalaya, University of Kalyani, Nadia District, West Bengal, India

Deepalok Banerjee Geoinformatics and Remote Sensing Cell, West Bengal State Council of Science and Technology, Government of West Bengal, Kolkata, West Bengal, India

Parama Bannerji Department of Geography, Nababarrackpore Prafulla Chandra Mahavidyalaya, Kolkata, West Bengal, India

Prabal Barua Department of Environmental Sciences, Jahangirnagar University, Dhaka, Bangladesh

Anindya Basu Department of Geography, Diamond Harbour Women's University, Sarisha, West Bengal, India

Dhiroj Kumar Behera Land Use and Cover Monitoring Division, National Remote Sensing Centre, ISRO, Hyderabad, India

Marie Asma Ben-Othmen INTERACT Research Unit–Innovation, Land Management, Agriculture, Agro-Industries, Knowledge, and Technology, UniLaSalle-France, Mont-Saint-Aignan, France

Radhika Bhanja Department of Geography, Presidency University, Kolkata, West Bengal, India

Siddhartha S. Biswas Crop Production Division, ICAR-National Research Centre for Orchids, Pakyong, Sikkim, India

Ashok Kumar Bora Department of Geography, Gauhati University, Guwahati, Assam, India

Amandeep Singh Brar Punjab Agricultural University (PAU), Krishi Vigyan Kendra (KVK), Moga, Punjab, India

Juliette Canchel Graduate school of Agronomy and Agroindustry, UnilaSalle-France, Mont-Saint-Aignan, France

Srijita Chakrabarty Urban Management Professional, Independent Researcher-Consultant, Kolkata, West Bengal, India

Jyotibrata Chakraborty Geoinformatics and Remote Sensing Cell, West Bengal State Council of Science and Technology, Government of West Bengal, Kolkata, West Bengal, India

Uday Chatterjee Department of Geography, Bhatter College, Dantan (Affiliated to Vidyasagar University), Paschim Medinipore, West Bengal, India

Jaspreet Kaur Chhabda Civil Engineering Department, College of Engineering, Pune, Maharashtra, India

Kishor Dandapat Department of Geography, Seva Bharati Mahavidyalaya, Kapgari (Vidyasagar University), Jhargram, West Bengal, India

Krishna Das Department of Geography, Pragjyotish College (Affiliated to Gauhati University), Guwahati, Assam, India

Sandipan Das Symbiosis Institute of Geo-Informatics (SIG), Symbiosis International (Deemed University), Pune, Maharashtra, India

Venkappayya R. Desai Department of Civil Engineering, Indian Institute of Technology (IIT) Kharagpur, Kharagpur, West Bengal, India
Director, Indian Institute of Technology (IIT) Dharwad, Dharwad, Karnataka, India

Lucie Devillers Graduate school of Agronomy and Agroindustry, UnilaSalle-France, Mont-Saint-Aignan, France

Merel D'Silva Department of Geography, Government College, Khandola, Goa, India

Shreyasee Dutta Symbiosis Institute of Geo-Informatics (SIG), Symbiosis International (Deemed University), Pune, Maharashtra, India

Subrata B. Dutta Department of Science and Technology and Biotechnology, Government of West Bengal, Kolkata, West Bengal, India

Dian Ekowati Department of Economics and Business, & Planning, Jawa Timur, Surabaya, Indonesia

Angela Chinelo Enweruzor Department of Sociology and Anthropology, University of Benin, Benin City, Edo State, Nigeria
Spirit Filled Women International (SFWI). A non-governmental Organization (NGO) for Women Development, Anambra State, Nigeria

Bonin Mahdavi Estalkhsari Department of Landscape Architecture, Shahid Beheshti University, Tehran, Iran

Alireza Faraji Department of Engineering, University of Basilicata, Potenza, Italy
Visiting Scholar, Clemson University, SC, Clemson, USA

Rakotoarisoa Maminirina Fenitra ADS Scholar, Faculty of Economics and Business, Universitas Airlangga, Surabaya Indonesia, Jawa Timur, Surabaya, Indonesia

Mohammed Firoz C Department of Architecture and Planning, National Institute of Technology Calicut, Kozhikode, Kerala, India

Kensuke Fukushi Institute of Future Initiatives, The University of Tokyo, Tokyo, Japan
United Nations University Institute for the Advanced Study of Sustainability, Tokyo, Japan

Siddhi Gaonkar Department of Geography, Government College, Khandola, Goa, India

Mohammadreza Goodarzi Department of Civil Engineering, Yazd University, Yazd, Iran

Anthony Hennart Graduate school of Agronomy and Agroindustry, UnilaSalle-France, Mont-Saint-Aignan, France

Marcin Pawel Jarzebski Tokyo College, The University of Tokyo, Tokyo, Japan

Ngozi Joe-Ikechebelu Department of Community Medicine and Primary Healthcare, Chukwuemeka Odumegwu, Ojukwu University (COOU) Teaching Hospital (COOUTH), Anambra, Nigeria
Nigeria Coalition on EcoSocial Health Research (NCEHR), Anambra, Nigeria

Justin George Kalambukattu Agriculture and Soils Department, Indian Institute of Remote Sensing (IIRS), Indian Space Research Organization (ISRO), Dehradun, Uttarakhand, India

Alireza Karimi Instituto Universitario de Arquitectura y Ciencias de la Construcción, Escuela Técnica Superior de Arquitectura, Universidad de Sevilla, Sevilla, Spain

Manpreet Kaur Department of Agricultural Economics, CCS Haryana Agricultural University, Hisar, Haryana, India

Sadaf Khan Symbiosis Institute of Geo-Informatics (SIG), Symbiosis International (Deemed University), Pune, Maharashtra, India

Kanchan Khare Department of Civil Engineering, Symbiosis Institute of Technology, Pune, Maharashtra, India

Arunima KT Department of Architecture and Planning, National Institute of Technology Calicut, Kozhikode, Kerala, India

Rajiv Kumar Land Use and Cover Monitoring Division, National Remote Sensing Centre, ISRO, Hyderabad, India

Suresh Kumar Agriculture and Soils Department, Indian Institute of Remote Sensing (IIRS), Indian Space Research Organization (ISRO), Dehradun, Uttarakhand, India

Aman Kumari Centre of Advanced Study in Geography, Panjab University, Chandigarh, India

Sakul Kundra Department of Social Sciences, Fiji National University, Suva, Fiji

Rajib Maity Department of Civil Engineering, Indian Institute of Technology (IIT) Kharagpur, Kharagpur, West Bengal, India

Gurdeep Singh Malhi Punjab Agricultural University (PAU), Krishi Vigyan Kendra (KVK), Moga, Punjab, India

D. P. Malik Department of Agricultural Economics, CCS Haryana Agricultural University, Hisar, Haryana, India

Santanu Mandal Department of Geography, Bankura Zilla Saradamani Mahila Mahavidyapith, Bankura University, Purandarpur, West Bengal, India

Sankar Mariappan Indian Institute of Soil and Water Conservation (IISWC), Indian Council of Agricultural Research (ICAR), Dehradun, Uttarakhand, India

Kalyani Mawale College of Engineering, Pune, Maharashtra, India

Binaya Kumar Mishra School of Engineering, Faculty of Science and Technology, Pokhara University, Pokhara, Kaski, Nepal

Contributors

Monalisha Mishra GIS Developer, Water Resource Department, Government of Odisha, Bhubaneswar, Odisha, India

Anisa Mitra Department of Zoology, Sundarban Hazi Desarat College, South 24 Parganas, West Bengal, India

Fuminari Miura Centre for Infectious Disease Control, National Institute for Public Health and the Environment, Bilthoven, The Netherlands
Center for Marine Environmental Studies (CMES), Ehime University, Matsuyama, Ehime, Japan

Mohit Modi Land Use and Cover Monitoring Division, National Remote Sensing Centre, ISRO, Hyderabad, India

Pir Mohammad Department of Earth Sciences, Indian Institute of Technology, Roorkee, Uttarakhand, India

Unaisi Nabobo-Baba Department of Education, Fiji National University, Suva, Fiji

F. M. Nadaf Department of Geography, DPM's Shree Mallikarjun ShriChetan Manju Desai College, Canacona, Goa, India

Kenebechukwu Jane Okafor Management Science, Accountancy Department, Nnamdi Azikiwe University, Awka, Anambra, Nigeria
Spirit Filled Women International (SFWI). A non-governmental Organization (NGO) for Women Development, Anambra State, Nigeria

Fredrick Aideniosa Omoruyi Department of Statistics, Nnamdi Azikwe University, Awka, Anambra, Nigeria

Chinenye Blessing Oranu Department of Geology, Faculty of Physical Sciences, University of Benin, Benin City, Nigeria
Spirit Filled Women International (SFWI). A non-governmental Organization (NGO) for Women Development, Anambra State, Nigeria

Mariia Ostapchuk INTERACT Research Unit–Innovation, Land Management, Agriculture, Agro-Industries, Knowledge, and Technology, UniLaSalle-France, Mont-Saint-Aignan, France

Gopal Krishna Panda KISS Deemed to be University, Bhubaneswar, Odisha, India

Arati Siddharth Petkar Civil Engineering Department, College of Engineering, Pune, Maharashtra, India

Venkatesh G. Prabhu Gaonkar Department of Geoinformatics, S. P. Chowgule College, Margao, Goa, India

Anu David Raj Agriculture and Soils Department, Indian Institute of Remote Sensing (IIRS), Indian Space Research Organization (ISRO), Dehradun, Uttarakhand, India

Muhammad Ishaq Asif Rehmani Department of Agronomy, Ghazi University, Dera Ghazi Khan, Pakistan

Lucie Rouyer Graduate school of Agronomy and Agroindustry, UniLaSalle-France, Mont-Saint-Aignan, France

Abira Dutta Roy Department of Geography, Bankura Zilla Saradamani Mahila Mahavidyapith, Bankura University, Purandarpur, West Bengal, India

Trisha Roy Human Resource Development and Social Science Division, ICAR-IISWC, Dehradun, Uttarakhand, India

Bimalesh Samanta Department of Science and Technology and Biotechnology, Government of West Bengal, Kolkata, West Bengal, India

Martiwi Diah Setiawati Research Center for Oceanography, The National Research and Innovation Agency (BRIN), Jakarta, Indonesia

Sumata Shetkar Department of Geography, Government College, Khandola, Goa, India

A. Simhachalam NIRD and PR, NERC, Guwahati, Assam, India

Sudhir Kumar Singh K. Banerjee Centre of Atmospheric and Ocean Studies, IIDS, Nehru Science Centre, University of Allahabad, Praygaraj, UP, India

Sudarshana Sinha Department of Humanities and Social Sciences, Indian Institute of Technology, Kharagpur, West Bengal, India

Aman Srivastava Department of Civil Engineering, Indian Institute of Technology (IIT) Kharagpur, Kharagpur, West Bengal, India

Fendy Suhariadi Department of Postgraduate Program, Jawa Timur, Surabaya, Indonesia

Shruti S. Tikhe Department of R & D (Research), DTK Hydronet Solutions, Pune, Maharashtra, India

Uche Marian Umeh Department of Community Medicine and Primary Healthcare, Chukwuemeka Odumegwu, Ojukwu University (COOU) Teaching Hospital (COOUTH), Anambra, Nigeria
Nigeria Coalition on EcoSocial Health Research (NCEHR), Anambra, Nigeria

Akanchha Vaishali Symbiosis Institute of Geo-Informatics (SIG), Symbiosis International (Deemed University), Pune, Maharashtra, India

Part I
Introduction

Chapter 1
Ecological Footprints in Changing Climate: An Overview

Suresh Kumar, Uday Chatterjee, and Anu David Raj

Abstract Human exploitation on the natural resources is continuing in an overwhelming rate. Nonrenewable natural resources are expected to deplete in the near future; in addition, humans are consuming the nonrenewable resources at a rate which is far above the time required for regeneration. The exponential growing population and global economic competency drive the overexploitation of natural resources. Apart from this, the climate change also possesses boundless threat for the natural resources as well as human habitats. Overexploitation of the natural fuels and other resources also amplifies the climate change and can act as a positive cyclic feedback mechanism. These activities drastically decrease the biocapacity and efficiency of the Earth which leads to higher ecological footprint for the products industrialized from the natural resources. The carbon emission is the one of the major contributors of ecological footprint which contributes to global warming- and climate change-related disasters as well as natural resource degradations. This demands the sustainability for land or soil, forest, and aquatic ecosystems as well as for human habitats. Sustainability is the quintessential solution which can supply the remedies for the abovementioned issues. The integrated approach of climate resilience acquiring from the adaptation and mitigation strategies, nature-based solutions, and UN sustainable development goals can deliver minimum ecological footprint generations in milieu of changing climate.

Keywords Biocapacity · Forest land · Crop land · Carbon footprint · Climate change · Adaptation and mitigation · Sustainability

S. Kumar · A. David Raj
Agriculture and Soils Department, Indian Institute of Remote Sensing (IIRS), Indian Space Research Organization (ISRO), Dehradun, Uttarakhand, India
e-mail: suresh_kumar@iirs.gov.in; sureshkumar100@gmail.com; anudraj@iirs.gov.in; anudraj2@gmail.com

U. Chatterjee (✉)
Department of Geography, Bhatter College, Dantan (Affiliated to Vidyasagar University), Paschim Medinipore, West Bengal, India
e-mail: raj.chatterjee459@gmail.com

Introduction: Concept of Ecology, Ecosystem, and Natural Resources

The term "ecology" is the relationship between the organisms and their environment or habitat. It includes a vast number of interrelationships between individuals, populations, and communities. The ecosystem can be classified into aquatic (freshwater, lentic, and marine ecology) and terrestrial (grassland, forest, and desert ecology). In addition, an ecosystem can provide room for performing these ecological functions and activities. Hence, a system resulting from these interactions can be considered as an ecosystem. These interactions provide numerous benefits to each organism which is part of the ecosystem. We can symbolize it as economy/asset of nature or biological capacity of the ecosystem. The ecosystem can be primarily classified into natural and artificial (manmade) ecosystem. Similar to the concept of ecology, the ecosystem also can be classified as aquatic and terrestrial ecosystems. Aquatic ecosystems include ocean, sea, pond, lake, and river, while terrestrial comprise of forest, grassland, and desert ecosystems. The artificial ecosystems are maintained or engineered by the humans for the purpose of enhanced food production and economic benefit. Here we can say that planned manipulation of some ecosystem can disturb the natural ecosystems. The crop land and plantations are the prime example of artificial ecosystem.

The components of ecosystem perform using various goods and services delivered by the ecosystem. The biotic and abiotic (climatic and edaphic factors) are the major components. The biotic can be further divided into producers, consumers, and decomposers. The climatic factors are precipitation, temperature, sunlight, and wind, while edaphic includes topography, soils, and minerals. These abiotic factors can limit the population of an ecosystem in certain level. The biotic factors produce the food/energy or simply by utilizing the resources (abiotic factors). Thus, energy in one trophic level is converted to another through various cycles (hydrologic, carbon, and other nutrients). This process used the whole environment for the completion of the process, i.e., lithosphere, hydrosphere, atmosphere, cryosphere, and biosphere. Hence, the goods and services provided from these vital processes and functions are the raw/natural resources for the utilization of human beings. We can consider that land/soil, forest, and water are the three major forms of natural resources and all other resources are derived from this. As mentioned above, climate, as a limiting factor, has play a vital role in natural resource development. Apart from this, human interventions also disturb the natural processes and cycles along with goods and services available for human beings.

Biological Capacity/Biocapacity and Ecological Footprint

The biological capacity, or biocapacity, is the natural ability of the ecosystem to regenerate the goods and services according to the demand of the humans. Hence, it includes the ability to produce the biological material and recycle the carbon emissions generated by humans. Biocapacity solely depends on the type of ecosystem, climate, and management and may change year to year and region to region. The land (forest, grass land, crop land) and water (marine and freshwater) contribute the photosynthetic activity and recycling the waste produced (CO_2 emissions) by the humans. According to Global Footprint Network, the total productive area (both land and water) comprises of 12.2 billion hectares. These can be referred as natural capital, which has been used for the yield of goods and services for humans. The ecological footprint (EF) is the consumption of goods and services exploited from the natural capital, i.e., from biological capacity, or biocapacity. This consumption involves food, shelter, mobility, and other goods and services extracted from the crop land, built-up land, forest, grassland, and fishing ground and carbon emission from various sources. According to Wackernagel and Rees (1996), the amount of land and water required to sustain the human population perpetually, expressed in terms of delivering all energy/natural resources used and absorbing all waste discharged, is described as ecological footprint. Environmental developments such as deforestation, the collapse of fisheries, and the building of carbon dioxide in the atmosphere indicate that human demand is plausible to exceed the biosphere's regenerative and absorptive capacity (Borucke et al. 2013). The carbon dioxide emission from fossil fuels and other sources increases the ecological footprint of particular product or region. Apart from this, increased population, climate change, land degradation, pollutions, ocean acidifications, etc. can increase the ecological footprint and also had the potential for reducing the biological capacity. On the other hand, the increased ecological footprint/exploitation also fuels the process of global warming (climate change).

Climate Change

While considering the efforts to combat climate change, sustainability, and poverty alleviation, the IPCC reported on the implications of global greenhouse gas emission, which made an increase in temperature of 1.5 °C above the pre-industrial levels (IPCC 2018). Climate change is altering the natural climatic condition of a region and adversely affecting the various vital processes around the globe. There will be less biodiversity loss and extinction on the land at 1.5 °C than at 2 °C increase. Allowing terrestrial, freshwater, and coastal ecosystems to continue to deliver more services to humans while limiting global warming to 1.5 °C rather than 2 °C is projected (IPCC 2018). It is also projected that drought and precipitation deficiency hazards are higher at 2 °C than 1.5 °C. High-latitude and/or high-elevation areas in the northern hemisphere, eastern Asia, and eastern North America face greater dangers from heavy precipitation events at 2 °C than 1.5 °C. Tropical cyclone-related heavy precipitation will be more common if global warming increases by 2 °C than if it does so at 1.5 °C. Flood hazards are expected to affect a bigger proportion of the worldwide land area at 2 °C than they are at 1.5 °C of global warming due to excessive precipitation. The IPCC emphasizes that the consequences of increased temperature may critically affect the weather patterns, species diversity, crop production, and land degradation processes all over the world. These projections indicated that slight rise in temperature may adversely affect the global climate dynamics. Hence, the mitigation for reducing the temperature rise below the 1.5 °C is the need for the future.

Global warming poses a growing threat to biodiversity and many ecosystems, due to its vast and rising effects. Individual species and their interactions with other creatures and their environments are affected due to climate change, which affects the character of ecosystems and also the products and services they provide to society (Díaz et al. 2019). Land serves as both a source and a sink of greenhouse gas emissions, facilitating the movement of these pollutants from the land surface to the atmosphere. Land ecosystems and biodiversity are under risk from climate change, as well as weather and temperature extremes, to varied degrees. The tundra has already been invaded by woody bushes, and this will only get worse as temperatures rise. Over several generations, it is predicted that keeping global warming below 2 °C will prevent a permafrost region of between 1.5 and 2.5 million square kilometers from thawing. Similarly, an increase in sea level of 0.8 millimeters per year was induced by a warming ocean. 226 Gt/year of glacial retreat occurred between 1971 and 2009, while 301 Gt/year occurred between 2005 and 2009. Even if global surface temperatures do not rise, glacial retreat will continue. Greenland's ice sheet has shrunk steadily during the past 20 years (Kamaruddin et al. 2016; IPCC-AR5-WG1 2013). Eastern Asia, some part of Europe, and the western Mediterranean have all seen large-scale flooding in the last five centuries. High sea level events have been anticipated to occur more frequently since 1970 (Nakicenovic et al. 2000). Increasing greenhouse gas concentrations are causing broad changes in the world's physical climate and ecosystems (IPCC 2014a, b). For more insight to the climate change issues observed and expected in various regions around the globe, geographic region-wise information is provided in the next section. We used World Meteorological Organization (WMO)'s State of the Global Climate (2020) in Asia, Central America, the Caribbean, Africa, Latin America, and South-West Pacific and European State of the Climate, Summary 2020, published by European Centre for Medium-Range Weather Forecasts (ECMWF) for the preparation of this section. It will provide more insight to the latest status of climate change around the globe.

Asia

The average land surface air temperature over Asia in 2020 was 1.42+/−0.13 °C which is higher than the last 30-year average and was the highest on the record. According to the IPCC, Asia's land surface air temperature has increased faster than the world average, especially in the Arctic. Siberia's yearly temperatures near the Arctic Ocean shore were higher than 5 °C above average in 2020. Precipitation is an important climate component since it provides essential resources for human activities; however, it causes major climatic phenomena like droughts and floods. In

2020, yearly precipitation levels in the South and East Asian summer monsoon regions were above normal, with the East Asian summer monsoon region receiving up to 200% of average annual precipitation. The temperature of the sea surface is a crucial physical indicator for the Earth's climate system. In 2020, the Asian oceanic region experienced an overall warming trend. Temperature anomalies on the sea surface as a whole in the Indian (>0.5 °C), Pacific (>0.6 °C), and Arctic (>0.6 °C) oceans recorded new highs in 2020. Globally, sea level rises at a pace of 3 millimeters per year, and this rate is increasing due to ocean warming and land ice melt. All four glaciers in the High Mountain Asian region with relative long-term measurements emphasized that a massive amount of land mass has been lost in the last 40 years, with an accelerated trend in the twenty-first century. It is estimated that climate change will leave 48.8 million Southeast Asians, 305.7 million South Asians, and 42.3 million West Asians to have been undernourished by 2020. Extreme weather events continue to pose a threat to long-term development, causing billions of dollars in damage, amplifying health hazards, and hastening the extinction of natural ecosystems (State of Climate in Asia-WMO 2021).

Central America and the Caribbean

Mexico and numerous Central American countries such as Panama, Belize, Nicaragua, etc. endured below-average rainfall. In numerous sectors of South America, including the central Andes, southern Chile, Amazon and Pantanal regions, and Southeastern South America, the annual precipitation levels in 2020 were likewise below the long-term normal. In southern Paraguay, sections of Peru, and the semiarid region of Northeast Brazil, as well as the Pacific coasts of Costa Rica, El Salvador, and Jalisco in Mexico, above-average rainfall was recorded. Between 1993 and 2020, sea level in the Caribbean rose at a slightly faster rate than the global average of 3.3 mm per year, averaging 3.6 mm per year. Throughout the year, the sea surface temperature in the tropical North Atlantic Ocean was much higher-than-average temperature. The year 2020 in the Caribbean witnessed the highest positive anomalies on record. Sea surface temperatures in the equatorial Pacific began to fall gradually in May 2020, and La Niña arise. This, combined with a warmer Atlantic Warm Pool (AWP), might lead to a more active hurricane season than usual. As a consequence of climate change, over 8 million people in Central America were hit by extreme weather events, aggravating food insecurity (State of Climate in Central America and the Caribbean-WMO 2021).

Africa

The trend of rising temperatures over Africa has persisted. The average rate of temperature rise in Africa is higher than the global average. The average near-surface air temperature over Africa in 2020 was 0.45–0.86 °C which is higher than the 1981–2010 average, making 2020 the third of eighth warmest year on record. Regional trends suggest that sea level rise rates on the Atlantic side of Africa are uniform and near to the global average, whereas rates on the Indian Ocean side are slightly higher than average. The strongest trend (4.1 mm/year) and significant inter-annual variability may be found around the Indian Ocean coast, which is likely due to the Indian Ocean Dipole (IOD), a pattern of internal climatic variability in the Indian Ocean. In case of glaciated region, if current retreat rates continue, the African mountains will be deglaciated by the 2040s, with Mount Kenya likely losing its glaciers sooner than that, making it one of the first complete mountain ranges to lose its glaciers as a result of anthropogenic climate change. Heavy rains were recorded in the northern Sahel region, the Rift Valley, the central Nile, and north-eastern Africa in 2020, as well as the Kalahari Basin and Congo River's lower course. The monthly flow of the two main rivers, Congo-Oubangui at the top and Niger at the bottom, greatly surpassed the typical flow values, notably during the peak in months of October and November, indicating higher-than-normal rainfall. In Africa, as a result of climate change, about 98 million people faced acute food insecurity and required humanitarian aid in 2020, which is nearly 40% rise from 2019 (State of Climate In Africa-WMO 2021).

Latin America and the Caribbean

With 1.0 °C, 0.8 °C, and 0.6 °C above the 1981–2010 period, 2020 was one of the three hottest on record in Central America and the Caribbean and the second warmest year in South America. Since 2010, the rate of glacier erosion has accelerated, owing to rising temperatures and a large decrease in precipitation. More than 27% of the population in Latin America and the Caribbean lives along the coast, with an estimated 6–8% residing in areas at high or very high risk of being affected by coastal hazards. Between 1993 and 2020, sea level in the Caribbean rose at a slightly faster rate than the global average with a rate of 3.6 mm per year. Throughout the year, the sea surface temperature in the tropical North Atlantic Ocean was much higher than typical. The highest positive anomalies in the Caribbean Sea were recorded in the year 2020. Drought conditions in Latin America and the Caribbean also had a substantial impact on crop productivity in 2020 (State of Climate in Latin America and the Caribbean-WMO 2021).

South-West Pacific

Interpreting the data source, 2020 was the second warmest year on record in the South-West Pacific area. The average near-surface temperature across land and sea in the region is 0.37–0.44 °C which is higher than the 1981–2010 average. The region had a wide range of precipitation anomalies. It was a relatively wet year in tropical areas. In contrast, several equatorial locations along the international date line and further east experienced a dry year. Up to 90% of the coral reefs in the Great Barrier Reef might be severely degraded under a scenario in which global mean temperature rises 2 °C over pre-industrial levels. Thus, building resilience to extreme climate disasters is critical to meeting the Sustainable Development Goals of the 2030 Agenda. This necessitates a greater awareness of the specific dangers that afflict individual regions and countries, as well as increased capacity to respond to them (State of Climate in South-West Pacific-WMO 2021).

Europe

The annual temperature in Europe in 2020 was the warmest on record, at least 0.4 °C more than the following 5 warmest years, all of which happened in the previous decade. With regard to the warmest winter and fall on record, the winter record was particularly impressive, with temperatures more than 3.4 °C above the 1981–2010 normal and roughly 1.4 °C higher than the previous warmest winter. While overall precipitation levels were ordinary, there were a wide range of anomalies between various locations and at different periods of the year. For example, February had higher precipitation than usual, but November had fewer precipitation than typical. Precipitation levels, river discharge, and vegetation cover all showed a significant change from wet to dry conditions in early spring. This indicates the changing trend of normal climatic events and its spatial as well as temporal distribution (European State of Climate, Summary-ECMWF 2021).

Components of Ecological Footprint and Climate Change Consequences

The six components of the ecological footprint formed from the five components of biocapacity are arable land, grazing land, forest land, urbanized land, and carbon emissions. As assessed by the forest product footprint, forest land supplies biocapacity for the use of wood and fiber as well as the assimilation of anthropogenic CO_2 waste (Borucke et al. 2013). Figure 1.1 presents a comprehensive illustration of the biocapacity and its five components, ecological footprints, and way of generation of ecological footprints and carbon emissions. The climate change, land degradation, and deforestation are the major threats to biocapacity and driving force for the increased ecological footprints from a region or product. The climate-resilient, sustainable development goals and nature-based solutions are key for the reducing ecological footprint and carbon emissions during the era of economic development. The carbon sequestration process occurring in the soil, forest, and ocean is absorbing

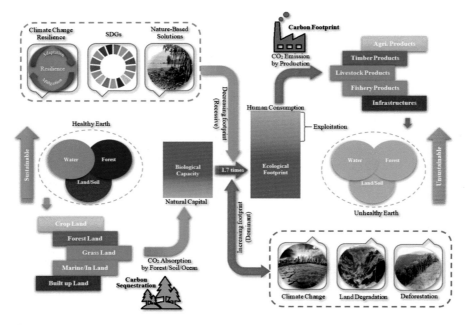

Fig. 1.1 Illustration of ecological footprint and components in changing climate and its solutions

the CO_2 emission from human activities and stored it in the form of soil carbon stock, biomass, and other non-harmful forms of carbon. The increased consumption leads to increased footprint from the products, which can also enhance the higher emission from a product and again result in climate change-related issues. Thus, it can act as a cyclic process which again and again fuels climate change. Thus, climate resilience in farming and urban sector is vital. It can assist all the basic requirements of humans and can alleviate the adverse effects of climate change by mitigation and adaptation. Hence, we believe this as a sustainable way for a prosperous Earth. It will enhance optimum and efficient use of natural resources instead of exploitation. In addition, it can provide more time for the regeneration of the natural resources. Currently, humans are exploiting the earth more than its regenerative capacity. This way of utilization leads to unsustainability. Henceforth, an integrated approach will be useful and efficient for the mitigation of the issues for long term effects. Here we are discussing the components of footprint from ecological/ecosystem point of view. Hence, we considered these five biocapacity/biological capacity subcomponents to land or soil, forest, and water.

Land/Soil Ecosystem

Land or soil is essential to sustainable agricultural development, key ecosystem functions, and food security. They are essential for the survival of life on Earth. Food, water, and a variety of ecosystem services are all dependent on land for their development and growth. Land is also a fundamental source of human well-being. Around 70% of the world's ice-free territory is directly impacted by human activity. Land also has a significant impact on the climate system and its regulation. The majority of the world's food is derived from soil, which is a valuable natural resource and an important part of the environment. Humans and the environment benefit from soil's ability to regulate water supply, protect biodiversity, and store carbon dioxide while simultaneously providing a range of essential ecological services. Soil is providing an anchoring medium for the plants, trees, and human infrastructure and is highly significant for crop production. Therefore, it is essential for the survival and growth of human existence on this planet that we support food and agriculture on a worldwide scale. Naturally occurring vegetation and other cultivated plants also rely on soil to grow. One of the largest carbon pools in the biosphere is soil organic matter, and it acts as an essential driver of climate change as well as a climate change response variable that can operate both as a source and sink of carbon. Sustainable and climate-resilient soil management is crucial for the climate system and the terrestrial carbon equilibrium.

However, land degradation emerged as a major threat for sustainability. Soil erosion, acidification, desertification, salinization, and compaction are the major land degradation issues for the land/soil ecosystem functions. The climate change and unsustainable land management practices pressing the land/soil leads to enhanced degradation. The loss of biological production and the complexity of terrestrial ecosystems are known as land degradation (Lal et al. 2012). Degradation of land occurs in all climate zones, with soils, vegetation, and water all included in the term "land" (Conacher 2009). Globally, roughly 1.5 billion people are affected by land degradation (Gnacadja 2012). Erosion induced by wind or water and loss of soil's physical, chemical, and biological qualities are examples of soil degradation (Lal et al. 2012). Since soils that have been degraded are less able to support vegetation growth, this has an impact on the environment (Gisladottir and Stocking 2005). As a result, in degraded ecosystems, vegetation hardly survives. The loss of natural vegetation has a direct impact on a wide range of ecological processes and services. Land productivity, hydrological cycles, drought occurrences, increased greenhouse gas (GHG) levels in the atmosphere, and biodiversity are all affected by agricultural activities (Stavi and Lal 2013).

Agricultural, forestry, and pastureland, as well as energy production and extraction of raw materials, have all taken a toll on land resources (UNCCD 2012). Land degradation processes reduce carbon sequestration and increase GHG emissions, making it more difficult to meet GHG reduction targets. In addition, unless land degradation is appropriately handled, global food security targets will be missed due to productivity losses and, as a result, a decrease in food supply (Lal et al. 2012). An estimated 3.5 billion hectares (23% of the Earth's total land surface) have been degraded, at an annual rate of 5–10 million hectares, according to contemporary estimates of land degradation (Lal 2012a). Soil organic carbon (SOC) depletion is another type of deterioration that reduces soil quality and fertility (Lal et al. 2012). After land-use conversion, some croplands have lost half to two-thirds of their initial SOC pool, culminating in a combined loss of 30–40 tons of carbon per hectare (Lal 2004). Erosion, in particular, removes substantial amounts of SOC from the topsoil of agro-ecosystems. Reduced soil quality and productivity can be attributed to lower concentrations of SOC on site, whether they are vented to the atmosphere as carbon dioxide (Lal and Pimentel 2008), buried in off-site terrestrial depressions, or deposited in aquatic bodies (Harden et al. 2008; Stavi and Lal 2013). Several factors can contribute to the damage of natural and semi-natural lands, including human interventions that produce severe ecological disruptions (Wilcox et al. 2012). These eventually lead to nonproductive crop land ecosystem and directly threaten the global food security and higher ecological footprint.

Soils contribute to climate change by emitting radiatively active gases such as carbon dioxide, methane, and nitrous oxide. Soil organic matter (SOM) depletion causes increased gaseous emissions into the atmosphere due to biological decomposition. Temperature rises related to global warming may hasten SOM mineralization, impair aggregation and aggregate stability, and increase the danger of soil erosion. Climate erosivity can be exacerbated by an increase in the frequency of extreme occurrences, such as heavy and strong rains and high winds. As a result, climatic warming could amplify climate erosivity while also increasing soil erodibility. Evaporation can lower available water capacity, diminish vegetation cover, and limit the quantity of carbon biomass returned to the soil as a result of increased evaporation. Thawing of permafrost (cryosols) and drainage of peat soils might result in a positive feedback loop, speeding up SOM mineralization and aggravating climate change (Lal 2012b). Recovery of degraded/desertified soils, on the other hand, can boost the ecosystem's carbon reservoir and mitigate certain anthropogenic emissions. Carbon sequestration potential is predicted to be 0.2–0.7 Pg C/year through desertification control and 0.3–0.7 Pg C/year by salt-affected soil regeneration. In addition, Reccomened Management Practices (RMPs) can sequester 0.4–1.2 Pg C/year in cropland soils, 0.3–0.5 Pg C/year in grassland/grazing land soils, and 1.4–1.9 Pg C/year through the establishment of forest plantations by the formation of afforestation. As a result, the total potential for C sequestration in the terrestrial environment is between 2.55 and 4.96 Pg C/year (Lal 2010). It emphasizes that the prime role of soil in mitigating the climate change consequences by storing the atmospheric carbon in the form of soil organic carbon stock.

As abovementioned, one of the main challenges to humanity's future is the degradation of agricultural areas (Oldeman 1998). Human well-being would be adversely affected by the loss of these lands' ability to provide important services like regional and global climate regulation or biodiversity habitats if they were degraded (Lee 2011). Thus, we require a sustainable plan to adapt and mitigate the adverse effect of land degradation issue. As a result, ZNLD focuses on agricultural operations that degrade environmental and ecosystem services. Under the current

ZNLD scope, only "agricultural" land types like crops, rangelands, and woodlands would be acceptable for conservation or restoration efforts (Stavi and Lal 2015). Similarly, for the recovery of degraded soils, transition to recuperative land use and adoption of soil/plant/animal/water management practices that provide a positive carbon budget and improved soil quality were required, that is, the optimum usage of land or soil resources through the minimum disturbance. Proper sustainable management framework can enhance the soil quality/health and reduce the degradation of land from climate-related threats.

Forest Ecosystem

The forest is considered as one of the richest ecosystems bearing a highly diverse variety of flora and fauna. In addition, it can provide more goods and services to human beings. More than half of Earth's animal and plant species live in forest ecosystems, which occupy over a third of the planet's land surface. The huge number of these species may be found in tropical forests. Shifting of forest land to agricultural land is a major danger to forest ecosystems in most parts of the world. An average of 130,000 km^2 of forest each year was lost between 2000 and 2010, according to the UN's third Global Biodiversity Outlook, compared with an average of 160,000 km^2 annually in the 1990s. Water regulation, wood supply, carbon storage, and biodiversity are all crucial functions provided by forest ecosystems. They will, however, face a significant variation in the future. Climate change and its repercussions for ecosystem services are numerous and intricately linked. Forest ecosystems, as well as the populations and industries that rely on them, are all vulnerable to climate change. Forests are important to humanity in a variety of ways. Climate change may have an impact on biodiversity, recreation, and cultural value, as well as timber and other material production. However, there are various sources of uncertainty, both inherent in projected future temperatures and in the comparatively lengthy timescales involved in forest development cycles. Hence, it influences the probable effects of climate change on forests and the ecological services they provide to the ecosystem and human beings. Furthermore, human interventions, which are impacted by sensitivities and involvements of the forest and climate, mediate forest responses to climate change.

Mountain ecosystems have a vital role in biodiversity protection, water resource management, global temperature regulation, and soil erosion avoidance, among other things. Most of the mountain regions are covered with forest or vegetation. They are referred to as "hot areas of biological diversity" at the genetic, species, and ecosystem levels, covering a wide range of ecosystem types that provide a variety of ecosystem services and advantages to society (Körner and Ohsawa 2005). Forest ecosystems are native to a significant portion of the world's biodiversity and have a significant responsibility in supplying goods and services that assist human well-being, including regulating weather and hydrology and producing timber (MEA 2005). Nearly half of Earth's terrestrial carbon storage are found in forest ecosystems, and forest biomass makes up around three by four of all land biomass. This information clarifies the significance of woodlands in the total carbon cycle and the role that environmentally sound forest management in alleviating global warming (Erb et al. 2018). There are few places in the world that have a greater impact on biodiversity than tropical rainforests (Giam 2017). More than two-thirds of the world's biodiversity is supposed to be supported by tropical rain forests, which cover less than 15% of the planet's surface (Gardner et al. 2009).

FAO (2015), Keenan et al. (2015), and MEA (2005) estimated that forest ecosystems covered roughly 4356 million hectares (30%) of Earth's land area in 2005; large areas were found in Asia (8.9%), Europe (6.9%), North and Central America (6.2%), South America (6.1%), and Africa (4.5%). Most of the world's forest area is covered by tropical forests (2027 million ha/13.6% of the planet's surface), which are accompanied by boreal forests (1258 million ha/8.4%). There are 697 million hectares of temperate forest and 353 million hectares of subtropical forest covering 4.7% and 2.4% of Earth's area, respectively (FAO 2015). Forest biomes have dissimilar forms in their terrestrial (portions of above the ground) net primary productivity (NPP) due to changes in climate, notably growing season length. As a general rule of thumb, tropical rainforests have the peak NPP of about 25 Mg ha^{-1} year^{-1}, accompanied by temperate forests of 22 Mg ha^{-1} year^{-1}, and sub-tropical periodic forests which have 13 Mg ha^{-1} year^{-1} (Keeling and Phillips 2007). A typical NPP of 7 Mg ha^{-1} year^{-1} is the smallest of all forest biomes in the boreal zone. Boreal forests are significant carbon sinks and thus crucial to preventing climate change, even with their low output. Among forest biomes, the soils of boreal forests are the most carbon-rich: as compared to tropical forests, boreal forests preserve 60% of the total carbon stock in soil, whereas tropical forests store 56% of the ecosystem's carbon in biomass (Pan et al. 2011). Hence, the combined process of land/soil and forest can sequestrate large amount of atmospheric carbon, which can act as sink/reservoir of carbon, while improper forest and soil management can release vast amount of carbon to the atmosphere, which can act as a further driving force for climate change. Thus, management of forest and soil ecosystem is critical for the climate change mitigation.

However, species are becoming extinct at an alarming rate as a result of human-induced environmental changes (Ceballos and Ehrlich 2018), and continuous biodiversity damage is projected to severely disrupt ecosystem performance, reducing the advantages that people derive from forests (Isbell et al. 2017). As a result, in the facet of global environmental change, biodiversity conservation is increasingly becoming a critical societal obligation (Griggs et al. 2013). The six main causes of biodiversity damage among forest types are deforestation, overexploitation, lack of safeguarding, climate change, pollution, and invasive species. These six factors must be addressed (Mazor et al. 2018; MEA 2005) for sustainable ecosystem. Biological resource use (especially wood harvesting and logging), agriculture, natural system alteration, residential and commercial advancement especially in urban development, pollution, and climate change are the global drivers of biodiversity loss that are decreasing in importance, according to IUCN (2019). These drivers destroy the natural characteristics of the forest/vegetation and lead to the unsustainability and decreased efficiency from the natural resources. Hence, the forest ecosystem becomes more vulnerable to the external threats like climate change and its consequences.

For the past three decades, forest ecosystems have been predominantly susceptible to the consequences of climate change (MEA 2005; IPCC 2013). Because temperature and precipitation are the two greatest underlying factors influencing variation and productivity characteristics, climate change will produce geographic alterations in species ranges, plant communities, and vegetation zones (MEA 2005). Most boreal-dependent tree, bird, and mammal species in North America's boreal forests have seen their ranges shrink and move northward as a result of climate change. Species ranges became progressively fragmented, with smaller typical sites and a greater divergence of environmentally suitable landscape pieces (Murray et al. 2017). This change in climate will lead to long-term alterations in the execution of ecosystem because forest biodiversity may also play a key role in reducing the negative effects of climate change on forest ecosystem performance (Ammer 2019). On a global scale, forests sequester and store tremendous quantities of carbon (Pan et al. 2011). However, biodiversity loss can change this role, as increased tree production in species-rich forests corresponds to more carbon deposited above and below ground in the ecosystem (Chen et al. 2018; Liu et al. 2018). In the interest of preserving biodiversity and reducing the effects of global warming, natural forest regeneration should take precedence over afforestation strategies such as plantation establishment, which fail to provide long-term environmental and communal benefits (Lewis et al. 2019). Hence, we demand proper sustainable forest management should be needed for mitigating the adverse effects of climate change especially in highly vulnerable forest ecosystems around the globe. Apart from this, high rate of the consumption of forest ecosystem-derived raw products (primarily timber) need to be controlled by the local- and regional-level administration with respect to the national and international forest policies.

Aquatic Ecosystem

Ecosystem services, such as agroecosystem water supply, protection of coastal and marine ecosystems, and forest water supply, all rely on water because of structural element. Water also serves as a support for other aquatic ecosystem services like food production, carbon sequestration, or species habitat (Guo and Xu 2019). Because of the ocean's abundance of life, Earth stands apart from the rest of the solar system. The oceans, which span 71% of the Earth's surface, have sustained life on Earth and serve to play an integral role in climate regulation. For as long as there is life on Earth, there is constant change. As a habitat for aquatic ecosystems or a set of services provided by other ecosystems, water is an irreplaceable and crucial ingredient for life in this sense (Diaz et al. 2018). As a result of the human society's reliance on ecosystem services, water plays a critical part in its growth (Manju and Sagar 2017). For example, food and raw material production is made easier with water, as it aids in storing nutrients and sequestering of carbon dioxide. Water also aids in supply and purification, biodiversity preservation, climate control, coastal and flood fortification, recreational opportunities, and tourism (Flávio et al. 2017; Wang et al. 2011). Water management and distribution, water filtration, flood mitigation, and aquatic fauna preservation are all given by healthy catchments through their landscapes and rivers (Hayat and Gupta 2016). Water control and nitrogen cycling can both occur in free-flowing rivers because they act as ecological networks for freshwater ecosystem fluxes across different areas (Schröter et al. 2018).

Multiple causes, such as population expansion, fluctuations in land-use area, and agrarian and urban development, are contributing to the deterioration and damage of marine and freshwater ecosystems and biodiversity at a greater rate than any other ecosystems (Feng et al. 2018; Paredes et al. 2019). Changes in long-term droughts and water supply imbalances are among the projected implications of global climate change (Mitrică et al. 2017). As a result, both wetlands and other water-dependent ecosystems are experiencing significant damage that may be irreversible (Chitsaz and Azarnivand 2017; Mitrică et al. 2017). Climate change projections are ambiguous, which leads to spatial and temporal ambiguity in consequences (Koutroulis et al. 2018; Salmoral et al. 2019), offering considerable adaptation problems for catchment supervision to recover or preserve freshwater ecosystem distribution in the future. In respect to freshwater ecosystem, process-based modeling approaches are often favored for assessing climate change consequences and defining adaptation decision-making strategies (Momblanch et al. 2019; Runting et al. 2017). Marine ecosystems' composition and function have undergone significant biological changes as a result of historical climate change (Pecl et al. 2017; Yasuhara and Danovaro 2016).

Changes in oceanographic and biological processes are predicted to occur at varying speeds throughout ocean basins as a result of climate change on a variety of time and geographic scales (Fossheim et al. 2015). Increased ocean temperatures, for example, cause distributional shifts in marine organisms; changes to colder, deeper, further offshore, or polar waters are likely in the near future with equatorial range contractions and global developments approaching to higher latitudes (Fossheim et al. 2015; Pinsky et al. 2013). Mild temperature rises boost the metabolic rates, which in turn influence life features, population expansion, and ecological processes (O'Connor et al. 2007).

Since many freshwater ecosystems are already under stress from a variety of other human-caused factors, the very rapid rate of warming that has occurred in recent decades poses a serious threat to their stability (Malmqvist et al. 2008). According to Scheurer et al. (2009), climate change persuaded spikes in river discharge and sediment loads throughout the winter and early spring which could be particularly harmful to the reproduction and development of numerous organisms at juvenile life stages. Erwin (2009) discussed the issues of wetland management and renovation in the aspect of climate change, emphasizing the importance of reducing non-climate stresses, such as monitoring, especially of invasive species that are affected by climate change. He predicted that, as a result of climate change, a lot of wetlands, particularly those on the drier end of the spectrum, will disappear. Similarly, water birds that rely on inland freshwater systems in semi-arid and dry regions are the most vulnerable to climate change, according to Finlayson et al. (2006). Henceforth, the studies indicate that climate concerns are numerous and how it affects each individual is unpredictable. The shift in rainfall and temperature has the potential for massive destruction of the aquatic ecosystems.

There has been at least a 6% yearly loss in ocean primary production since the early 1980s, with over 70% of the decline occurring at higher latitudes and substantial relative drops appearing in the Pacific and Indian ocean gyres (Gregg et al. 2003; Polovina et al. 2008). The richness and density of coral reef fishes and other species are already being reduced as a result of mass coral bleaching and mortality caused by rising temperatures (Hoegh-Guldberg et al. 2007). Both local and global pressures are posing increasing risks to coastal angiosperms such as mangroves, sea grass, and saltmarsh populations. Although mangrove deforestation (1–2% each year) poses a bigger immediate hazard, sea level rise poses a higher long-term concern, with 10–20% of mangroves estimated to be lost by 2100 (Alongi 2008). A crucial role in the organization of polar ocean biodiversity is played by sea ice, along with coral reefs and kelp forests. The current krill fall of 75–21% every decade could be explained by the reduction in food webs that rely on sea-ice algae due to the loss of sea ice (Atkinson et al. 2004; Gradinger 2009). Antarctic species like penguins and seals are in decline and, in certain cases, face an increasing threat of extinction due to current projections for Antarctic warming (Barbraud et al. 2008). These enormous rising threat over various organisms in aquatic ecosystem will hamper the food chain as well as the functions and provided by them to the ecology. Hence, the conservation of these organisms should also be taken into consideration.

The changing climate may press the normal functioning of the aquatic ecosystems. The precipitation and temperature change are the major climate-induced driver which can act adversely to the marine and freshwater ecosystems. The aquatic ecosystem is a major sector which provides various kinds of food, medicinal products to human being and especially carbon sink. The reduced functioning of aquatic ecosystem as a consequence of climate change increased footprint from the aquatic sector. Henceforward, an optimum usage of aquatic resources along with procedures and actions to mitigate and adapt the climate change for aquatic ecosystems is very much necessitated.

Carbon Footprint, Climate Change, and Sustainability

During the last two decades, people have been increasingly concerned about the impact of anthropogenic activities on ecological systems, biodiversity, and people's health (IPCC 2007; Steffen et al. 2004). An increased focus on the field of sustainability science is a result of this awareness, which aims to better understand the connections between nature and society and to enable the growth of tailored management solutions and tools to achieve environmental quality, economic strength, and social equity, which are the three pillars of sustainable development (Chapin et al. 2010; Clark 2007; Kates et al. 2001). Environmental management must be flexible to deal with the rapid and nonlinear changes that are predicted to characterize most ecosystems in the coming years. A reduction in greenhouse gas emissions, which would not only reduce the massive costs of adaptation but also the growing crisis of driving our planet into an unpredictable and extremely dangerous situation, is still a top concern for the future of our planet (Hoegh-Guldberg and Bruno 2010). To attain sustainability, climate change should be addressed as it possesses enormous threat day by day. In addition, carbon footprint is the quantification of carbon emission, which should be mitigated for sustainability.

Given the term's popularity over the last decade, the concept of "carbon footprint" is surprisingly ambiguous. The term "ecological footprint" (Wackernagel et al. 1999) is found in the literature, which explains the whole area of land required to produce goods for human consumption at a certain level. This means that the ecological footprint is inherently a full lifecycle assessment because the land utilized to produce most consumer products is far away from the final user. However, the term's new counterpart, the carbon footprint, appears to be different. Wiedmann and Minx (2007) found that there is a wide range of classifications that differ in the inclusion of gases, where the analyses' boundaries are placed, and a number of other factors. The carbon footprint is considered as one component of ecological footprint. Also, it poses more emphasis than the remaining ecological footprints. The prime reason associated with this is the highest contributor to the total ecological footprint which is also directly driving the climate change.

The term "carbon footprint" (CFP) is commonly used to refer to greenhouse gas accounting, which is a common way for corporations and governments to track the effects on climate change. In addition to climate change, environmental sustainability includes challenges such as chemical pollution and depletion of natural resources and a priority on CFP risk crisis shifting if CFP reductions are accomplished at the price of other environmental impacts (Laurent et al. 2012). Many enterprises and organizations are implementing "carbon footprint" projects in an effort to measure their own carbon emissions and their impact on global climate change. Carbon registries let businesses calculate their carbon footprints by defining protocol definitions. The scope of these protocols varies, but in general, only straightforward emissions and procured energy emissions are recommended for measurement, with less attention devoted to emissions along the supply chain. Approaches based on extensive ecologic lifecycle assessment methods for tracking total emissions across the entire supply chain and findings show that following narrowly focused valuation protocols will normally result in large decrement of carbon emissions for delivering goods and services (Matthews et al. 2008).

In recent decades, the predominance of power generation from fossil fuels (coal, oil, and gas) has led to an increase in energy demand and problems associated with fast increases in carbon dioxide emissions globally (Asumadu-Sarkodie and Owusu 2016). Climate change is one of the most pressing challenges of our day. If efforts are made to reform current energy systems, it may still be possible to avert its terrible repercussions. Reducing greenhouse gas emissions from fossil fuel-based power generation is the primary goal of renewable energy sources (Edenhofer et al. 2011).

Many nations' national policies, goals, and planning processes have recently placed a higher emphasis on sustainable development than ever before. At the United Nations in New York, the Open Working Group presented 17 global Sustainable Development Goals (SDGs) with 169 specific targets. A preliminary set of 330 indicators was introduced in March 2015 as well (Lu et al. 2015). As compared to the Millennium Development Goals, the Sustainable Development Goals place a higher value on and have higher expectations of scientists. Food, health, and water security are all dependent on worldwide monitoring and modeling of a wide range of socioeconomic and environmental challenges due to climate change (Hák et al. 2016; Owusu et al. 2016). Hence, the Sustainable Development Goals have a vital responsibility in conserving the ecosystem by optimum usage.

The natural resources (natural capital or biocapacity), ecological and carbon footprint, and climate change are deeply interconnected with each other. The natural capital is providing building blocks for growth and development of organisms and human being. The ecological footprint represents how much resources we are using/exploiting for a given period of time for a particular region or product. Carbon emission occurs during the production of particular product and is called as carbon footprint. The exponential population growth (above the carrying capacity) demands more resources from the natural capital. This exploitation further enhances the emission and eventually climate change. In addition, climate change enhances the ecological footprint, and this process continuous to occur. Thus, if we are reducing the ecological footprint, we can able to reduce the climate change and related adverse effects. The climate resilience, Sustainable Development Goals, and nature-based solutions together with a strong administration can resist the climate change through strong adaptation and mitigation measures. Table 1.1 shows a country-wise comparison of the biocapacity (biodiversity index, share in total GDP), footprint (ecological and carbon), climate risk index, and human development index. It provides an overview of major countries which having the highest natural resources and economic condition and carbon emission and climate risk.

Ecosystem and Sustainability

The "landscape" is the most appropriate, integrated conceptual, and practical tool for long-term sustainability. A landscape includes various ecosystems and its processes and functioning. Ecosystem service degradation, pollution, and global climate change have led to the formation of the ideas of sustainability and sustainable development (WCED 1987). As a scientific basis for sustainable development, sustainability science was first introduced by Kates et al. (2001) and has progressed rapidly since then (Fang et al. 2018; Wu 2013). Wu (2013) proposed the landscape sustainability science framework, which fosters the convergence of the natural and communal sciences, primarily landscape ecology, sustainability research, and landscape planning and design (Liao et al. 2020; Opdam et al. 2018; Wu 2019). Human activities are having enormous and rapid impacts on the Earth's temperature,

Table 1.1 Status of biocapacity, ecological footprint, and biodiversity

Country	Biocapacity m gha (rank)[a]	Ecological footprint m gha (rank)[a]	Carbon footprint -Mt CO_2 (rank)[b]	Deficit (−)/reserve (+) − % (no. of Earths required to meet demand on nature)[a]	Share of agriculture, forestry, and fishing value added in total GDP (%)[c]	Climate Risk Index (Rank)[d]	National Biodiversity Index[e]	Human Development Index (Rank)[f]
Brazil	1800 (1)	588 (6)	467 (12)	+206 (1.8)	4.8	33.67 (27)	0.877[g]	0.765 (84)
China	1330 (2)	5350 (1)	10668 (1)	−302 (2.3)	7.6	42.83 (32)	0.839[g]	0.761 (85)
USA	1120 (3)	2610 (2)	4713 (2)	−1570 (5)	1.0	–	0.677[g]	0.926 (17)
Russian Federation	1000 (4)	788 (4)	1577 (4)	+27 (3.4)	3.7	50.67 (39)	0.447	0.824 (52)
India	577 (5)	1600 (3)	2442 (3)	−177 (0.7)	15.1	16.67 (7)	0.732[g]	0.645 (131)
Canada	549 (6)	296 (12)	536 (11)	+85 (5.1)	2.0	65.67 (62)	0.299	0.929 (16)
Indonesia	322 (7)	439 (7)	590 (10)	−36 (1)	12.8	24.83 (14)	1.0[g]	0.718 (107)
Australia	309 (8)	178 (24)	392 (15)	+74 (4.6)	2.0	28.0 (19)	0.853[g]	0.944 (8)
Germany	127 (15)	386 (8)	644 (7)	−204 (2.9)	0.5	61.33 (56)	0.365	0.947 (6)
Japan	75.5 (25)	593 (5)	1031 (5)	−685 (2.9)	1.0	14.50 (4)	0.638	0.919 (19)
United Kingdom	71.8 (28)	278 (14)	330 (17)	−287 (2.6)	0.6	90.83 (102)	0.320	0.932 (13)
Norway	37.9 (57)	30.6 (73)	41 (60)	+24 (3.6)	1.5	92.33 (106)	0.297	0.957 (1)
Niger	27.6 (65)	33.6 (67)	1.7 (158)	−22 (1)	36.1	18.17 (9)	0.412	0.394 (189)
Singapore	0.32 (167)	33.5 (68)	46 (57)	−10,300 (3.7)	0.0	118.0 (130)	–	0.938 (11)
Bhutan	3.95 (135)	3.53 (150)	1.9 (156)	+12 (2.6)	13.9	118.0 (130)	0.607	0.654 (129)
World	12060	20926	34807	−(1.7)	4.2	–	–	–

[a]National Footprint and Biocapacity Accounts, 2021 Edition, Global Footprint Network
[b]Global Carbon Atlas 2021
[c]Statistical Yearbook, World Food and Agriculture, 2021
[d]Global Climate Risk Index, 2021, Germanwatch
[e]Global Biodiversity Outlook 1, Convention on Biological Diversity
[f]Human Development Index: Human Development Report, 2020, United Nations Development Programme
[g]Mega biodiversity countries

environment, and ecosystems (Haberl et al. 2007; Foley et al. 2005), resulting in the degradation of numerous ecosystem services (MEA 2005). This unsustainable trend necessitates a significant shift in human relationships with the planet's environment and life-support system (Foley et al. 2005; MEA 2005). Increasingly, ecologists and social scientists who are working to develop policies to promote the provision of goods and services for an expanding human population focused on ecological sustainability (Goodland 1995).

Climate change, diverse disturbances, soil or water resources, and functional biota groupings all interact with each other to alter natural ecosystems' structure and functions. In order to counteract the positive feedbacks linked with exponential population growth and cumulative changes in rates of nutrient cycling, negative-feedback processes such as predation, successive disturbances/recoveries, and the use of the same limiting resource by many competing species are necessary. The long-term survival of an ecosystem is dependent on the complicated interconnections and positive and negative feedbacks generated by interactive controls. Due to negative feedback, an ecosystem can only be sustained for a limited amount of time under specific conditions (Chapin et al. 1996). We believe that successful management of human-modified ecosystems requires an understanding of the general principles of ecological sustainability. Knowledge of interactive controls, in particular, may aid in the management of natural resources and biological variety. The most critical stage in ensuring a managed ecosystem's long-term viability is to preserve or regulate interaction controls so that they establish negative feedback loops that maintain desirable ecological traits. Negative feedbacks can be increased by preserving historic links across ecosystems and employing laws to create negative feedbacks where human involvement has a significant impact on natural ecosystems' interaction controls (Chapin et al. 1996). Climate change (warming), soil and water resources (deposition), perturbation regime (land-use change, fire control), and the functional groups of organisms (species introductions and extinctions) are all threat for sustainability. Thus, all ecosystems are experiencing directional changes in ecosystem controls. This results in novel conditions and the formation of new types of ecosystems in many cases as a positive feedback mechanism. Because of these developments, it is becoming increasingly challenging to successfully manage natural resources and natural ecosystems around the world (Chapin et al. 1996).

Soils are a key component of ecosystems, and understanding their processes requires multidisciplinary and transdisciplinary approaches. Soil is created when the lithosphere, biosphere, atmosphere, and hydrosphere collide. It is home to a huge part of the world's biodiversity and manages the bulk of ecological processes in landscapes, as well as provides the physical underpinning for numerous human activities. Human influences on soil have gotten worse since the first agricultural revolution (Barrios 2007; Brevik et al. 2015). Soil ecosystem (ES) are being severely degraded as a result of climate change and human pressure, putting future generations' food security at risk. Changes in the soil-water-gas equilibrium, as well as a decrease in soil organic carbon, may reduce soil fertility. Furthermore, soil is the greatest terrestrial carbon sink, but it may also be a significant source of greenhouse gas emissions if improperly managed. Agriculture, unsustainable practices, changing natural ecosystems in rural areas, and urban sprawl are all contributing to soil degradation at an astonishing rate and lowering soil quality (Pereira et al. 2016, 2017). Soils are an essential natural capital since they determine a country's economic standing (Dominati et al. 2010).

Apart from this, for a given land use, soil's ability to supply the ES is characterized as soil natural capital, given that sustainable practices are adopted (Hewitt et al. 2015). The importance of soil processes to ES has been overlooked by earlier study, despite the fact that soil components are crucial to ES (Robinson et al. 2013, 2014). The methods we choose to manage our fields can have a significant impact on soil ES. For instance, non-sustainable use can lead to several negative consequences, but it can also have an adverse effect on the land's environmental characteristics, soil qualities and functions, as well as its ability to provide these services. It is possible to protect and improve these services through sustainable approaches. Using non-sustainable ways accelerates the effects of climate change on soil ES, whereas using sustainable methods reduces them. Our activities will have an effect on our society and economy if they hinder or help the soil's long-term viability (Pereira et al. 2018).

Hence, it is clear that the natural ecosystem itself can act as a tool for sustainability. The adverse effect of climate change can be adapted according to proper management of land/soil, forest, and aquatic ecosystem. The soil ecosystem has enormous potential for storing the carbon dioxide from the atmosphere. The proper/sustainable soil and land-use management can increase the efficiency of soil ecosystem goods and services. The forest and soil ecosystem together can store vast amounts of carbon from the atmosphere, and land degradation and deforestation are the major threats for proper functioning of these ecosystems. The planktons in the ocean also store carbon from the atmosphere. However, human interventions decrease the efficiency of carbon-storing capacity of various ecosystems. The nature-based solution is one good example of mitigating various environmental issues in a non-harmful way. These solutions are eco-friendly and also highly effective in terms of cost required for them. Hereafter, the climate resilience of various sectors, sustainable development goals and its outcomes, and nature-based solutions deliver a sustainable ecosystem which can resist the consequence of changing climate to greater extent.

There are lot of global frameworks initiated for the well-being of natural ecosystem as well as human habitat. These initiatives are based on the basic principles of conserving the nature while utilizing for economic development. The Ramsar Convention is intended for wetland conservation and the Convention on International Trade in Endangered Species of Wild Fauna and Flora (CITES) for flora and fauna. The Montreal Protocol is preventing the overuse of ozone layer-depleting substances. The Rio Summit, also called Earth Summit, emphasizes the need for implementing the United Nations Convention to Combat Desertification (UNCCD), UNCBD, and UNFCCC. The United Nations Framework Convention on Climate Change (UNFCCC) is a universal treaty to combat climate change through the adaptation and mitigation measures. The Convention on Biological Diversity (UNCBD) is for conserving the biological diversity all around the world. UNCCD

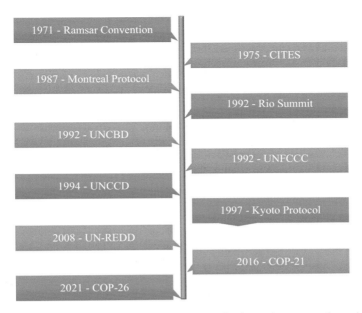

Fig. 1.2 The timeline of major environmental initiatives for the environment and sustainability

aims to combat the desertification and drought through a strategic framework. The Kyoto Protocol aims to reduce greenhouse gas emissions from various sources to reduce the global temperature (global warming). Reducing Emissions from Deforestation and forest Degradation (REDD+) is an initiative against the adverse effects of climate change by mitigating and reducing greenhouse gas emissions, sequestrating carbon along, and preventing forest degradation as well as sustainable use of biodiversity by proper conservation. The Conference of Parties (COP) is a global meeting of the UNFCCC conducted every year. Similarly, during the UNCCD COP13 in 2017, countries devoted to accomplish Land Degradation Neutrality by 2030 (Fig. 1.2).

Summary and Conclusion

The human population is growing exponentially with no sign of decreasing in the near future. Together with this concern, the natural resource exploitation is increasing day by day to meet the food and economic requirement for humans. Most of the natural resources (fossil fuels) are nonrenewable in nature. However, human utilization is very high. In addition, renewable natural resources are also under threat of overconsumption by human beings. Human consumption rate is far above the natural regeneration capacity. The natural resources we obtained or available are considered as natural capital or biocapacity/biological capacity. The land/soil, forest, and aquatic ecosystems are the basic building blocks which provide sources for human consumption. The crop, grass, forest, aquatic land, and built-up land are the major components of the biocapacity. The process of production derived from these components which are consuming biological productive land produces biological footprint. The carbon emission/footprint is considered as one of the major components of ecological footprint. It represents the amount of carbon emission to the atmosphere due to production process of a product. Higher ecological footprint leads to higher carbon emission, which means more pressure to natural resources and results higher warming which eventually leads to climate change. Climate change also possesses threat to biocapacity by decreasing its efficiency. It may lead to higher ecological footprint from a product due to nonfavorable situation. Hence, both of these (ecological footprint and climate change) act as a fuel for each other.

We consider nature itself as the best solution for these kinds of concerns. It depends upon how we are utilizing the natural resources and ecosystem services. Optimum usage is always the better way to increase the Earth's regenerative capacity. The increased footprint has long-lasting consequences for the present and future scenarios. Climate resilience in all sector is the better way to manage climate change-related adverse effects (flood, drought, extreme weather events, etc.). The resilience acquired by the adaptation and mitigation can also help maintain the natural resources. Climate resilience in agricultural field and urban sector can provide quality food and suitable habitat for the human beings. The 17 UN Sustainable Development Goals are a proper framework for attaining sustainability in all sector while considering the nature/ecosystem. It enhances the optimum ecological footprint for each region or product. Equity is one of the prime concepts tangled in sustainable development goals. In addition, nature-based solution provides an eco-friendly solution for the environment and other related issues. The multifaceted unified approach along with optimum utilization and proper administration can counterattack the issues emerging in the era of changing climate.

> Earth provides enough to satisfy every man's need, but not every man's greed. (Mahatma Gandhi)

References

Alongi DM (2008) Mangrove forests: resilience, protection from tsunamis, and responses to global climate change. Estuar Coast Shelf Sci 76(1):1–13

Ammer C (2019) Diversity and forest productivity in a changing climate. New Phytol 221(1):50–66

Asumadu-Sarkodie S, Owusu PA (2016) Feasibility of biomass heating system in Middle East Technical University, Northern Cyprus campus. Cogent Eng 3. https://doi.org/10.1080/23311916.2015.1134304

Atkinson A, Siegel V, Pakhomov E, Rothery P (2004) Long-term decline in krill stock and increase in salps within the Southern Ocean. Nature 432(7013):100–103

Barbraud C, Weimerskirch H, Bost CA, Forcada J, Trathan P, Ainley D (2008) Are king penguin populations threatened by Southern Ocean warming? Proc Natl Acad Sci 105(26):E38–E38

Barrios E (2007) Soil biota, ecosystem services and land productivity. Ecol Econ 64(2):269–285

Borucke M, Moore D, Cranston G, Gracey K, Iha K, Larson J, Lazarus E, Morales JC, Wackernagel M, Galli A (2013) Accounting for demand and supply of the biosphere's regenerative capacity: the National Footprint Accounts' underlying methodology and framework. Ecol Indic 24:518–533

Brevik EC, Cerdà A, Mataix-Solera J, Pereg L, Quinton JN, Six J, Van Oost K (2015) The interdisciplinary nature of SOIL. Soil 1:117–129

Ceballos G, Ehrlich PR (2018) The misunderstood sixth mass extinction. Science 360(6393): 1080–1081

Chapin FS, Torn MS, Tateno M (1996) Principles of ecosystem sustainability. Am Nat 148(6): 1016–1037

Chapin FS, Carpenter SR, Kofinas GP, Folke C, Abel N, Clark WC et al (2010) Ecosystem stewardship: sustainability strategies for a rapidly changing planet. Trends Ecol Evol 25(4): 241–249

Chen S, Wang W, Xu W et al (2018) Plant diversity enhances productivity and soil carbon storage. Proc Natl Acad Sci U S A 115:4027–4032

Chitsaz N, Azarnivand A (2017) Water scarcity management in arid regions based on an extended multiple criteria technique. Water Resour Manag 31(1):233–250

Clark WC (2007) Sustainability science: a room of its own. Proc Natl Acad Sci 104(6):1737–1738

Conacher A (2009) Land degradation: a global perspective. N Z Geogr 65(2):91–94

Díaz ME, Figueroa R, Alonso MLS, Vidal-Abarca MR (2018) Exploring the complex relations between water resources and social indicators: the Biobío Basin (Chile). Ecosyst Serv 31:84–92

Díaz S, Settele J, Brondízio ES, Ngo HT, Guèze M, Agard J et al (2019) Summary for policymakers of the global assessment report on biodiversity and ecosystem services of the Intergovernmental Science-Policy Platform on Biodiversity and Ecosystem Services. Intergovernmental Science-Policy Platform on Biodiversity and Ecosystem Services, Bonn

Dominati E, Patterson M, Mackay A (2010) A framework for classifying and quantifying the natural capital and ecosystem services of soils. Ecol Econ 69(9):1858–1868

Edenhofer O, Pichs-Madruga R, Sokona Y, Seyboth K, Matschoss P, Kadner S et al (2011) Renewable energy sources and climate change mitigation. Cambridge University Press, Cambridge. https://doi.org/10.1017/CBO9781139151153

Erb KH, Kastner T, Plutzar C, Bais ALS, Carvalhais N, Fetzel T et al (2018) Unexpectedly large impact of forest management and grazing on global vegetation biomass. Nature 553(7686): 73–76

Erwin KL (2009) Wetlands and global climate change: the role of wetland restoration in a changing world. Wetl Ecol Manag 17(1):71–84

European State of The Climate, Summary 2020 (2021) European Centre for Medium-Range Weather Forecasts (ECMWF) Available online at https://climate.copernicus.eu/esotc/2020

Fang X, Zhou B, Tu X, Ma Q, Wu J (2018) "What kind of a science is sustainability science?" An evidence-based reexamination. Sustainability 10(5):1478

FAO – Food and Agriculture Organisation of the United Nations (2015) Global forest resources assessment. FAO, Rome, pp 1–253

Feng T, Wang C, Hou J, Wang P, Liu Y, Dai Q et al (2018) Effect of inter-basin water transfer on water quality in an urban lake: a combined water quality index algorithm and biophysical modelling approach. Ecol Indic 92:61–71

Finlayson C, Gitay H, Bellio M, van Dam R, Taylor I (2006) Climate variability and change and other pressures on wetlands and waterbirds: impacts and adaptation. In: Waterbirds around the world: a global overview of the conservation, management and research of the world's waterbird flyways. The Stationery Office, Edinburgh, pp 88–97

Flávio HM, Ferreira P, Formigo N, Svendsen JC (2017) Reconciling agriculture and stream restoration in Europe: a review relating to the EU Water Framework Directive. Sci Total Environ 596:378–395

Foley JA, DeFries R, Asner GP, Barford C, Bonan G, Carpenter SR et al (2005) Global consequences of land use. Science 309(5734):570–574

Fossheim M, Primicerio R, Johannesen E, Ingvaldsen RB, Aschan MM, Dolgov AV (2015) Recent warming leads to a rapid borealization of fish communities in the Arctic. Nat Clim Chang 5(7): 673–677

Gardner TA, Barlow J, Chazdon R et al (2009) Prospects for tropical forest biodiversity in a human-modified world. Ecol Lett 12:561–582

Giam X (2017) Global biodiversity loss from tropical deforestation. Proc Natl Acad Sci 114(23): 5775–5777

Gísladóttir G, Stocking M (2005) Land degradation and mitigation. Land Degrad Dev 16(2):97–97

Gnacadja L (2012) Moving to zero-net rate of land degradation. In: Statement by Executive Secretary. UN Convention to Combat Desertification. (Rio de Janeiro). http://www.unccd.int/Lists/SiteDocumentLibrary/secretariat/2012/UNCCD%20ES%20Statement%20at%20PR%20in%20NY%20on%2026%20March%202012.pdf

Goodland R (1995) The concept of environmental sustainability. Annu Rev Ecol Syst 26(1):1–24

Gradinger R (2009) Sea-ice algae: major contributors to primary production and algal biomass in the Chukchi and Beaufort Seas during May/June 2002. Deep Sea Res Part II Top Stud Oceanogr 56(17):1201–1212

Gregg WW, Conkright ME, Ginoux P, O'Reilly JE, Casey NW (2003) Ocean primary production and climate: global decadal changes. Geophys Res Lett 30(15):OCE 3-1–OCE 3-4

Griggs D, Stafford-Smith M, Gaffney O, Rockström J, Öhman MC, Shyamsundar P et al (2013) Sustainable development goals for people and planet. Nature 495(7441):305–307

Guo C, Xu H (2019) Use of functional distinctness of periphytic ciliates for monitoring water quality in coastal ecosystems. Ecol Indic 96:213–218

Haberl H, Erb KH, Krausmann F, Gaube V, Bondeau A, Plutzar C et al (2007) Quantifying and mapping the human appropriation of net primary production in earth's terrestrial ecosystems. Proc Natl Acad Sci 104(31):12942–12947

Hák T, Janoušková S, Moldan B (2016) Sustainable development goals: a need for relevant indicators. Ecol Indic 60:565–573. https://doi.org/10.1016/j.ecolind.2015.08.003

Harden JW, Berhe AA, Torn M, Harte J, Liu S, Stallard RF (2008) Soil erosion: data say C sink. Science 320(5873):178–179

Hayat S, Gupta J (2016) Kinds of freshwater and their relation to ecosystem services and human well-being. Water Policy 18:1229–1246. https://doi.org/10.2166/wp.2016.182

Hewitt A, Dominati E, Webb T, Cuthill T (2015) Soil natural capital quantification by the stock adequacy method. Geoderma 241:107–114

Hoegh-Guldberg O, Bruno JF (2010) The impact of climate change on the world's marine ecosystems. Science 328(5985):1523–1528

Hoegh-Guldberg O, Mumby PJ, Hooten AJ, Steneck RS, Greenfield P, Gomez E et al (2007) Coral reefs under rapid climate change and ocean acidification. Science 318(5857):1737–1742

Intergovernmental Panel on Climate Change (2007) Climate Change 2007: the physical science basis. Contribution of Working Group I to the Fourth Assessment Report of the Intergovernmental Panel on Climate Change (IPCC), Solomon S (eds); Cambridge University Press, Cambridge

IPCC (2014a) Summary for policymakers Climate Change 2014: impacts, adaptation, and vulnerability. Part A: global and sectoral aspects. Contribution of Working Group II to the Fifth Assessment Report of the Intergovernmental Panel on Climate Change ed C B Field, Cambridge University Press, Cambridge/New York, p 1–32

IPCC (2014b) Climate Change, 2013: the physical science basis. Contribution of Working Group I to the Fifth Assessment Report of the Intergovernmental Panel on Climate Change. Cambridge University Press, Cambridge

IPCC (2018) Summary for Policymakers. In: Global Warming of 1.5°C. An IPCC Special Report on the impacts of global warming of 1.5°C above pre-industrial levels and related global greenhouse gas emission pathways, in the context of strengthening the global response to the threat of climate change, sustainable development, and efforts to eradicate poverty

[Masson-Delmotte V, Zhai P, Pörtner H-O, Roberts D, Skea J, Shukla PR, Pirani A, Moufouma-Okia W, Péan C, Pidcock R, Connors S, Matthews JBR, Chen Y, Zhou X, Gomis MI, Lonnoy E, Maycock T, Tignor M, Waterfield T (eds)]. World Meteorological Organization, Geneva, Switzerland, 32 pp

IPCC – Intergovernmental Panel on Climate Change (2013) Climate change 2013: the physical science basis. Contribution of working group I to the fifth assessment report of the intergovernmental panel on climate change. Cambridge University Press, Cambridge, MA

IPCC-AR5-WG1 (2013) Climate Change 2013: the physical science basis – Working Group 1 (WG1) contribution to the Intergovernmental Panel on Climate Change (IPCC) 5th assessment report (AR5). Cambridge University Press, UK

Isbell F, Gonzalez A, Loreau M, Cowles J, Díaz S, Hector A et al (2017) Linking the influence and dependence of people on biodiversity across scales. Nature 546(7656):65–72

IUCN – International Union for Nature Conservation (2019) IUCN spatial data. http://www.iucnredlist.org/techical-documents/spatial-data

Kamaruddin AH, Din AHM, Pa'suya MF, Omar KM (2016, August) Long-term sea level trend from tidal data in Malaysia. In: 2016 7th IEEE control and system graduate research colloquium (ICSGRC), IEEE, p 187–192

Kates RW, Clark WC, Corell R, Hall JM, Jaeger CC, Lowe I, McCarthy JJ, Schellnhuber HJ, Bolin B, Dickson NM, Faucheux S, Gallopin GC, Gruebler A, Huntley B, Jager J, Jodha NS, Kasperson RE, Mabogunje A, Matson P, Mooney H, Moore B III, O'Riordan T, Svedin U (2001) Sustainability science. Science 292:641–642

Keeling HC, Phillips OL (2007) The global relationship between forest productivity and biomass. Glob Ecol Biogeogr 16(5):618–631

Keenan RJ, Reams GA, Achard F, de Freitas JV, Grainger A, Lindquist E (2015) Dynamics of global forest area: results from the FAO Global Forest Resources Assessment 2015. For Ecol Manag 352:9–20

Körner C, Ohsawa M (2005) Mountain systems. In: Hassan R, Scholes R, Ash N (eds) Ecosystems and human well-being: current state and trends, vol 1. Island Press, Washington, DC, pp 687–716

Koutroulis AG, Papadimitriou LV, Grillakis MG, Tsanis IK, Wyser K, Caesar J, Betts RA (2018) Simulating hydrological impacts under climate change: implications from methodological differences of a Pan European assessment. Water 10(10):1331

Lal R (2004) Soil carbon sequestration to mitigate climate change. Geoderma 123(1–2):1–22

Lal R (2010) Managing soils and ecosystems for mitigating anthropogenic carbon emissions and advancing global food security. Bioscience 60:708–721

Lal R (2012a) Land degradation and pedological processes in a changing climate. Pedologist 55(3):315–325

Lal R (2012b) Restoring degraded lands and the flow of its provisioning services. In: Proceedings of the 4th international conference on drylands, deserts and desertification, p. 65

Lal R, Pimentel D (2008) Soil erosion: a carbon sink or source? Science 319:1040–1042. https://doi.org/10.1126/science.319.5866.1040

Lal R, Safriel U, Boer B (2012, May) Zero net land degradation: a new sustainable development goal for Rio+ 20. In: United Nations Convention to Combat Desertification (UNCCD). http://www.unccd.int/Lists/SiteDocumentLibrary/secretariat/2012/Zero%20Net%20Land%20Degradation%20Report%20UNCCD%20May%202012%20background.pdf

Laurent A, Olsen SI, Hauschild MZ (2012) Limitations of carbon footprint as indicator of environmental sustainability. Environ Sci Technol 46(7):4100–4108

Lee DK (2011) Land and soil in the context of a green economy for sustainable development, food security and poverty eradication. The submission of the UNCCD Secretariat to the Preparatory Process for Rio+ 20 UNCCD, Bonn http://www.unscd2012.org/content/documents/462unccd.pdf

Lewis SL, Wheeler CE, Mitchard ET et al (2019) Restoring natural forests is the best way to remove atmospheric carbon. Nature 568:25–28

Liao C, Qiu J, Chen B, Chen D, Fu B, Georgescu M et al (2020) Advancing landscape sustainability science: theoretical foundation and synergies with innovations in methodology, design, and application. Landsc Ecol 35(1):1–9

Liu X, Trogisch S, He JS et al (2018) Tree species richness increases ecosystem carbon storage in subtropical forests. Proc R Soc B 285:20181240

Lu Y, Nakicenovic N, Visbeck M, Stevance A-S (2015) Policy: five priorities for the UN sustainable development goals. Nature 520:432–433. https://doi.org/10.1038/520432a

Malmqvist B, Rundle SD, Covich AP, Hildrew AG, Robinson CT, Townsend CR (2008) Prospects for streams and rivers: an ecological perspective. In: Polunin N (ed) Aquatic systems: trends and global perspectives. Cambridge University Press, Cambridge, pp 19–29

Manju S, Sagar N (2017) Renewable energy integrated desalination: a sustainable solution to overcome future fresh-water scarcity in India. Renew Sust Energ Rev 73:594–609

Matthews HS, Hendrickson CT, Weber CL (2008) The importance of carbon footprint estimation boundaries. Environ Sci Technol 42:5839–5842

Mazor T, Doropoulos C, Schwarzmueller F, Gladish DW, Kumaran N, Merkel K, Di Marco M, Gagic V (2018) Global mismatch of policy and research on drivers of biodiversity loss. Nat Ecol Evol 2:1071–1074

MEA – Millennium Ecosystem Assessment (2005) Ecosystems and human well-being: synthesis. Island Press, Washington, DC

Mitrică B, Mitrică E, Enciu P, Mocanu I (2017) An approach for forecasting of public water scarcity at the end of the 21st century, in the Timiş Plain of Romania. Technol Forecast Soc Change 118: 258–269

Momblanch A, Holman IP, Jain SK (2019) Current practice and recommendations for modelling global change impacts on water resource in the Himalayas. Water 11(6):1303

Murray DL, Peers MJ, Majchrzak YN, Wehtje M, Ferreira C, Pickles RS et al (2017) Continental divide: predicting climate-mediated fragmentation and biodiversity loss in the boreal forest. PLoS One 12(5):e0176706

Nakicenovic N, Alcamo J, Davis G, Vries BD, Fenhann J, Gaffin S et al (2000) Special report on emissions scenarios. A Special Report of Working Group III of the Intergovernmental Panel on Climate Change. Cambridge University Press, Cambridge

O'Connor MI, Bruno JF, Gaines SD, Halpern BS, Lester SE, Kinlan BP, Weiss JM (2007) Temperature control of larval dispersal and the implications for marine ecology, evolution, and conservation. Proc Natl Acad Sci 104(4):1266–1271

Oldeman LR (1998) Soil degradation: a threat to food security? Report 98/01. International Soil Reference and Information Centre, Wageningen

Opdam P, Luque S, Nassauer J, Verburg PH, Wu J (2018) How can landscape ecology contribute to sustainability science? Landsc Ecol 33(1):1–7

Owusu PA, Asumadu-Sarkodie S, Ameyo P (2016) A review of Ghana's water resource management and the future prospect. Cogent Eng 3. https://doi.org/10.1080/23311916.2016.1164275

Pan Y, Birdsey RA, Fang J et al (2011) A large and persistent carbon sink in the world's forests. Science 333:988–993

Paredes I, Ramírez F, Forero MG, Green AJ (2019) Stable isotopes in helophytes reflect anthropogenic nitrogen pollution in entry streams at the Doñana World Heritage Site. Ecol Indic 97:130–140

Pecl GT, Araújo MB, Bell JD, Blanchard J, Bonebrake TC, Chen IC et al (2017) Biodiversity redistribution under climate change: impacts on ecosystems and human well-being. Science 355(6332)

Pereira P, Ferreira A, Pariente S, Cerda A, Walsh RPD, Keesstra S (2016) Preface: urban soils and sediments. J Soils Sediments 16:2493–2499

Pereira P, Brevik E, Munoz-Rojas M, Miller B, Smetanova A, Depellegrin D, Misiune I, Novara A, Cerda A (2017) Soil mapping and process modelling for sustainable land management. In: Pereira P, Brevik E, Munoz-Rojas M, Miller B (eds) Soil mapping and process modelling for sustainable land use management. Elsevier, pp 29–60

Pereira P, Bogunovic I, Muñoz-Rojas M, Brevik EC (2018) Soil ecosystem services, sustainability, valuation and management. Curr Opin Environ Sci Health 5:7–13

Pinsky ML, Worm B, Fogarty MJ, Sarmiento JL, Levin SA (2013) Marine taxa track local climate velocities. Science 341(6151):1239–1242

Polovina JJ, Howell EA, Abecassis M (2008) Ocean's least productive waters are expanding. Geophys Res Lett 35(3):L03618

Robinson DA, Hockley N, Cooper DM, Emmett BA, Keith AM, Lebron I et al (2013) Natural capital and ecosystem services, developing an appropriate soils framework as a basis for valuation. Soil Biol Biochem 57:1023–1033

Robinson DA, Fraser I, Dominati EJ, Davíðsdóttir B, Jónsson JOG, Jones L et al (2014) On the value of soil resources in the context of natural capital and ecosystem service delivery. Soil Sci Soc Am J 78(3):685–700

Runting RK, Bryan BA, Dee LE, Maseyk FJ, Mandle L, Hamel P et al (2017) Incorporating climate change into ecosystem service assessments and decisions: a review. Glob Chang Biol 23(1): 28–41

Salmoral G, Rey D, Rudd A, de Margon P, Holman I (2019) A probabilistic risk assessment of the national economic impacts of regulatory drought management on irrigated agriculture. Earth Future 7(2):178–196

Scheurer K, Alewell C, Bänninger D, Burkhardt-Holm P (2009) Climate and land-use changes affecting river sediment and brown trout in alpine countries—a review. Environ Sci Pollut Res 16(2):232–242

Schröter M, Koellner T, Alkemade R, Arnhold S, Bagstad KJ, Erb KH et al (2018) Interregional flows of ecosystem services: concepts, typology and four cases. Ecosyst Serv 31:231–241

State of the Climate in Africa 2020 (2021) World Meteorological Organization, (WMO-No. 1275) Available online at https://library.wmo.int/index.php?lvl=notice_display&id=21973#.Ye7uzf5Bzb0

State of the Climate in Asia 2020 (2021) World Meteorological Organization, (WMO-No. 1273) Available online at https://library.wmo.int/index.php?lvl=notice_display&id=21977#.Ye7tlv5Bzb1.

State of the Climate in Latin America and the Caribbean 2020 (2021) World Meteorological Organization, (WMO-No. 1272) Available online at https://library.wmo.int/index.php?lvl=notice_display&id=21926#.Ye7ue_5Bzb0

State of the Climate in South-West Pacific 2020 (2021) World Meteorological Organization, (WMO-No. 1276) Available online at https://library.wmo.int/index.php?lvl=notice_display&id=21990#.Ye7vJf5Bzb0

Stavi I, Lal R (2013) Agriculture and greenhouse gases, a common tragedy. A review. Agron Sustain Dev 33(2):275–289

Stavi I, Lal R (2015) Achieving zero net land degradation: challenges and opportunities. J Arid Environ 112:44–51

Steffen W, Sanderson A, Tyson PD, Jager J, Matson PA, Moore B III, Oldfield F, Richardson K, Schellnhuber HJ, Turner BL, Wasson RJ (2004) Global change and the earth system: a planet under pressure. Springer, Berlin

UNCCD Zero Net Land Degradation, a Sustainable Development Goal for Rio+20 (2012). http://www.unccd.int/Lists/SiteDocumentLibrary/Rio+20/UNCCD_PolicyBrief_ZeroNetLandDegradation.pdf

Wackernagel M, Rees WE (1996) Our ecological footprint: reducing human impact on the earth. New Society, Gabriola Island

Wackernagel M, Onisto L, Bello P, Linares AC, Falfan ISL, Garcia JM, Guerrero AIS, Guerrero CS (1999) National natural capital accounting with the ecological footprint concept. Ecol Econ 29(3):375–390

Wang MH, Li J, Ho YS (2011) Research articles published in water resources journals: a bibliometric analysis. Desalin Water Treat 28(1-3):353–365

WCED (1987) Our common future. Oxford University Press, Oxford

Wiedmann T, Minx J (2007) A definition of 'carbon footprint': integrated sustainability analysis UK. pp 1–11

Wilcox BP, Fox WE, Prcin LJ, McAlister J, Wolfe J, Thomas DM et al (2012) Contour ripping is more beneficial than composted manure for restoring degraded rangelands in Central Texas. J Environ Manag 111:87–95

Wu J (2013) Landscape sustainability science: ecosystem services and human well-being in changing landscapes. Landsc Ecol 28:999–1023

Wu J (2019) Linking landscape, land system and design approaches to achieve sustainability. J Land Use Sci 14:173–189

Yasuhara M, Danovaro R (2016) Temperature impacts on deep-sea biodiversity. Biol Rev 91(2): 275–287. https://doi.org/10.1111/brv.12169

Chapter 2
Assessing Global-Scale Synergy Between Adaptation, Mitigation, and Sustainable Development for Projected Climate Change

Aman Srivastava ⓘ, Rajib Maity ⓘ, and Venkappayya R. Desai ⓘ

Abstract The theoretical idea of a "greenhouse effect" has existed for centuries; however, the human influence via greenhouse gas (GHG) emissions is realized post-mid-twentieth century. The resultant – "global warming" – has been evidently observed in causing intensified hydrological cycle and recurrent extreme climatic events. Consequently, unprecedented rise in glacier melting at a rate of 3.5–4.1% per decade, ocean warming due to >90% energy accumulation, ocean acidification due to falling in pH by 0.1, and mean sea-level rise of ~19 cm due to aforesaid changes are observed globally, which are further endangering associated livelihoods, ecosystem, biodiversity, and coastal environment. This study evaluates key drivers of long-term future climate change, risk, and impacts at a global scale. The GCM data (a subset of the CMIP5 multi-model ensemble snapshot) has been used, comprising monthly means of climate variables from 1850 to 2100. The assessment is conducted considering two distinct future periods (2046–2065 and 2081–2100) under RCP2.6, RCP4.5, RCP6.0, and RCP8.5 scenarios. Major findings indicated irreversible and pervasive impacts of continued GHG emission across all components of the climate system. As projected climate change will largely be driven by CO_2 emissions, mere reduction of GHGs to zero will not prevent climate change during and post-twenty-first century. To combat future climate change, substantial and sustained GHG reductions by exploiting co-benefits of adaptation and mitigation are needed. The findings presented here can be useful for decision-making, climate policy drafting, and decentralized governance for enhancing mitigation shifts and adaptive capacity across all spatial scales.

A. Srivastava (✉) · R. Maity
Department of Civil Engineering, Indian Institute of Technology (IIT) Kharagpur, Kharagpur, West Bengal, India
e-mail: amansrivastava1397@kgpian.iitkgp.ac.in; rajib@civil.iitkgp.ac.in

V. R. Desai
Department of Civil Engineering, Indian Institute of Technology (IIT) Kharagpur, Kharagpur, West Bengal, India

Director, Indian Institute of Technology (IIT) Dharwad, Dharwad, Karnataka, India
e-mail: venkapd@civil.iitkgp.ac.in

Keywords Climate change · Global warming · Mean sea level · Greenhouse gas (GHG) emissions · Climate-resilient pathways

Introduction to Climate Change

The observational evidence of human influence on climate change is clear, given the record emission of greenhouse gases (GHGs) and widespread impacts of climate change on the human, earth, and natural systems (IPCC 2014). As compared to the past 1400 years, the period between 1983 and 2012 likely[1] remained the warmest period across the globe (Beevers 2019). The comparative linear trend analysis of the globally averaged combined ocean surface and land temperature datasets between 1850–1900 and 2003–2012 indicated a sharp increase in overall temperature by 0.78 °C (IPCC 2014; Rohde and Hausfather 2020; Smith et al. 2008). Rising global temperature, especially between 1971 and 2010, caused the ocean to warm and store more energy (~90%) as compared to 1% stored energy in the atmosphere (Johnson and Lyman 2020). Consequently, the twentieth century witnessed an unprecedented surge in precipitation intensity (Allan et al. 2020) which significantly intensified the natural hydrological cycle in the twenty-first century. One of the impacts of change in precipitation has been reflected in the uneven salinity of the ocean, wherein the regions receiving higher rainfall are observed to be less saline than otherwise (Durack 2015; IPCC 2014). Contrarily, at the same time, one of the negative influences of industrialization caused greater carbon dioxide (CO_2) uptake resulting in acidification of oceans (Hönisch et al. 2012; Riebesell and Gattuso 2015). For the period 1992–2011, studies observed that the colder regions, such as Greenland and the Arctic, have significantly lost ice sheet mass, which likely occurred at a rate greater than 2002–2011 (IPCC 2014). As a combined influence of the aforementioned changes, the global mean sea level experienced a rapid rise of ~19 cm during the twentieth century (Gregory et al. 2013; Watson et al. 2015). As far as the rate is concerned, the sea-level rise remained more profound since 1850 as compared to the last two millennia (IPCC 2014). Given the rapidly increasing global climate change impacts, it becomes imperative to investigate the shreds of evidence that are behind these occurrences. The aforesaid climate change evidence has been further detailed in Table 2.1.

Studies across the world have identified true evidence of human influence on the ongoing climate change issues. The large increase in the concentrations of CO_2, methane (CH_4), and nitrous oxide (N_2O) (categorized in the group of GHG) in the atmosphere has been primarily driven by GHG emissions from anthropogenic

---

[1] Each finding or evidence presented in this study is based on IPCC (2014) and works published thereafter. The evidences are represented in terms of limited or medium or robust evidences, while agreements are represented in terms of low- or medium- or high-level agreements. In addition, confidence levels are represented in terms of very low, low, medium, high, and very high levels, whereas the assessed likelihood of results is represented as virtually certain (99–100% probable), very likely (90–100% probable), likely (66–100% probable), or not likely (33–66% probable), and unlikely (0–33% probable).

Table 2.1 Evidence of global climate change and their assigned confidence between pre-industrial period and 2012

Phenomena	Study period	Evidence	Confidence
Earth's surface warming	1983–2012	Northern Hemisphere likely witnessed the warmest climate as compared to the historic 1400 years (back in the past)	Medium
	1880–2012	Warming of combined land and ocean surface by 0.85 [0.65–1.06] °C	Medium
Ocean warming	1971–2010	>90% of the energy is accumulated in oceans with merely 1% stored in the atmosphere	High
	1971–2010	The warming effect remained more profound near the ocean surface such that the temperature for the upper 75 m increased at the rate of 0.11 °C per decade	High
	1870s–1971	Upper ocean (0–700 m) likely warmed	High
Precipitation intensification	1901–1951	Precipitation has increased over the midlatitude land areas of the Northern Hemisphere	Medium
	1951–2012	Precipitation continued increasing trend in the aforementioned locations	High
	1951–2012	Regions experiencing greater evaporation and higher salinity very likely become more saline, whereas regions experiencing greater precipitation and lower salinity very likely become fresher	Medium
Anthropogenic GHG emission	1750–2011	~40% [880 ± 35 gigatons of CO_2 equivalent ($GtCO_2$)] of the total emissions (2040 ± 310 $GtCO_2$) remained in the atmosphere	High
	1970s–2012	~50% of the aforesaid emissions have occurred recently	High
	1970–2010	Heavy fossil fuel combustion and industrial processes caused total GHG emissions to increase by 78% in these 30 years	High
	2000–2010	These 10 years alone witnessed a rise of ~78% of the total GHGs	High
Oceanic uptake of CO_2	1750–2011	Absorbed ~30% of the emitted anthropogenic CO_2	High
	1950s–2012	pH has decreased by 0.1, corresponding to a 26% increase in acidity causing ocean acidification	High
Increased glacier melting	1992–2011	Greenland and Antarctic ice sheets have been losing mass; sustained decrement in spring snow cover observed across Northern Hemisphere	High
	1979–2012	Annual mean Arctic sea-ice extent decreased with a very likely rate of 3.5 to 4.1% per decade	High
	1980s–2012	Most regions experienced increased temperature for permafrost; mostly attributed to increased surface temperature and changing snow cover	High
Rising mean sea level	1850s–2012	With respect to the previous two millennia, the rate at which sea-level rise occurs has been observed comparatively higher during this period	High
	1901–2010	An increment of ~19 cm observed in the global mean sea level	High

Source: IPCC (2014)

sources (constituting economic sectors) since the pre-industrial era (Dijkstra et al. 2012; Tian et al. 2015). About 60% of the cumulative emissions [i.e., 1160 ± 275 gigatons of CO_2-equivalent (GtCO_2)] between 1750 and 2011 in the atmosphere got sequestrated by land and oceans. Specifically, by 2010, sectors such as building, transport, industry, and energy were observed directly contributing 6.4%, 14%, 21%, and 35% to the net annual anthropogenic GHG emissions, respectively, such that between 2000 and 2010, an increase of 10 GtCO_2 equivalents (eq) was recorded (Fig. 2.1a). On the contrary, despite an increase in emissions from all sectors, "agriculture, forestry, and other land uses" (AFOLU) observed a decline, which recorded 24% of the net emissions in 2010 majorly from land-based CO_2 emission sources. Emissions based on 100-year global warming potential (GWP100), 20-year global warming potential (GWP20), and 100-year global temperature change potential (GTP100) have also been estimated for the aforesaid sectors and are shown in Fig. 2.1b (IPCC 2014).

Despite implementing mitigation policies globally, the GHG emissions from anthropogenic sources uninterruptedly increased since the 1970s, wherein the rate remained far greater between 2000 and 2010 and reached 49 ± 4.5 GtCO_2-eq/y (IPCC 2014). Besides, the population explosion (since the last three decades) coupled with economic growth (risen sharply in the twenty-first century) remained a consistent driver of CO_2 emissions in the twenty-first century. Furthermore, as coal-based applications, such as electricity generation, are thriving post-industrialization, the age-old process of unceasing decarbonization has been affected negatively (Farquharson et al. 2017). In general, anthropogenic increase in GHG concentrations and other anthropogenic forcings together have extremely likely increased global average surface temperature by more than 50% between 1951 and 2010 (IPCC 2014). In fact, anthropogenic forcings have likely resulted in the misbalancing of the global water cycle and retreating of glaciers post-1960, increasing global upper ocean heat content and mean sea-level rise post-1970s, and increasing melting of the Arctic sea-ice and Greenland ice sheet surface since 1979 and 1993, respectively (Tables 2.1 and 2.2).

In the context of the negative influence of the observed climate change, profound implications have been reported on human and natural systems. Within hydrological systems, alterations in precipitation intensity and the number of days during the monsoon season are affecting water resources systems and their quality and quantity (Sun et al. 2018; Elbeltagi et al. 2022; Vystavna et al. 2021). In the terrestrial system, factors such as patterns of migration, spatial changes, specific activities related to seasons, and other associated factors in response to ongoing climate change are bringing drastic alterations in freshwater and marine species (Campbell et al. 2017; Nolan et al. 2018). While in the ocean systems, aquatic lives have been drastically influenced due to ocean acidification (Cavicchioli et al. 2019; Gattuso et al. 2018). In agricultural systems, crop yields have also been negatively affected due to climate change vulnerabilities (Eyhorn et al. 2019; Kumar et al. 2022; Fellmann et al. 2018). The impacts of climate change on human systems across continents are apparent (e.g., Orru et al. 2017), as documented in Table 2.2.

Extreme climatic events have substantially increased the susceptibility and undesirable interaction of many human systems and some ecosystems with the climate change impacts (Mora et al. 2018; Reichstein et al. 2021). For example, the number

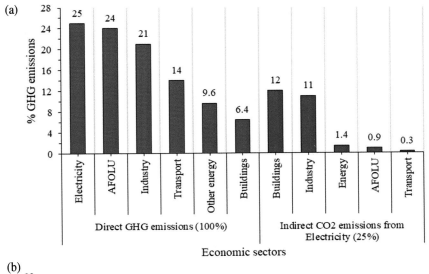
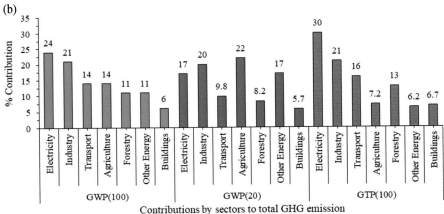

Fig. 2.1 Sector-wise anthropogenic GHG emissions. (Source: Developed by authors after IPCC (2014, 2019))
(**a**) Total anthropogenic greenhouse gas (GHG; in % of total anthropogenic GHG emissions) emissions from economic sectors in 2010; left-hand bars shows the shares of direct GHG emissions from five economic sectors in 2010 and the right-hand bars shows how shares of indirect CO_2 emissions from "electricity and heat production" are attributed to sectors of final energy use. (Note: AFOLU stands for agriculture, forestry, and other land uses; "Other energy" refers to all GHG emission sources in the energy sector other than "electricity and heat production")
(**b**) Contributions from different sectors to total metric-weighted global GHG emissions in the year 2010, calculated using 100-year global warming potential (GWP100, left), 20-year GWP (GWP20, middle), and 100-year global temperature change potential (GTP100, right)

Table 2.2 Region-wise relative impacts (major (M_a) or minor (M_i) with confidence interval – very low (*VL*), low (*L*), medium (*M*), high (*H*), or very high (*VH*) – attributed to climate change since the beginning of the twenty-first century

Particulars	Polar region	North America	Central-South America	Europe	Africa	Asia	Australasia	Small Islands
Physical systems								
Glaciers, snow, ice and/or permafrost	M_a (M to H)	M_a (H)	M_a (H)	M_a (M to H)	M_a (H)	M_a (M to H)	M_a (M)	-
Rivers, lakes, floods and/or drought	M_a (L)	M_i (M) to M_a (H)	M_a (M to H)	M_i (VL)	M_a (L to H)	M_i (L to M) to M_a (M to H)	M_i (L) to M_a (H)	M_i (VL)
Coastal erosion and/or sea-level effect	M_a (M)	M_a (M)	-	-	-	M_a (L)	-	M_i (L)
Biological systems								
Terrestrial ecosystems	M_a (M to H)	M_i (L) to M_a (M)	M_i (L)	M_a (L to H)	M_a (M)	M_a (L to H)	M_a (L to H)	M_i (L) to M_a (M)
Wildfire	-	M_i (M) to M_a (M)	M_i (L)	M_a (H)	M_a (L)	-	-	-
Marine ecosystems	M_a (M)	M_a (M)	M_i (L) to M_a (H)	M_a (M to H)	M_a (H)	M_a (L to H)	M_a (L to H)	M_i (VL) to M_a (H)
Human managed systems								
Food production	-	-	M_a (M)	M_i (M)	M_i (L) to M_a (L)	M_i (L to M)	M_i (L) to M_a (M)	-
Livelihoods, health and/or economics	M_a (M)	M_a (M)	M_a (M)	M_a (L to M)	M_i (VL) to M_a (L)	M_i (L) to M_a (L)	M_a (L)	M_i (L)

Source: Developed by authors after IPCC (2014)
[a]The absence of additional impacts (−) attributed to climate change does not imply that such impacts have not occurred. For more information, readers are directed to refer to literature assessed in the report of Working Group–II of Assessment Report–5 (IPCC 2014)

of cold days and nights has decreased against the rising number of warm days and nights; human mortality due to heat waves has increased in some regions than the decrease in cold-related human mortality in other regions; the events of heavy precipitation in some regions have increased against the cases of decrement in other regions; etc. Considering the widespread negative impressions of climate change on several primary natural phenomena, it becomes pertinent to evaluate their causes, as of now impacts, and future impacts, thereby reviewing effective, sustainable, and green technologies and scope for development within the adaptation and mitigation framework.

This study is primarily aimed to review and assess (1) key drivers of long-term future climate change, risk, and impacts; (2) complementary strategies for adaptation and mitigation against future climate change impacts; (3) various dimensions toward achieving sustainable development in the context of future challenges, limits, and benefits; and (4) common enabling factors, constraints, and integrated responses while devising effective strategies to combat future climate change. The General

Circulation Model (GCM) data, a subset of the Fifth Coupled Model Intercomparison Project (CMIP5) multi-model ensemble snapshot, has been used, given the discussion from the IPCC Working Group–I Fifth Assessment Report (AR5). This dataset consists of monthly means of climate variables from 1850 to 2100. The assessment is conducted considering three distinct 20-year periods, viz., a baseline period (1986–2005) and two future periods (2046–2065 and 2081–2100), and four different scenarios, RCP2.6, RCP4.5, RCP6.0, and RCP8.5. In the next section, the risks and impacts of projected climate change are assessed along with the factors attributing the same to the ecosystem and human systems. It ends with the assessment of changes beyond the twenty-first century in view of irreversible and abrupt future climate changes. Section 2.3 evaluates adaptation and mitigation strategies considering their aids, jeopardies, abrupt and incremental changes, and probable transformations. Furthermore, it discusses choices and prospects that influence mitigation and adaptation. Section 2.4 presents short-term global-scale integrated response options in terms of exploiting co-benefits of adaptation and mitigation in synergy with sustainable development. Finally, Sect. 2.5 concludes the study by summarizing key issues from and emphasizing the scope of the aforementioned scientific literature investigation.

Key Drivers of Projected Climate Changes, Risks, and Impacts

The key drivers of projected climate and associated risks and impacts are mostly observed as economic development, demography, lifestyle and behavioral changes, energy consumption, alterations in land use/land cover, advancement in technologies, and policy regulations in climate change control. The carbon cycle simulating models and climate models, such as comprehensive general circulation models (GCMs) and Earth system models (ESMs), are being deployed for determining and quantifying future climate changes given simulating anthropogenic GHG emissions, large-scale patterns of precipitation, continental-scale surface temperature, thermal expansion of oceans, glacial covers and ice sheets, and global mean sea level (Seneviratne et al. 2021; IPCC 2014).

In order to set future scenarios and assess the costs associated with emission reductions, several approaches have been proposed from time to time, such as Special Report on Emissions Scenarios (SRES), Representative Concentration Pathways (RCPs), and more recently Socioeconomic Pathways (SSPs). The present work has focused on RCPs for understanding the impacts under projected climate change. RCPs designate four different twenty-first-century pathways of GHGs and atmospheric concentrations, air pollutant emissions, and land use. This includes (1) stringent mitigation scenario (RCP2.6), (2) intermediate scenario (RCP4.5), (3) another intermediate scenario (RCP6.0), and (4) very high GHG emissions scenario (RCP8.5). Among this, RCP2.6 objects to retain global Earth temperature likely less than 2 °C as compared to pre-industrial levels. While pathways from RCP6.0 to RCP8.5 are regraded for constraining emissions without additional efforts and thus also termed as baseline scenarios. In addition, under the forcing levels comparable to RCP2.6, scenarios are characterized by substantial net negative emissions by the twenty-first century, whereas in the case of RCPs for land use scenarios, the possible futures range between net reforestation and further deforestation. While for the case of air pollutants [e.g., sulfur dioxide (SO_2)], a consistent decrease in emissions is assumed, given the positive outcomes from air pollution control and GHG mitigation policy, if implemented effectively. It is to be noted that the natural forcings, such as the natural greenhouse effect, emissions due to volcanic activities or other natural sources, etc., are not accounted for in the aforesaid future scenarios (IPCC 2014; van Vuuren et al. 2011).

As compared to the previously assessed scenarios of SRES, the future scenario range is significantly wider in RCPs than in SRES. Considering overall forcings, RCP4.5 is approximately analogous to the SRES B1, RCP6.0 to B2, and RCP8.5 to A2/A1FI, while no equivalent scenarios exist for RCP2.6 (IPCC 2014; Pedersen et al. 2021). Besides RCPs and SRES, Kriegler et al. (2012) and Van Vuuren et al. (2012) proposed the design of the socioeconomic dimension of the scenario framework defined as SSPs. These scenarios consider a set of alternative reference assumptions about future socioeconomic development and don't consider policy regulations for climate change. Kriegler et al. (2014) emphasized that the combination of SSPs with different climate policy assumptions and climate change projections can enable the comprehensive analyses of impacts, vulnerability, mitigation, and adaptation. Further discussions on SSPs are beyond the scope of present work; however, readers are directed to also refer to Van Vuuren and Carter (2014) and O'Neill et al. (2017).

Contemplating projected climate changes, associated risks, vulnerability, and impacts, past studies have used experiments (Claessens et al. 2012), analogies (Raimi et al. 2017), and model-based methods (Yuan et al. 2015) for their precise estimation. First, in *experiments*, one or more climate system factors are deliberately changed, which affect a "subject of interest," against the other factors (which are kept unchanged) that affect the "subject constant." Second, in *analogies*, existing variations in the climate system factors are used. It is employed when it becomes impractical to rely on controlled experiments, which may be due to high system complexity or ethical constraints. Third, in *models*, numerical simulation is conducted for real-world systems, and the aforesaid two methods (viz., experiments and analogies) are employed for calibration and validation. Besides, the interrelationship between future changes in the climate system and diverse dimensions of vulnerability in societies and ecosystems allows evaluating risks alongside their severity, magnitude, and rate of occurrence. All these aforementioned approaches reflect anticipated future scenarios. Coherent to the aforesaid understanding of the characteristics of key drivers of future climate and associated methods of simulating the same, it is imperative to discuss the findings from recent past researches on the projected climate.

Projected Changes, Risks, and Impacts Under Climate Change

Past studies have identified future climate changes in the diverse climate system components and variables such as air temperature (Orru et al. 2017; Khan and Maity 2022), water cycle (Tapley et al. 2019), ocean (Bronselaer et al. 2018), cryosphere, sea level (IPCC 2019, Naren and Maity 2018), carbon cycle (Mitchard 2018), biogeochemistry (Crowther et al. 2019), and climate system responses (Zscheischler et al. 2018). This section will follow a discussion on findings from IPCC (2014) on global-scale projected changes in the future period 2081–2100 (unless otherwise indicated) by considering 1986–2005 as the base period.

In the case of global mean surface temperature, the change will be likely in the range between 0.3 and 0.7 °C against all four RCPs for the period 2016–2035. In view of global warming, three factors, viz., past anthropogenic emissions, future anthropogenic emissions, and natural climate variability, will warrant the warming scenarios. Figure 2.2 depicts the magnitude of the projected climate change post-2050s which will substantially be influenced by the choice of emissions. For example, considering the targets of the recently organized 26th Conference of Parties (COP26) to the United Nations Framework Convention on Climate Change (UNFCCC), it has been jointly decided upon securing global net-zero emissions by the mid-twenty-first century and not allowing the warming to increase further from 1.5 °C above pre-industrial levels (Arora and Mishra 2021). Such initiatives will very likely positively reduce future emissions, thereby global mean surface temperature in the near future. However, in the distant future (2081–2100), there is high confidence that the global temperature will likely exceed 1.5 °C in RCP4.5, RCP6.0, and RCP8.5 scenarios. In spatial terms, studies have found mean warming to be larger on land than on ocean bodies, and importantly the warming on land will be greater than the global mean warming (Scheffers et al. 2016; Sutton et al. 2007). Therefore, it is virtually certain that the frequent hot and fewer cold temperature

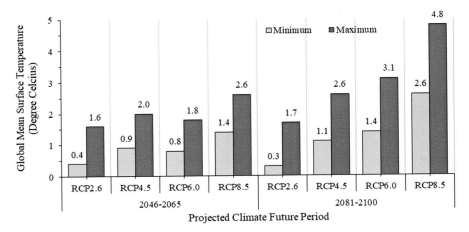

Fig. 2.2 Future changes in global mean surface temperature for the future period 2046–2065 and 2081–2100, relative to the base period 1986–2005 under all RCPs. (Source: Developed by authors after IPCC (2014, 2019))

extremes will increase drastically on land bodies from the near to the distant projected period. This is very likely to trigger the heat waves having more intensity than ever with greater duration coupled with occasional cold winter extremes.

In the case of the water cycle, abrupt changes in the precipitation are likely to influence the water balance. In spatial terms and under RCP8.5 scenarios, the heavy precipitation events are likely to increase in high latitudes and the equatorial Pacific regions. The projections for many midlatitude wet regions are similar. This may become the primary reason for floods in the near future. On the contrary, a decrease in precipitation is likely to observe in most parts of the midlatitude and subtropical dry regions, which may further trigger recurrent drought issues in the near future. In fact, under warming conditions, it is expected that the events of heavy precipitation will very likely become recurrent and intense. In addition, under all RCPs, both monsoon system and precipitation events related to El Niño-Southern Oscillation (ENSO) are very likely to witness intensification that can subsequently result in regional imbalances across the globe (Sarkar and Maity 2021).

In the case of oceans, the surface of subtropical regions in the Northern Hemisphere alongside the tropical regions is likely to witness the strongest ocean warming due to the warming climate. Also, the likely weakening of Atlantic Meridional Overturning Circulation (AMOC) will occur by ~11% in RCP2.6 and 34% in RCP8.5 across the twenty-first century. In the case of the cryosphere, year-round decrement in ice sheet volume is projected with higher and lower confidence in the Arctic Ocean and Antarctica regions, respectively. Furthermore, by 2100, the Northern Hemisphere is likely to witness declining spring snow cover by 7% in RCP2.6 and by 25% in RCP8.5. Besides, amidst rising global mean temperature, it is virtually certain that near-surface permafrost will decrease up to 37% in RCP2.6 and 81% in RCP8.5. As a consequence, the global glacier volume is likely to decrease in the range between 15% and 55% in RCP2.6 and between 35% and 85% in RCP8.5. One of the direct impacts of the weakening of AMOC and declining ice sheet cover and glacier volume is reflected as a rise in the global mean sea level. As compared to the rate of mean sea-level rise of ~2.0 mm/y during 1971–2010, the twenty-first century will very likely experience an increment in the rate under all RCPs (Fig. 2.3). The rate can reach as high as 8–16 mm/y for the period 2081–2100 under RCP8.5. Furthermore, in spatial terms, global mean sea-level rise is expected to be nonuniform. More specifically, the rise will follow the CO_2 emissions pathway which may cause a rise far greater in one region than another. For example, the impact of sea-level rise will likely be more in ocean areas, more precisely along coastlines across the globe.

In the case of the carbon cycle and biogeochemistry, ESM models have predicted, with high confidence, that carbon uptake will increase under all RCPs along the twenty-first century and its feedback with climate change will magnify global warming. In fact, by 2100, surface ocean pH will decrease as a result of increased acidification of oceans, which will be reflected as decreased pH in the range between 0.06 and 0.07 under RCP2.6, 0.14 and 0.15 under RCP4.5, 0.20 and 0.21 under RCP6.0, and 0.30 and 0.32 under RCP8.5. Increased CO_2 uptake with increasing ocean acidity will likely to further result in a decline of the dissolved oxygen content of the ocean, predominantly in the subsurface midlatitude oceans.

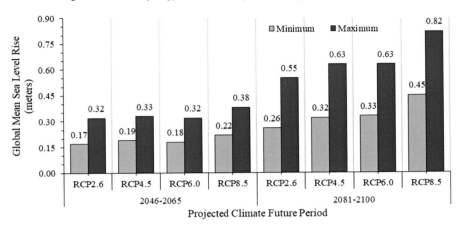

Fig. 2.3 Projected change in global mean sea-level rise for the future period 2046–2065 and 2081–2100, relative to the 1986–2005 base period under all RCPs. (Source: Developed by authors after IPCC (2014, 2019))

In the case of climate system responses, climate models have estimated equilibrium climate sensitivity (ECS) to be likely in the range between 1.5 and 4.5 °C, which will very unlikely be exceeding 6 °C. However, the global mean peak surface temperature will likely be in the range of 0.8–2.5 °C for the change per 1000 GtC (gigatons of carbon) or ~ 3650 $GtCO_2$ emitted as CO_2. It is critical to note that the warming due to CO_2 emissions will become irreversible if adequate measures in terms of adaptation and mitigation for removing excess CO_2 are not enforced. For example, in order to attain warming below 2 °C, the cumulative CO_2 emissions from all anthropogenic sources must be maintained at less than ~3650 $GtCO_2$ (Fig. 2.4). Unfortunately, more than 50% of the maximum limit to emissions is already emitted. Also, limiting total human-induced warming to less than 2 °C by considering 1861–1880 as the base period requires maintaining the anthropogenic and non-anthropogenic emissions between 2550 $GtCO_2$ and 3150 $GtCO_2$.

The projected changes in the climate system provide a basis to understand the interaction between climate hazards (such as extreme climatic events) and the revelation of the human system, ecosystem, and natural systems to climate change, which is perceived as a risk (defined in IPCC (2014)). Based on the aforementioned findings on projected climate, it can be inferred that the rising rates and magnitudes of warming and thereby extreme climatic events will amplify the risk of severe and irreversible detrimental impacts. There are primarily four risks that can significantly trigger climate change and directly affect the human system, ecosystem, and biodiversity, viz., risk of (1) storms, sea-level rise, and floods that are known to severely disrupt livelihoods, thereby resulting in ill-health, (2) extreme weather and climatic events destroying infrastructure-based essential services, (3) insecurity due to food inadequacy and water scarcity resulting in loss of livelihoods and income, and (4) losing ecosystem and biodiversity, thereby associated goods, functions, and services. To combat the total risks of future climate change impacts, there is an urgency to limit the rate and magnitude of climate change (discussed in Sect. 2.4).

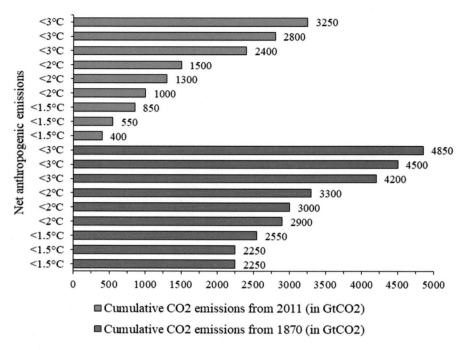

Fig. 2.4 Cumulative carbon dioxide (CO_2) emission in gigaton of CO_2 ($GtCO_2$) corresponding to limiting warming against the specified limits of temperature. (Source: Developed by authors after IPCC (2014, 2019))

Even though adaptation-based measures can markedly reduce the risks of climate change impacts, failure to control risks and impacts increases the likelihood of surpassing adaptation potential. Coherent to the aforesaid reasoning, Table 2.3 enlists region-wise representative risk with the potential of risk reduction through adaptation for the long-term future period of 2080–2100.

Scenarios Beyond Twenty-First Century

Under all the RCP scenarios, studies have predicted continuous warming of the Earth beyond 2100 (except for RCP2.6). Only after reaching net-zero anthropogenic CO_2 emissions, the surface temperatures will likely get nearly constant at elevated levels for many centuries. Except for the condition of a large net removal of CO_2 from the atmosphere over a sustained period, the climate change impact due to anthropogenic emission will, unfortunately, remain irreversible for many centuries. Furthermore, mere stabilization of global mean surface temperature cannot stabilize other facets of the climate system, such as sea-level rise, re-equilibrating soil carbon, etc., because every climate system follows its own intrinsic long timescales of achieving stabilization. For example, ocean acidification will continue beyond the

Table 2.3 Region-wise major risks identification with the likelihood of risk reduction using adaptive pathways; assessed at very low (*VL*), low (*L*), medium (*M*), high (*H*), or very high (*VH*); presented for the long-term future period 2080–2100 and two possible future temperature increased scenarios, viz., 2 °C and 4 °C, above pre-industrial levels

Region	Key risks	Risk level with high adaptation		Risk level with current adaptation	
		2°C	4°C	2°C	4°C
Polar region (Arctic and Antarctic)	Risks for ecosystems	H	VH	H	VH
	Risks for health and well-being	H	H	VH	VH
	Unprecedented challenges, especially from the rate of change	H	H	VH	VH
North America	Increased damages from wildfires	H	VH	VH	VH
	Heat-related human mortality	L	M	H	VH
	Increased damages from the river and coastal urban floods	M	H	H	VH
Central and South America	Reduced water availability and increased flooding and landslide	H	H	VH	VH
	Reduced food production and quality	M	M	VH	VH
Europe	Increased damages from river and coastal floods	VL	L	L	M
	Increased water restrictions	M	M	H	VH
	Increased damages from extreme heat events and wildfires	M	M	H	VH
Africa	Compounded stress on water resources	M	H	H	VH
	Reduced crop productivity and livelihood and food security	M	VH	VH	VH
	Vector- and waterborne diseases	M	M	VH	VH
Asia	Increased flood damage to infrastructure, livelihoods, and settlements	M	H	H	VH
	Heat-related human mortality	H	VH	VH	VH
	Increased drought-related water and food shortage	L	M	M	H
Australasia	The significant change in composition and structure of coral reef systems	H	VH	H	VH
	Increased flood damage to infrastructure and settlements	L	M	M	H
	Increased risks to coastal infrastructure and low-lying ecosystems	L	M	M	H

(continued)

Table 2.3 (continued)

Small Islands	Loss of livelihoods, settlements, infrastructure, ecosystem services, and economic stability	M	H	H	VH
	Risks for low-lying coastal areas	H	H	VH	VH
The Oceans	Distributional shift and reduced fisheries catch potential at low latitudes	L	M	M	H
	Increased mass coral bleaching and mortality	VH	VH	VH	VH
	Coastal inundation and habitat loss	M	H	H	VH

Source: Developed by authors after IPCC (2014, 2019)

twenty-first century even if the emission level stops increasing, and over to that, the impact will exacerbate if global warming continues to increase. However, with virtual certainty, the global mean sea-level rise will continue to increase even after 2100 and is subjected to increase up to 1 m under RCP2.6 and more than 3 m in RCP8.5 (Hoegh-Guldberg et al. 2007; IPCC 2014; Lyon et al. 2022; Meehl et al. 2012, 2013; Nazarenko et al. 2015).

There is high confidence that near-complete loss of the Greenland ice sheet can occur as a consequence of exceeding global warming limits, which can further cause sea level to rise by 7 m (van den Broeke et al. 2017). However, it is virtually certain that there is a reduction in the extent of permafrost by 180–920 $GtCO_2$ in the RCP8.5 scenario due to sustained increment in the global mean temperature (IPCC 2014; Wang et al. 2019). Besides this, the composition, structure, and function of marine, terrestrial, and freshwater ecosystems are likely to witness a high threat of abrupt and irreversible changes from RCP4.5 to RCP8.5 at a regional scale. Nevertheless, due to the limitation of studies being conducted for post-2100 scenarios, there is comparatively lower confidence for the persistence of (already discussed) hazards, beyond the twenty-first century.

Future Adaptation and Mitigation Strategies

The rate and magnitude at which the changes are occurring and likely to occur, along with the exposure and vulnerability of humans, biodiversity, and other natural systems, will decide how the severity of climate change impacts can alter human livelihood their economy and health, ecosystem, food chain and web system, critical infrastructure, and urban, rural, and coastal regulations. To combat climate change impacts, "adaptive responses" that require actions from incremental to transformational changes and "mitigation strategies" that regulate human societies for appropriate use of energy services and land or natural resources can comprehensively provide solutions. In general, adaptation is complementary to mitigation and vice versa (Landauer et al. 2019). Adaptation is a process to adjust with the ongoing or future climate system in such a way that either the possible harm or climate disaster is prevented or possible benefits are exploited, while the processes intended to reduce the emissions or increase the sink for emission (especially the GHGs) define mitigation. This section will discuss the strategic responses to climate change under co-benefits of enhancing adaptative capacity and mitigation shifts in synergy with sustainable development.

Basics of Decision-Making

Contribution to GHGs in the past and current varies from country to country and thereby the associated climate change risks and impacts (Tables 2.2 and 2.3). Given the issues being raised by adaptation and mitigation, in terms of equity, justice, and fairness, their co-benefits provide the necessity to achieve sustainable development and poverty eradication. However, a delay in implementing mitigation shifts and adaptive responses increases burdens from current to future and thus endangers sustainable development (Di Gregorio et al. 2017). Furthermore, the mitigation will not be considered effective in view that the agent (such as individual, community, company, and country) independently advances their private interests by risking global interests. In the larger perspective, it is apparent that GHGs, being emitted by diverse agents, accumulate over time and mix globally. Therefore, international cooperation is required for devising strategies for addressing climate change issues. This demands collective actions and decision-making on a global scale (Sarkodie and Strezov 2019).

Decision-making is driven by valuation and mediation such that normative disciplines and analytical methods may further aid in their drafting. For example, ethics helps in answering political questions on who is responsible for the effects of emissions, while economics help in determining the social cost of carbon, cost-benefit ratio, effectiveness, optimization-related aspects, etc. One of the critical limitations of these analytical methods is their inability to provide the single appropriate combination between "mitigation and adaptation measures" and "risks, impacts, and vulnerabilities of residual climate." This is primarily attributed to the complex social processes and natural phenomena, large disagreement on valuation, and varying distributional effects of adaptation and mitigation approaches. However, certain key inputs are being generated from these methods, such as alternatives to emission pathways, that more or less can positively direct the decision-making process (Frantzeskaki et al. 2019; IPCC 2014; McNamara et al. 2018; Tiernan et al. 2019).

In general, it can be inferred that decision-making is an iterative process, as strategies require adjustment after every new information and associated new understanding. It is worthy to note that mitigation governs the prospects for climate-resilient pathways, which means that delay in their enforcement will leave decision-makers with limited duration and opportunities to exploit advantages of positive synergies between "adaptation and mitigation" and "sustainable development." In addition, the process of perceiving risk and uncertainties varies from individuals to the organization. One example could be overestimating or underestimating climate change risks based on status quo bias. Due to the varying relative importance of such measures and inherent complexities, decision-making is influenced in both short-term and long-term frameworks. Hence, the formal approach, while designing decision-making, should consider uncertainty for risk and incline more attention to both near-time and distant-future consequences. Hence, it becomes imperative to understand the characteristics of mitigation and adaptation (discussed in the following section) individually so as to devise innovative approaches through their integration.

General Aspects of Mitigation and Adaptation Approaches

There is high confidence that without enforcing appropriate mitigation efforts, high to very high risk of climate change impacts, concerning severity and irreversibility, will occur by 2100 such that the presence of the adaptative measures will not be of much significance (IPCC 2014). Although there exist adverse side effects in mitigation shifts, they are not as severe and irreversible as impacts of climate change and can effectively mitigate near-term challenges. Hence, by understanding the individual characteristics of both adaptation and mitigation, a scope is generated to integrate adaptation and mitigation to exploit the resulting co-benefits. Coherently, this section will emphasize general pathways of first, mitigation, followed by adaptation in the context of reducing the risk of ongoing and future climate change impacts.

Mitigation Pathways

The rising population and economic growth are, more or less, the primary factors leading to global emission rise. This rise of emission (in terms of GHGs) is likely to persist in the absence of additional efforts to reduce them beyond the level in ongoing times. Without additional mitigation, the global GHG emission under RCP6.0 to RCP8.5 for projected 2100 may reach between 75 $GtCO_2$-eq/y and 140 $GtCO_2$-eq/y. In the absence of additional mitigation and under all RCPs by 2030, global GHG emission will increase to 450 ppm CO_2-eq, which by 2100, will further exacerbate between 750 ppm CO_2-eq and > 1300 ppm CO_2-eq. Likewise, global mean surface temperature will likely increase between 3.7 and 4.8 °C by 2100, as compared to the mean for the 1850–1900 period (IPCC 2014; Robinson 2020; Sharma and Ravindranath 2019).

In one of the studies by IPCC (2014), ~900 mitigation scenarios were collected to assess possible pathways to long-term climate goals. The study concluded with high confidence (findings are documented in Table 2.4) that by 2100, varying groupings of technological, behavioral, and policy options can maintain emission concentrations between 430 ppm CO_2-eq and > 720 ppm CO_2-eq. The study further added that the global mean warming across the twenty-first century is likely to be maintained below 2 °C (as compared to pre-industrial levels) if the aforesaid scenarios can limit the CO_2-eq concentrations below 450 ppm. Continuing these propositions for 2021, more likely than not, the concentration levels reaching 500 ppm CO_2-eq will cause warming to be less than 2 °C, while if concentration levels exceed 530 ppm CO_2-eq, warming is as likely as not to be less than 2 °C; exceeding the concentration from 650 ppm CO_2-eq is unlikely to limit warming to less than 2 °C; and on the contrary, concentration levels below 430 ppm CO_2-eq can result in limiting the warming below 1.5 °C by 2100. Further details on the projected scenarios, concentration level, change in CO_2-eq emission relative to 2010, and the likelihood of continuing less than specific temperature level over the current century (as compared to the 1850–1900 period) are provided in Table 2.4.

Considering the projections limiting the warming to below 2 °C, there is a need to substantially reduce GHG emissions from anthropogenic sources by 2050. To achieve this, appropriate advancement in the prevailing energy systems and green development of land use applications is required at a global scale. These actions are

Table 2.4 Key characteristics of the RCP scenarios given change in emission and likelihood of maintaining to less than a stated temperature (1.5 °C, 2 °C, 3 °C, and 4 °C) in the twenty-first century

Representative Concentration Pathways (RCPs)	ªCO₂-eq concentration in 2100 (ppm CO₂-eq)	% Change in CO₂-eq. emission relative to 2010				Likelihood of staying below a specific temperature level over the 21ˢᵗ century (relative to 1850–1900)			
		2050		2100		1.5°C	2°C	3°C	4°C
		ªMin	ªMax	Min	Max				
RCP2.6	430 to 480	-72	-41	-118	-78	More unlikely than likely	Likely		
	480 to 530	-57	-42	-107	-73		More likely than not		
		-55	-25	-114	-90		About as likely as not	Likely	
	530 to 580	-47	-19	-81	-59				Likely
		-16	7	-183	-86	Unlikely	More unlikely than likely		
RCP4.5	580 to 650	-38	24	-134	-50				
	650 to 720	-11	17	-54	-21			More likely than not	
RCP6.0	720 to 1000	18	54	-7	72		Unlikely	More unlikely than likely	
RCP8.5	>1000	52	95	74	178			Unlikely	More unlikely than likely

Source: Adapted from IPCC (2014, 2019)
ªppm CO_2-eq, parts per million of carbon dioxide equivalent; Min, minimum; Max, maximum

required to be strategized in such a way that for the greater limiting of warming, large-scale changes should occur less quickly and for the smaller limiting of warming, large-scale changes should occur more quickly. In similar lines, Table 2.4 shows that the reduction in GHG emission from 40% to 70% by 2050 (relative to 2010) for the concentration level of 450 ppm CO_2-eq is likely to limit warming to less than 2 °C and emission to zero-level by 2100. Further, the reduction in GHG emission from 25% to 55% by 2100 (relative to 2010) for the concentration level of 500 ppm CO_2-eq is about as likely as not to limit warming below 2 °C. It can be inferred that emission reduction is more rapid in limiting warming to 2 °C than 3 °C and 4 °C, while limited findings indicate warming more likely than not to be 1.5 °C by 2100. In general, achieving concentration levels <430 ppm CO_2-eq by 2100 will likely result in emission reduction from 70% to 95% (relative to 2010), which is desirable to be targeted globally.

As discussed, over the next few decades, natural and anthropogenic GHG emissions along with agents of climate forcings will certainly be increasing the climate change impacts and their rate and magnitude. Although CO_2-based emissions are primarily driving the long-term warming, controlling the emission from climate forcing agents, which are known to cause short-lived impacts, can drastically reduce short-term warming. Recent studies have found that until 2010, the non-CO_2-based emissions contributed ~27% to the total emissions of Kyoto gases (Gerber et al. 2013; IPCC 2014; Kuleszo et al. 2010). To quickly combat climate change impacts due to such emissions, reductions in climate forcing agents, particularly the ones that are short-lived, are required urgently. Nevertheless, the recent decade is experimenting with near-term, low-cost options for reducing emissions from non-CO_2 gases. However, due to the implications in the radiative properties of non-CO_2 gases [primary sources are livestock (emitting CH_4) and fertilizers (emitting N_2O gases)], their concentrations are difficult to reduce to zero even under stringent mitigation scenarios (IPCC 2014). Since the reduction in emissions from such short-lived climate forcing agents has profound near-term advantages, they have been found to have a fairly immediate impact on climate change. Thus, opportunities related to their reduction should be investigated and promoted globally (Reisinger and Clark 2018; Vermont and De Cara 2010).

Notably, it can be reiterated with evidence that, at a global scale, delays in implementing additional mitigation by 2030 will markedly lessen the likelihood of maintaining the warming less than 2 °C, as compared to pre-industrial levels. IPCC (2014) estimated that under cost-effective scenarios that limit the warming to below 2 °C by 2100, GHG emissions should be maintained between 30 $GtCO_2$-eq/y and 50 $GtCO_2$-eq/y by 2030. However, in the case of GHG emissions exceeding 55 $GtCO_2$-eq/y during 2030–2050, a ~200% higher rate of emission reduction will be required as compared to cost-effective scenarios. This will require urgent scale-up and tripling of net-zero carbon-emitting options relative to 2010, which is likely to exceed the mitigation limits. To achieve this, immediate implementations of mitigation strategies are required such that a common carbon price at a global scale can be fixed and every possible appropriate technology can be made accessible at all spatial scales that are aiming to restrict warming to less than 2 °C by 2100. If this is assumed to be systematically followed in near, mid, and future time, future mitigation scenarios will require reducing global consumption from 1% to 4% by 2030, 2% to 6% by 2050, and 3% to 11% by 2100. Failing to achieve this will increase mitigation costs substantially depending upon (1) the type of technologies, such as CO_2 capture and storage (CCS), nuclear, wind, and solar, as shown in Fig. 2.5a, and (2) further delay in bringing additional mitigation by 2030, as shown in Fig. 2.5b. Also, it can be noticed that the mitigation cost increases from medium-term (2030–2050) to long-term (2050–2100) under immediate mitigation scenarios. In general, considering the ongoing situation, there is an urgency concerning the action to be framed and time to be taken to bring them into practice at a global scale.

Adaptation Pathways

Considering the ongoing and future climate change impacts, adaptation measures have the potential to enhance the well-being of current and future generations along with moderating and maintaining the ecosystem in terms of goods, function, and services. The effective and strategic adaptation measures consider interlinkages between "climate change, sustainable development, and driving socioeconomic processes" and "vulnerability and exposure" (IPCC 2014; Fazey et al. 2016; Magnan et al. 2020). However, it is to be stressed that the adaptation measures have no stringent approach; instead, it is site-specific as well as context-specific. This means that a successful approach at one location, say for a remote village in the desert, may not be suitable at another location, say for a developed city in a frigid zone.

The government at the local, regional, and central levels considers societal values, objectives, and risk perceptions when planning and implementing adaptation measures. In fact, having such information provides frameworks for effective decision-making (Wellstead and Stedman 2015; York et al. 2021). For example, on the one hand, via a top-down approach, the Central Government passes an order on reducing GHG emissions via infrastructure-based construction activities, and on the other hand, via a down-top approach, local indigenous methods of constructing bamboo houses that cut emissions to more than half (when compared with cement), brought to the notice of the Central Government; this becomes an opportunity to scale up a local solution to combat the global issue of rising emission due to conventional cement manufacturing at least at a regional scale. Such initiatives may provide scope to integrate modern and local indigenous knowledge from individual level to government level, thereby increasing the practice of adaptation, decision-making, and policy drafting on a complementary basis.

Sub-graded decision-making in view of poor planning, lack of stringent implementation, and overemphasizing short-term outcomes are very likely to result in maladaptation (Magnan et al. 2016). Such practices amplify the future exposure and vulnerability of individuals, communities, companies, and countries. As a primary step, it is necessarily required to reduce the aforesaid to current climate change impacts by promoting (1) integration of adaptation with decision-making and policymaking and (2) collaborations between developmental agencies and

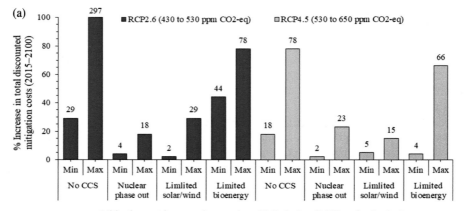

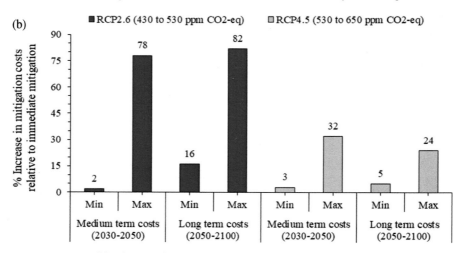

Fig. 2.5 Mitigation costs at a global scale relative to (**a**) specific technologies that are limited in their availability or accessibility, (**b**) cost-effective scenarios under delayed additional mitigation (Abbreviations: *CCS* carbon dioxide capture and storage, *RCP* Representative Concentration Pathway, *ppm CO₂-eq* parts per million of carbon dioxide equivalent; Min, minimum; Max, maximum). (Source: Developed by authors after IPCC (2014, 2019))

organizations on disaster risk reduction. Furthermore, constraints that can severely fail effective adaptation planning and implementation require due consideration. This includes limitations on financial and human resources utilization, limitation on coordination of governance, varying discernments of risks, limited adaptation leaders and advocates, difficulties in monitoring the effectiveness of adaptation, inherent uncertainties about projected climate impacts, and insufficient research, monitoring, and observation for ongoing, twenty-first century, and beyond periods.

In coherence to the effective adaptation design and strategies, it is imperative to emphasize that not controlling rising rates and magnitudes of climate change is very likely to exceed the aforesaid adaptation limits. Such situations attain when certain adaptation measures are either not available at a location or yet not have been devised against intolerable risks. Lack of synergies between adaptation and mitigation may also be aggravated under limiting interaction between climate change and socioeconomic constraints (Sharifi 2021). Therefore, in order to enhance sustainable development (detailed in the next subsection), appropriate transformations in social, technological, economic, and political aspects are essentially required by considering context, decisions, and actions. This can be achieved by developing adaptive capacity and through effective "adaptation options" selection and implementation. Thus, improvement in the exploitation of co-benefits, due to synergies between mitigation and adaptation and adaptation responses, can result in greater security for ecosystems in terms of ecosystem services and carbon storage, a significant reduction in energy and water consumption via greening cities, enhanced sustainability in agriculture and forestry, and improved energy efficiency coupled with greener and cleaner technologies for exploitation (IPCC 2014).

Sustainable Development with Adaptation and Mitigation

Until Sect. 2.3, the deliberations majorly focused on integrating the frameworks of adaptation and mitigation. Since both these approaches are closely related to sustainable development, a need has been felt for establishing synergies between sustainable development plans and adaptation-mitigation approaches. It is required to combat increasing climate change threats to equitable and sustainable development. Furthermore, as it is known that amidst climate change impacts, associated risks and vulnerabilities can aggravate threats to social and natural systems, additional complications can occur especially in the underdeveloped and developing countries, thereby constraining their inclusive development. However, it is imperative to highlight that risks of higher costing to social and environmental developments could be imposed due to some responses to climate change and through adverse distributional effects of adaptive and mitigation-based approaches. To reduce such threats and risks, climate policy is required to be drafted in coherence to sustainable development ideology alongside exploiting complementary advantages of integrated adaptation and mitigation strategies.

Maximizing the co-benefits by improving the interaction of adaptation, mitigation, and sustainable development can result in improved food-air-energy-water nexus (Di Gregorio et al. 2017). It can be observed in terms of sustainable agriculture and forestry, enhanced energy security, optimal energy utilization, reduced water consumption, and improved air quality resulting in protected ecosystems. Strategies' development in lines with aforesaid interaction for ongoing climate change yields climate-resilient pathways. Although integrating adaptation, mitigation, and sustainable development provides climate-resilient pathways, considering the sequence, it is the mitigation-based approaches in the first place that target to lessen the global warming rate and magnitude. However, as a result of mitigation, the time available for adaptation to a particular climate change level significantly increases in the second place. In summary, it can be realized that adaptation alone may not be sufficient to combat climate disasters because delay in implementing mitigation actions likely reduces climate-resilient options for both the near and distant future.

In addition to understanding the aspects of adaptation and mitigated-based approaches, it is imperative to assess constraints and enabling factors that are common for their responses. Also, aspects such as livelihoods, lifestyles, behavior, ethics, and culture of human societies comprehensively influence the adaptative capacity and mitigation potential (IPCC 2014). For example, the effectiveness of climate policies can be influenced by incentivizing at a regional scale by bringing changes in lifestyles or individual choices that, in general, create social acceptability. Furthermore, promoting green innovation and appropriate investments in environmentally sound technologies at local to global scale can drastically reduce GHG emissions, thereby improving climate change resilience and adaptive capacity (Pauleit et al. 2017). It can be inferred that technologies, development, and policies in the direction of improving capacity for adaptation and mitigation should be considered as a foundation to managing climate change risks. This can be achieved by positive coordination and cooperation via institutional-based decentralized governance. Therefore, there is a need for extending this discussion to further understand the response options, given cooperation at all scales and due consideration to societal objectives.

Response Options for Mitigation and Adaptation

Starting with exploring response options for mitigation first, studies have found that in comparison to individual technologies and sectors, mitigation strategies that are appropriately designed and cross-sectoral are comparatively substantial in reducing GHG emissions (England et al. 2018). Due to the characteristics of intersecting with other societal goals, mitigation creates a greater likelihood of both co-benefits and adverse side effects. Nevertheless, if these intersections are handled sustainably, climate action endorsement can be strengthened. For example, as per IPCC (2014), in order to limit the warming within 2 °C above pre-industrial levels, considering the scenario of <450 ppm CO_2-eq concentrations by 2100, there is an urgency to decarbonize electricity and heat generation coupled with reducing energy demand via behavioral changes. While limiting the same warming, considering the scenario of >450 ppm CO_2-eq concentrations by 2100, it is expected that the energy sector will decline by >90% between 2040 and 2070 as compared to 2010 levels. However, for scenarios reaching ~450–~500 ppm CO_2-eq concentration, it is expected that the share of low-carbon electricity supply increases from the current share of

~30% to >80% by 2050, such that by 2100 fossil fuel power generation will phase out in its entirety. Besides cutting down emission-based electricity generation, there are yet other cost-effective and green mitigation options, which include sustainable forest management via afforestation and reforestation (McDowell et al. 2020), cropland management, grazing land management (Smith et al. 2020), and restoration of organic soils (Amelung et al. 2020). Considering the behavioral, lifestyle-related, and cultural aspects of human society, there are ample opportunities to lower down global emissions, especially via bringing changes in consumption patterns of any natural resources and switching to energy-saving measures (Brown et al. 2021; Jorgenson et al. 2019; Vita et al. 2019).

As far as response options for adaptation are concerned, recent studies have found that the adaptive capacity is enhancing globally and within communities across both public and private sectors (Thomas et al. 2019). Along with considering the social value, local and indigenous knowledge base, institutional governance, and biodiversity and ecosystem-based measures, constraints to adaptation and related experience are being considered as the inherent and recognized component of the planning process. Given the virtual certainty that climate change challenges will surge as a consequence of delay in mitigation shift which will eventually be crossing the adaptation limit, context-wise adaptation options require to be documented in terms of their potential to reduce climate change vulnerability and associated disaster risks. This demand for accounting adaptation-based approaches and opportunities for climate change risk management within broader strategic goals in an overlapping manner (Table 2.5).

The synergy between vulnerability reduction, adaptation, and transformation include *human development* in view of improving access to education, nutrition, health facilities, energy, safe housing, and social support structures; *poverty alleviation* via improving access to and control of local and natural resources and social safety framework and protection; *livelihood security* through livelihood diversification alongside income and asset; *disaster risk management* in view of mapping hazard and vulnerability, advancing early warning systems, and developing disaster-resilient infrastructure; *ecosystem management* via wetland green spaces' management, large-scale afforestation, integrated watershed development, and

Table 2.5 Adaptation-based approaches for climate change risk management; figure showing approaches in an overlapping manner, as they are pursued simultaneously

Category	Human development	Poverty alleviation	Livelihood security	Disaster risk management	Ecosystem management	Spatial planning	Structural	Institutional	Social/spheres of change
Vulnerability reduction	9	8	7	6	5	4	3	2	1
Adaptation	2	3	5	6	8	9	7	6	4
Transformation	1	2	3	4	5	6	7	8	9

Note: shades of green; numbers against each category are merely indicating the strength of the shade
Source: Developed by authors after IPCC (2014, 2019)

community-based decentralized natural resource management; *spatial planning* through urban planning and managing development in high-risk areas; *structural-based* adaptation including engineered and built-environment options (e.g., flood and cyclone shelters), technological options (e.g., revisiting traditional knowledge, technologies, and methods), ecosystem-based options (e.g., afforestation and reforestation), and general services (e.g., essential public health services and vaccination programs); *institutional-based* adaptation including economic options, framing laws and regulation, and policymaking and program making at all governmental levels; *social and sphere of changes* achieved in view of educational, informational, and behavioral options. Readers are directed to refer to IPCC (2014) for further information on response options for mitigation and adaptation.

Integrated Response Options

From the evaluation of mitigation, adaptation, and sustainable development, it is now apparent that integrated responses are required to be framed to address linkages between them so as to achieve success while drafting climate policy. This section will discuss the growing evidence base on the integrated responses between adaptation and mitigation in the sustainable development context.

Mitigating and adapting to changing climate eventually cumulate complexities of interaction among human systems, ecosystems, natural resources, and biodiversity. For example, achievements due to societal goals (such goals include sustainable development, clean and green energy access, livelihood generation, food and water security, health and nutrition, and quality environment) can be well supported by mitigation-based approaches, whereas adaptation, in addition, can also support trade-offs along with societal goals while delivering mitigation co-benefits. As discussed, both adaptation and mitigation are complementary; hence, integrated frameworks are required to consider for planning, decision-making, and climate policy development in view of synergizing with sustainable development. A multi-objective approach to policymaking must be adopted to account for these synergies and trade-offs such that appropriate multi-goal policies can be attracted for global support. Furthermore, given the challenges in managing synergies and trade-offs, there is a need for developing a suitable coordination structure across scales and sectors such that appropriate institutions can provide support for achieving effective integrated responses (IPCC 2014; Shukla et al. 2019).

Sectors such as energy, water, food, agriculture, etc. in rural and urban settings can be explored for understanding integrated response options. The potential for complementary benefits of adaptation and mitigation with due consideration to their adverse side effects in the case of energy planning and implementation (Cronin et al. 2018) can be assessed by an integrated approach and by capturing complementarities across multiple climates and social and environmental objectives. As discussed by IPCC (2014), analytical tools such as cost-benefit analysis, cost-effectiveness analysis, multi-criteria analysis, etc. can be useful in determining the correlation of energy policy objectives with social, environmental, and climate objectives. While considering the interaction between water, food, energy (Märker et al. 2018), and biological carbon sequestration (Rosenzweig et al. 2020), the integrated responses in this interaction can yield effective decisions for climate-resilient pathways. For example, to mitigate climate change, one of the options can be large-scale afforestation which can additionally check high-discharge floods amidst maintenance of ecosystem function and services; as a result, alternative water sources such as surface and groundwater storage can be recharged that can improve water use efficiency for agriculture, thereby controlling food inadequacy. On the contrary, increased water consumption for agriculture may also result in the exploitation of surface and groundwater resources, thereby a decline in water availability for other uses (Srivastava et al. 2022). However, in the case of urban and urbanizing settlements (Frantzeskaki et al. 2019; Sharifi 2021), integrated responses can be foundational to generating opportunities for sustainable development by enhancing resilience and reducing anthropogenic GHG emissions. Since it is known that more than 50% of global primary energy use and the associated emissions from CO_2-based energy sources are caused by urban belts (IPCC 2014), there is a need for developing mitigation strategies amidst an increasing population and rising economic activities. Several integrated solutions have been proposed and have been cited from time to time in this study, in view of reducing direct and indirect energy use across sectors, such as mixed-use zoning, transport-oriented development, and co-located jobs and homes. Besides preserving land carbon stocks and land for agriculture and bioenergy, concepts of greening cities and recycling water can provide opportunities to reduce water and energy consumption. In general, developing resilient infrastructure systems and governance structures at all scales can substantially check the vulnerability of any type of settlement vis-à-vis which can monitor and control climate-induced stresses.

Conclusions

This study critically evaluates the confirmation that there is an apparent indication of human influence on the climate system whose impacts are not just limited to the present time but are likely to endanger the future climate as well. As far as real-life observation-based studies are concerned, more or less, it has been concluded that changes since the mid-twentieth century are extraordinary and perilous. In fact, there is a 95% likelihood that there exists a tremendous role of anthropogenic actions behind the rapid surge in global mean surface temperature. Consequently, the risks of adverse, persistent, and irreparable impacts on various components of the climate system are substantially rising given their long-lasting impacts. Studies have comprehensively realized the significance of the synergy and integrated responses between mitigation and adaptation in the sustainable development context that have the potential to markedly stabilize warming to less than 2 °C as compared to pre-industrial levels. However, establishing synergy between them, thereby achieving this goal, will require a pressing departure from the customary practice approach. Coherently, this study has concluded the following:

- Since the pre-industrial time, a rapid rise in anthropogenic greenhouse gas (GHG) emissions has been observed as the dominant cause behind observed global warming; impacts of observed changes are widespread on human systems, ecosystems, biodiversity, and natural resources across all continents and oceans.
- The projected change in foreseeable climate will largely be driven by cumulative emissions of CO_2 such that the changes, risks, and magnitude of climate disasters will intensify across the twenty-first century. Mere reducing GHGs to zero by the twenty-first century will not prevent climate change post-twenty-first century, as each climate system has its own time of achieving stability.
- Despite risks due to a few limitations of mitigation and adaptation, severe, widespread, and irreversible impacts of climate change can be reduced by exploiting co-benefits of mitigation shifts and adaptation measures. However, in a longer-term perspective, mere dependence only on mitigation or adaptation and delaying in bringing mitigation effort will limit their effectiveness.
- Given the context of sustainable development, diverse opportunities can be generated to establish linkages between mitigation, adaptation, and societal objectives through an integrated response approach. The findings can be a great input for decision-making, drafting climate policy, and suitable governance structures for enhancing mitigation shifts and adaptive capacity across all scales.

Acknowledgments The authors are grateful to the contributors of the three working groups – "The Physical Science Basis," "Impacts, Adaptation, and Vulnerability," and "Mitigation of Climate Change" – to the Fifth Assessment Report (AR5) of the Intergovernmental Panel on Climate Change (IPCC). The authors appreciate the Indian Institutes of Technology (IIT) Kharagpur authorities for providing technical infrastructure and a research environment for conducting this review research.

Authors' Contribution All authors contributed to the study's conception and design. AS collected the data and conducted the analysis, while RM and VRD guided the material preparation, data collection, and analysis. The first draft of the manuscript was written by AS, while RM and VRD commented on previous versions of the manuscript. All authors read and approved the final manuscript.

Competing Interest The authors declare that they have no known competing financial interests or personal relationships that could have appeared to influence the work reported in this study.

Funding No funding was received.

References

Allan RP, Barlow M, Byrne MP, Cherchi A, Douville H, Fowler HJ et al (2020) Advances in understanding large-scale responses of the water cycle to climate change. Ann N Y Acad Sci 1472(1):49–75. https://doi.org/10.1111/nyas.14337

Amelung W, Bossio D, de Vries W, Kögel-Knabner I, Lehmann J, Amundson R et al (2020) Towards a global-scale soil climate mitigation strategy. Nat Commun 11(1):1–10. https://doi.org/10.1038/s41467-020-18887-7

Arora NK, Mishra I (2021) COP26: more challenges than achievements. Environ Sustain 4:585–588. https://doi.org/10.1007/s42398-021-00212-7

Beevers MD (2019) Geoengineering: a new and emerging security threat? In: Gueldry M, Gokcek G, Hebron L (eds) Understanding new security threats. Routledge, Milton, pp 32–43. https://doi.org/10.4324/9781315102061

Bronselaer B, Winton M, Griffies SM, Hurlin WJ, Rodgers KB, Sergienko OV et al (2018) Change in future climate due to Antarctic meltwater. Nature 564(7734):53–58. https://doi.org/10.1038/s41586-018-0712-z

Brown MA, Dwivedi P, Mani S, Matisoff D, Mohan JE, Mullen J et al (2021) A framework for localizing global climate solutions and their carbon reduction potential. Proc Natl Acad Sci 118(31). https://doi.org/10.1073/pnas.2100008118

Campbell JE, Berry JA, Seibt U, Smith SJ, Montzka SA, Launois T et al (2017) Large historical growth in global terrestrial gross primary production. Nature 544(7648):84–87. https://doi.org/10.1038/nature22030

Cavicchioli R, Ripple WJ, Timmis KN, Azam F, Bakken LR, Baylis M et al (2019) Scientists' warning to humanity: microorganisms and climate change. Nat Rev Microbiol 17(9):569–586. https://doi.org/10.1038/s41579-019-0222-5

Claessens L, Antle JM, Stoorvogel JJ, Valdivia RO, Thornton PK, Herrero M (2012) A method for evaluating climate change adaptation strategies for small-scale farmers using survey, experimental and modeled data. Agric Syst 111:85–95. https://doi.org/10.1016/j.agsy.2012.05.003

Cronin J, Anandarajah G, Dessens O (2018) Climate change impacts on the energy system: a review of trends and gaps. Clim Chang 151(2):79–93. https://doi.org/10.1007/s10584-018-2265-4

Crowther TW, Van den Hoogen J, Wan J, Mayes MA, Keiser AD, Mo L et al (2019) The global soil community and its influence on biogeochemistry. Science 365(6455). https://doi.org/10.1126/science.aav0550

Di Gregorio M, Nurrochmat DR, Paavola J, Sari IM, Fatorelli L, Pramova E et al (2017) Climate policy integration in the land use sector: mitigation, adaptation and sustainable development linkages. Environ Sci Pol 67:35–43. https://doi.org/10.1016/j.envsci.2016.11.004

Dijkstra FA, Prior SA, Runion GB, Torbert HA, Tian H, Lu C, Venterea RT (2012) Effects of elevated carbon dioxide and increased temperature on methane and nitrous oxide fluxes: evidence from field experiments. Front Ecol Environ 10(10):520–527. https://doi.org/10.1890/120059

Durack PJ (2015) Ocean salinity and the global water cycle. Oceanography 28(1):20–31. https://www.jstor.org/stable/24861839

Elbeltagi A, Raza A, Hu Y, Al-Ansari N, Kushwaha NL et al (2022) Data intelligence and hybrid metaheuristic algorithms-based estimation of reference evapotranspiration. Appl Water Sci 12(7):1–18. https://doi.org/10.1007/s13201-022-01667-7

England MI, Dougill AJ, Stringer LC, Vincent KE, Pardoe J, Kalaba FK et al (2018) Climate change adaptation and cross-sectoral policy coherence in southern Africa. Reg Environ Chang 18(7):2059–2071. https://doi.org/10.1007/s10113-018-1283-0

Eyhorn F, Muller A, Reganold JP, Frison E, Herren HR, Luttikholt L et al (2019) Sustainability in global agriculture driven by organic farming. Nat Sustain 2(4):253–255. https://doi.org/10.1038/s41893-019-0266-6

Farquharson D, Jaramillo P, Schivley G, Klima K, Carlson D, Samaras C (2017) Beyond global warming potential: a comparative application of climate impact metrics for the life cycle assessment of coal and natural gas based electricity. J Ind Ecol 21(4):857–873. https://doi.org/10.1111/jiec.12475

Fazey I, Wise RM, Lyon C, Câmpeanu C, Moug P, Davies TE (2016) Past and future adaptation pathways. Clim Dev 8(1):26–44. https://doi.org/10.1080/17565529.2014.989192

Fellmann T, Witzke P, Weiss F, Van Doorslaer B, Drabik D, Huck I et al (2018) Major challenges of integrating agriculture into climate change mitigation policy frameworks. Mitig Adapt Strateg Glob Chang 23(3):451–468. https://doi.org/10.1007/s11027-017-9743-2

Frantzeskaki N, McPhearson T, Collier MJ, Kendal D, Bulkeley H, Dumitru A et al (2019) Nature-based solutions for urban climate change adaptation: linking science, policy, and practice communities for evidence-based decision-making. Bioscience 69(6):455–466. https://doi.org/10.1093/biosci/biz042

Gattuso JP, Magnan AK, Bopp L, Cheung WW, Duarte CM, Hinkel J et al (2018) Ocean solutions to address climate change and its effects on marine ecosystems. Front Mar Sci 5:337. https://doi.org/10.3389/fmars.2018.00337

Gerber PJ, Henderson B, Makkar HP (2013) Mitigation of greenhouse gas emissions in livestock production: a review of technical options for non-CO2 emissions. Food and Agriculture Organization of the United Nations (FAO), Rome

Gregory JM, White NJ, Church JA, Bierkens MF, Box JE, Van den Broeke MR et al (2013) Twentieth-century global-mean sea level rise: is the whole greater than the sum of the parts? J Clim 26(13):4476–4499. https://doi.org/10.1175/JCLI-D-12-00319.1

Hoegh-Guldberg O, Mumby PJ, Hooten AJ, Steneck RS, Greenfield P, Gomez E et al (2007) Coral reefs under rapid climate change and ocean acidification. Science 318(5857):1737–1742. https://doi.org/10.1126/science.1152509

Hönisch B, Ridgwell A, Schmidt DN, Thomas E, Gibbs SJ, Sluijs A et al (2012) The geological record of ocean acidification. Science 335(6072):1058–1063. https://doi.org/10.1126/science.1208277

IPCC (2014) Climate change 2014: synthesis report. Contribution of working groups I, II, and III to the fifth assessment report of the Intergovernmental Panel on Climate Change. IPCC, Geneva, p 151

IPCC (2019) The ocean and cryosphere in a changing climate. A special report of the Intergovernmental Panel on Climate Change

Johnson GC, Lyman JM (2020) Warming trends increasingly dominate global ocean. Nat Clim Chang 10(8):757–761. https://doi.org/10.1038/s41558-020-0822-0

Jorgenson AK, Fiske S, Hubacek K, Li J, McGovern T, Rick T et al (2019) Social science perspectives on drivers of and responses to global climate change. Wiley Interdiscip Rev Clim Chang 10(1):e554. https://doi.org/10.1002/wcc.554

Khan MI, Maity R (2022) Hybrid deep learning approach for multi-step-ahead prediction for daily maximum temperature and heatwaves. Theor Appl Climatol 149(3–4):945–963. https://doi.org/10.1007/s00704-022-04103-7

Kriegler E, O'Neill BC, Hallegatte S, Kram T, Lempert RJ, Moss RH, Wilbanks T (2012) The need for and use of socio-economic scenarios for climate change analysis: a new approach based on shared socio-economic pathways. Glob Environ Chang 22(4):807–822. https://doi.org/10.1016/j.gloenvcha.2012.05.005

Kriegler E, Edmonds J, Hallegatte S, Ebi KL, Kram T, Riahi K et al (2014) A new scenario framework for climate change research: the concept of shared climate policy assumptions. Clim Chang 122(3):401–414. https://doi.org/10.1007/s10584-013-0971-5

Kuleszo J, Kroeze C, Post J, Fekete BM (2010) The potential of blue energy for reducing emissions of CO2 and non-CO2 greenhouse gases. J Integr Environ Sci 7(S1):89–96. https://doi.org/10.1080/19438151003680850

Kumar P, Vishwakarma DK, Markuna S, Ali R, Kumar D, Jadhav N et al (2022) Evaluation of Catboost method for predicting weekly pan evaporation: case study of subtropical and subhumid regions of India. PREPRINT (version 1) available at Research Square. https://doi.org/10.21203/rs.3.rs-1538970/v1

Landauer M, Juhola S, Klein J (2019) The role of scale in integrating climate change adaptation and mitigation in cities. J Environ Plan Manag 62(5):741–765. https://doi.org/10.1080/09640568.2018.1430022

Lyon C, Saupe EE, Smith CJ, Hill DJ, Beckerman AP, Stringer LC et al (2022) Climate change research and action must look beyond 2100. Glob Chang Biol 28(2):349–361. https://doi.org/10.1111/gcb.15871

Magnan AK, Schipper ELF, Duvat VKE (2020) Frontiers in climate change adaptation science: advancing guidelines to design adaptation pathways. Curr Clim Change Rep 6:166–177. https://doi.org/10.1007/s40641-020-00166-8

Magnan AK, Schipper ELF, Burkett M, Bharwani S, Burton I, Eriksen S et al (2016) Addressing the risk of maladaptation to climate change. Wiley Interdiscip Rev Clim Chang 7(5):646–665. https://doi.org/10.1002/wcc.409

Märker C, Venghaus S, Hake JF (2018) Integrated governance for the food–energy–water nexus–the scope of action for institutional change. Renew Sust Energ Rev 97:290–300. https://doi.org/10.1016/j.rser.2018.08.020

McDowell NG, Allen CD, Anderson-Teixeira K, Aukema BH, Bond-Lamberty B, Chini L et al (2020) Pervasive shifts in forest dynamics in a changing world. Science 368(6494). https://doi.org/10.1126/science.aaz9463

McNamara KE, Bronen R, Fernando N, Klepp S (2018) The complex decision-making of climate-induced relocation: adaptation and loss and damage. Clim Pol 18(1):111–117. https://doi.org/10.1080/14693062.2016.1248886

Meehl GA, Hu A, Tebaldi C, Arblaster JM, Washington WM, Teng H et al (2012) Relative outcomes of climate change mitigation related to global temperature versus sea-level rise. Nat Clim Chang 2(8):576–580. https://doi.org/10.1038/nclimate1529

Meehl GA, Washington WM, Arblaster JM, Hu A, Teng H, Kay JE et al (2013) Climate change projections in CESM1 (CAM5) compared to CCSM4. J Clim 26(17):6287–6308. https://doi.org/10.1175/JCLI-D-12-00572.1

Mitchard ET (2018) The tropical forest carbon cycle and climate change. Nature 559(7715): 527–534. https://doi.org/10.1038/s41586-018-0300-2

Mora C, Spirandelli D, Franklin EC, Lynham J, Kantar MB, Miles W et al (2018) Broad threat to humanity from cumulative climate hazards intensified by greenhouse gas emissions. Nat Clim Chang 8(12):1062–1071. https://doi.org/10.1038/s41558-018-0315-6

Naren A, Maity R (2018) Modeling of local sea level rise and its future projection under climate change using regional information through EOF analysis. Theor Appl Climatol 134(3–4): 1269–1285. https://doi.org/10.1007/s00704-017-2338-8

Nazarenko L, Schmidt GA, Miller RL, Tausnev N, Kelley M, Ruedy R et al (2015) Future climate change under RCP emission scenarios with GISS ModelE2. J Adv Model Earth Syst 7(1): 244–267. https://doi.org/10.1002/2014MS000403

Nolan C, Overpeck JT, Allen JR, Anderson PM, Betancourt JL, Binney HA et al (2018) Past and future global transformation of terrestrial ecosystems under climate change. Science 361(6405): 920–923. https://doi.org/10.1126/science.aan5360

O'Neill BC, Kriegler E, Ebi KL, Kemp-Benedict E, Riahi K, Rothman DS et al (2017) The roads ahead: narratives for shared socioeconomic pathways describing world futures in the 21st century. Glob Environ Chang 42:169–180. https://doi.org/10.1016/j.gloenvcha.2015.01.004

Orru H, Ebi KL, Forsberg B (2017) The interplay of climate change and air pollution on health. Curr Environ Health Rep 4(4):504–513. https://doi.org/10.1007/s40572-017-0168-6

Pauleit S, Zölch T, Hansen R, Randrup TB, Konijnendijk van den Bosch C (2017) Nature-based solutions and climate change – four shades of green. In: Kabisch N, Korn H, Stadler J, Bonn A (eds) Nature-based solutions to climate change adaptation in urban areas. Theory and practice of urban sustainability transitions. Springer, Cham. https://doi.org/10.1007/978-3-319-56091-5_3

Pedersen JST, Santos FD, van Vuuren D, Gupta J, Coelho RE, Aparício BA, Swart R (2021) An assessment of the performance of scenarios against historical global emissions for IPCC reports. Glob Environ Chang 66:102199. https://doi.org/10.1016/j.gloenvcha.2020.102199

Raimi KT, Stern PC, Maki A (2017) The promise and limitations of using analogies to improve decision-relevant understanding of climate change. PLoS One 12(1):e0171130. https://doi.org/10.1371/journal.pone.0171130

Reichstein M, Riede F, Frank D (2021) More floods, fires and cyclones—plan for domino effects on sustainability goals. Nature 592:347–349. https://doi.org/10.1038/d41586-021-00927-x

Reisinger A, Clark H (2018) How much do direct livestock emissions actually contribute to global warming? Glob Chang Biol 24(4):1749–1761. https://doi.org/10.1111/gcb.13975

Riebesell U, Gattuso JP (2015) Lessons learned from ocean acidification research. Nat Clim Chang 5(1):12–14. https://doi.org/10.1038/nclimate2456

Robinson SA (2020) Climate change adaptation in SIDS: a systematic review of the literature pre and post the IPCC Fifth Assessment Report. Wiley Interdiscip Rev Clim Chang 11(4):e653. https://doi.org/10.1002/wcc.653

Rohde RA, Hausfather Z (2020) The Berkeley Earth land/ocean temperature record. Earth Syst Sci Data 12(4):3469–3479. https://doi.org/10.5194/essd-12-3469-2020

Rosenzweig C, Mbow C, Barioni LG, Benton TG, Herrero M, Krishnapillai M et al (2020) Climate change responses benefit from a global food system approach. Nat Food 1(2):94–97. https://doi.org/10.1038/s43016-020-0031-z

Sarkar S, Maity R (2021) Global climate shift in 1970s causes a significant worldwide increase in precipitation extremes. Sci Rep 11(1):11574. https://doi.org/10.1038/s41598-021-90854-8

Sarkodie SA, Strezov V (2019) Economic, social and governance adaptation readiness for mitigation of climate change vulnerability: evidence from 192 countries. Sci Total Environ 656:150–164. https://doi.org/10.1016/j.scitotenv.2018.11.349

Scheffers BR, De Meester L, Bridge TC, Hoffmann AA, Pandolfi JM, Corlett RT et al (2016) The broad footprint of climate change from genes to biomes to people. Science 354(6313). https://doi.org/10.1126/science.aaf7671

Seneviratne SI, Zhang X, Adnan M, Badi W, Dereczynski C, Luca AD et al (2021) Weather and climate extreme events in a changing climate. In: Masson-Delmotte V, Zhai P, Pirani A, Connors SL, Péan C, Berger S, Caud N, Chen Y, Goldfarb L, Gomis MI, Huang M, Leitzell K, Lonnoy E, Matthews JBR, Maycock TK, Waterfield T, Yelekçi O, Yu R, Zhou B (eds) Climate Change 2021: the Physical Science Basis. Contribution of Working Group I to the Sixth Assessment Report of the Intergovernmental Panel on Climate Change. Cambridge University Press. In press. https://www.ipcc.ch/report/ar6/wg1/downloads/report/IPCC_AR6_WGI_Chapter_11.pdf

Sharifi A (2021) Co-benefits and synergies between urban climate change mitigation and adaptation measures: a literature review. Sci Total Environ 750:141642. https://doi.org/10.1016/j.scitotenv.2020.141642

Sharma J, Ravindranath NH (2019) Applying IPCC 2014 framework for hazard-specific vulnerability assessment under climate change. Environ Res Commun 1(5):051004. https://doi.org/10.1088/2515-7620/ab24ed

Shukla PR, Skea J, Calvo Buendia E, Masson-Delmotte V, Pörtner HO, Roberts DC et al (2019) IPCC, 2019: climate change and land: an IPCC special report on climate change, desertification, land degradation, sustainable land management, food security, and greenhouse gas fluxes in terrestrial ecosystems. A special report of the Intergovernmental Panel on Climate Change

Smith TM, Reynolds RW, Peterson TC, Lawrimore J (2008) Improvements to NOAA's historical merged land–ocean surface temperature analysis (1880–2006). J Clim 21(10):2283–2296. https://doi.org/10.1175/2007JCLI2100.1

Smith P, Calvin K, Nkem J, Campbell D, Cherubini F, Grassi G et al (2020) Which practices co-deliver food security, climate change mitigation and adaptation, and combat land degradation and desertification? Glob Chang Biol 26(3):1532–1575. https://doi.org/10.1111/gcb.14878

Srivastava A, Jain S, Maity R, Desai VR (2022) Demystifying artificial intelligence amidst sustainable agricultural water management. In: Zakwan M, Wahid A, Niazkar M, Chatterjee U (eds) Water resource Modeling and computational technologies. Current directions in water scarcity research, vol 7. Elsevier. https://doi.org/10.1016/B978-0-323-91910-4.00002-9

Sun Q, Miao C, Duan Q, Ashouri H, Sorooshian S, Hsu KL (2018) A review of global precipitation data sets: data sources, estimation, and intercomparisons. Rev Geophys 56(1):79–107. https://doi.org/10.1002/2017RG000574

Sutton RT, Dong B, Gregory JM (2007) Land/sea warming ratio in response to climate change: IPCC AR4 model results and comparison with observations. Geophys Res Lett 34(2). https://doi.org/10.1029/2006GL028164

Tapley BD, Watkins MM, Flechtner F, Reigber C, Bettadpur S, Rodell M et al (2019) Contributions of GRACE to understanding climate change. Nat Clim Chang 9(5):358–369. https://doi.org/10.1038/s41558-019-0456-2

Thomas K, Hardy RD, Lazrus H, Mendez M, Orlove B, Rivera-Collazo I et al (2019) Explaining differential vulnerability to climate change: a social science review. Wiley Interdiscip Rev Clim Chang 10(2):e565. https://doi.org/10.1002/wcc.565

Tian H, Chen G, Lu C, Xu X, Ren W, Zhang B et al (2015) Global methane and nitrous oxide emissions from terrestrial ecosystems due to multiple environmental changes. Ecosyst Health Sustain 1(1):1–20. https://doi.org/10.1890/EHS14-0015.1

Tiernan A, Drennan L, Nalau J, Onyango E, Morrissey L, Mackey B (2019) A review of themes in disaster resilience literature and international practice since 2012. Policy Design Pract 2(1): 53–74. https://doi.org/10.1080/25741292.2018.1507240

van den Broeke M, Box J, Fettweis X, Hanna E, Noël B, Tedesco M et al (2017) Greenland ice sheet surface mass loss: recent developments in observation and modeling. Curr Clim Chang Rep 3(4):345–356. https://doi.org/10.1007/s40641-017-0084-8

van Vuuren DP, Carter TR (2014) Climate and socio-economic scenarios for climate change research and assessment: reconciling the new with the old. Clim Chang 122(3):415–429. https://doi.org/10.1007/s10584-013-0974-2

van Vuuren DP, Edmonds J, Kainuma M et al (2011) The representative concentration pathways: an overview. Clim Chang 109:5. https://doi.org/10.1007/s10584-011-0148-z

van Vuuren DP, Riahi K, Moss R, Edmonds J, Thomson A, Nakicenovic N et al (2012) A proposal for a new scenario framework to support research and assessment in different climate research communities. Glob Environ Chang 22(1):21–35. https://doi.org/10.1016/j.gloenvcha.2011.08.002

Vermont B, De Cara S (2010) How costly is mitigation of non-CO2 greenhouse gas emissions from agriculture? A meta-analysis. Ecol Econ 69(7):1373–1386. https://doi.org/10.1016/j.ecolecon.2010.02.020

Vita G, Lundström JR, Hertwich EG, Quist J, Ivanova D, Stadler K, Wood R (2019) The environmental impact of green consumption and sufficiency lifestyles scenarios in Europe: connecting local sustainability visions to global consequences. Ecol Econ 164:106322. https://doi.org/10.1016/j.ecolecon.2019.05.002

Vystavna Y, Matiatos I, Wassenaar LI (2021) Temperature and precipitation effects on the isotopic composition of global precipitation reveal long-term climate dynamics. Sci Rep 11(1):1–9. https://doi.org/10.1038/s41598-021-98094-6

Wang C, Wang Z, Kong Y, Zhang F, Yang K, Zhang T (2019) Most of the northern hemisphere permafrost remains under climate change. Sci Rep 9(1):1–10. https://doi.org/10.1038/s41598-019-39942-4

Watson CS, White NJ, Church JA, King MA, Burgette RJ, Legresy B (2015) Unabated global mean sea-level rise over the satellite altimeter era. Nat Clim Chang 5(6):565–568. https://doi.org/10.1038/nclimate2635

Wellstead A, Stedman R (2015) Mainstreaming and beyond: policy capacity and climate change decision-making. Mich J Sustain 3. https://doi.org/10.3998/mjs.12333712.0003.003

York AM, Otten CD, BurnSilver S, Neuberg SL, Anderies JM (2021) Integrating institutional approaches and decision science to address climate change: a multi-level collective action research agenda. Curr Opin Environ Sustain 52:19–26. https://doi.org/10.1016/j.cosust.2021.06.001

Yuan X, Wood EF, Ma Z (2015) A review on climate-model-based seasonal hydrologic forecasting: physical understanding and system development. Wiley Interdiscip Rev Water 2(5):523–536. https://doi.org/10.1002/wat2.1088

Zscheischler J, Westra S, Van Den Hurk BJ, Seneviratne SI, Ward PJ, Pitman A et al (2018) Future climate risk from compound events. Nat Clim Chang 8(6):469–477. https://doi.org/10.1038/s41558-018-0156-3

Chapter 3
Global Warming Impacts on the Environment in the Last Century

Sankar Mariappan 🄳, Anu David Raj 🄳, Suresh Kumar 🄳, and Uday Chatterjee 🄳

Abstract Global warming is continuing to occur globally as result of fossil fuel burning and no signs of decreasing concentration of greenhouse gases. The environment is highly dependent on the climate of a particular region. Any variation may negatively impact the proper functioning and processes of the ecosystem. Every sector is severely under threat of global warming and its associated impacts. The consequences include reduced agricultural productivity, forest degradation, biodiversity loss, species shift, sea level rise, habitat loss, enhanced land degradation, increased occurrences of cyclones, floods and heat waves, and so on. In recent decades the impact of climate change is observed more frequently. The preparedness against the adverse impact is vital. It indicates the development of a climate-resilient sustainable community for tackling the climate change and associated issues. Policies and proper implementation generated through site-specific impact studies will help us achieve the targets. A case study conducted at various watersheds of Himalayan region showed increase in temperature and rainfall under various IPCC emission scenarios. Hence, the fragile Himalayas necessitate the development of climate-resilient sustainable communities.

Keywords Climate change · Radiative forcing · Ecosystem · Adaptation · Himalayas

---

S. Mariappan (✉)
Indian Institute of Soil and Water Conservation (IISWC), Indian Council of Agricultural Research (ICAR), Dehradun, Uttarakhand, India
e-mail: M.Sankar1@icar.gov.in

A. David Raj · S. Kumar
Agriculture and Soils Department, Indian Institute of Remote Sensing (IIRS), Indian Space Research Organization (ISRO), Dehradun, Uttarakhand, India
e-mail: anudraj@iirs.gov.in; anudraj2@gmail.com; suresh_kumar@iirs.gov.in; sureshkumar100@gmail.com

U. Chatterjee
Department of Geography, Bhatter College, Dantan (Affiliated to Vidyasagar University), Paschim Medinipore, West Bengal, India
e-mail: raj.chatterjee459@gmail.com

Introduction

The temperature of the earth planet is rising continuously and the main cause for this is global warming. Global warming is initiated based on the sun's radiation as it reaches the earth. The atmosphere particles, clouds, sea surface, and ground surface re-emit nearly certain portion of radiation to the space, and the remaining is absorbed by oceans, air, and land. This heats the surface of the planet and atmosphere and makes life comfortable for living beings. After warming up the earth, this thermal radiation and infrared rays passed onto space that favor cooling the earth. But the portion of outgoing radiation is reabsorbed by gases in the atmosphere such as carbon dioxide (CO_2), water vapors, methane, CFCs, ozone, etc. and radiate back to the earth's surface. These gases are called greenhouse gases due to their heat-trapping capacity. Maintaining the ideal temperature at earth reabsorption of radiation by greenhouse gases is necessary; otherwise, the earth would be very cold.

However, the concentration of greenhouse gases caused by human activities increased at an alarming rate for the last two centuries. Up to the end of the twentieth century, about 8 billion tons of CO_2 was released into the atmosphere, and an increased level of greenhouse gases caused by humans enhanced the global warming effect. Over the last 100 years (between 1906 and 2006), the Earth has experienced the highest increase in temperature that is between 0.6 and 0.9 °C per year. Since the industrial evolution of 1750, CO_2 and CH_4 levels raised by 35% and 148%, respectively. For a clear understanding of global warming, there is a need to understand the greenhouse effect. There are two types of the greenhouse effect: the first one is the natural greenhouse gas effect and another one is the human-enhanced greenhouse effect. The natural greenhouse effect normally traps the portion of the heat and favor safe life on Earth, but human-enhanced greenhouse gas effect causes global warming due to fossil fuels which increase the release of greenhouse gases into the atmosphere (CO_2, CH_4, oxides of nitrogen, etc.) (Marc 2015). Gas molecules that absorb thermal infrared radiation, and are in significant enough quantity, can force the climate system are called greenhouse gases. These gases act like a mantle, absorbing infrared radiation and restricting it to escape outer space. The net effect causes the regular heating of the earth's surface. This is called the greenhouse effect, combined with further increases of greenhouse gases resulting in global warming. If it endures unabated, it will result in major climate change.

As a result of this warming, abnormal events such as droughts and floods, cold and heat waves, forest fires, landslides, and other natural disasters regularly have a negative effect on economic growth. Even if not associated with meteorological events, natural catastrophes such as earthquakes, tsunamis, and volcanic activities could alter the chemical constitution of the atmosphere. Because of the forest degradation that naturally blocks rainfall and permits it to be consumed by the soil, rainfall spreads throughout the ground, transporting top soil and causing floods and droughts. In dry years, this removal of trees, ironically, worsens drought by causing the soil to dry up faster. Drought-influenced areas are more probable to be spread out. Heavier rainfall patterns are extremely likely to become more common, posing a greater danger of flooding. In this context, this chapter tries to shed some light on the global warming impact in the last century in various sectors. This chapter is focusing on basic reasons of global warming and how it affects various components and issues of environment (agriculture, soil, land degradation, water resources, forest, biodiversity, air quality, sea and coastal region). Separate sections were provided to each component for general understanding. In addition, future possible climate change projections over Himalayas (Shiwalik, Lesser, and Middle) are provided in this chapter.

Radiative Forcing to Climate Change

The phrase "radiative forcing" is used in IPCC assessments to describe an outwardly inflicted disruption in the Earth climate system's radiative energy budget. The notion was first created in the scope of one-dimensional radiative-convective models, which looked at the equilibrium state global, yearly average surface temperature sensitivities to radiative disturbances generated by variations in radiatively active component levels (Manabe and Wetherald 1967; Ramanathan and Coakley 1978). In the era of climate change, the phrase forcing refers to superficially enforced changes in the surface troposphere system's radiation balance, without a variation in stratospheric kinetics, no surface and tropospheric responses, and no dynamically prompted quantity and dissemination of atmospheric water in diverse states or forms. Because of the long life span and relatively consistent spatial disseminations of well-mixed greenhouse gases like CO_2, CH_4, N_2O, and halocarbons, a few measurements combined with a solid understanding of their radiative characteristics will be enough to produce a relatively precise idea of the radiative forcing (Shine and Forster 1999). Radiative forcing remains an important term, as it provides a simple first-order assessment of the comparative climatic relevance of many drivers. Because the sun is the primary basis of all energy in the climate system, variations in solar output can cause climate change through radiative forcing. Between the minimum and maximum of the 11-year solar cycle, satellite measurements imply a fluctuation in yearly mean total solar irradiance (TSI) of the order of 0.08% (or around 1.1 Wm^2). Global warming potentials (GWPs) are a form of concise index that relies on radiative characteristics which could be utilized to assess the possible future effects of diverse gases on the climate system in a comparative sense, similar to how radiative forcing offers a simplified method of evaluating the multiple aspects that are known to affect the climate system. The IPCC (1994) went into great lengths about how GWPs are calculated, why different time horizons are chosen, and the influence of clouds, scenarios, and several other factors on GWP numbers.

While comparing global and annual mean radiative forcing (1750–2000), it was identified that greenhouse gases such as CO_2, CH_4, N_2O, and halocarbons are the major components of forcing. They contribute +2.43 W/m^2: carbon with a value of 1.46 W/m^2, CH_4 with 0.48, N_2O with 0.15, and halocarbon with 0.34. Apart from this, tropospheric O3 contributes 0.35 W/m^2, and black carbon release due to fossil fuel burning contributes 0.20 W/m². It's vital to note that variations in tropospheric water vapor are treated as a feedback factor instead of a forcing factor. The distinction is less evident in the case of the second indirect aerosol forcing. Anthropogenic emissions (e.g., aircraft, fossil fuels) or antecedents to water vapor are insignificant. The involvement of well-mixed greenhouse gases to global mean radiative forcing has been growing steadily throughout time. These facts explicitly indicate the need for regulating (primarily reducing) the radiative forcing sources because it drives the climate change and associated consequences. As mentioned earlier, carbon dioxide is the main source of radiative processing; hence, during the twenty-first century, the carbon reduction measures, technologies, and frameworks are very much necessitated. Net zero emission, carbon neutrality, REDD+, etc. will help to achieve reduced forcing from greenhouse gases. Apart from this, natural mitigation measures also can be used for effective reduction of carbon. For example, healthy soil can store vast amounts of carbon, if we manage it sustainably. Hence, it can play a major role in mitigation of climate change by carbon sequestration while decreasing the greenhouse gas emission to the atmosphere.

Global Response to 1.5 °C Global Warming

The IPCC reports emphasize that the anthropogenic activities contributed a 1 °C warming of the global atmosphere while comparing it with preindustrial periods. The projections indicate that it will reach to 1.5 °C in the near future (2030–2050) considering the current developmental rate. The increase in temperature may contribute vigorous hydrological cycle and thereby extreme weather events as well as other related disasters. In this context, the Anthropocene is a term that can be associated with the Paris Agreement. It highlights the importance of bolstering the global response to 1.5 °C warming. Many scientists have called for a recognition that perhaps the Earth has created a different geological era: the Anthropocene, based on overwhelming experimental signs of the astonishing degree and global size of anthropological involvement on the Earth system (Crutzen 2002; Waters et al. 2016). Increase in concentration of the greenhouse gases (especially CO_2) links the connection between Anthropocene and global warming. Around 2000, global CO_2 levels have risen at a rate of roughly 20 ppm/decade; this is around ten times considerably faster than any other prolonged CO_2 growth in the preceding 800,000 years (Bereiter et al. 2015). During 1970, the global average temperature has risen at a pace of 1.7 °C per the last hundred years while comparing to a long-term drop of 0.01 °C per century for the previous 7000 years (NOAA 2016). Figure 3.1 portrays the link between increased concentration of CO_2 and the warming temperature observed and projected at a global scale. The projections indicate that the global mean temperature will reach 3.8 °C relative to the increased concentration of CO_2 of 790 ppm during the end of the twenty-first century. Hence, the evidences demand our action for mitigating and adapting the consequences of climate change while projections warn us to prepare for the corollaries.

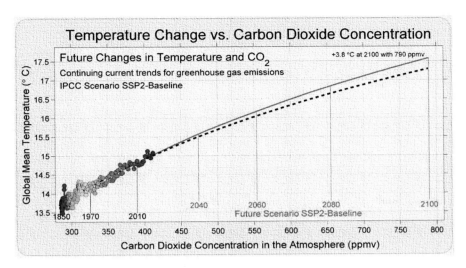

Fig. 3.1 Global CO_2 concentration and global mean temperature. (Available at http://berkeleyearth.org/)

Equality, sustainable development, and alleviation of unstable economy are viewed as simultaneously supporting and attainable well within the face of climate change action and are supported by a variety of additional global firm and delicate law frameworks (as per AR5). The UNDP currently utilizes a Multifaceted Poverty Index to determine that 1.5 billion people are living in multifaceted poverty around the world, primarily in rural parts of South Asia and Sub-Saharan Africa, with another billion at danger of sliding into poverty (UNDP 2016). Restricting global warming to 1.5 °C would necessitate significant communal and scientific advancement changes, which would be contingent on worldwide and local sustainable development routes. The consolidation of instant growth co-assistances from mitigation and adaptation to resolve daily development inadequacies like organizations, market dynamics, and democratic processes (Pelling et al. 2018) which could improve the competence of communities of critical components at threat from 1.5 °C climate change effects to water, fuel, food, ecosystems, cities, local, and coastal systems. The question is whether connecting 1.5 °C paths with the Sustainable Development Goals (SDGs) might well garner reinforcement for a rapid transformation and new development cycle (Stern 2015). Because without achieving the SDG on metropolises and sustainable urbanization in emerging economies (Revi 2016) or reforming the global economic transitional system, it is challenging to envisage how a 1.5 °C world might be realized. In order for the world community to acclimate to 1.5 °C dependable paths, the water management is very much required. To increase water availability, utilization, and efficacy, as well as to assure hydrologic sustainability, comprehensive management involving a wide range of stakeholders would be essential. Controlling global warming to 1.5 °C over pre-industrial stages would necessitate radical progressive change combined with long-term growth. Upscaling and accelerating the deployment of further, multilevel, and cross-sectoral climate mitigation and barrier removal would be required for such transformation. The transition to a low-carbon energy system would have been forced to maintain global warming to a minimum. Electrification, hydrogen, bio-based bio-fuels, and replacements and in some other situations, carbon dioxide acquisition, use, and storage would all help to achieve the severe emissions reductions needed in energy-exhaustive businesses to keep global warming under control. Switching in global and regional land use and ecosystems, as well as accompanying behavioral changes, are needed to keep global warming below 1.5 °C, which can help future adaptation and land-based farming and forestry mitigation capability. Modifying agricultural techniques can be an efficient climate adaptation approach, but such changes may have ramifications for people's income that rely on agriculture and natural capitals. Substantial and comprehensive transitions, especially in urban and rural regions are required for minimizing warming. It can be enabled through a combination of mitigation and adaptation strategies conducted in a participative and coherent way. To make this investment possible, a variety of policy instruments must be mobilized and better integrated. If the conversion to a 1.5 °C world turn out to be an actuality, knowledge gaps regarding executing and bolstering the global reaction to climate change must be addressed immediately (IPCC 2018).

Global Warming and the Environment

This section primarily assembles the IPCC Fifth Assessment report by evaluating new scientific data of climate system changes and their implications on social-ecological systems, with a particular focus on the scale and nature of risks resulting from climate change of 1.5 °C above pre-industrial averages. Figure 3.2 presents the linkage of global warming and the components and process in the environment. It provides more perception about the global warming consequences and responses against it. Global warming is experienced the around globe. Various global datasets indicate the increased temperature trend from 1850 to the present time period (Fig. 3.3). It is also indicated that year by year the magnitude of the warming trend is also rising. Hence, the changes occurring to the environment need to be monitored and addressed in proper seriousness. Consequences of climate change are caused by a variety of environmental factors, including growing atmospheric CO_2, altering precipitation patterns (Lee et al. 2018), sea level rise, escalating ocean acidity, and extreme occurrences including floods, droughts, and heat waves, in line with rising temperatures (IPCC 2014). Growing concentrations of greenhouse gases caused changes in plant productivity but increased ocean acidification, which has a number of consequences (Forkel et al. 2016; Hoegh-Guldberg et al. 2007). The positive effects of the global warming will be diminished by its own negative effects.

With 1.5 °C of global warming, exposed to various and cumulative climate-related risks is expected to grow, with greater amounts of people simultaneously displayed and prone to poverty in Africa and Asia. A 1.5 °C increase in global

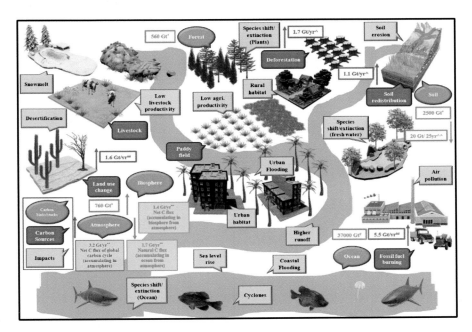

Fig. 3.2 Global warming (carbon source and sink) and the environment (# Melieres and Marechal, 2015; * Lal, (2004); ^ Lal, (2007); ^^FAO Soil portal ## Courtesy of UNEP/GRID-Arendal; ** Matthews et al. (2003)

Melieres, M. and Marechal, C. "Warming in the 20th century," in Climate Change: Past, Present and Future 1st ed., U.K.: Wiley, 2015, ch.29, sec.1, pp. 298–301. (Ocean Carbon Reservoir)
Courtesy of UNEP/GRID-Arendal
* Lal, R. (2004). Soil carbon sequestration impacts on global climate change and food security. science, 304(5677), 1623–1627. (Atmospheric and Biosphere, soil carbon stock)
** Matthews, R. W., Broadmeadow, M. S. J., Mackie, E., Benham, M. W. S., & Harris, K. (2003). Survey methods for Kyoto Protocol monitoring and verification of UK forest carbon stocks. UK Emissions by Sources and Removals by Sinks due to Land Use, Land Use Change and Forestry Activities
^ Lal, R. (2007). Carbon management in agricultural soils. Mitigation and adaptation strategies for global change, 12(2), 303–322
^^ FAO, FAO Soil portal. Available at https://www.fao.org/soils-portal/soil-management/soil-carbon-sequestration/en/#:~:text=It%20is%20estimated%20that%20soils,the%20enhancement%20of%20bio%2Ddiversity. Accessed on [15 February 2022]

warming could result to an increase of the global land area, with large surges in runoff. While global warming is 2 °C, the likelihood of a sea-ice-free Arctic Ocean during the summer is significantly higher than when global warming is 1.5 °C. It indicates a small warming from the normal condition will drive larger adverse impacts in biosphere. In a 1.5 °C warmer world, global mean sea level rise is expected to be roughly 0.1 m inferior by the end of the century than in a 2 °C warmer one. Local species damage and, as a result, extinction threats are much lower in a 1.5 °C warmer world than in a 2 °C warmer world. Under 1.5 °C of warming, the occurrence and severity of floods and droughts are expected to be lower than under

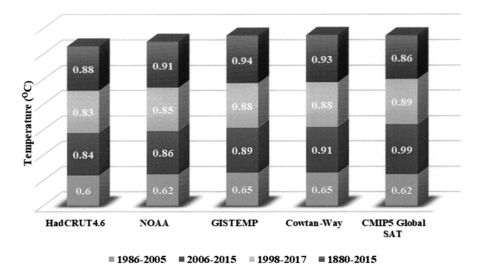

Fig. 3.3 Detected surge in global average surface temperature as observed. HadCRUT4.6, Hadley Centre/Climatic Research Unit Temperature; NOAAGlobalTemp, National Oceanic and Atmospheric Administration Global Temperature; GISTEMP, Goddard Institute for Space Studies Surface Temperature Analysis; Cowtan-Way, Global temperature reconstructions version 2; CMIP5 global SAT, Coupled Model Intercomparison Project phase 5 surface air temperature

2 °C of warming in some regions. Confining global warming to 1.5 °C rather than 2 °C is expected to develop in slighter net decrease in maize, rice, wheat, and possibly other cereal crop yields, especially in Sub-Saharan Africa, Southeast Asia, and Central and South America. Hence, the adverse effect of the global warming has the potential to alter our food security. The global warming responses indicated that climate change is a multifaceted phenomenon which has enormous capability to affect various components of the environment (both natural and artificial). Detailed descriptions of various sectors are described below.

Global Warming Impacts on Agriculture

Global warming (i.e., increased greenhouse gas) is predicted to have major impacts on agriculture. Due to global warming, frequency of droughts and flooding increased in many tropical countries and affected several countries' reduction in food production. Therefore, there is more concern about climate change and its direct/indirect effect on the agriculture sector (IPCC 2001; Aggarwal 2003). CO_2 concentration was increased from 280 ppm in 1850 to 379 ppm in 2005; after that, it was rising at the rate of 1.5–1.8 ppm per year. CO_2 concentration is likely to be doubled at the end of the twenty-first century (Keeling et al. 1995). Kimball (1983) and Uprety (2003) found that increased levels of CO_2 concentration result in the increased photosynthetic rate and crop yield. Uprety et al. (2003) found that elevated CO_2 concentration resulted in increased grain yield of paddy. Few studies reported that elevated temperature in the growing season could drop 40% of the crop harvest in major crops at the end of this century (Pathak et al. 2003; Mall et al. 2006). Climate change will have different consequences on farming. It's difficult to predict how climate change will disturb farming; a wide range of consequences are possible. Climate change will have a significant impact on world cultivation, particularly in tropical areas, temperature fluctuations, precipitation patterns, and the consequent rise in CO_2 levels. Crop yield is projected to shift as a result of such climate changes, as well as weather events and variations in pest and disease regimes. Climate change may cause geographic adjustments in the acceptable land areas for the farming of essential major crops. Due to predicted temperature rise, tropical and subtropical countries are more affected compared to temperate regional countries. The effect of environmental change is expected to be higher in India because of extreme dependence on the agriculture sector. Global warming impacts agriculture by a change in availability of water for irrigation and frequency and intensity of seasonal droughts, floods, soil organic matter formation, soil erosion intensity, etc., all together impacting agricultural production and national and global food security (Fig. 3.4).

The greenhouse effect of global warming is expected to impact the hydrological cycle (rainfall pattern, soil moisture, evapotranspiration, etc.) which causes a new challenge for agriculture (Selvaraju 2003). Particularly in countries like India, the agriculture sector plays an important role in the economy (Kumar and Parikh 1998), and around 60% of the population depends on climate-sensitive sectors (agriculture, forest, fisheries, and natural resources) for their subsistence and livelihood. A small change in climate causes a major impact on the agricultural sector and its dependence. Farming sector's climate sensitivity is uncertain due to regional differences in precipitation, temperature, cultivars and farming patterns, soils, and managing approaches. Inter-annual rainfall and temperature fluctuations were far greater than projected changes in temperature and rainfall. However, yield reduction may worsen if climate variability rises as a result of expected climate change. Deforestation as a result of agrarian development and land acquisition was a significant source of carbon emissions in the past. While in situ vegetation is transformed to agricultural land, trees, plants, and dead organic matter are eradicated, and a major amount of the soil carbon is displaced. This activity is responsible for around a third of worldwide CO_2 emissions.

For certain soils in certain climatic zones, variations in soil attributes such as depletion of soil organic matter, loss of soil nutrients, saltwater intrusion, and erosion are expected consequences of climate change. On a worldwide basis, the localized rises and declines linked to climate change are going to occur in significant impact on food productivity during the next century. As a result, in some low-latitude regions, influences on local or regional food supply could result in significant variations in present production. Climate change might cost some industries a lot of money. Furthermore, warming above the levels predicted by existing literature could result in higher costs in terms of overall food supply. Several economic estimates project significant economic losses when temperature rises

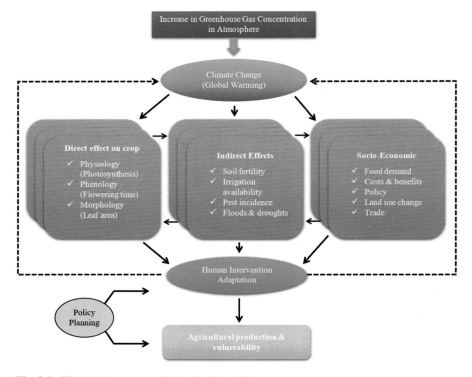

Fig. 3.4 Climate change on agricultural vulnerability

above the equivalent of twice CO_2. This emphasizes the need of determining the extent of heating that may occur as a result of the existing CO_2 increase in the atmosphere and adaptation and mitigation for this concern.

Global Warming Impact on Soil Resources

The IPCC reports show that the global average temperature may rise between 1.1 and 6.4 °C by 2090–2099 compared to the 1980–1999 temperature rise of between 1.8 and 4.0 °C. This rise in temperature causes changes in the climate system and will cause effects in the environment including the soil (Brevik 2012). Soils are the most important natural resources to food security (Lal 2010), and global warming-induced climate change has major potential to threaten food security through its effect on soil properties and processes (Brevik 2013). Climate change and soil conservation can alter soils' ability to fulfil soil functions. Numerous researches have looked at how climate change affects soil functions (Coyle et al. 2016; Xiong et al. 2014). Climate is one of the important factors controlling soil formation and soil properties. Any change in climate affects soil properties such as SOC content, soil biota, moisture, temperature, erosion, and all other physical, chemical, and biological fertility. Soil formation is controlled by many factors but important are climate factors such as temperature and precipitation. The change in climate will alter the energy requirement for weathering minerals and increase mineral tolerance to weathering. It will cause loss of soil function particularly fertility maintenance to create greater dependence on mineral fertilizer (Pareek 2017). Water is essential for chemical weathering, and increased rainfall due to climate change accelerates weathering. The same primary minerals undergo different weathering conditions that result in the production of secondary minerals. Climate change (change in the pattern of rainfall and temperature) influences soil organic matter. This turn affects soil properties mainly aggregate formation and stability, water-holding capacity, cation exchange capacity, soil nutrient content, availability, etc.

Soil functioning can be affected by climate change both directly and indirectly. Variations in organic carbon conversions and nutrient cycling in the soil due to the changing moisture and temperature or higher soil erosion thresholds attributed to the escalated occurrence of extreme intensity rainfall events, which are examples of direct effects. Soils are affected by and cause climate change and also act as a carbon sink and source to the atmosphere (Mariappan et al. 2022). Soils regulate greenhouse gas emission consequently causing global warming and climate change. Those produced by climate change adaptation choices are among the secondary consequences of climate change on soil functioning. Agricultural governance can help to alleviate the consequences of climate change by increasing soil organic carbon (SOC) sequestration, for example (Rumpel et al. 2020). In addition, the significant proportion of the planet's soil resources are in fair, poor, or very bad condition, according to the latest United Nations (UN) assessment on the condition of global soil resources. It emphasizes that soil erosion is a serious agro-ecological concern around the world (FAO, ITPS 2015). The consequences of climatic abnormalities may exacerbate soil concerns such as soil erosion, compaction, degraded soil fertility, and poorer grain production, threatening nutritional safety and environmental sustainability in the long run (Lal et al. 2011). The future function of soils as a viable source for food production is challenged by these climate-related threats.

Excess soil erosion is the major threat that affects the natural ability of the soil functioning. The primary drivers of human-induced soil erosion are tillage, improper agricultural operations, forest destruction, and overgrazing (Amundson et al. 2015). This has a cumulative impact on the planet, resulting in nutrient loss, decreased carbon storing, dwindling biodiversity, and soil and ecological instability (Robinson et al. 2017). The fundamental croplands are both major causes of soil and ecological destruction and a key contributor of biogenic greenhouse gas emissions (Montanarella 2015; Paustian et al. 2016). Soil erosion could be reduced by the use of sustainable management approaches and the implementation of appropriate legislative subsidies. Conservation agriculture (CA) is predicted to span 11–14% of agricultural land globally or 1.42 billion ha (FAOSTAT 2019).

Climate change is a big concern which might cause agricultural productivity to drop in many parts of the world. To control the risks and reap the advantages of climate change, adaptability is critical. If the principal goal is to improve food production, soils and ecosystem services may suffer. Therefore, it is critical to comprehend the conceivable future implications of agricultural adaptation strategies to combat probable soil degradation threats. Climate change, with maximal effects on soils, poses a severe danger to worldwide food security. Transformed precipitation trends can have a big effect on the organic matter and mechanisms in soils and the plants and crops that grow. Agriculture and land management methods must experience major changes in order to solve the connected issues of global food security and climate change. Organic farming, conservation agriculture, agro-ecology, and agroforestry are examples of better agronomic and soil management approaches that build up soil organic carbon (FAO 2015).

Global Warming Impact on Water Resources

For practically all community and financial activity, water is among the most important natural resource. It is mostly used for agronomic operations, but, as a result of socioeconomic and population developments, household water requirement is increasing. Mostly with steep increase of inhabitants and development, global water supplies per capita have decreased considerably. From 13,406.92 m^3 in 1962 to 5724.52 m^3 in 2017, the world's renewable water sources per capita have dropped (World Bank 2022). Mankind, intentionally or not, are influencing the earth's climate and, as a result, the hydrological cycle. The consequences of global warming on freshwater supplies and governance will be observed in the coming years. Due to the decline in rainfall in many parts of the world, global climate change may increase water scarcity (Cramer et al. 2014). Conversely, increased evaporation and transpiration by plants might enhance water requirements as a result of global warming (Alkama et al. 2010). According to a recent study, an increase in air temperature might result in a 3.78 L per capita per day surge in household water requirement (Al-Zubari et al. 2018). Higher evaporation degree from the marine and an upsurge in moisture delivery to the landmass are projected to boost the monsoon as temperatures rise in the Indo-Pacific Oceans.

Climate change is just one of several factors influencing prospective water availability and consumption. Climate change will have an impact on the supply of water but also its purity, delivery, and management systems. All water-related problems are being exacerbated by climate change as follows: floods such as coastal floods, flash floods, riverine floods, and droughts; landslides and erosion, particularly in highlands and hot dry areas under which soils are subjected to aggressive rainfall events; sedimentation and siltation, which minimize groundwater rejuvenation and storage capacity, curtailing dry season irrigation and hydropower development; invasion of saltwater into upstream aquatic environments instigated by water increase and/or decline of river reaches; and intrusion of saltwater into interior aquatic environments. Groundwater is the primary supply of water in numerous aquifers around the world, particularly in arid and semi-arid areas. As a consequence of climate change, early summer recharge reverts at roughly the similar rates from summer to winter in many aquifers around the world, but summer recharge falloffs intensely. Moreover, as the climate warms, the demand for water for particular crops is increasing. Water deficits for such crops are exacerbated by higher evaporation losses related to surface water reservoirs and decreased groundwater recharge related to enhanced siltation in alluvial plains.

Climate change affects the severity, occurrence, and periodicity of these water-related concerns in various ways. A surge in atmospheric moisture as a result of rising temperatures might contribute to higher relative humidity and more clouds, reducing solar radiation and the usable energy for evaporation. Quite extreme and focused rainfall causes stronger and even more violent flash floods, notably in highlands, as well as higher erosion and floodplain siltation, leading to diminished groundwater recharge, and decrease in water storage capacity owing to mutually changed rainfall patterns and siltation of water bodies. A change in the percentage composition of snow and rain, as well as the period of snowmelt and runoff, could almost certainly come from global warming. The rise in temperature is also causing significant impacts. Glacial lake outburst floods (GLOFs) are more likely as snowmelt and ice melt surge, and higher flood peaks in the springtime in snow-fed/ice-fed rivers do not counterpart the greatest requirement for agricultural water in the summer, jeopardizing food security.

Within high latitudes, increasing rainfall will certainly result in much more yearly runoff. On the contrary, a blend of higher evaporation and decreasing rainfall may lead to huge declines in runoff and exacerbated water crisis in certain low latitude basins. Droughts may become extra prevalent in some locations as a result of lower overall rainfall, further recurrent dry spells, and increased evapotranspiration. With increasing air temperature, potential evapotranspiration increases. As a result, even in locations with more rainfall, increased evapotranspiration rates may result in less runoff, thus reducing renewable water supplies. The hydrology of arid and semiarid regions is especially vulnerable to climate change. In certain places, even minor variations in temperature and rainfall can lead to huge variation in runoff, creating a risk and frequency of droughts and floods. If higher precipitation occurs as rain rather than snow and the period of the snow storage season is limited, seasonal interruptions in hilly areas' water supplies may follow. In places where there is less flow to dilute toxins brought from natural and human sources, water quality issues may worsen. Water quality is a serious issue that is being aggravated by climate change; the possibility of pathogenic intrusion increases as temperature and hydrologic extremes rise.

Global Warming and Land Degradation

Human existence is entirely dependent on land resources. Land is a finite resource that serves as a vital physical basis and a safe haven for human growth (Tong et al. 2018). The vegetations, organisms, and water are the major components which entirely depend on the land or soil. Carbon dioxide levels in the atmosphere are growing due to the combustion of fossil fuels and deforestation, causing substantial climate change and having extensive effects for the terrestrial ecosystem (Douville et al. 2000). Over the next century, this warming is predicted to accelerate, potentially causing severe climate changes, particularly in the hydrological cycle (Melillo et al. 1995). In many regions of the world, land degradation is becoming increasingly severe and widespread, with more than 20% of all agricultural areas, 30% of forests, and 10% of grasslands are being degraded (Bai et al. 2008). Every year, millions of hectares of land are degraded across all climate zones. Land degradation and desertification are predicted to harm 2.6 billion people in over a hundred countries, affecting about 33% of the earth's land surface (Adams and Eswaran 2000). The UNCCD, CBD, Kyoto Protocol, and Millennium Development Goals all emphasize this as a global environment, economic, and social issue (UNCED 1992; UNEP 2008). It endangers ecology, food and energy safekeeping, land deterioration, long-term socioeconomic growth, and individual living conditions (Reed et al. 2011; Wieland et al. 2019). Soil is considered as the major component of the land. Soil is valuable in terms of public, environmental, financial, ethnic, and spiritual aspects. In ecosystem services, it also serves as an assistance, provision, governance, and cultural service. Human existence and development are inextricably linked to soil. Tackling global concerns, such as food availability, water management, climate change, and biodiversity, require effective soil management (Xie et al. 2020).

Land and climate coexist in complicated ways across geographical and temporal dimensions related to variation in forcing and several biophysical and biogeochemical feedbacks. The soil organic carbon (SOC) stock could be used as a general indicator of soil deterioration. Because of its pivotal involvement in many soil activities, the amount of SOC may be one of the most relevant soil indicators (Stockmann et al. 2015). Hence, the land degradation can be easily identified with the presence of organic matter in the soil. Apart from this SOC is directly linked to climate change through the carbon dynamics between the soil and atmosphere. Deterioration of land elements, such as soils, rocks, river systems, and plants, can all contribute to land degradation. Land degradation caused by natural and human activities such as loss of organic matter, fall in soil health, erosion, and toxic chemical impacts is a severe global environmental concern that could be exacerbated by climate change (Lal 1997). Land degradation also refers to fluctuations in soil qualities that disturb crop output, infrastructural preservation, and the quality of natural resources. Acidity, alkalization, deterioration of soil nutrients, decline in soil organic matter (SOM), compactness, soil erosion, and biodiversity loss are all possible consequences (Acharya and Kafle 2009). Land degradation is caused by a variety of physical, biological, and chemical processes that are either direct or indirect driven by human activity. Land degradation, especially soil erosion, affects around 60% of the world's land surface, making it one of the most difficult concerns for land managers (Pimentel 2006). Soil erosion, in instance, is a critical issue that concerns people worldwide (Guerra et al. 2017). Increased rates of soil erosion, forest degradation, reduced agronomic productivity, and biodiversity loss have resulted from the demand for increasing production associated with unsustainable land management practices. These kinds of pressures will be compounded as climate change (Millennium Ecosystem Assessment 2005; IPCC 2013).

The erosivity of rainfall-runoff is among the most important climatic elements that determine soil erosion due to water (Gupta and Kumar 2017). Beneath the domain of climate change, rainfall amount and intensity comprised of the runoff erosivity is regarded as the most dominant parameters for soil loss (Teng et al. 2018). The disquieting influence of climate change on the development of soil salinity has previously garnered considerable attention in the scientific community. The increase in atmospheric greenhouse gas (GHG) concentrations, the subsequent rise in air temperature and decrease in relative humidity, and intense rainfall occurrences are likely markers of climate change that seem to have a significant impact on the rate at which soil salinity develops (IPCC 2013). Due to rising sea levels, climate change may hasten saltwater incursion into rich soils, and excessive groundwater extraction in arid places of the world may further exacerbate soil and groundwater salinity (Dasgupta et al. 2015). Apart from this, the desertification in arid region is also a major concern around the globe. Since it is widely agreed that desertification is triggered by a mixture of inadequate land management and climate change (UNCCD 1994), climate change plays a critical role in desertification having a significant impact on local water availability, soil moisture, and vegetation cover variance (Costa and Soares 2012; Sivakumar 2007). Land degradation contributes to climate change by releasing greenhouse gases (GHGs) and carbon absorption rates. Land degradation is a severe and pervasive issue, but there are still many unknowns about its scope, severity, and connections to climate change. To solve the global land concerns of opposing climate change, adaptation, combating land degradation and desertification, and guaranteeing food security, there is a clear need for revolutionary modification in the land managing and food manufacturing sectors (Smith et al. 2020). They deliver a quantitative measure of the potential of several approaches for resolving all of these land concerns owing to land deterioration in the context of climate change on a universal scale.

Global Warming and the Forest Ecosystem

Forests are important components of mountain communal and environmental systems, providing a variety of products and ecosystem services to mankind (Costanza et al. 2007; Pandey and Jha 2012). Forest ecosystems provide a range of ecosystem services that relate to social well-being. Such services include plenty of supplying wood products and freshwater to regulatory services such as flood mitigation and climate stabilization, cultural services such as entertainment and mental amenities, supporting services such as transportation, and other related services (de Groot et al. 2002). Variations in phenological properties, tree line shifting, shifts in forest-type distribution, and forest productivity are all indicators of a forest's vulnerability to climate change (Bertin 2008; Devi et al. 2018; Pan et al. 2015; UAPCC 2014). Due to various concurrent impacts that forests have on human welfare, assessing for anticipated future shifts in forest ecosystem services is challenging, for instance, the concern on how trees might contribute most effectively to climate regulation and mitigation. On the one hand, enhancing forest area and standing volume mitigates anthropogenic climate change by increasing carbon (C) sequestration and storage (Vass and Elofsson 2016). Hence, the forest ecosystems play a vital role in adapting and mitigating the adverse of climate change in diverse pathways. The survival of forest ecosystems is mostly dependent on the development and implantation of tree sprouts, while climate change has the potential to disrupt these processes (Clark et al. 2016). Water availability is a vital determinant of tree seedling sprouting in forests, and climate change threatens this activity (Pérez-Ramos et al. 2013). Warmer temperatures and fewer water availability (rainfall) volumes can increase seed water loss and trigger embryo mortality, limiting seedling emergence (Jot et al. 2013). This is especially important for tree species that generate refractory seeds. Forest ecosystems occupy about 4.1 billion hectares which cover a large portion of the Earth's surface and play a critical responsibility in the global carbon cycle as carbon storage reservoirs (Pan et al. 2011).

Since the Industrial Revolution, global carbon dioxide (CO_2) levels and mean surface temperature have risen, and these patterns may exist in the future (IPCC 2013). On a global scale, these variations are likely to have an impact on the form and function of vegetation in terrestrial ecosystems (Grimm et al. 2013). Forest net primary production (NPP) is highly responsive in CO_2 levels and climate and has received considerable attention in climate change research. As a result, forest NPP can be an important indicator for studying forest ecosystem responses to future climate change (Li et al. 2017; Wang et al. 2018). Dormancy management, bud burst, and early growth may be affected by extended growth periods and a more unpredictable climate (Chmura et al. 2011). Hence, it poses lot of uncertainties with respect to the effect of climate change. Climate change could also cause successional activities, resulting in shifting plant groups (Bolte et al. 2014). The delivery of ecosystem services from forests, such as wood supply, natural disaster prevention, water supply, and biodiversity, may be severely harmed at some point (Jandl et al. 2019). Vicissitudes in climate can have an impact on forest ecosystem disturbance regimes (Seidl et al. 2017). However, perturbations are a normal part of ecosystem evolution, and they can reset potential change trajectories at irregular intervals. Climate change has been connected to abiotic perturbations including the occurrences and magnitude of wildfires and rain events (Abram et al. 2021). Climate change has the potential to affect tree species distributions and forest habitats, resulting in novel and unexpected species mixes (O'Hara 2016). Hickler et al. (2012) found that climatic change has a significant impact on natural vegetation capacity. These climatic and perturbation pattern variations may lead to significant tree death, pushing ecosystems into alternate stable states and consequently diminishing forest health, variety, and even forest land cover (Beisner et al. 2003). Climate change may have an influence on worldwide commercial forestry by reducing yields and affecting market pricing for wood products, lowering the economic worth of forests (Hanewinkel et al. 2013). It is explicit that the climate change as a consequence of rising temperature poses serious threats to forest ecosystems, although they have potential to withstand the adverse effect to a certain extent. Hence, a proper forest management will help increase the tolerance level of forest ecosystems to climate change.

Global Warming and Impact on Biodiversity

Diversity is the one important component of ecosystem, which has a role in coexistence of various species, genus, family, and classes. This existence is sensitive to any perturbations in their habitat. Forest habitats also have a high amount of biodiversity, which is important for ecological processes and resilience to its inherent worth (Thompson et al. 2009). Climate change, which includes warming, fluctuating precipitation, altering extreme event patterns, and altering perturbation regimes, has a significant impact on forests. Even as warming trend continues, the occurrence of negative climate change effects is growing (Steffen et al. 2015). Landscape change is mostly driven by perturbations, and climate change is expected to create more ubiquitous and extreme disruption across a variety of ecosystems (Hicke et al. 2012a, b). Climate change has had a substantial impact on forest ecosystem processes and biodiversity (Sala et al. 2000). Due to anthropocentric influences, environmental destruction and climate change are swiftly contributing to a considerable decline in world species variety (Hooper et al. 2012). Climate change has the potential to cause biome shifts on a worldwide scale, and it is expected to affect roughly half of the world's vegetation by the end of the century (Bergengren et al. 2011). Warming and reduced water supply have resulted in the loss of regional plant species and a decline in species diversity at the taxonomic level (Thuiller et al. 2005). Climate change has an impact on biodiversity because climate factors play a key part in defining species distribution ranges (Pearson and Dawson 2003). As a result, species expand their geographic extents and become extinct in places where the environment is no longer adequate, relying on their distribution abilities (Bellard et al. 2012).

Fortunately, the use of biodiversity to counteract the effects of climate change on ecosystem processes has recently been established (Hisano et al. 2018). Ecosystem functionality and biodiversity are both affected by global climate changes and vice versa. Biodiversity loss can have a direct impact on ecosystem functioning (e.g., carbon stock reductions), hastening global change (Hisano et al. 2018). Biodiversity has the ability to reduce two major aspects of climate change impacts: biodiversity and ecosystem functioning. More diversified systems may be more resilient to climate change impacts (Chapin et al. 2000; Ruiz-Benito et al. 2016). Furthermore, because of the direct link among biodiversity and ecosystem functioning, enhanced biodiversity might reduce the detrimental effects of climate change on ecological processes by enhancing ecosystem performance (Sakschewski et al. 2016). By minimizing consequences of climate change on carbon sequestration, management techniques that rely on our approach will boost the performance of "the climate change abatement capacity of forests" (Canadell and Raupach 2008). Biodiversity could be a technique to offset the adverse financial consequences of climate change on biological processes even while allowing for an effective forest management strategy. Hence, an integrated forest and biodiversity management approach can withstand the adverse effects. It is explicit that nature itself can act as a tool for mitigating the climate change. Henceforth, the act to identifying and utilizing its inherent potential needs to be implemented.

Global Warming and Sea and Coastal Region

Climate change consequences on coastal areas can have a distinct character that is subject to local climate change as well as regional landform, biological, geochemical, environmental, and social elements that influence climate sensitivity. Climate change consequences in coastal regions are likely to be a major concern in this century, with millions of people and trillions of dollars in assets at risk on a worldwide scale (Hanson et al. 2011). Climate change, particularly faster SLR and possibly altered wave climates, is probable to cause the erosion issue (Bray and Hooke 1997). As a result of climate change, coastal habitats may face a number of consequences. A decline in sediment load; differences in the severity, frequency, and strength of storms and cyclones; and fluctuations in sea level and wave climate are only a few of the impacts linked to erosional processes (Mase et al. 2015; Wong et al. 2014). As shown in the Intergovernmental Panel on Climate Change, Fifth Assessment Report, which analyzed global forecasts, the mean sea level might rise around 0.26 m and 0.55 m in an ideal scenario (optimal) and among 0.45 m and 0.82 m in a worst-case scenario in the twenty-first century (IPCC 2013).

Merlis et al. (2016) employed globally uniform SST Aquaplanet computations to illustrate that higher SST resulted in greater severity but reduced number. Water temperature is a fundamental driver of ecosystem functions in aquatic environments, as well as one of the major components controlling aquatic species variation and life histories (Vannote and Sweeney 1980). Climate change is also being connected to shifts in species distribution, phenotypic expression, and development that have been seen and anticipated (Parmesan and Yohe 2003). Watershed carbon cycles rely heavily on aquatic environments, which recycle a significant quantity of carbon from the terrestrial environment (Raymond et al. 2013). However, there are significant gaps in our knowledge of how aquatic ecosystem metabolism will adapt to temperature and hydrologic alterations caused by climate change (Jankowski and Schindler 2019). Apart from this, simulation studies imply that recent surge in the frequency of extremely strong tropical cyclones in the post-monsoon season is likely attributable to manmade factors in the Arabian Sea (Murakami et al. 2017). With rising sea surface temperatures, climate change is expected to exacerbate the strength of tropical cyclonic storms in some ocean basins (Gupta et al. 2019). According to Sun et al.'s (2017) numerical simulation study, tropical cyclones will get faster and larger as the world warms. Hence, the warming temperature is adversely affecting the optimum conditions for aquatic organisms and increasing SST intensifies the hydrological cycle which is increases the exposure of coastal region through the intense and frequent storms and cyclones.

Global Warming and Air Quality/Pollution and Urban regions

Air quality is highly reliant on the weather and thus vulnerable to climate change. By modifying solar and planetary radiation regimes, air pollution and their antecedents can drive the climate system, whose dispersion is in response highly reliant on local climate (Fiore et al. 2012). Latest studies have been utilizing correlations of air quality with meteorological factors and variation analyses using different models governed by general circulation model predictions of twenty-first-century climate change to quantify this climate influence (Jacob and Winner 2009). Ozone and its manmade predecessors, which are vented from sources and distributed in the open troposphere, add a major foundation to surface ozone, which is becoming a growing concern for fulfilling air quality regulatory limits (Holloway et al. 2003). As a result of their interactions involving solar and terrestrial radiation, ozone and PM have been identified as key climatic driving factors (Forster et al. 2007). Because of such a multi-function, climate change's impact on surface air quality is frequently discussed on the basis of chemistry-climate interconnections (Giorgi and Meleux 2007). Indeed, in prevailingly hot climes, such as Egypt, a significant association of high ozone with temperature is a common feature of findings in polluted places (Elminir 2005). In the southwestern United States, Wise and Comrie (2005) show a negative association between PM and relative humidity, indicating the relevance of dust as a PM source in that region. As mentioned by Liao et al. (2006), a general result across models seems to be that the ozone rise from climate change is greatest in metropolitan regions where current ozone levels are already high (Bell et al. 2007; Jacobson 2008; Nolte et al. 2008). Pye et al. (2009) reported that high water vapor in the future climate contributes to improved levels of H_2O_2, the primary SO_2 oxidant, and correspondingly high sulfate concentrations, according to Hong et al. (2019). According to our findings, climate change and more violent extremes are likely to increase the danger of devastating pollution events in China. Aerosols in the atmosphere have a variety of effects on climate (Isaksen et al. 2009). The aerosol straightforward effect is defined as the scattering and absorption of shortwave and longwave radiation by aerosols (Haywood and Boucher 2000).

By 2050, urbanization is anticipated to occur in a population of 6 billion people. Cities will be subjected to climate change as a result of greenhouse gas-induced radiative forcing as well as regional consequences of urbanization, including the urban heat island. Existing data, on the other hand, suggests developments in urban heat islands in certain sites that are comparable to or greater than those caused by greenhouse gas-induced climate change (Stone 2007; Fujibe 2009). Urban form and physical features, urban expanse, waste heat output, and regional climate variables all influence the extent of the urban heat island (Arnfield 2003). As a result of enhanced susceptibility resulting through urbanization process, wealth, and infrastructural facilities to fewer regions, urbanization necessarily enhances flood risk (Ishigaki et al. 2009). Flood hazard is additionally exacerbated by hydrological and hydro-climatological factors ranging from land use changes to microclimatic modifications made on by urbanization (WMO/GWP 2008). Climate change and variability have a major effect on worldwide flood risk (Milly et al. 2008). Climate change-driven increases in rainfall events are partly to blame for the rising prevalence of flood disasters around the world (IPCC 2007). The global water cycle is projected to be accelerated as a result of global warming, leading to higher flood severity and phase in several areas.

Climate Change Projections over Indian Himalayas: A Case Study

We are describing some selected projections of future climate change scenario in various regions of Himalayas. Here we are focusing diverse IPCC scenarios in various future time periods of the twenty-first century (three 30-year periods). A brief description of study area, methodology, and relevant findings are given below.

Locations

Three study area locations were selected from Indian Himalayas: outermost Shiwaliks, Lesser Himalayas, and Mid-Himalayas (Fig. 3.5). The average annual rainfall in the study area (watersheds) varies from approximately 1300 mm (Shiwaliks), 2200 mm (Lesser Himalayas), and 1200 mm (Mid-Himalayas). In Shiwaliks, January is the coldest, with an average temperature of 11.4 °C and the hottest month is June, with an average temperature of 31 °C. The average annual temperature ranges from 15.8 to 33.3 °C in Lesser Himalayas. The average annual maximum and minimum temperatures are 9.9 °C and 19.2 °C, respectively, in Mid-Himalayas. Soils were characterized as dominantly sandy loam texture in Shiwalik and sandy loam to loam in Lesser Himalayas and Mid-Himalayas.

Materials and Methods

There are various sources and methodologies available for predicting future climate change. The projected climate data obtained for Shiwalik Himalayas (David Raj et al. 2022; Raj 2020), Lesser Himalayas (Sooryamol 2020; Sooryamol et al. 2022), and Mid-Himalayas (Gupta and Kumar 2017) from the literature. Here we applied both MarkSim weather generator and Statistical Downscaling Model (SDSM) for

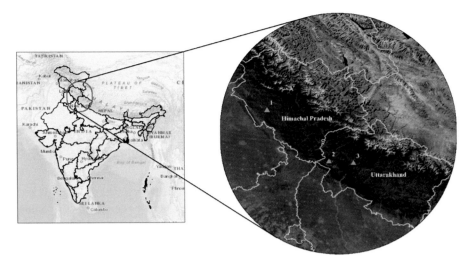

Fig. 3.5 Study locations: Shiwalik (1); Lesser Himalayas (2); Mid-Himalayas (3)

future climate change projections. MarkSim provides downscaled future projections of climate scenario (both temperature and precipitation). We employed SRES A2 and B2 scenarios for SDSM and RCP 4.5 and 8.5 scenarios for MarkSim weather generator to project the future climate change. MarkSim requires the latitude and longitude of the area of interest for the projected climate. SDSM uses observed climate data and predictor variables for predicting the future climate. More details on SDSM are discussed by Wilby et al. (2002) and whether MarkSim is obtainable by Jones and Thornton (2000).

Salient Findings

Climate change and its associated effects are already visible in various regions of Himalayas. Himalayas are very fragile and sensitive to various perturbations due to its geological formation, young and steep slopes, and also unique habitat for various endemic organisms. Hence, the socioeconomic sector is also very fragile and the community livelihood is moreover vulnerable to climate change. Therefore, watershed-based measures are vital in purview of climate change and associated consequences. Agriculture and livestock are the major economic sources of the Himalayan villages. The high land agriculture is already under threat of lower soil quality due to higher erosion from landscapes (Fig. 3.6).

Various IPCC scenario (SRES A2 and B2 and RCP 4.5 and 8.5) projections for the Shiwalik region indicates that the minimum temperature may increase up to 1.2–1.6 °C, 2–3.4 °C, and 2.7–5.4 °C while maximum temperature may surge up to

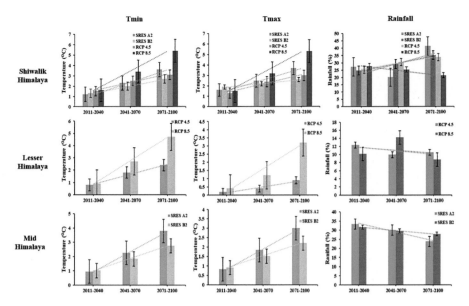

Fig. 3.6 Projected future temperature and rainfall over various regions of the Himalayas (IPCC scenarios)

1.3–1.9 °C, 2.2–3.2 °C, and 2.6–5.3 °C during 2011–2040 (near future), 2041–2070 (mid-future), and 2071–2100/2095 (far future), respectively. Hence, it is explicit that the minimum and maximum temperature over Shiwalik Himalaya is increasing under all IPCC emission scenarios. From near future to far future, it showed an increasing tendency, indicating the need for mitigation and adaptation measures. SRES A2 and RCP 8.5 (worst-case scenario) scenarios exhibited a more increasing tendency in temperature than SRES B2 and RCP 4.5 (intermediate/optimum scenario).

Similar to watershed in Shiwalik region, watershed in Lesser Himalaya also showed the same trend in case of minimum and maximum temperature. The minimum temperature may increase up to 0.8–0.9 °C, 1.8–2.7 °C, and 2.4–4.7 °C, while maximum temperature may surge up to 0.2–0.4 °C, 0.4–1.2 °C, and 0.9–3.2 °C during 2011–2040 (near future), 2041–2070 (mid-future), and 2071–2100/2095 (far future), respectively, under various scenarios (RCP 4.5 and 8.5). The minimum temperature showed a higher increase than the maximum temperature. It indicates an increase in the night-time temperature in the watershed of Lesser Himalayas. Considering the warming trend of mid-Himalayas, an increase was observed as per various emission scenarios (SRES A2 and B2). The minimum temperature may increase up to 0.9–1.0 °C, 1.8–2.3 °C, and 2.7–3.4 °C, while maximum temperature may surge up to 0.8–0.9 °C, 1.5–1.9 °C, and 2.2–3.0 °C during 2011–2040 (near future), 2041–2070 (mid-future), and 2071–2100/2095 (far future), respectively. In Mid-Himalayas both minimum and maximum temperature showed a similar magnitude of increase. While generalizing the warming trend over the various watersheds in Himalayas, all IPCC scenarios projected an increase in average annual temperature. It indicates the need for the climate resilience in the highly sensitive hilly and mountainous terrains of Himalayas. Apart from this, these warming trends are also harmful for the glaciers in Himalayas, which may result in increased melt of glaciers, thereby resulting in higher runoff through Himalayan rivers and their down streams.

The increase in temperature may cause vigorous hydrological cycle. It has the potential to remove the nutrient-rich top soil from the high land agriculture. Thus, it has serious consequences due to increased rainfall occurrence owed to climate change. The future rainfall may increase up to 20–40% during the twenty-first century under various IPCC scenarios (SRES A2 and B2 and RCP 4.5 and 8.5) over Shiwalik Himalayas. In addition, it may surge up to 8–14% during the twenty-first century under various scenarios (RCP 4.5 and 8.5) over lesser Himalayas. In Mid-Himalayas, the rainfall may increase up to 23–33% under SRES A2 and B2 scenarios over the end of the twenty-first century. Consequently, the temperature and rainfall over various watersheds in the Himalayan regions may be severely attributed by climate change impacts. Hence, it demands site-specific management practices and mitigation measures. Acquiring climate resilience through adaptation and mitigations can resist the adverse effects. Thus, the response to the consequence of global warming is required from the administrators especially in fragile Himalayas. Hence, stringent actions over the Himalayas are vital.

Conclusions

The concentration of greenhouse gases is increasing following the industrialization, and in recent decades it is growing rapidly. Consequently, warming of global atmosphere is observed around the globe. It may disturb the normal or optimal habitat conditions for various organisms and human beings. The increased warming has the potential to exacerbate the hydrological cycle and climate dynamics. The IPCC models suggest that the warming trend continues to occur until if any necessary action has not been taken. This warming has the potential to perturb our daily lives in the form of various climate abnormalities. Each segment of the environment is affected directly to the exposure of climate change. It demands the adaptation and mitigation measures globally by emphasizing site-specific measures. Climate change will affect forest, agriculture, water resources, soil, biodiversity, coastal regions and seas, and urban and rural human habitat, and moreover worsen various kinds of land degradation issues. Climate resilience achieved through adaptation and mitigation is the efficient way to tackle this issue. Hence, the global response against climate change is vital. Various international treaties and conferences (Conference of Parties, Convention on Biological Diversity, Convention to Combat Desertification, Reducing Emissions from Deforestation and forest Degradation, etc.) help us to generate a better framework and identify critical issues for fighting against this. However, the implementation and finding uncertainties are the major challenges for the site-specific implementation of various measures. The sustainable way of development will be a key tool which can produce a resilient community against the adverse impact of climate change. Henceforth, site-specific impact studies are required to identify the vulnerable locations and processes in a watershed.

Thus, we performed a case study in various segments of Himalayas: Shiwalik, Lesser and Middle. To identify the future climate change impact, we employed various IPCC emission scenarios over some selected watershed. In both optimal and worst-case scenarios, the temperature (both minimum and maximum) over three watersheds showed an increase during the twenty-first century. In addition, the rainfall amount over the three watersheds showed a rising trend during the twenty-first century. This increase in temperature has the potential to disturb the mountain community in the form of agricultural productivity. The enhanced rainfall also delivers more threat to mountain region and may cause flash floods and removal of nutrient-rich top soil and result in reduced productivity from high land farming. It demands the proper measures to tackle this issue especially in the hilly and mountainous terrain of highly fragile Himalayas.

References

Abram NJ, Henley BJ, Sen Gupta A, Lippmann TJ, Clarke H, Dowdy AJ et al (2021) Connections of climate change and variability to large and extreme forest fires in southeast Australia. Commun Earth Environ 2(1):1–17

Acharya AK, Kafle N (2009) Land degradation issues in Nepal and its management through agroforestry. J Agric Environ 10:133–143

Adams CR, Eswaran H (2000) Global land resources in the context of food and environmental security. In: Gawande SP (ed) Advances in land resources management for the 20th century. New Delhi, Soil Conservation Society of India, pp 35, 655 pp–50

Aggarwal PK (2003) Impact of climate change on Indian agriculture. J Plant Biol 30(2):189–198

Alkama R, Decharme B, Douville E, Becker M, Cazenave A, Sheffield J, Voldoire A, Tyteca S, Lemoigne P (2010) Global evaluation of the ISBA-TRIP continental hydrological system. Part 1: A twofold constraint using GRACE terrestrial water storage estimates and in situ River discharges. J Hydrometeorol 11(3):583–600. https://doi.org/10.1175/2010JHM1211.1

Al-Zubari WK, El-Sadek AA, Al-Aradi MJ, Al-Mahal HA (2018) Impacts of climate change on the municipal water management system in the Kingdom of Bahrain: vulnerability assessment and adaptation options. Clim Risk Manag 20(2006):95–110. https://doi.org/10.1016/j.crm.2018.02.002

Amundson R, Berhe AA, Hopmans JW, Olson C, Sztein AE, Sparks DL (2015) Soil and human security in the 21st century. Science 348(6235):1261071

Arnfield AJ (2003) Two decades of urban climate research: a review of turbulence, exchanges of energy and water, and the urban heat island. Int J Climatol 23:1–26. https://doi.org/10.1002/joc.859

Bai ZG, Dent DL, Olsson L, Schaepman ME (2008) Global assessment of land degradation and improvement 1. Identification by remote sensing. Report 2008/01. ISRIC, Wageningen

Beisner BE, Haydon DT, Cuddington K (2003) Alternative stable states in ecology. Front Ecol Environ 1(7):376–382

Bell ML, Goldberg R, Hogrefe C, Kinney PL, Knowlton K, Lynn B et al (2007) Climate change, ambient ozone, and health in 50 US cities. Clim Chang 82(1):61–76

Bellard C, Bertelsmeier C, Leadley P, Thuiller W, Courchamp F (2012) Impacts of climate change on the future of biodiversity. Ecol Lett 15(4):365–377

Bereiter B, Eggleston S, Schmitt J, Nehrbass-Ahles C, Stocker TF, Fischer H et al (2015) Revision of the EPICA Dome C CO2 record from 800 to 600 kyr before present. Geophys Res Lett 42(2): 542–549. https://doi.org/10.1002/2014gl061957

Bergengren JC, Waliser DE, Yung YL (2011) Ecological sensitivity: a biospheric view of climate change. Clim Chang 107:433–457

Bertin RI (2008) Plant phenology and distribution in relation to recent climate change. J Torrey Bot Soc 135:126–146

Bolte A, Hilbrig L, Grundmann BM, Roloff A (2014) Understory dynamics after disturbance accelerate succession from spruce to beech-dominated forest – the Siggaboda case study. Ann For Sci 71(2):139–147

Bray MJ, Hooke JM (1997) Prediction of soft-cliff retreat with accelerating sea-level rise. J Coast Res:453–467

Brevik EC (2012) Soils and climate change: gas fluxes and soil processes. Soil Horiz 53. https://doi.org/10.2136/sh12-04-0012

Brevik EC (2013) Climate change, soils, and human health. In: Brevik EC, Burgess LC (eds) Soils and human health. CRC Press, Boca Raton, pp 345–383

Canadell JG, Raupach MR (2008) Managing forests for climate change mitigation. Science 320: 1456–1457

Chapin FS III, Zavaleta ES, Eviner VT, Naylor RL, Vitousek PM, Reynolds HL, Hooper DU, Lavorel S, Sala OE, Hobbie SE (2000) Consequences of changing biodiversity. Nature 405: 234–242

Chmura DJ, Anderson PD, Howe GT et al (2011) Forest responses to climate change in the Northwestern United States: ecophysiological foundations for adaptive management. For Ecol Manag 261:1121–1142

Clark JS, Iverson L, Woodall CW, Allen CD, Bell DM, Bragg DC et al (2016) The impacts of increasing drought on forest dynamics, structure, and biodiversity in the United States. Glob Chang Biol 22(7):2329–2352

Costa AC, Soares A (2012) Local spatiotemporal dynamics of a simple aridity index in a region susceptible to desertification. J Arid Environ 87:8–18

Costanza R, Fisher B, Mulder K, Liu S, Christopher T (2007) Biodiversity and ecosystem services: a multi-scale empirical study of the relationship between species richness and net primary production. Ecol Econ 61(2–3):478–491

Coyle C, Creamer RE, Schulte RPO, O'Sullivan L, Jordan P (2016) A functional land management conceptual framework under soil drainage and land use scenarios. Environ Sci Pol 56:39–48. https://doi.org/10.1016/j.envsci.2015.10.012

Cramer W, Yohe G, Auffhammer M, Huggel C, Leemans R (2014) Detection and attribution of observed impacts, impacts, adaptation, and vulnerability. Part A: Global and sectoral aspects. Working Group II contribution to the fifth assessment report of the Intergovernmental Panel on Climate Change. Clim Chang 6(2):13–36. https://doi.org/10.1017/CBO9781107415379.005

Crutzen PJ (2002) Geology of mankind. Nature 415(6867):23. https://doi.org/10.1038/415023a

Dasgupta S, Hossain M, Huq M, Wheeler D (2015) Climate change and soil salinity: the case of coastal Bangladesh. Ambio 44(8):815–826

David Raj A, Kumar S, Sooryamol KR (2022) Modelling climate change impact on soil loss and erosion vulnerability in a watershed of Shiwalik Himalayas. Catena 214:106279

de Groot RS, Wilson MA, Boumans RM (2002) A typology for the classification, description and valuation of ecosystem functions, goods and services. Ecol Econ 41:393–408

Devi RM, Patasaraiya MK, Sinha B, Saran S, Dimri AP, Jaiswal R (2018) Understanding the linkages between climate change and forest. Curr Sci:987–996

Douville H, Planton S, Royer JF, Stephenson DB, Tyteca S, Kergoat L et al (2000) Importance of vegetation feedbacks in doubled-CO2 climate experiments. J Geophys Res-Atmos 105(D11): 14841–14861

Elminir HK (2005) Dependence of urban air pollutants on meteorology. Sci Total Environ 350(1–3):225–237

FAO (2022) Soils help to combat and adapt to climate change by playing a key role in the carbon cycle. https://www.fao.org/documents/card/en/c/39058c2c-991b-4a24-a0bd-ccd0544e8320/#:~:text=Healthy%20soils%20provide%20the%20largest,gas%20emissions%20in%20the%20 atmosphere. Accessed 4 Mar 2022

FAO, ITPS (2015) The status of the world's soil resources (Main report). FAO

FAOSTAT (2019) Crops. Crop. Natl. Prod. Food and Agriculture Organization of the United Nations. http://www.fao.org/faostat/en/#data/QC. Accessed 4 Mar 2022

Fiore AM, Naik V, Spracklen DV, Steiner A, Unger N, Prather M et al (2012) Global air quality and climate. Chem Soc Rev 41(19):6663–6683

Forkel M, Carvalhais N, Rödenbeck C, Keeling R, Heimann M, Thonicke K et al (2016) Enhanced seasonal CO2 exchange caused by amplified plant productivity in northern ecosystems. Science 351(6274):696–699

Forster P, Ramaswamy V, Artaxo P, Berntsen T, Betts R, Fahey DW, Van Dorland R (2007) Chapter 2: Changes in atmospheric constituents and in radiative forcing. In: Climate change 2007. The Physical Science Basis

Fujibe F (2009) Detection of urban warming in recent temperature trends in Japan. Int J Climatol 29:1811–1822. https://doi.org/10.1002/joc.1822

Giorgi F, Meleux F (2007) Modelling the regional effects of climate change on air quality. Compt Rendus Geosci 339(11–12):721–733

Grimm NB, Chapin FS III, Bierwagen B, Gonzalez P, Groffman PM, Luo Y et al (2013) The impacts of climate change on ecosystem structure and function. Front Ecol Environ 11(9): 474–482

Guerra AJT, Fullen MA, Jorge MDCO, Bezerra JFR, Shokr MS (2017) Slope processes, mass movement and soil erosion: a review. Pedosphere 27:27–41

Gupta S, Kumar S (2017) Simulating climate change impact on soil erosion using RUSLE model – a case study in a watershed of mid-Himalayan landscape. J Earth Syst Sci 126(3):1–20

Gupta S, Jain I, Johari P, Lal M (2019) Impact of climate change on tropical cyclones frequency and intensity on Indian coasts. In: Proceedings of international conference on remote sensing for disaster management. Springer, Cham, pp 359–365

Gutiérrez JM, Jones RG, Narisma GT, Alves LM, Amjad M, Gorodetskaya IV, Grose M, Klutse NAB, Krakovska S, Li J, Martínez-Castro D, Mearns LO, Mernild SH, Ngo-Duc T, van den Hurk B, Yoon J-H (2021) Atlas. In: Masson-Delmotte V, Zhai P, Pirani A, Connors SL, Péan C, Berger S, Caud N, Chen Y, Goldfarb L, Gomis MI, Huang M, Leitzell K, Lonnoy E, Matthews JBR, Maycock TK, Waterfield T, Yelekçi O, Yu R, Zhou B (eds) Climate change 2021: the physical science basis. Contribution of Working Group I to the sixth assessment report of the Intergovernmental Panel on Climate Change. Cambridge University Press. Interactive Atlas available from Available from http://interactive-atlas.ipcc.ch/ (Atlas). Accessed 12 Feb 2022. (in press)

Hanewinkel M, Cullmann DA, Schelhaas M-J, Nabuurs G-J, Zimmermann NE (2013) Climate change may cause severe loss in the economic value of European forest land. Nat Clim Chang 3: 203–207

Hanson S, Nicholls R, Ranger N, Hallegatte S, Corfee-Morlot J, Herweijer C, Chateau J (2011) A global ranking of port cities with high exposure to climate extremes. Clim Chang 104(1):89–111

Haywood J, Boucher O (2000) Estimates of the direct and indirect radiative forcing due to tropospheric aerosols: a review. Rev Geophys 38(4):513–543

Hicke JA, Allen CD, Desai AR, Dietze MC, Hall RJ, Hogg EH et al (2012a) Effects of biotic disturbances on forest carbon cycling in the United States and Canada. Glob Chang Biol 18(1): 7–34

Hicke JA et al (2012b) Effects of biotic disturbances on forest carbon cycling in the United States and Canada. Glob Chang Biol 18:7–34

Hickler T, Vohland K, Feehan J, Miller PA, Smith B, Costa L et al (2012) Projecting the future distribution of European potential natural vegetation zones with a generalized, tree species-based dynamic vegetation model. Glob Ecol Biogeogr 21(1):50–63

Hisano M, Searle EB, Chen HY (2018) Biodiversity as a solution to mitigate climate change impacts on the functioning of forest ecosystems. Biol Rev 93(1):439–456

Hoegh-Guldberg O et al (2007) Coral reefs under rapid climate change and ocean acidification. Science 318(5857):1737–1742. https://doi.org/10.1126/science.1152509

Holloway T, Fiore A, Hastings MG (2003) Intercontinental transport of air pollution: will emerging science lead to a new hemispheric treaty? Environ Sci Technol 37(20):4535–4542

Hong C, Zhang Q, Zhang Y, Davis SJ, Tong D, Zheng Y et al (2019) Impacts of climate change on future air quality and human health in China. Proc Natl Acad Sci 116(35):17193–17200

Hooper DU, Adair EC, Cardinale BJ, Byrnes JE, Hungate BA, Matulich KL, Gonzalez A, Duffy JE, Gamfeldt L, O'Connor MI (2012) A global synthesis reveals biodiversity loss as a major driver of ecosystem change. Nature 486:105–108

Intergovernmental Panel on Climate Change [IPCC], Climate Change 2013 (2013) The Physical Science Basis

IPCC (1994) Climate change 1994. In: Houghton JT, Filho LGM, Bruce J, Lee H, Callander BA, Haites EF, Harris N, Maskell K (eds) Radiative forcing of climate change and an evaluation of the IPCC IS92 emission scenarios. Cambridge University Press, Cambridge/New York

IPCC (2013a) Sea level change supplementary material. In: Church JA et al (eds) Climate change 2013: the physical science basis. Contribution of Working Group I to the fifth assessment report of the Intergovernmental Panel on Climate Change, Cambridge, pp 1–8

IPCC (2013b) Climate change 2013. In: Stocker TF, Qin D, Plattner G-K, Tignor M, Allen SK, Boschung J, Nauels A, Xia Y, Bex V, Midgley PM (eds) The Physical Science Basis. Contribution of Working Group I to the fifth assessment report of the Intergovernmental Panel on Climate Change. Cambridge University Press, Cambridge/New York, p 1535

IPCC (2014) Summary for policymakers. In: Field CB, Barros VR, Dokken DJ, Mach KJ, Mastrandrea MD, Bilir TE, Chatterjee M, Ebi KL, Estrada YO, Genova RC, Girma B, Kissel ES, Levy AN, MacCracken S, Mastrandrea PR, White LL (eds) Climate change 2014: impacts, adaptation, and vulnerability. Part A: Global and sectoral aspects. Contribution of Working Group II to the fifth assessment report of the Intergovernmental Panel on Climate Change. Cambridge University Press, Cambridge/New York, pp 1–32

IPCC (2018) Global warming of 1.5 °C. In: Masson-Delmotte V, Zhai P, Pörtner HO, Roberts D, Skea J, Shukla PR, Pirani A, Moufouma-Okia W, Péan C, Pidcock R, Connors S, Matthews JBR, Chen Y, Zhou X, Gomis MI, Lonnoy E, Maycock T, Tignor M, Waterfield T. An IPCC Special Report on the impacts of global warming of 1.5 °C above pre-industrial levels and related global greenhouse gas emission pathways, in the context of strengthening the global response to the threat of climate change, sustainable development, and efforts to eradicate poverty (in press)

IPCC Climate change 2007: synthesis report – an assessment of the Intergovernment Panel on Climate Change, p 2007

IPCC (Intergovernmental Panel for Climate Change) (2001) Climate change 2001 – The Scientific Basis. In: Houghton JT, Ding Y, Griggs DJ, Noguer M, van der Linden PJ, Dai X, Maskell K, Johnson CA (eds) Contribution of Working Group I to the third assessment report of the Intergovernmental Panel on Climate Change. Cambridge University Press, Cambridge, p 881

IPCC (Intergovernmental Panel on Climate Change) (2013) Summary for policymakers. In: Stocker TF, Qin D, Plattner GK, Tignor M, Allen SK, Boschung J, Nauels A, Xia Y, Bex V, Midgley PM (eds) Climate change 2013: the Physical Science Basis. Contribution of Working Group to the fifth assessment report of the Intergovernmental Panel on Climate Change. Cambridge University Press, Cambridge/New York, p 1535

Isaksen IS, Granier C, Myhre G, Berntsen TK, Dalsøren SB, Gauss M et al (2009) Atmospheric composition change: climate-chemistry interactions. Atmos Environ 43(33):5138–5192

Ishigaki T, Kawanaka R, Onishi Y, Shimada H, Toda K, Baba Y (2009) Assessment of safety on evacuating route during underground flooding. In: Advances in water resources and hydraulic engineering. Springer, Berlin/Heidelberg, pp 141–146

Jacob DJ, Winner DA (2009) Effect of climate change on air quality. Atmos Environ 43(1):51–63

Jacobson MZ (2008) On the causal link between carbon dioxide and air pollution mortality. Geophys Res Lett 35(3)

Jandl R, Spathelf P, Bolte A, Prescott CE (2019) Forest adaptation to climate change – is non-management an option? Ann For Sci 76(2):1–13

Jankowski KJ, Schindler DE (2019) Watershed geomorphology modifies the sensitivity of aquatic ecosystem metabolism to temperature. Sci Rep 9(1):1–10

Joët T, Ourcival JM, Dussert S (2013) Ecological significance of seed desiccation sensitivity in Quercus ilex. Ann Bot 111(4):693–701

Jones PG, Thornton PK (2000) MarkSim: software to generate daily weather data for Latin America and Africa. Agron J 92(3):445–453

Keeling CD, Whorf TP, Wahlen M, Van der Plicht J (1995) Interannual extremes in the rate of rise of atmospheric carbon dioxide since 1980. Nature 375:666–670

Kimball BA (1983) Carbon dioxide and agricultural yield: an assemblage and analysis of 430 prior observations. Agron J 75:779–786

Kumar KS, Parikh J (1998) Climate change impacts on Indian agriculture: the Ricardian approach. In: Dinar et al (eds) Measuring the impacts of climate change on indian agriculture, World Bank technical paper no. 402. World Bank, Washington, DC

Lal R (1997) Deforestation effects on soil degradation and rehabilitation in western Nigeria. IV. Hydrology and water quality. Land Degrad Dev 8(2):95–126

Lal R (2010) Managing soils and ecosystems for mitigating anthropogenic carbon emissions and advancing global food security. Bioscience 60:708–721

Lal R, Delgado JA, Groffman PM, Millar N, Dell C, Rotz A (2011) Management to mitigate and adapt to climate change. J Soil Water Conserv 66:276–285. https://doi.org/10.2489/jswc.66.4.276

Lee D, Min SK, Fischer E, Shiogama H, Bethke I, Lierhammer L, Scinocca JF (2018) Impacts of half a degree additional warming on the Asian summer monsoon rainfall characteristics. Environ Res Lett 13(4):044033

Li P, Peng C, Wang M, Li W, Zhao P, Wang K et al (2017) Quantification of the response of global terrestrial net primary production to multifactor global change. Ecol Indic 76:245–255

Liao H, Chen WT, Seinfeld JH (2006) Role of climate change in global predictions of future tropospheric ozone and aerosols. J Geophys Res-Atmos 111(D12)

Mall RK, Singh R, Gupta A, Srinivasan G, Rathore LS (2006) Impact of climate change on Indian agriculture. A review. Clim Chang 78:445–478

Manabe S, Wetherald R (1967) Thermal equilibrium of the atmosphere with a given distribution of relative humidity. J Atmos Sci 24:241–259

Marc L (2015) What is the greenhouse effect. http://www.livescience.com/37743-greenhouse-effect.html. Accessed 5 Mar 2022

Mariappan S, Hartley IP, Cressey EL, Dungait JAJ, Quine TA (2022) Soil burial reduces decomposition and offsets erosion-induced soil carbon losses in the Indian Himalaya. Glob Change Biol 28(4):1643–1658. https://doi.org/10.1111/gcb.15987

Mase H, Tamada T, Yasuda T, Karunarathna H, Reeve DE (2015) Analysis of climate change effects on seawall reliability. Coast Eng J 57(03):1550010

Melillo JM, Prentice IC, Farquhar GD, Schulze ED, Sala OE (1995) Terrestrial biotic responses to environmental change and feedbacks to climate. In: Houghton JT, Filho LGM, Callander BA, Harris N, Kattenberg A, Maskell K (eds) Climate change. Cambridge University Press, New York

Merlis TM, Zhou W, Held IM, Zhao M (2016) Surface temperature dependence of tropical cyclone-permitting simulations in a spherical model with uniform thermal forcing. Geophys Res Lett 43: 2859–2865

Millennium Ecosystem Assessment (2005) Ecosystems and human well-being. Synthesis Island Press, Washington, DC

Milly PCD, Betancourt J, Falkenmark M, Hirsch RM, Kundzewicz ZW, Lettenmaier DP, Stouffer RJ (2008) Stationarity is dead: whither water management? Science 319:573–574. https://doi.org/10.1126/science.1151915

Montanarella L (2015) Agricultural policy: govern our soils. Nature 528(7580):32–33

Murakami H, Vecchi GA, Underwood S (2017) Increasing frequency of extremely severe cyclonic storms over the Arabian Sea. Nat Clim Chang 7(12):885

NOAA (2016) State of the climate: global climate report for annual 2015. National Oceanic and Atmospheric Administration (NOAA) National Centers for Environmental Information (NCEI). Available at: www.ncdc.noaa.gov/sotc/global/201513. Accessed 9 Feb 2022

Nolte CG, Gilliland AB, Hogrefe C, Mickley LJ (2008) Linking global to regional models to assess future climate impacts on surface ozone levels in the United States. J Geophys Res-Atmos 113 (D14)

O'Hara KL (2016) What is close-to-nature silviculture in a changing world? For Int J For Res 89(1): 1–6

Pan Y, Birdsey RA, Fang J, Houghton R, Kauppi PE, Kurz WA, Phillips OL, Shvidenko A, Lewis SL, Canadell JG (2011) A large and persistent carbon sink in the world's forests. Science 333: 988–993

Pan S, Tian H, Dangal SR, Ouyang Z, Lu C, Yang J et al (2015) Impacts of climate variability and extremes on global net primary production in the first decade of the 21st century. J Geogr Sci 25(9):1027–1044

Pandey R, Jha S (2012) Climate vulnerability index-measure of climate change vulnerability to communities: a case of rural Lower Himalaya, India. Mitig Adapt Strateg Glob Chang 17(5): 487–506

Pareek N (2017) Climate change impact on soils. Adapt Mitigat 2(3):2–5. https://doi.org/10.15406/mojes.2017.02.00026

Parmesan C, Yohe G (2003) A globally coherent fingerprint of climate change impacts across natural systems. Nature 421(6918):37

Pathak H, Ladha PK, Aggarwal S, Peng D, Singh Y, Singh B, Kamra SK, Mishra B, Sastri ASRAS, Aggarwal PK, Das DK, Gupta RK (2003) Trends of climatic potential and on-farm yields of rice and wheat in the Indo-Gangetic Plains. Field Crop Res 80:223–234

Paustian K, Lehmann J, Ogle S, Reay D, Robertson GP, Smith P (2016) Climate-smart soils. Nature 532(7597):49–57

Pearson RG, Dawson TP (2003) Predicting the impacts of climate change on the distribution of species: are bioclimate envelope models useful? Glob Ecol Biogeogr 12(5):361–371

Pelling M, Leck H, Pasquini L, Ajibade I, Osuteye E, Parnell S et al (2018) Africa's urban adaptation transition under a 1.5 climate. Curr Opin Environ Sustain 31:10–15. https://doi.org/10.1016/j.cosust.2017.11.005

Pérez-Ramos IM, Rodríguez-Calcerrada J, Ourcival JM, Rambal S (2013) Quercus ilex recruitment in a drier world: a multi-stage demographic approach. Perspect Plant Ecol Evol Systematics 15(2):106–117

Pimentel D (2006) Soil erosion: a food and environmental threat. Environ Dev Sustain 8:119–137

Pye HOT, Liao H, Wu S, Mickley LJ, Jacob DJ, Henze DK, Seinfeld JH (2009) Effect of changes in climate and emissions on future sulfate-nitrate-ammonium aerosol levels in the United States. J Geophys Res-Atmos 114(D1)

Raj AD (2020) Modelling climate change impact on surface runoff and sediment yield in a watershed of Shivalik region. M.Sc. Thesis, Academy of Climate Change Education and Research, Kerala Agricultural University, Vellanikkara

Ramanathan V, Coakley J (1978) Climate modeling through radiative-convective models. Rev Geophys Space Phys 16:465–490

Raymond PA, Hartmann J, Lauerwald R, Sobek S, McDonald C, Hoover M, Butman D et al (2013) Global carbon dioxide emissions from inland waters. Nature 503(7476):355–359

Reed MS, Buenemann M, Atlhopheng J, Akhtar-Schuster M, Bachmann F, Bastin G, Bigas H, Chanda R, Dougill AJ, Essahli W et al (2011) Cross-scale monitoring and assessment of land degradation and sustainable land management: a methodological framework for knowledge management. Land Degrad Dev 22:261–271

Revi A (2016) Afterwards: habitat III and the Sustainable Development Goals. Urbanisation 1(2):x–xiv. https://doi.org/10.1177/2455747116682899

Robinson DA, Panagos P, Borrelli P, Jones A, Montanarella L, Tye A, Obst CG (2017) Soil natural capital in Europe; a framework for state and change assessment. Sci Rep 7(1):1–14

Ruiz-Benito P, Ratcliffe S, Jump AS, Gómez Aparicio L, Madrigal-González J, Wirth C, Kändler G, Lehtonen A, Dahlgren J, Kattge J (2016) Functional diversity underlies demographic responses to environmental variation in European forests. Glob Ecol Biogeogr 26:128–141

Rumpel C, Amiraslani F, Chenu C, Garcia Cardenas M, Kaonga M, Koutika LS et al (2020) The 4p1000 initiative: opportunities, limitations and challenges for implementing soil organic carbon sequestration as a sustainable development strategy. Ambio 49(1):350–360

Sakschewski B, von Bloh W, Boit A, Poorter L, Peña-Claros M, Heinke J, Joshi J, Thonicke K (2016) Resilience of Amazon forests emerges from plant trait diversity. Nat Clim Chang 6: 1032–1036

Sala OE, Chapin FS, Armesto JJ, Berlow E, Bloomfield J, Dirzo R, Huber-Sanwald E, Huenneke LF, Jackson RB, Kinzig A (2000) Global biodiversity scenarios for the year 2100. Science 287: 1770–1774

Secretary General's Report on Land Chapter of Agenda 21 to Commission on Sustainable Development (CSD8, UN, New York 2000) (1994) UNCCD Agenda 21, Rio de Janeiro, 1992 and UNCCD Paris

Seidl R, Thom D, Kautz M, Martin-Benito D, Peltoniemi M, Vacchiano G et al (2017) Forest disturbances under climate change. Nat Clim Chang 7(6):395–402

Selvaraju R (2003) Impact of El Nino-Southern oscillation on Indian food grain production. Int J Climatol 23:187–206

Shine KP, PM de F Forster (1999) The effects of human activity on radiative forcing of climate change: a review of recent developments. Glob Planet Chang 20:205–225

Sivakumar MVK (2007) Interactions between climate and desertification. Agric For Meteorol 142(2-4):143–155

Smith P, Calvin K, Nkem J, Campbell D, Cherubini F, Grassi G et al (2020) Which practices co-deliver food security, climate change mitigation and adaptation, and combat land degradation and desertification? Glob Chang Biol 26(3):1532–1575

Steffen W, Richardson K, Rockström J, Cornell SE, Fetzer I, Bennett EM et al (2015) Planetary boundaries: guiding human development on a changing planet. Science 347(6223):1259855

Stern N (2015) Economic development, climate and values: making policy. Proc R Soc B Biol Sci 282(1812):20150820. https://doi.org/10.1098/rspb.2015.0820

Stockmann U, Padarian J, McBratney A, Minasny B, de Brogniez D, Montanarella L et al (2015) Global soil organic carbon assessment. Glob Food Secur 6:9–16. https://doi.org/10.1016/j.gfs.2015.07.001

Stone B (2007) Urban and rural temperature trends in proximity to large US cities: 1951–2000. Int J Climatol 27:1801–1807. https://doi.org/10.1002/joc.1555

Sun Y, Zhong Z, Li T, Yi L, Hu Y, Wan H et al (2017) Impact of ocean warming on tropical cyclone size and its destructiveness. Sci Rep 7(1):1–10

Sooryamol K R (2020) Potential impact of climate change on surface runoff and sediment yield in a watershed of lesser Himalayas (M.Sc. Thesis, Academy of Climate Change Education and Research, Kerala Agricultural University, Vellanikkara)

Sooryamol KR, Kumar S, Regina M, David Raj A (2022) Modelling climate change impact on soil erosion in a watershed of north-western lesser Himalayan region. J Sedimen Environ 7 (2):125–146

Teng H, Liang Z, Chen S, Liu Y, Rossel RAV, Chappell A et al (2018) Current and future assessments of soil erosion by water on the Tibetan Plateau based on RUSLE and CMIP5 climate models. Sci Total Environ 635:673–686

Thompson I, Mackey B, McNulty S, Mosseler A (2009) Forest resilience, biodiversity, and climate change. A synthesis of the biodiversity/resilience/stability relationship in forest ecosystems, vol 43. Secretariat of the Convention on Biological Diversity, Montreal, pp 1–67

Thuiller W, Lavorel S, Araùjo MB, Sykes MT, Prentice IC (2005) Climate change threats to plant diversity in Europe. Proc Natl Acad Sci U S A 102:8245–8250

Tong S, Zhiming F, Yanzhao Y, Yumei L, Yanjuan W (2018) Research on land resource carrying capacity: progress and prospects. J Resour Ecol 9(4):331–340

UAPCC (2014) Uttarakhand action plan on climate change 'Transforming crisis into opportunity'

UNCCD (1994) Elaboration of an International Convention to Combat Desertification in countries experiencing serious drought and/or desertification, particularly in Africa. A/AC.241/27

UNDP (2016) Human development report 2016: human development for everyone. United Nations Development Programme (UNDP), New York, 286 pp

UNEP (2008) Africa: Atlas of our changing environment. Division of Early Warning and Assessment (DEWA) United Nations Environment Programme (UNEP), Nairobi

Uprety DC (2003) Rising atmospheric carbon dioxide and crops: Indian studies. Souvenir: 2nd international congress of plant physiology, New Delhi, pp 87–93

Uprety DC, Dwivedi N, Jain V, Mohan R, Saxena MJ, Paswan G (2003) Response of rice cultivars to the elevated CO. Biologia Plantarum, 2 46(1):35–39

Vannote RL, Sweeney BW (1980) Geographic analysis of thermal equilibria: a conceptual model for evaluating the effect of natural and modified thermal regimes on aquatic insect communities. Am Nat 115(5):667–695

Vass MM, Elofsson K (2016) Is forest carbon sequestration at the expense of bioenergy and forest products cost-efficient in EU climate policy to 2050? J For Econ 24:82–105

Wang H, Liu G, Li Z, Wang P, Wang Z (2018) Assessing the driving forces in vegetation dynamics using net primary productivity as the indicator: a case study in Jinghe river basin in the Loess Plateau. Forests 9(7):374

Waters CN, Zalasiewicz J, Summerhayes C, Barnosky AD, Poirier C, Gałuszka A et al (2016) The Anthropocene is functionally and stratigraphically distinct from the Holocene. Science 351(6269):aad2622–aad2622. https://doi.org/10.1126/science

Wieland R, Lakes T, Yunfeng H, Nendel C (2019) Identifying drivers of land degradation in Xilingol, China, between 1975 and 2015. Land Use Policy 83:543–559

Wilby RL, Dawson CW, Barrow EM (2002) SDSM – a decision support tool for the assessment of regional climate change impacts. Environ Model Softw 17(2):145–157

Wise EK, Comrie AC (2005) Meteorologically adjusted urban air quality trends in the Southwestern United States. Atmos Environ 39(16):2969–2980

WMO/GWP Associate Program on Flood Management: Urban Flood Risk Management – a tool for Integrated Flood Management, 2008

Wong PP, Losada IJ, Gattuso JP, Hinkel J, Khattabi A, McInnes KL et al (2014) Coastal systems and low-Lying areas. In: Field CB et al (eds) Climate change 2014: impacts, adaptation, and vulnerability. Part A: Global and sectoral aspects. Contribution of Working Group II to the fifth assessment report of the Intergovernmental Panel on Climate Change. Cambridge University Press, Cambridge/New York, pp 361–409

World Bank (2022) Renewable internal freshwater resources per capita (cubic meters). Available at http://data.worldbank.org/indicator/ER.H2O.INTR.PC. Accessed 22 Aug 2022

Xie H, Zhang Y, Wu Z, Lv T (2020) A bibliometric analysis on land degradation: current status, development, and future directions. Land 9(1):28

Xiong X, Grunwald S, Myers DB, Ross CW, Harris WG, Comerford NB (2014) Interaction effects of climate and land use/land cover change on soil organic carbon sequestration. Sci Total Environ 493:974–982. https://doi.org/10.1016/j.scitotenv.2014.06.088

Chapter 4
Analysis of Low-Flow Indices in the Era of Climate Change: An Application of CanESM2 Model

Mohammadreza Goodarzi and Alireza Faraji

Abstract Low-flow hydrology shifts can change streamflow criteria such as dissolved oxygen content, nutrient concentration, and terrestrial and aquatic qualities, as well as different water and wastewater functions, including water allocation, power plant generation, navigation, and waste load allocation. Climate changes play a significant role in the low-flow hydrology regime among various factors. Therefore, sustainable water resources planning and restoration of ecosystems depend on low-flow extremes and their consequent impacts. This research aimed to assess the consequences of climate change on the low-flow indices (7Q10, Q80) in a subbasin so-called Gharesou located in Iran employing the HEC-HMS rainfall-runoff modeling and the general population circulation outputs (CanESM2). For this purpose, the precipitation, temperature, and streamflow datasets in 1970–2000 were applied in the current chapter. The results revealed that the subbasin would experience a reduced average flow in all scenarios for the low-flow seasons. The 7Q10 index approached 0.008 m^3/s and zero for the SDSM model and change factor, respectively, for the future period; these values were 0.724 and 1.429 m^3/s in the corresponding historical periods, respectively. Furthermore, Q_{80} decreased from 4.27 to 0.1 for SDSM and from 5.3 to 0.3 m^3/s for the change factor method in future projection studies.

Keywords CanESM2 · Climate change · Extreme events · HEC-HMS model · Low-flow

M. Goodarzi (✉)
Department of Civil Engineering, Yazd University, Yazd, Iran
e-mail: Goodarzimr@yazd.ac.ir

A. Faraji
Department of Engineering, University of Basilicata, Potenza, Italy

Visiting Scholar, Clemson University, SC, Clemson, USA
e-mail: Afaraji@g.clemson.edu

© The Author(s), under exclusive license to Springer Nature Switzerland AG 2022
U. Chatterjee et al. (eds.), *Ecological Footprints of Climate Change*, Springer Climate, https://doi.org/10.1007/978-3-031-15501-7_4

Introduction

Low-flow regimes can affect dissolved oxygen contents, the level of nutrients, and aquatic habitat quality which are important surface water factors (Hagg and Melvin 2008). Therefore, assessing the low-flow regime is one of the most vital watershed management and planning components. They highlight the necessity of profound scientific studies on low-flow indices, especially through dry seasons. Several factors influence and control the low-flow hydrology, including soil characteristics, vegetation cover, watershed hydraulics, topography, evapotranspiration trend, and climatic change (Smakhtin 2001). Guzha et al. (2018) studied the land-use changes and land cover losses, especially vegetation cover and their impacts on the characteristics of runoff and low-flow regimes. Their results highlighted that loss of forest cover results in an increased annual flow, surface runoff, and high flows (16%, 45%, and 10%, respectively), while a mean of 7% reduces low flows. However, the authors discussed that considering just the land cover changes cannot represent hydrological regimes in the studied catchments because of weak correlations among land cover and runoff, mean flow, and peak flow (Guzha et al. 2018). In another research by Li et al. (2018), the change in topography and its possible effects on low-flow regimes were performed over a basin with snow-dominated character in British Columbia, Canada, including 28 watersheds. They concluded that topography acts as a major factor in low-flow index ($\leq Q_{75}\%$) than high flows, and they discovered five significant topography indices (TIs): perimeter, slope length factor, surface area, openness, and terrain characterization index (Li et al. 2018). Besides, Széles et al. (2018) studied the effects of spatial changes in runoff generation processes on the daily discharges during the low-flow duration. Their results indicated that a split of scales in transpiration affects low-flow temporal and spatial levels. It means that the daily streamflow changes are generated by transpiration from the riparian vegetation (Széles et al. 2018). Among the factors mentioned above, coming climate change may intensify extreme hydrological occurrences such as low and high flows. Additionally, it interrupts the hydrological cycle process, including changes in precipitation patterns, evaporation rates, sea level rises, and late/early snowmelt (IPCC 2014; Pathak et al. 2016). The previous literature discussed that likely changes in hydrological variables such as precipitation and temperature could disturb rainfall patterns (IPCC 2014). It can lead to periodic, intense, long-term, and continuous drought in many catchments and basins (Frederick and Major 1997).

Nevertheless, several works of literature have concentrated on the outcomes of climate change aspects on flooding events and runoff variation worldwide. There is still a lack of comprehensive and specific studies on low-flow regimes affected by climate change impacts, especially in Iran, where droughts and extreme events have affected aquatic and human life. For example, Nasonova et al. (2019) assessed the possible effects of climate on river runoff variations within the twenty-first century using five different general circulation models (GCMs) accompanied by four representative concentration pathway (RCP) scenarios in the Northern Dvina, Indigirka, and Taz River basins (Nasonova et al. 2019). GCMs are a primary step to project future climate change and are commonly employed to project or predict local climate parameters under emission scenarios. In a study conducted on Europe by Marx et al. (2018), the low-flow indices affected by climate change were analyzed by applying five various GCMs under three different RCPs. They outlined a decrease in low-flow indices in the Mediterranean region. In contrast, the Alpine and Northern regions observed an increase in low-flow values (Marx et al. 2018). In another study regarding GCM models, Jin et al. (2018) assessed the changes in hydrologic characteristics of the Yellow River, such as precipitation and temperature, to observe the runoff changes by employing seven GCMs from the Coupled Model Intercomparison Project-Phase 5 (CMIP5) under various RCPs. They demonstrated that the amount of streamflow might rise in the future. Also, several droughts and floods may occur in China's Yellow River in the future (Jin et al. 2018).

The coarse horizontal resolution of GCMs imposes limitations on the accuracy of future projections. From a practical standpoint, hydrological models need more local-scale resolution (Xu 1999). Operating the full GCM at a higher resolution can solve the problem, but it takes longer to perform a simulation. Also, it needs a powerful computer or simulation for a much shorter period (e.g., 5 years). The GCM-RCMs are numerical coupled simulations that model the interactions between atmosphere, oceans, land surface, sea ice, and earth systems (Fowler et al. 2007). GCM-RCM outputs present the scales between 12.5 and 50 km, while hydrology modeling needs finer data resolutions. Owing to this limitation, an obvious demand for higher resolution climate outputs for climate change studies has become an attractive topic in scientific communities (Cohen 1990). The significant gap between the climate model outputs and required hydrological modeling resolutions causes a substantial challenge. In many studies (e.g., Trzaska and Schnarr 2014), dynamical and statistical techniques establish the main downscaling methods. RCMs or fine spatial-scale numerical atmospheric models represent the dynamic downscaling such as limited area models (LAM) (Feser et al. 2011; Fowler et al. 2007).

In contrast, statistical downscaling methods evaluate the large-scale climate factors and the documented data correlation in the studied area. Several studies focused on the climate change impacts on low-flow indices. For example, Ryu et al. (2011) studied hydrology and low-flow frequency in the Geum river basin in South Korea exposed to climate change factors and concluded a decrease in the 7Q10 index. This research used five regional climate scenarios and a coupled method of change factor statistical downscaling by adding daily temperature increases and multiplying precipitation by a factor for each scenario. They found an 8–12% reduction in streamflow and a sharp decrease in the low-flow index (7Q10) to about 0.03 m^3/s (Ryu et al. 2011). The current periods of low-flow (August to September related to the five river basins located in Germany) may extend through late autumn by 2100 based on Huang et al. (2013) research. They investigated low-flow conditions in Danube, Elbe, Ems, Rhine, and Weser basins in Germany under the impact of climate change by coupling three RCMs and using the SWIM hydrological model. RCMs, including CCLM and REMO with the spatial resolution of 0.2° and 0.088°, were applied to overcome the coarse resolution of ECHAM5 (as based on GCM) (Huang et al. 2013). Chen et al. (2019) analyzed the impacts of land use and climate change on extreme events employing SWAT (the soil and water assessment tool) as well as a statistical downscaling method named DBC. This technique considers each quantile's equal deviations between future and recorded climate. The results revealed that land-use change causes a slight impact on low and high flows.

In contrast, climate change played one of the major changes in the extreme hydrological situation (Chen et al. 2019). In another study in the Great Miami River watershed, Shrestha et al. (2019) used several GCMs accompanied by RCP 4.5 and 8.5 to evaluate the change of water resources and low-flow indices caused by changing climates over the twenty-first century as well as by using daily bias correction constructed analogs (BCCA) statistical downscaled technique. This method eliminates biases between the GCM output and recorded data by establishing a linear interaction of selected days to generate an analog, relatively resembling the GCM (Hwang and Graham 2013). This study argued that the low-flow ensemble in the twenty-first century was slightly higher than historical indices (Shrestha et al. 2019).

In this chapter, the Canadian Earth System Model (CanESM2) from the fifth assessment report (AR5) of IPCC was evaluated using the new scenarios of the RCPs (RCP2.6, RCP4.5, and RCP8.5). The statistical downscaling model (SDSM) and change factor (CF) followed by a frequency analysis of low-flow periods were employed to consider the uncertainty of downscaling methods. Besides, the hydrologic modeling system (HEC-HMS) was applied to calculate the different future low-flow indices and compare their values with the historic period. This chapter aimed to evaluate the low-flow indices' changes over the future period (2040–2069). The results in this chapter may present useful data on future low-flow conditions for water policy-makers toward resolving climate change issues at a basin scale.

The Rationale of the Study

As mentioned in the introduction, several pieces of literature have been conducted on the effects of climate change on high flows and the flooding phenomena in recent years. In contrast, climate changes also affect low-flow indices, emphasizing the necessity of evaluating different low-flow criteria at the subbasin scale. However, less is known recently about the frequency of low flows in the presence of climate change, especially in the selected subbasin in Iran, where longer and severe droughts have made it difficult for water resources planning and water allocation strategies. Therefore, the rationale of this chapter is to evaluate the changes in low flows by projecting climate change scenarios in a subbasin exposed to severe fluctuations in rainfalls and temperature patterns. Besides, considering the uncertainties in downscaling methods is another aspect of this chapter that compares different emission scenario outputs.

Materials and Methods

Study Area

The Karkheh catchment basin placed in the center and southwest of the Zagros Mountains (46°06′ and 49°10′ E and 30°58′ and 34°56′ N) covers an area of 50,764 km² in western Iran. The Gharesou subbasin, in the northwest of the Karkheh basin (Fig. 4.1), selected for this chapter, covers 5354 km² and has a maximum and minimum altitude of 3364 m and 1180 m, respectively. The annual average precipitation of this subbasin changes from 300 to 800 mm. Also, three major rivers flow into the basin (the Merck, the Gharesou, and the Razavar rivers). Several synoptic and rain gage stations are in the basin, as listed in Table 4.1. In Iran, the Karkheh River is one of the longest rivers, approximately 900 kilometers, formed by joining

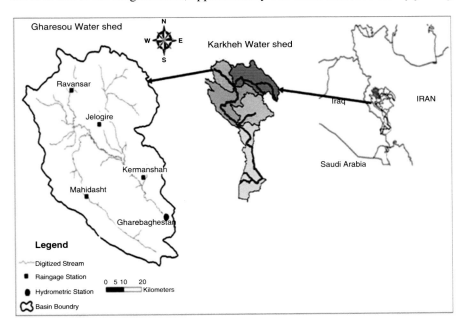

Fig. 4.1 Map of stations in Gharesou basin

Table 4.1 Station characteristics

Station	Station type	Elevation (m)	Latitude	Longitude	Annual precipitation (mm)	Data range
Kermanshah	Synoptic	1318.6	34°21′	47°09′	447	1971–2000
Mahidasht	Rain gage	1415	34°16′	46°49′	403	1988–2002
Jelogireh	Rain gage	1180	34°35′	46°51′	490	1988–2000
Ravansar	Synoptic	1379.7	34°43′	46°39′	525	1988–2000
Gharebaghestan	Hydrometric	1230	34°14′	47°15′	–	1971–2000

the Gamasiab and Gharesou rivers and finally falling into the Persian Gulf. These two sub-rivers are significant in this basin and vital for the agriculture sector and watershed restoration. Therefore, any changes in the Gharesou flow regime, especially within dry seasons recognized as low-flow periods, will lead to a lack of mentioned functions.

GCMs and Emission Scenarios

GCMs present the initial tool for interpreting past climate variations and projecting future climate situations (Wilby and Harris 2006). The CanESM2 used in this chapter is a new version of CanESM1 presented by the CCCma (Canadian Centre for Climate Modelling and Analysis). CanESM2 incorporates the CanCM4 model and the terrestrial carbon cycle presented by the CTEM (Canadian Terrestrial Ecosystem Model) (Arora and Boer 2010) to model the land-atmosphere carbon interaction. Another feature of CanESM2 is taking the greenhouse gas emissions and solar variability into account based on CMIP5 instructions and adding the impacts of volcanic eruptions (Chylek et al. 2011). The atmospheric part of CanCM4, including 35 layers from the land up to 1 hPa in the atmosphere, is the fourth-generation atmospheric GCM. The ocean part of CanCM4 involves 40 levels with roughly 10 m resolution above the ocean.

It is worth mentioning that the CanESM2 is the only current AR5 GCM series for SDSM downscaling model. Given that hydrological modelers require more accurate spatial and temporal scale than the ability of GCMs, the outputs of these GCMs need further analysis in climate change studies (Xu 1999). As mentioned, dynamic and statistical downscaling methods (Wilby and Harris 2006) are used to overcome the problem.

SDSM Downscaling Model

SDSM is a decision support tool that helps users project the effects of global climate change by employing spatial-scale modification of data produced by large-resolution GCMs (Wilby et al. 2002). The regressions and weather generators are valuable features of SDSM, which allow users to simulate a series of daily climatic data based on observed records for the present and coming future. The stochastic element of SDSM enables the generation of 100 simulations (Gagnon et al. 2005). SDSM modeling includes several steps. The primary screening of potential downscaling predictor variables is the first step where large-scale predictors based on the correlation with observed (predictand) data are defined. This step may include several variables obtained from mean sea-level pressure ranges. Next, the large-scale predictors classified in the previous step facilitate the calibration of the SDSM. Various linear regression relationships are used between predictors and predictands using almost half of the available dataset. The establishment of statistical models is monthly, seasonal, or annual. Then, employing the rest of the observed data helps validate the performance of downscaling models. The stochastic element of SDSM generates up to 100 datasets with similar statistical features but changing daily. After that, the model generates future weather data using the statistical relationships calculated in step 2. Finally, it is possible in the SDSM model to measure the statistical features of both the observed and modeled data (Wilby and Harris 2006). After completing all the stages, the downscaled temperature and precipitation datasets are presented to the hydrologic model (HEC-HMS) to simulate the streamflow hydrographs.

Emission Scenarios

In general, a non-climate scenario contains data from socioeconomic status and the level of greenhouse gas emission in the atmosphere of earth, which is also referred to as emission scenarios. The World Meteorological Organization (WMO) and Intergovernmental Panel on Climate Change (IPCC) are responsible for identifying all aspects of the climate change phenomenon. In this regard, new scenarios of RCPs were released in 2010, aiming to provide a collection of information that can be applied to climate models (IPCC 2014). Climate models use the results of emission scenarios to indicate the concentration and emission of greenhouse gases and the level of pollution and changes in land use. The RCP family scenario includes four different scenarios, RCP2.6, RCP4.5, RCP6.0, and RCP8.5, based on the various specifications of the technology level, socioeconomic status, and future policies. It can lead to a different emission level of greenhouse gases and climate changes in any condition. In these scenarios, greenhouse gas emissions are classified into four categories of 8.5, 6, 4.5, and 2.6 w/m^2 based on their impact on the radiation forces surface by the end of the twenty-first century. In addition, these scenarios cover from 1850 up to 2100, and they are formulated till 2300. Details on RCPs specifications are provided in many sources (e.g., (Riahi et al. 2011; Van Vuuren et al. 2011; Collins et al. 2013). Special Reports on Emissions Scenarios (SRES) are other published scenarios of this association in 2000. Greenhouse gas emission scenarios given in the mentioned report provide an image of future climate change on earth. SRES scenarios were employed in the Third (TAR) and Fourth (AR4) reports published in 2001 and 2007. These scenarios include A1B, A1FI, A1T, A2, B1, and B2 families (IPCC 2007).

The RCPs presented in the AR5 of IPCC describe diverse possible radiative forcing scenarios. Greenhouse gas, aerosol concentrations, land-use change, and climate factors employed by the climate modeling community are a part of RCPs. Radiative forcing as the additional heat that the lower atmosphere will hold due to extra greenhouse gases is another part of RCPs. These scenarios include one moderate outline with a very low forcing level (RCP2.6) and two medium stabilization scenarios (RCP4.5 and RCP6). Furthermore, one extreme baseline emission

Table 4.2 The prospect of variation in mean surface temperature and mean sea-level rise over the mid-twenty-first, comparable to 1986–2005

2046–2065	Global mean surface temperature change (°C)		Global mean sea-level rise (m)	
Scenario	Mean	Range	Mean	Range
RCP2.6	1.0	0.4 to 1.6	0.24	0.17 to 0.32
RCP4.5	1.4	0.9 to 2.0	0.26	0.19 to 0.33
RCP6.0	1.3	0.8 to 1.8	0.25	0.18 to 0.32
RCP8.5	2.0	1.4 to 2.6	0.30	0.22 to 0.38

IPCC (2014)

(RCP8.5) (Meinshausen et al. 2011) arises from the slight action to cut emissions. It illustrates a breakdown to stop warming by 2100, same as A1FI, the highest-emission scenario in the IPCC Fourth Assessment Report (AR4). The RCP6.0 upholds total radiative forcing quickly after 2100, considering the use of cutting-edge technologies and guidelines for cutting greenhouse gas emissions. RCP4.5 is considered the lowest emission scenario, as its counterpart B1 in the IPCC AR4.

On the other hand, RCP 2.6 is the most promising pathway. It indicates emission peak early and then drops due to operational elimination of atmospheric CO_2. RCP 2.6 highly demands early collaboration between all the top emitters, specifically developing countries. It has no counterpart in IPCC AR4. The RCPs estimate a wider range of likelihoods than the SRES scenarios in the IPCC third and AR4 modeling. However, RCPs start with atmospheric greenhouse gas concentrations instead of socioeconomic processes. This is because of overcoming the uncertainty in socio-economic scenarios under climate change impacts. Unlike SRES scenarios, some RCPs also involve mitigation and adaptation schemes (Van Vuuren et al. 2011).

Furthermore, RCP8.5 projects the severe concentration of GHGs by 2100. On the other hand, RCP2.6 estimates an increase in CO_2 and NO_2 by almost 2050; then, the emissions will fall slightly till 2100. Also, CH4 will experience a decrease during this time in all RCPs except 8.5. Table 4.2 presents the changes in mean surface temperature and mean sea-level rise between 2046–2065 and 1986–2005. According to Table 4.2, the RCP8.5 presents a 2 °C increase in future temperature and 0.3 m sea level higher than 1986–2005. Overall, the changes in mean temperature are between 1 and 2 °C, and the sea levels vary in a range of 0.24 and 0.3 m in all RCPs (Clarke et al. 2009; Meinshausen et al. 2011; Van Vuuren et al. 2011; IPCC 2014).

HEC-HMS Modeling and Automatic Calibration with Optimization Trials

The HEC-HMS is a conceptual and numerical model for simulating the precipitation-runoff processes of watershed systems and places in the event and continuous models (Bennett and Peters 2012). This model simulates the rainfall-runoff processes using the SMA (soil moisture accounting) algorithm. It assumes the

4 Analysis of Low-Flow Indices in the Era of Climate Change: An... 103

Table 4.3 Final values of optimized parameters for HEC-HMS simulation

Parameter	Optimized value	Parameter	Optimized value
Base-flow initial flow (m³/s)	25.986	GW2 initial storage (%)	30
Base-flow threshold flow (m³/s)	44.34	GW2 percolation rate (mm/hr)	0.01035
Canopy capacity (mm)	141.66	GW2 storage coefficient (hr)	388.24
Canopy initial storage (%)	0.43816	Recession constant (cte)	0.96637
Clark storage coefficient (hr)	159.93	Soil capacity (mm)	114.5
Clark time of concentration (hr)	81.424	Soil infiltration rate (mm/hr)	4.361
GW1 capacity (mm)	66.112	Soil initial storage (%)	5
GW1 initial storage (%)	20	Soil percolation rate (mm/hr)	3.638
GW1 percolation rate (mm/hr)	0.6429	Surface capacity (mm)	13.12
GW1 storage coefficient (hr)	401.33	Surface initial storage (%)	0.801
GW2 capacity (mm)	9.9773	Tension zone capacity (mm)	67.32

soil moisture condition of the watershed over a long time for projecting discharge daily, monthly, and seasonal. The SMA algorithm considers complete runoff elements, including direct runoff (surface flow) and indirect runoff (interflow and groundwater flow). The model needs daily rainfall, evaporation, and observed discharge data. The SMA algorithm represents the watershed with five storage layers (canopy-interception, surface-depression, soil profile, and two groundwater storages). In addition to precipitation, the SMA model needs potential evapotranspiration data (HEC 2021). The first step in the optimization method considers the initial parameter and modifies them to match simulated and observed flows. Optimization trials make it possible to assess subbasin parameters and reach components automatically. For this purpose, a dataset of observed flow is needed in one component before the beginning of the optimization process. Seven objective functions can evaluate the goodness of fit between the simulated results and observed flow. In this research, the coefficient of determination (R^2) and Nash-Sutcliffe (E_{NS}) are employed to do so. Various parameters have been selected for model calibration following previous research on the calibration of the HEC-HMS model. Table 4.3 shows the value obtained for the optimized parameters at the calibration step.

Low-Flow Indices and Selection

An FDC (flow-duration curve) indicates the particular flow value and the percentage of time the specific flow equals or exceeds a certain limit. Annual, monthly, and daily datasets can establish the FDC figures (Smakhtin 2001). The most important indices of low flows derive from FDC, such as Q_{75}, Q_{80} (Ouyang 2012), Q_{90}, and

Q_{95}. In addition to FDC, low-flow frequency curve, so-called LFFC, displays the ratio of years when the discharge exceeds (in other words, the average interval in years ("return period" or "recurrence interval") that the river drops below a given discharge) (Smakhtin 2001). 7Q10 and the 7Q2 are the most commonly used indices in the USA. These indices introduce the lowest average flows in the recurrence interval of 2 and 10 years that occur for a consecutive 7-day period (Pyrce 2004). Some other studies point out other indices, for example, 3Q20 (the 3 days 20-year low flow), 30Q10 (the 30 days 10-year low flow), and 90Q10 (the 90 days 10-year low flow). The low-flow selection is based on 7Q10 (Laaha and Blöschl 2007) and FL (frequent low) (Ouyang 2012). The FL approach is one of the five low-flow managing techniques used in St. Johns River in Florida. Minimum flows or levels, including five groups of low flows, are assumed to evade harsh damage to water resources by water withdrawals (Neubauer et al. 2008). The researchers have argued that low flow in FDC varies from 70% to 99%. The authors used the FL method for the following reasons. Firstly, the 7Q10 values fit intense low-flow situations occurring 95–99% of the FDC time, which happens over severe short-term drought and has a wide recurrence interval. Besides, the FL approach is within the low-flow selection range, and FL levels have considered low-flow effects on the floodplain, flora and fauna, and ecosystems (Ouyang 2012).

Distribution Function

The observed available flow data are inadequate for the reliability of the low-flow event frequency. Accordingly, different theoretical distribution functions facilitate extrapolating the recorded data and enhancing the precision of low-flow analysis. However, probability functions are unknown to fit low-flow analysis. Determining an optimal distribution and quantitating its parameters are common problems (Smakhtin 2001). This process involves fitting many theoretical distribution functions into the recorded data of the low flow and determining based on graphical and statistical tests which relate the best distribution to the data. Several articles have pointed out the low flows and forms of distribution functions among the distribution functions, including a variety of Weibull, Gumbel, and Pearson type III and log-normal.

Environmental Water Requirements (EWR)

Water requirements are indispensable for every inland or coastal ecosystem to maintain its structure and functions. For river basins, these are defined as environmental flows. However, water allocation for environmental purposes has low priority for water resources management (Revenga et al. 2000). In a report, Smakhtin et al. (2004) assessed environmental flow requirements on a global scale. They divided EWR into two components, (a) the environmental low-flow requirement (LFR) and (b) the environmental high-flow requirement (HFR), which are described in Table 4.4. It should be noted that LFR was obtained from monthly FDC and equaled to Q_{90} of this curve. Furthermore, MAR defines mean annual runoff, and the EWR is a sum of LFR and HFR (Smakhtin et al. 2004).

Overall Methodology

This chapter includes four main steps to evaluate future low-flow indices. The first step was the preparation of GCM accompanied RCPs. Then, downscaling the GCM output by the SDSM model and CF method was the second step. The third part of this chapter was calibrating the HEC-HMS model and generating future flow series. The low-flow indices form the final part. Figure 4.2 displays a schematic view of this chapter, including climate models, downscaling process, hydrological modeling, and low-flow indices.

Table 4.4 The estimation of high-flow requirement (HFR)

Low-flow requirement (Q_{90})	High-flow requirement (HFR)
If $Q_{90} < 10\%$ MAR	Then HFR = 20% MAR
If 10% MAR $\leq Q_{90} < 20\%$ MAR	Then HFR = 15% MAR
If 20% MAR $\leq Q_{90} < 30\%$ MAR	Then HFR = 7% MAR
If $Q_{90} \geq 30\%$ MAR	Then HFR = 0

Smakhtin et al. (2004)

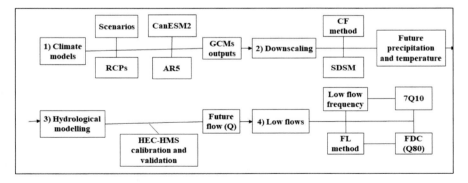

Fig. 4.2 Schematic view of steps of this chapter

Results and Discussion

Calibration Results

To calibrate (period 1993–1996) and validate (period 1988–1990), the model employed data collected from the precipitation in four available stations, Kermanshah station temperature, as well as the observed flow data by Gharebaghestan hydrometric station. Table 4.5 presents the daily, monthly, and seasonal average values of R2 and ENS coefficients. Also, Fig. 4.3 displays a time-series analysis of daily simulated and observed flow basin. Fig. 4.2 shows that the simulations for the Gharesou River capture the daily variability of flows and their seasonality quite well. However, the simulated hydrograph recession is more crucial than observed and extreme low flows have a substantial low bias. The timing of severe low flows, however, closely matches the observations.

Table 4.5 HEC-HMS calibrated (1993–1996) and validated (1988–1990) results for Gharesou basin

Time period	Criteria	Daily	Monthly	Seasonal
Calibration	R^2	0.70	0.72	0.80
	E_{NS}	0.66	0.65	0.74
Validation	R^2	0.60	0.67	0.71
	E_{NS}	0.57	0.64	0.65

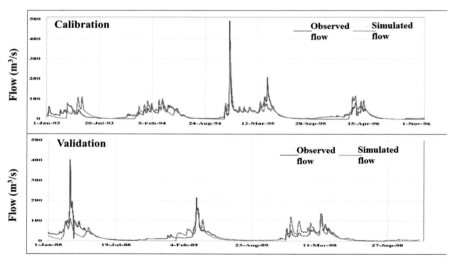

Fig. 4.3 Observed and simulated flow (m^3/s) at Gharebaghestan station. (**a**) Calibration period (1993–1996), (**b**) validation period (1988–1990)

Performance of CanESM2 in the Historic Data Simulating

Table 4.6 presents the performance of the CanESM2 simulation in the historical data for the SDSM model and CF method. Two stations, Ravansar and Jelogireh, have been excluded because of the low values of R^2 and E_{NS} in the validation period. This exclusion leads to a better future projection of HEC-HMS with the SDSM model.

Future Climate Change Outputs

Figure 4.4 compares simulated historical and projected future monthly flows for the Gharesou basin by downscaling methods and using CanESM2 and three RCPs. The precipitation and temperature outputs of the CanESM2 model resulting from the SDSM statistical model were entered into the HEC-HMS calibrated precipitation-runoff model. Besides, the flow of the future period (2040–2069) was generated at the Gharebaghestan hydrometric station. Figure 4.4 shows that the timing of maximum future streamflow will shift from March to February in the SDSM model. The number of figures will not experience any changes in the CF method as both historical and future projections occur in March. All RCPs estimate an extension in the low-flow period from August to October to the end of the year (August–December) by the SDSM model (2040–2069). However, in the CF method, the lowest monthly streamflow figures will occur 1 month earlier, from July to October.

RCP outputs project a monthly flow increase in January, February, and June between 12% and 180% and a decrease in the year from 1% to 100% in the SDSM model. On the other hand, the figures for the CF method show a flow reduction from March to October, approximately 10–90%. The average monthly flow will increase in November and December by three times higher than historical values in RCP2.6 and RCP 8.5 as two extreme scenarios. Moreover, a significant reduction occurred in the low-flow seasons (summer and autumn) for all scenarios, confirming previous findings in Ryu et al. (2011) and Huang et al. (2013) studies. It is worth mentioning that RCP8.5 was an extreme scenario. It has predicted more flow increase within November and December. On the other hand, RCP4.5 has shown a substantial

Table 4.6 The performance of CanESM2

SDSM					CF method	
Precipitation	R^2 (calibration)	E_{NS} (calibration)	R^2 (validation)	E_{NS} (validation)	R^2	E_{NS}
Kermanshah	0.90	0.90	0.80	0.70	0.94	0.92
Mahidasht	0.51	−2.09	0.71	0.56	0.86	0.79
Ravansar	0.49	−2.66	0.19	−7.40	0.91	0.75
Jelogireh	0.01	−96	0.001	−143	0.89	0.78
Temperature						
Kermanshah	0.99	0.99	0.99	0.98	0.99	0.99

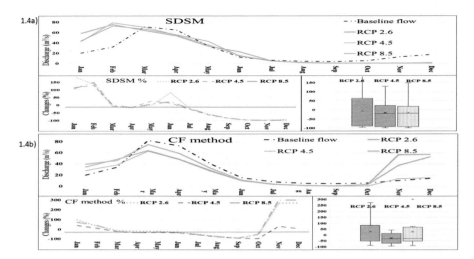

Fig. 4.4 Comparison of baseline versus simulated monthly average flows (SDSM model)

annual flow reduction in the future period (−30%). The annual streamflow changes were between −17 and − 6.5% by SDSM model (−6.5, −17, and − 16% for RCP2.6, RCP4.5, and RCP8.5 respectively) and between −30 and + 40% in the CF method (+27.5, −30, and + 40% for RCP2.6, RCP4.5, and RCP8.5 respectively).

Q_{80} Index (FL Approach)

Figure 4.5 presents the FDCs derived from the FL approach by two different downscaling methods and daily data. Also, low-flow sections have been shown for all FDCs to provide more details. The results illustrated that Q_{80} would reduce in the future period. The arrows shown in the low-flow sections demonstrate the differences between future and basic Q_{80}. Table 4.7 shows the Q_{80} index for two downscaling methods. As expected, the results of the Q_{80} index prove that low-flow conditions will intensify in the future period. This index was close to 0.1 m^3/s for all RCPs using the SDSM model. Also, it was about 0.3 for RCP2.6 and RCP8.5 and 0.1 for RCP4.5. The historical Q_{80} index was 4.27 m^3/s for 1971–2000 and 5.4 m^3/s for 1971–1983 by the SDSM model and CF method.

Low-Flow Frequency Analysis in the Future Period

Figure 4.6 displays the resulting changes in 7Q low flows with several recurrence intervals for each RCP scenario. The results show reductions in all RCPs of about

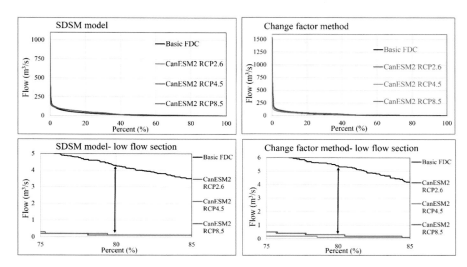

Fig. 4.5 FDC and low-flow section

Table 4.7 Q_{80} index (m³/s) for future period

Q_{80}	RCP 2.6	RCP 4.5	RCP 8.5	Historical value
SDSM model	0.1	0.1	0.1	4.27
CF method	0.3	0.1	0.3	5.4

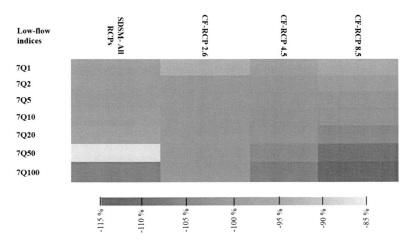

Fig. 4.6 Changes in 7Q low flows; negative quantities indicate a reduction in the magnitude of 7Q

85–115%. The reductions in RCP 8.5 are more notable than other RCPs. Also, as the recurrence interval increases, the changes become more critical. For example, 7Q flow with a 2-year recurrence interval was 2.33 m³/s in the base period. This index

(7Q2) reached about 0.01 m³/s using the SDSM model, demonstrating a severe and crucial condition for the Gharesou river basin. Also, 7Q10, as one of the most common low-flow indices, reaches 0.008 using the SDSM model, while its historical value was 0.724 m³/s. This index has reduced dramatically to zero for the CF method in the future, with the value of 1.493 m³/s in 1971–1983. These results represent the severe conditions and the extreme drought of this basin in the future period.

Environmental Water Requirement and Other Low-Flow Characteristics

Table 4.8 presents the EWR, Vmax, and Vmin (maximum and mean deficit volume as the sum of daily flow below Q_{80}) and d_{max} and d_{mean} (maximum and mean low-flow prolongation) of days with flows below Q_{80} in the historical and future period. Obviously, the maximum and mean deficit volume declined significantly in the future. Conversely, the number of days below Q_{80} increased remarkably from 56 days to 68, 70, and 69 days for RCP2.6, RCP 4.5, and RCP 8.5, respectively. It is noteworthy that the environmental water requirements rose from 22.3 mm in the historical period to nearly 30 mm, emphasizing that the environmental situation will be intensified in the future, and more flows are needed to protect the river ecosystem. Also, the change factor method data showed the same trend except for RCP 4.5. The low-flow duration (d_{mean}) fell by 3 days to 65 days, and the minimum water requirement remained stable to around 24 mm in RCP 4.5.

Limitations of the Study

There were some limitations in this chapter while conducting the research. The main restriction was the length of recorded data (precipitation and temperature) in some stations. Nevertheless, it cannot hamper the whole way of modeling. However, it can

Table 4.8 The changes in the values of EWR and other low-flow indices

SDSM model					
Index	V_{max} (million m³)	V_{mean} (million m³)	d_{max} (day)	d_{mean} (day)	EWR (mm)
Historical	5546	2522	115	56	22.3
RCP 2.6	52.5	49.07	92	68	31.3
RCP 4.5	52.5	51.03	95	70	30
RCP 8.5	52.5	50.97	100	69	30.2
Change factor					
Historical	6049	3613	129	68	24.7
RCP 2.6	151	137	109	76	35.5
RCP 4.5	52.5	44.5	101	65	24.4
RCP 8.5	149.5	136.5	123	75	34.9

change the quality of the calibration and validation of rainfall-runoff models. Besides, another important matter was related to downscaling methods, where there is more room for dynamic ones that provide more exact data for future projection. These models are not dependent on recorded datasets.

Recommendations

For future studies:

- It is recommended to evaluate multivariate analysis for low-flow criteria (e.g., copulas as commonly employed for multivariate analysis).
- Using the various GCM outputs and models can cover a full uncertainty study if the GCM models are provided for the SDSM.
- Adding one dynamical downscaling model can be a great option for observing the uncertainty in different downscaling methods.

Conclusions

This chapter provides useful data regarding future low-flow conditions for water policy-makers. The present findings might help better manage existing water resources. It also shows the importance of future protection or reducing the environmental and human societies from severe damages. Future studies should examine flow simulation uncertainty using two or more precipitation-runoff models. Additionally, an important matter to resolve is finding more operational low-flow indices. The results of this chapter indicated that all the indices would approach zero: a situation that indicates severe conditions in basin catchments.

This chapter has evaluated the effects of climate change on the hydrology of basin flow, likewise low-flow frequency. The HEC-HMS precipitation-runoff model measured hydrological responses to coming climate change. The HEC-HMS model was calibrated and validated to measure the performance of the precipitation-runoff simulation. Nash-Sutcliff coefficient (ENS) and the coefficient of determination (R2) were calculated to assess the performance of the flow simulation. The outputs of climate models using different scenarios were routed to the calibrated precipitation-runoff model, producing simulations of the future flow series. This chapter determined the projected climate change impact in the Gharesou basin using the CanESM2 model as the output of the newest report of IPPC (AR5). Various RCP scenarios were measured using the SDSM downscaling model and change factor downscaling. The indices of low-flow frequency and FL approach were analyzed on the future flows. Additionally, various probability functions were used to analyze the frequency. The findings suggest the following:

- That future flow will reduce during low-flow periods (summer and autumn) due to climate change.
- A severe decline in the 7Q10 index to nearly 0.008 in the SDSM downscaling method and the value of zero in the change factor method, as well as the Q_{80} index, will change between 0.1 and 0.3 m³/s.
- The findings highlight a role for climate change that can help water management policy-makers to apply in their management and sustainable scenarios. These changes in basin flow regimes will impose intense challenges on the reliability and quantity of water resources.
- The FL method, based on protecting animal life and the environment and based on the Q_{80} index, implies the appropriate management measures to mitigate or eliminate the damage to life and ecosystems.
- Climate change will affect the minimum water requirements to conserve the environment. The level of required water will increase to 31 mm in SDSM outputs. In contrast, it was around 22 mm in the historical period. Increasing from 24 to about 35 mm, the amount of water required will alter the change factor method.
- Furthermore, the discharge sum below a specific threshold (Q_{80}) will plummet. The average low-flow duration will prolong in most scenarios in both methods. That is to say, the future low-flow condition will worsen considerably.
- This chapter provides useful data regarding future low-flow conditions for water policy-makers. It can address problems of climate change on water resources planning and management, water stress, and the safety of vulnerable communities at a basin scale.

However, a missing link between climate change and water resources management remains. It involves communicating these two subjects and interpreting this sentence: "what information do we have and what should we do now?" Although there are limitations due to fewer recorded data for some rain gage stations, this chapter has selected the newest GCM to project future precipitation. The present findings might help manage existing water resources better and provide appropriate planning for the future. It can protect or reduce environmental hazards and human societies from severe damages. Future studies can examine flow simulation uncertainty using two or more precipitation-runoff models. Additionally, an important matter to resolve is broadening the knowledge of conceptional low flow. This chapter shows that all the indices will approach zero: a situation that indicates severe conditions in basin catchments.

References

Arora VK, Boer GJ (2010) Uncertainties in the 20th century carbon budget associated with land use change. Glob Change Biol 16:3327–3348. https://doi.org/10.1111/j.1365-2486.2010.02202.x

Bennett TH, Peters JC (2012) Continuous soil moisture accounting in the Hydrologic Engineering Center Hydrologic Modeling System (HEC-HMS) 1–10. https://doi.org/10.1061/40517(2000)149

Chen Q, Chen H, Wang J, Zhao Y, Chen J, Xu C (2019) Impacts of climate change and land-use change on hydrological extremes in the Jinsha river basin. Water 11:1398. https://doi.org/10.3390/w11071398

Chylek P, Li J, Dubey MK, Wang M, Lesins G (2011) Observed and model simulated 20th century Arctic temperature variability: Canadian earth system model CanESM2. Atmos Chem Phys Discuss 11:22893–22907. https://doi.org/10.5194/acpd-11-22893-2011

Clarke L, Edmonds J, Krey V, Richels R, Rose S, Tavoni M (2009) International climate policy architectures: overview of the EMF 22 international scenarios. Energy Econ 31:S64–S81. https://doi.org/10.1016/j.eneco.2009.10.013

Cohen SJ (1990) Bringing the global warming issue closer to home: the challenge of regional impact studies. Bull Am Meteorol Soc 71:520–526. https://doi.org/10.1175/1520-0477(1990)071<0520:BTGWIC>2.0.CO;2

Collins M, Knutti R, Arblaster J, Dufresne J-L, Fichefet T, Friedlingstein P, Gao X, Gutowski WJ, Johns T, Krinner G, Shongwe M, Tebaldi C, Weaver AJ, Wehner M (2013) In: IPCC (ed) Chapter 12 – Long-term climate change: projections, commitments and irreversibility. Cambridge University Press, Cambridge

Feser F, Rockel B, von Storch H, Winterfeldt J, Zahn M (2011) Regional climate models add value to global model data: a review and selected examples. Bull Am Meteorol Soc 92:1181–1192. https://doi.org/10.1175/2011BAMS3061.1

Fowler HJ, Blenkinsop S, Tebaldi C (2007) Linking climate change modelling to impacts studies: recent advances in downscaling techniques for hydrological modelling. Int J Climatol 27:1547–1578. https://doi.org/10.1002/joc.1556

Frederick KD, Major DC (1997) Climate change and water resources. In: Frederick KD, Major DC, Stakhiv EZ (eds) Climate change and water resources planning criteria. Springer Netherlands, Dordrecht, pp 7–23. https://doi.org/10.1007/978-94-017-1051-0_2

Gagnon S, Singh B, Rousselle J, Roy L (2005) An application of the statistical down scaling model (SDSM) to simulate climatic data for streamflow modelling in Québec. Can Water Resour J Rev Can Ressour Hydr 30:297–314. https://doi.org/10.4296/cwrj3004297

Guzha AC, Rufino MC, Okoth S, Jacobs S, Nóbrega RLB (2018) Impacts of land use and land cover change on surface runoff, discharge and low flows: evidence from East Africa. J Hydrol Reg Stud 15:49–67. https://doi.org/10.1016/j.ejrh.2017.11.005

Hagg W, Melvin L, J.W. (2008) Effects of severe drought on freshwater mussel assemblages. Am Fish Soc 137:1165–1178

HEC user manual: Hydrologic Engineering Center. HEC-HMS User's Manual version 4.7. Retrieved February 18, 2021., from https://www.hec.usace.army.mil/confluence/hmsdocs/hmsum/4.7

Huang S, Krysanova V, Hattermann FF (2013) Projection of low-flow conditions in Germany under climate change by combining three RCMs and a regional hydrological model. Acta Geophys 61:151–193. https://doi.org/10.2478/s11600-012-0065-1

Hwang S, Graham WD (2013) Development and comparative evaluation of a stochastic analog method to downscale daily GCM precipitation. Hydrol Earth Syst Sci 17:4481–4502. https://doi.org/10.5194/hess-17-4481-2013

IPCC (2007) Climate change 2007: synthesis report. In: Core Writing Team, Pachauri RK, Reisinger A (eds) Contribution of working groups I, II and III to the fourth assessment report of the intergovernmental panel on climate change. IPCC, Geneva, 104 pp

IPCC (2014) Climate change 2014: synthesis report. In: Core Writing Team, Pachauri RK, Meyer LA (eds) Contribution of working groups I, II and III to the fifth assessment report of the intergovernmental panel on climate change. IPCC, Geneva, 151 pp

Jin J, Wang G, Zhang J, Yang Q, Liu C, Liu Y, Bao Z, He R (2018) Impacts of climate change on hydrology in the Yellow River source region. China J Water Clim Change 11:916–930. https://doi.org/10.2166/wcc.2018.085

Laaha G, Blöschl G (2007) A national low-flow estimation procedure for Austria. Hydrol Sci J 52:625–644. https://doi.org/10.1623/hysj.52.4.625

Li Q, Wei X, Yang X, Giles-Hansen K, Zhang M, Liu W (2018) Topography significantly influencing low flows in snow-dominated watersheds. Hydrol Earth Syst Sci 22:1947–1956. https://doi.org/10.5194/hess-22-1947-2018

Marx A, Kumar R, Thober S, Rakovec O, Wanders N, Zink M, Wood EF, Pan M, Sheffield J, Samaniego L (2018) Climate change alters low flows in Europe under global warming of 1.5, 2, and 3 °C. Hydrol Earth Syst Sci 22:1017–1032. https://doi.org/10.5194/hess-22-1017-2018

Meinshausen M, Smith SJ, Calvin K, Daniel JS, Kainuma MLT, Lamarque J-F, Matsumoto K, Montzka SA, Raper SCB, Riahi K, Thomson A, Velders GJM, van Vuuren DPP (2011) The RCP greenhouse gas concentrations and their extensions from 1765 to 2300. Clim Chang 109:213. https://doi.org/10.1007/s10584-011-0156-z

Nasonova ON, Gusev YM, Kovalev EE, Ayzel GV, Panysheva KM (2019) Projecting changes in Russian northern river runoff due to possible climate change during the 21st century: a case study of the northern Dvina, Taz and Indigirka Rivers. Water Resour 46:S145–S154. https://doi.org/10.1134/S0097807819070145

Neubauer CP, Hall GB, Lowe EF, Robison CP, Hupalo RB, Keenan LW (2008) Minimum flows and levels method of the St. Johns river water management district, Florida, USA. Environ Manage 42:1101–1114. https://doi.org/10.1007/s00267-008-9199-y

Ouyang Y (2012) A potential approach for low-flow selection in water resource supply and management. J Hydrol 454–455:56–63. https://doi.org/10.1016/j.jhydrol.2012.05.062

Pathak P, Kalra A, Ahmad S, Bernardez M (2016) Wavelet-aided analysis to estimate seasonal variability and dominant periodicities in temperature, precipitation, and streamflow in the midwestern United States. Water Resour Manag 30:4649–4665. https://doi.org/10.1007/s11269-016-1445-0

Pyrce R (2004) Hydrological low-flow indices and their uses. Trent University, Watershed Science Centre, Peterborough, p 37

Revenga C, Brunner J, Henninger N, Kassem K, Payne R (2000) Pilot analysis of global ecosystems: freshwater systems. World Resources Institute, Washington, DC

Riahi K, Rao S, Krey V, Cho C, Chirkov V, Fischer G, Kindermann G, Nakicenovic N, Rafaj P (2011) RCP 8.5—a scenario of comparatively high greenhouse gas emissions. Clim Change 109:33. https://doi.org/10.1007/s10584-011-0149-y

Ryu JH, Lee JH, Jeong S, Park SK, Han K (2011) The impacts of climate change on local hydrology and low-flow frequency in the Geum River Basin, Korea. Hydrol Process 25:3437–3447. https://doi.org/10.1002/hyp.8072

Shrestha S, Sharma S, Gupta R, Bhattarai R (2019) Impact of global climate change on stream low flows: a case study of the great Miami river watershed, Ohio, USA. Int. J. Agric. Biol. Eng 12:84–95. https://doi.org/10.25165/ijabe.v12i1.4486

Smakhtin VU (2001) Low-flow hydrology: a review. J Hydrol 240:147–186. https://doi.org/10.1016/S0022-1694(00)00340-1

Smakhtin V, Revenga C, Döll P (2004) A pilot global assessment of environmental water requirements and scarcity. Water Int 29:307–317. https://doi.org/10.1080/02508060408691785

Széles B, Broer M, Parajka J, Hogan P, Eder A, Strauss P, Blöschl G (2018) Separation of scales in transpiration effects on low flows: a spatial analysis in the hydrological open air laboratory. Water Resour Res 54:6168–6188. https://doi.org/10.1029/2017WR022037

Trzaska S, Schnarr E (2014) A review of downscaling methods for climate change projections (technical report, United States Agency for International Development). Tetra Tech ARD, Burlington

Van Vuuren DP, Edmonds J, Kainuma M, Riahi K, Thomson A, Hibbard K, Hurtt GC, Kram T, Krey V, Lamarque J-F, Masui T, Meinshausen M, Nakicenovic N, Smith SJ, Rose SK (2011) The representative concentration pathways: an overview. Clim Chang 109:5. https://doi.org/10.1007/s10584-011-0148-z

Wilby RL, Harris I (2006) A framework for assessing uncertainties in climate change impacts: low-flow scenarios for the River Thames, UK. Water Resour Res 42. https://doi.org/10.1029/2005WR004065

Wilby RL, Dawson CW, Barrow EM (2002) sdsm—a decision support tool for the assessment of regional climate change impacts. Environ Model Softw 17:145–157. https://doi.org/10.1016/S1364-8152(01)00060-3

Xu C (1999) Climate change and hydrologic models: a review of existing gaps and recent research developments. Water Resour Manag 13:369–382. https://doi.org/10.1023/A:1008190900459

Part II
Climate Change and Contemporary Issues, Challenges and Sustainability

Chapter 5
Climate Change Impact on Land Degradation and Soil Erosion in Hilly and Mountainous Landscape: Sustainability Issues and Adaptation Strategies

Suresh Kumar, Anu David Raj, Justin George Kalambukattu, and Uday Chatterjee

Abstract Soil degradation declines the inherent soil quality, resistance and stability which leads to less productive and highly vulnerable soil. The world's population is increasing in an exponential rate and through this pressure on land is increasing rapidly. Consequently, to meet the food requirement, cultivation is shifting towards hilly and mountainous terrain where the susceptibility to soil erosion is very high. Soil erosion is a major land degradation issue, which decreases the soil productivity and causes soil carbon loss, which in turn accelerate various climate change processes. The studies reported that, in the context of climate change and due to its ancillary effects, soil erosion is anticipated to intensify at a higher rate in future periods than the baseline. Hence, it was evidenced that climate change has an enormous potential to increase the soil erosion rates. The hilly and mountainous regions are most and first affected regions due to changes in climate. The unsustainable land use practices, steep slopes and climate change are the prime factors which enhance the soil erosion in hilly and mountainous region. Sustainability is the prime key, which can solve the concern of soil erosion. Soil erosion identification, monitoring and quantification using remote sensing and GIS coupled with various erosion models can provide a comprehensive information about the soil erosion processes and its quantity. Studying impact of climate change on soil erosion provides an in-depth knowledge about possible future erosion scenarios. These kinds of impact studies at local scale can generate more beneficial and site-specific measures to overcome soil erosion. Adoption of various site-specific adaptation

---

S. Kumar (✉) · A. David Raj · J. G. Kalambukattu
Agriculture and Soils Department, Indian Institute of Remote Sensing (IIRS), Indian Space Research Organization (ISRO), Dehradun, Uttarakhand, India
e-mail: suresh_kumar@iirs.gov.in; sureshkumar100@gmail.com; anudraj@iirs.gov.in; anudraj2@gmail.com; justin@iirs.gov.in

U. Chatterjee
Department of Geography, Bhatter College, Dantan (Affiliated to Vidyasagar University), Paschim Medinipore, West Bengal, India
e-mail: raj.chatterjee459@gmail.com

© The Author(s), under exclusive license to Springer Nature Switzerland AG 2022
U. Chatterjee et al. (eds.), *Ecological Footprints of Climate Change*, Springer Climate, https://doi.org/10.1007/978-3-031-15501-7_5

and mitigation strategies based on Sustainable Development Goal (SDG) framework, nature-based solutions and appropriate local government policies can weaken the soil erosion driving forces through a more viable and environment-friendly pathway.

Keywords Soil degradation · Soil erosion · Climate change · Adaptation and mitigation · Soil sustainability · Sustainable Development Goals (SDG)

Introduction

Land degradation is the term which simply represents as the name indicates the process- degradation of land. Although it is not simple, different researchers characterise the land degradation as per their perception. The term "land" represents a comprehensive entities and processes in the surface and subsurface of the Earth, while "degradation" epitomises deterioration of functions and vital processes. According to UNCCD (1994), it is the decline in productivity of various terrestrial bio-productive systems including vegetation, soil, biological and hydrological processes due to change/loss in vegetation, soil properties and soil erosion, while FAO (2002) stated that land degradation is the loss in productivity and bio-diversity of soil with depletion of natural resources. It has a direct relationship with the food productivity and economic stability of a country. It eventually declines the soil quality and results in loss of soil productivity. Land degradation may occur due to human-induced, climate change or other natural processes. The geographic location and climatic factors can also determine the type of degradation processes. For example, desertification may occur in arid regions, rain-abundant regions severely affected by soil erosion, low-lying plains severely affected by waterlogging and some coastal and flood plains affected by salinisation or alkalisation.

The improper management of arable lands, unsustainable forest management, deforestation, mining, industrial wastes, overgrazing and climatic extremes enhance land degradation in various ways. Land degradation has a great significance among the major global concerns due to its potential to alter quality of life and global food security. There are numerous research and evaluations that prove that land degradation has an impact on all aspects of livelihood. Soil degradation is a universal challenge in the twenty-first century that is predominantly severe in the tropics and subtropics (Lal 2015). The land degradation negatively affects the soil quality. Thus, it may further alter the proper functioning of soil ecosystem services and processes. Many difficulties confront people working in the field of land resources, including methods for reducing degradation as well as strategies for assessing and monitoring it. Thus, understanding the land degradation based on different characteristics of the processes is necessary. The cause of soil erosion also may vary depending upon the driving forces predominant in the geological region. In majority of geographical regions, climate plays a prime role as a driving force for soil erosion. Hence, climate change is expected to greatly influence various erosion processes. The population pressure, shifting cultivation, improper land use management, deforestation, grazing and mining are other major driving forces which drastically affect soil erosion. According to FAO, globally in most of the regions, the erosion condition is poor and the status of soil is deteriorating (Table 5.1). These above-stated reasons verify why the soil erosion is considered as a global concern.

For a long time, it has been recognised that soil erosion reduces crop production, resulting in decreased agricultural potential (Pimentel 1987). Accelerated erosion rates are known to significantly contribute towards the destruction of natural resources (Kovda 1977). The soil erosion is assumed to have permanently ruined 430 million hectares of land in various countries. This associates to roughly 30% of the world's existing farmed land area. Globally, natural/geological erosion is estimated to cause loss of 9.9 billion tonnes of soil annually, while human-induced accelerated erosion is expected to cause loss of nearly 26 billion tonnes a year (more than 2.5 times) (Brown 1984a, b; Holdgate et al. 1982). In general, soil erosion is observed to be much severe in hilly terrain than in flat areas. Previous studies have estimated that nearly 24–75 billion tonnes of fertile soil are lost each year from agricultural fields (Crosson 1997; Myers 1993). Each year, India loses 6.6 billion tonnes of soil, according to Lal and Stewart (1990). India's major land degradation problem is due to water erosion, which results in topsoil loss and soil displacement. In India 146.8 Mha of total land are degraded, with soil erosion by water accounting for the majority (93.7 million hectares) (NBSS&LUP 2004). The average rate of soil erosion was calculated to be 16.4 tonnes per hectare per year, resulting in an annual loss of 5.3 billion tonnes of soil across the country (Narayan and Babu 1983).

Erosion is a very serious problem in the Himalayan region, and the rivers flowing through the region transport a massive amount of debris and thus enormous quantities of soil nutrients (Sharma et al. 1991). Due to excessive soil erosion, the hydrological system, crop productivity, water quality and the ecology are all impacted (Lal 1998). The Northwestern Himalayan region is particularly more vulnerable to soil erosion. The Himalayan-Tibetan mountain range is one of the world's most deteriorated ecosystems (Dent 1984). Due to Nepal's rugged landscape and high population density, particularly in the mountainous areas of the Shivalik highlands in the Himalayas, it is very prone to erosion (Dregne 1982). 336 million tonnes of soil are transported by India's rivers to the main river system (Brown 1981). The Terai rivers' bed levels are reported to be rising by 15 to 30 centimetres every year as a result of sediment deposition (Dent 1984). Each year, the Ganges River in India deposits 1.6 billion tonnes of soil into the Bay of Bengal (Brown 1984a, b). To tackle this global concern, a well-equipped integrated soil erosion prevention framework is very much necessitated. The proper implementation of UN's 17 Sustainable Development Goals (SDGs) will help us to adapt and mitigate this concern.

The United Nations' implementation of the 17 Sustainable Development Goals (SDGs) as part of the 2030 Agenda for Sustainable Development placed a burden on the scientific community to generate sound data to aid in the development and scrutinising of socio-economic development that is environmentally sustainable (UN 2015). The SDG stands for food security and safety, the nutrient pollution from land to sea, urban development and sustainable survival of the terrestrial

Table 5.1 Global water erosion zones and major causes

Latitude zone	Continent	Subzone	Erosion condition (FAO)	Current erosion trend (FAO)	Causes
Tropics	Africa	Ethiopian highlands	Poor	Deteriorating ⬇	Population density
		East Africa	Poor	Deteriorating ⬇	Extreme climate
		Southern Africa and Madagascar	Poor	Deteriorating ⬇	Intensive farming
		Tropical wet-dry regions	Poor	Deteriorating ⬇	Uncontrolled grazing
		Equatorial Africa	Poor	Deteriorating ⬇	Excessive stocking rate
		Semi-arid and arid Africa	Poor	Deteriorating ⬇	Soil profile characteristics
	Asia	Southeast Asia	Poor	Deteriorating ⬇	Cultivation in steep slopes
		South Asia	Poor	Deteriorating ⬇	Deforestation
					Shifting cultivator in highlands
					Production of upland crops (maize, cassava)
					Extreme climate
	America	America and Caribbean	Poor	Deteriorating ⬇	Conversion of natural vegetation for cropland
	Australia	Australia	Fair	Improving ⬆	Abandoned mines
					Permeable vertisols
					Heavy harvesting
					High-intensity tropical rain
Subtropics	Asia	East Asia	Poor	Deteriorating ⬇	Geologically unstable
		The Himalayan-Tibetan Mountain ecosystem	Poor	Deteriorating ⬇	Soils and their parent materials are vulnerable
		China	Poor	Deteriorating ⬇	Population density
					Intensive land use
					Monsoonal climate (torrential rain)
					Cultivation in steep hills
					Natural resource degradation

(continued)

Table 5.1 (continued)

Latitude zone	Continent	Subzone	Erosion condition (FAO)	Current erosion trend (FAO)	Causes
	Africa	West Asia and North Africa	Very poor	Deteriorating ⬇	Extreme rainfall Denuded hills
	Europe	Southern Europe	Fair	Improving ⬆	Flysch rock types Farming in steep slopes Extreme rainfall
	America	Midlatitude regions in America	Fair	Improving ⬆	Grazed land Intensive cultivation
	Australia	Australia	Fair	Improving ⬆	Intensive cultivation Deforestation and logging Grazing
Temperate	Europe	Europe	Fair	Improving ⬆	Orchards and croplands Steep lands
	America	North America	Fair	Improving ⬆	Stream instability Logging and mining Climatic extreme

ecosystem in which all these are directly related to soil sustainability. There are several connections between the Sustainable Development Goals and soil sustainability management (Keesstra et al. 2016). The SDGs aid in rationalising soil monitoring programmes, which can supply the requisite data and information when paired with novel bases of soil knowledge and widespread use of contemporary technical resolutions (Tóth et al. 2018). SDG 13 also demands the sustainable climate action which can aid in soil sustainability. Hence, the SDG explicitly covers most of the fields which comes under the human livelihood and which can enhance the quality of life and ecosystem through aiding the proper functioning of ecosystem. Soil erosion and climate change are found to be interdependent of each other according to the studies conducted globally. Thus, it is very necessary to study the impact of climate change on soil erosion and other associated processes. Here we are describing the soil erosion (soil degradation) and its quantification techniques, climate change and climate models, downscaling, sustainability and approaches in a comprehensive way. Apart from this, we are describing two case studies conducted in the mountainous environment of Himalayas for providing more insight to climate change impact on soil erosion and its effects.

Land Degradation

Degradation of land is induced by physical, chemical and biological factors (Lal 1994a, b, c). Land degradation is demarcated by old geological formations, juvenile soil profile development or the presence of heavily eroded soil cover, and the occurrence of dissected geomorphological landforms (Mahala 2020). The physical degradation includes the erosion of soil induced by water and wind, crusting, sealing, desertification and compaction. It may reduce the soil's physical stability and structural features. In agricultural fields, the heavy agricultural machineries cause the deformation of soil through the stress applied by them (Blum 2011). The chemical degradation is primarily described by salinisation, acidification, leaching and nutrient exhaustion. It directly increases toxicity of soil which reduces the productivity of soil. The biological land degradation reduces the soil biodiversity and carbon storing capacity, resulting in higher emission of greenhouse gases from the soil. It eventually causes the decline in soil organic matter content. Comprehensively, these three factors affect the ecological balance, thereby disrupting hydrological and nutrient cycle and resulting in decreased net biome productivity (Lal 2015).

Land degradation is driven by a range of natural processes, but it is often hastened by human activity, resulting in a loss of proper soil functioning. Land that is degrading usually goes through three stages. Because a steady state exists between soil development and soil degradation, natural degradation is often gradual. While human-induced degradation occurs as a result of poor land use and management. Natural deterioration takes longer than human-induced degradation. Desertification occurs when a level of deterioration has occurred to the point where the land's resilience has been compromised (Fitzpatrick 2002). However, generally, the degradation in dry land is represented as desertification. The biophysical processes and features that regulate the type of degradative processes are known as land degradation factors. Climate, topography, vegetation and biodiversity, particularly soil biodiversity, and intrinsic properties are included. The factors responsible for determining the rate of land degradation are referred to as the causes of land degradation. These are the biophysical, social and political aspects that influence how land deteriorations and factors affect the environment (USDA-NRCS n.d.). There is no specific boundary between these land degradation factors. Thus, in some geographical region, overlapping of this process is visible. Identifying the exact processes and assessing their quantity and extent are very vital in identifying risks and reversing it. As mentioned, the soil erosion is considered as the most threatened and extended land degradation issue in the world. Hence, the soil erosion, processes, measurement and modelling are described in the following sections.

Soil Erosion

Soil erosion, the accelerated removal of soil from land surface by various erosive agents such as water, wind, etc., is considered as a major threat to soil resources across the globe (FAO 2015). Various weathering forces in the form of physical, chemical as well as biological processes result in disintegration of rocks and minerals to form soil material, which gets eroded in due course of time. The formation rate and soil characteristics are heavily dependent on various factors such as rock characteristics, climate, vegetation, etc. Erosion is a three-stage process comprising of detachment, transport as well as deposition of soil material by the action of various erosive agents. Water, wind, gravity as well as glaciers are the predominant erosive agents acting at various geographical locations across the globe. The nature of predominant erosive agent varies across geographical landscapes as well as climatic regions as indicated by higher erosion rates in humid tropics caused by water, whereas wind plays the major role in dry arid environments.

Factors Affecting Water Erosion

Water erosion being the most dominant form of erosion accounts for nearly 80% of soil loss. Rainfall and the associated runoff play major roles in detachment as well as transportation of soil material, thus causing water erosion. The loss of material by water erosion is governed by various factors related to characteristics of rainfall, topography, soil, nature of ground cover, etc. Intensity and duration are the major rainfall characteristics playing pivotal roles in water erosion. Soil aggregates are mainly broken down due to the direct impact of rainfall. Detached particles from broken aggregates get splashed and disrupt other aggregates and cause sealing of soil pores by fine particles, thus reducing the infiltration rate. In addition, raindrop characteristics such as mass, diameter as well as velocity also significantly influence erosion rates. These various rainfall characteristics have been considered and assimilated during the formulation of a single factor called rainfall kinetic energy. The rainfall kinetic energy is used for estimating the rainfall's potential ability to cause erosion, defined as rainfall erosivity. Field-based long-term continuous rainfall observations, climate records, remote sensing-derived rainfall products, etc. are widely being used for estimation of rainfall erosivity values.

The ability of soils to withstand or resist erosion processes, defined as soil erodibility, is entirely dependent on various soil characteristics. The major soil characteristics influencing erosion rates are textural composition, stability of aggregates, infiltration capacity, amount of organic matter present, etc. The interplay of these different characteristics improves soil's ability to withstand erosion by inhibiting the easy breakdown of aggregates, slowing down the particle transportation as well as runoff/overland flow by improved infiltration rates, etc. Soil erodibility estimation involves use of various relationships, incorporating the values of different easily measurable soil properties. In general, presence of high amount of organic matter, increased infiltration rates and good structure improves the resistance to erosion processes. Similarly, particles like silt and very fine sand are found to be more susceptible to erosion compared to other soil types/particles. Among the various topographic factors, slope length and gradient play most important roles in influencing erosion processes. The effect of these factors is predominant especially in hilly as well as undulating terrains. A disproportionate increase in erosion risk has been observed in case of steeper and longer slopes compared to flatter terrains mainly due to the increased runoff volume and velocity. This causes an increased capacity of runoff to carry sediments resulting in higher erosion rates. Apart from these, slope shapes are also known to affect erosion rates by influencing flow and deposition characteristics.

Vegetation, both natural and cultivated, along with their residues greatly influences the erosion rates. The vegetation mainly influences the risk as well as erosion rate by various interventions comprising of rainfall interception by canopy, runoff velocity reduction by physical obstruction, surface coverage by litter/residues, soil structure and strength improvement, improved porosity, protection from higher temperatures, etc. Canopy absorbs the impact of rainfall and significantly lowers the kinetic energy, thus reducing the particle detachment at the surface. Similarly, the runoff slowdown due to physical obstruction improves the infiltration, thus influencing the transportation of eroded materials. Various vegetation characteristics such as crop and canopy type, planting distance, above-ground height of canopy, rooting pattern, residues, etc. determine the influence of vegetation on erosion during different periods. Crops such as cereals and row-planted crops results in increased risk as well as rates of erosion, as they leave the soil bare during major periods of a year, which receive high erosive rainfall events such as summer as well as monsoon. This is in contrast to the effect of legumes and many cover crops which reduces erosion by providing improved protective cover on soil surface. Various crop and residue management, tillage practices as well as cultural operations when planned and executed scientifically are known to significantly reduce amount of soil erosion.

Soil Erosion Types and Characteristics

Soil erosion is categorised into various types, namely, splash, sheet, rill as well as gully erosion based on the mechanism of occurrence. Splash erosion mainly occurs due to the detachment of soil particles in response to direct impact of raindrops on soil surface or shallow water surfaces. This mainly happens as a result of soil aggregates' breakdown under the impact of high-energy rainfall, which releases finer soil particles in the form of a "splash". The splash erosion rates are mainly dependent on various rainfall characteristics like mass/size of raindrops, velocity, rainfall direction/intensity as well as other factors such as canopy cover, previous soil moisture status, retention of residues, etc. Sheet erosion indicates the uniform removal of detached or splashed soil particles by surface runoff, especially in case of gently sloping lands. As the removal of soil particles occurs in a uniform fashion, such losses remain unnoticed for quite some time until land surface is lowered due to loss of a good amount of productive topsoil. Such soil loss in fields could be easily identified in the form of muddy runoff. Soil loss rates due to sheet erosion are influenced by slope and other factors similar to splash erosion.

Rill erosion indicates the loss of soil particles along with concentrated runoff in small channels that are few millimetres wide and deep, referred to as rills. Runoff concentration and subsequent flow in rills lead to particle removal by scouring action and result in widening and deepening of rills in due course of time, if ignored. Rills do not interfere with field operations generally and can be eliminated by various tillage operations. Gully erosion considered as the most advanced form of soil erosion indicates removal of soil particles by concentrated runoff in comparatively permanent channels (gullies) with high width and depth dimensions and steeper side slopes. Generally, channels designated as gullies are known to have a cross-sectional area greater than 1 m^2 (Poesen 1993). Due to high width and depth dimensions reaching to several metres, gullies are known to cause physical obstruction of various field operations like tillage as well as natural movement of personal and livestock. The runoff volume and sediments transported via gullies are much higher compared to rills, and if left unattended, the gullies can result in land dissection due to widening as well as deepening. The land dissection may result in the formation of ravenous lands by the interconnection and networking of gullies.

Field and Laboratory Estimation Methods

Measurement and estimation of soil erosion rates in field as well as laboratory conditions is a vital component in studying erosion processes. The selection of measuring methods depends primarily on the objective. Field measurements are observed to be the most reliable methods for obtaining realistic soil loss data, whereas laboratory-based experiments having ability to control many factors and their effects are adopted to determine the major erosion causes as well as understanding various processes.

Lal (1994a, b, c) as well as FAO (1993) has given detailed description of various soil erosion estimation and measuring methods. The various erosion estimation methods can be broadly classified as reconnaissance methods, field plots, stream flow estimation, sediment transport estimation, estimation using various models, etc. Reconnaissance methods are comparatively cheap and simple and provide ways to get first approximation of amount of soil eroded. It also provides us with the opportunity to have repetitive measurements, thus improving the reliability and representative nature of erosion estimation. The various reconnaissance methods can be categorised as (i) measuring change of surface level and (ii) volumetric methods. The methods which measure surface level changes are suitable in locations characterised by high erosion rate, and we are able to predict/identify the position of erosion such as deforested steep lands, cattle track in rangelands, etc. Surface level changes could be measured in single dimension at a point, whereas multidimensional measurements are possible in case of profile (two dimension), rills and gullies (three-dimensional/volumetric measurement). The various instruments/methods for point measurement include use of erosion pins, paint collars, bottle tops, pedestals, tree mounds and tree roots, as well as profile metres whereas the volumetric measurements involve the measurement of length as well as cross-sectional area and their changes of rills, roads, gullies and streambanks; use of catch pits, landslips and landslides; etc. The volume measurements could be effectively conducted using measuring tapes, scanners, etc.

Another category of erosion measuring techniques is the erosion plot/runoff plots. Erosion plots are generally used for demonstration purpose in field conditions as well as comparative assessment of the effect of various conditions/treatments such as mulching, tillage, slope positions, etc. They also form a vital component in the data collection for validation as well as construction of runoff/soil loss prediction models/equations. They are mainly created by separating a part of slope by rectangular partition, thus protecting it from runoff originating from surrounding. Runoff generated and the associated sediments from the plot will be collected and measured at the lowest portion of plot, using certain devices. The collection as well as storage tank, sampling devices, different types of recorders, etc. forms the essential components of erosion measurement using erosion plots. Runoff plots are considered expensive, in terms of initial construction, maintenance and operating costs.

Hydrological methods form another widely adopted category of erosion measurement, which involve measurement of stream flow/runoff in channels as well as the sediment transport. Discharge in channels could be measured by making use of various approaches such as volumetric method (direct measurement in case of small channels), velocity/area method (using average velocity of flow and channel's cross-sectional area), rating gauging stations (using rating curves), using empirical velocity estimation formulas and making use of various types of gauging weirs, measuring flumes as well as water level recorders. Simultaneously, the sediment load (sediment quantity in unit flow quantity) needs to be estimated by collecting water samples periodically using various sampling techniques. The sample collection includes hand sampling or the use of various types of automatic samplers. Once the flow volume and sediment load are known, total sediment transport in the form of suspension could be easily estimated. The sediment lost in the form of surface movement on riverbeds (bed load) could be estimated using dug holes, bed load traps, samplers, radioactive tracers, etc. These measurements play a vital role in the development, calibration as well as validation of various erosion models, which helps us to assess the soil erosion scenario at system levels such as watershed level, river basins, etc. Recently, radioactive 137Cs isotope-based long-term soil erosion rate estimation has been gaining more importance especially in areas under the Northern Hemisphere. The method could provide us with much higher insights into the long-term impact of changing climates on past as well as current erosion rates (FAO & IAEA 2017).

Geospatial technology involving the use of remote sensing data at varying spatial as well as temporal resolutions provides vital information regarding vegetation, topography as well as climatic parameters, which are known to be the most important drivers of soil erosion process at varying scales and geographies. Various remote sensing-derived information when integrated in a GIS platform could aid us in spatial assessment of soil erosion rates at different scales (local, regional, global) with considerable accuracies using different empirical, semi-empirical as well as process-based soil erosion models (Kalambukattu and Kumar 2017; Kalambukattu et al. 2021; Borrelli et al. 2017).

Soil Erosion Models

Soil erosion is the major threat for soil quality, health, agricultural productivity and water quality. The planning and implementing of the appropriate adaptation and mitigation measures (conservation practices and appropriate land use plans) need the explicit quantification of spatial and temporal distribution as well as critical source area for generating the soil erosion. Soil erosion models are capable of capturing the soil erosion processes, based on the various factors directly or indirectly contributing soil erosion. These models differ in their capability, application and spatial and temporal scales of soil erosion simulation. The erosion models are primarily classified into four, viz. (i) empirical, (ii) physically based, (iii) conceptual and (iv) hybrid models.

Empirical Model

The empirical erosion models are developed based on simple linear statistical relationship between the dependent (soil loss) and independent variables (erosion-causing factors). Hence, this kind of model does not carry large uncertainties. For example, the USLE (Universal Soil Loss Equation), established in the mid-1960s, was developed for cropland (Wischmeier and Smith 1965). Later Renard et al. (1994) developed Revised Universal Soil Loss Equation (RUSLE), which has enhanced capability and widely used integrated with GIS. RUSLE model is considered as the most widely used and tested soil erosion model which provides annual average soil loss from small field to large catchment scale. Primarily, USLE was a point-scale field (plot)-based model. But nowadays, researchers are using USLE for larger catchment with the aid of geographic information system (GIS) (Mahapatra et al. 2018).

Physically Based Model

These kinds of models commonly represent the physical processes which occur during the erosion processes. They simulate erosion processes based on the principles of law of conservation of mass, energy and momentum. Hence, they use equation of continuity to represent the erosion processes. Most of these models are capable of simulating erosion on event or daily basis up to hundreds of thousands of years if sufficient data is available for the simulation. In physically based models, numerical articulations which are derived from particular processing have several assumptions that need not be important in much of the natural environment (Beven 1989). Calibrated and properly parameterised physically based models provide more realistic results than the empirical models. In the mid-1980s for the Limburg Provincial Government (De Roo et al. 1996), the Limburg Soil Erosion Model (LISEM) was established. Smith et al. (1995) developed an event-oriented, deterministic and physically dependent model, which is the Kinematic Erosion (KINEROS) model for the USDA Agricultural Research Service. Water Erosion Prediction Project (WEPP) (Flanagan and Nearing 1995) is a reliable process-based model created by the USDA-ARS for hillslopes and watersheds. The GIS interface of WEPP model (GeoWEPP) is also available, which enhances the input parameterisation and output visualisation to be very user-friendly.

Conceptual Models

Conceptual models are a mixture of empirical and physical models and are increasingly important to address general queries of erosion (Beck 1987). These models have been created based on spatially lumped types of water and the soil condition (Lal 1994a, b, c). Erosion-Productivity Impact Calculator (EPIC) model, developed by Williams et al. (1984), is used to study the impact of soil erosion on soil productivity. The model's hydrology module predicts the surface runoff of daily precipitation with an equation-curve number. Another conceptual model that is frequently under processing and mostly used around the world is the Soil and Water Assessment Tool (SWAT) model. The model is a global, continuous, semi-disseminated river basin-centred model based on processes (Arnold et al. 1998; Williams et al. 2008).

Hybrid Models

Hybrid models represent a combination of methods to measure soil erosion. The framework of the hybrid model is usually physical or logical in nature, while the model is largely based on empirical assumptions and relies on proven regression relationships in space and time scales. The hybrid models apply physical and measured soil erosion procedures and models of sedimentation systems. For example, a promising method for gauging source and transport of sediment at catchment scales is given by a Sediment River Network Model (SedNet) (Prosser et al. 2001). SedNet is a mean annual model that produces slope, soil, ravine and bank erosion on an annual basis. The Morgan-Morgan-Finney model (Morgan et al. 1984; Morgan 2001) is a concept that is being used in a wide variety of different circumstances of soil erosion in the fields of land, hillslopes and small-field areas.

Soil Carbon Models

Carbon dynamics is a vital component which is affected by soil erosion processes and also has direct impact on global climate change. The soil degradation (primarily soil erosion) reduces the carbon content in the soil. The increased carbon loss from the soil leads to decreased aggregate stability. It further increases the vulnerability of soil erosion during torrential rain. Intense and torrential precipitation along with steep slopes is the major threat for carbon loss from hilly and mountainous region. Hence, soil erosion-induced carbon loss carries an import role in climate change and soil erosion processes in a cyclic way. The carbon dynamic model provides an in-depth insight to the processes which affect carbon loss from the soil. There are plenty of carbon dynamic models that are available to simulate carbon dynamics. The CENTURY simulation model enables long-term soil organic carbon (SOC) trends to be predicted using mathematical techniques by simulating carbon cycling processes in soil-plant systems. Colorado State University and the USDA-ARS collaborated to build the model, which is a multicompartmental ecosystem model and is used to assess carbon dynamics in Great Plains grasslands (Parton et al. 1987). Many researchers studied the soil organic carbon changes using the CENTURY model around the world at various geographical, topographical and climatological conditions (Gupta and Kumar 2017a, b; Kelly et al. 1997; Nicoloso et al. 2020; Voroney and Angers 2018). Few models are Carbon, Organisms, Rhizosphere, and Protection in the Soil Environment (CORPSE) (Sulman et al. 2014), Microbial Enzyme-Mediated Decomposition (MEND) (Wang et al. 2015), MIcrobial-MIneral Carbon Stabilization (MIMICS) (Wieder et al. 2014) and DAYCENT (Parton et al. 1998) which is a daily version of the CENTURY model. Rothamsted carbon (RothC) models are widely used to predict the long-term dynamics of soil organic carbon (SOC) and carbon dioxide (CO_2) emissions in topsoil for a variety of plant types and various land uses (Coleman and Jenkinson 2014). These kinds of model enhance the perception about the soil carbon dynamics in the soil, plant and atmosphere in the global climate change context. Hence, it will aid in helping proper management of soil with minimum or no emission of carbon from the soil biota.

Climate Change

The United Nations Framework Convention on Climate Change (UNFCCC) defines climate change as "a change in climate which is attributed directly or indirectly to human activity that alters the composition of the global atmosphere and which is in addition to natural climate variability observed over comparable periods of time" (UNFCCC 1992). The outcomes of climate change nowadays are rapid, frequent and devastating. Extreme weather events can have the capability to destroy a stable system/process abruptly. As mentioned in the UNFCCC definition, human-induced climate change is very rapid because of increased greenhouse gas emission from the fossil fuels. CO_2, CH_4, N_2O and water vapour are the major greenhouse gases which contribute to global warming. Among them, CO_2 is considered to have the highest potential capability to primarily contribute to global warming through anthropogenic activities. The soil is considered as one of the major sinks of carbon, while soil degradation (primarily soil erosion) process can convert it as global soil carbon source. The climate models are used to simulate complex mechanisms which drive the climate over a region. The land, atmosphere and ocean are considered as the major components of climate system. Soil also plays a major role in controlling the gaseous cycles, which include soil-plant-atmospheric continuum. For the impact studies (soil erosion), the past, present and future changes in climate with respect to radiative forcing are very much required. The complex models help to achieve reliable future physically based condition of atmosphere which is discussed in the following sections.

Climate Models

Climate models are the numerical representation of physical and chemical processes which occur in the land, atmosphere and oceanic continuum. These models obey all the physical laws which are applicable to them. However, these models have different capabilities and drawbacks according to the processes they represent. The World Climate Research Group (WCRP) is controlling and maintaining the climate modelling approaches through the Coupled Model Intercomparison Project (CMIP). It provides proper guidelines and comparison of various climate models around the world. The energy balance model, one-dimensional radiative convective model, earth system model of intermediate complexity and global circulation models (GCMs) are the major types of climate model. These models are used based on the various applications of atmospheric and oceanic phenomena. The GCMs are more complex and majority of the models are used as the combination of the oceanic and atmospheric general circulation model (AOGCM). The IPCC provides various emission and socio-economic pathways for future climate scenario. These models simulate the future climate and their predictor variables (variables which control the rainfall and temperature) according to the various radiative forces and socio-economic conditions.

Emission Scenarios

Due to the integration of diverse natural courses into climate models and the socio-economic aspects incorporated in IPCC scenarios, downscaled IPCC climate scenarios are the most frequently employed for more credible assessments (Garbrecht and Zhang 2015, Zhang et al. 2004). The SRES (Special Report Emissions Scenarios) family of climate scenarios is the most frequently employed in present impact studies (Nakicenovic et al. 2000). It has mainly four storylines or scenarios, viz. A1, A2 and B1, B2; A represents economically rapid growth and B represents environment-friendly growth, while 1 represents a global scenario and 2 represents a locally emphasised scenario. However, given the new set of climate scenario, Representative Concentration Pathways (RCPs) include short-lived emissions and land use change, as well as climate policies for mitigation, and additional impact assessments are planned to employ this new collection of climate scenarios in the future. The RCP scenarios are 2.6, 4.5, 6 and 8.5 representing the future possible radiative forcing; 2.6 is low-emission scenario and 8.5 is high-emission scenario. The very latest forms of scenarios are developed under the CMIP6 project named shared socio-economic pathway (SSP). It combines socio-economy as well as radiative forcing. It has five scenarios from SSP 1 to SSP 5, and detailed description can be found from WCRP documentation. Based on the application and feasibility researchers, various scenarios can be selected.

Downscaling

Downscaling is the commonly used term for a procedure that uses existing data at large scales to make predictions at smaller scales. Global or general circulation models (GCMs) replicate the Earth's atmosphere using scientific assumptions that define the atmosphere, ocean and biotic processes, associations and contributions. In any case, due to the spatial resolution difference between GCMs and hydrological models, the GCM cannot be used directly at the regional level, but can be used to detect the hydrological effects of climate change at the watershed and local levels. These models have a resolution of approximately 2° x 2° and a grid dimension of approximately 100–500 kilometres. Additionally, they are frequently compact at the month-to-month temporal scale, which is insufficient for basin-scale climate change evaluations (Buytaert et al. 2010; Mora et al. 2014). The two vital methods used for downscaling atmospheric data are dynamic and statistical. Dynamic downscaling necessitates the employment of high-resolution atmosphere models in territorial subspace, as well as the use of model performance at low resolution as a confined zone (Wilby et al. 2004; Karamouz et al. 2010).

Regional climate models (RCM) incorporate enormous amounts of atmospheric data from parallel GCM production and increasingly complex geology, land-ocean differentiation, surface heterogeneity and point-by-point physical methods to produce realistic atmospheric data with a spatial resolution of approximately 20–50 kilometres. Dynamic downscaling requires significant computational resources, expertise and a large amount of data as inputs. It typically applies to evaluations at the state or regional level that get substantial government aid and possessions (Fowler et al. 2007; Wilby et al. 2009). Statistical downscaling can provide site-specific projections of the atmosphere. It is based on experimental links between observed present-day huge scale atmospheric variables and factors in the surrounding environment. Once a relationship has been established and recognised, future atmospheric factors are used to forecast future local atmospheric factors using GCMs. The Statistical Downscaling Models are based on the premise that regional climate is largely influenced by the method used in global-scale circulation models (von Storch 1999) and that the relationship between predictor and predictand factors is invariant under future atmospheric conditions. Statistical Downscaling Model (SDSM) (Wilby et al. 2002) and Long Ashton Research Station Weather Generator (LARS-WG) (Racksko et al. 1991) are the models which are used to downscale the GCM climate data to local–/station-scale data. Nowadays, artificial intelligence is used to downscale the future climate scenario at station scale. Examples of artificial intelligence methods are machine learning (ML) and deep learning (DL) methods which are usually used based on the applications. They help to capture large amount of data and try to understand the statistical non-linear relationship between predictor and predictand. The artificial neural network (ANN) (Nourani et al. 2019), recurrent neural network (RNN) (Wang et al. 2020) and convolutional neural network (CNN) (Tu et al. 2021) are the common AI used to downscale and predict future climate scenario based on observed station data.

Soil Erosion and Climate Change

Soil erosion is a significant ecological issue in many regions around the world (Mondal et al. 2016). Water erosion is the dominant cause of soil degradation, which contributes approximately 55% of all eroded land (Bridges and Oldeman 1999). According to García-Ruiz et al. (2015), locations with high rates of erosion or sediment yield are primarily found in semi-arid and subhumid climates. Thus, it is plausible to assert that soil erosion is already severe in several regions of the world, including the United States, Australia, China, India and portions of Europe, Africa and South America, and may become worse as a consequence of climate change (Li and Fang 2016). According to Borrelli et al. (2020), the global hydrological cycle is becoming more vigorous and it could have the potential to increase soil erosion by water. Before the 1950s, the impacts of climate change on soil erosion were recognised (Bryan and Albritton 1943; Raeside 1948). The climate change affects the soil erosion in both direct and indirect ways (Fig. 5.1). The direct impacts are amount, intensity and spatio-distribution of rainfall. The indirect consequences are related to increase in global warming potential. Climate change has a significant effect on soil erosion, mostly via changes in plant cover and soil moisture (Nearing et al. 2004). A more indirect effect is associated with human-caused or socio-economic issues (Li and Fang 2016).

Indirectly, soil erosion can occur in a variety of ways such as increased temperatures. When the atmospheric CO_2 concentration and temperature increase, evapotranspiration rates rise and soil moisture levels decrease, enhancing the higher soil infiltration capacity and thus, reducing runoff and soil erosion (Xu 2003). Additionally, increasing the CO_2 content in the atmosphere may result in enhanced plant biomass production (Rosenzweig and Hillel 1998), which aids in enhancing canopy

5 Climate Change Impact on Land Degradation and Soil Erosion in Hilly...

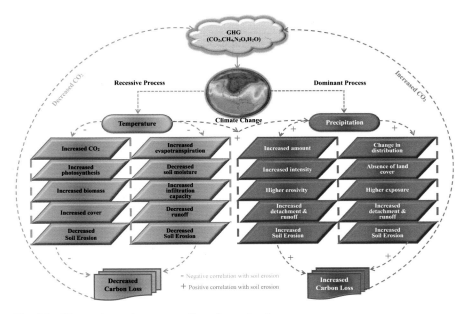

Fig. 5.1 Climate change impact on soil erosion and pathways

interception and reducing runoff and soil erosion (Fig. 5.1). Soil erosion, in particular, may contribute to the intensification of the effects of climate change by drastically lowering soil quality and fertility. Erosion disrupts the soil and degrades the aggregates, resulting in the partial mineralisation of soil organic carbon and the release of CO_2 into the atmosphere (Lal 2004; Lal and Pimentel 2008). Carbon (C) interactions among land ecosystems and the atmosphere are being altered by global processes such as increase in temperature and atmospheric CO_2 concentrations (Bond-Lamberty et al. 2018). Warming stimulates biological activities and breakdown at first, resulting in a reduction in SOC stocks (Melillo et al. 2017; Li et al. 2016).

Terrestrial C is the third major pool of C, consisting of soil and vegetation constituents. The soil carbon pool is composed of two components: 1550 Pg soil organic carbon pool (SOC) and 750 Pg soil inorganic carbon (SIC) pool. Batjes (1996) and Eswaran et al. (1995) estimated a total soil C pool of around 2300 Pg to a depth of 1 metre. At the moment, the atmospheric pool contains 760 Pg and is growing at a rate of 3.2 Pg C each year. Thus, the soil carbon pool of 2300 Pg is approximately four times the size of the biotic/vegetative carbon pool and approximately three times the larger size of the atmospheric carbon pool. Soil erosion occurs in four stages: sediment detachment, breakdown, transport/redistribution and depositing. Each of the four stages has an effect on the pool of soil organic carbon (SOC). Gross water erosion may total 75 billion Mg, depending on the delivery ratio or fraction of material delivered to the river system, of which 15–20 billion Mg is delivered via rivers through aquatic ecosystems and eventually into the ocean. The quantity of total carbon displaced worldwide by erosion is estimated to be between 4.0 and 6.0 Pg/year, assuming a 10% delivery ratio and a SOC concentration of 2–3%. With a 20% contribution from mineralisation of displaced carbon, erosion-induced emission on the earth might range between 0.8 and 1.2 Pg C/year (Lal 2003). Hence, this figure indicates how soil erosion enhances the climate change through the increased concentration of soil organic carbon to atmosphere.

Erosion also degrades land, reducing the amount of vegetation capable of absorbing climate-warming carbon dioxide. In a single year, soils can trap enough greenhouse gases to account for approximately 5% of all human-caused GHG emissions. Improved land management can help preserve soils by allowing them to grow more carbon dioxide-fixing vegetation. This is already practised in China, where the Grain-for-Green initiative has conserved soil and water while lowering carbon emissions in the Yellow River basin. According to the Intergovernmental Panel on Climate Change (IPCC), soil is presently eroding at a rate of up to 100 times quicker than it is generated when it is not conserved. Erosion danger will increase in the future as a result of temperature changes caused by emissions, resulting in declines in agricultural production, land value and human health.

Climate change has been documented to have an effect on soil erosion throughout the world, and numerous studies indicated that rainfall is the primary influencing element. These studies used empirical and process-based model to simulate long-term soil erosion and downscaling methods to predict future climate scenario for the study area. Various climate models provide diverse picture and methods to represent the future climate scenario, and most of the studies projected that an increase in precipitation in the future period expects a few of this scenario. IPCC emission scenarios are commonly used for future radiative forcing representation and rainfall prediction. There are a variety of downscaling methods and bias correction methods and tools available for local climate projection from GCM. Most studies reported that with the increase in precipitation, increase in soil erosion will occur in the future period. The increase in precipitation is attributed by amount, intensity and spatial distribution. Hence, it is obvious that climate change has a significant potential to enhance the present soil erosion unless appropriate amelioration measures are not adopted. Thus, site-specific impact studies will help us to create more reliable and sustainable practices to adapt and mitigate these issues (Table 5.2).

Sustainability Issues

The most widely accepted perspective of soil sustainability as a necessity is "soil management that meets the needs of the present without compromising the ability of future generations to meet their own needs from that soil" (Abbott and Murphy 2007). Sustaining soil management requires mitigating or resolving such threats. Thus, in the near future, "it will be essential to minimise soil erosion, increase soil organic matter content, promote soil nutrient balance and cycles, prevent, minimise, and mitigate soil salinization and alkalinisation, prevent and minimise soil

Table 5.2 Global climate change impact on soil erosion studies and the methods used

Country	Soil erosion model	Climate model	Climate scenario	Downscaling	Future status Rainfall	Erosion	References
United States	WEPP	HadCM3	A2a, B2a and GGa1	CLIGEN	Increased frequency of large storms	Increase	Zhang and Nearing (2005)
Austria	WEPP	ECHAM4 and ECHAM5	A1B and A2	Pattern scaling technique	Increased intensity	Increase and decrease	Klik and Eitzinger (2010)
Germany	EROSION 3D	ECHAM5-OPYC3	A1B	WETTREG2010	Increased intensities of extreme rainfall events	Increase and decrease	Routschek et al. (2014)
Kenya	USLE	ECHAM5	A1B, A2 and B1	Monte Carlo simulation and PDF	Increased erosivity	Increase	Maeda et al. (2010)
Thailand	RUSLE	HadCM3, NCAR CSSM3 and PRECIS RCM	A1b, A2, B1 and B2	Delta change approach	Increase/decrease	Increase and decrease	Plangoen et al. (2013)
East Africa (Tanzania, Malawi)	RUSLE	HadGEM2-ES and CP4A	RCP 8.5	Bias correction	Increases in rainfall intensity	Increase	Chapman et al. (2021)
India	RUSLE	MIROC5	RCP 2.6, 4.5, 6.0 and 8.5	Statistical	Increases in rainfall	Increase	Pal and Chakrabortty (2019)
China	WEPP	HadCM3	A2, B2 and GGa	CLIGEN	Increases in rainfall	Increase	Zhang et al. (2009)
Spain	GeoWEPP	MarkSim (avg. of 17 models)	RCP 2.6 and 6.0	CLIGEN	Decreases in rainfall	Decrease	Luetzenburg et al. (2020)

(continued)

Table 5.2 (continued)

Country	Soil erosion model	Climate model	Climate scenario	Downscaling	Future status Rainfall	Erosion	References
India	RUSLE	HadCM3	A2a and B2a	SDSM	Increases in rainfall	Increase	Gupta and Kumar (2017a, b)
Switzerland	D-RUSLE	ECHAM6 and CCSM4	RCP 2.6, 4.5 and 8.5	Stochastic time random cascades	Increase and decrease in rainfall	Increase and decrease	Gianinetto et al. (2020)
Italy	D-RUSLE	EC-Earth3.0, CESM2 and ECHAM6.3	SSP 2.6, 4.5, 7.0 and 8.5 (RCP)	Stochastic space random cascade model	–	Increase	Maruffi et al. (2022)
India	RUSLE	CanESM5, Access-ESM1–5, IPSL-CM6A-LR and MRI-ESM2–0	SPP 2–4.5 and 5–8.5	SDSM	Higher rainfall erosivity	Increase	Kumar et al. (2021)
Madagascar	SWAT	NorESM2 and CanESM5	SSP 3–7.0 and 5–8.5	ANN	Decreases in rainfall	Decrease	Tanteliniaina et al. (2021)
China	InVEST	MPI-ESM-LR	RCP 2.6 and 4.5	Bilinear interpolation	Increase and decrease	Increase and decrease	Hu et al. (2020)

contamination, preserve and enhance soil biodiversity, minimise soil sealing, prevent and mitigate soil compaction, and improve soil water management" (FAO 2017, p. 26) as cited by Dazzi and Papa (2021). Continuous soil erosion is an environmental damage as a result of enormous development, particularly in mountainous areas. However, soil erosion is a particular concern on arable lands and agriculture is simply one of the numerous expansion activities that significantly increases the process of soil erosion. Additionally, road development, mining, extraction and construction can result in significant soil erosion. Agricultural experts have showed substantial research on soil erosion, and preventing soil erosion is extensively acknowledged as a critical component of sustainable agriculture (Harden 2001).

Soil loss reduces crop yield, alters on- and off-site hydrology and causes potentially harmful off-site deposits (Pimentel et al. 1995). Global climate and land use variations have the capability to expressively impact the soil erosion and the ability of soils to sustain agriculture, hence jeopardising regional or global food security. Soil is a critical component of the food chain and of our culture. Soil production is a gradual process, while soil deterioration can occur quickly. As a result, soil is regarded as a non-renewable resource, and its sustainability is critical. Erosion is one of the major hazards to soil, as it is a natural occurrence that can be altered by global change (Paroissien et al. 2014; Walker 1994). Changes in rainfall patterns and land use patterns as a result of human causes (e.g. macroeconomics, technical advancements, social trends and governance systems) may have an effect on soil erosion and, consequently, soil sustainability. The models, which represent the soil ecosystem, can simulate the processes and can identify the issues with the natural soil ecosystem. For example, Paroissien et al. (2014) examined the effects of climate and land use change on soil erosion and sustainability in a Mediterranean watershed using a model. These kinds of studies provide more insight to the ecosystem processes and measures to mitigate the undesirable effect of climate change.

Sustaining a sustainable use of soil properties is reliant on a wide variety of interdependent elements (Fig. 5.2). Several of these are significant, including soil resilience, soil stability and soil quality. These properties are mainly reliant on the tillage techniques used and the maintenance of the soil surface, i.e. the nature and intensity of tillage. The soils have the potential to introduce a variety of processes that affect the soil's stability, resilience, quality and sustainability of land use. These interacting effects have an effect on management decisions about tillage methods, land usage and farming/cropping systems (Lal 1993). Soil structure is a critical soil feature because it affects a number of processes that are critical for soil productivity, ecological quality and agricultural sustainability. Degradation of soil structure can result in the initiation of deterioration processes such as compaction, increased erosion, soil salinity and loss of soil fertility. Multiple factors contribute to the degradation of soil structure on a local, regional and global scale, affecting economic, environmental and resource sustainability (Lal 1991a, b). They are widely acknowledged as key components of ecosystem services and sustainability, land use planning, climate change mitigation and food security (Bampa et al. 2019; Hou et al. 2020).

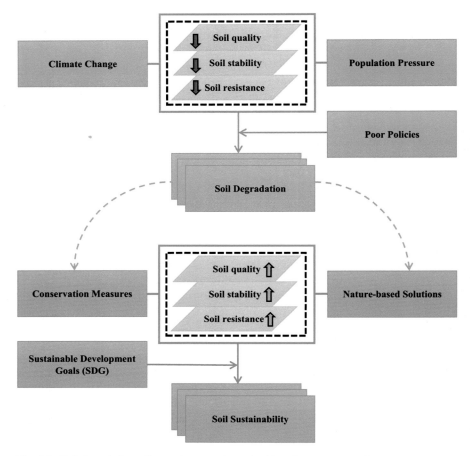

Fig. 5.2 Soil degradation, climate change and sustainable soil management flow

Global soils are being degraded at an alarming rate as a result of population increase, financial growth and climate change (Montanarella et al. 2016). Soil erosion is a significant source of soil deterioration, affecting over 1 billion hectares globally (e.g. water, wind and gully) (Lal 2003). Hence the integrated policy approaches can check the soil degradation and provide soil sustainability for the long-term benefits. The Sustainable Development Goals (SDGs) constructed by UN can be considered as the key for soil sustainability. These goals represent the importance of soil and its ecosystem services and functions and how much they are vital to humans as well as nature. Soils have a critical role in achieving SDGs 1, 2, 3, 6, 7, 11, 13, 14 and 15 (Bouma & Montanarella 2016; Keesstra et al. 2016). Table 5.3 shows the major SDG goals which are related to soil and its processes and functions. SDGs 2 and 6 state the food security, SDG 3 comprises food safety, SDG 14 concerns about the land-based nutrient pollution to the sea, SDG 11 concentrates on development of urban region and SDG 15 states about the long-term survivability

Table 5.3 SDG and relationship with soil

SDG no.	SDG icon	Goal	Relationship with soil
1	1 NO POVERTY	No poverty	People's incomes can be raised by appropriate land utilisation, mostly in rural areas of the world. Different soils require distinct management strategies and land uses. As a result, extensive knowledge of the characteristics and capacities of the soil in each location will aid in achieving this goal
2	2 ZERO HUNGER	Zero hunger	Maintaining healthy soils and managing them responsibly enables increased food quality and quantity, assuring food safety and accessibility for an expanding population
3	3 GOOD HEALTH AND WELL-BEING	Good health and well-being	The standard of living for humanity is reliant upon the soil's quality on which individuals reside and employ
6	6 CLEAN WATER AND SANITATION	Clean water and sanitation	Soils are key components of the water cycle. The soil directly or indirectly affects the majority of freshwater quality for human consumption, mostly through its cation exchange capacity
7	7 AFFORDABLE AND CLEAN ENERGY	Affordable and clean energy	Utilising fewer yield soils to cultivate energy crops has the potential to make a significant contribution to reducing energy consumption from fossil or non-renewable sources
11	11 SUSTAINABLE CITIES AND COMMUNITIES	Sustainable cities and communities	Metropolitan soils contribute significantly to the aesthetics of cities through the vegetation they support, to lessening the "heat island" issue that exemplifies urban areas, and to the well-being and social pleasure associated with urban garden care
13	13 CLIMATE ACTION	Climate action	Soil is the world's greatest reservoir of organic carbon and plays a critical role in the global carbon cycle
15	15 LIFE ON LAND	Life on land	Soil is the world's greatest reservoir of organic carbon and plays a critical role in the global carbon cycle

After Dazzi and Papa (2021)

of terrestrial ecosystem, all relying on soil ecosystem services in which soil plays a very critical role. SDG 15.3 on land degradation neutrality contains aims for preventing desertification; rehabilitating degraded land and soil, particularly land afflicted by desertification, drought and flooding; and attaining a world devoid of land degradation by 2030. Additionally, soils play a key role in the adaptation and mitigation of climate change (SDG 13). Similarly, SDGs 7 and 12 will depend on the accessibility of healthy soil resources in some capacity (Mihály et al. 2021; Tóth et al. 2018). Apart from this, the UN Statistical Commission has recognised the various global SDG indicators during its 48th session (UN 2017). There is currently no soil-based indicator attached to any of the soil-associated SDGs. In circumstances when disaggregation of indicators is relevant, the option of incorporating such indicator is expressly presented (UN 2017). The importance of adding soil indicators from the signalling through implementation stages is obvious, to accomplish the SDG targets based on soil resources (Bouma and Montanarella 2016).

Adaptation Strategies/Approaches

Soil and water are critical natural resources that must be maintained in conjunction with the surrounding environment in order to maintain sustainable agroecosystems. Political, economic and technical actions are required to avert soil erosion and sedimentation (Okalp 2005). Population growth and shrinking arable land deliver a significant resistance to modern agriculture. Agriculture production must be well-adjusted against the ever-increasing population in order to cover food supply demands. These changes have resulted in agricultural intensification, culminating in the conversion of natural vegetation to arable land. In combination with environmental variables, the persistent overexploitation of land resources results in the elimination of the top productive layer of soil. On a global scale, the initial substantial change in land use correlates to the beginning of the first wave of soil erosion. Human-affected areas experience a high rate of soil erosion. Soil erosion control becomes a critical component in striking a balance between agricultural productivity and conservation. To control and prevent soil erosion, an integrated soil erosion control system must be developed, including engineering, new technologies, law enforcement, biological approaches and land management. Structures for soil conservation, as well as enhanced soil loss models, would be necessary for land management (Bhat et al. 2019). To estimate the risk of increased sediment yield in a watershed, several sediment yield threshold values can be used to classify the risk of sedimentation as high, medium or low. The threshold values are determined by the basin and the amount of sediment generated (Shrestha et al. 2020). This approach can be very useful in determining the type of soil and water conservation measures used.

For soil erosion control, a variety of biological, mechanical and agronomic management approaches are existing. There are distinctions between these techniques in terms of their erosion control mechanisms and efficiency. Crop residues, manure application and conditioner application all come into direct interaction with the soil surface and act as buffers or thin films that protect the soil. In comparison, standing vegetation such as cover crops can also contribute to soil erosion prevention by the shielding nature of its canopy cover, which interrupts rainwater above the soil surface. Cover crops are novel conservation methods that are planted particularly for the purpose of preventing soil erosion, enhancing soil characteristics, increasing soil fertility and controlling weeds. Crop residues are significant resources on agricultural soils, providing a variety of ecosystem services such as lowering soil erosion and increasing the physico-chemical and biological aspects of the soil. They also help to enhance crop productivity and the environmental conditions. Crop residues, in particular, are crucial for minimising runoff and erosion, enhancing the hydraulic characteristics of the soil, enhancing storage of soil water, controlling soil temperature, adding or sustaining soil organic matter and enriching soil fertility. Manure is a rich resource of organic matter as well as macro- and micronutrients required for plant development. Fertilisers are manufactured using both solid and liquid animal manure. Manuring enhances soil quality and decreases soil erosion while also increasing crop production. Manuring also contributes to soil erosion prevention by improving the growth, strength and stability of aggregates through the incorporation of organic matter.

Runoff water and soil erosion can be controlled using a variety of biological and agronomic management practices, for example, crop rotations, no-till, reduced tillage, agroforestry, vegetative filter strips and riparian buffers. Buffers are continuous vegetated strips or passageways that are used to minimise erosion caused by water. These protection buffers are meant to limit runoff water and its velocity, filter the sediment and eradicate pollutants such as fertilisers and pesticides transported through sediment from upland regions. Grassed waterways must be carefully maintained to avoid runoff and soil damage. Agroforestry is a type of land management that involves growing trees and/or shrubs with agricultural crops and livestock on the same piece of land. It is a novel approach that has the potential to increase soil and water conservation vividly. Agroforestry, in a broad sense, refers to a variety of practices that involve the intentional planting and management of trees adjacent or within agricultural land and pasturelands with the goal of reducing soil erosion, emerging sustainable agronomic production systems, improving rural and urban landscapes, mitigating environmental pollution and improving farm income through the harvesting of products (Blanco and Lal 2008).

Mechanical or engineering structures are crucial in the management of soil erosion. While biological solutions are more cost-efficient and eco-friendly than constructed structures, soils prone to severe erosion require the installation of structures capable of intercepting massive volumes of runoff and silt. Mechanical structures decrease runoff, collect runoff water, transfer runoff water to a non-erosive speed, check sediment and nutrients, protect adjacent lands from flooding and reduce sedimentation. Permanent engineering structures include terraces, drop structures, spillways, culverts, gabions, etc. which are suitable for preventing excessive soil erosion due to torrential rainfall and runoff (Blanco and Lal 2008). Their selection is influenced by erosion intensity, soil slope and climate.

The climate change and increased population pressure put enormous pressure on the soil ecosystem. The poor practices and policies implemented by the farmers and local government also lead to the soil degradation. It will diminish the natural ability of soil such as quality, resistivity and stability. These unsustainable practices lead to decreased productivity and services from the soil ecosystem. An alternative for this issue is soil sustainability, which can be achieved through improving the soil by the conservation measures explained above and nature-based solutions appropriate for each site/location. The SDG framework will help us to improve the soil ecosystem from degraded state to sustainable state. It will increase the natural ability of soil and provide soil sustainability (Fig. 5.2).

Modelling Climate Change Impact on Soil Erosion in Himalayan Watershed: Case Study

The climate change and its after-effects possess very less predictability because of complexities involved in the processes. Hence, direct impact estimating studies at local/regional scales are very vital in policy making for government and other bodies. The more straightforward method of climate change impact assessment on soil erosion is through soil erosion modelling with downscaling future climate scenario (mostly precipitation). Here, we are describing two studies carried out at Himalayan region which used various erosion models with Statistical Downscaling Model (SDSM) for quantifying potential impact of climate change on soil erosion. Various climate scenarios developed by IPCC were used to identify the various rainfall scenarios for the future period (twenty-first century) (Fig. 5.3).

Chamba (Mid-Himalayas)

Experimental Site

The study was carried in the watershed of Tehri-Garhwal district of Uttarakhand and elevation ranges from 500 m to 2500 m. The Central Himalayan rocks characterise the study region parent material. The area is classified physiographically as moderately sloping hill, steep slope hill, and very steep sloping hill (Gupta and Kumar 2017a, b).

Method of Approach

The objective of this study is to quantify the possible influence of climate change on soil erosion in a Himalayan watershed using the RUSLE model. The future climate projection for the study area was downscaled using GCM (global circulation model)-derived scenarios (HadCM3 A2a and B2a). The future rainfall was downscaled for 2011–2040, 2041–2070 and 2071–2099 at large time scale. The RUSLE model was parameterised for the current time using the satellite data, field data collection and soil sample analysis data. To quantify yearly soil loss in the watershed, geographic input parameters such as soil erodibility, slope length and steepness, cover management factor and conservation practice factor were generated using the CARTOSAT DEM and IRS LISS III. For the future soil loss estimation period, the rainfall erosivity factor derived from future rainfall was used.

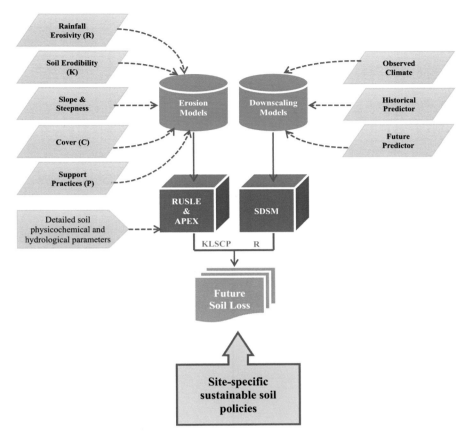

Fig. 5.3 Common methodology adopted for case study

Salient Findings

The coefficient of determination indicated that downscaling was preferable in the case of precipitation ($r^2 = 0.71$–0.80). The study concluded that precipitation may increase until 2020, but then decline between 2050 and 2080 relative to 2020. Additionally, the study discovered that the average annual rainfall in the area might increase by up to 33.3 per cent under A2 emission scenarios and by up to 31.67 per cent under B2 emission scenarios during the twenty-first century (Fig. 5.4). The RUSLE model predicted that 39.97% of the area is at very low to low risk of erosion this year, 33.15 per cent is at moderate to moderately high risk of erosion and 36.84 per cent is at high to very high risk of erosion. Annual rainfall erosivity was predicted to be 546 m t ha^{-1} cm^{-1} for the 30-year period, from 1985 to 2013. The rainfall erosivity value was projected to be up to 701 million tonnes ha^{-1} cm^{-1} in the A2 scenario and 693.8 million tonnes ha^{-1} cm^{-1} in the B2 scenario. According to the findings, the average yearly soil loss in the twenty-first century might rise to

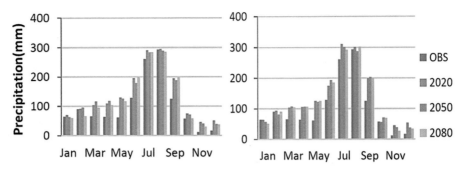

Fig. 5.4 Monthly future rainfall under A2 and B2 scenarios

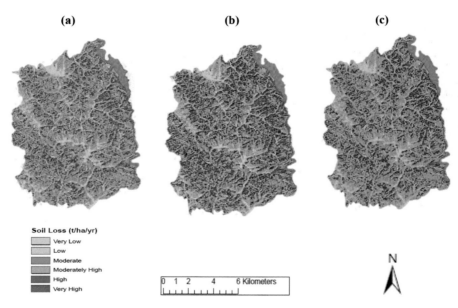

Fig. 5.5 Spatial distribution of soil loss during base period (**a**) and during 2080 under A2 (**b**) and B2 (**c**) scenarios

28.38 per cent under the A2 scenario. Similarly, in the B2 scenario, average yearly soil loss in the twenty-first century might increase by 27.06 per cent from the baseline period (1985–2013) (Fig. 5.5). The study predicted very high soil loss from the study area during the future period under the influence of climate change. The study provides comprehensive possible future soil erosion scenarios for the mid-Himalayan landscape especially for government and policy makers.

Hamirpur (Shivalik Himalayas)

Experimental Site

The study was conducted in a hilly watershed in the Shivalik Himalayan region of Himachal Pradesh, India, to model the climate change impact on soil erosion (Raj 2020; David Raj et al. 2021). Nearly the entire Hamirpur is lying under the tertiary formations. The terrain of the study area is typically hilly and undulating. The major physiographic units are structural hills, upland (600–900 m) and valley/alluvial plain (400–600 m). The study area comprises of the humid subtropical zone.

Method of Approach

The calibrated APEX model and MarkSim weather generator were used to assess the climate change impact on soil erosion. The model was calibrated using the observed surface runoff and sediment yield data received from watershed gauging station. The terrain parameters such as slope, aspect and drainage lines were generated using CARTOSAT DEM 30 m resolution. IRS LISS-IV sensor data was used to prepare land use/land cover map of the watershed. The landform and land use/land cover map were intersected to generate physiographic soil map for the watershed. Transect sampling approach was used for the collection of soil samples during the monsoon period. APEX model was parameterised using these data as well as field and soil analysis data. The future climate scenario for the study area was downloaded from the MarkSim weather generator tool. The base period soil erosion is compared with the future period for the change in soil erosion as a consequence of climate change.

Salient Findings

The APEX model generates hydrologic land use unit (HLU)-based soil loss for the area of interest. The HLU is characterised as unique soil, land use, topographical and climatic conditions. The model was calibrated and validated successfully for low to medium rainfall events. After calibration and validation, the APEX model was run based on a yearly scale to identify surface runoff and sediment yield from each subarea. The study found that the future rainfall in the study area is increasing (Fig. 5.6). The increase in rainfall is more in RCP 4.5 which is up to 33.71% and relatively less in RCP 8.5 which is up to 27.63%. As a result of this, the future soil erosion is also showing an increasing trend. Further soil erosion was predicted for future periods, and the results showed increases up to 44.0% under RCP 4.5 scenario, while 28.3% for RCP 8.5 scenario (Fig. 5.7). Higher soil loss was observed in RCP 4.5 scenario than in RCP 8.5 scenario. The APEX model identified scrubland as the land use which contributes the highest amount of surface runoff as well as soil loss followed by open forest and maize cropped field. The dense forest in the

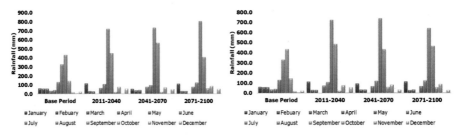

Fig. 5.6 Monthly future rainfall under RCP 4.5 and RCP 8.5 scenarios

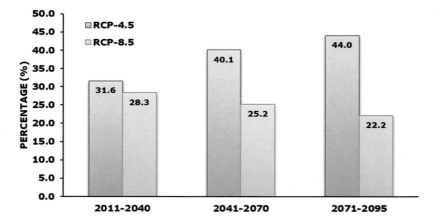

Fig. 5.7 Percentage increase in soil loss during various future periods

watershed was predicted to have the lowest surface runoff and soil loss. Because of cover factor, a dense forest has a good land cover which reduces the direct contact of rainfall with soil surface. Thus, slope factor is diminished by cover factor. A very less land cover and absence of conservation practices contribute to a higher surface runoff and sediment yield from the scrubland. It was also identified that only forest can resist future extreme soil erosion due to torrential rainfall and scrubland has the least.

Conclusion

Soil erosion is considered as the major land degradation issue which is a global threat for food security as well as economic development. Each year countries spend billions of dollars to tackle or overcome the soil erosion problem. High-intensity rainfall, sloping terrain, absence of soil cover and highly erodible soils are the major driving forces which enhance soil erosion. The soil erosion and its risk vary according to the geographical position, topography and economic condition of the country. The deforestation for agriculture, logging and cultivation in highlands without proper support practices are also severely affecting extent of soil erosion. The climate change is one of the major factors which directly and indirectly enhance the soil erosion risk of an area. The torrential high-intensity rainfall and associated runoff carry million tonnes of soils every year which are deposited in low-lying regions, reservoirs, dams and some parts of seas. The increase in temperature and carbon di oxide in the atmosphere helps to reduce the soil erosion through the higher vegetative growth (cover factor). But the rainfall effect is dominant and thus temperature effect becomes recessive in most cases. Various researches around the world identified that the climate change has a huge potential to increase the soil erosion rate. Similarly, higher soil erosion rate lessens the soil organic carbon content from the soil and acts as a carbon source to the atmosphere. It further enhances the global warming and eventually leads to climate change. Hence, climate change and soil erosion are acting as cyclic processes and one process enhances the other process and vice versa. Although soil erosion is a global threat, the mitigation and adaptation measures should be site-specific. The Sustainable Development Goals (SDGs) help to achieve the sustainable soil ecosystem and services from the soil. The nature-based solutions also enhance the soil erosion mitigation and adaptation measures in a more comprehensive and eco-friendly pathway. The huge potential of climate change on soil erosion can only be prevented through the integrated approach of site-specific sustainable framework, nature-based solutions and better policies (site-specific policies according to UN) from the local government. It will help us to build a sustainable climate-resilient ecosystem for adapting and mitigating land degradation issues.

References

Abbott LK, Murphy DV (2007) What is soil biological fertility? In: Abbott LK, Murphy DV (eds) Soil biological fertility, a key to sustainable land use in agriculture. Springer, Dordrecht, pp 1–15. https://doi.org/10.1007/978-1-4020-6619-1_1

Arnold JG, Srinivasan R, Muttiah RS, Williams JR (1998) Large area hydrologic modeling and assessment part I: model development. J Am Water Resour Assoc 34(1):73–89. https://doi.org/10.1111/j.1752-1688.1998.tb05961.x

Bampa F, O'sullivan L, Madena K, Sandén T, Spiegel H, Henriksen CB, Ghaley BB, Jones A, Staes J, Sturel S, Trajanov A (2019) Harvesting European knowledge on soil functions and land management using multi-criteria decision analysis. Soil Use Manag 35(1):6–20

Batjes NH (1996) Total carbon and nitrogen in the soils of the world. Eur J Soil Sci 47(2):151–163

Beck MB (1987) Water quality modeling: a review of the analysis of uncertainty. Water Resour Res 23(8):1393–1442

Beven K (1989) Changing ideas in hydrology—the case of physically-based models. J Hydrol 105(1–2):157–172

Bhat SA, Dar MUD, Meena RS (2019) Soil erosion and management strategies. In: Sustainable management of soil and environment. Springer, Singapore, pp 73–122

Blanco H, Lal R (2008) Principles of soil conservation and management, vol 167169. Springer, New York

Blum A (2011) Plant water relations, plant stress and plant production. In: Plant breeding for water-limited environments. Springer, New York, pp 11–52

Bond-Lamberty B, Bailey VL, Chen M, Gough CM, Vargas R (2018) Globally rising soil heterotrophic respiration over recent decades. Nature 560(7716):80–83

Borrelli P, Robinson DA, Fleischer LR, Lugato E, Ballabio C, Alewell C et al (2017) An assessment of the global impact of 21st century land use change on soil erosion. Nat Commun 8(1):1–13

Borrelli P, Robinson DA, Panagos P, Lugato E, Yang JE, Alewell C, Wuepper D, Montanarella L, Ballabio C (2020) Land use and climate change impacts on global soil erosion by water (2015-2070). Proc Natl Acad Sci 117(36):21994–22001

Bouma J, Montanarella L (2016) Facing policy challenges with inter-and transdisciplinary soil research focused on the UN Sustainable Development Goals. Soil 2(2):135–145

Bridges EM, Oldeman LR (1999) Global assessment of human-induced soil degradation. Arid Soil Res Rehabil 13(4):319–325

Brown LR (1981) World population growth, soil erosion, and food security. Science:995–1002

Brown LR (1984a) The global loss of top soil. I Soil Water Conserv 36(5):255–260

Brown LR (1984b) The global loss of top soil. J Soil Water Conserv 39:162–165

Bryan K, Albritton CC (1943) Soil phenomena as evidence of climatic changes. Am J Sci 241(8):469–490

Buytaert W, Vuille M, Dewulf A, Urrutia R, Karmalkar A, Célleri R (2010) Uncertainties in climate change projections and regional downscaling in the tropical Andes: implications for water resources management. Hydrol Earth Syst Sci 14(7):1247–1258

Chapman S, Birch CE, Galdos MV, Pope E, Davie J, Bradshaw C, Eze S, Marsham JH (2021) Assessing the impact of climate change on soil erosion in East Africa using a convection-permitting climate model. Environ Res Lett 16(8):08400

Coleman K, Jenkinson DS (2014) RothC—a model for the turnover of carbon in soil: model description and users guide (updated June 2014). Lawes Agricultural Trust, Harpenden. Accessed 27 Nov 2021 https://www.rothamsted.ac.uk/sites/default/files/RothC_guide_WIN.pdf

Crosson P (1997) Will erosion threaten agricultural productivity? Environment 39(8):4–9 and 29–31

David Raj A, Kumar S, Regina M, Sooryamol KR, Singh AK (2021) Calibrating APEX model for predicting surface runoff and sediment loss in a watershed – a case study in Shivalik region of India. J Hydrol Sci Technol, Int. https://doi.org/10.1504/IJHST.2021.10041820

Dazzi C, Papa GL (2021) A new definition of soil to promote soil awareness, sustainability, security and governance. International Soil and Water Conservation Research

De Roo APJ, Wesseling CG, Ritsema CJ (1996) LISEM: a single-event physically based hydrological and soil erosion model for drainage basins. I: theory, input and output. Hydrol Process 10(8):1107–1117

Dent FR (1984) Land degradation: present status, training and education needs in Asia and Pacific. UNEP investigations on environmental education and training in Asia and Pacific: FAO, Regional Office, Bangkok.

Dregne HE (1982) Historical perspective of accelerated erosion and effect on world civilization. Determinants of soil loss tolerance 45:1–14

Eswaran H, Van den Berg E, Reich P, Kimble J (1995) Global soil carbon resources. In: Lal R, Kimble JM, Levine E, Stewart BA (eds) Soils and global change. CRC/Lewis, Boca Raton, pp 27–43

FAO (2002) Land Degradation Assessment in Drylands (LADA) Project: Meeting Report, 23–25 January 2002 (World Soil Resources Reports)

FAO (2015) Global soil status, processes and trends. Status of the World's Soil Resources (SWSR)—Main Report of the Food and Agriculture Organization, New York

FAO (2017) Voluntary Guidelines for sustainable soil management. Food and Agriculture Organization of the United Nations, Rome, p 26

FAO/IAEA. (2017). Use of 137Cs for soil erosion assessment. Fulajtar, E., Mabit, L., Renschler, C. S., Lee Zhi Yi, A., Food and Agriculture Organization of the United Nations, Rome, 64 p

Fitzpatrick RW (2002) Land degradation processes. ACIAR Monograph Series 84:119–129

Flanagan DC, Nearing MA (1995) USDA-Water Erosion Prediction Project: hillslope profile and watershed model documentation. Nserl Rep 10:1–123

Fowler HJ, Blenkinsop S, Tebaldi C (2007) Linking climate change modelling to impacts studies: recent advances in downscaling techniques for hydrological modelling. Int J Climatol 27(12): 1547–1578

Garbrecht JD, Zhang XC (2015) Soil erosion from winter wheat cropland under climate change in Central Oklahoma. Appl Eng Agric 31(3):439–454

García-Ruiz JM, Beguería S, Nadal-Romero E, González-Hidalgo JC, Lana-Renault N, Sanjuán Y (2015) A meta-analysis of soil erosion rates across the world. Geomorphology 239:160–173

Gianinetto M, Aiello M, Vezzoli R, Polinelli FN, Rulli MC, Chiarelli DD et al (2020) Future scenarios of soil erosion in the Alps under climate change and land cover transformations simulated with automatic machine learning. Climate 8(2):28

Gupta S, Kumar S (2017a) Simulating climate change impact on soil erosion using RUSLE model—a case study in a watershed of mid-Himalayan landscape. Journal of Earth System Science 126(3):43

Gupta S, Kumar S (2017b) Simulating climate change impact on soil carbon sequestration in agro-ecosystem of mid-Himalayan landscape using CENTURY model. Environ Earth Sci 76(11): 1–15

Harden CP (2001) Soil erosion and sustainable mountain development. Mt Res Dev 21(1):77–83

Holdgate MW, Kassas M, White GF (1982) The world environment: 1972–1982. Tycool International, Dublin

Hou D, Bolan NS, Tsang DC, Kirkham MB, O'Connor D (2020) Sustainable soil use and management: an interdisciplinary and systematic approach. Sci Total Environ 729:138961

Kalambukattu J, Kumar S (2017) Modelling soil erosion risk in a mountainous watershed of Mid-Himalaya by integrating RUSLE model with GIS. Eurasian Journal of Soil Science 6(2): 92–105

Kalambukattu JG, Kumar S, Hole RM (2021) Geospatial modelling of soil erosion and risk assessment in Indian Himalayan region—a study of Uttarakhand state. Environ Adv 4:100039

Karamouz M, Nazif S, Fallahi M (2010) Rainfall downscaling using Statistical Downscaling Model and Canonical Correlation Analysis: a case study. Proceedings of the World Environmental and Water Resources Congress 2010, Providence, Rhode Island, USA, 4579–4587

Keesstra SD, Bouma J, Wallinga J, Tittonell P, Smith P, Cerdà A, Montanarella L, Quinton JN, Pachepsky Y, van der Putten WH, Bardgett RD, Moolenaar S, Mol G, Jansen B, Fresco LO (2016) The significance of soils and soil science towards realization of the United Nations Sustainable Development Goals. Soil 2(2):111–128. https://doi.org/10.5194/soil-2-111-2016

Kelly RH, Parton WJ, Crocker GJ, Graced PR, Klir J, Körschens M, Poulton PR, Richter DD (1997) Simulating trends in soil organic carbon in long-term experiments using the century model. Geoderma 81(1–2):75–90

Klik A, Eitzinger J (2010) Impact of climate change on soil erosion and the efficiency of soil conservation practices in Austria. J Agric Sci 148(5):529–541

Kovda VA (1977) Biosphere, soil cover and their changes. In: Meldinger fra Norges Landbrukshogskele, 56(4). Moscow University, Moscow

Kumar N, Singh SK, Dubey AK, Ray RL, Mustak S, Rawat KS (2021) Prediction of soil erosion risk using earth observation data under recent emission scenarios of CMIP6. Geocarto Int:1–24

Lal D (1991a) Cosmic ray labeling of erosion surfaces: in situ nuclide production rates and erosion models. Earth Planet Sci Lett 104(2–4):424–439

Lal R (1991b) Soil structure and sustainability. J Sustain Agric 1(4):67–92

Lal R (1993) Soil erosion and conservation in West Africa. World soil erosion and conservation:7–25

Lal R (1994a) Tillage effects on soil degradation, soil resilience, soil quality, and sustainability. Soil Tillage Research 27:1–8

Lal R (1994b) Soil erosion research methods. CRC Press, Boca Raton, 352p

Lal R (1994c) Soil erosion research methods, soil and water conservation society (U. S.). CRC Press, 352p

Lal R (1998) Soil erosion impact on agronomic productivity and environment quality. Crit Rev Plant Sci 17(4):319–464

Lal R (2003) Soil erosion and the global carbon budget. Environ Int 29(4):437–450

Lal R (2004) Soil carbon sequestration impacts on global climate change and food security. Science 304(5677):1623–1627

Lal R (2015) Restoring soil quality to mitigate soil degradation. Sustainability 7(5):5875–5895

Lal R, Pimentel D (2008) Soil erosion: a carbon sink or source? Science 319(5866):1040–1042

Lal R, Stewart BA (1990) Soil degradation. Springer, New York

Li Z, Fang H (2016) Impacts of climate change on water erosion: a review. Earth Sci Rev 163:94–117

Li Y, Zhou G, Huang W, Liu J, Fang X (2016) Potential effects of warming on soil respiration and carbon sequestration in a subtropical forest. Plant Soil 409:247–257. https://doi.org/10.1007/s11104-016-2966-2

Luetzenburg G, Bittner MJ, Calsamiglia A, Renschler CS, Estrany J, Poeppl R (2020) Climate and land use change effects on soil erosion in two small agricultural catchment systems Fugnitz–Austria, Can Revull–Spain. Sci Total Environ 704:135389

Maeda EE, Pellikka PK, Siljander M, Clark BJ (2010) Potential impacts of agricultural expansion and climate change on soil erosion in the Eastern Arc Mountains of Kenya. Geomorphology 123(3–4):279–289

Mahala A (2020) Land degradation processes of Silabati River Basin, West Bengal, India: a physical perspective. In: Gully Erosion Studies from India and Surrounding Regions. Springer, Cham, pp 265–278

Mahapatra SK, Reddy GO, Nagdev R, Yadav RP, Singh SK, Sharda VN (2018) Assessment of soil erosion in the fragile Himalayan ecosystem of Uttarakhand, India using USLE and GIS for sustainable productivity. Curr Sci 115(1):108–121

Maruffi L, Stucchi L, Casale F, Bocchiola D (2022) Soil erosion and sediment transport under climate change for Mera River, in Italian Alps of Valchiavenna. Sci Total Environ 806:150651

Melillo JM, Frey SD, DeAngelis KM, Werner WJ, Bernard MJ, Bowles FP, Pold G, Knorr MA, Grandy AS (2017) Long-term pattern and magnitude of soil carbon feedback to the climate system in a warming world. Science 358(6359):101–105

Mihály S, Remetey-Fülöpp G, Kristóf D, Czinkóczky A, Palya T, Pásztor L, Rudan P, Szabó G, Zentai L (2021) Earth observation and geospatial big data management and engagement of stakeholders in Hungary to support the SDGs. Big Earth Data 5(3):306–351

Mondal A, Khare D, Kundu S (2016) Impact assessment of climate change on future soil erosion and SOC loss. Nat Hazards 82(3):1515–1539

Montanarella L, Pennock DJ, McKenzie N, Badraoui M, Chude V, Baptista I, Mamo T, Yemefack M, Singh Aulakh M, Yagi K, Young Hong S, Vijarnsorn P, Zhang GL, Arrouays D, Black H, Krasilnikov P, Sobocká J, Alegre J, Henriquez CR, de Lourdes M-SM, Taboada M, Espinosa-Victoria D, AlShankiti A, AlaviPanah SK, Elsheikh EAEM, Hempel J, Camps Arbestain M, Nachtergaele F, Vargas R (2016) World's soils are under threat. Soil 2:79–82

Mora D, Campozano L, Cisneros F, Wyseure G, Willems P (2014) Climate changes of hydrometeorological and hydrological extremes in the Paute basin. Ecuadorean Andes, Hydrology and Earth System Sciences 18(2):631–648

Morgan RPC (2001) A simple approach to soil loss prediction: a revised Morgan–Morgan–Finney model. Catena 44:305–322

Morgan RPC, Morgan DDV, Finney HJ (1984) A predictive model for the assessment of erosion risk. J Agric Eng Res 30:245–253

Myers N (1993) Gaia: an atlas of planet management. Anchor/Doubleday, Garden City (NY)

Nakicenovic, N., Alcamo, J., Davis, G., Vries, B.D., Fenhann, J., Gaffin, S., Gregory, K., Grubler, A., Jung, T.Y., Kram, T. & La Rovere, E.L., (2000). Special report on emissions scenarios

Narayana DV, Babu R (1983) Estimation of soil erosion in India. J Irrig Drain Eng 109(4):419–434

National Bureau of Soil Survey & Land Use Planning (NBSS&LUP) (2004) Soil Map (1:1 Million Scale). NBSS&LUP, Nagpur, India

Nearing MA, Pruski FF, O'neal, M. R. (2004) Expected climate change impacts on soil erosion rates: a review. J Soil Water Conserv 59(1):43–50

Nicoloso RS, Amado TJ, Rice CW (2020) Assessing strategies to enhance soil carbon sequestration with the DSSAT-CENTURY model. Eur J Soil Sci 71(6):1034–1049

Nourani V, Paknezhad NJ, Sharghi E, Khosravi A (2019) Estimation of prediction interval in ANN-based multi-GCMs downscaling of hydro-climatologic parameters. J Hydrol 579:124226

Okalp K (2005) Soil erosion risk mapping using geographic information systems: a case study on Kocadere Creek Watershed, Izmir (doctoral dissertation). Middle East Technical University, Turkey

Pal, S.C. & Chakrabortty, R., (2019). Simulating the impact of climate change on soil erosion in sub-tropical monsoon dominated watershed based on RUSLE, SCS runoff and MIROC5 climatic model. Adv Space Res, 64(2), pp. 352–377.6

Paroissien JB, Darboux F, Couturier A, Devillers B, Mouillot F, Raclot D, Le Bissonnais Y (2014) A method for modeling the effects of climate and land use changes on erosion and sustainability of soil in a Mediterranean watershed (Languedoc, France). J Environ Manag 150:57–68

Parton WJ, Schimel DS, Cole CV, Ojima DS (1987) Analysis of factors controlling soil organic matter levels in Great Plains grasslands. Soil Sci Soc Am J 51(5):1173–1179

Parton WJ, Hartman M, Ojima D, Schimel D (1998) DAYCENT and its land surface submodel: description and testing. Glob Planet Change 19:35–48. https://doi.org/10.1016/S0921-8181(98)00040-X

Pimentel D (1987) Erosion and soil productivity: proceedings of the National Symposium on Erosion and Soil Productivity, American Society of Agricultural Engineers, St. Joseph, MI, 1985, 289 pp.

Pimentel D, Harvey C, Resudarmo P et al (1995) Environmental and economic costs of soil erosion and conservation benefits. Science 267:1117–1123

Plangoen P, Babel MS, Clemente RS, Shrestha S, Tripathi NK (2013) Simulating the impact of future land use and climate change on soil erosion and deposition in the Mae Nam Nan sub-catchment. Thailand Sustainability 5(8):3244–3274

Poesen J (1993) Gully typology and gully control measures in the European loess belt. In: Wicherek S (ed) Farmland erosion in temperate plains environment and hills. Elsevier, Amsterdam, pp 221–239

Prosser IP, Young B, Rustomji P, Hughes A, Moran CA (2001) Model of river sediment budgets as an element of river health assessment. Proceedings of the International Congress on Modelling and Simulation (MODSIM'2001), Canberra, Australia, 10–13

Racksko P, Szeidl L, Semenov M (1991) A serial approach to local stochastic weather models. Ecol Model 57:27–41

Raeside J (1948) Some post-glacial climatic changes in Canterbury and their effect on soil formation Trans. R Soc N Z:153–171

Raj AD (2020) Modelling climate change impact on surface runoff and sediment yield in a watershed of Shivalik region (M.Sc. Thesis, Academy of Climate Change Education and Research, Kerala Agricultural University, Vellanikkara)

Renard KG, Foster GR, Yoder DC, McCool DK (1994) RUSLE revisited: status, questions, answers, and the future. J Soil Water Conserv 49(3):213–220

Rosenzweig C, Hillel D (1998) Climate change and the global harvest: potential impacts of the greenhouse effect on agriculture. Oxford University Press

Routschek A, Schmidt J, Kreienkamp F (2014) Impact of climate change on soil erosion—a high-resolution projection on catchment scale until 2100 in Saxony/Germany. Catena 121:99–109

Sharma PD, Goel AK, Minhas RS (1991) Water and sediment yields into the Sutlej river from the high Himalaya. Mt Res Dev 11(2):87–100

Shrestha S, Sattar H, Khattak MS, Wang G, Babur M (2020) Evaluation of adaptation options for reducing soil erosion due to climate change in the Swat River Basin of Pakistan. Ecol Eng 158: 106017

Smith RE, Goodrich DC, Quinton JN (1995) Dynamic, distributed simulation of watershed erosion: the KINEROS2 and EUROSEM models. J Soil Water Conserv 50(5):517–520

Sulman BN, Phillips RP, Oishi AC, Shevliakova E, Pacala SW (2014) Microbe-driven turnover offsets mineral-mediated storage of soil carbon under elevated CO_2. Nat Clim Chang 4:1099–1102. https://doi.org/10.1038/nclimate2436

Tanteliniaina RMF, Rahaman M, Zhai J (2021) Assessment of the future impact of climate change on the hydrology of the Mangoky River. Madagascar Using ANN and SWAT Water 13(9):1239

Tóth G, Hermann T, da Silva MR, Montanarella L (2018) Monitoring soil for sustainable development and land degradation neutrality. Environ Monit Assess 190(2):1–4

Tu T, Ishida K, Ercan A, Kiyama M, Amagasaki M, Zhao T (2021) Hybrid precipitation downscaling over coastal watersheds in Japan using WRF and CNN. J Hydrol Regional Stud 37: 100921

UN (2015) Transforming our world: the 2030 Agenda for Sustainable Development. Resolution adopted by the General Assembly on 25 September 2015. United Nations, p 35

UN (2017) Report of the inter-agency and expert group on sustainable development goal indicators. United Nations, Economic and Social Council. E/CN.3/2017/2. United Nations, New York

UNCCD (1994) United Nations convention to combat desertification. United Nations General Assembly, New York City, p 54p

UNFCCC (1992) United Nations framework convention on climate change: text. UNEP/WMO Information Unit on Climate Change, Geneva

USDA-NRCS (n.d.) USDA-NRCS home page [online]. Available: https://www.nrcs.usda.gov/wps/portal/nrcs/detail/soils/use/?cid=nrcs142p2_054028. [16 July 2021]

von Storch H (1999) The global and regional climate system. In: von Storch H, Floscr G (eds) Anthropogenic climate change. Springer, Berlin, pp 3–36

Voroney RP, Angers DA (2018) Analysis of the short-term effects of management on soil organic matter using the CENTURY model. In: Soil management and greenhouse effect. CRC Press, pp 113–120

Walker BH (1994) Global change strategy options in the extensive agriculture regions of the world. In: Climate change: significance for agriculture and forestry. Springer, Dordrecht, pp 39–47

Wang G, Jagadamma S, Mayes MA, Schadt CW, Steinweg JM, Gu L et al (2015) Microbial dormancy improves development and experimental validation of ecosystem model. ISME J 9: 226–237

Wang Q, Huang J, Liu R, Men C, Guo L, Miao Y, Jiao L, Wang Y, Shoaib M, Xia X (2020) Sequence-based statistical downscaling and its application to hydrologic simulations based on machine learning and big data. J Hydrol 586:124875

Wieder WR, Grandy AS, Kallenbach CM, Bonan GB (2014) Integrating microbial physiology and physio-chemical principles in soils with the MIcrobial-MIneral Carbon Stabilization (MIMICS) model. Biogeosciences 11(14):3899–3917. https://doi.org/10.5194/bg-11-3899-2014

Wilby RL, Dawson CW, Barrow EM (2002) SDSM—a decision support tool for the assessment of regional climate change impacts. Environ Model Softw 17(2):145–157

Wilby RL, Charles SP, Zorita E, Timbal B, Whetton P, Mearns LO (2004) Guidelines for use of climate scenarios developed from statistical downscaling methods. Supporting material of the Intergovernmental Panel on Climate Change, Available at DDC of IPCC TGCIA, 27

Wilby RL, Troni J, Biot Y, Tedd L, Hewitson BC, Smith DM, Sutton RT (2009) A review of climate risk information for adaptation and development planning. Int J Climatol 29(9): 1193–1215

Williams JR, Jones CA, Dyke PT (1984) A modelling approach to determining the relationship between erosion and soil productivity. Transactions of the ASAE 27:129–0144

Williams JR, Arnold JG, Kiniry JR, Gassman PW, Green CH (2008) History of model development at Temple. Texas Hydrol Sci J 53(5):948–960. https://doi.org/10.1623/hysj.53.5.948

Wischmeier WH, Smith DD (1965) Predicting rainfall erosion losses from cropland east of the Rocky Mountains: guide for selection of practices for soil and water conservation, Agriculture handbook, 282. USDA, Washington, 47p

Xu J (2003) Sedimentation rates in the lower Yellow River over the past 2300 years as influenced by human activities and climate change. Hydrol Process 17(16):3359–3371

Zhang XC, Nearing MA (2005) Impact of climate change on soil erosion, runoff, and wheat productivity in Central Oklahoma. Catena 61(2–3):185–195

Zhang XC, Nearing MA, Garbrecht JD, Steiner JL (2004) Downscaling monthly forecasts to simulate impacts of climate change on soil erosion and wheat production

Zhang XC, Liu WZ, Li Z, Zheng FL (2009) Simulating site-specific impacts of climate change on soil erosion and surface hydrology in southern Loess Plateau of China. Catena 79(3):237–242

Chapter 6
Vulnerability Assessment of the Inherent Hazards of Climate Change on the Coastal Environment of the Mahanadi Delta, East Coast of India

Monalisha Mishra, Gopal Krishna Panda, Kishor Dandapat, and Uday Chatterjee

Abstract This study makes a vulnerability assessment of the coastal hazards of the Mahanadi delta across Bay of Bengal using coastal hazard wheel model which is a multi-hazard assessment and management tool in a changing global climate. This paper analyses the coastal environment of the Mahanadi delta to evaluate the vulnerability profile by assessing its inherent hazards such as ecosystem disruption, gradual coastal inundation, intrusion of salt water into the freshwater aquifers, coastal flooding and beach erosion. The study uses the bio-geophysical parameters such as geomorphological layout, nature of wave exposure, range of tidal amplitude, coastal vegetation, balance and deficit of coastal sediments and presence or absence of storm climate which are also used in assessing hazards of climate change which are inherent to a coastal zone. The model has used remote sensing data from the open sources and a GIS framework to assess the problem from the management perspective. The study has generated data and maps about the degree to which a particular location along the coast is vulnerable to the hazards inherent to the coast dividing the coastline into several morphological segments. The study reveals that the deltaic and pro-grading shoreline of the Mahanadi delta has undergone erosion and accretion in a phased manner with erosion exceeding the rate of accretion. The rate of erosion has varied from 4 m to 15 m per year at different sites along the delta front. The study has

M. Mishra (✉)
GIS Developer, Water Resource Department, Government of Odisha, Bhubaneswar, Odisha, India

G. K. Panda
KISS Deemed to be University, Bhubaneswar, Odisha, India

K. Dandapat
Department of Geography, Seva Bharati Mahavidyalaya, Kapgari (Vidyasagar University), Jhargram, West Bengal, India

U. Chatterjee
Department of Geography, Bhatter College, Dantan (Affiliated to Vidyasagar University), Paschim Medinipore, West Bengal, India

generated vulnerability data and maps for the different coastal hazards inherent to the coast, i.e. saltwater intrusion, saline inundation, coastal erosion, flooding, ecosystem disruption and associated risk levels at the regional level for the 279 km of coastline of the delta. The study reveals that 28.6% of Mahanadi delta's coastline has a very high risk, 47.6% has high risk, 16.99% has moderate risk and 6.81% has a lower risk in respect of different inherent hazards. The study reveals an increasing magnitude of vulnerability to the coastal environment from the inherent hazards than previous assessments by different researchers. The study has also identified a number of possible environmentally adoptable management interventions as a part of its sustainable solutions. The findings of this study are significant in understanding the retardation in the growth of the delta and environmental problems arising out of the inherent hazards of the coastal zone in a deltaic region. The study also suggests the significance of using open-source satellite data and geo-informatics-based CHW and DSAS in assessing the vulnerability of environmental degradation for integrated coastal zone management.

Keywords Biophysical environment · Climate change · Coastal hazard wheel · Geographic information system · Integrated coastal zone management

Introduction

Environments by definition are organizations or networks of biotic and abiotic components connected through food chain and energy cycles. In the deltaic zone, the environment usually provides services such as provisioning and production of food and water, regulating the nutrient cycle and helping in crop pollination and recreation. Environmental degradation arises out of the exploitation of resources (air, water, soil), destruction of the environment and extinction of flora and fauna. The primary causes can be ascribed to a blend of natural and man-made causes. Natural causes contain complex features like El Niño-La Niña and extreme events such as tropical cyclones, droughts, tsunami, heat and cold waves and climate change. Environmental change impacts speed up the ocean-level ascent, changes in the breeze environment, temperature and precipitation (Bakker et al. 2016). Anthropogenic causes include an increase in the density of population and urbanization, alternation of river flows and sediment loads, land-use change and habitat destructions, industrial and developmental activities and consequent pollution (Nichols et al. 2019). Rising population and associated activities build pressures on the environment which often leads to deterioration of key environmental parameters and biological system capacities (Balmford and Bond 2005). Along the coastal zone, 20% of the earth's surface remains 200 m above sea level (Pernetta and Milliman 1995), and it houses a huge portion of the world's human population (NRC 1990; Cohen et al. 1997). There is a significant increase in population of the coastal zone due to migration (Curran et al. 2002) and with the result that 70% of the world's megacities (>1.6 million) are now within the coastal zone (LOICZ 2002). These growing anthropogenic pressures have come about in an enduring worldwide misfortune of coral reefs, mangrove forests, brackish swamps and seagrass glades over the past 50 years. The predicted climate change of IPCC includes additional hazards for human activities in the coastal ranges. (Ellison and Stoddart 1991; Jackson et al. 2001; Adam 2002; Duarte 2002; Bellwood et al. 2004; Lotze et al. 2006; Orth et al. 2006). IPCC has observed that these are likely to alter with present-day climate change regimes (IPCC 2013). Coastlines will respond to movements in different ways and at various speeds depending upon their profile geological qualities, yet by and large, society should perceive that past beachfront pattern can't be straightforwardly projected into the future. All things being in the rise, it is vital to consider how different coastal environments will react to predicted climate change and associated coastal hazards to recognize appropriate strategies, improve coastal communications and take important administrative decisions as a part of integrated coastal zone management (Appelquist and Balstrorm 2014). While the coast responds to the present trend of global warming and sea-level rise, there is a need to understand that the climate-induced inherent hazards are significant in causing environment degradation and ecosystem disruption along the coast.

Review of Literature

Coastal zones, the world's most densely populated regions, are increasingly threatened by climate change stressors – rising and warming seas, intensifying storms and droughts and acidifying oceans. Although coastal zones have been affected by local human activities for centuries, how local human impacts and climate change stressors may interact to jeopardize coastal ecosystems remains poorly understood. Here we provide a review of literature on climate-induced hazards and their environmental impacts on the coastal zone. We highlight how these interactions can impair and, at times, decimate a variety of coastal ecosystems and examine how understanding and incorporating these interactions can reshape theory on climate change impacts and ecological resilience. We further discuss implications of the interactions between climate change and local human impacts for coastal conservation and elucidate the context when and where local conservation is more likely to buffer their impacts. Our review underscores that an enhanced understanding of interactions between climate change and its impact on the environment are of profound importance for improving predictions of climate change impacts, devising climate-smart conservation measures and enhancing adaption of coastal societies to climate change in the Anthropocene (Qiang H and Silliman R).

Climate change, reflected by significant environmental changes such as warming, sea-level rise, changes in salinity, oxygen and other ocean conditions, is expected to affect the coastal environment and associated ecosystem services. Studies on assessment of the potential impacts and vulnerability of marine biodiversity and fishery catches in the Persian Gulf have used three separate niche modelling approaches under a 'business-as-usual' climate change scenario. The model results suggest a high rate of local extinctions, up to 35% of initial specific wealth by 2090 compared to 2010. Spatially, predicted local extinctions are higher in the South-Western part of the Gulf region, off the coast of Saudi Arabia, Qatar and UAE. Although the predicted models provided useful indicators of the potential impact of climate change, on the diversity of the region, the extent of changes in habitat quality is more uncertain (Wabnitz CCC, Lam VWY, Reygondeau G, Teh LCL, Al-Abdulrazzak D, Khalfallah M, et al. 2018).

A study on 'threats of coastal communities of Mahanadi delta due to consecutive erosion' has been done through semi-automated Tasselled Cap technique. The study revealed that coastal communities will face a significant threat of erosion in the near future especially in surrounding areas of Puri and Paradip (Mukhopadhyay A, Ghosh P, Chanda A, Ghosh A, Ghosh S, Das S, Ghosh T, Hazra, 2018). Based on the most recent Intergovernmental Panel on Climate Change (IPCC) forecasts, mangrove forests along arid coasts, in subsiding river deltas, and on many islands are predicted to decline in area, in structural complexity and/or in functionality, but mangroves will continue to expand polewards. It is highly likely that they will survive into the foreseeable future as sea level, global temperatures and atmospheric CO_2 concentrations continue to rise (Daniel 2015).

Study Area and Its Coastal Hazards

The state of Odisha along the Bay of Bengal extends over 480 km of coastline along the east coast of India. The physiography of the Odisha coast embraces the combined delta of the Mahanadi, the Brahmani, and the Baitarani besides the flood plains of the Subarnarekha, the Budhabalanga and the Rushikulya. The entire coast covers six districts with 22 development blocks (Fig. 6.1). Among them this study is focussing on the Mahanadi delta which extends over three districts, i.e. Kendrapara (83 km), Jagatsinghpur (59 km) and Puri (137 km) (Fig. 6.2). The deltaic zone is characterized by tidal flats, mudflats and mangroves. The delta development of the Brahmani is obliterated by the encroachment of the distributaries of the Mahanadi and Baitarani from the south and north, respectively. In this portion, the coast is convex, while from Chandabali to the Subarnarekha River mouth, the coastline is concave (Murali and Vethamony 2014).

A large amount of sediment transport along the coast through the littoral drift as well as the inflow of tidal sediments and materials from the longshore drift to the coast by the waves and tides had created an ideal depositional environment along the coast forming extensive sandy beaches, dunes, spits, bars and hooks leaving behind lagoons and backwaters. But now this trend is changing slowly from an accretional environment to an erosional environment. Coastal erosion and submergence had already displaced some of the human habitations along the coast, while several villages and beaches are on the verge of submergence. This is becoming more conspicuous during the southwest monsoon season due to the impact of the storm surges and high tides induced by the low-pressure systems over the Bay of Bengal and the landfall of cyclones along the coast.

6 Vulnerability Assessment of the Inherent Hazards of Climate Change on... 161

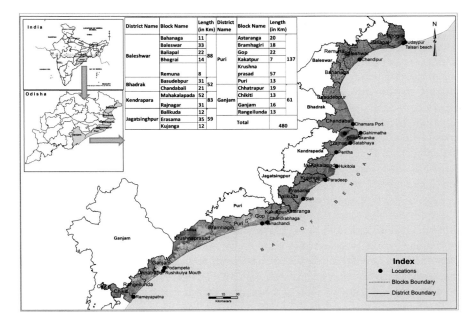

Fig. 6.1 Map of the Mahanadi delta along Odisha coast and its administrative blocks

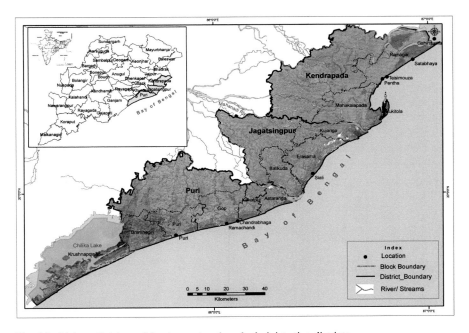

Fig. 6.2 Mahanadi delta and its river network and administrative districts

Mahanadi delta of Odisha coast is endowed with unique environmental heritages like the Chilika Lake, the largest lagoon ecosystem of Asia with a bird sanctuary; the second-largest patch of mangroves along India's east coast, i.e. Bhitarkanika; world-famous rockeries of the Olive Ridley tortoise at Gahirmatha; sanctuaries of a saltwater crocodile at Bhitarkanika; and spotted deer of Balukhand sanctuary along Puri-Konark coast. Landfall of the supercyclone of 1999, Phailin of 2013, Hudhud of 2014, Fani of 2019 and Yaas of 2021 along the Odisha coast had devastated the coastal environment with loss of vegetation, saltwater intrusion, coastal flooding and erosion besides damage to the coastal infrastructure along Odisha coast. As a hot spot of global warming and climate change, Odisha coast has been under the scanner and support of the World Bank working towards its integrated coastal zone management.

Odisha's Chilika Lake is one of India's focal points of biodiversity and one of the biggest brackish water lakes in Asia. The lake is home to the uncommon Irrawaddy dolphin. The Bhitarkanika wetland contains the second-largest mangrove biological system in Asia. Both these regions are likewise home to large vulnerable populations dependent on coastal resources. A range of port development activities is also underway. Consistently, a huge number of endangered Olive Ridley sea turtles come to nest on some of Odisha's beaches. The turtles are, however, at risk from uncontrolled mechanized fishing in prohibited areas, the non-use of turtle excluder devices and insensitive tourism. Moreover, their mass settling site on the Gahirmatha coast has been steadily moving towards the north throughout the most recent 20 years because of waterfront disintegration. Portions of the mangroves have been recovered for development, fuelwood and timber, just as for huge scope in shrimp cultivation. Bhitarkanika's sensitive environment is confronting a significant danger from the modification of freshwater inflows because of the development of hydrological structures upstream. Sea levels have been known to ascend by around 4 m in certain stretches, immersing up to 3 km of beachfront land. Besides, continuous tropical storms cause substantial misfortunes to agribusiness and fisheries, whereas withdrawing life for an enormous number of helpless ranchers and anglers. The coast comes under the impact of a strong littoral drift, making an anticipated 1.5 million tons of sand move from the southwest towards the upper northeast in a year, whereas certain coastal regions face pollution, contamination of municipal sewage from the coastal towns and effluents from the coastal industries (Banerjee 2012).

Materials and Methods

Methodology and Database

Over the last 50 years, the science of coastal mapping has evolved considerably due to advances in technology, satellite data and high-speed geospatial data processing tools. These procedures are valuable and offer an advantage over conventional field

Table 6.1 Datasets used in the CHW application for mapping of bio-geophysical parameters of Odisha's coast and their data sources

Sl. no.	Bio-geophysical parameters	Data used and sources
1.	Geological layout	Fundamental geological outline map of the study area, Google Earth's satellite images and Google Earth's terrain elevation function
2.	Wave exposure	Google Earth and essential data information on the local wind climate. Global wave environments (Davies 1980, modified by Masse link and Hughes 2003 and Rosendahl Appelquist, 2014)
3.	Tidal range	Global tidal environment maps (Davies 1980, modified by Masse link and Hughes 2003). Local data from the tide table of Paradip, Gopalpur and Chandabali port
4.	Coastal vegetation	Google Earth imagery, latitude information of the study area and the UNEP-WCMC global coral reef database
5.	Sediment Balance	The satellite imagery of Google Earth and the functionality timeline of Google Earth
6.	Storm climate	Basic information on the local wind climate and Google Earth images. Global wave environments (Davies 1980, modified by Mass link and Hughes 2003 and Rosendahl Appelquist, 2014)

strategies because of their cost-effectiveness, lessening of manual mistakes, the need for subjective choice and tedious coverage. The coastal hazard wheel (CHW) decision support system of the Danish Institute of Hydrology and Danish Technology University (Appelquist and Halsnaes 2015) is used here to evaluate the presence, spatial extent and intensity of coastal ecosystem disruption and vulnerability towards environmental impacts along Odisha coast. In the present study, assessment of ecosystem disruption is being done by the use of geospatial technology, GIS-based coastal hazard wheel framework and remote sensing data from the public domain. Spatial information utilized for this appraisal are topographical outline of the Odisha coast and an administrative map showing the boundaries of the coastal districts, blocks and villages. Data on wave characteristics, tidal range and presence of storm climate are obtained from the model which are included in the original CHW (Masselink and Hughes 2003). Satellite data used in this study are Google Earth images with timeline and ground elevation functions (Google Earth), Bing maps and Rapid Eye images (GRAS 2012). Diverse spatial data and non-spatial information are used for each of the coastal classification circles of CHW (Table 6.1).

Coastal Hazard Wheel (CHW) Model

Coastal hazard wheel (CHW, 3.0 version) is a particularly versatile and standardized means to objectively assess the degree of ecosystem disruption of the coastal stretches. The CHW has been recommended by the United Nations Environment Programme (UNEP) as a risk assessment tool to help coastal managers, planners and policymakers assess how coastal areas may be affected relative to different levels of risk induced by climate change (Appelquist 2012). This tool was primarily designed to improve decision-making in developing states (Appelquist and Balstrom 2014), as it can be applied without the need for the wide availability of digital data (UNEP 2012). In its most basic form, CHW can be applied to assess the vulnerability profile where data availability is limited. This option facilitates a preliminary assessment of the types and locations of presence of the hazard and possible management options to address the problem. Although a more accurate (intermediate level) result can be achieved with the inclusion of field observed data (in addition to remotely sensed data), high accuracy and locally focused hazard assessment is possible with the use of higher resolution in situ studies in support of the above two categories. The method chosen for this study was performed with an intermediate level of accuracy using in situ field observations and satellite imagery.

The coastal hazard wheel consisted of successive layers representing a range of optional configurations for six coastal traits (Fig. 6.3) beginning from the coastline as much as 200 metre inland followed by five rows describing various 'hazard levels' for five different climate change-sensitive coastal hazards. The centre of the wheel provides the starting point with eight possible geological layouts for the coastal surroundings being studied (sloping hard rock, sloping soft rock coast, flat hard rock coast, coral island, sedimentary plain, barrier island, delta/low estuary island, tidal inlet/sand spit/river mouth (Micallef et al. 2017)). Fom centre of CHW, the Subsequent coastal traits take into consideration the extent of wave exposure (expressed as exposed, moderately exposed or protected) followed by the tidal range (such as micro, <2 m; meso, 2.4 m and macro >4 m tidal), the flora and fauna band (intermittent marsh, intermittent mangrove, marsh/tidal flat, mangrove, marsh/mangrove, vegetated, not vegetated, coral or 'any'), the sediment balance band (balance/deficit), surplus, beach/no beach and 'any' and the storm climate recorded as presence or absence of tropical cyclones with regard to sediment balance coastal characteristics. In this model, its moves from the centre of the CHW address, in turn, geological layout, wave exposure, tidal range, vegetation, sediment balance and storm climate to the external edge where the inherent hazard levels (starting from low, moderate, high and very high) are identified for ecosystem disruption, gradual inundation, saltwater intrusion, erosion and flooding (Micallef et al. 2017).

Data Layers and Digital Analysis

The coastal classification and hazard evaluation is carried out in ArcGIS based totally on a Hybrid Bing Map. As the resolution of the satellite images is generally better in the Quick Bird image than in Bing Maps, the entire classification is done in this image. In the primary stage, a geo-database is created in ArcGIS to comprise the coastal classification data in addition to facts on environment disruption levels. An updated digitized shoreline map of Odisha is created with inside the geo-database referencing with the WGS1984, UTM zone 45 coordinate system. The polyline feature is then used for generating a digitized shoreline of Odisha by manually

6 Vulnerability Assessment of the Inherent Hazards of Climate Change on...

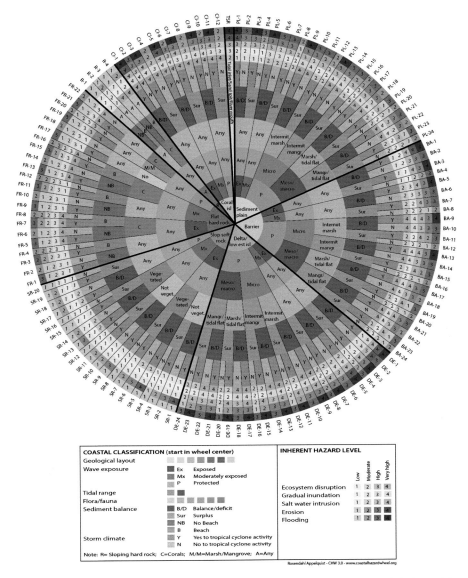

Fig. 6.3 The coastal hazard wheel 3.0 comprises six coastal type circles, five hazard circles and the coastal type codes. It is used initially in the wheel centre moving outwards through the coastal classification. (Modified from Rosendahl Appelquist and Halsnæs 2015 and Rosendahl Appelquist 2012)

digitizing the coast on the approximate suggest mean sea level (MSL) with a zoom level of 2–4 km. Since the satellite images are taken at different times during the tidal cycle, the polyline features will most likely deviate from the actual MSL. This is considered of minor importance for this assessment, as it only requires a relatively accurate and up-to-date coastline. The digitizing is accomplished with an accuracy of about 5–10 m leaving gaps for river mouths and tidal inlets. Islands are digitized as separate units. This digitized line feature was the foundation for further coastal classification and the creation of hazard maps.

To facilitate the evaluation of the coastal slope and sediment balance, separate line features are created with the Quick Bird image. The line feature for facilitating the slope evaluation consists of a range of shore parallel line sections that are drawn landwards of the coastline in all coastal areas with a slope greater than 3–4%. This enables the user to quickly determine whether a particular coastal area is sloping or not when carrying out the coastal classification. The slope of a particular coastal section is determined by manually placing the cursor over the first 200 m landwards of the coastline in the image, taking note of elevation levels given in the button of the image window. In this way it's carried out for each approximately 200 m length of shoreline on a Quick Bird satellite image at zoom level 2–4 km. The assessment of the sediment balance is done by a non-continuous line drawn on the approximate coastal vegetation line. When coastal classification is performed, sediment stability is assessed by evaluating satellite images taken at different times through the satellite image timeline function, observing the evolution of the coast associated with the most recent digitized coast. Since satellite images are acquired at different tide levels and at different times of the year, the width of the beach cannot be reliably used to determine sediment balance, but the vegetation line is considered an indicator of a balance, relatively good for overall assessment. Polylines are used to divide the original linear feature into sections, each representing a different type of shoreline defined in the CHW framework. In this process, the CHW-3 has mapped the important bio-geophysical parameters of the coast related to geological layout, wave exposure, tidal range, flora/fauna, sediment balance, storm climate and their subgroups for characterizing a specific generic coastal environment (Table 6.1).

Subsequently, the coastal environment type is entered in the attribute table of the respective polyline in the ID field, since the attribute table only accepts numbers, and the coastal environments in the CHW frame are assigned values between 1 and 131, with 1 given a CHW type PL-1 (Fig. 6.3). Following the classification of each coastal stretch according to CHW and also the input data listed, the user must adopt an appropriate type of coast and its extension before finalizing each polyline. Occasionally, a coastline can support equivalent assets over long distances, which means that the length of a polygon can vary from less than 50 m to several kilometres. The polylines are then used to divide the initial digitized coastline into sections, each representing a specific coastal category (Fig. 6.3). The hazard levels are specified in the CHW document (Appelquist 2012) and then entered in a separate attribute table which is attached to the Coastal Classification File's attribute table. From this, five different hazard maps are generated for the respective hazard types, and different hazard levels are encoded and derived.

Results and Discussion

Bio-geophysical Characterization of Odisha Coast Based on CHW Model

Based on the CHW application framework, the coastline of Odisha brings out an outline table of the prevalence of the coastal ecosystem disruption along with their risk levels and a range of subregional and regional hazard maps and corresponding data tables. Table 6.2 describes the four common coastal categories and their 21 subtypes according to the geological and geomorphological characteristics of the Mahanadi delta. Each zone is characterized by a set of bio-geophysical parameters giving a distinct generic characteristic; it also presents its nature of vulnerability towards ecosystem disruption and level of risk along with the length of these coastal stretches under different categories (Fig. 6.4) as per the coastal hazard wheel. The use of bio-geophysical parameters and their characterization in the coastal hazard wheel along with the geological layout for identifying the vulnerability and risk are discussed below for understanding the nature of ecosystem disruption along the Mahanadi delta.

Geological and Geomorphological Layout

Classification of the Odisha coastline about the geological and geomorphological outline is conducted based on the published maps of the Geological Survey of India. The 'geological layout' types presented by the CHW reveal in terms of four important types of generic coastal stretches such as delta/low estuary island (DE), sediment plain (PL), sloping soft rock (SR) and tidal inlet/sand spit/river mouth (TSR). These four classes are further classified into 15 different subgroups in combination with different micro-geomorphic features related to the above four major groups. Figure 6.4 shows the 15 coastal types and their spatial distribution along the Mahanadi delta. It can be noticed that 56% of the coastline falls under sediment plain followed by 24% deltaic low estuary islands and 15% tidal inlet/sand spit/river mouths and 5% is under soft sloping rocks. Details of the coastal types and their associated parameters, i.e. wave exposure, tidal range, coastal vegetation, sediment balance and storm climate, are given in Table 6.2. Based on these 15 coastal types and their above-mentioned five bio-geophysical parameters, the application of CHW has revealed the following hazard profile of the ecosystem disruption along the Mahanadi delta (Table 6.2).

Wave Exposure

CHW applies the perspective wave environment to distinguish between exposed bank, moderately exposed bank and protected shoreline or bank. The wave duration is delineated by the wave chart from the original CHW article (Rosendahl

Table 6.2 Different regions of the Mahanadi Delta coast with their biophysical characteristics and their hazard risk are described as low (1), moderate (2), high (3) and very high (4)

Sl no	CHW	Geological composition	Exposure to wave	Tidal range	Flora/fauna	Sediment balance	Storm climate	Length in km	% of coastline
1	DE-13	Delta/low estuary island	Protected	Micro	Intermittent mangrove	Balance/deficit	Yes	3	1
2	DE-15	Delta/low estuary island	Protected	Micro	Intermittent mangrove	Surplus	Yes	9	3
3	DE-16	Delta/low estuary island	Protected	Micro	Intermittent mangrove	Surplus	No	28	10
4	DE-21	Delta/low estuary island	Protected	Meso/micro	Mangrove/tidal flat	Surplus	Yes	6	2
5	DE-5	Delta/low estuary island	Moderately exposed	Any	Any	Balance/deficit	Yes	22	8
6	PL-1	Sediment plain	Exposed	Any	Any	Balance/deficit	Yes	5	2
7	PL-11	Sediment plain	Protected	Micro	Intermittent mangrove	Surplus	Yes	6	2
8	PL-17	Sediment plain	Protected	Meso/macro	Marsh/tidal flat	Balance/deficit	Yes	7	3
9	PL-20	Sediment plain	Protected	Meso/macro	Marsh/tidal flat	Surplus	No	12	4
10	PL-21	Sediment plain	Protected	Meso/macro	Mangrove	Balance/deficit	Yes	17	6
11	PL-5	Sediment plain	Moderately exposed	Any	Any	Balance/deficit	Yes	99	35
12	PL-8	Sediment plain	Moderately exposed	Any	Any	Surplus	No	4	2
13	PL-9	Sediment plain	Protected	Micro	Intermitted marsh	Balance/deficit	Yes	6	2
14	SR-13	Slope soft rock	Moderately exposed	Any	Vegetated	Balance/deficit	Yes	14	5
15	TSR	Tidal inlet/sand spit/river mouth						41	15
Total								279	

6 Vulnerability Assessment of the Inherent Hazards of Climate Change on...

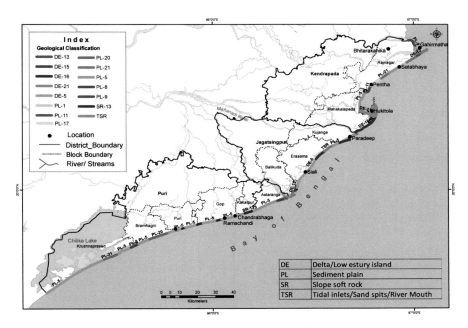

Fig. 6.4 Coastal morphological types based on CHW for Mahanadi delta

Appelquist 2012; Masselink and Huges 2003). Since Odisha is a part of the monsoon climate zone, exposure levels depend on free fetch and wind speed. In this assessment, Google Earth imagery was used to measure out whether the free fetch for a particular coastal area was less than or equal to 10–100 km or greater than 100 km. This is the defined boundary of the protected, moderately exposed shore and exposed coastline in the coastal hazard wheel structure. In general, the outer expanses of Mahanadi delta are classified as exposed because it is directly exposed to waves in the Bay of Bengal and moderately exposed, while the shore of the inner estuary is protected (Fig. 6.3).

Range of Tide

Tides can have a significant impact on coastal morphology and changes in the coastal landscape. They are an indicator of the gravitational pull of the moon and sun as they act on the Earth's hydrosphere, emerging as ocean waves with wavelengths of thousands of miles that cause periodic fluctuations in coastal water levels (Davis Jr and Fitzgerald 2004). Tides fluctuate daily in a diurnal, midday and mixed tidal cycle (Davis Jr and Fitzgerald 2004; Rosendahl Appelquist 2012). The tide range is determined based on the tide range maps contained in the original CHW documents (Rosendahl Appelquist 2012; Masselink and Hughes 2003). However, the Odisha falls somewhere between the micro and meso tide types, and more local tidal range data has been collected. These data show that the Mahanadi delta remains in the microbial category as the tidal range increases towards the northern part of the state. Therefore, it was decided to use this category for coastal classification (Fig. 6.3). Mesotidal conditions may be the current north side of the delta. However, the tidal range at these locations should remain close to the boundary between the micro and meso tide levels. Therefore, the categories of micro and meso tides apply uniformly to the entire delta.

Flora and Fauna

Flora and fauna are important parameters of the coastal environment, influencing coastal sediments, erosion and the coastal ecosystem. The CHW identifies nine different categories of natural vegetation, namely, intermittent wetlands, intermittent mangroves, swamp/tidal plate, mangrove/mudflats, swamp/mangrove, vegetation, no vegetation, coral and all. Intermittent swamps and swamps/high tides belong to coasts whose geological configuration falls into the categories of sedimentary plains. Salty, moist herbaceous plants are found along the coast in safe areas with less energy. Wetlands gradually form from successive sedimentary deposits, which often prevent coastal flooding and saline inundation. The flora and fauna are determined by a visual estimate of the coast on Google Earth combined with information on the geographic location and general signal of the reef. As Odisha is in a tropical climate, it is protected by coastal lowlands such as coastal lowlands and generally has mangrove vegetation barriers in the area, but the mangrove area is incomplete due to a low degree of suffocation. Coral rocks generally do not exist in the delta region. Therefore, the reef option did not apply to any part of the delta area.

Sediment Balance

Sediment availability plays a major role in coastal tract to develop various types of coastal landscape. The balance of sediment in CHW model is divided into two (2) main classes, such as balance/deficit and excess, associated with soft sedimentary rock, and two special categories other than beach and shore, associated with hard rock coast. For the soft sedimentary rock coast, the sediment balance can be determined based on coastal vegetation line changes. Sediment budget assessment uses remote sensing information from Google Earth satellite images and the timeline feature to compare images of the coast over the past decade. Excess sediment in coast expanses has been generally accepted to overestimate certain levels of danger. As for the exposed shores of the Odisha, Google Earth images from the past 5–10 years have been able to provide a reliable statement on the sediment balance, as changes in the vegetation line are visible in most cases. However, for protected coasts, it was difficult to estimate small variations over time from satellite images, and therefore these coasts were classified as balance/deficit in many cases. In addition to the general challenge of estimating the sediment budget, Google Earth has some shortcomings in its temporal function, such that some areas are only covered by satellite imagery, making it impossible to estimate the weather. This is the case of the coast south of the Vamsadhara River to the south of the Devi River (PC22), south of the mouth of the Devi River to Bhitarkanika Spit (PC 23), spit from Bhitarkanika to the mouth of Paga Nadi (PC24), mouth of Paga Nadi south of the Subarnarekha River (PC25) and south of Subarnarekha north of Rasulpur (PC26), and therefore these coasts have been classified as balance/deficit. Sometimes only two images are available on Google Earth after a few years, which also leads to uncertainties in the estimates. In this case, the available LISS III images are considered to supplement the data requirements.

Storm Climate

Landfall of tropical cyclones in coastal areas and its strong winds, waves and rainfall have a major impact on coastal morphology and adverse impact on the coastal environment. The model distinguishes between locations with and without tropical cyclone activity, regardless of their frequency. The influence of tropical cyclones on coastal areas (Masselink and Hughes 2003) is shown in Fig. 6.3. In areas designated as 'tropical cyclone', the categorization applies 'yes' to cyclone movement and 'no' to locations outside these areas. It is determined by wave/storm maps from the original CHW publication (Rosendahl Appelquist 2012; Masselink and Hughes 2003). Since Mahanadi delta appears to be under the influence of tropical storms, the entire coast is classified as a place prone to tropical cyclones.

The Inherent Hazard Levels of Ecosystem Disruption and Environmental Degradation

As a general rule, coastal areas with poorly shorted sedimentary characteristics tend to erode faster than those with a hard rock composition. Furthermore, low-lying shores are likely to erode faster, although this largely depends on the composition of the rock (IPCC 2007). Along the delta area, 28.60% of Mahanadi delta's coastline has a very high risk (VHR), 47.60% has high risk (HR), 16.99% has moderate risk (MR) and 6.81% has a lower risk (LR) of delta environment (see Table 6.3). In the coastal delta tract, the most risk able or very high inherent hazard is flooding (68.82%). The high, medium and low inherent hazards are saltwater intrusion (58.42%), inundation (26.52%)-ecosystem disruption (26.52%) and coastal erosion

Table 6.3 Hazard level (in percentage) along the Mahanadi delta

Mahanadi delta	Hazard level (km)			
	Very high (VH)	High	Moderate	Low
Level of hazards in percentage (%)	28.60	47.60	16.99	6.81

(43%), respectively. The very high-risk zone ecosystem disruption is correlated to the mangrove area of Bhitarkanika and Gahirmatha beach, which is the landing site of Olive Ridley turtles and mouth area of Chilika lagoon. The very-high risk of ecosystem forbearance is specially related to the existing barrier coasts and protected estuary coasts in the long-term change of climate. They are likely to carry additional risks due to sea-level rise because of their low-lying topography adjacent to the coastline. These sites are environmentally rich natural heritages that provide valuable ecosystem services as well as the breeding ground for Olive Ridley turtles, estuarine fishes and migratory birds. The moderate areas of ecosystem disruption cover part of the coast which are low-lying and frequently flooded due to storm surges and inundation from spring tides, and the delta is free from any kind of dense vegetation. A major part of the delta from Kendrapara, Jagatsinghpur and Puri belong to this risk zone. The magnitude of moderate risk indicates long-term possible damage to the delta ecosystem following climate change and sea-level rise unless suitable intervention measures are adopted (Fig. 6.5).

The mouth region of Chilika between Satapada and Humma is identified in the application model as having a high risk for ecosystem disruption. This zone is the 25 km length and 1 km width sand spit with *Casuarina* plantation (Fig. 6.6). This part of the coast is coming under sediment plain and moderately exposed and reveals deficit or sediment balance with the presence of storm climate and a tidal range of nearly 1–2 m (Table 6.4).

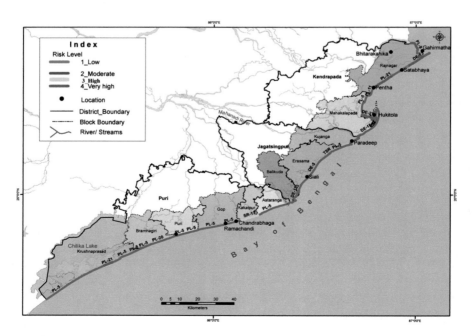

Fig. 6.5 Degree of vulnerability of the coastal stretches of Mahanadi delta to the hazard of ecosystem disruption

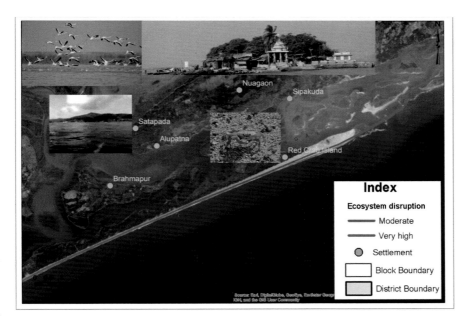

Fig. 6.6 The red portion of the Chilika coast facing the Bay of Bengal is being identified as an important site of ecosystem disruption along the Odisha coast. This may affect the lagoon ecosystem as well as its inherent biodiversity which includes one of the largest migratory birds landing and nesting sites, estuarine crabs, prawns, home to Irrawaddy dolphins and its rich saltwater weeds

Table 6.4 Hazard types and their level of risk as % of total length for the Mahanadi delta

Hazard type	Hazard level				Total length of coast line (in km)
	Very high	High	Moderate	Low	
Saltwater intrusion	44 (15.77%)	163 (58.42%)	59 (21.15%)	13 (4.66%)	279 (100%)
Flooding	192 (68.82%)	54 (19.35%)	22 (7.89%)	11 (3.94%)	279 (100%)
Inundation	43 (15.41%)	148 (53.05%)	74 (26.52%)	14 (5.02%)	279 (100%)
Coastal erosion	77 (27.60%)	151 (54.12%)	08 (2.87%)	43 (15.41%)	279 (100%)
Ecosystem disruption	43 (15.41%)	148 (53.05%)	74 (26.52%)	14 (5.02%)	279 (100%)

Hazard Assessment of Coastal Environment at Regional Level

A detailed regional analysis of delta environment is being done at the district and block level identifying the vulnerable localities and corresponding coastal stretches. Analysis at this level indicates the model efficacy in assessing the hazard at the

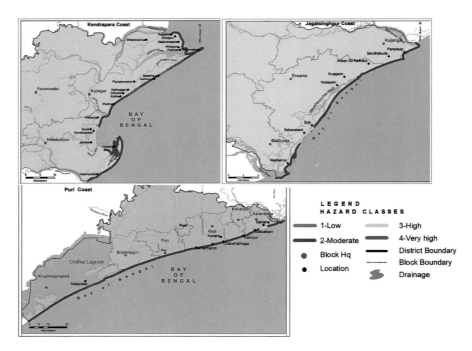

Fig. 6.7 A district-level vulnerability and risk profile of environment disruption along the Odisha coast using CHW model

regional level as well as locating the hazard presence at a regional level for ease of interventions. Figure 6.7 shows the ecosystem disruption assessment of the three coastal districts of Odisha, namely, Kendrapara, Jagatsinghpur and Puri. From the hazard vulnerability maps, it can be observed how the coastal hazard wheel (CHW) framework can be used for regional hazard mapping. These maps provide a relatively good overview of risk zones and related hazards, which require special attention and provide the basis for planning and management decisions on climate change and related processes.

Puri district has 137 km of coastline. Along this coastline, the Chilika coast is very highly vulnerable to ecosystem disruption, whereas all other sections are moderately vulnerable. High dunes with high-energy waves and beach plantations at different locations along the coastline give some amount of protection from storm surges and flooding. The ecosystem degradation of Puri coast is very high in which only 3.65% covers only 5 km length of the tract, and high, medium and low are 70.80%, 15.33% and 10.22%, respectively (see Table 6.5). The environment disruption map for the delta zone of Kendrapara district is 83 km long. This delta depicts the very high hazard levels, almost 20 km which is 24.10% (see Table 6.5), at four different locations, i.e. Bhitarkanika mangrove and estuarine crocodile sanctuary region, Gahirmatha coast which is the world's largest Olive Ridley tortoise nesting site, highly eroded coasts of Satabhaya and Pentha villages and Hukitola

Table 6.5 Coastline length (in km) and their level of risk to different hazard

District	Coastline length (in km) and their level of risk to flooding				Total coastline length
	Very high	High	Moderate	Low	
Kendrapara	39	38	6	–	83
Jagatsinghpur	47	12	–	–	59
Puri	106	4	16	11	137
Delta total	192	54	22	11	279
District	Coastline length (in km) and their level of risk to saltwater intrusion				Total coastline length
	Very high	High	Moderate	Low	
Kendrapara	15	35	33	–	83
Jagatsinghpur	25	24	10	–	59
Puri	4	104	16	13	137
Delta total	44	163	59	13	279
District	Coastline length (in km) and their level of risk to gradual inundation				Total coastline length
	Very high	High	Moderate	Low	
Kendrapara	20	27	36	–	83
Jagatsinghpur	18	24	17	–	59
Puri	5	97	21	14	137
Delta total	43	148	74	14	279
District	Coastline length (in km) and their level of risk to coastal erosion				Total coastline length
	Very high	High	Moderate	Low	
Kendrapara	38	31	–	14	83
Jagatsinghpur	35	08	4	12	59
Puri	04	112	4	17	137
Delta total	77	151	8	43	279
District	Coastline length (in km) and their level of risk to ecosystem disruption				Total coastline length
	Very high	High	Moderate	Low	
Kendrapara	20	27	36	–	83
Jagatsinghpur	18	24	17	–	59
Puri	5	97	21	14	137
Delta total	43	148	74	14	279

bay region to the north of the Mahanadi River mouth. Vulnerability of this zone is related to the dominant action of waves and tides together, cyclonic passages, presence of many estuarine mouths, lot of creeks and loss of mangroves as well as coastal plantations. Besides these areas, other parts of the coast depict the presence of a high hazard level of nearly 27 km in length which is 32.53% (see Table 6.5) of Kendrapara's coastline. This entire coast is under the moderate hazard zone. This zone is mostly wave-dominated and the tide is relatively lower in amplitude. In this zone, accretion is visible at multiple locations although there are few sites where the presence of coastal erosion is visible. Landfall of cyclones usually brings saline inundation and flooding with high storm surges in this district which makes this coast vulnerable to ecosystem degradation and 36 km (45.37%) fall in moderate zone (see Table 6.5). The Jagatsinghpur delta depicts the very high risk to ecosystem disruption almost 18 km, high 24 km and moderate 17 km which is 30.51%, 40.68% and 28.81%, respectively (see Table 6.5). Among all the inherent hazards in Mahanadi delta, i.e., flooding, saltwater intrusion, gradual inundation and coastal erosion, Puri district is more vulnerable, then Kendrapara and then Jagatsinghpur (see Table 6.5).

Summary and Conclusion

The findings of the present study provide relevant maps and datasets about the coastal stretch of the Mahanadi delta to guide coastal planners, stakeholders and policymakers in adopting appropriate mitigation and protection measures to mitigate the risks of the ecosystem disruption. In the future, additional parameters may be considered in the characterization of coastal environments that may be of particular regional importance when selecting a particular method and strategy for a specific site. Furthermore, the CHW-based decision support must consider the 'human interventions in the coastal environment which is a limitation in this study'. The current study of coastal vulnerability assessment and ecosystem disruption of the Mahanadi delta in relation to the impact of climate change reveals the suitability of the application of the CHW model to depict a broad picture of the inherent hazards and their impact on the coastal environment. The input of local data based on observations and the use of high-resolution satellite data can bring greater accuracy to the output of the model. The management options emerging out of the study can be first-hand information for the decision-making process which can be further improved while considering the results along with the human interventions along the delta and socioeconomic vulnerability of the delta communities. Based on the coastal hazard wheel model, the ecosystem disruption along the Mahanadi delta can be managed by adopting some of the soft solutions like beach nourishment, beach plantation, dune stabilization, beach replenishment and plantations restoring the ethics of the coastal environment. Limited hard solutions can be an alternative for the protection of the infrastructure existing in the towns like Puri, Gopalpur and Paradip. While the present data provides an exhaustive database on the presence and degree of vulnerability of the deltaic coast, field-level data generation will be suitable for implementation of the findings of this study.

References

Adam P (2002) Saltmarshes in a time of change. Environ Conserv 29:39–61
Appelquist LR (2012) A generic framework for mesoscale assessment of climate change hazards in coastal environments. J Coast Conserv 17(1):59–74. https://doi.org/10.1007/s11852-012-0218-z
Appelquist L, Balstrorm T (2014) Application of the Coastal Hazard Wheel methodology for coastal multi-hazard assessment and management in the state of Djibouti. Clim Risk Manag 3:79–95
Appelquist L, Halsnaes K (2015) The Coastal Hazard Wheel system for coastal multi-hazard assessment and management in a changing climate. J Coast Conserv 19(2):157e179
Bakker JP et al (2016) Environmental impacts – coastal ecosystems. In: Quante M, Colijn F (eds) North Sea region climate change assessment. regional climate studies. Springer, Cham. https://doi.org/10.1007/978-3-319-39745-0_9
Balmford A, Bond W (2005) Trends in the state of nature and their implications for human well-being. Ecol Lett 8:1218–1234
Bellwood DR, Hughes TP, Folke C, Nystrom M (2004) Confronting the coral reef crisis. Nature 429:827–833
Cohen JE, Small C, Mellinger A, Gallup J, Sachs J (1997) Estimates of coastal populations. Science 278:1211–1212
Curran SR, Kumar AA, Lutz W, Williams M (2002) Interactions between coastal and marine ecosystems and human population systems: perspectives on how consumption mediates this interaction. Ambio 31:264–268
Davis RA Jr, Fitzgerald DM (2004) Beaches and coasts. Blackwell Publishing
Duarte CM (2002) The future of seagrass meadow. Environ Conserv 29:192–206
Ellison JC, Stoddart DR (1991) Mangrove ecosystem collapse during predicted sea-level rise: Holocene analogs and implications. J Coast Res 7:151–165
IPCC (2007) In: Parry ML, Canziani OF, Palutikof JP, van der Linden PJ, Hanson CE (eds) Climate change 2007: impacts, adaptation, and vulnerability: contribution of Working Group II to the fourth assessment report of the Intergovernmental Panel on Climate Change. Cambridge University Press, Cambridge, p 976
IPCC (2013) Summary for policymakers. In: Stocker TF, Qin D, Plattner-K TM, Allen SK, Boschung J, Nauels A, Xia Y, Bex V, Midgley PM (eds) Climate change 2013: the physical science basis. Contribution working Group I to the fifth assessment report of the Intergovernmental Panel on Climate Change. Cambridge University Press, Cambridge
Jackson JBC, Kirby MX, Berger WH, Bjorndal KA, Botsford LW, Bourque BJ, Bradbury RH, Cooke R, Erlandson J, Estes JA, Hughes TP, Kidwell S, Lange CB, Lenihan HS, Pandolfi JM, Peterson CH, Steneck RS, Tegner MJ, Warner RR (2001) Historical overfishing and the recent collapse of coastal ecosystems. Science 293:629–638
LOIC (2002) Report of the LOICZ synthesis and futures meeting 2002: coastal change and the Anthropocene. LOICZ International Project Office (IPO), Netherlands Institute for Sea Research (NIOZ), Texel
Lotze HK, Lenihan HS, Bourque BJ, Bradbury RH, Cooke RG, Kay MC, Kidwell SM, Kirby MX, Peterson CH, Jackson JBC (2006) Depletion, degradation, and recovery potential of estuaries and coastal seas. Science 312:1806–1809
Masselink G, Hughes MG (2003) Introduction to coastal processes and geomorphology. Oxford University Press
Micallef S et al (2017) Application of the Coastal Hazard Wheel to assess erosion on the Maltese coast. Ocean Coast Manag 156:209–222

Murali RM, Vethamony P (2014) Morpho-dynamic evolution of Ekakula spit of Odisha coast, India using satellite data. Indian J Mar Sci 43(7):1157–1161

Nichols CR, Zinnert J, Young DR (2019) Degradation of coastal ecosystems: Causes, impacts and mitigation efforts. In: Wright L, Nichols C (eds) Tomorrow's coasts: complex and impermanent. Coastal research library, vol 27. Springer, Cham. https://doi.org/10.1007/978-3-319-75453-6_8

Orth RJ, Carruthers TJB, Dennison WC, Duarte CM, Fourqurean JW, Heck KL, Hughes AR, Kendrick GA, Kenworthy WJ, Olyarnik S, Short FT, Waycott M, Williams SL (2006) A global crisis for seagrass ecosystems. Bioscience 5612:987–996

Pernetta JC, Milliman JD (1995) Land-ocean interactions in the coastal zone – implementation plan. IGBP global change report no. 33. 215. International Geosphere-Biosphere Programmed, Stockholm

Chapter 7
Assessment of Groundwater Vulnerability to Climate Change of Jalgaon District (M.S.), India, Using GIS Techniques

Kalyani Mawale, Jaspreet Kaur Chhabda, and Arati Siddharth Petkar

Abstract Jalgaon District falls in the northwestern region of Maharashtra state in India. Groundwater is a primary source for drinking and irrigation in this region. The study aims at identifying the various parameters affecting groundwater of Jalgaon District and demarcates vulnerable zones using weighted overlay analysis with the application of geographic information system (GIS) and remote sensing (RS). This study has primarily understood groundwater depletion trends, and eight parameters such as distance from major residential area, distance from major roads, LULC, slope, drainage density, rainfall, net groundwater recharge, and average groundwater level which have been identified are responsible for groundwater vulnerability. The subordinate stage was to assign weights to the parameters by the Delphi technique.

Further, using weighted overlay analysis, eight thematic maps were generated and superimposed to demarcate vulnerable areas. The results show that 17% of the total study area is under high or more vulnerability zones. Vulnerability within the region increases while moving toward the north of the study region. The southern part is under very low to low vulnerability, the central part is under low to moderate vulnerability, and the northern part of the district comes under the high vulnerability zone. Suggestions have been provided for the immediate improvement of the highly vulnerable groundwater zones.

Keywords Groundwater vulnerability index · Geographic informatics system (GIS) · Remote sensing (RS) · Delphi technique · Urbanization

---

K. Mawale (✉)
College of Engineering, Pune, Maharashtra, India

J. K. Chhabda · A. S. Petkar
Civil Engineering Department, College of Engineering, Pune, Maharashtra, India
e-mail: asp.civil@coep.ac.in

© The Author(s), under exclusive license to Springer Nature Switzerland AG 2022
U. Chatterjee et al. (eds.), *Ecological Footprints of Climate Change*, Springer Climate, https://doi.org/10.1007/978-3-031-15501-7_7

Introduction

On our planet, seawater shares 97.6%, whereas freshwater occupies only 2.4% share of all water resources (Tillery et al. 1987; Mapani 2005; Palenzuela et al. 2015). Of the total fresh water available, 79.1% is held in ice caps, 20.83% is groundwater, 0.83% in rivers and lakes, and 0.042% in clouds and aerosols (Mapani 2005). Furthermore, precipitation in the ocean is 77% since the oceans cover substantial areas on the planet, while the land receives only 23% (Mapani 2005; Mapani and Schreiber 2008). This concerns the policymakers for conservation and proper usage of the water.

The changes in the urban environment are constantly impacting the natural environment (Heynen et al. 2006; Ulrich et al. 1991). Today, humankind is dealing with environmental depletion and its consequences on the natural and built environment. Urban and rural environments are affecting groundwater vulnerability. Urbanization has four repercussions on the hydrological cycle: water shortage (e.g., rising population and consumption), flooding, water pollution, and alterations in the river regime and groundwater activities (Wakode et al. 2014). These changes make long-term impacts on groundwater resources, quality, and levels in the underlying aquifers.

Throughout the world, groundwater is considered a primary source for drinking and irrigation (Rose and Krishnan 2009; Kemper 2004; Takal and Quaye-Ballard 2018). Climate change is being felt in the form of unprecedented changes in almost all major ecosystems in India, including adverse effects on the water balance in different parts of the country resulting from changes in precipitation and evapotranspiration (Mohan and Sinha 2009; Hallegatte et al. 2010; Cronin et al. 2014). Especially in arid and semiarid zones, the rural areas in the vicinity of urban centers are experiencing groundwater depletion due to overexploitation (Singh et al. 2011). Over the last five decades, technological advancements in India have made groundwater more accessible for utilization by pumping water through deeper tube wells. Along with this, factors such as increased water demand due to the ever-rising population paved surfaces due to development, and low infiltration rates of rainfall into the soil make urban areas vulnerable to groundwater.

Groundwater vulnerability to climate change refers to its sensitivity to current and potential threats from climatic stressors (Aslam et al. 2018). In semi-humid regions, especially in urban areas, a lower infiltration rate increases groundwater vulnerability (Starke et al. 2010). It is generally agreed that global climate change poses a great challenge to human and natural systems (Solomon 2007; Aslam et al. 2018). As a result, there has been an increasing demand for dependable methods to assess the relative vulnerability of systems to likely impacts of climate change (Carter et al. 2007; Aslam et al. 2018).

Vulnerability mapping is a method to identify the susceptibility of the resource to the environment, making it an accountable technique for future research (Rahman 2008). The overlay analysis results by juxtaposing and further assigning appropriate weights and indexing the parameters (Nobre et al. 2007; Basharat et al. 2016;

Sharma 2021). The identification of vulnerable areas through this study helps determine effective measures to counter groundwater recharge.

Jalgaon District of Maharashtra has been identified as the study area. Tapi River and its tributaries, along with groundwater reserves, are the various sources of drinking and irrigation. The increasing demand for groundwater in the last 20 years has adversely affected groundwater within the district. Observations of groundwater level (GWL) from the study for the period of post-monsoon and seasonal fluctuations show that the district is vulnerable to groundwater in recent years.

Materials and Methods

Study Area: Jalgaon District

Jalgaon District lies in the Maharashtra state of India. It lies in the northwestern part of the state. It shares its northern border with the state of Madhya Pradesh. Nashik lies to the west of Jalgaon, Buldhana and Dhule districts on the east, and Aurangabad District on the south. It lies between east longitudes 74° 55′ and 76° 28′ and north latitudes 20° 15′ and 21° 25′. The Jalgaon District occupies a total area of 11,765 km^2. The district headquarter is located in the Jalgaon Municipal Corporation area. Figure 7.1 shows that the district has been divided into 15 blocks: Jalgaon, Amalner, Bhusawal, Bhadgaon, Bodwad, Chopda, Chalisgaon, Dharangaon Erandol, Jamner, Muktainagar, Palora, Raver, Pachora, and Yawal for administrative convenience. Along with the 15 blocks, the district has 16 towns and 1519 villages. On average, the study area falls under the semiarid region (Arjun et al.

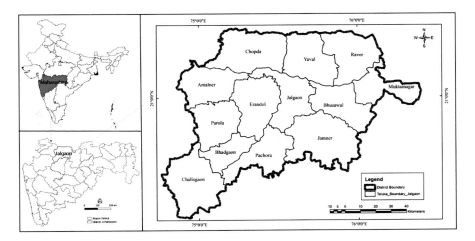

Fig. 7.1 Location map of Jalgaon District in Maharashtra

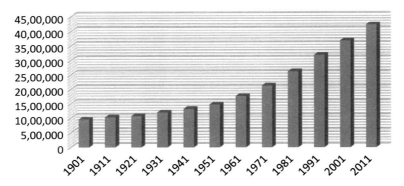

Fig. 7.2 Population chart of Jalgaon District from 1901 to 2011

2017; ENVIS Centre: Maharashtra 2015). A large part of the district falls under the Tapi Basin (Fig. 7.2).

The population of the district has shown gradual increment over the decades. From around 900 thousand people in 1901 to around 4.2 million people in 2011, the population has increased. The percentage rate of population change has decreased in recent decades. As per the 2011 census, the district population is around 4.2 million with a population density of 930 people per m^2. The population growth rate from 2001 to 2011 was 14.71% (Census of India 2011). Agriculture is the major land use, followed by forest, in the district. The average annual rainfall of the district from the 110 years of data is 721.58 mm (Ingle et al. 2018).

The highest usage of water in Jalgaon District has been for agriculture and livestock, then industrial, followed by domestic (Patil and Suryawanshi 2017).

In the study region, 9% area is underhill and uplands. Also, uplands constitute a large portion of cultivable area, which are not bounded and thus generate a huge amount of runoff (Arjun et al. 2017).

Objectives

The study's primary objective is to understand groundwater depletion trends within the study area of the Jalgaon District and identify parameters responsible for groundwater vulnerability. These parameters include the distance from the major residential area, distance from major roads, LULC, slope, drainage density, rainfall, net groundwater recharge, and average groundwater level. The study's secondary objective is to assign weights to these parameters and derive the groundwater vulnerable index of the Jalgaon District using weighted overlay analysis. Finally, measures have been suggested to overcome the vulnerability of groundwater.

Methodology

The study has used satellite data for the analysis of different parameters acting on groundwater. Satellite data help to obtain data related to soil classification, land slope, geomorphology, land use, land cover, and drainage (Mallick et al. 2015; Jha and Peiffer 2006; Jha et al. 2007). Remote sensing (RS) and geographic information systems (GIS) have been recently applied widely to access natural hazards. Several GIS tools are used in the current study. A geographic information system (GIS) is a set of tools used in gathering, managing, and analyzing information. Remote sensing is mainly used to survey and gather information through satellites about any remote object. Data for the following study have been obtained from different sources such as WorldClim and USGS in the form of a raster. It has been then processed in the GIS application. For this study, ArcGIS 10.3 has been used. The study has used tools in ArcGIS 10.3 such as raster calculator, weighted sum, unsupervised classification, vector to raster tool, reclassification, flow direction, and flow accumulation.

Figure 7.3 shows the detailed methodology for data collection and processing. The relevant parameters and their weighted overlay analysis are vital for identifying the vulnerable areas of groundwater in the district. For analyzing these parameters, primary task is to collect the data for 8 selected parameters finalized using a literature review.

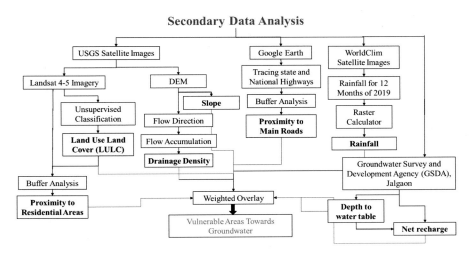

Fig. 7.3 Detailed methodology for the chapter

Sources of Data

Groundwater Surveys and Development Agency (GSDA), Jalgaon

Groundwater Surveys and Development Agency (GSDA) is an organization by the Government of Maharashtra working at the district level in the field of groundwater. It is one of the nation's leading institutions concerned with groundwater surveys, monitoring, development, management, exploration, and regulation of groundwater resources for irrigation, drinking, and industrial needs. GSDA, Jalgaon, is the primary department working on data collection regarding groundwater. GSDA department collects groundwater level data for 4 months every year: January, March, May, and October. There is a total of 178 monitoring stations (Fig. 7.4) whose data have been taken for the chapter. Data collected for GWL are in the form of excel. For using data in the ArcGIS application, excel has been imported to ArcGIS, and processing on the data has been done for the years 1991, 1996, 2001, 2006, 2011, 2016, and 2019.

WorldClim

WorldClim is a set of global climate data that can be used for mapping and spatial modeling, and it is present in GeoTiff format. Data of monthly average precipitation

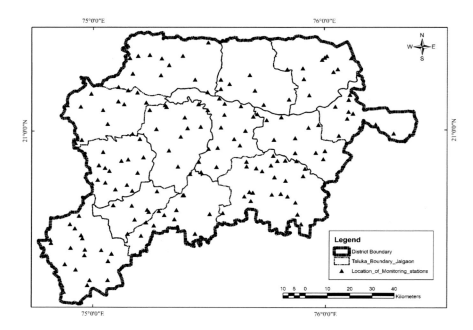

Fig. 7.4 Location of groundwater monitoring stations by GSDA Jalgaon

(mm) have been collected from WorldClim with a spatial resolution of 2.5 min. Each download is in a "zip" file format which contains GeoTiff (.tif) files for each month of the years 1991, 1996, 2001, 2006, 2011, 2016, and 2019.

Landsat 8

Landsat 8 is a joint mission launched by the U.S. Geological Survey mission and NASA (National Aeronautics and Space Administration). It is the eighth satellite in the Landsat series. Landsat 8 has been used to download a multispectral image which has then been classified into land uses using ArcGIS.

Results and Discussion

Observation of Groundwater Over 2 Decades

Figure 7.4 shows the locations of groundwater monitoring stations that the GSDA department of Jalgaon monitors. These monitoring stations are a combination of dug wells and bore wells. The data of 178 monitoring stations have been used for the chapter. In the inverse distance weighted (IDW) method, it is substantially considered that similarities between neighbors and the rate of correlations are proportional to the distance between them, which can be defined as a distance reverse function of every point from neighboring points (Setianto and Triandini 2013). IDW method has been selected over the Kriging model, as the data are of weak spatial correlation (Almodaresi et al. 2019; Jie et al. 2013). The IDW formula is given as follows (Li et al. 2018):

$$z_p = \frac{\sum_{i=1}^{n} \left(\frac{z_i}{d_i^p}\right)}{\sum_{i=1}^{n} \left(\frac{1}{d_i^p}\right)}$$

where Z is the estimated value for the prediction point, Z_i is the measured value for the sample point, d_i is the Euclidean distance between the sample point and prediction point, p is a power parameter, and n represents the number of sample points. The main factor affecting IDW interpolation result is the p value. When the p value increases, the smoothness of IDW output surface increases (Li et al. 2018).

The "IDW interpolation" tool has been used to generate each of the maps shown below in Figs. 7.5, 7.6, and 7.7. Maps are generated with the interval of 5 years for pre-monsoon GWL, post-monsoon GWL, and groundwater recharge.

Figure 7.5 shows GWL variation of the pre-monsoon period for the region. Pre-monsoon is the period before the monsoon. Hence, the groundwater level for

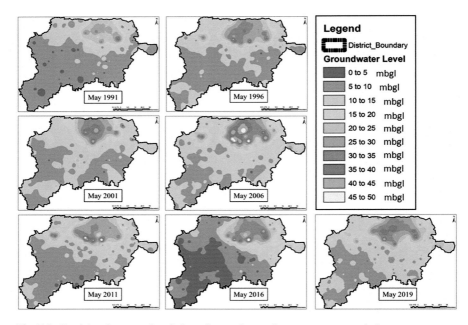

Fig. 7.5 Spatial and temporal variation of groundwater for pre-monsoon period

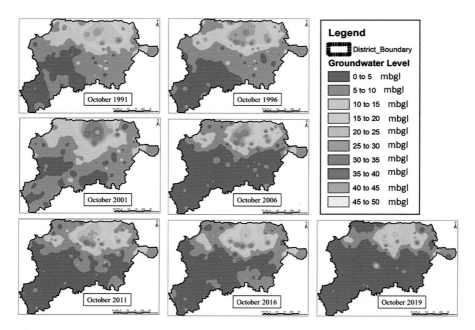

Fig. 7.6 Spatial and temporal variation of groundwater for post-monsoon period

7 Assessment of Groundwater Vulnerability to Climate Change of...

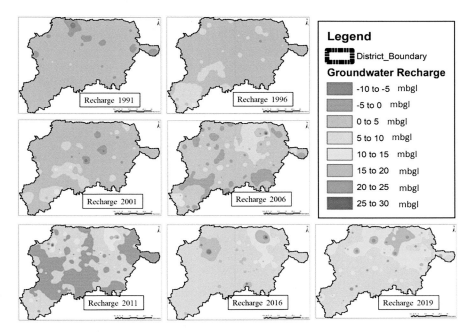

Fig. 7.7 Spatial and temporal variation of groundwater recharge

May has been chosen for spatial and temporal analysis of the pre-monsoon groundwater level. To make maps easily comparative and identify vulnerable areas, 5-meter (m) interval ranges are defined starting from 0 to 50 m deep. It has been observed in Fig. 7.5 that overall GWL is getting deeper over the period. For the year 2006, GWL for May is critical as compared to GWL of other years. In May 1991, most parts of the district fell under the category of 5–10 m level, whereas in May 2019, most parts of the district fell under the category of 10–15 m level. The maps show that the northern portion of the district is vulnerable to groundwater levels for May. The groundwater of the northern part is 30–45 m deep and at some portions is even deeper as 50 m. For the whole district, the earlier groundwater level was majorly falling in the range of 5–10 m range, whereas now it has been shifted to 10–15 m range. Comparing the findings with the CGWB report of the Jalgaon district, the findings of the following study seem correct. The northern portion of the district seems highly vulnerable.

Figure 7.6 shows GWL variation of the post-monsoon period for the region. Post-monsoon is the period after the monsoon withdraws from the Indian subcontinent. Groundwater level from October has been chosen for the post-monsoon analysis. Similar to the pre-monsoon groundwater level maps, post-monsoon GWL maps are generated with the 5 m interval. From Fig. 7.6, it is observed that the northern part of the district has a deeper GWL in October and is again the vulnerable part of the district. GWL has increased comparatively over the past 30 years. The reason can be the improvement in rainfall, but the northern portion remains vulnerable. For the year 2001, GWL for October is critical as compared to other years. The year 2000 observed to be declared as a drought year (Bhuiyan et al. 2006; Bhardwaj and Mishra 2021).

Figure 7.7 shows the variation of net groundwater recharge spatially and temporally. Temporally, it varies from 1991 till 2019 with a 5-year interval. Spatially, it varies with 5 m intervals and in the limits of the district boundary. The expression for net groundwater recharge is as follows (Narasimha Prasad 2003):

$$\text{Net groundwater rechrge} = \text{Change in groundwater storage} + \text{Groundwater extraction}$$

The above equation can be simplified and written as follows:

$$\text{Net goundwater recharge} = \text{Postmonsoon groundwater level} - \text{Premonsoon groundwater level}$$

A general expression for the groundwater balance equation is as follows (Muthuwatta et al. 2010):

$$(P + Gin) - (Q + ET + Gout) = \Delta Sgw$$

where P = precipitation, Gin = groundwater inflow, Q = discharge, ET = evapotranspiration, $Gout$ = groundwater outflow, and ΔS = change is storage. In the case of a typical unconfined aquifer, the major factors that contribute to the inflow and outflow components are recharged from rain, canals, irrigation, tanks, influent recharge from rivers, inflow from other basins, draft from groundwater, discharge to rivers, outflow to other basins, etc. (Muthuwatta et al. 2010). ΔSgw is the change in groundwater storage.

$$\Delta Sgw + \text{Groundwater Extraction} = \text{Postmonsoon GWL} - \text{Premonsoon GWL}$$

Net groundwater recharge has been increasing continuously from 1991 till 2019. More areas are getting vulnerable. Figure 7.7 shows that spatial variation of net recharge is higher in the northern portion and temporally it is deeper in recent years. The net recharge till 2001 is spatially varying from 0 to 5 m, whereas temporally it is getting deeper till 10–15 m in the year 2019 for most of the district.

Parameter Identification and Expert Opinion Survey

Through the literature review, parameters affecting groundwater vulnerability have been identified. The literature studied for identifying parameters for this chapter includes (Wakode et al. 2014; Ghosh et al. 2015; Nobre et al. 2007; Janipella et al.

7 Assessment of Groundwater Vulnerability to Climate Change of... 189

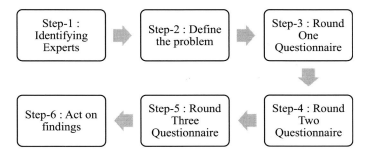

Fig. 7.8 Methodology for Delphi method

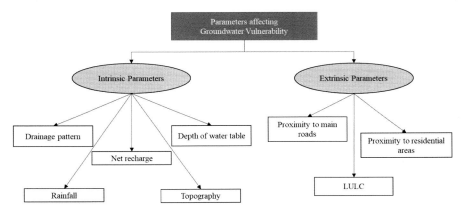

Fig. 7.9 Selected parameters for weighted overlay analysis

2020; Stempvoort et al. 1993; Aydi 2018; Singh et al. 2011). These parameters are then divided into two categories which are (i) environment-related (intrinsic vulnerability) parameters and (ii) human activity-related (extrinsic vulnerability) parameters.

The Delphi method has been used to finalize and assign weights to these parameters. Figure 7.8 shows the methodology of the Delphi method. In step 3, the round one questions involve the precursory questions to experts asking them about their area of interest, work experience, etc., for gaining a general understanding of experts along with the open-ended questions for gaining their views on groundwater vulnerability and parameters affecting the groundwater vulnerability. In step 4, the second round of questions was where the experts were asked to compare all the parameters through a questionnaire and rank them as per their effect on groundwater. The highest affecting parameter will be given the first rank and so forth. Out of all the parameters highest, weighted eight parameters were finalized for the study. Figure 7.9 shows the selected 8 parameters under the two groups (Fig. 7.10).

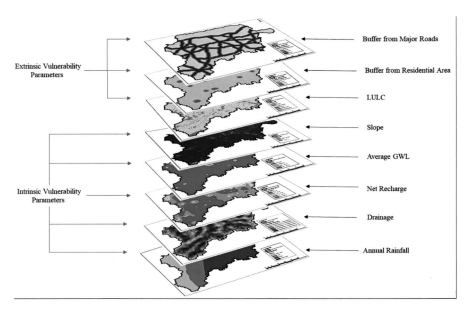

Fig. 7.10 Layer pyramid of intrinsic and extrinsic vulnerability parameters

Step 5 is the final round of questions and is conducted to assign weights to each parameter as compared to the other seven parameters. In step 6, the mean weight based on expert ratings is given out of 100. Table 7.1 shows the relative weight of each parameter along with their ranks out of 4. Relative weight is the importance of one parameter over the other, whereas the ranks are given as per their adversity and impact of each parameter on groundwater. The formula for assigning relative weight is as follows. Using this formula, the relative weight for each parameter has been calculated assigned based on weights assigned by 15 experts.

$$\text{Mean Weight (A)} = \frac{\sum n = 0 \text{ to } n = 15 \text{ (A)}}{15} * 100$$

where A is the parameter. As per the values of relative weights, the depth of the water table has been given importance over the other seven parameters, and its relative weight is highest among others. Proximity to roads has been given the least importance for its effect on groundwater. Its relative weight is 3.4 and is the least amongst others.

LULC Map

The land-use/land cover (LULC) map refers to the classification of natural elements on the globe as per their usage and human activities. A multispectral Landsat 8 imagery of the year 2015 from the USGS satellite series is used to carry out

Table 7.1 Parameters, weights, and ranking criteria

Sr. no.	Research papers	Study area	Parameters	
			Intrinsic vulnerability	Extrinsic vulnerability
I	*International literature*			
1	Groundwater vulnerability and risk mapping using GIS, modeling, and a fuzzy logic tool	Northeastern Brazil	1. Depth of water table, 2. net recharge, 3. aquifer media, 4. topography, 5. hydraulic conductivity, and 6. impact of vadose zone	7. LULC
2	Evaluation of groundwater vulnerability to pollution using a GIS-based multicriteria decision analysis	Agareb area, Southern Tunisia	1. Aquifer type, 2. hydraulic conductivity, 3. depth to water table, 4. slope, and 5. proximity to rivers,	1. Proximity to concentrated land use, 2. proximity to residential areas, and 3. proximity to main roads
II	*National literature*			
3	Evaluation of groundwater vulnerability to pollution using GIS-based DRASTIC method	Nangasai River Basin, India	1. Observation wells, 2. geology 3. rainfall, and 4. slope	–
4	Assessment of impact of urbanization on groundwater resources using GIS techniques	Hyderabad, India	1. Mean annual groundwater level	1. Land-use activities in the recharge area
5	A GIS-based DRASTIC model for assessing groundwater vulnerability of Katri Watershed, Dhanbad, India	Katri Watershed, Dhanbad, India	1. Depth of water table, 2. net recharge, 3. aquifer media, 4. soil media, 5. topography, and 6. hydraulic conductivity	–
6	Application of GWQI to assess effect of land-use change on groundwater quality in lower Shiwaliks of Punjab: remote sensing and GIS-based approach	Lower Shiwaliks of Punjab	1. Soil map, 2. hydrology map, and 3. chemical parameters	1. LULC

unsupervised classification. Using the "unsupervised classification" tool, similar spectral characteristics were clustered together. Through this classification, land uses are segregated into five categories: forest, agriculture, built-up, scrubland, and water bodies. Table 7.2 shows the rank for each land use from 1 to 4 for weighted overlay analysis. 1 rank has been given to the land use, which causes a very less negative impact on groundwater. These land uses are forest and water bodies. On the

Table 7.2 Parameters, weights, and ranking criteria

	Parameters for weighted overlay analysis					
		Rank				
Sr. no.	Parameters	1 Best	2 Good	3 Bad	4 Worst	Mean weights out of 100
1.	Depth of water table	<10 m	10–20 m	20–30 m	>40 m	22.7
2.	Net groundwater Recharge	<10 m	10–15 m	15–20 m	>25 m	10.3
3.	Rainfall	550–630 mm	631–711 mm	712–792 mm	793–873 mm	15
4.	Proximity to residential areas	>4 km	2–3 km	1–2 km	<1 km	9.2
5.	LULC	Forest, water bodies	Agriculture	Scrub land	Built-up	11.8
6.	Drainage density	0.00076–0.0012	0.00058–0.00076	0.0004–0.00058	0.000016–0.0004	11.4
7.	Slope	<10 m	10–20 m	20–30 m	>40 m	13.2
8.	Proximity to main roads	>4 km	2–3 km	1–2 km	<1 km	6.4

other hand, 4 ranks have been given to the land use which causes serious impact on groundwater, and hence, built-up area has been given rank 4.

Buffer from Built-Up Area

The built-up area has been extracted from the LULC map through query building, and then, the layer was created from the selected feature. The "buffer tool" has been used to create a buffer from the built-up area boundary. The four buffers have been created—1 km, 2 km, 3 km, and 4 km above buffer. Table 7.2 shows the ranks for the built-up buffer. The area under 4 km above category is safe and is causing a negligible impact on groundwater; hence, rank 1 has been given. On the other hand, 1 km buffer from the built-up area is harmful to groundwater, and hence, rank 4 has been given.

Buffer from Major Roads

Similar to that of built-up area buffer, buffer for roads was created till 4 km and above buffer is the least harmful to groundwater as it is away from paved surfaces. Table 7.2 shows the ranks of major roads buffer.

Slope

The lower slope values show flat terrain, and the higher slope value indicates more steep terrain. The slope varies from 0 to 66 value. The slope values are then classified into 4 classes—less than 10%, 10–20%, 20–30%, and 30% above. Table 7.2 shows the ranks allotted to each class of the slope. The steeper slope has been given a lower rank as rainwater will drain off from the surface and will not penetrate inside the soil. The extreme north side shows the steepest slope. This shows the presence of mountainous regions there.

Average Groundwater Level

The groundwater level is the level below which the rock masses and subsoil of the earth are fully saturated with water. The formula for calculating average GWL is as follows:

$$\text{Average GWL} = \frac{\text{Premonsoon GWL} + \text{Postmonsoon GWL}}{2}$$

The data regarding GWL of the May and October months have been collected from the GSDA department, Jalgaon. As GSDA, Jalgaon has recent groundwater data for 2019, and GWL of the year 2019 has been taken to carry out this study. The northern portion of the district has a deeper groundwater level, and the groundwater level is increasing as moving toward the southern portion of the district.

Net Groundwater Recharge

Net groundwater recharge shows the improvement of GWL in the post-monsoon period due to rainfall. Higher values of net recharge show that the deeper GWL has been refiled by rainfall to a higher extent. For weighted overlay analysis, the net groundwater recharge of 2019 has been taken. Table 7.2 shows the ranks given to each class of net groundwater recharge.

Drainage Density

Drainage density is a measurement of the sum of the channel lengths serving per unit area around that channel. The drainage density values are then classified into 4 classes as shown in Table 7.2.

Rainfall

For creating a total rainfall map of the district, monthly rainfall data have been downloaded from the WorldClim satellite in the form of the raster. Using the raster calculator, the rainfall data for all the 12 months of 2019 have been added, and a new raster file with the total rainfall of 2019 has been generated.

$$\text{Total Rainfall} = \sum (\text{January to December}) \text{ Rainfall}$$

The generated raster file is then subdivided into 4 classes and ranked from 1 to 4 as shown in Table 7.2.

Groundwater Vulnerability Index

Figure 7.12 shows the groundwater vulnerability index of the region and its spatial extent. The eight maps shown in Fig. 7.11 were weighted among each other using the relative weights calculated by the Delphi method. It is observed that very high and high groundwater vulnerable zones are located in the northern parts of the study area. The southern side of the district shows very low and low vulnerability, which shows that groundwater usage in these areas is low. The comparison of Fig. 7.12 with the LULC map of Jalgaon (Fig. 7.11c) shows that agricultural land use is concentrated in southern parts of the study area where the groundwater vulnerability is coming low. In contrast, northern parts of the study area, where the groundwater vulnerability is high, the hills and concentration of built-up area are observed.

There is an urgent need to address the highly vulnerable areas and implement solutions to reduce groundwater vulnerability. Major parts of the Raver, Jalgaon, Chopda, and Yawal blocks are under high to very high vulnerability zones. Blocks such as Amalner, Bhusawal, and Muktainagar fall under high to medium groundwater vulnerability zone. Blocks such as Parola, Erandol, Bhadgaon, and Chalisgaon fall under low to moderate vulnerability zones. In contrast, most of the portion of Pachora and Jamner blocks are occupied with low to very low vulnerability zones. Figure 7.12 shows that Raver, Jalgaon, Chopda, and Yawal blocks need immediate attention.

Around 17% of the study area falls under high and very high vulnerability zones. 35% of the district comes under moderate vulnerability zones. A very low vulnerability zone has occupied the area of 14%, followed by the low vulnerability zone occupying the area of 34%. Even though there is a river and its arteries flowing from the northern side of the study area, the high and very high vulnerability zones are present here. The reason behind this situation can be the high slope and concretization in built-up areas.

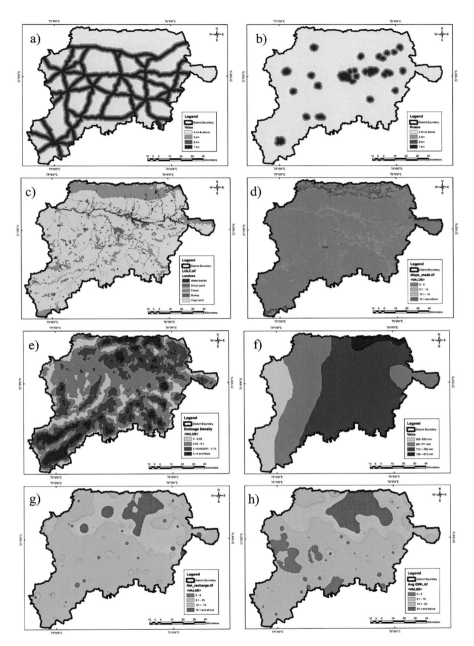

Fig. 7.11 (**a**) Distance from major residential area, (**b**) distance from major roads, (**c**) LULC, (**d**) slope, (**e**) drainage density, (**f**) rainfall, (**g**) net groundwater recharge, and (**h**) average groundwater level

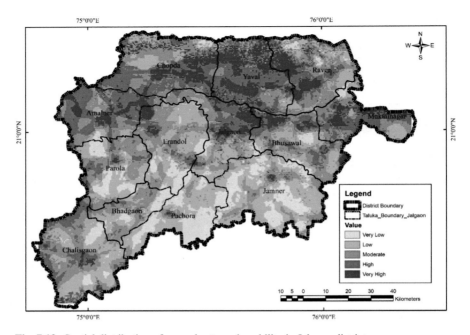

Fig. 7.12 Spatial distribution of groundwater vulnerability in Jalgaon district

Limitations

The data used for the study are limited to secondary data. An expert opinion survey has been carried out through online mode by creating a Google Form. Hence, the conversations with the experts were restricted.

Recommendations

Blocks with high to very high vulnerability are not recommended for further groundwater development except for drinking purposes. Rather, these blocks are recommended to introduce artificial recharge to increase the groundwater level. Parola, Erandol, Bhadgaon, Chalisgaon, Pachora, and Jamner blocks are recommended for further groundwater development.

There are several techniques of artificial recharge of water both under and above the surface throughout the world. Techniques such as rooftop rainwater harvesting, building reservoirs, and check dams are some of the techniques used in India from past times. There are two types of artificial groundwater recharge technologies, which are direct and indirect recharge methods. Artificial recharge through injection wells is a new technique initiated by many countries recently. Today, injection wells are gaining much attention from the government of India. This technique is a direct recharge technique (Ghazavi et al. 2018). It is recommended as the solution for improving groundwater levels in high and very high vulnerability zones of the study area. Injection wells are similar to a tube well with the function opposite to a tube well. It injects water inside the ground from nearby water resources. This method increases groundwater storage by pumping in treated surface water under pressure (Bhattacharya 2010; David and Pyne 2017). This method is highly suitable for the current study. Tapi River and its tributaries, such as Purna and Girna, flow in the region (Nagamallesh and Mathur 2019). Existing dug wells within the study area can also be used as injection wells after the proper filtration of the source water. The site location of the dug well is such that a sufficient thickness of desaturated/unsaturated aquifer exists, and the water level is more than 5 m deep (Central Ground Water Board (CGWB) 2015).

Conclusion

In this chapter, the groundwater vulnerability index of the Jalgaon District region has been assessed by studying intrinsic and extrinsic vulnerability parameters. Relative weights to these parameters were assigned by the Delphi technique. Delphi's method helped carry out expert opinion surveys. Using these relative weights and ranks for all the eight parameters, a weighted overlay analysis has been done to prepare the groundwater vulnerability index. It shows that the northern part of the region is highly vulnerable. This portion needs the immediate attention of policymakers, nongovernmental organizations (NGOs), and government officials working for the groundwater sector. If the groundwater condition of highly vulnerable zones of the region is not taken care of in the present situation, the situation will get worse in the near future. Complete dryness in these zones and the neighboring areas around these zones of the region can be seen in the future. To overcome this situation, the chapter has also given recommendations that can be implemented. Past trend of groundwater in the region has shown that net recharge has increased drastically within the span of two decades which shows high dependency on rainfall.

Acknowledgement All the authors would like to thank the Groundwater Surveys and Development Agency (GSDA) of Jalgaon for making the groundwater-related data available. The authors also would like to thank all the experts for their quick response to the online questionnaire survey.

References

Almodaresi SA, Mohammadrezaei M, Dolatabadi M, Nateghi MR (2019) Qualitative analysis of groundwater quality indicators based on Schuler and Wilcox diagrams: IDW and Kriging models. J Environ Health Sustain Dev 4:903

Arjun PN, Shivaji SD, Madhukar KA (2017) Strategies for water balance and deficit in drought-prone areas of Jalgaon District (MS) India. Int J Eng Res Techno (IJERT) 6:167–171

Aslam RA, Shrestha S, Pandey VP (2018) Groundwater vulnerability to climate change: a review of the assessment methodology. Sci Total Environ 612:853–875

Aydi A (2018) Evaluation of groundwater vulnerability to pollution using GIS-based multi-criteria decision analysis. Groundw Sustain Dev 7:204–211

Basharat M, Shah HR, Hameed N (2016) Landslide susceptibility mapping using GIS and weighted overlay method: a case study from NW Himalayas, Pakistan. Arab J Geosci 9(4):1–19

Bhardwaj K, Mishra V (2021) Drought detection and declaration in India. Water Sec 14:100104

Bhattacharya AK (2010) Artificial groundwater recharge with a special reference to India. Int J Res Rev Appl Sci 4(2):214–221

Bhuiyan C, Singh RP, Kogan FN (2006) Monitoring drought dynamics in the Aravalli region (India) using different indices based on ground and remote sensing data. Int J Appl Earth Obs Geoinf 8(4):289–302

Carter TR, Jones RN, Lu X, Bhadwal S, Conde C, Mearns LO et al (2007) New assessment methods and the characterisation of future conditions. In: Climate change 2007: impacts, adaptation and vulnerability. Contribution of working group II to the fourth assessment report of the intergovernmental panel on climate change. Cambridge University Press, Cambridge, pp 133–171

Central Ground Water Board (CGWB) (2015) Maharashtra district profile. Retrieved from cgwb.gov.in. http://cgwb.gov.in/District_Profile/Maharashtra/Jalgaon.pdf

Cronin AA, Prakash A, Priya S, Coates S (2014) Water in India: situation and prospects. Water Policy 16(3):425–441

David R, Pyne G (2017) Groundwater recharge and wells: a guide to aquifer storage recovery. CRC Press, America

ENVIS Centre: Maharashtra (2015) Climate. Retrieved from mahenvis. http://mahenvis.nic.in/Climate.aspx

Ghazavi R, Babaei S, Erfanian M (2018) Recharge wells site selection for artificial groundwater recharge in an urban area using the fuzzy logic technique. Water Resour Manag 32(12):3821–3834

Ghosh A, Tiwari AK, Das S (2015) A GIS-based DRASTIC model for assessing groundwater vulnerability of Katri Watershed, Dhanbad, India. Model Earth Syst Environ 1(3):1–14

Hallegatte S, Ranger N, Bhattacharya S, Bachu M, Priya S, Dhore K, Farhat Rafique F, Mathur P, Naville N, Henriet F, Patwardhan A, Narayanan K, Ghosh S, Karmakar S, Patnaik U, Abhayankar A, Pohit S, Corfee-Morlot J, Herweijer C (2010) Flood risks, climate change impacts and adaptation benefits in Mumbai: an initial assessment of socio-economic consequences of present and climate change induced flood risks and of possible adaptation options, OECD environment working papers, no. 27. OECD Publishing, Paris

Heynen N, Kaika M, Swyngedouw E (eds) (2006) In the nature of cities: urban political ecology and the politics of urban metabolism, vol 3. Taylor & Francis, Hoboken

INDIA, P (2011) Census of India 2011 provisional population totals. Office of the Registrar General and Census Commissioner, New Delhi

Ingle ST, Patil SN, Mahale NK, Mahajan YJ (2018) Analysing rainfall seasonality and trends in the North Maharashtra region. Environ Earth Sci 77(18):1–12

Janipella R, Quamar R, Sanam R, Jangam C, Jyothi V, Padmakar C, Pujari PR (2020) Evaluation of groundwater vulnerability to pollution using GIS-Based DRASTIC method in Koradi, India—a case study. J Geol Soc India 96(3):292–297

Jha MK, Peiffer S (2006) Applications of remote sensing and GIS technologies in ground water hydrology: past, present, and future. BayCEER, Bayreuth, p 201

Jha MK, Chowdhury A, Chowdary VM, Peiffer S (2007) Groundwater management and development by integrated remote sensing and geographic information systems: prospects and constraints. Water Resour Manag 21(2):427–467

Jie C, Hanting Z, Hui Q, Jianhua W, Xuedi Z (2013, June) Selecting proper method for groundwater interpolation based on spatial correlation. In: 2013 Fourth international conference on digital manufacturing & automation, IEEE, pp 1192–1195

Kemper KE (2004) Groundwater—from development to management. Hydrogeol J 12(1):3–5

Li Z, Wang K, Ma H, Wu Y (2018, November) An adjusted inverse distance weighted spatial interpolation method. In: Proceedings of the 2018 3rd international conference on communications, information management and network security (CIMNS 2018), pp 128–132

Mallick J, Singh CK, Al-Wadi H, Ahmed M, Rahman A, Shashtri S, Mukherjee S (2015) Geospatial and geostatistical approach for groundwater potential zone delineation. Hydrol Process 29(3):395–418

Mapani BS (2005) Groundwater and urbanisation, risks and mitigation: the case for the city of Windhoek, Namibia. Phys Chem Earth Parts A/B/C 30(11–16):706–711

Mapani BS, Schreiber U (2008) Management of city aquifers from anthropogenic activities: Example of the Windhoek aquifer, Namibia. Phys Chem Earth Parts A/B/C 33(8–13):674–686

Mohan D, Sinha S (2009) Vulnerability assessment of people, livelihoods and ecosystems in the Ganga Basin. World Wildlife Fund, Delhi

Muthuwatta LP, Ahmad MUD, Bos MG, Rientjes THM (2010) Assessment of water availability and consumption in the Karkheh River Basin, Iran—using remote sensing and geo-statistics. Water Resour Manag 24(3):459–484

Nagamallesh G, Mathur RR (2019) Sustainable groundwater resources for environmental management in watershed areas–a case study. Int J Sci Eng Sci 3:68–75

Narasimha Prasad NB (2003) Assessment of groundwater resources in Nileshwar Basin. J Appl Hydrol XVI:52–60

Nobre RCM, Rotunno Filho OC, Mansur WJ, Nobre MMM, Cosenza CAN (2007) Groundwater vulnerability and risk mapping using GIS, modeling and a fuzzy logic tool. J Contam Hydrol 94(3–4):277–292

Palenzuela P, Alarcón-Padilla DC, Zaragoza G (2015) Concentrating solar power and desalination plants: engineering and economics of coupling multi-effect distillation and solar plants. Springer, Cham

Patil N, Suryawanshi D (2017) Water demand in drought-prone areas of Jalgaon district, Maharashtra state: past, present and future. Int J Recent Adv Multidiscip Res 04

Rahman A (2008) A GIS-based DRASTIC model for assessing groundwater vulnerability in shallow aquifer in Aligarh, India. Appl Geogr 28(1):32–53

Rose RS, Krishnan N (2009) Spatial analysis of groundwater potential using remote sensing and GIS in the Kanyakumari and Nambiyar basins, India. J Ind Soc Remote Sen 37(4):681–692

Setianto A, Triandini T (2013) Comparison of kriging and inverse distance weighted (IDW) interpolation methods in lineament extraction and analysis. J Appl Geol 5(1):21–29

Sharma M (2021) Groundwater analytics for measuring quality and quantity. In: Geospatial technology and smart cities. Springer, Cham, pp 415–430

Singh CK, Shashtri S, Mukherjee S, Kumari R, Avatar R, Singh A, Singh RP (2011) Application of GWQI to assess the effect of land-use change on groundwater quality in lower Shiwaliks of Punjab: remote sensing and GIS-based approach. Water Resour Manag 25(7):1881–1898

Solomon S (2007, December) IPCC (2007): climate change the physical science basis. In: Agu fall meeting abstracts, vol. 2007, pp U43D-01

Starke P, Göbel P, Coldewey WG (2010) Urban evaporation rates for water-permeable pavements. Water Sci Technol 62(5):1161–1169

Stempvoort DV, Ewert L, Wassenaar L (1993) Aquifer vulnerability index: a GIS-compatible method for groundwater vulnerability mapping. Can Water Resour J 18(1):25–37

Takal JK, Quaye-Ballard JA (2018) Bacteriological contamination of groundwater in relation to septic tanks location in Ashanti Region, Ghana. Cog Environ Sci 4(1):1556197

Tillery BW, Day JA, Hawkins GS, Picker L, Ridky RW (1987) Fresh water. Heath Earth Science. D.C. Heath and Company, Lexington

Ulrich RS, Simons RF, Losito BD, Fiorito E, Miles MA, Zelson M (1991) Stress recovery during exposure to natural and urban environments. J Environ Psychol 11(3):201–230

Wakode HB, Baier K, Jha R, Ahmed S, Azzam R (2014) Assessment of the impact of urbanisation on groundwater resources using GIS techniques-case study of Hyderabad, India. Int J Environ Res 8(4):1145–1158

Chapter 8
Impact of Climate Change on Water Crisis in Gujarat (India)

Nairwita Bandyopadhyay

Abstract Gujarat, an economically developed state of India, is under severe threat of water scarcity due to rising population, rapid urbanisation, growing water demand and overextraction of its limited groundwater resources. Recurrent droughts have further aggravated the situation as the aquifers are under intense water stress due to inadequate recharge owing to change in meteorological dynamics. With more than 60 million population, the state has already utilised 68% of its groundwater resource, and the trend is ever increasing. Rapid urbanisation, prolific industrialisation and multiple cropping practice have increased the water demand and consumption manifold over the years. At the same time, frequency of heatwaves and intensity of droughts have increased resulting in severe water scarcity. The state suffers from water scarcity every year, particularly during the summer. In the present study, spatiotemporal variations of groundwater recharge and quality have been examined in a GIS environment. Pre- to post-monsoon groundwater storage change in various districts of Gujarat has been analysed to evaluate aquifer recharge. Besides, the impact of meteorological drought on groundwater has been assessed. The study has revealed that water table in Northern Gujarat is falling steadily and has declined by 70 m below ground level (BGL) over the last 30 years. In many parts of the state, the groundwater level has reached 200 m BGL, leading to an irreversible risk of salinisation of aquifers. A significant increase in total dissolved solids (TDS) during droughts have made groundwater unsuitable for drinking, resulting in a different type of water crisis. A correlation analysis between population change and groundwater fluctuation has revealed a direct adverse impact of rising population on groundwater resource. Findings of this study predict that water supply in Gujarat may be in jeopardy in the near future and sustainable use of water and adaptation to climate change is the only way forward.

Keywords Drought · Groundwater · Gujarat · India · Climate change

---

N. Bandyopadhyay (✉)
Department of Geography, Haringhata Mahavidyalaya, University of Kalyani, Nadia District, West Bengal, India

Introduction

Inaccessibility of water over the long run possesses threat to food security in any country. Agricultural production based on groundwater is unsustainable. This fact threatens India's ability to maintain growth rates of food production. Since the introduction of the green revolution, technology and output have become significantly more sensitive to rainfall (Rao et al. 1988). Groundwater resources provide a reliable drought buffer (Calow et al. 2007), and in a country like India, groundwater systems have become the "lender of last resort", and depletion of renewable groundwater stocks is taken as the first indicator of water scarcity (Shah and Indu 2004).

Gujarat, an economically developed state of India (Fig. 8.1), is under severe threat of water scarcity due to rising population, rapid urbanisation, growing water demand and overextraction of its limited groundwater resources. With more than 60 million population, the state has already utilised 68% of its groundwater resource, and the trend is ever increasing (Gujarat Water Resources Development Corporation 2018). Recurrent droughts have further aggravated the situation as the aquifers are under intense water stress due to inadequate recharge. Frequency of heatwaves and intensity of droughts have increased in the last decade resulting in severe water scarcity.

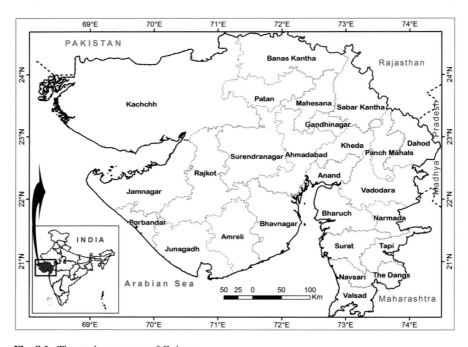

Fig. 8.1 The study area map of Gujarat

Water scarcity is more intense during the summer and is recurrent. The impact of drought on groundwater needs attention as groundwater plays a vital role in buffering the impacts of rainfall variability (Calow et al. 2005).

It has been observed in study conducted by CGWB in 2006 that the alluvial aquifer of northern part of Gujarat is one of the most exploited aquifers of India as the demand for groundwater and its abstraction has exceeded the net demand for water in several administrative blocks. The withdrawal of groundwater owing to human demand has exceeded more than 90% of annual recharge in the northern part of Gujarat, and for the groundwater levels in arid Kachchh Peninsula, the estimation goes beyond 80%. This has led to pollution of the groundwater table and resulted in high fluoride content and salt ingression in the zone of north Gujarat and Saurashtra Peninsula aquifer.

Drought Scenario in Gujarat

Gujarat witnesses drought in every 3 or 4 years. Gujarat gets rainfall only in the monsoon season (June to September). The frequency and intensity of droughts have also increased in the state over the years, resulting in serious drinking water scarcity (Planning Commission Report 2007). Gujarat was hit by 15 major droughts during the period from 1982 to 2010 (1982, 1985, 1986, 1987, 1990, 1991, 1993, 1995, 1998, 1999, 2000, 2001, 2002, 2004 and 2009). In major parts of Northern Gujarat, a significant rise in temperature coupled with deficient rainfall is observed during 2001–2010. Out of every 5 years, 2–3 years are drought years in Gujarat, and it is severe and widespread in 2–3 out of every 10 years (Bandyopadhyay et al. 2015).

Groundwater Resources in Gujarat

The diverse terrain conditions have given rise to different groundwater situations in the state. The rock formations ranging in age from Archaean to recent include gneisses, schists, phyllites, intrusive, medium- to coarse-grained sandstones, basalts and recent alluvium. The high relief area in the eastern and north-eastern part occupied by Archaean and Deccan Trap has steep gradient allowing high run-off and therefore has little groundwater potential. The yield of wells in these formations ranges from 5 to 10 m3/hr. The yield in sandstones varies from 50 to 170 m3/hr. The yield of wells tapping Quaternary alluvium in Cambay basin ranges between 75 and 150 m3/hr. There are five major aquifers in alluvial sediments out of which the top one has dried up due to overexploitation (CGWB 2016). It has been found that due to the varied water-yielding capacity of the soil, there is a contrast in the north to south groundwater extraction. In the northern part of Gujarat, the soils have better infiltration capacity owing to high permeability. The rainfall that is required to recharge this shallow aquifer is also less compared to the southern aquifers of the state which comprises of hard rock basalts.

Due to surface water accumulated in the southern part of Gujarat, the irrigated agriculture is dependent on groundwater, apart from industrial and domestic sector water demands. The mining of groundwater has thus resulted in areas becoming overexploited.

Gujarat's groundwater development is in the porous hydrogeological formations comprising of consolidated and semi-consolidated sedimentary rocks (Gupte et al. 2009). In the past two decades, the state has lost about 27% of its groundwater resources, and the loss is 50% in North Gujarat alone. About 87% area of the state has become "non-white" in groundwater, implying unsafe withdrawal of groundwater in these areas (Hirway 2005).

Being located in the semiarid region, precipitation is not evenly distributed and ranges from high rainfall in the south (2000 mm) to very low rainfall (300 mm) in the northwest. The mode of groundwater extraction in Gujarat is mainly tube wells. Most of the wells in Gujarat are individually owned, whereas the rest are owned by a group of farmers to overcome the costs required to drill a deep well. The refusal of farmers to comply with electricity rationing regulations and illegal consumption of electricity led to higher consumption and as a result of it led to higher abstraction of groundwater (IWMI Annual Report 2018). According to Dubash (2002), the ownership and control over groundwater are highly polarised as the benefits and access to extraction are unevenly distributed among different social classes.

Objectives

The main objectives of the study are to:

- Assess aquifer recharge by analysing pre- to post-monsoon groundwater storage change in various districts of Gujarat from 1981 to 2010
- Assess the impact of meteorological drought on the groundwater regime
- Examine the spatiotemporal variations in groundwater quality in the state

Data Source

The study has been conducted using mainly secondary data sources that are mentioned below:

Rainfall Data

- The CGIAR-CSI CRU TS 2.1 Climate Database (1981–2010)
- India Meteorological Department (IMD), Pune (1981–2010)

Groundwater Data

- District-wise data of average groundwater level and total dissolved salts (TDS) from Gujarat Water Resources Development Corp. Ltd (GWRDC) and Central Ground Water Board, Government of India (1981–2010).
- Data of groundwater levels were collected twice a year, i.e. pre-monsoon (May) and post-monsoon (October) periods, and are available with 5-year intervals (1981–2010).

Methodology

Trend in the change in groundwater level is measured by calculating the deviation of mean groundwater level in observation wells in respect of previous year data. The previous year is taken as the base year for calculation.

In similar manner, the change in rainfall is calculated by measuring the rainfall deviation of the present year in respect to last year data of the particular station.

The data for TDS to measure groundwater quality was also analysed on a time scale from 1981 to 2010.

Correlation between change in rainfall and change in groundwater has been performed to understand the impact of one factor on the other. Impact of rainfall was also analysed by correlating rainfall with change in TDS level. For statistical analysis, SPSS and EasyFit professional software have been used. A correlation coefficient (R) analysis was done to explain relationship between two or more random variables or observed data values. Pearson product-moment correlation coefficient, also known as R, is a measure of the strength and direction of the linear relationship between two variables.

Results

There were altogether 123 talukas in white category, i.e. in safe groundwater development, which reduced to 97 in 1997, and they further reduced to 85 in 2004. Similar reduction in talukas considered to be safe was noticed in case of non-DDP and DPAP talukas. The reduced numbers were 61, 45 and 20 in 1991, 1997 and 2004, respectively. However, DDP and DPAP talukas have registered slight improvement in white talukas in 2004. There were only seven talukas in white category in 1991, which reduced to three in the year 1997 across DDP areas. But it had consolidated in 2004, with 14 talukas coming into white category. However, this is not the gloomy picture about drought and drought-prone area of Gujarat. There has also been increase in the member of 54 talukas in grey, dark, OE and saline categories during the same period. In 2015, out of the 184 Talukas, there are 31 talukas that fall in the overexploited category, 12 talukas that fall in the critical and 69 talukas that fall in the semi-critical category. The stage of groundwater development is 76% according to Central Ground Water Board of Gujarat.

Change in Groundwater Level

During the pre-monsoon season, the decadal changes in water table fluctuation indicate change in groundwater recharge (in unconfined aquifer).

In the pre-monsoon period, particularly in May, over a period of 30 years, the decline in water level is observed in Ahmedabad, Banaskantha, Jamnagar, Kutch, Patan, Porbandar, Rajkot, Surendranagar and Dangs with more than 50 m decline of groundwater level found on an average (Fig. 8.2). Figure 8.3 shows the districts which are water-scarce during pre-monsoon and how the situation changes in post-monsoon period. During the post-monsoon season, particularly in October, over a period of 30 years, Kachchh is worst hit with more than 70% decline in average water level as this region falls under rainfall deficit and arid zone. It was also observed that Bhavnagar recorded more than 50% decline, followed by Ahmedabad, Bharuch, Kheda, Patan, Sabarkantha and the Dangs with more than 20% decline in average water level. The districts that recorded groundwater deficit in both the seasons are Ahmedabad, Bharuch, Gandhinagar, Kachchh, Kheda, Patan, Sabarkantha, Surendranagar and the Dangs. These nine districts require immediate attention as the groundwater level has declined more than 20% on an average over a period of 30 years.

Temporal and Spatial Change

The groundwater data from 1980 to 2010 indicates an interesting scenario of district-wise variation with a gap of 5-year interval. Figure 8.4 shows during which years these districts were more affected compared to the other years. It was found that the temporal change in the level of groundwater is very pronounced after each and every drought situation.

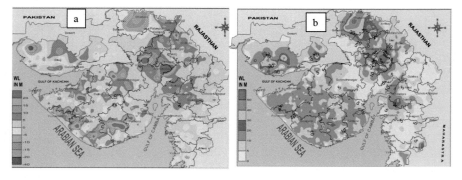

Fig. 8.2 (a) Pre- (May) and (b) post-monsoon (November) groundwater scenario in Gujarat (2003–2013). (Source: Central Ground Water Board, Gujarat, 2015)

8 Impact of Climate Change on Water Crisis in Gujarat (India)

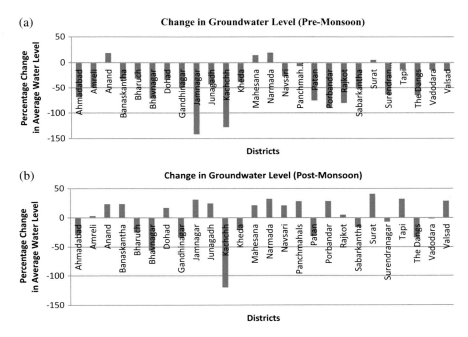

Fig. 8.3 Groundwater level fluctuation (1980–2010) pattern in (**a**) pre-monsoon and (**b**) post-monsoon

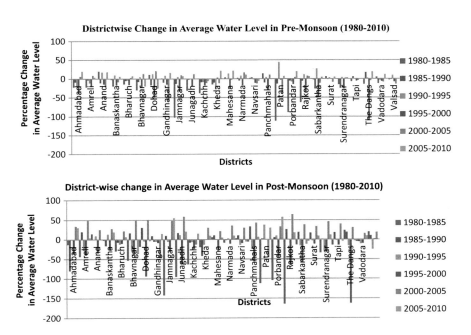

Fig. 8.4 Average groundwater fluctuation during 1981–2010

During the consecutive drought years in 1985, 1986 and 1987, water table fell below normal, and 18 out of 26 districts were not able to recover from the depletion in groundwater in 1990. In periods 1990–1995 and 1995–2000, frequency and intensity of drought were low, so the fall in groundwater was also not that alarming. It was found that since the year 2000, temperature has risen, with severe and frequent heatwaves leading to evaporation loss and overextraction of groundwater in drought years (Bandyopadhyay et al. 2015). From 2000 to 2005, four droughts in 2000, 2001, 2002 and 2004 led to sharp depletion of groundwater affecting more than 88% of the total area of the state. In the year 2009, groundwater level fell alarmingly affecting 25 out of 26 districts covering 96% area of Gujarat during intense drought.

From Fig. 8.4, it can be stated that in all the districts, the fall in water level did not happen at every time interval. It was the period of 1980–1985 during which Gandhinagar, Panchmahals, Patan, Porbandar were affected in both pre- and post-monsoon periods. It can be also observed that during the period of 1995–2000, the districts of Dohad, Junagadh, Rajkot, Sabarkantha and the Dangs were affected in comparison to the other periods. It can be thus observed that the districts are observing groundwater fluctuation from pre-monsoon to post-monsoon season during the drought years and deficit in rainfall is directly affecting the average groundwater level.

To establish the relation between rainfall and change in groundwater level, correlation analysis was done to estimate the impact. The correlation analysis (Fig. 8.5) shows a positive correlation between rainfall extreme and decline in groundwater indicating that a decrease in rainfall results in a fall in groundwater level. Maximum strength of correlation is for Kutch $R = 0.96$ explaining 94% of variation followed by Bhavnagar and Junagadh with $R = 0.95$ explaining 91% of variation, Porbandar $R = 0.93$ explaining 88% of variation and Surat $= 0.91$ explaining 83% of variation. Districts with high R values are more dependent on rainfall for groundwater recharge, i.e. more sensitive to rainfall departure, and hence are more vulnerable to drought and water scarcity. Low correlation between rainfall and recharge in Ahmedabad is most likely due to land cover. Ahmedabad being an urban area, the scope for infiltration and recharge of groundwater is less.

It can also be found that on a gap of every 5 years from 1980 to 2010, the percentage of change in groundwater due to change in rainfall has increased after 2005 and the rise in change is quite high (more than 50%) compared to the previous year's 1990, 1995 and 2000 (Fig. 8.6).

Impact of Rainfall on Groundwater Level and Groundwater Quality

The drought-prone and desert areas of Gujarat experience gross loss of surface and ground water due to deficit precipitation, as evapotranspiration exceeds precipitation in these regions. Perpetual and frequent occurrence of drought in these areas of the

8 Impact of Climate Change on Water Crisis in Gujarat (India)

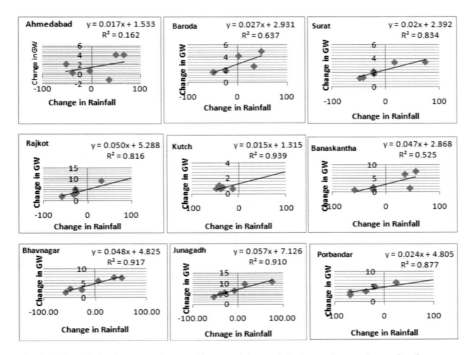

Fig. 8.5 Correlation between changes (departure) in precipitation and groundwater level

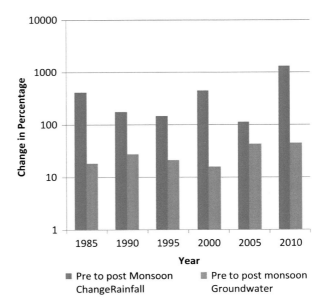

Fig. 8.6 Change in percentage of groundwater level due to change in rainfall from 1980 to 2010

state not only affects the quantity but also the quality of groundwater. The impact of drought on groundwater further adds to the severity of water crisis and leads to more serious situation where both the groundwater accessibility is hampered and pollution gets escalated.

The Bureau of Indian Standards (BIS) has set specifications in IS 10500 and subsequently the revised edition of IS 10500: 2012 in Uniform Drinking Water Quality Monitoring protocol which states that the desirable quality of drinking water is that which has total dissolved solid (TDS) content of 500 ppm or less (ppm stands for parts of the salt present in a million parts of water). Where water of this quality is not easily available, the compromise level is water having up to 2000 ppm. The presence of TDS beyond 500 mg/L in drinking water decreases palatability and may cause gastrointestinal irritation.

TDS is the term used to describe the inorganic salts and small amounts of organic matter present in solution in water. These are usually calcium, magnesium, sodium and potassium cations and carbonate, hydrogen carbonate, chloride, sulphate and nitrate anions.

Groundwater Quality Problem

The salinity intrusion/seepage is a problem that is invariably linked to excessive withdrawal of groundwater especially in area along the coastline and in the Rann area of Kutch. According to the Central Ground Water Board (CGWB) of Gujarat, 2016, the districts affected due to contaminants in groundwater are as follows (Table 8.1).

Drinking Water Scenario

According to NSSO, 59% of the rural population of Gujarat has to pay to access water. More than 29% of the total number of household receives untreated (though piped) drinking water, reveals the data collected during Census 2011.

A simple calculation will indicate that the total number of households in the state (including those that do not have access to piped drinking water) is 1,21,81,718. If every household has, on an average, five members, it means more than 6 lakhs people are getting untreated water for drinking and cooking through their taps. This comes to nearly 30% of the total population of Gujarat.

The data further reveals that the state government has undoubtedly been successful in increasing the number of households provided with piped drinking water. The number of such households has gone up from 46.5% in 2001 census to 64% in 2011 census.

More than 35 lakh households in the state have to make do with untreated (though piped) drinking water. This forms 29.2% of the total number of households in Gujarat. The number of households receiving purified tap water as their main source

Table 8.1 Groundwater scenario in Gujarat

Contaminants	Districts affected (in part)
Salinity (EC > 3000 μS/cm at 25 °C)	Ahmedabad, Amreli, Anand, Bharuch, Bhavnagar, Banaskantha, Dohad, Porbandar, Jamnagar, Junagadh, Kachchh, Mehsana, Navsari, Patan, Panchmahals, Rajkot, Sabarkantha, Surendranagar, Surat, Vadodara
Fluoride (>1.5 mg/l)	Ahmedabad, Amreli, Anand, Banaskantha, Bharuch, Bhavnagar, Dohad, Junagadh, Kachchh, Mehsana, Narmada, Panchmahals, Patan, Rajkot, Sabarkantha, Surat, Surendranagar, Vadodara
Chloride (>1000 mg/l)	Ahmedabad, Amreli, Bharuch, Bhavnagar, Banaskantha, Porbandar, Jamnagar, Junagadh, Kachchh, Dohad, Patan, Panchmahals, Sabarkantha, Surendranagar, Surat, Vadodara, Rajkot
Iron (>1.0 mg/l)	Ahmedabad, Banaskantha, Bhavnagar, Kachchh, Mehsana Narmada
Nitrate (>45 mg/l)	Ahmedabad, Amreli, Anand, Banaskantha, Bharuch, Bhavnagar, Dohad, Jamnagar, Junagadh, Kachchh, Kheda, Mehsana, Narmada, Navsari, Panchmahals, Patan, Porbandar, Rajkot, Sabarkantha, Surat, Surendranagar, Vadodara

Source: Central Ground Water Board, Gujarat, 2016. (http://cgwb.gov.in/gw_profiles/st_Gujarat.htm)

of water for drinking and cooking comes to only 39.8% of the total number of households in the state.

Further, there exists a rural-urban divide in availability of untreated and treated drinking water which is significantly wide.

Out of more than 54.16 lakh urban households in the state, 68.8% get treated tap water and 16.8% get untreated drinking water. The situation in rural Gujarat is the very opposite of this. There are more than 67.65 lakh households in the rural areas of the state, of which only 16.7% get treated water, while 39.7% households have to manage with untreated drinking water. In a study carried out by NGO, Pravah, which works in the area of availability of drinking water, it had been noticed that government schemes are introduced mainly in urban areas because of the high density of population. Further, city people are more vocal about their needs, as is apparent from incidents of women causing ruckus in municipality or ward offices whenever there is a breakdown in supply of drinking water. Such awareness is missing in rural areas. People in rural areas are mostly unaware of the fact that that they have a right to clean drinking water.

In the past two decades of 1981–1991 and 1991–2001, the state has lost about 27% of its groundwater resources, the loss being 50% in North Gujarat. Unsafe withdrawal of ground water in these areas has resulted in about 87% area of the state becoming "non-white" in ground water. The per capita availability of water supply has declined from 1322 cubic meters in 1991 to 1137 cubic meters in 1999–2000 against the norm of 1700 M3 at satisfactory level. This availability is 427 M3 in North Gujarat, 734 cubic meters in Saurashtra and 875 cubic meters in Kachchh (IRMA 2001), which indicates serious "water stress" situation. Innumerable incidents of public rallies and demonstrations (sometimes resulting in violence) were reported in local newspapers to be taken out to protest against water shortage in the different towns and cities as well as in villages in the state.

Deterioration of Water Quality

The decline in groundwater can be found to be accompanied by fluctuating rainfall. On an average, the value of R is 0.43 in case of relationship between rainfall and groundwater level, whereas it is 0.71 in case of rainfall and TDS (Fig. 8.7). In Baroda and Rajkot, there exists a negative correlation in case of rainfall and TDS. Rainfall has direct impact on raising the level of water in the aquifer, and this is most pronounced in the districts of Kutch, Bhavnagar, Junagadh, Surat, Porbandar and Baroda. They are most vulnerable to departure in rainfall, and any drought situation will have more impact on the availability of groundwater than on the water quality (Table 8.2).

It could be seen that there exists a negative correlation between groundwater fluctuation and chlorine content. As the water level increases from pre-monsoon to post-monsoon, the TDS concentration in groundwater becomes less. The year 1995 was the year of extreme heatwave (Bandyopadhyay et al. 2015). According to the

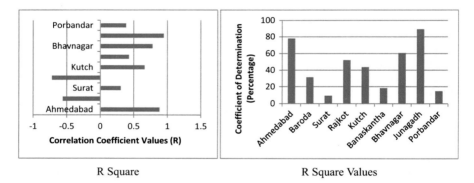

R Square   R Square Values

Fig. 8.7 Correlation coefficient showing how much variation in water quality is explained by variation in rainfall

Table 8.2 Correlation between groundwater fluctuation and TDS in groundwater for Gujarat

Year	R value
1980	−0.08
1985	−0.19
1990	−0.28
1995	0.24
2000	−0.28
2005	−0.41
2010	−0.09

8 Impact of Climate Change on Water Crisis in Gujarat (India) 213

table, the percentage of groundwater fluctuation was low and extreme heat has resulted in increase of TDS level.

Correlation Between Rainfall and TDS Level

The relationship between rainfall and TDS level was studied for nine stations. The relationship is positive for all the seven districts except Rajkot and Baroda. In Rajkot and Baroda, the gross cultivated area is more than 80%, with crops being sown more than once. In both in these areas, more rainfall results in more agricultural run-off, and more rainfall leads to more leaching of topsoil resulting in infiltration and deposition of salts in the soil which migrates to the groundwater increasing the TDS value. Irrigation water supplies also contain a substantial amount of salt, and irrigation can contribute a substantial amount of salts to the field in a season. This is the reason that in extensive agricultural areas of two stations of Rajkot and Baroda, instead of rainfall, irrigation water plays an important role in increasing the TDS value. In areas like Ahmedabad, Junagadh and Bhavnagar, a negative departure in rainfall also brings about negative increase in TDS level (Fig. 8.7). This may be due to the fact that due to negative departure of rainfall, more and more demand for water is met through extraction of groundwater which increases the level of TDS affecting the water quality. The districts of Ahmedabad, Junagadh, Bhavnagar and Kutch are extremely dependent on rainfall for water quality as a negative departure in rainfall affects the quality of water in an adverse way putting the people of these districts under threat of water-borne diseases and water contamination in the future. Groundwater recharge program for groundwater is the need of the hour in these districts which should be implemented on an emergency basis to protect the people.

Correlation Between (Departure in GW) Groundwater Depletion and Deterioration of Water Quality (Departure in TDS)

The correlation between groundwater depletion and deterioration of water quality is mainly positive except in three districts. The places where it shows very high positive correlation are Kutch, Junagadh and Bhavnagar. In these three areas, the more the level of groundwater falls, the more the water quality falls. In three places of Baroda, Rajkot and Banaskantha, the correlation is negative, but the R square value is very low and not significant which does not explain the impact of one variable on the other (Fig. 8.8). These three areas are urbanised areas with Baroda having 16 urban towns, Rajkot having ten towns and Banaskantha having six towns. It can be assumed after CAG published a report in 2012 that in these areas, due to industrial pollution, the water quality is not related to extraction of water for agriculture but to more effluents getting mixed in water which raises the TDS level due to the presence of industries which discharge its effluents without proper treatment. CAG also said in the report that there was an increase in the incidence of water-borne diseases due to heavily polluted water sources and about 32% of

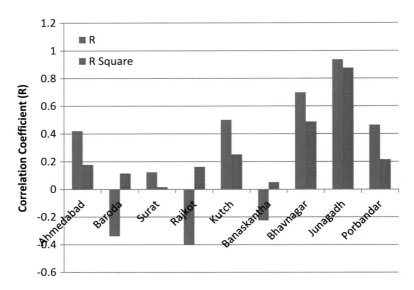

Fig. 8.8 Impact of (departure in GW) groundwater depletion on water quality (departure in TDS)

Gujarat's drinking water sources were found to be contaminated in a pre-monsoon survey about which the villagers were not alerted.

The positive values in the graph shows the impact of groundwater extraction due to agriculture as these are dominantly rural and agricultural areas where the level of groundwater has a direct impact on its quality.

Impact of Increase in Population on Groundwater

In a report published by the Government of Gujarat in 2005, it was revealed that the population of Gujarat has increased from 26.7 million in 1971 to over 50 million in 2001 which is at a rate of more than 20% per decade. At the same time, it has been observed that the demand for water in irrigated area has increased which primarily uses groundwater for irrigation. The dependency on groundwater has doubled in the last 30 years, and the pressure of water for domestic use has increased with rising population in the state as groundwater alone is responsible for 80% of the domestic water requirement of the state (Government of Gujarat 2005).

It could be seen from the table (Table 8.3) below that there exists a positive correlation on an average in the state of Gujarat between increase in population and change in groundwater level showing the direct adverse impact of a rising population. With increase in population, the departure in groundwater has increased putting more and more pressure on the resource where supply became less than demand for more consumption of water with time leading to more and more depletion. Except

Table 8.3 Correlation between population change and change in groundwater level

District	R	R square
Ahmedabad	0.84	0.70
Baroda	0.92	0.85
Surat	0.84	0.70
Rajkot	0.61	0.36
Kutch	−0.98	0.95
Banaskantha	0.78	0.61
Bhavnagar	0.50	0.25
Junagadh	0.95	0.90
Porbandar	0.57	0.33
Average	0.56	0.31

Table 8.4 Impact of population change on change in groundwater level (mm)

Decade	Change in groundwater level (mm)	Percentage decadal variation in population
1981–1991	0.62	21.19
1991–2001	1.38	22.66
2001–2011	−2.00	19.17

Kutch, in every district, population played a major role in diminishing the groundwater. In Kutch, the correlation possibly doesn't exist because it is a desert area with very scarce population, and such a small amount of population cannot influence the level of groundwater which is generally very low due to its geographic setting, and demand of water for agriculture there is nil as no water-intensive agriculture can be practised in the dry barren land of Kutch.

In the decade of 2001 to 2011, it was observed that the influence of increase in population on groundwater was more severe than the last two decades (1981–1991 and 1991–2001) as due to an increase in population of 19.17%, the groundwater level fell less than 2 mm (Table 8.4).

It could be observed that the increase in population in 2001–2010 compared to the last decade of 1991–2001 was lesser (less than 3%) but the decrease in groundwater level was more severe which may be attributed to the intensification of pumping groundwater by drilling deeper wells, installing more powerful pumps and using more electricity in Gujarat according to a study by CGIAR in 2014.

A decrease in population percentage could not replenish the water deficit caused in the earlier decades. Continuation of drought-like situation leads to rapidly falling water tables, and intensification of pumping groundwater has worsened the situation.

Discussion

This study proposes that impact of meteorological drought on groundwater in the backdrop of climate change is resulting in an unsustainable water crisis. This study indicates the fact that, spatially and temporally, the groundwater has declined in

Gujarat and it started declining more after the year 2000. The extraction of groundwater in addition to low rainfall and recurrent droughts has led to a situation of no return where the situation could not be countered by rainfall or aquifer recharge. The objective of the paper was to assess the groundwater trends in pre- and post-monsoon scenario to observe the response to climate change-induced rainfall patterns before and after the monsoon. The time series of groundwater levels and correlation analysis provided the possible linkages of declining water table with the extremes of key climatic variable like rainfall. The scope of the study is limited in examining recent trends in water table fluctuations as this paper aims at a time-bound analysis of groundwater data. Due to data gaps in accessing real-time groundwater data, the latest scenario could not be portrayed, but from secondary sources the findings have been validated that the depletion of groundwater table is posing a threat to the water security of Gujarat. The future studies on this can explore how the agricultural sector is also under pressure with the growing demand for water coupled with population explosion. Integrating GIS data with ground-level data can be a way forward in drought preparedness and managing water scarcity.

Conclusion

In this paper, analysis of groundwater quality and storage change has helped to identify zones and periods severely affected due to drought. Groundwater recharge and quality in Gujarat deteriorated in the previous decade (2001–2010) in a much faster pace compared to the earlier decades causing acute water scarcity in various parts of the state owing to climate change dynamics in the state. The study has revealed that water tables in Northern Gujarat are falling steadily and have declined by 70% BGL over the last 30 years. A significant increase in TDS during droughts has made groundwater unsuitable for drinking. Rising pressure of population on an already dwindling water table will eventually multiply the impact from extreme weather events like heatwaves and meteorological drought which will affect the vegetation and ultimately lead to hydrological drought. The conclusion drawn from this paper indicates that climate change has accelerated the deficit of water and has led to a deterioration in quality. Increasing intensity of drought events accompanied by heatwave events has already led to an increased demand for water. The study advocates for an incentive-based approach towards water conservation, zero water wastage, water treatment and reuse to combat drought and water scarcity in Gujarat taking into account the change in climate pattern.

The solution to curb this problem lies in effective policy management incorporating sustainable development goals, climate change adaptations and regulation of formal and informal water markets.

References

Bandyopadhyay N, Bhuiyan C, Saha AK (2015) Temperature extremes, moisture deficiency and their impacts on dryland agriculture in Gujarat, India. In: Andreu J, Solera A, Paredes-Arquiola-J, Haro-Monteagudo D, van Lanen H (eds) Drought: research and science-policy interfacing. CRC Press, Balkema, pp 119–124. ISBN 978-1-138-02779-4

Calow R, Macdonald D (eds) (2005) Community Management of Groundwater Resources in Rural India: Research Report. British Geological Survey Commissioned Report CR/05/36N

Calow RC, Robins NS, Macdonald AM, Macdonald DMJ, Gibbs BR, Orpen WRG, Mtembezeka P, Andrews AJ, Appiah SO (2007) Groundwater management in drought –prone- areas of Africa. Int J Water Resour Dev 13(2):241–262

Central Ground Water Board Report (2016). http://cgwb.gov.in/gw_profiles/st_Gujarat.htm

CGIAR (2014). https://wle.cgiar.org/thrive/2014/06/10/when-wells-fail-farmers%E2%80%99-response-groundwater-depletion-india

Dubash NK (ed) (2002) Power politics: equity and environment in electricity reform. World Resources Institute, Washington, DC

GoG (Government of Gujarat) (2005) Socio-economic review. Directorate of Economics and Statistics, Gujarat

Groundwater Resource Development Corporation (2018) Gujarat: https://timesofindia.indiatimes.com/city/ahmedabad/state-has-guzzled-68-of-groundwater/articleshow/63404908.cms. Accessed on 1/2/2019

Gupte P, Ngar A, Jain PK (2009) Importance of Surface Water Bodies for Groundwater Recharge, Ahmedabad Urban Areas, Gujarat State, India. Conference paper: International Conference on Water Harvesting, Storage and Conservation. (WHSC), 2009

Hirway I (2005) Ensuring drinking water to all: a study in Gujarat. 4th IWMI-TATA annual partners research meet, 1–46pp

International Water Management Institute (IWMI) (2018) IWMI annual report 2017. International Water Management Institute (IWMI), Colombo, 36p. https://doi.org/10.5337/2018.209

Planning Commision (2007) Groundwater management and ownership: Report of the expert group. Government of India, New Delhi

Rao HCH, Ray SK, Subbarao K (1988) Unstable agriculture and droughts: implications for policy. Vikas Publishing House Pvt. Ltd, New Delhi

Shah T, Indu R (2004) Fluorosis in Gujarat: a disaster ahead. International Water Management Institute-Tata Programme Annual Partner's Meet, Anand

Chapter 9
Factors Affecting Governance Aspect of Disaster Management: Comparative Study of the Sundarbans in India and Bangladesh

Srijita Chakrabarty

Abstract Climate change is closely associated with disaster risk reduction. More extreme weather events are predicted to increase the scale and number of disasters in the days to come. Frequency of extreme events like tidal surges and super cyclones in the Bay of Bengal are predicted to go up as per climatological simulations. Given that natural disasters transcend national boundaries, transboundary issues of disaster management become critical and need to be addressed through enhanced regional cooperation.

This research's main objective was to study the factors affecting governance aspect of disaster management. Though disaster management is a multi-faceted interdisciplinary subject by itself, this research, based on mixed method literature review, tries to focus on the governance aspect of the same. Even though transboundary ecosystems are not limited only to the case of the Sundarbans, this particular region is considered here as an overarching topic to understand the governance aspect of disaster management for both the countries in concern – India and Bangladesh – in the Bay of Bengal region, particularly in the context of climate change.

A gap in the literature exists, such that it does not detail out the result of enhanced binational cooperation in better management of this region. Based on the outcomes of this research, a more detailed study can be conducted on field to analyse the role of various stakeholders in effective management of transboundary ecosystems for the case of the Sundarbans and also utilise the findings of the same elsewhere.

Keywords Disaster risk governance · Disaster management policy · Transboundary cooperation · Climate change adaptation · Sundarbans

---

S. Chakrabarty (✉)
Urban Management Professional, Independent Researcher-Consultant, Kolkata, West Bengal, India

Introduction

The National Institute of Disaster Management notes that in the Indian context, 58.6% of the country's total geographical area is earthquake-prone, while 75% of the coastal line is prone to cyclone and tsunami, 68% of the cultivable area is prone to droughts and 12% of the land area is vulnerable to river erosion and floods. This study also noted that as per the CRED International Disaster Database, India was struck by 1291 disaster events between 1980 and 2018, killing over one hundred thousand people and affecting nearly two billion people, with total economic losses for the said period being estimated at around 100 billion USD (Harshan 2021). Over the last three decades, India's policy towards disaster management has shifted towards holistic management from mere rehabilitation and relief efforts. With increasing intensity and frequency of storm surges, landslides, flash floods, droughts and cyclones, climate change is expected to see far-fetched implications for disaster risk management in India (Dhar Chakrabarti 2011). Disaster-induced direct economic losses have shot up by more than 150% over the past two decades, and vulnerable developing countries are bearing such losses disproportionately (United Nations Office for Disaster Risk Reduction 2021b). Both adaptation to climate change and disaster risk management seek to modify human and environmental contexts and to reduce factors contributing to climate-related risk, which thus support and promote sustainability in economic and social development (Lavell et al. 2012).

Rationale of the Study

This section discusses the rationale of the study undertaken.

Problem Statement

Climate change is closely associated with disaster risk reduction. More extreme weather events are predicted to increase the scale and number of disasters in the days to come (United Nations International Strategy for Disaster Reduction 2008). Although some argue that the impacts of climate change are two-dimensional, that is, positive as well as negative (Bates et al. 2008), in the long term, climate change only has negative impacts, as it destabilises the climatic system. According to different estimates of the Intergovernmental Panel on Climate Change, by 2080 around 20% of the world's population (Filho et al. 2020) will be prone to increased flooding, while people residing in severely stressed river basins could increase to

4.3–6.9 billion by 2050 as compared to 1.4–1.6 billion people in 1995 (Pachauri and Reisinger 2007). Extreme weather events such as heavy flooding and severe droughts will create various challenges for resource managers and for environmentalists to understand the emerging conditions that have an effect on ecosystem processes (Bates et al. 2008; Murdoch et al. 2007).

With increase in frequency of extreme events (Sen 2019), such as tidal surges and super cyclones in the Bay of Bengal, as predicted by climatological simulations, homes are expected to be flooded, fields are expected to turn salty and villages are expected to be washed away. Scientists and policymakers are working towards devising climate adaptation strategies, and testing them in the Sundarbans, which is simultaneously preparing the Bengal delta for a climate-secure future (Bhattacharyya and Mehtta 2020). Some parts of the Sundarbans are witnessing advancement of the sea at about 200 yards annually. Kolkata and Dhaka that are not very far away are also expected to be adversely exposed to storm surges and cyclones (Schwartzstein 2019). Each year the combined river system of the Ganga, Brahmaputra and Meghna rivers, feeding the Sundarbans, carries about 40 billion cubic feet of silt to the Bay of Bengal, which is the highest silt load in the world at about 25% of the world total (Bhattacharyya and Mehtta 2020). Coastal zones like the Sundarbans are anticipated to become more and more prone to natural disasters with intensifying results of climate change (International Union for Conservation of Nature 2014).

Enhanced regional cooperation is very important in addressing transboundary issues of disaster management as natural disasters do not adhere to national boundaries (DharChakrabarti 2011). Furthermore, all countries are expected to be adversely affected as climate change will significantly impact the water resources and allied ecosystems. Impacts of climate change are predicted to have snowball effects on human health and also parts of the economy and society. Climate change adaptation, with special attention to shared water networks and ecosystems, needs to be a central element in any country's adaptation strategy as both water and climate changes do not follow international borders, thus making transboundary cooperation in climate change adaption not only necessary so as to prevent possible conflicts in the future due to unilateral adaptation measures but also to enable mutually beneficial and effective adaptation (Kirby and Edgar 2009).

Research Objectives

This research's main objective was to study the factors affecting governance aspect of disaster management. Though disaster management is a multi-faceted interdisciplinary subject by itself, this research will try to focus on the governance aspect of the same.

Significance of the Research

This research is significant from both professional and academic perspectives. The professional side, including the public sector, can benefit from the review of the existing scenario while also gaining from the learnings of the academia. The academia, on the other hand, can take the insights for furthering their work.

There has not been much research in the field of disasters and hazards, particularly focusing on the governance aspect, though a considerable amount of relevant research is available on disaster management, risk-reduction policies and programs, legislation and disaster management (Tierney 2012), thus making this study all the more significant. Moreover, the author could not come across any concrete comparative study on governance aspect of disaster management for shared ecosystems for India and Bangladesh, making this research all the more challenging and significant.

Materials and Methods

This section discusses the research design.

Research Strategy

To analyse the different factors affecting governance aspect of disaster management, more breadth than depth was deemed desirable in order to get a broader insight. Desk research was selected as the research strategy (van Thiel 2014) in this respect.

Site Selection

Along the Bay of Bengal, the Sundarbans spans roughly across 4000 square miles of India and Bangladesh. Only 52 are inhabited out of the 102 islands that make up the Sundarbans. A UNESCO World Heritage Site and the world's largest continuous mangrove forest (Rainforest Action Network 2022), along with its river creeks, Sundarbans hosts many wildlife, along with endemic cyclones and storms. For nearly 7.5 million people inhabiting this region, this forest is a natural barrier against cyclones and tides. In this region, the loss of land due to global warming surpasses the global average (Bhattacharyya and Mehtta 2020; Schwartzstein 2019). Figure 9.1 shows the contextual site location.

9 Factors Affecting Governance Aspect of Disaster Management:...

Fig. 9.1 Contextual site location. (Source: Schwartzstein 2019)

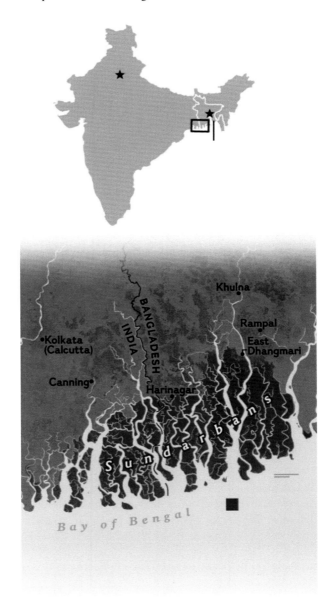

Data Collection Methods

The research methodology was based on secondary data – qualitative and quantitative – collected over the Internet. The research instrument is comprised of reports, etc. A database was created on Microsoft Excel. A mixed method literature review was performed on the final database after errors for duplication and appropriateness of classification were checked (vanThiel 2014; Heyvaert et al. 2011).

Data Analysis Methods

Content analysis was done on the collected data. Computer-based programs like Microsoft Word and ATLAS.ti were used for coding the quotes and reports. Primarily, Microsoft Excel and Microsoft Word were used in sorting out the data, and then ATLAS.ti was used for more complex analyses.

Validity and Reliability

A research is said to be valid if it measures what it is supposed to measure and is considered reliable if the same result is arrived at consistently (van Thiel 2014; Harris 2001). Content analysis allows for transparent analysis of data collected in a reproducible manner (Harris 2001), aiding in the validity and reliability of the research conducted. Furthermore, desk research being assimilation of already conducted research has in-built provision for further checks and analyses of the data used. Furthermore, caution was taken to cross-check all sources gathered for this study for making this study as valid and reliable as possible (Olabode et al. 2019).

Limitations

Disaster management is a multi-faceted interdisciplinary subject by itself, but primarily due to time limitation, this research tried to focus on the governance aspect of the same. Even though transboundary ecosystems are not limited only to the case of the Sundarbans, this particular region is considered here as an overarching topic to understand the governance aspect of disaster management for the both the countries in concern – India and Bangladesh – in the Bay of Bengal region, particularly in the face of climate change.

Due to constraints set by the coronavirus pandemic, field visit could not be undertaken. Hence, this study had to rely only on desk research. Thus, though the study will try to capture more breadth on the topic, it might miss out on some depth which would have otherwise been taken into detailed consideration via interviews or surveys and observations in the field.

Results and Discussion

This section discusses the significant findings with respect to the research objectives.

Site Description

The largest delta and the largest mangrove forest in the world (Baitalik and Majumdar 2015), Sundarbans covers around 10200 sq. km, out of which approximately 4200 sq. km of reserved forest falls in India (Dutta and Mukhopadhyay 2016) and the remaining 6000 sq. km of reserved forest falls in Bangladesh. The Indian part of the Sundarbans region has an additional 5400 sq. km of non-forested inhabited region, thus making the total area of the Sundarbans region in India – the Sundarbans Biosphere Reserve – 9600 sq. km (Baitalik and Majumdar 2015). In 1989, the Government of India constituted the Sundarbans Biosphere Reserve, and UNESCO, in November 2001, recognised the region under its Man and Biosphere Programme. In 1987, UNESCO also recognised the Sundarbans National Park, which forms the core area of the Sundarbans Tiger Reserve, as a World Heritage Site. The Government of India has nominated the region as Ramsar site. In 1973, under "Project Tiger" scheme, the Government of India constituted the Sundarbans Tiger Reserve. A project of biodiversity conversation through two-country approach was approved by the Ministry of Environment, Forest and Climate Change and Department of Economic Affairs, Government of India, with funding aid from the United Nations Foundation routed through the United Nations Development Programme (UNDP) – UNDP India and UNDP Bangladesh. Under the said project, the National Project Coordinators of both India and Bangladesh had prepared a joint project document, and the same was submitted to the respective UNDP centres for further funding support towards implementation of the project (Sundarban Biosphere Reserve 2010; Government of West Bengal 2020). Figures 9.2, 9.3 and 9.4 give the physical location of the Sundarbans region.

With some major distributaries of river Ganga that feed the Sundarbans being moribund, the region is experiencing severe freshwater shortage, particularly during the dry season. Anticipated changes in the Himalayan riverine flows along with increase in sea level due to climate change are expected to intensify salinity intrusion in the region. This is slated to alter the hydrological regime and the forest ecology, which would in turn significantly affect the forest-based livelihoods of a large number of poor inhabitants of the region and its adjacent areas (Dasgupta et al. 2016a, b).

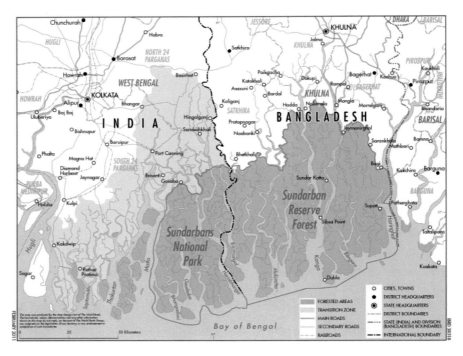

Fig. 9.2 Sundarbans region in India and Bangladesh. (Source: Ortolano et al. 2016)

Disaster Management

A serious disruption to community functioning causing losses to humans, material, economy and environment beyond a community's ability to cope is known as a disaster. Disaster management, defined as management and organisation of responsibilities and resources, so as to deal with all humanitarian aspects during emergencies, aims to reduce disaster impacts, with particular focus on preparedness, response and recovery, minimising loss of life and property (International Federation of Red Cross and Red Crescent Societies 2021; United Nations Office for Outer Space Affairs 2021). Disaster management comprises mainly of five phases, in that order – prevention, preparedness, response, mitigation and recovery (Ha et al. 2015). The recovery phase is yet to be understood in detail, while significant knowledge is available on the first four phases (United Nations Development Programme 2016; Pal and Shaw 2018a, b). Historically, environmental issues were seen as single issues. However, this notion evolved over time wherein ecological issues started being considered interlocked with development issues focused on management of resources across networks at local, national, transboundary and global levels (Gupta and Vegelin 2016). Disaster risk management is a multi-sectoral field requiring well-defined contributions from private sector and civil society, along with building

9 Factors Affecting Governance Aspect of Disaster Management:...

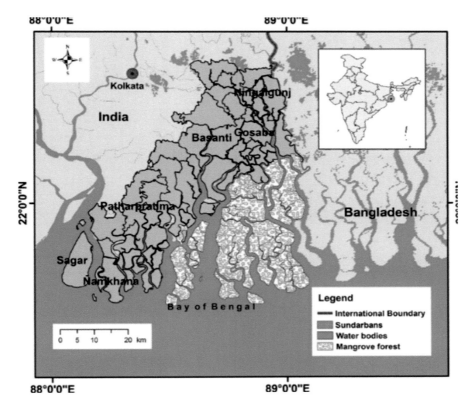

Fig. 9.3 Map showing location of the Sundarbans region in India and its physical surroundings. (Source: Pramanik et al. 2021)

vertical linkages amongst local, regional, national and local levels (United Nations Development Programme 2007). The approach of mostly practising disaster risk reduction as disaster management has been ineffective in attaining the policy goals of disaster risk reduction for multiple reasons, like lack of proper integration into development programming (United Nations Office for Disaster Risk Reduction 2021c).

Governance Aspect of Disaster Management

An emerging concept in the field of disaster, disaster governance, is closely associated with environmental governance and risk governance (Tierney 2012). The United Nations Development Programme (2012) views the mode by which public authorities, private sector, civil servants, civil society and media coordinate at national, regional and community levels so as to manage and reduce disaster- and

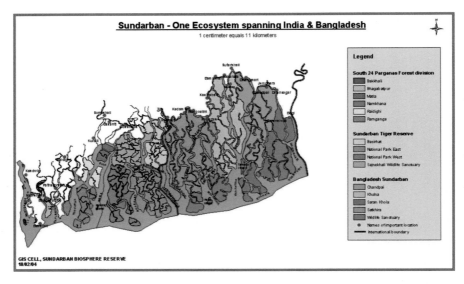

Fig. 9.4 Sundarbans ecosystem across India and Bangladesh. (Source: Sundarban Biosphere Reserve 2010)

climate-related risks (DuryogNivaran Secretariat 2020), as disaster risk governance. Governance is a core aspect of disaster risk reduction at the broader policy level and also in the context of individual warning systems (Ahrens and Rudolph 2006). Disaster governance is built within the overarching existing governance system (Tierney 2012). Governance is about interactions amongst structures, processes, role-players and traditions determining the exercise of power and decision-making. The disaster risk governance framework has a tendency to inform the actors involved about how to deal with possible future risks. Hence, disaster risk governance also ensures availability of sufficient legally binding instruments, resources and capacity levels in a decentralised manner so as to prevent disaster, prepare for, manage and recover from the same in the future (Van Niekerk 2015; United Nations 2022). Often seen as a form of collaborative governance, disaster governance usually brings together multiple organisations in order to solve problems extending beyond boundaries of a single organisation. Cross-border collaboration and complex governance arrangements determine many disaster risk-reduction efforts, for example, as in the case of the Rhine river basin, the management of which involves collaboration amongst the European Union and nine European nations (Tierney 2012). How the society as a whole manages disaster risks across scales and levels, emerging as networks, irrespective of clear-cut entities, can thus be seen as disaster risk governance (Lassa 2010).

Sendai Framework

The Sendai Framework for Disaster Risk Reduction 2015–2030 (Sendai Framework), which provides member states concrete actions for protecting development gains from disaster risk, resulted out of the first major agreement from the post-2015 development agenda. This works alongside the other 2030 Agenda agreements. This framework gives the state the primary role in disaster risk reduction, but other stakeholders like local government and private sector should share that responsibility too (United Nations Office for Disaster Risk Reduction 2021d; Government of India 2022; United Nations Office for Disaster Risk Reduction 2015). Figure 9.5 gives a snapshot of the Sendai Framework targets for 2030.

In the post-2015 Sendai Framework era, multi-stakeholder forums, known as regional platforms, form a key component in tackling transboundary issues. These reflect the commitment of governments towards improved coordination and implementation of disaster risk-reduction activities and help build resilient communities and nations (United Nations Office for Disaster Risk Reduction 2021a). Interconnected social and economic processes are considered in both Sustainable Development Goals and Sendai Framework. Thus, these two policy instruments are synergistic in nature (United Nations Office for Disaster Risk Reduction 2021b).

Factors Affecting Governance Aspect of Disaster Management

Institutional structure of national disaster management system determines a country's disaster governance, and effective early warning depends on the institutional frameworks, policies and laws and also on the officers' capacities (Fakhruddin and Chivakidakarn 2014). Horizontal and vertical coordination amongst the stakeholders, integrated multi-sectoral institutional framework and knowledgeable and committed institutional stakeholders across levels are critical determinants of effectiveness of early warning systems in reducing disaster risk and vulnerabilities. Gaps in coordination can lead to misunderstandings and errors. Clarity of roles and responsibilities of the stakeholders is also important (Tierney 2012; Sakalasuriya et al. 2018; Spahn et al. 2010; Sakalasuriya et al. 2020).

Disaster governance is influenced by forces such as social inequality, world-system dynamics, globalisation, sociodemographic trends and relationship amongst various actors (Tierney 2012). At a global level, the author observes, countries participate in negotiations only when their interests are met, often leading to some countries not participating and taking responsibility of their actions, often at a cost of others, who often are the weakest entities in the society – individuals and countries (Gupta 2014). Though globalisation can thus contribute to socioeconomic inequalities, it can also bolster the depth of participation amongst diverse actors shaping environmental governance. A shift towards alternative forms of government with higher levels of citizen participation in governance is also observed (Lemos and

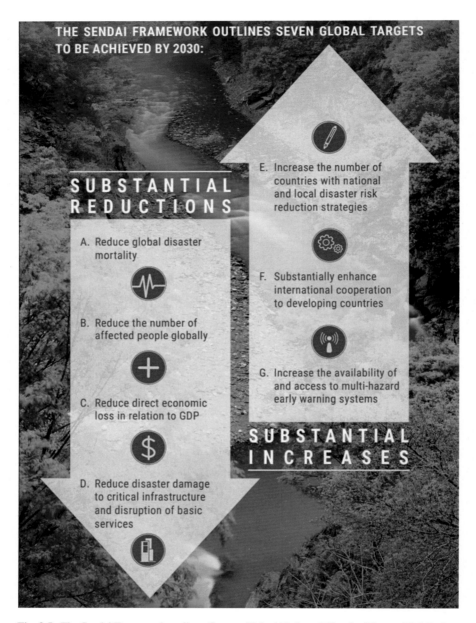

Fig. 9.5 The Sendai Framework outline. (Source: United Nations Office for Disaster Risk Reduction 2021d)

Agrawal 2006). Institutional innovation also is a key indicator of risk reduction across levels (Lassa 2010). However, government performance alone is not sufficient for assessing effectiveness of governance systems, and accountability within and across various organisations and stakeholders is needed along with assessment of all stakeholders from local to global scales (Dahiya 2012; Gupta and Sharma 2006; Gall et al. 2014). This contrasts research findings in developing countries, where decentralisation and devolution of power to local levels is generally considered as desirable (Uddin 2013). Climate change, disaster risks and population growth present a combined challenge for achieving sustainable development (Pal and Shaw 2018a, b). Therefore, disasters are litmus tests for good governance. While good governance is an important prerequisite for disaster risk management, it forms a core good governance practice in disaster-prone countries. At the operational level, with regard to disaster risk reduction, participation, transparency, rule of law, consensus orientation, responsiveness, efficiency, effectiveness, equity, strategic vision and accountability translate to good governance (Tun 2018). Governance is the exercise of political, administrative and economic authority in managing a country's affairs across levels, and interplay of governance attributes happens before, during and after any situation in case of disasters (Pal and Shaw 2018a, b). Dominant influential actors pursuing market-based narratives, like donors, multinationals and nongovernmental organisations, are often seen to treat ecological and social issues as separate, externalising certain impacts, thus missing out on local contexts and aspirations (Gupta and Lebel 2020). A variety of gaps across capacity, funding, policy, administration, objective, accountability, information, etc., may translate to global-level problems found at the national level. This further highlights the influence of global-level actors in national policies, though controversial (Gupta and Pahl-Wostl 2013).

Context of the Sundarbans Across India and Bangladesh

It might be noted here that this region in context – the Sundarbans across international borders – is culturally, ecologically and socioeconomically unified and was one until 1947, from historical and political perspectives too. However, planning and management of the region currently being undertaken by respective agencies in both countries have made limited progress in coordinating their efforts, despite various unifying factors. Signed in 2011, two binational agreements have created opportunities for cooperation. These have established aspirations, but formal action has hardly come as rapidly as planned (Ortolano et al. 2016). This memorandum of understanding signed in 2011 fails to consider the issues regarding effective rise of sea level, land loss, coastal flooding and intensifying cyclonic storms. It also does not consider that a common understanding of capacity building and climate change impacts is needed for effective adaptation. It is unlikely to work out without a cooperation mechanism within the countries and between the two, mainly in terms of external assistance pertaining to finance and technology (Danda 2019).

To help people in this delta adapt better against climate change, governments, nongovernmental organisations and foreign bodies have been working on identifying innovative approaches (Rahman et al. 2020). Some of the guiding documents prepared by the Government of Bangladesh are the National Adaptation Programme of Action 2005, Bangladesh Climate Change Strategy and Action Plan 2009 and Bangladesh Climate Change and Gender Action Plan 2013 (Haq et al. 2018; Government of the People's Republic of Bangladesh 2005, 2009, 2013), while the Indian part had initiated the National Action Plan on Climate Change in 2008 at the national level, and the state of West Bengal had prepared the State Action Plan on Climate Change in 2012, with regular updates of both thereon (Dey et al. 2016; Government of India 2021; Government of West Bengal 2017). Plans and sectoral policies in both countries have shown a tendency to be sector-specific and short term, instead of focussing on cross-cutting issues (Haq et al. 2018; Dey et al. 2016).

India and Bangladesh both mainly have disaster-focused adaptation policies, which lack in sectoral coherence and with less focus on gender sensitivity and migration as components of climate change adaptation. Policies and plans started incorporating gender awareness post 2009. The Bangladesh Climate Change and Gender Action Plan (2013) is the only dedicated document for gender-specific needs in this context (Haq et al. 2018, Government of the People's Republic of Bangladesh 2013; Juran and Trivedi 2015; Rashid and Michaud 2002). The Indian side has fared better in this regard, though gender sensitivity and inclusivity need mainstreaming in the context of disaster management and climate change (Rahman et al. 2020). The Bangladesh Delta Plan 2100 developed by the Government of Bangladesh aims at integrating global, national and sectoral plans and targets coherent long-term strategies across ministries towards adaptive delta management while considering issues of climate change, population growth and economic development (Government of the People's Republic of Bangladesh 2018).

Treaties and international cooperation like the Sendai Framework and Sustainable Development Goals have influenced policy choices for deltas (Lwasa 2015). Transformations focussing on substantial long-term change may question the effectiveness of existing systems (Lonsdale et al. 2015), but such transformational policies are usually undertaken when incremental changes do not work (Kates et al. 2012). This region, despite having a long history of adaption to environmental changes, any specific detail of this adaption is hardly known. However, the global stocktaking mandated by the Paris Agreement should enable focus on documenting such adaptation practices (Tompkins et al. 2018). Imperfect implementation does not help attain the desired social consequences despite having adaptation policies in place (Mimura et al. 2014). However, emergence of delta management plans is a positive sign, though this involves judicial understanding of trade-offs and the direction to take (Tompkins et al. 2020).

Context of India

Poor coordination at the local level, lack of capacity (Smith et al. 2010), poor community empowerment, lack of early warning systems and lack of search and rescue facilities, amongst other factors, have led to poor response to disasters previously. The National Disaster Response Plan marks the first formal step in

India towards policy development regarding disaster care, which was initially formulated for natural disasters only but included man-made disasters too by a later amendment. The National Disaster Management Authority led by the Prime Minister of India aims at providing national guidelines highlighting roles and responsibilities of various departments involved in disaster management. Respective disaster management authorities and district disaster management committees need to be set up by each state. The Environment Protection Act is the only central legislation regarding disaster management, which is not covered directly by either of the central, state or concurrent lists of the Indian Constitution. Primary responsibility concerning rescue, relief and rehabilitation efforts rests on the state government, with the centre supporting by means of financial and physical resources (Kaur 2006). The Department of Sundarbans Affairs, Government of Bengal, is the ministry of primary concern in the Indian part of the Sundarbans, which implements its work programme through the Sundarbans Development Board (Government of West Bengal 2020).

Split of responsibilities between the central and state governments is one of the major criticisms in the Indian part of the Sundarbans. The central government has designated the entire Sundarbans (the forested and the inhabited parts) as a biosphere reserve, and demarcated areas for tiger conservation and coastal zone protection as protected forest, while the state government governs the inhabited areas. Overlapping responsibilities amongst central and state governments and different departments concerned have often adversely affected the work (Mahadevia and Vikas 2012). Structural integrity of many embankments in this region has adversely been affected by anthropogenic activities along with upstream water diversions, cyclonic storms and creekside erosion. Growing population has led to human-animal conflict resulting out of habitat loss of tigers. Pollution from upstream Kolkata and increased salinity, amongst other factors, have led to out-migration. Coordination and cooperation amongst different organisations, like the Department of Sundarbans Affairs, Department of Fisheries and Aquaculture, Forest Department, Irrigation and Waterways Department and Department of Panchayat and Rural Development, is of utmost importance (Sen 2019; Sánchez-Triana et al. 2018).

Context of Bangladesh

Many people in Bangladesh cannot interpret warnings despite improvement in early warning systems (Mallick et al. 2017). Local people find it difficult to interpret hazard map as people are hardly aware of reading maps. Moreover, local government uses the hazard map, not the local people. Additionally, people mostly migrate for the months of inundation period (Fujita et al. 2017). Many people decide to evacuate as the weather condition turns worst (Parvin et al. 2019). However, apart from emergencies, there is hardly any support for migrants, and the linkage with local government is weak (Martin et al. 2013; Fujita and Shaw 2019).

Bangladesh has made significant progress regarding cyclone, with the concept of total disaster management replacing the concept of acting only post-disaster occurrence after the devastating 1991 cyclone. Progress came in terms of technological advancements, developing evacuation plans, constructing coastal embankments and cyclone shelters, modernising early warning systems, maintenance and improvement of coastal forest cover and community awareness through collaboration across levels. Fewer than expected deaths occurred due to the 2007 Cyclone Sidr when the Bangladesh Meteorological Department issued instant early warning signals directly to the National Coordination Committee, which then relayed it to the at-risk community (Habiba and Shaw 2018).

Bangladesh has worked out the policy document – Climate Change Strategy and Action Plan 2009 (Government of the People's Republic of Bangladesh 2009) – proposing six main action areas (Islam and Nursey-Bray 2017), and a climate change trust fund has been set up in 2009–2010 with the government's own fund to implement the said action plan (Ayers et al. 2013, Government of the People's Republic of Bangladesh 2022). In Bangladesh, the Ministry of Local Government, Ministry of Water Resources and Ministry of Environment (Islam and Amstel 2021) got maximum fund allocation, though local government institutions performed better in adaption work with lesser funding compared to the central agencies (Vij et al. 2018). Pending rehabilitation of many dykes and polders since being damaged in cyclones of 2007 and 2009 has stretched community suffering majorly owing to lack of freshwater supply and continued flooding (Sadik et al. 2018). The Bangladesh Climate Change Resilience Fund Management Committee noted salinity intrusion in coastal Bangladesh as a key element to climate change adaptation (Dasgupta et al. 2016a, b, 2017).

Conclusions and Recommendations

The conclusion summarises the research. Some recommendations have been noted towards the end, which may pave the path for further research in the domain.

Conclusion

A robust methodology is needed for vulnerability assessment for climate adaption in the Sundarbans – a region facing development challenges due to lack of infrastructure, health, education and modern energy services (Jana et al. 2013). Improved accountability mechanisms held in place by legislation, social audit and free press is needed. Local planning and financing along with civil society partnerships are essential in scaling up community initiatives. The role of political leaders is also crucial in effective integration of disaster risk reduction across levels (United Nations Office for Disaster Risk Reduction 2021c).

A gap in the literature exists, such that it does not detail out on the result from enhanced binational cooperation in better management of this region. Though both the Indian state of West Bengal and Bangladesh have disaster management programmes, the effectiveness of Bangladesh is considered superior. Early warning system in Bangladesh incorporates indigenous knowledge in its community-based programme. In the Indian side, disaster management framework is more centralised and vertically integrated than its Bangladeshi counterpart. Door-to-door communication and volunteer structure is also more developed in Bangladesh. Forest management problems with regard to wildlife poaching, overfishing and increased water and soil salinity are faced by both countries. Corruption and lack of trust on the Bangladesh Forest Department has been observed. West Bengal, on the other hand, has been the first Indian state to adopt a joint government-community forest management strategy wherein people dependent on forest resources are inducted into forest management activities, and they get to share forest benefits in return (Ortolano et al. 2016). Though most of the activities in Bangladesh came as reactive response to the 2009 Cyclone Aila, anticipatory approaches are yet to be explored for effective and adaptive delta management (Government of the People's Republic of Bangladesh 2018; Parven et al. 2021; Seijger et al. 2017).

Governance actors have uneven capacities. Empowering the ones at the lowest rank is critical for sustainably addressing long-term challenges for spatial justice and better risk reduction (Fuentealba et al. 2020). Cross-border learning and collaborative research on joint issues are required to work out replicable solutions (Lumbroso et al. 2017). Innovation in tidal river management is also crucial with rising sea level (Rahman et al. 2020). Sustainable development and economic growth is to incorporate reconceptualised disaster risk reduction (Gall et al. 2014) along with focus on science and technology for informed decision-making and enhanced stakeholder partnership (Shaw et al. 2018). Equity and active citizen participation is also important for success of governance on disaster risk management (United Nations Development Programme 2007).

Finally, both India and Bangladesh would benefit by viewing the Sundarbans as a unified system by means of greater bilateral collaboration and management (Ortolano et al. 2016).

Recommendations

New and emerging technologies can play a pivotal role in effective disaster management having a combination of pre-, during and post-disaster activities (Shaw 2019). Protection of mangroves from anthropogenic threats is very important for encouraging resilience to global climate change (McLeod and Salm 2006). Promoting science-policy interface by means of enhanced legal resources and government engagement is required to fulfil goals of the Sendai Framework for Disaster Risk Reduction 2015–2030, particularly for nations to establish state targets and local indicators (Lian et al. 2018). An integrated long-term approach is required for effectiveness of environmental policies (Murdoch et al. 2007). Responsibilities need to be delegated at the local level along with strong coordination across sectors

(United Nations Office for Disaster Risk Reduction 2021c; Parvin et al. 2018). To address disaster risk reduction at a regional level, intergovernmental collaboration is key, as risks and vulnerabilities transcend national boundaries. Regional offices of United Nations Office for Disaster Risk Reduction aid in the process (United Nations Office for Disaster Risk Reduction 2021a). For decision-making in complex situations, iterative risk management can be undertaken as an effective tool (Intergovernmental Panel on Climate Change 2014).

Unless local ecological conditions are improved through hydrological regime changes, the Sundarbans, despite being treaty-protected, will not be saved from the growing concerns of increased salinity (Potkin 2004). Maintenance of upland freshwater catchments and connectivity between nearby river sources and mangrove systems, along with identification of areas likely to survive seal-level rise, need to be part of the management protocols (Dasgupta et al. 2016a, b). Developing nonstructural measures is important for communities residing in hazardous areas (Sanyal and Routray 2016). Good governance also needs to have inclusive and transparent decision-making (Tun 2018). Even though India has initiated adoption of holistic multidisciplinary methods in disaster management, focus has majorly been on logistical and technical aspects. Subtler aspects of mental health need to be incorporated into disaster care (Kaur 2006). Strengthening supply chains is urgently required at the village level to improve the resilience of economic activities. Empowering local communities is also critical for making the Sundarbans more resilient (Pramanik et al. 2021). Many innovations are also coming up that link disaster risk reduction to Sustainable Development Goals, which include numerous games to involve children in disaster risk reduction and to create awareness (Izumi et al. 2020).

To sum it up, stakeholders share a broad consensus regarding protecting the ecosystem and people's livelihoods in the Sundarbans. A balance needs to be sought between total flexibility and absolute rigidity, with institutionalisation of ecological partnership between India and Bangladesh along with active stakeholder engagement, in order to overcome barriers to transboundary settings, for adaptive disaster management in the context of climate change (Danda 2019).

Further Research

Based on the outcomes of this research, a more detailed study can be conducted on field to analyse the role of various stakeholders in effective management of transboundary ecosystems for the case of the Sundarbans and also utilise the findings of the same elsewhere. Future research could possibly also be directed towards evaluation of net benefits from bilateral cooperation in the Sundarbans across India and Bangladesh that can possibly be studied in the future (Ortolano et al. 2016).

Acknowledgement The author is greatly thankful to Dr. Senthamizh Kanal for his support and to Mr. Ujjwal Sardar for his inspirational work in the Sundarbans.

References

Ahrens J, Rudolph PM (2006) The importance of governance in risk reduction and disaster management. J Contingencies Crisis Manage 14(4):207–220 Available at: https://doi.org/10.1111/j.1468-5973.2006.00497.x. Available at: https://onlinelibrary.wiley.com/doi/abs/10.1111/j.1468-5973.2006.00497.x. Accessed 1 July 2021

Ayers JM, Huq S, Faisal AM, Hussain ST (2013) Mainstreaming climate change adaptation into development: a case study of Bangladesh. WIREs Clim Change 5(1):37–51 Available at: https://doi.org/10.1002/wcc.226. Available at: https://onlinelibrary.wiley.com/doi/10.1002/wcc.226. Accessed 1 July 2021

Baitalik A, Majumdar S (2015) Coastal tourism destinations in West Bengal: historical background and development. Int J Social Sci Manage 2(3):267–272. Available at: https://doi.org/10.3126/ijssm.v2i3.12910. Available at: https://nepjol.info/index.php/IJSSM/article/view/12910. Accessed 31 Jan 2022

Bates B, Kundzewicz ZW, Wu S, Palutikof J (2008) Climate change and water. Intergovernmental Panel on Climate Change, Geneva. Available at: https://www.ipcc.ch/publication/climate-change-and-water-2/. Accessed 1 July 2021

Bhattacharyya D, Mehtta M (2020) More than rising water: living tenuously in the sundarbans. In: Diplomat. Available at: https://thediplomat.com/2020/08/more-than-rising-water-living-tenuously-in-the-sundarbans/; manage.thediplomat.com; https://thediplomat.com/. Accessed 1 July 2021

Dahiya B (2012) Cities in Asia, 2012: demographics, economics, poverty, environment and governance. Cities 29:S44–S61. Available at: https://www.academia.edu/5459983/Cities_in_Asia_2012_Demographics_economics_poverty_environment_and_governance. Accessed 1 July 2021

Danda AA (2019) Environmental security in the sundarban in the current climate change era: strengthening India-Bangladesh cooperation, vol 220. Observer Research Foundation, New Delhi. Available at: https://www.orfonline.org/research/environmental-security-in-the-sundarban-in-the-current-climate-change-era-strengthening-india-bangladesh-cooperation-57191/; https://www.orfonline.org/. Accessed 1 July 2021

Dasgupta S, Huq M, Golam M, Istiak S et al (2016a) Impact of climate change and aquatic salinization on fish habitats and poor communities in southwest coastal Bangladesh and Bangladesh sundarbans, vol 7593. World Bank, Washington, DC. Available at: https://openknowledge.worldbank.org/handle/10986/24135. Accessed 1 July 2021

Dasgupta S, Sobhan I, Wheeler D (2016b) Impact of climate change and aquatic salinization on mangrove species and poor communities in the Bangladesh sundarbans, vol 7736. World Bank, Washington, DC. Available at: https://openknowledge.worldbank.org/handle/10986/24653; https://documents1.worldbank.org/curated/en/452761467210045879/pdf/WPS7736.pdf; documents.worldbank.org. Accessed 1 July 2021

Dasgupta S, Huq M, Mustafa MG, Sobhan MI et al (2017) The impact of aquatic salinization on fish habitats and poor communities in a changing climate: evidence from southwest coastal Bangladesh. Ecol Econ 139:128–139. Available at: https://doi.org/10.1016/j.ecolecon.2017.04.009. Available at: https://www.sciencedirect.com/science/article/pii/S0921800916312137?via%3Dihub. Accessed 10 Jan 2022

Dey S, Ghosh AK, Hazra S (2016) Review of West Bengal state adaptation policies, Indian Bengal delta, vol 107642. Collaborative Adaptation Research Initiative in Africa and Asia, Southampton. Available at: https://www.researchgate.net/publication/330134934_Review_of_West_Bengal_State_Adaptation_Policies_Indian_Bengal_Delta. Accessed 1 July 2021

Dhar Chakrabarti PG (2011) Challenges of disaster management in India: implications for the economic, political, and security environments. In: DharChakrabarti PG, Joseph M (eds) Nontraditional security challenges in India: human security and disaster management. The National Bureau of Asian Research, Washington, DC. Available at: https://www.nbr.org/publication/challenges-of-disaster-management-in-india-implications-for-the-economic-political-and-security-environments/; https://www.nbr.org/publication/nontraditional-security-challenges-in-india-human-security-and-disaster-management/. Accessed 1 July 2021

DuryogNivaran Secretariat (2020) DuryogNivaran. Available at: http://duryognivaran.org/. Accessed 2022

Dutta MK, Mukhopadhyay SK (2016) Reviews and syntheses: methane biogeochemistry in Sundarbans mangrove ecosystem, NE coast of India; a box modeling approach. Biogeosciences. Available at: https://doi.org/10.5194/bg-2016-58. Available at: https://bg.copernicus.org/preprints/bg-2016-58/bg-2016-58.pdf. Accessed 31 Jan 2022

Fakhruddin SHM, Chivakidakarn Y (2014) A case study for early warning and disaster management in Thailand. Int J Disaster Risk Reduc 9:159–180. Available at: https://doi.org/10.1016/j.ijdrr.2014.04.008 Available at: https://www.sciencedirect.com/science/article/abs/pii/S2212420914000375. Accessed 1 July 2021

Filho WL, Wall T, Azul AM, Brandli L et al (2020) Good health and well-being. Springer, Cham. Available at: https://doi.org/10.1007/978-3-319-95681-7. Accessed 31 Jan 2022

Fuentealba R, Verrest H, Gupta J (2020) Planning for exclusion: the politics of urban disaster governance. Polit Govern 8(4):244–255. Available at: https://doi.org/10.17645/pag.v8i4.3085. Available at: https://www.researchgate.net/publication/344612481_Planning_for_Exclusion_The_Politics_of_Urban_Disaster_Governance. Accessed 1 July 2021

Fujita K, Shaw R (2019) Preparing international joint project: use of Japanese flood hazard map in Bangladesh. Int J Disaster Risk Manag 1(1):62–80. Available at: https://doi.org/10.18485/ijdrm.2019.1.1.4. Available at: https://www.researchgate.net/publication/332261211_Preparing_International_Joint_Project_Use_of_Japanese_Flood_Hazard_Map_in_Bangladesh. Accessed 1 July 2021

Fujita K, Shaw R, Nakagawa H (2017) Social background of riverbank erosion areas in Bramaptra and Ganges rivers: implication for Japanese hazard mapping technology in Bangladesh. DPRI Annuals 60:701–710. Available at: https://www.dpri.kyoto-u.ac.jp/nenpo/no60/ronbunB/a60b0p39.pdf. Accessed 1 July 2021

Gall M, Cutter SL, Nguyen KH (2014) Governance in disaster risk management, vol 3. Integrated Research on Disaster Risk, Beijing. Available at: https://www.researchgate.net/publication/273449927_Governance_in_Disaster_Risk_Management#fullTextFileContent. Accessed 1 July 2021

Government of India (2021) Frequently Asked Questions (FAQs): National Action Plan on Climate Change (NAPCC) – Ministry of Environment, Forest and Climate Change. RU-49-03-0004-011221/FAQ). Press Information Bureau, Ministry of Information and Broadcasting, Government of India, New Delhi. Available at: https://static.pib.gov.in/WriteReadData/specificdocs/documents/2021/dec/doc202112101.pdf. Accessed 10 Jan 2022

Government of India (2022) Sendai framework. Available at: https://ndma.gov.in/Global/sendai-framework. Accessed 2022

Government of the People's Republic of Bangladesh (2005) National Adaptation Programme of Action (NAPA). Ministry of Environment and Forest, Government of the People's Republic of Bangladesh, Dhaka. Available at: https://unfccc.int/resource/docs/napa/ban01.pdf. Accessed 10 Jan 2022

Government of the People's Republic of Bangladesh (2009) Bangladesh Climate Change Strategy and Action Plan 2009. Ministry of Environment and Forest, Government of the People's Republic of Bangladesh, Dhaka. Available at: https://www.iucn.org/content/bangladesh-climate-change-strategy-and-action-plan-2009. Accessed 1 July 2021

Government of the People's Republic of Bangladesh (2013) Bangladesh Climate Change and Gender Action Plan (ccGAP: Bangladesh). Ministry of Environment and Forest, Government of the People's Republic of Bangladesh, Dhaka. Available at: http://nda.erd.gov.bd/files/1/Publications/CC%20Policy%20Documents/CCGAP%202009.pdf. Accessed 10 Jan 2022

Government of the People's Republic of Bangladesh (2018) Bangladesh delta plan 2100 (Abridged Version). General Economics Division, Bangladesh Planning Commission, Government of the People's Republic of Bangladesh, Dhaka. Available at: http://plancomm.portal.gov.bd/sites/default/files/files/plancomm.portal.gov.bd/files/dc5b06a1_3a45_4ec7_951e_a9feac1ef783/BDP%202100%20Abridged%20Version%20English.pdf. Accessed 1 July 2021

Government of the People's Republic of Bangladesh (2022) Bangladesh Climate Change Trust. Available at: http://www.bcct.gov.bd/. Accessed 2022

Government of West Bengal (2017) State action plan on climate change 2017–2020. Saraswaty Press Limited, Kolkata. Available at: http://www.environmentwb.gov.in/pdf/WBSAPCC_2017_20.pdf; http://www.environmentwb.gov.in/. Accessed 10 Jan 2022

Government of West Bengal (2020) Know sundarban. Available at: https://sundarbanaffairswb.in/home/page/sundarban_biosphere. Accessed 2021

Gupta J (2014) Normative issues in global environmental governance: connecting climate change, water and forests. J Agric Environ Ethics 28:413–433. Available at: https://doi.org/10.1007/s10806-014-9509-8 Available at: https://link.springer.com/article/10.1007/s10806-014-9509-8. Accessed 1 July 2021

Gupta J, Lebel L (2020) Access and allocation in earth system governance: lesson learnt in the context of the Sustainable Development Goals. Int Environ Agreements Polit Law Econ 20: 393–410. Available at: https://doi.org/10.1007/s10784-020-09486-4. Available at: https://link.springer.com/article/10.1007/s10784-020-09486-4. Accessed 1 July 2021

Gupta J, Pahl-Wostl C (2013) Editorial on global water governance. Ecol Soc 18(4). Available at: https://doi.org/10.5751/ES-06115-180454. Available at: http://www.ecologyandsociety.org/vol18/iss4/art54/; https://www.ecologyandsociety.org/. Accessed 1 July 2021

Gupta M, Sharma A (2006) Coumpounded loss: the post tsunami recovery experience of Indian island communities. Disaster Prev Manag 15(1):67–78. Available at: https://doi.org/10.1108/09653560610654248. Available at: https://www.researchgate.net/publication/241338274_Compounded_loss_The_post_tsunami_recovery_experience_of_Indian_island_communities. Accessed 1 July 2021

Gupta J, Vegelin C (2016) Sustainable development goals and inclusive development. Int Environ Agreements Polit Law Econ 16(3):433–448. Available at: https://doi.org/10.1007/s10784-016-9323-z. Available at: https://link.springer.com/article/10.1007%2Fs10784-016-9323-z. Accessed 1 July 2021

Ha H, Fernando RLS, Mahmood A (eds) (2015) Strategic disaster risk management in Asia. Springer, New Delhi. Available at: https://link.springer.com/book/10.1007/978-81-322-2373-3. Accessed 31 Jan 2022

Habiba U, Shaw R (2018) 24 – Improvement of responses and recovery approaches for cyclone hazards in coastal Bangladesh. In: Shaw R, Shiwaku K, Izumi T (eds) Science and technology in disaster risk reduction in Asia. Academic, New York, pp 409–430. Available at: https://doi.org/10.1016/B978-0-12-812711-7.00024-9; https://www.sciencedirect.com/science/article/pii/B9780128127117000249?via%3Dihub; https://keio.pure.elsevier.com/en/publications/improvement-of-responses-and-recovery-approaches-for-cyclone-haza. Accessed 1 July 2021

Haq MI, Omar MAT, Zahra QA, Jahan I (2018) Evaluation of adaptation policies in GBM delta of Bangladesh, vol 107642. Collaborative Adaptation Research Initiative in Africa and Asia, Southampton. Available at: http://generic.wordpress.soton.ac.uk/deccma/wp-content/uploads/sites/181/2017/02/Evaluation-of-adaptation-policies-in-GBM-1.pdf. Accessed 1 July 2021

Harris H (2001) Content analysis of secondary data: a study of courage in managerial decision making. J Bus Ethics 34:191–208. Available at: https://doi.org/10.1023/A:1012534014727. Available at: https://link.springer.com/article/10.1023%2FA%3A1012534014727. Accessed 1 July 2021

Harshan TP (2021) Chapter 28 – Agricultural production and income in a disaster year: findings from the study of Melalinjippattu village affected by cyclone Thane. In: Chaiechi T (ed) Economic effects of natural disasters: theoretical foundations, methods, and tools. Academic, New York, pp 477–492. Available at: https://www.sciencedirect.com/science/article/pii/B9780128174654000285; https://doi.org/10.1016/B978-0-12-817465-4.00028-5. Accessed 1 July 2021

Heyvaert M, Maes B, Onghena P (2011) Mixed methods research synthesis: definition, framework, and potential. Qual Quant 47:659–676. Available at: https://doi.org/10.1007/s11135-011-9538-6. Available at: https://link.springer.com/article/10.1007/s11135-011-9538-6. Accessed 1 July 2020

Intergovernmental Panel on Climate Change (2014) Summary for policymakers. In: Field CB, Barros VR, Dokken DJ, Mach KJ, Mastrandrea MD, Bilir TE, Chatterjee M, Ebi KL, Estrada YO, Genova RC, Girma B, Kissel ES, Levy AN, MacCracken S, Mastrandrea PR, White IL (eds) Climate change 2014: impacts, adaptation, and vulnerability. Part A: global and sectoral aspects. Contribution of Working Group II to the Fifth Assessment Report of the Intergovernmental Panel on Climate Change. Cambridge University Press, New York, pp 1–32. Available at: https://www.researchgate.net/publication/272150376_Climate_change_2014_impacts_adaptation_and_vulnerability_-_IPCC_WGII_AR5_summary_for_policymakers. Accessed 1 July 2021

International Federation of Red Cross and Red Crescent Societies (2021) About disaster management. Available at: https://www.ifrc.org/en/what-we-do/disaster-management/about-disaster-management/. Accessed 2021

International Union for Conservation of Nature (2014) Natural hazard regulation: Sundarbans National Park (India) and The Sundarbans (Bangladesh). Available at: https://worldheritageoutlook.iucn.org/benefits/benefits-case-studies/2014/natural-hazard-regulation-sundarbans-national-park-india-and. Accessed 2021

Islam MN, Amstel AV (eds) (2021) Bangladesh II: climate change impacts, mitigation and adaptation in developing countries. Springer, Cham. Available at: https://link.springer.com/book/10.1007%2F978-3-030-71950-0. Accessed 31 Jan 2022

Islam MT, Nursey-Bray M (2017) Adaptation to climate change in agriculture in Bangladesh: the role of formal institutions. J Environ Manag 200:347–358. Available at: https://doi.org/10.1016/j.jenvman.2017.05.092. Available at: https://www.sciencedirect.com/science/article/pii/S0301479717305777?via%3Dihub. Accessed 1 July 2021

Izumi T, Shaw R, Ishiwatari M, Djalante R et al. (2020) 30 innovations linking Disaster Risk Reduction with Sustainable Development Goals. Tokyo. Available at: https://www.researchgate.net/publication/339712814_30_innovations_linking_Disaster_Risk_Reduction_with_Sustainable_Development_Goals. Accessed 1 July 2021

Jana R, Mohapatra S, Gupta AK (2013) Integrating climate change adaptation and disaster resilience: issues for sundarbans. Disaster Dev 7. Available at: https://nidm.gov.in/PDF/Journal/2_Dec_2013.pdf. Accessed 1 July 2021

Juran L, Trivedi J (2015) Women, gender norms, and natural disasters in Bangladesh. Geogr Rev 105(4):601–611. Available at: https://doi.org/10.1111/j.1931-0846.2015.12089.x. Available at: https://www.tandfonline.com/doi/full/10.1111/j.1931-0846.2015.12089.x. Accessed 1 July 2021

Kates RW, Travis WR, Wilbanks TJ (2012) Transformational adaptation when incremental adaptations to climate change are insufficient. Proc Natl Acad Sci U S A 109(19):7156–7161. Available at: https://doi.org/10.1073/pnas.1115521109. Available at: https://pubmed.ncbi.nlm.nih.gov/22509036/. Accessed 1 July 2021

Kaur J (2006) Administrative issues involved in disaster management in India. Int Rev Psychiatry 18(6):553–557. Available at: https://doi.org/10.1080/09540260601038449. Available at: https://www.tandfonline.com/doi/full/10.1080/09540260601038449. Accessed 1 July 2021

Kirby A, Edgar C (2009) Guidance on water and adaptation to climate change. United Nations Economic Commission for Europe, Geneva. Available at: https://unece.org/environment-policy/publications/guidance-water-and-adaptation-climate-change. Accessed 1 July 2021

Lassa JA (2010) Institutional vulnerability and governance of disaster risk reduction: macro, meso and micro scale assessment (With case studies from Indonesia). PhD. Rheinische-Friedrich-Wilhelm Universität Bonn, Bonn. Available at: https://d-nb.info/1016181752/34. Accessed 1 July 2021

Lavell A, Oppenheimer M, Diop C, Hess J et al (2012) Climate change: new dimensions in disaster risk, exposure, vulnerability, and resilience. In: Moser S, Takeuchi K (eds) Managing the risks of extreme events and disasters to advance climate change adaptation. Cambridge University Press, New York, pp 25–64. Available at: https://www.ipcc.ch/site/assets/uploads/2018/03/SREX-Chap1_FINAL-1.pdf; https://www.ipcc.ch/. Accessed 1 July 2021

Lemos MC, Agrawal A (2006) Environmental governance. Annu Rev Environ Resour 31:297–325. Available at: https://doi.org/10.1146/annurev.energy.31.042605.135621. Available at: https://www.annualreviews.org/doi/abs/10.1146/annurev.energy.31.042605.135621; https://www.annualreviews.org/. Accessed 1 July 2021

Lian F, Lu L, Shaw R (2018) 2 – priority actions for science and technology to implement the Sendai Framework for Disaster Risk Reduction. In: Shaw R, Shiwaku K, Izumi T (eds) Science and technology in disaster risk reduction in Asia. Academic, New York, pp 17–27. Available at: https://www.sciencedirect.com/science/article/pii/B978012812711700002X?via%3Dihub. Accessed 1 July 2021

Lonsdale K, Pringle P, Turner B (2015) Transformative adaptation: what it is, why it matters & what is needed. University of Oxford, Oxford. Available at: https://ukcip.ouce.ox.ac.uk/wp-content/PDFs/UKCIP-transformational-adaptation-final.pdf. Accessed 1 July 2021

Lumbroso DM, Suckall NR, Nicholls RJ, White KD (2017) Enhancing resilience to coastal flooding from severe storms in the USA: international lessons. Nat Hazards Earth Syst Sci 17(8): 1357–1373. Available at: https://doi.org/10.5194/nhess-17-1357-2017. Available at: https://nhess.copernicus.org/articles/17/1357/2017/. Accessed 1 July 2021

Lwasa S (2015) A systematic review of research on climate change adaptation policy and practice in Africa and South Asia deltas. Reg Environ Chang 15:815–824. Available at: https://doi.org/10.1007/s10113-014-0715-8. Available at: https://link.springer.com/article/10.1007/s10113-014-0715-8. Accessed 1 July 2021

Mahadevia K, Vikas M (2012) Climate change – impact on the sundarbans: a case study. Int Sci J Environ Sci. Available at: https://www.researchgate.net/publication/311607858_Climate_Change_-_Impact_on_the_Sundarbans_A_case_study. Accessed 1 July 2021

Mallick B, Ahmed B, Vogt J (2017) Living with the risks of cyclone disasters in the south-western coastal region of Bangladesh. Environment 4(1). Available at: https://doi.org/10.3390/environments4010013. Available at: https://www.mdpi.com/2076-3298/4/1/13/htm. Accessed 1 July 2021

Martin M, Kang YH, Billah M, Siddiqui T et al (2013) Policy analysis: climate change and migration Bangladesh, vol 4. Migrating out of Poverty, Sussex. Available at: http://www.migratingoutofpoverty.org/files/file.php?name=wp4-ccrm-b-policy.pdf&site=354. Accessed 1 July 2021

McLeod E, Salm RV (2006) Managing mangroves for resilience to climate change. The World Conservation Union (IUCN), Gland. Available at: https://portals.iucn.org/library/sites/library/files/documents/2006-041.pdf. Accessed 1 July 2021

Mimura N, Pulwarty RS, Duc DM, Elshinnawy I et al (2014) Adaptation planning and implementation. In: Field CB, Barros VR, Dokken DJ, Mach KJ, Mastrandrea MD, Bilir TE, Chatterjee M, Ebi KL, Estrada YO, Genova RC, Girma B, Kissel ES, Levy AN, MacCracken S, Mastrandrea PR, White LL (eds) Climate change 2014: impacts, adaptation, and vulnerability. Part A: Global and Sectoral Aspects. Contribution of Working Group II to the Fifth Assessment Report of the Intergovernmental Panel on Climate Change. Cambridge University Press, New York, pp 869–898. Available at: https://www.ipcc.ch/report/ar5/wg2/. Accessed 1 July 2021

Murdoch PS, Baron JS, Miller TL (2007) Potential effects of climate change on surface-water quality in North America. J Am Water Resour Assoc 36(2):347–366. Available at: https://doi.org/10.1111/j.1752-1688.2000.tb04273.x. Available at: https://onlinelibrary.wiley.com/doi/abs/10.1111/j.1752-1688.2000.tb04273.x. Accessed 1 July 2021

Olabode SO, Olateju OI, Bakare AA (2019) An assessment of the reliability of secondary data in management science research. Int J Business Manage Rev 7(3):27–43. Available at: https://www.eajournals.org/wp-content/uploads/An-Assessment-of-the-Reliability-of-Secondary-Data-in-Management-Science-Research.pdf. Accessed 1 July 2021

Ortolano L, Sánchez-Triana E, Paul T, Ferdausi SA (2016) Managing the Sundarbans region: opportunities for mutual gain by India and Bangladesh. Int J Environ Sustain Dev 15(1):16–31. Available at: https://doi.org/10.1504/IJESD.2016.073331. Available at: https://www.researchgate.net/publication/286856798_Managing_the_Sundarbans_region_Opportunities_for_mutual_gain_by_India_and_Bangladesh; https://www.inderscience.com/info/inarticle.php?artid=73331. Accessed 1 July 2021

Pachauri RK, Reisinger A (2007) Climate change 2007: Synthesis report. Contribution of Working Groups I, II and III to the Fourth Assessment Report of the Intergovernmental Panel on Climate Change. Synthesis Report. Intergovernmental Panel on Climate Change, Geneva. Available at: https://www.ipcc.ch/report/ar4/syr/. Accessed 1 July 2021

Pal I, Shaw R (2018a) Disaster governance and its relevance. In: Pal I, Shaw R (eds) Disaster risk governance in India and cross cutting issues. Disaster risk reduction (Methods, approaches and practices). Springer, Singapore. Available at: https://link.springer.com/chapter/10.1007/978-981-10-3310-0_1. Accessed 1 July 2021

Pal I, Shaw R (eds) (2018b) Disaster risk governance in India and cross cutting issues. Springer, Singapore. Available at: https://link.springer.com/book/10.1007/978-981-10-3310-0. Accessed 31 Jan 2022

Parven A, Pal I, Hasan MS (2021) Chapter 12 – Ecosystem for disaster risk reduction in Bangladesh: a case study after the Cyclone "Aila". In: Pal I, Shaw R, Djalante R, Shrestha S (eds) Disaster resilience and sustainability. Elsevier, New York, pp 277–300. Available at: https://www.sciencedirect.com/science/article/pii/B9780323851954000032. Accessed 1 July 2021

Parvin GA, Rahman R, Fujita K, Shaw R (2018) Overview of flood management actions and policy planning in Bangladesh. Int J Public Policy 14. Available at: https://doi.org/10.1504/IJPP.2018.096700. Available at: https://www.inderscienceonline.com/doi/abs/10.1504/IJPP.2018.096700. Accessed 1 July 2021

Parvin GA, Sakamoto M, Shaw R, Nakagawa H et al (2019) Evacuation scenarios of cyclone Aila in Bangladesh: investigating the factors influencing evacuation decision and destination. Progr Disaster Sci 2. Available at: https://doi.org/10.1016/j.pdisas.2019.100032 Available at: https://www.sciencedirect.com/science/article/pii/S2590061719300328. Accessed 1 July 2021

Potkin A (2004) Watering the Bangladeshi Sundarbans. In: Mirza MMQ (ed) The Ganges water diversion: environmental effects and implications. Springer, Dordrecht, pp 163–176. Available at: https://link.springer.com/chapter/10.1007%2F978-1-4020-2792-5_8. Accessed 1 July 2021

Pramanik M, Szabo S, Pal I, Udmale P et al (2021) Twin disasters: tracking COVID-19 and Cyclone Amphan's impacts on SDGs in the Indian sundarbans. Environ Sci Policy Sustain Dev 63(4):20–30. Available at: https://doi.org/10.1080/00139157.2021.1924575. Available at: https://www.researchgate.net/publication/352976339_Twin_Disasters_Tracking_COVID-19_and_Cyclone_Amphan's_Impacts_on_SDGs_in_the_Indian_Sundarbans. Accessed 1 July 2021

Rahman MM, Ghosh T, Salehin M, Ghosh A et al (2020) Ganges-Brahmaputra-Meghna Delta, Bangladesh and India: a transnational mega-delta. In: Nicholls RJ, Adgar NW, Hutton CW, Hanson SE (eds) Deltas in the anthropocene. Palgrave Macmillan, Cham, pp 23–51. Available at: https://link.springer.com/chapter/10.1007/978-3-030-23517-8_2. Accessed 1 July 2021

Rainforest Action Network (2022) Rainforest action network. Available at: https://www.ran.org/. Accessed 2022

Rashid SF, Michaud S (2002) Female adolescents and their sexuality: notions of honour, shame, purity and pollution during the floods. Disasters 24(1):54–70. Available at: https://doi.org/10.1111/1467-7717.00131. Available at: https://onlinelibrary.wiley.com/doi/10.1111/1467-7717.00131. Accessed 1 July 2021

Sadik MS, Nakagawa H, Rahman R, Shaw R et al (2018) A study on cyclone Aila recovery in Koyra, Bangladesh: evaluating the inclusiveness of recovery with respect to predisaster vulnerability reduction. Int J Disaster Risk Sci 9:28–43. Available at: https://doi.org/10.1007/s13753-018-0166-9. Available at: https://link.springer.com/article/10.1007%2Fs13753-018-0166-9. Accessed 1 July 2021

Sakalasuriya M, Amaratunga D, Haigh R, Hettige S (2018) A study of the upstream-downstream interface in end-to-end Tsunami early warning and mitigation systems. Int J Adv Sci Eng Inf Technol 8(6):2421–2427. Available at: https://doi.org/10.18517/ijaseit.8.6.7487. Available at: https://www.researchgate.net/publication/330343998_A_Study_of_The_Upstream-downstream_Interface_in_End-to-end_Tsunami_Early_Warning_and_Mitigation_Systems. Accessed 1 July 2021

Sakalasuriya M, Haigh R, Hettige S, Amaratunga D et al (2020) Governance, institutions and people within the interface of a Tsunami early warning system. Polit Govern 8(4):432–444. Available at: https://doi.org/10.17645/pag.v8i4.3159. Available at: https://www.cogitatiopress.com/politicsandgovernance/article/view/3159. Accessed 1 July 2021

Sánchez-Triana E, Ortolano L, Paul T (2018) Managing water-related risks in the West Bengal Sundarbans: policy alternatives and institutions. Int J Water Resour Dev 34(1):78–96. Available at: https://doi.org/10.1080/07900627.2016.1202099. Available at: https://www.tandfonline.com/doi/full/10.1080/07900627.2016.1202099. Accessed 1 July 2021

Sanyal S, Routray JK (2016) Social capital for disaster risk reduction and management with empirical evidences from Sundarbans of India. Int J Disaster Risk Reduc 19:101–111. Available at: https://doi.org/10.1016/j.ijdrr.2016.08.010. Available at: https://www.researchgate.net/publication/307610696_Social_capital_for_disaster_risk_reduction_and_management_with_empirical_evidences_from_Sundarbans_of_India. Accessed 1 July 2021

Schwartzstein P (2019) This vanishing forest protects the coasts – and lives – of two countries. National Geographic Magazine. Available at: https://www.nationalgeographic.com/magazine/article/sundarbans-mangrove-forest-in-bangladesh-india-threatened-by-rising-waters-illegal-logging; https://www.nationalgeographic.com/. Accessed 1 July 2021

Seijger C, Douven W, van Halsema G, Hermans L et al (2017) An analytical framework for strategic delta planning: negotiating consent for long-term sustainable delta development. J Environ Plan Manag 60(8):1485–1509. Available at: https://doi.org/10.1080/09640568.2016.1231667. Available at: https://www.tandfonline.com/doi/full/10.1080/09640568.2016.1231667. Accessed 1 July 2021

Sen HK (ed) (2019) The Sundarbans: a disaster-prone eco-region. Springer, Cham. Available at: https://www.springerprofessional.de/en/the-sundarbans-a-disaster-prone-eco-region/16442288. Accessed 31 Jan 2022

Shaw R (2019) Chapter 1 – global coasts in the face of disasters. In: Krishnamurthy RR, Jonathan MP, Srinivasalu S, Glaeser B (eds) Coastal management. Academic, New York, pp 1–4. Available at: https://www.sciencedirect.com/science/article/pii/B9780128104736000017?via%3Dihub. Accessed 1 July 2021

Shaw R, Izumi T, Shiwaku K (2018) 1 – Science and technology in disaster risk reduction in Asia: post-Sendai developments. In: Shaw R, Shiwaku K, Izumi T (eds) Science and technology in disaster risk reduction in Asia. Academic, New York, pp 3–16. Available at: https://www.sciencedirect.com/science/article/pii/B9780128127117000018?via%3Dihub. Accessed 1 July 2021

Smith SM, Gorski J, Vennelakanti HC (2010) Disaster preparedness and response: a challenge for hospitals in earthquake-prone countries. Int J Emerg Manag 7(3–4):209–220. https://doi.org/10.1504/IJEM.2010.037006 . Available at: https://www.indersicenceonline.com/doi/abs/10.1504/IJEM.2010.037006. Accessed 31 Jan 2022

Spahn H, Hoppe M, Vidiarina HD, Usdianto B (2010) Experience from three years of local capacity development for tsunami early warning in Indonesia: challenges, lessons and the way ahead. Natural Hazards Earth Syst Sci 10(7):1411–1429. Available at: https://doi.org/10.5194/nhess-10-1411-2010. Available at: https://www.researchgate.net/publication/45492192_Experience_from_three_years_of_local_capacity_development_for_tsunami_early_warning_in_Indonesia_Challenges_lessons_and_the_way_ahead. Accessed 1 July 2021

Sundarban Biosphere Reserve (2010) Sundarban Biosphere Reserve. Available at: http://sundarbanbiosphere.org/index.htm; http://www.sundarbanbiosphere.org/. Accessed 2021

Tierney K (2012) Disaster governance: social, political, and economic dimensions. Annu Rev Environ Resour 37:341–363. Available at: https://doi.org/10.1146/annurev-environ-020911-095618. Available at: https://www.annualreviews.org/doi/full/10.1146/annurev-environ-020911-095618; https://www.annualreviews.org/. Accessed 1 July 2021

Tompkins EL, Vincent K, Nicholls RJ, Suckall N (2018) Documenting the state of adaptation for the global stocktake of the Paris Agreement. WIREs Clim Change 9(5). Available at: https://doi.org/10.1002/wcc.545. Available at: https://onlinelibrary.wiley.com/doi/10.1002/wcc.545. Accessed 1 July 2021

Tompkins EL, Vincent K, Suckall N, Rahman R et al (2020) Adapting to change: people and policies. In: Nicholls RJ, Adgar NW, Hutton CW, Hanson SE (eds) Deltas in the anthropocene. Palgrave Macmillan, Cham, pp 201–222. Available at: https://link.springer.com/chapter/10.1007/978-3-030-23517-8_9. Accessed 1 July 2021

Tun SS (2018) Challenges of good governance in disaster management in Myanmar: case studies of emergency response to the 2015 flood and landslide in the two townships of the Ayeyarwaddy Region. Masters. Kobe University, Kobe. Available at: https://www.researchgate.net/publication/333566438_Challenges_of_Good_Governance_in_Disaster_Management_in_Myanmar_Case_Studies_of_Emergency_Response_to_the_2015_Flood_and_Landslide_in_the_two_townships_of_the_Ayeyarwaddy_Region_Faculty_Graduate_Schoo. Accessed 1 July 2021

Uddin MN (2013) Towards good urban local governance in Bangladesh: lessons learnt from Japanese local government system to overcome the challenges. Lex Localis J Local Self-Govern 11(4):933–953 Available at: https://doi.org/10.4335/11.4.933-953. Available at: https://www.researchgate.net/publication/260872060_Towards_Good_Urban_Local_Governance_in_Bangladesh_Lessons_learnt_from_Japanese_Local_Government_System_to_Overcome_the_Challenges. Accessed 1 July 2021

United Nations (2022) United Nations Bosnia and Herzegovina. Available at: https://bosniaherzegovina.un.org/. Accessed 2022.

United Nations Development Programme (2007) Governance for disaster risk management: 'how to' guide. United Nations Development Programme, New York. Available at: https://www.eird.org/wiki/images/Governance_for_DRM_How_to_Guide_June_07_GPDR_Conference_Draft.pdf. Accessed 1 July 2021

United Nations Development Programme (2012) Issue brief: disaster risk governance – crisis prevention and recovery. United Nations Development Programme, New York. Available at: http://www.undp.org/content/dam/undp/library/crisis%20prevention/20121112_Issue_brief_disasterriskgovernance.pdf. Accessed 1 July 2021

United Nations Development Programme (2016) A guidance note: national post-disaster recovery planning and coordination. United Nations Development Programme, New York. Available at: https://reliefweb.int/sites/reliefweb.int/files/resources/UNDP_Guidance_Note_Disaster%20Recovery_final.pdf. Accessed 1 July 2021

United Nations International Strategy for Disaster Reduction (2008) Climate Change and disaster risk reduction: weather, climate and climate change. Briefing note 1. United Nations International Strategy for Disaster Reduction, Geneva. Available at: https://eird.org/publicaciones/Climate-Change-DRR.pdf; https://www.preventionweb.net/; https://drupal.undrr.org/. Accessed 1 July 2021

United Nations Office for Disaster Risk Reduction (2015) Sendai framework for disaster risk reduction 2015–2030. 1. United Nations Office for Disaster Risk Reduction, Geneva. Available at: https://www.undrr.org/publication/sendai-framework-disaster-risk-reduction-2015-2030; https://www.preventionweb.net/files/43291_sendaiframeworkfordrren.pdf; https://www.preventionweb.net/sendai-framework/sendai-framework-for-disaster-risk-reduction; https://www.preventionweb.net/; https://drupal.undrr.org/. Accessed 10 Jan 2022

United Nations Office for Disaster Risk Reduction (2021a) Regional platforms. Available at: https://www.undrr.org/implementing-sendai-framework/regional-platforms; https://www.undrr.org/; https://www.preventionweb.net/; https://drupal.undrr.org/. Accessed 2021

United Nations Office for Disaster Risk Reduction (2021b) The Sendai framework and the SDGs. Available at: https://www.undrr.org/implementing-sendai-framework/sf-and-sdgs; https://www.undrr.org/; https://www.preventionweb.net/; https://drupal.undrr.org/. Accessed 2021

United Nations Office for Disaster Risk Reduction (2021c) Weak governance. Available at: https://www.preventionweb.net/disaster-risk/risk-drivers/weak-governance/; https://www.undrr.org/; https://www.preventionweb.net/; https://drupal.undrr.org/. Accessed 2021

United Nations Office for Disaster Risk Reduction (2021d) What is the Sendai framework for disaster risk reduction? Available at: https://www.undrr.org/implementing-sendai-framework/what-sendai-framework; https://www.undrr.org/; https://www.preventionweb.net/; https://drupal.undrr.org/. Accessed 2021

United Nations Office for Outer Space Affairs (2021) Disaster management. Available at: https://www.unoosa.org/oosa/en/ourwork/topics/disaster-management.html; https://www.preventionweb.net/. Accessed 2021

Van Niekerk D (2015) Disaster risk governance in Africa: a retrospective assessment of progress against the Hyogo Framework for Action (2000-2012). Disaster Prev Manage 24(3):387–416. Available at: https://doi.org/10.1108/DPM-08-2014-0168. Available at: https://www.researchgate.net/publication/277142908_Disaster_risk_governance_in_Africa_A_retrospective_assessment_of_progress_against_the_Hyogo_Framework_for_Action_2000-2012. Accessed 1 July 2021

vanThiel S (2014) Research methods in public administration and public management: an introduction. Routledge, New York

Vij S, Biesbroek R, Groot A, Termeer K (2018) Changing climate policy paradigms in Bangladesh and Nepal. Environ Sci Policy 81:77–85 Available at: https://doi.org/10.1016/j.envsci.2017.12.010. Available at: https://www.sciencedirect.com/science/article/pii/S1462901117311425?via%3Dihub. Accessed 1 July 2021

Chapter 10
Application of Geospatial Technology in Seasonal Flood Hazard Event in Dhemaji District of Assam

Krishna Das , A. Simhachalam , and Ashok Kumar Bora

Abstract Flood is a recurring natural event in Assam and Dhemaji district in particular. As per the flood hazard atlas of Assam prepared by the National Remote Sensing Centre of ISRO, Dhemaji is one of the worst flood-affected districts in the state with 46% of flood inundating land area. The repeated flood events in Dhemaji district cause havoc to the people of the district by damaging properties and agricultural lands. To understand the seasonal flood event in Dhemaji district of Assam, this study has been carried out using geospatial technology. This study focuses on the seasonal trend of rainfall, accumulated surface water extent, transitions and seasonality, seasonal flood water extent and affected land use/land cover by recent seasonal flood events in Dhemaji district. The geospatial datasets incorporated in this study are extracted from the sources such as *GPM IMERG*-derived data product using the *Giovanni* platform, the *Global Surface Water* dataset, the *EOS Land Viewer* platform and the *ESA CCI Land Cover* data. According to the results of the study, the study area receives more than 400 mm of mean monthly rainfall during monsoon season in June, July and August. The district has about one-fifth of accumulated surface water extent area, and about 10% of new seasonal water area is added to the district during 1984–2019. This study will help policy- and decision-makers in mitigating the frequent flood events in the study area and also provide a basis for similar studies in the coming days.

Keywords Seasonal flood · Rainfall · Surface water transitions · Affected land use/land cover · Geospatial technology

---

K. Das (✉)
Department of Geography, Pragjyotish College (Affiliated to Gauhati University), Guwahati, Assam, India

A. Simhachalam
NIRD and PR, NERC, Guwahati, Assam, India

A. K. Bora
Department of Geography, Gauhati University, Guwahati, Assam, India

© The Author(s), under exclusive license to Springer Nature Switzerland AG 2022
U. Chatterjee et al. (eds.), *Ecological Footprints of Climate Change*, Springer Climate, https://doi.org/10.1007/978-3-031-15501-7_10

Introduction

Flood is a hydrological event that occurs especially in the riverine floodplains, coastal lands and other low-lying areas of the Earth surface with a maximum accumulation of surface water especially due to heavy rainfall (Dartmouth Flood Observatory 2014). Flood events are frequent and recurring phenomena, and flood events cause severe damage to property, infrastructure and crops with human and animal casualties when flooding becomes a hazard or a disaster (Jonkman and Kelman 2005). Over the period 1998–2017, there were 3148 flood disaster events that occurred globally that affected 2 billion people with 747,234 deaths and led to an economic loss of US$ 656 billion (CRED and UNISDR 2018). India alone recorded an absolute economic loss of US$ 79.5 billion due to flood disasters between the period 1998 and 2017 (CRED and UNISDR 2018). Over the period 1985–2010, there were a total of 257 large flood events that occurred in India especially during the monsoon season (Brakenridge 2016). The state of Assam situated in the north-eastern part of India is also a severely flood-prone area with severe flood risks. The heterogeneous topography, high monsoonal rainfall and unique geographical location crisscrossed by a vast network of perennial rivers originating from the surrounding highlands are largely responsible for the recurring floods in the state (Bhattachaiyya and Bora 1997). As per the Flood Hazard Atlas of Assam (2016), the flood-prone area of the state is 31,500 km^2 (as assessed by the Rashtriya Barh Ayog) which shares about 39.58% of the state's total land area. This represents 9.40% of the total flood-prone area of the whole country. Every year the recurring flooding in Assam damages croplands, settlements, infrastructure, etc., and poses a great loss to the state's economy. Physiographically, the Brahmaputra River valley and the Barak River valley of the state are highly susceptible to flood hazards (Fig. 10.1).

The Dhemaji district of Assam is situated in the north-eastern corner of the state and is physiographically located to the eastern edge of Brahmaputra Valley (right bank to Brahmaputra River). It is also facing the recurring flood hazard especially during the rainy monsoon season. Dhemaji district is one of the worst flood-affected districts of Assam with about 33% of the district's area having low, moderate to very high flood severity as per the flood hazard index calculated by the National Remote Sensing Centre, Hyderabad (Flood Hazard Atlas for Assam state 2016). During 2016, flood hazard affected around 462 villages in the district with around 185,550 persons affected (Flood Report of Dhemaji district, 2016). In this study, an attempt has been made to understand the seasonal flood hazard in Dhemaji district of Assam using geospatial technology.

Geospatial technology mainly refers to the integrated approach of remote sensing, geographical information system (GIS) and global navigational satellite system (GNSS) to make spatial decisions based on geographic data of the Earth's environment. Nowadays, geospatial technology is the most widely used scientific tool for Earth observation and disaster management. The availability of spatial data especially in the form of satellite data at global to local scales has made the application of

10 Application of Geospatial Technology in Seasonal Flood Hazard Event...   249

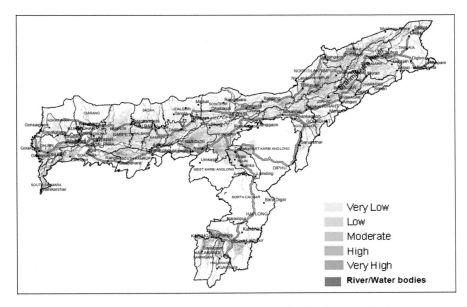

Fig. 10.1 Flood hazard map of Assam. (After Flood Hazard Atlas of Assam 2016)

geospatial technology an efficient, timely and cost-effective approach for researchers to examine the spatiotemporal nature of disasters including floods for sustainable monitoring and management. Various scholars utilized the geospatial technology in flood hazard studies and observed effective research findings (Bhatt et al. 2013; Albano et al. 2015; Aggarwal 2016; Samanta et al. 2018; Das 2020). Besides, Sanyal and Lu (2004) addressed the flood hazard problem in the monsoon regime of Asia utilizing remote sensing data and GIS. Mishra et al. (2019) assessed the extent of heavy flooding in Assam and Bihar by integrating multi-temporal satellite imageries in GIS. Surampudi and Yarrakula (2020) examined the flood event over the Brahmaputra River basin of Assam effectively utilizing microwave remote sensing data. Shivaprasad Sharma et al. (2018) analysed the flood risk in Kopili River basin of Assam through a GIS-based multi-criteria decision approach. Salunkhe et al. (2018) observed the flood inundation model-based flood susceptibility using GIS in the Lakhimpur-Majuli region of Assam with the help of pre-monsoon and on-monsoon satellite imageries of the region. These studies found sound effectiveness in the application of geospatial technology in meaningfully understating the flood hazard event in the monsoonal lands of Assam. Besides, there is a lack of scientific studies on flood hazard issues in Dhemaji district of Assam using geospatial technology. Therefore, this study also tries to adopt the geospatial approach for understanding the spatiality of recent flood hazard events in Dhemaji district of Assam. Moreover, this study will provide the stepping stone to further research and studies on recent flood hazard phenomenon in Dhemaji district of Assam as well as similar surrounding region.

Study Area

This study is confined to Dhemaji district of Assam located between 27.274° north to 27.869° north latitudes and 94.250° east to 95.516° east longitudes (Fig. 10.2). The district is bounded by Arunachal Pradesh to the north, Brahmaputra River to the

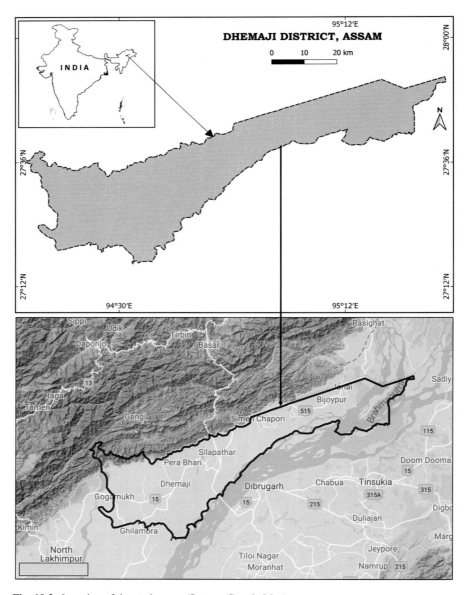

Fig. 10.2 Location of the study area. (Source: Google Map)

south, Arunachal Pradesh, Brahmaputra River and Sadiya plain to the east and Lakhimpur district in the west. The Dhemaji district of Assam has a GIS-derived shaped area of 2576.78 km^2 in WGS84/UTM 46 north projected coordinate system. Physiographically, Dhemaji district is a narrow river valley (with an average elevation of 160 meter) and a part of Brahmaputra plain, and it is surrounded by Arunachal Himalaya to its north and Brahmaputra River to its south. Geologically, older alluvium and newer alluvium sediment deposits of Quaternary age occupy the entire Dhemaji district of Assam. The main soil types found in the district are alluvial soil, sandy soils and silty and loamy soils. As per National Remote Sensing Centre's land use and land cover assessment through *Bhuvan* geo-platform during 2015–2016, the major land use categories of the districts are agriculture (47%), forests and grazing (32%), waterbodies (15%) and built-up (2%) lands. The climate of the district falls under the tropical monsoon climate regime with high rainfall during summer with dry winter. The maximum and minimum temperatures are 39.9 °C and 5.9 °C, respectively (NEDFI Databank 2014). The temperature begins to rise from the later part of March and falls in the latter part of October. January is the coldest month in the district with an average temperature below 11 °C. The annual rainfall of the district ranges from 1900 to 3900 mm. Dhemaji district receives maximum rainfall from May to July. As per the 2011 population census by the Census of India, the district has 129,504 households with a total population of 686,133 persons (334,884 females) sharing around 2.20% of the state's total population with a population density of 212 persons per sq. km.

Rationale of the Study

The Dhemaji district of Assam experiences recurring seasonal flood every year. With the help of geospatial technology, the flood hazard can be addressed at spatiotemporal scale. This study of understanding the flood hazard in Dhemaji district using geospatial technology would provide sound information on the spatiality of seasonal flood in a river valley region. This study is conducted using open-source geospatial tools in such a way that the findings would effectively depict the application advantages of geospatial technology in flood hazard study which will certainly provide significant inputs for the nonstructural management of flood hazards to the researchers and policymakers.

Materials and Methods

The objective of this study mainly focuses on the four aspects of the seasonal flood hazard in the study area – firstly, visualizing and estimating the spatiotemporal surface water pattern based on Global Surface Water dataset; secondly, understanding the main causal factor of seasonal flood using satellite-derived precipitation trend

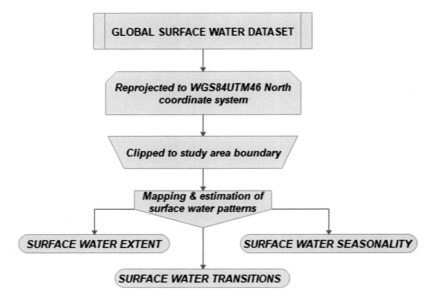

Fig. 10.3 GIS procedure as adopted for mapping surface water extent

statistics and secondary rainfall records; thirdly, mapping the extent of seasonal floodwater inundation for the recent year flood and estimation of last 3-year consecutive seasonal flood extent pattern on four sampled areas of the district using web-GIS-derived radar satellite data; and, fourthly, observing the impact of flood hazard by assessing the secondary flood damage statistics and identifying the flood-susceptible land use/cover on the sampled areas of the district based on satellite-derived land use/cover products.

The Global Surface Water dataset developed by the European Commission's Joint Research Centre has been used for mapping and estimating the spatiotemporal pattern of surface water in the study area using GIS (Fig. 10.3). The Global Surface Water dataset (https://global-surface-water.appspot.com/map) was created by the Joint Research Centre of European Commission which provides global-scale open surface water and its dynamics between 1984 and 2019. This dataset displays water surfaces that are visible from space, including natural as well as artificial waterbodies in approximately 30 meter pixel size (Pekel et al. 2016).

The European Space Agency's (ESA) Sentinel-1 Synthetic Aperture Radar (SAR) active remote sensing imagery was incorporated as the main database for delineating seasonal flood extents in the study area. They were freely downloaded from Land Viewer geoportal in VV (vertical transmit and vertical receive) polarization at 100 meter spatial resolution. The advantage of active remote sensing data is that these data can be obtained in cloud-cover conditions as the cloud cover is a common phenomenon during the rainy monsoon season, and SAR sensors can detect surface water through penetrating clouds (Uddin et al. 2019). The preprocessed ESA's Sentinel-1 SAR imageries which are geometrically and topographically corrected can be visualized in different polarization methods in the Earth Observation System's (EOS) Land Viewer geoportal (https://eos.com/landviewer/).

The satellite-derived ESA CCI land cover datasets are also incorporated in this study to visualize the flood-susceptible land use/cover in the district. The ESA CCI global land cover maps are analysis-ready datasets and have a spatial resolution of 300 meter with more than 70% of overall accuracy (ESA 2014; Liu et al. 2018). The ESA CCI global land cover maps are open-source products, and these maps describe the Earth's terrestrial surface in 37 discrete land cover classes based on the United Nations Land Cover Classification System (Li et al. 2018).

The open-source Quantum GIS 3.16 software package (https://www.qgis.org/en/site/forusers/download.html) has been used as the GIS platform in this study for geoprocessing, analysis and geo-visualization of GIS datasets.

Results and Discussion

In Dhemaji district of Assam, the seasonal flooding events create havoc to the people by damaging transportation networks, settlements, croplands, etc. The flooding occurs mainly during the monsoon season in the district as the entire state receives maximum rainfall in monsoon. To understand the spatiotemporal nature of flood hazard events in the district, a geospatial application approach is taken in this study by using primarily the open-source geospatial data products and software for a meaningful interpretation.

The Spatiotemporal Extent of Surface Water

The spatiotemporal surface water extent pattern of Dhemaji district is delineated and mapped using a satellite-derived surface water dataset. Three components of the surface water extent are represented for the study area based on the satellite-derived datasets. These three components include the maximum extent of surface water from 1984 to 2019, surface water transitions during 1984–2019 and surface water seasonality patterns during 2019 in the study area. These three surface water components are mapped for the study area in QGIS, and the map statistics of these components are extracted using raster layer unique value report tools in QGIS based on 30 meter spatial resolution of the surface water datasets.

The surface water extent refers to the maximum surface water cover on all the locations ever detected as water from 1984 to 2019 in the study area (Fig. 10.4). It represents all inland surface water pixels. This has been revealed that the maximum surface water extent from 1984 to 2019 in the study area is 542.18 km^2 which represents 21.05% of the district's total area.

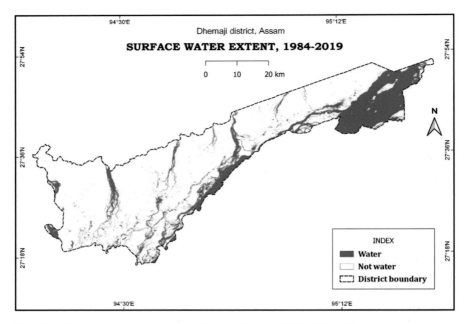

Fig. 10.4 Surface water extent in the study area. (Source: EC JRC/Google)

The surface water transitions provide information on the change in seasonality between the first and last years and capture changes between the three classes of not water, seasonal water and permanent water from 1984 to 2019 in the study area (Fig. 10.5). This has been observed in respect to surface water transitions in the district that the district records 12.61 km^2 of unchanging permanent water surfaces, 38.77 km^2 of new permanent water surfaces (conversion of land into permanent water), 6.73 km^2 of lost permanent water surfaces (conversion of permanent water into land), 32.80 km^2 of unchanging seasonal water surfaces, 217.11 km^2 of new seasonal water surfaces (conversion of land into seasonal water), 51.47 km^2 of lost seasonal water surfaces (conversion of seasonal water into land), 1.40 km^2 of conversion of seasonal water into permanent water, 27.74 km^2 of conversion of permanent water into seasonal water, 5.19 km^2 of ephemeral permanent and 138.81 km^2 of ephemeral seasonal water (Table 10.1).

The surface water seasonality provides information concerning the intra-annual behaviour of water surfaces for a single year (2019) and shows permanent and seasonal water as well as the total number of months when water was present over the surface. A permanent water surface is underwater throughout the year, while a seasonal water surface is underwater for less than 12 months of the year. In Dhemaji district of Assam, the seasonal surface water is mostly observed during the rainy monsoon season. The surface water seasonality pattern during 2019 of the district represents 52.78 km^2 of area (2.05% to district's total area) perennial under water which comprise of perennial, rivers, streams and wetlands, whereas seasonal water

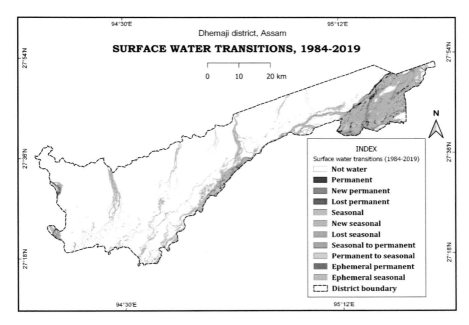

Fig. 10.5 Surface water transitions in the study area. (Source: EC JRC/Google)

Table 10.1 Surface water transition statistics in Dhemaji district, 1984–2019

Transition class	Area (km²)	%
Not water	2044.14	79.33
Permanent	12.61	0.49
New permanent	38.77	1.50
Lost permanent	6.73	0.26
Seasonal	32.80	1.27
New seasonal	217.11	8.43
Lost seasonal	51.47	2.00
Seasonal to permanent	1.40	0.05
Permanent to seasonal	27.74	1.08
Ephemeral permanent	5.19	0.20
Ephemeral seasonal	138.81	5.39

inundated area is estimated at 277.66 km² (10.78% to district's total area) (Figs. 10.6 and 10.7).

The Main Flood-Causing Factor

The spatiotemporal patterns in the seasonality, transitions and surface water extent in Dhemaji district signify that the district experiences the seasonal flooding

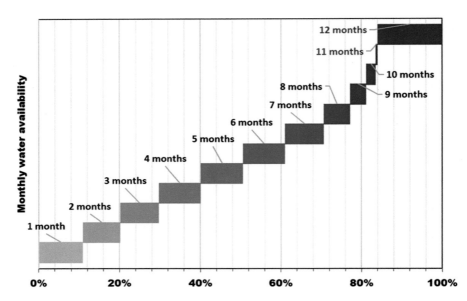

Fig. 10.6 Surface water availability (monthly) in the study area, 2019

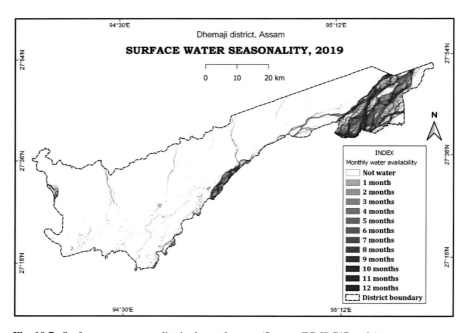

Fig. 10.7 Surface water seasonality in the study area. (Source: EC JRC/Google)

Table 10.2 Season-wise rainfall in Dhemaji district of Assam and in Lower Dibang Valley, East and West Siang districts of Arunachal Pradesh, 2015

Season	Rainfall figures in millimetre				
District	Winter	Pre-monsoon	Monsoon	Post-monsoon	Annual total
Dhemaji (Assam)	30.00	563.00	1694.00	11.00	2298.0
Lower Dibang Valley (Arunachal Pradesh)	105.10	1118.40	4110.80	287.70	5622.00
East Siang (Arunachal Pradesh)	170.10	829.50	3551.20	79.40	4630.20
West Siang (Arunachal Pradesh)	31.00	475.5	1311.9	172.2	1990.60

Source: Indian Meteorological Department (2015)
Note: Lower Dibang Valley, East Siang and West Siang districts of Arunachal Pradesh are located in the northern part of Dhemaji district from where the all perennial rivers flow through the district originate

phenomenon every year. Flood hazard occurring in the district is mainly rain-induced seasonal flood. The narrow-valley location with monsoonal climate condition and the enormous monsoonal rainfall in and around Dhemaji district may be considered as the main causes of seasonal flooding in the district. Dhemaji district receives rainfall almost the entire year, but maximum rainfall occurs during the monsoon season especially in June, July and August (Table 10.2). High and torrential rainfall in the district and also in the surrounding river catchment areas leads to the accumulation of enough amount of rainwater as surface water. These maximum amounts of accumulated water cause the rise of river beds and trigger flood inundation in the district (Bhatt et al. 2013). To understand the trend of precipitation in the Dhemaji district of Assam including nearby areas, the satellite-based precipitation time series of NASA's *Global Precipitation Measurement* mission-based *Integrated Multi-satellite Retrievals* (GPM-IMERG) products are analysed and visualized in the NASA's *Giovanni* web-geospatial interactive portal (https://giovanni.gsfc.nasa.gov/giovanni/) for the period 2001–2019 (Liu et al. 2017; Bhuiyan et al. 2020). From this assessment, it is revealed that the district records maximum precipitation trends during June, July and August with more than 400 millimetre precipitation (Fig. 10.8). Whereas, December, January, and February have the least precipitation in the district.

Visualizing the Recent Seasonal Flood Events

To address the seasonal flood water extent in Dhemaji district of Assam, for the recent flood event of 2020, geo-visualization was done using active satellite data in the GIS platform for monsoon (rainy) and non-monsoon seasons. The Sentinel-1 SAR imageries can also be freely obtained from the Land Viewer platform at various resolutions in georeferenced raster file formats (Kavats et al. 2019). To visualize the flood water extent of 2020 in the study area, the Sentinel-1 SAR imageries for

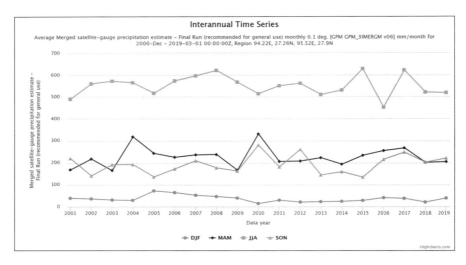

Fig. 10.8 A 3-monthly precipitation trend in and nearby Dhemaji district, 2001–2019 (*DJF* December, January, February, *MAM* March, April, May, *JJA* June, July, August, *SON* September, October, November)

monsoon and non-monsoon seasons for the entire Dhemaji district. In Sentinel-1 VV polarization the dark black to black hues, and VV polarization is suitable for surface water delineation in SAR imageries (Cao et al. 2019). The downloaded Sentinel-1 imageries are geo-visualized in QGIS to delineate the seasonal surface water extents during 2020. After identifying the seasonal flood water extent in the district, four sample areas within the district are selected to observe the seasonal flood events there for the last 3 years (2020, 2019 & 2018) (Fig. 10.9). These sampled areas are selected based on the maximum flood water extent during monsoon season through visual assessment (Fig. 10.10). These four sampled areas are the *Galighat area (51.54 km^2), Lamajan area (37.92 km^2), Amguri area (55.39 km^2)* and *Bordoibam area (31.47 km^2)* of Dhemaji district. These sampled areas are basically inland wetland areas of the district that include a group of villages within each sampled area. All these four sampled areas are located in the southern floodplain of the district nearby the Brahmaputra River. The monsoon season Sentinel-1 IW SAR imagery of Dhemaji district for 2020 is obtained on 29 July 2020, and non-monsoon season Sentinel-1 IW SAR imagery is obtained on 7 January 2020. Similarly, the recent years rectified Sentinel-1 IW SAR imageries for the sampled areas are also obtained for monsoon and non-monsoon seasons from the Land Viewer geoportal at 20 meter spatial resolutions with VV polarization. 2018, 2019 and 2020 non-monsoon imageries for the sampled areas are obtained on the same week of January, and 2018, 2019 and 2020 monsoon imageries are obtained on the same week of July. The visual assessment of the flood water extent during the 2020 monsoon season imagery in Dhemaji district of Assam shows flood inundated areas in almost all parts of the district except the northern foothills, and the maximum floodwater areas are seen mainly along the major rivers of the district and also in the southern parts of the

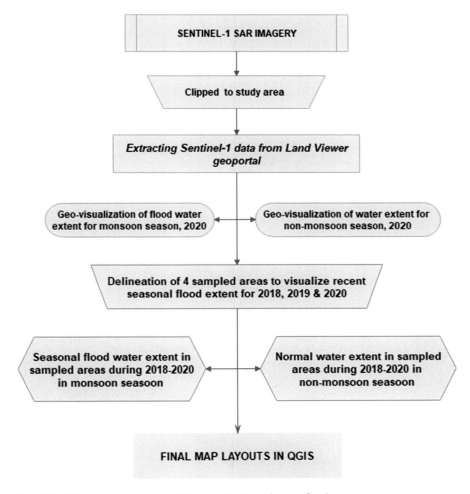

Fig. 10.9 GIS procedure as adopted for visualization of recent flood events

district along the Brahmaputra floodplains, whereas the non-monsoon imagery shows the same flood inundated areas of the district as non-water cover areas or with limited normal surface water cover areas in the forms of river water and wetland water (Fig. 10.10). This signifies that the flood inundation in the study area is a seasonal phenomenon that mainly occurs during the monsoon season due to excessive precipitation in and around the district. Similarly, the recent flood observations in the sampled areas from visual analysis reveal that the Galighat area of the district has shown an increasing trend of seasonal flood water extent since 2018 with maximum flood inundation in the sampled area during 2020. The Lamajan area has also shown an increasing trend with maximum flood inundation during 2020; the Amguri area has shown an increasing trend with maximum flood inundation during

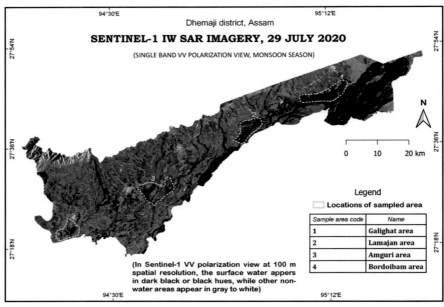

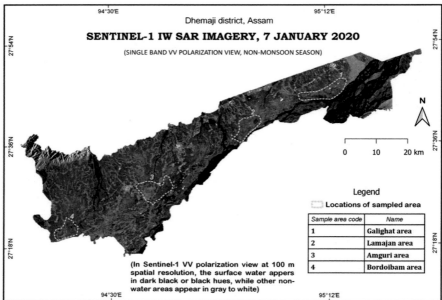

Fig. 10.10 Seasonal SAR imagery visualization of surface water extent in Dhemaji district. (Source: EOS)

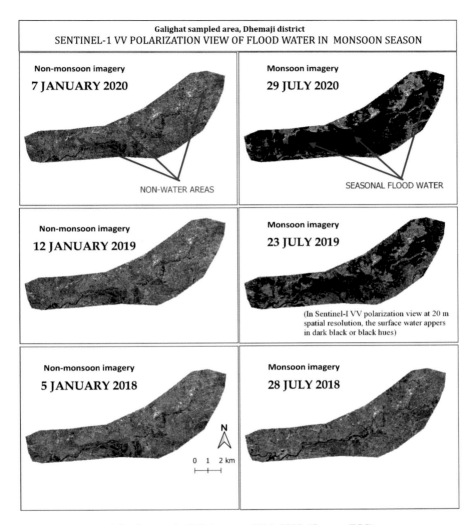

Fig. 10.11 Seasonal flood extent in Galighat area, 2018–2020. (Source: EOS)

2019, whereas the Bordoibam area has shown relatively decreasing trend with maximum flood inundation during 2018 (Figs. 10.11, 10.12, 10.13 and 10.14). These geo-visualization-based observations of recent seasonal flood events in the sampled areas of Dhemaji district during 2018–2020 indicate the fact that during the monsoon season there is flood inundation in the district and this seasonal recurring phenomenon shows an increasing occurrence as its areal extent includes new land areas in recent years.

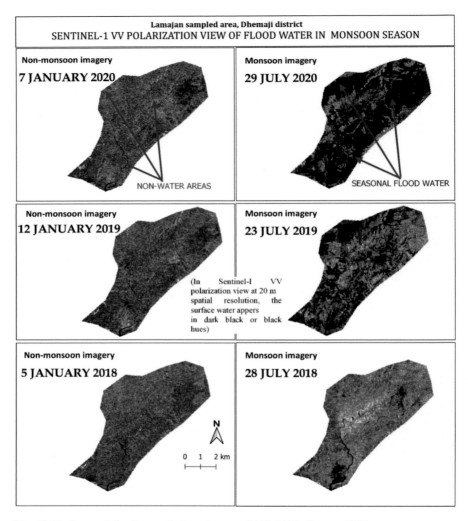

Fig. 10.12 Seasonal flood extent in Lamajan area, 2018–2020. (Source: EOS)

Assessing the Impact of Recent Seasonal Flood Events

The impact of recent flood hazard events in Dhemaji district of Assam has been assessed in this study by considering the flood damage statistics from a secondary source and geo-visualizing the flood-susceptible land use/land cover within the sampled areas with the estimation of susceptible croplands based on satellite-derived land use/land cover datasets. The flood damage statistics of Dhemaji district are obtained from the flood situation reports published by Assam State Disaster Management Authority (ASDMA), Government of Assam. The flood reports of Dhemaji

10 Application of Geospatial Technology in Seasonal Flood Hazard Event... 263

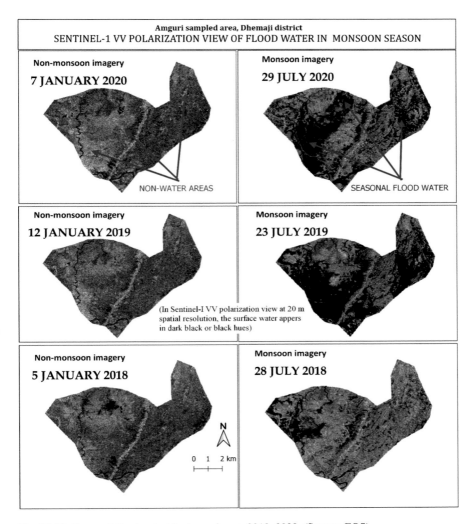

Fig. 10.13 Seasonal flood extent in Amguri area, 2018–2020. (Source: EOS)

district on 1 August 2018, 23 July 2019 and 29 August 2020 have been chosen to know the flood-affected status in terms of affected villages, affected croplands, affected population and affected animals in the study area. This has been found that Dhemaji district has recorded maximum flood-affected population in 2020 (45,457 persons), maximum affected croplands in 2019 (3896 hectares), maximum affected villages in 2019 (155 villages) and maximum affected animals in 2019 (15,198 numbers) as per the flood damage reports by ASDMA on the selected dates during 2018–2020 (Table 10.3). The satellite-derived land use/land cover data are obtained from the European Space Agency's Climate Change Initiative (CCI)-provided global land cover maps of 2018. These ESA CCI global land cover maps

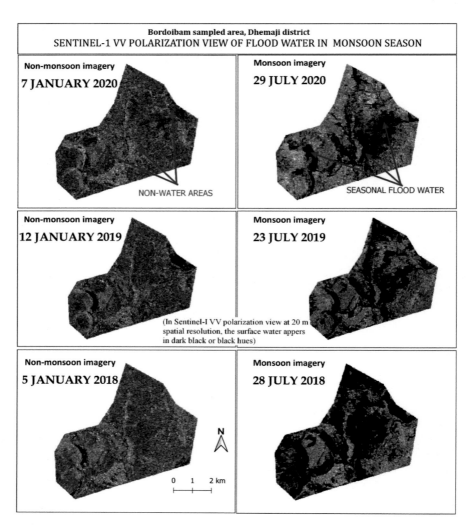

Fig. 10.14 Seasonal flood extent in Bordoibam area, 2018–2020. (Source: EOS)

Table 10.3 Flood-affected statistics of Dhemaji district

Flood events	Affected villages (numbers)	Affected croplands (hectare)	Affected population (persons)	Affected animals (numbers)
1 August 2018	41	1758	15,352	11,169
23 July 2019	155	3896	39,347	15,198
29 July 2020	122	3247	45,457	6959

Source: Assam State Disaster Management Authority (2020)

Fig. 10.15 GIS procedure for geo-visualization of flood-susceptible land use/land cover in the sampled areas of Dhemaji district

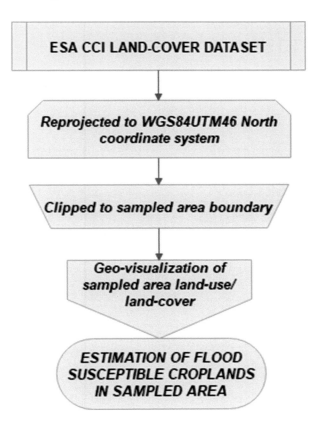

are first downloaded from the ESA CCI geoportal (https://www.esa-landcover-cci.org/) and then extracted these data to sampled area boundary with a projected coordinate system for geo-visualization and estimation of land use/land cover in GIS platform (Fig. 10.15). As seasonal floodwater inundation is observed to the maximum extent of each of the sampled areas in Dhemaji district during 2018–2020, therefore, it is evident to know the land use/land cover during non-monsoon season within each sampled area for understanding the seasonal vulnerability of different land use/land cover categories. Here, the geo-visualization of the sampled area land use/land cover during 2018 has shown that the Galighat area has nine land use/land cover categories, Lamajan area has seven land use/land cover categories, Amguri area has seven land use/land cover categories and Bordoibam area has eight land use/land cover categories. Among these four sampled areas, each sample area has the cropland (rainfed) land use/land cover category which is common and prominent with the maximum areal extent within the sampled area (Fig. 10.16). The GIS-based estimation of the croplands within these sampled areas shows 35.80 km^2 of croplands in Galighat area (*69.46% to total sampled area*), 22.50 km^2 of croplands in Lamajan area (*59.33% to total sampled area*), 39.18 km^2 of croplands in Amguri

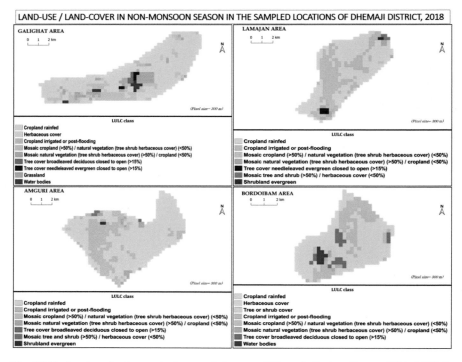

Fig. 10.16 Land use/land cover in the sampled areas of Dhemaji district. (Source: ESA CCI-LC)

area (*70.73% to total sampled area*) and 19.91 km² of croplands in Bordoibam area (*63.26% to total sampled area*).

Limitations of the Study

This study also has certain limitations as follows: the spatial resolutions of various geospatial datasets are not uniform, there is the unavailability of timely datasets on the research problem, the seasonal flood extent is assessed over the sampled locations using visual interpretation only and field data are not collected on seasonal flood in these sampled locations. These limitations might indirectly influence the accuracy of the results.

Recommendations

Applications of diverse open-source geospatial datasets to map the near real time as well as trend of seasonal flooding for efficient hazard mitigation, GIS-based assessment of flood damage extent in terms of land utilization pattern and spatial planning of nonstructural flood mitigation for a river valley region at local scale using sample areas' seasonal flood extent trend maps are some significant recommendations drawn from this study.

Conclusion

The study has applied open-source geospatial technology in understanding the seasonal flood hazard event in Dhemaji district of Assam. The study reveals the advantages of the open-source dataset with an open-source GIS platform for the spatial assessment of flood hazard events with no cost and at time-effective manner. The study has observed the spatiotemporal surface water extent, the main flood-causing factor, the geo-visualization of recent flood events and the impact of seasonal flooding to understand the seasonal flood hazard in Dhemaji district of Assam. It has been found that about one-fifth of the area in Dhemaji district has accumulated surface water extent during 1984–2019, and within the surface water extent areas, about the one-tenth area is under seasonal water extent. The study area receives the highest rainfall during the monsoon season especially in June, July and August, and this excessive monsoonal rainfall triggers the seasonal flooding in the district. The seasonal flood inundated areas are observed mainly in the southern parts of the district, and the sampled areas in flood inundation maps during the last 3 years have shown the same results. The croplands are found to be highly susceptible to seasonal flood hazards in Dhemaji district as the land use/land cover distribution in sampled areas indicates this result. Another significant finding of this study is that the seasonal flood hazard event in Dhemaji district of Assam is in an increasing trend as the recent flood inundation maps in the sampled areas of the district reveal this fact. As this study has successfully implemented the open-source geospatial inputs in effectively understanding the recurring flood hazard events in Dhemaji district of Assam, therefore, this study will provide the basis for further research in the field of flood hazard management and will also offer valuable inputs to the planners and policymakers for nonstructural management of rainfall-accumulated seasonal flood hazards using geospatial tools.

References

Aggarwal A (2016) Exposure, hazard and risk mapping during a flood event using open source geospatial technology. Geomat Nat Haz Risk 7(4):1426–1441

Albano R, Mancusi L, Sole A, Adamowski J (2015) Collaborative strategies for sustainable EU flood risk management: FOSS and geospatial tools – challenges and opportunities for operative risk analysis. ISPRS Int J Geo Inf 4(4):2704–2727

Assam State Disaster Management Authority (2020) Assam flood reports. http://www.asdma.gov.in/reports.html

Bhatt CM, Rao GS, Begum A, Manjusree P, Sharma SVSP, Prasanna L, Bhanumurthy V (2013) Satellite images for extraction of flood disaster footprints and assessing the disaster impact: Brahmaputra floods of June–July 2012, Assam, India. Curr Sci:1692–1700

Bhattachaiyya NN, Bora AK (1997) Floods of the Brahmaputra river in India. Water Int 22(4): 222–229

Bhuiyan MAE, Yang F, Biswas NK, Rahat SH, Neelam TJ (2020) Machine learning-based error modeling to improve GPM IMERG precipitation product over the brahmaputra river basin. Forecasting 2(3):248–266

Brakenridge GR (2016) Global active archive of large flood events. Dartmouth Flood Observatory, University of Colorado, USA. https://floodobservatory.colorado.edu/Archives/index.html

Cao H, Zhang H, Wang C, Zhang B (2019) Operational flood detection using Sentinel-1 SAR data over large areas. WaterSA 11(4):786

Centre for Research on the Epidemiology of Disasters & UN Office for Disaster Risk Reduction (2018) Economic losses, poverty and disasters 1998–2017. https://reliefweb.int/sites/reliefweb.int/files/resources/61119_credeconomiclosses_0.pdf

Dartmouth Flood Observatory (2014) Global flood hazard atlas. Space-based measurement of surface water. University of Colorado

Das S (2020) Flood susceptibility mapping of the Western Ghat coastal belt using multi-source geospatial data and analytical hierarchy process (AHP). Remote Sens Appl Soc Environ 20: 100379

European Space Agency (2014) CCI land cover product user guide version 2.4. ESA CCI LC project

Flood Hazard Atlas for Assam state (2016) Assam State Disaster management Authority. http://sdmassam.nic.in/pdf/publication/Flood_Hazard_Atlas2016.pdf

https://databank.nedfi.com/content/dhemaji-district

Indian Meteorological Department (2015) Rainfall statistics of India. Ministry of Earth Sciences, Government of India

Jonkman SN, Kelman I (2005) An analysis of the causes and circumstances of flood disaster deaths. Disasters 29(1):75–97

Kavats O, Khramov D, Sergieieva K, Vasyliev V (2019) Monitoring harvesting by time series of Sentinel-1 SAR data. Remote Sens 11(21):2496

Li W, MacBean N, Ciais P, Defourny P, Lamarche C, Bontemps S, Peng S (2018) Gross and net land cover changes in the main plant functional types derived from the annual ESA CCI land cover maps (1992–2015). Earth Syst Sci Data 10:219–234

Liu Z, Ostrenga D, Vollmer B, Deshong B, Macritchie K, Greene M, Kempler S (2017) Global precipitation measurement mission products and services at the NASA GES DISC. Bull Am Meteorol Soc 98(3):437–444

Liu X, Yu L, Si Y, Zhang C, Lu H, Yu C, Gong P (2018) Identifying patterns and hotspots of global land cover transitions using the ESA CCI Land Cover dataset. Remote Sens Lett 9(10):972–981

Mishra AK, Meer MS, Nagaraju V (2019) Satellite-based monitoring of recent heavy flooding over north-eastern states of India in July 2019. Nat Hazards 97(3):1407–1412

NEDFI Databank (2014) Dhemaji District

Pekel JF, Cottam A, Gorelick N, Belward AS (2016) High-resolution mapping of lobal surface water and its long-term changes. Nature 540(7633):418–422

Salunkhe SS, Rao SS, Prabu I, Venkataraman VR, Murthy YK, Sadolikar C, Deshpande S (2018) Flood inundation Hazard modelling using CCHE2D hydrodynamic model and geospatial data for embankment breaching scenario of Brahmaputra River in Assam. J Indian Soc Remote Sens 46(6):915–925

Samanta RK, Bhunia GS, Shit PK, Pourghasemi HR (2018) Flood susceptibility mapping using geospatial frequency ratio technique: a case study of Subarnarekha River Basin, India. Model Earth Syst Environ 4(1):395–408

Sanyal J, Lu XX (2004) Application of remote sensing in flood management with special reference to monsoon Asia: a review. Nat Hazards 33(2):283–301

Shivaprasad Sharma SV, Roy PS, Chakravarthi V, Srinivasa Rao G (2018) Flood risk assessment using multi-criteria analysis: a case study from Kopili River Basin, Assam, India. Geomat Nat Haz Risk 9(1):79–93

Surampudi S, Yarrakula K (2020) Mapping and assessing spatial extent of floods from multitemporal synthetic aperture radar images: a case study on Brahmaputra River in Assam State, India. Environ Sci Pollut Res 27(2):1521–1532

Uddin K, Matin MA, Meyer FJ (2019) Operational flood mapping using multi-temporal sentinel-1 SAR images: a case study from Bangladesh. Remote Sens 11(13):1581

Chapter 11
Geospatial Approach in Watershed Vulnerability to Climate Change and Environmental Sustainability

Anu David Raj, Justin George Kalambukattu, Suresh Kumar, and Uday Chatterjee

Abstract Watershed is considered as a natural unit for the resource management, planning and adoption of soil and water conservation practices. Hence, it plays a vital role in social, economic and ecological functioning of a system. The hydrological processes occurring in the watershed drive major vital services to various organisms and human beings. In recent years, climate change delivers chaos to various sectors around the globe both directly and indirectly. Climate change also poses vulnerability to the watershed and its various components. The watershed vulnerability assessment is one approach which can identify the exposure, sensitivity and adaptive capacity of watershed and adopt the solutions to counter the changing climate. The spatial extent of vulnerability also showed an increase in amount with respect to the future climate change scenario. The chapter deals with various aspects of watershed vulnerability-related indicators, adaptive approaches, assessment methods and sustainability. A case study demonstrates geospatial approach in characterising natural resources and watershed vulnerability analysis in future climate scenario in the catchment of lesser Himalaya, India, in detail.

Keywords Watershed vulnerability · Geospatial modelling · Environmental sustainability · Climate change

---

A. David Raj (✉) · J. G. Kalambukattu · S. Kumar
Agriculture and Soils Department, Indian Institute of Remote Sensing (IIRS), Indian Space Research Organization (ISRO), Dehradun, Uttarakhand, India
e-mail: anudraj@iirs.gov.in; anudraj2@gmail.com; justin@iirs.gov.in; suresh_kumar@iirs.gov.in; sureshkumar100@gmail.com

U. Chatterjee
Department of Geography, Bhatter College, Dantan (Affiliated to Vidyasagar University), Paschim Medinipore, West Bengal, India
e-mail: raj.chatterjee459@gmail.com

Introduction

Watershed primarily serves as an ideal natural planning unit for soil and water conservation to encourage sustainable development of diverse natural resources distributed unevenly within a defined geographical area (Knox and Gupta 2000; Sheng 1990). Watershed management provides immense scope for optimal utilisation of natural resources and participatory involvement of the people. It is a continuous process, which is related to soil and water resources and integrated the active expansion of socio-economic and ecological concerns. Condition of the watershed denote to the state of the physical and biological characteristic as well as the soil-hydrological function which support the sustainable development of natural resources. The watershed supports the important ecosystem services such as supply of clean water, recharge of aquifers and streams, soil quality and offsetting climate change impact through soil carbon sequestration, improving crop productivity and high biomass production. Functions of the watershed vary with the type of parent material (geology), soils, land use/land cover, typical climate and topography. Human influences also affect watershed functions and resilience which are reliant on the management-related activities. Indicators related to sensitivity of watershed can be selected to assess vulnerability of watershed by combining information of resource values, exposure and sensitivity.

Emerging geospatial approaches such as remote sensing (RS), geographic information system (GIS) and geographic positioning system (GPS) are providing reliable and up-to-date information of natural resources of the Earth. Currently, several RS satellites provide Earth observation data with spatial resolution ranging from 30 cm to 20 m. The RS satellite such as IRS Resourcesat-1 and Resourcesat-2, Landsat series, Sentinel 1 and 2, SPOT, Rapid Eye and World View are being used to derive information of natural resources of the watershed. The satellite-borne digital elevation model (DEM) such as Carto DEM, SRTM, ASTER, ALOS and GDEM is available to characterise topographical characteristics of the watershed. DEMs are used for precise watershed delineation and characterisation of other terrain variables for better computation of surface runoff, sediment loss, water quality simulation and watershed resource management. Physically based semi-distributed and distributed hydrological models are used to simulate watershed processes and linked responses.

Climate change has emerged as the biggest challenge to environment and it's sustainability. It is a global phenomenon and is caused by anthropogenic activities and poses serious challenge to humanity and sustainable development. It isn't evidenced in the same way throughout the region. Therefore, all watersheds are not equally vulnerable to climate change. Its impact varies across regions due to variances in the level of exposure and vulnerability of various ecosystems. Identifying most susceptible ecosystem is one of the significant ways to suggest suitable adaptation measures to reduce the climate change impact and their risk. Climate change scenarios are analysed by downscaling projected climate data from GCMs that use either empirical-statistical or dynamic models for assessing its future impact on water availability, water quality, sediment and nutrient losses from the watershed/catchments.

Vulnerability to climate change pertains to a system and its inability to cope with the negative effects of climate variability and extremes. It depends on adaptive capacity, sensitivity and exposure dimensions (Aleksandrova et al. 2014; Engle 2011). Watershed development programmers have the potential to make a significant role in reducing the vulnerability, improve resilience and develop adaptive capabilities to local societies to cope with climate adversaries. There is an immediate necessity to integrate climate change-related adaptation strategies into ongoing watershed development programmes to minimise climate change risk. There is utmost need to fine-tune the policies and the programmes to improve the watershed resilience to climate variability and risk with the aid of technology and institutional supports. Hahn et al. (2009) established a livelihood vulnerability assessment based on the general perspective of Intergovernmental Panel on Climate Change (IPCC) (IPCC 2007).

Watershed is viewed as ecosystem that includes ecology, human health and their sustainability. Sustainability is defined as "to create and maintain conditions, under which human and nature can exist in productive harmony, that permit fulfilling the social, economic, and other requirements of present and future generations" (NRC 2011). Improved ecosystem resilience is needed for efficient watershed development preparation and adaptive management measures (Furniss et al. 2010). The better approach to adapt to climate change is to sustain and enhance resilience, according to Williams and Jackson (2007). There is need to identify vulnerable watershed for effective allocation resources and to implement management activities. Climate change is adversely affecting both biophysical and socio-economic systems (IPCC 2014). Its impact varies across space and time depending primarily on topography, climatic condition, vegetation and human activities. Himalayan regions are characterised by friable ecosystems, unstable geology with convoluted topography and high biodiversity deemed as extremely subtle to the minor vicissitudes in the climate condition. Agrarian communities in the region are highly affected due to their poor adaptive capacity and limited access to alternative means of livelihood (Ives et al. 2000). The communities of the watershed in the region are vastly reliant on natural resources for their daily life and considered as most susceptible. Climate change also poses serious problem to biodiversity and socio-economic development of the region. Watershed development programmes are required to evolve a holistic approach to manage natural resources.

Climate vulnerability comprises of several components such as hydrological, vegetation, soil ecosystem in association with sociodemographic profile, food securities and natural disasters. The ecological and economic circumstances differ mostly by the hydrology and the people living in the watershed (Cabello et al. 2015). Quantification of vulnerability and identification of most vulnerable watershed are required for the prioritisation and preparation of the actions to halt with the impact of climate change (Krishnamurthy et al. 2014). This chapter deals with various aspects of watershed vulnerability-related indicators, adaptive approaches, assessment methods and sustainability. A case study demonstrates geospatial approach in characterising natural resources and watershed vulnerability analysis in future climate scenario in the catchment of lesser Himalaya, India, in detail.

Watershed: Natural Unit for Management of Natural Resources

Watershed is defined as a natural geohydrological unit, which collects water/runoff and drains to a common point or outlet. It is also being referred to as catchment area or drainage basin. The runoff generated within the watershed area is mainly contributed by rainfall as well as snowmelt, depending on the geographical locations and the predominant form of precipitation in the area. The ridges or hills, which act as a boundary between adjacent watersheds, are the highest points within a watershed and are commonly referred to as water divide lines/ridge lines. The common point or outlet to which the entire water drains may be a small stream, river, lake, dam, wetlands, estuary, sea or ocean. As watershed is a unit with geographical dimensions, it encompasses various physical-biological components such as water resources, soil resources, mineral resources, livestock, land use/land cover, pasture, human resources, biomass, etc., along with socio-political features. These various features as well as components need to be integrated during the planning as well as implementation of various watershed management activities. It is impossible to exaggerate the significance of watersheds as ecological units in natural resource protection and management. They are hydrologic units that have been frequently utilised as biogeochemical, economical, communal or political units in natural utilisation and management (Brooks et al. 1991). The watersheds could be as large as a major river basin (covering thousands of square kilometres) or as small as a farm micro-watershed (covering few hectares), with the smaller watersheds always forming a part of the larger one.

Characterisation of watershed mainly includes the detailed description and analysis of various components present in the watershed area. This provides us with vital information necessary for assessing the general behaviour of watershed with respect to runoff generation, groundwater recharge, water availability, soil erosion, sediment transport, nutrient loss, etc., that can be made use of with various other details for sustainable use of resources. It mainly describes the various biophysical as well as socio-economic features prevalent in a watershed and can be broadly categorised into climate, geology and physiography, soils, land use and land cover conditions, watershed hydrology and socio-economic features. Various aspects of precipitation, evaporation, wind, relative humidity, etc., are described in case of climate characterisation, whereas the watershed size and its shape, elevation, slope, aspect, etc., are included in physiography. The nature and type of parent rock as well as different drainage features such as drainage density, pattern, etc., are included in geology characterisation, whereas various soil properties such as depth, type, textural composition, infiltration capacity, erosiveness, etc., are detailed during characterisation

Table 11.1 List of freely available DEMs along with their specifications and source

Sl. No	Name of DEM	Agency	Resolution (m)	Source
1	Cartosat DEM for India	ISRO	30 and 90	http://bhuvan.nrsc.gov.in/data/download/index.php
2	Space Shuttle Radar Topography Mission (SRTM)	NASA	30 and 90	USGS Earth Explorer
3	Advanced spaceborne thermal emission and reflection radiometer (ASTER) Global DEM	NASA	1 arc second (30 m)/3 arc second (90 m)	USGS Earth Explorer
4	Advanced Land Observing Satellite (ALOS) Global Digital Surface Model	JAXA	30	http://www.eorc.jaxa.jp/ALOS/en/aw3d30/
5	GOTOPO 30	NASA	30 arc second (approx. 1 km)	USGS Earth Explorer
6	Global Multi-resolution Terrain Elevation Data (GMTED)	NASA	0 arc second (1 km), 15 arc second (450 m) and 7.5 arc second (225 m)	USGS Earth Explorer

of soil component. Another most important component, i.e. land use/land cover, is described in terms of predominant land use types (forest, grassland, agriculture, urban, etc.), existing ownership pattern (government, private, industrial, etc.), condition of forest lands in the watershed, dominant forest types, type and condition of rangelands, prevalent agricultural practices, existence of road networks and their conditions, presence of wildlife and other livestock resources, etc. With the advent of geospatial technologies, characterisation of these diverse components of a watershed could be easily and increasingly done using various RS products such as optical images, DEMs, microwave data, etc.

RS data interpreted by various visual as well as digital techniques aids us in identification and mapping of various components. The digital terrain analysis using DEMs could easily generate drainage/stream network, quick estimation of watershed area, slope, stream order, shape, drainage pattern, drainage density and a lot more parameters, which help us in faster characterisation and further modelling of watershed processes. A list of freely available DEMs widely used for watershed studies at various scales ranging from regional to local levels is given in Table 11.1.

Human resources, residing within the boundary, are an integral component of watershed. As they are dependent on various biophysical components, their livelihood is related to the socio-economic condition of the watershed. The socio-economic condition defines the land tenure/usage, capital expenditure, labour utilisation, etc., that play a crucial role in effective utilisation of resources and thus sustainable watershed management. The optimum availability of land and its quality, labour availability and its cost as well as capital determine the livelihood of people residing in the watershed by greatly influencing various activities such as construction of soil conservation and water harvesting structures, agricultural operations, livestock production and management of forests, pasturelands and other resources. A vicious cycle of low productivity, low income, low savings and low investment could be observed as a major characteristic in most of the poverty-stricken watersheds, thus adversely affecting the livelihood of its inhabitants. This invites special attention, as most of the poverty-stricken people are critically dependent on environment for their livelihood needs such as food, medicine, shelter, fuel, fruits, etc.

The concept of watershed management is primarily focused on the concept of optimal utilisation and conservation of natural resources within the area for prolonged and sustainable availability. Watersheds have been identified as the basic unit for developmental planning across the globe as well as in the country since many years as it encompasses a holistic development approach aimed at the conservation as well as improvement of various components, thus bringing favourable changes in the livelihood, socio-economic conditions of the inhabitants and environment within the watershed boundary as well as downstream areas. In the 1970s and 1980s, ICAR and subsequently the Ministry of Rural Development and Ministry of Environment and Forest initiated a series of integrated initiatives coping with rural area advancement, work opportunities, eradicating poverty and watershed development. The fundamental notion of watershed management was centred on resource management in medium and large river valleys. Using emerging watershed treatment technology and community participatory methodologies, the Integrated Watershed Development (Hills II) Project intends to boost India's productive potential in five states. In the inhabited areas of the Shivalik Hills in the five project states, the initiative contributes greatly to reducing soil erosion, boosting water security and eliminating poverty. Watershed conservation and redevelopment, as well as building capacity, are the two key aspects of the project. Watershed treatments, feed and livestock development and rural infrastructure development are all part of the first component. New policies, investigations and human capital development are included in the second component, as well as consumer capacity development, earning initiatives for women, data management, surveillance, assessment and planning and coordination support (World Bank n.d.). Accordingly, watersheds are perhaps the most acceptable management and planning units for natural resources since soil or land, forest and water resources can be handled at the watershed level. Watersheds are not too vast as a hydrological unit in order for them to function as a communal and economical unit as well.

Watershed Processes

As watershed is primarily a natural hydrological unit, the most common functions/processes in a watershed are defined as various aspects involving water movement. The main hydrological functions of watershed are (i) water capture/collection, (ii) water storage and (iii) water release. Apart from the above-mentioned physical processes, two ecological functions, i.e. (i), provide diverse sites for the occurrence of chemical reactions and (ii). Providing habitat for various flora and fauna is also defined for watershed. Further discussions will be focused only on hydrological function as ecological functions are out of the scope of this chapter. These major processes and their interactions will result in various sub-functions and characteristics, namely, erosion and soil processes, nutrient cycling, transport of pollutants, riparian habitat characteristics, stream network and their characteristics, downstream water quality, etc.

Runoff Generation Processes

Runoff generation in a watershed is a direct result of the mutual linkages and interaction between the three hydrological functions indicated earlier. Rainfall as well as snowmelt forms the major sources of water in most of the watersheds across the globe. Moisture received on the soil/land surface normally undergoes the processes of infiltration at the surface as well as percolation into the deeper layers at varying rates depending on different factors. Runoff, which is the movement of excess water across the surface through channels, streams, gullies or rivers, originates by the concentration of overland flow. Overland flow and subsequent runoff generation is initiated once the amount of water received is in excess after meeting the requirements of infiltration, evapotranspiration and other losses. Numerous fixed as well as controllable factors influence the collection of water and infiltration rates such as soil depth, texture, rainfall (duration, intensity, etc.), climate, topography, type and nature of vegetation cover, soil organic matter and soil compaction. The water infiltrated will be stored in the pore space between soil particles, the maximum amount of moisture a soil could hold being determined by various soil properties as well as the management practices adopted in the region. Once the soil gets saturated, additional water will either percolate into deeper layers or appear as surface runoff (referred to as direct runoff). The percolated water may join the groundwater table and continue to flow along the slope direction to other places (indirect runoff).

Soil Erosion Processes, Sediment and Nutrient Loss

The flow of water in the form of surface runoff is primarily responsible for the transportation as well as deposition of sediments as well as pollutants into streams or channels, via various erosion processes. Soil erosion starts with the detachment of soil particles resulting from the breakdown of aggregates caused by the beating action of rainfall. The individual particles released will cause sealing of soil pores, thus adversely affecting infiltration and percolation processes, thus resulting in runoff generation. The detached particles will be transported by the runoff in varying quantities depending on various characteristics related to rainfall, soil, topography, vegetation, terrain, etc. Different soil erosion processes could be categorised into four types mainly, i.e. splash, sheet, rill and gully erosion, depending on the mechanism of occurrence. The splash erosion is caused by the direct impact of rainfall on bare soil or shallow water surfaces, which results in the breakdown of aggregates, thus releasing finer soil particles in the form of a "splash". The rate of soil loss via splash erosion is dependent on various characteristics of rainfall, canopy cover, residue retention, previous moisture condition of soil, etc. The overland flow and surface runoff could remove the "splashed" as well as detached soil particles in a uniform thin-layer fashion, especially in gently sloping lands, which is commonly referred to as sheet erosion. The uniform soil loss many at times goes unnoticed until a considerable amount of productive topsoil is lost resulting in the lowering of land surface. The factors influencing sheet erosion rates are more or less the same as in case of splash erosion. Rill erosion is defined as the process of soil loss happening in few millimetres wide and deep channels (referred to as rills) due to the concentrated runoff. Prolonged unattended soil loss via rills could result in their widening as well as deepening, due to the scouring action performed by moving soil particles in the concentrated runoff. Gully erosion, which is considered as the most advanced stage/process of soil erosion, is characterised by the soil transportation caused by concentrated runoff happening in much wider, deeper and steeper channels (referred to as gullies) of permanent nature. Gullies, unlike the rills, are known to cause physical obstruction of different tillage and other field operations, owing to their larger dimensions. The gullies if left unattended could lead to the dissection of land resulting in the formation of ravenous landscapes.

Watershed Services

Watershed plays an important role in determining the quality of human life as well as environmental quality by virtue of performing various ecosystem services. Watershed not only influences the quality of life and livelihood of people residing within its boundary but also affects the people as well as environment outside, especially the downstream area. Watershed and its numerous components perform various ecosystem services ranging from supporting, provisioning, regulating and cultural services as mentioned in and not limited to Table 11.2. Various categories of services indicated in the table are purely dependent on the three processes/services of soil formation, nutrient cycling and primary production, which are collectively referred to as supporting services (services necessary for the production of all other watershed services).

Majority of the provisioning services have a direct impact within the watershed boundary in general, whereas the regulating services play a crucial role in influencing the quality of life both within and outside the watershed boundary, even in far-off places. Regulating services such as water purification and quality improvement by filtering various organic and inorganic pollutants helps us in minimising the ill effects of point as well as nonpoint sources of pollution in downstream areas.

Table 11.2 Different categories of watershed services

Provisioning services (products obtained from watershed)	Regulating services (benefits obtained from regulation of ecological processes)	Cultural services (nonmaterial benefits obtained from watershed)	Supporting services
Food and fibre	Water regulation – both surface flow and groundwater	Aesthetic	
Fuelwood	Water purification and quality control	Inspirational and spiritual	
Fresh water	Erosion control and sediment retention	Educational	
Timber	Air quality maintenance	Recreational and ecotourism	
Biochemicals, natural medicines and pharmaceuticals	Climate regulation – both local and global	Cultural heritage	
Genetic resources	Storm protection	Sense of home	

Similarly, the watersheds with adequate and healthy natural land cover and soil resources have the ability to sequester carbon in good amounts, thus helping us in offsetting the adverse impact of greenhouse gas emissions, up to a certain extent (Hansen et al. 2013). Watersheds also play key roles in controlling erosion processes, sediment loss within the area and stream bank stabilisation, thus preserving the soil fertility for sustainable agricultural production and ecosystem productivity as well as reducing the issues of reservoir sedimentation in downstream catchment area. Healthy watersheds with well-functioning riparian areas and floodplains are known to exhibit increased resilience to the threats posed by climate change due to their improved adaptability to extreme weather events (EPA 2012).

Watershed Vulnerability

Vulnerability is defined as a system's or a unit's incapability to prevent the occurrence of a stressful atmosphere. Using the IPCC's insights, "the degree to which a system is susceptible to, and unable to cope with, adverse effects of climate change, including climate variability and extremes. Vulnerability is a function of the character, magnitude and rate of climate change and the variation to which a social-ecological system is exposed, its sensitivity and its adaptive capacity" (IPCC 2007). Hence, the watershed vulnerability is considered as the magnitude in which a watershed disposed and disabled to the adverse effect of a hazard (climate change). It can be defined as an efficacy of scale, characteristics and its frequency of occurrence to which a watershed is exposed (in this context the watershed is exposed to climate change). Besaw et al. (2009) stated the vulnerability as a system's sensitivity to stress and the extent to which it will be damaged by exposure to a

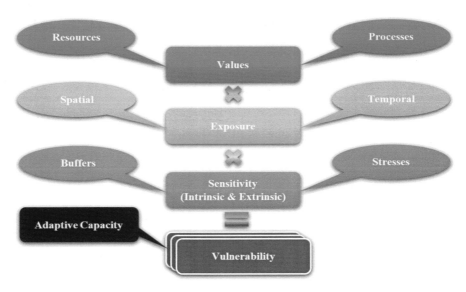

Fig. 11.1 Concept of watershed vulnerability

source of stress or perturbation. Watershed condition is one of the most important factors in determining a system's or watershed's vulnerability. The status of a watershed's physical and biological traits and processes as they relate to the soil and hydrologic activities that sustain aquatic ecosystems is referred to as its "watershed condition". Aquatic and riparian species rely on healthy, well-functioning watersheds for their survival. System (watershed) resilience is the ability of a system to absorb and reorganise even though facing challenges while maintaining its functions, structure and individuality (Walker et al. 2004) or refers to a system's ability to bounce back after a stressful event. The term "watershed health" or "watershed condition" is often used interchangeably, which refers to the state of a watershed's ability to withstand stress and restore itself (Furniss et al. 2010). A watershed assessment typically examines the watershed's vulnerability to future deterioration since watershed health is a dynamic quality that can alter with future changes in climate and anthropological activities. The potential of an ecosystem to defence variation and sustain it's structure and function even in the face of mounting stressors is called resistance (Vieira et al. 2004).

Because of the complexity, interconnectedness and dimensionless, vulnerability cannot be assessed or evaluated directly. As a result, we deem it to be a function of exposure, sensitivity and adaptive capacity (Fig. 11.1), which can be measured numerically or subjectively (Glick et al. 2011). The size and extent (by representing the spatial and temporal scales) of climate change impacts are measured by exposure. Sensitivity is a measure of how a system will react if it is subjected to a stress caused by climate change. The values/resources represented as the natural resources being affected due to various threats such as climate change. As per IPCC, the greatest pressures on, and challenges for, societies and the environment under climate change would be water and its accessibility and quality. The hydrological process such as interception, infiltration, saturation, surface and sub-surface runoff and sediment transfer directly affects the natural environment and human habitat. Hence, majority of studies focused on the aspect of water resource values such as water quantity and quality. These resource values are affected by the adverse consequences of climate change called the exposure. Assessing the exposure needs the historical as well as projected climate data. It will give an idea about how the resources are behaving with respect to the historical climate, while the projected climate will give the response of values with respect to baseline or historic climate data. It will provide the exposure of the water resource value in accordance with a changing climate.

The purpose of the sensitivity assessment was to categorise the watersheds based on how they would react to projected climate-induced changes in hydrological processes. Watershed sensitivity to changes is influenced in part by parent geology, soils, normal climate, terrain and vegetative cover. Depending on the scope and location of management-related activities, human factors also have an impact on watershed resilience. While we must determine the management, factors influence the hydrological processes of interest, for example, construction of highways and reservoirs. The sensitivity indicators chosen were those that had the greatest influence on the hydrological process and the value of the water resource in question. Some indications tend to reduce (commonly called buffers) effects, while others increase them (commonly called stressors). Adaptive capacity is a measure of a system's ability, potential or possibilities to reduce exposure or susceptibility to a climate-driven stress or simply adaptation. Adaptive capacity is the other aspect of vulnerability which can control the degree of vulnerability through management. To put it another way, adaptive capacity refers to a system's ability to deal with the impacts of climate change by either reducing its exposure or increasing its sensitivity as described above (Glick et al. 2011). Preparation for stresses or changes and tolerance and response to stressors are examples of resilience (Smit and Pilifosova 2001).

While we are considering about the vulnerability, the hydrological process is strongly affected by the climate because precipitation is the major source of water. The surface runoff is one of the major water resource values which serve as the dynamic entity in a watershed which control the hydrologic response in a catchment (Hendriks 1993). The availability of water (surface runoff) in the stream/river depends on the rainfall availability. Thus, any change in the precipitation pattern and depth may affect the watershed condition. At this moment, the historical climate data provide the current exposure of climate on the surface runoff, while the projected runoff will provide the future change of surface runoff. The exposure can be measured using various kinds of hydrological, soil and nutrient erosion and other models (lump sum, distributed, process-based model, etc.), and more information on this is provided in Section 8. After this, the sensitivity of the indicator needs to be selected for identifying its influence. For example, an increase in the number of roads could increase the peak flow response and the risk of flooding in areas with sensitive infrastructure near waterways. Investment in road enhancements, such as the separation of road surfaces from streams, would, on the other hand, lessen the impact. Thus, indicator-based approach provides fast and reliable estimation of the watershed vulnerability.

Climate change is expected to have a significant impact on global hydrologic systems, according to current models and predictions. Anthropogenic-induced greenhouse gas emission is the prime reason for faster global warming than the usual. The fast-growing concentration of these gases increases the warming potential year to year. The vigorous hydrological cycle as a result of warming drives excess precipitation and higher surface runoff and flooding. The increased water temperature also affects the aquatic ecosystem and improper functions of their vital processes. Apart from this, anthropogenic intervention adversely affects the watershed functioning. The construction of roads and infrastructures in unstable and fragile slopes of hilly and mountainous watershed is increasing vulnerability of watershed. It can fuel the climate change-induced disasters occurring in the watershed. Road building has been linked to an increase in landslides in the Himalayan region (Ives and Messerli 1989; Singh et al. 2014). This kind of unsustainable activities increases the vulnerability of a watershed significantly. Watershed variables such as climate, landform, geology, type of soil, land use/land cover and socio-economics all play a role in determining the amount of runoff and soil erosion that can be anticipated (Kang et al. 2001). Soil erosion causes the relocation of suspended sediment particles, which has negative effects on indigenous soil quality such as topsoil separation, soil organic material depletion and nutrient depletion (Creamer et al. 2010). When nutrients are removed from the soil through sediments and runoff water, they not only deplete soil fertility but also cause environmental problems as they travel deeper into valleys, lakes and reservoirs (Kunimatsu 1986; Kin-Che et al. 1997). This indicates monitoring and evaluation methods are very much required for reducing the watershed vulnerability. Hence, the watershed vulnerability assessment is a vital tool for the adaptation and mitigation policies.

Watershed Vulnerability Evaluation Methods

The process of analysing, quantifying and/or describing the exposure, sensitivity and adaptive capability of a natural or man-made system to perturbations is known as vulnerability assessment. Vulnerability can be assessed using a variety of methods. The tools given are able to be utilised in any of the several vulnerability assessment methodologies. The methodology used will be determined by the information and ideas available, technical competency, capability terms of individuals, time and finances and kind of information required by decision-makers in a given situation. The spectrum of techniques to vulnerability assessment includes creating the theoretical foundation for vulnerability assessment, analysing effects of climate change, establishing metrics or indicators to estimate vulnerability and consuming the outcomes for climate change management and adaptation (Nelitz et al. 2013).

The importance of specifying the water resources to be included in the overall assessment could be emphasised. Choosing what to cover is one of the most difficult aspects of performing a large-scale analysis. The land regions under question are vast, and the ecological and societal systems they support are exceedingly complex (Furniss et al. 2013). Thus, we can be able to narrow the scope to the most severe perturbations or climate change-related issues. Identifying the indicator which has more potential to explain or reveal the vulnerability is vital in vulnerability assessment. Other valuable tools that aren't numerical models are included in this category, and they provide options for organisations who don't have the capacity to run complex numerical models. Because each indicator, index and statistical model is different, this tool class contains a wide range of methodologies, frameworks, input demands and output features. A vulnerability indicator is a hydrological and climatologic physical, chemical and biological property that is susceptible to perturbation and can represent the functioning of hydrological processes and climate system functions in the watershed of interest. These indicators are dynamic and extremely sensitive to outside influences. Precipitation, temperature, solar radiation (climate), soil temperature, soil moisture, nutrient availability (soil), water level, stream flow, discharge (hydrology) and so on are some examples of variables (Tiburon et al. 2010). Indicators and indices are often simple to use, highly transferrable and cost-effective due to the origination of extensive database.

Schultz (2001) expresses an explicit criticism of the Environmental Protection Agency's (EPA) index of watershed indicators. Some of the challenges in applying indicator-based vulnerability assessment include the provision of adequate data on some attributes of vulnerability, uncertainty regarding vulnerability indicators, indicator screening process, inferences employed in weighing indicators and the arithmetic of aggregation (Eakin and Luers 2006; Hinkel 2011; Tapia et al. 2017). These difficulties arise when determining appropriate indicators for exposure, sensitivity and adaptive ability from global databases that are appropriate for national and local contexts (Tessema et al. 2021). Terrain characteristics, landforms, hydrology, water quality, habitats and biological conditions, according to a report by the US EPA (2012), are six critical indicators for determining the condition of watersheds. It was found that numerous indicators were established to assess the climate change vulnerability of regional water resources, comprising stream flow, climate sensitivity indicators, water balance dryness ratios and groundwater depletion ratios which consider yearly recharge rates into account (Hurd et al. 1999). In general, a collection of indicators is chosen to measure and assemble the subindices of vulnerability that are most important to the watersheds under consideration. Then, by combining these subindices, a single vulnerability index can be created, which can be used to identify prospective threats such as future climate, land and water use change. Statistical methodologies such as the Monte Carlo analysis, principal component analysis and factor analysis or participative methods such as household surveys, cognitive mapping, expert opinion, discourse analysis, semi-structured interviews and thought experiments can be used to produce a set of meaningful indicators. Table 11.3 describes vulnerability dimensions of some selected indicators which can be useful for the watershed vulnerability assessments.

Table 11.3 Selected indicators for dimension of vulnerability (assessment)

Dimension of vulnerability	Indicators
Exposure	Rainfall, temperature, wind speed, soil moisture and temperature, water level, discharge, nutrient and sediment loss, water temperature, snow depth/cover, etc.
Sensitivity	Land use/land cover, growing degree days, water quality and quantity, stream size, groundwater recharging, natural disorders, erosion features, species density and shifts, infrastructures, stream density, groundwater availability, etc.
Adaptive capacity	Sources of income, technological accessibility, education, availability of skilled labours, availability health services, income distribution and inherent adaptive capacity of the ecosystem, etc.

The statistical models are primarily user-built; they have a lot of flexibility in terms of structure, approach and data requirements. They're also usually based on historical data, which means they don't recognise altering physical processes. In a watershed close to Assateague Island National Seashore in the USA, LaMotte and Greene (2007) employed geospatial statistical methods to investigate land use and groundwater susceptibility. Greene et al. (2005) used geographic probability models to assess groundwater sensitivity to nitrate intrusion at several levels in the mid-Atlantic region. Weil et al. (2019) employed two statistical methods to uncover physical and biological and watershed land cover characteristics that influence stream susceptibility to urbanisation: Kruskal-Wallis analysis and logistic regression. Furthermore, these models can be resource intensive, necessitating high-performance computing when applied to larger areas, limiting a complete evaluation of ambiguity in climate change scenarios (Elsner et al. 2014). To get over these restrictions, statistical models with a small number of input variables might be employed as a substitute tool to examine the spatial aspects of land hydroclimate processes (Abatzoglou and Ficklin 2017). Apart from this, GIS-based tools and approaches provide more insights as well as reduction of knowledge gap in vulnerability assessment. For example, Swansburg et al. (2004) used regression-based statistical downscaling of GCM estimates to create location-oriented future climate scenarios for regions over New Brunswick. Beier et al. (2008) used a geospatial decision-support tool in controlled ecosystems to study ecosystem services and emergent vulnerability. The geospatial approach of the occurrence, health and vulnerability of Wyoming's wetlands is described by Copeland et al. (2010). In certain sections of the Upper Subarnarekha river basin in India, a geospatial analysis of soil erosion susceptibility at the watershed level was conducted by Chatterjee et al. (2014). Tiburan Jr et al. (2013) used a geospatial-oriented ecological vulnerability index represented as the Geospatial-based Regional Environmental Vulnerability Index (GeoREVIEW) for ecosystems and watersheds to estimate the vulnerability of watersheds. A Watershed-Based Geospatial Groundwater Specific Vulnerability Assessment Tool was created by Baloch and Sahar (2014). Using geospatial approaches, Bhattacharjee et al. (2021) applied hydrodynamic modelling and vulnerability assessment to quantify flood risk in a populous Indian metropolis. David Raj et al. (2022) generated vulnerability map of the Himalayan watershed in terms of climate change and soil erosion potential.

Hence, the applicability of statistical, geospatial and their combined approach has already proven the capability of watershed-based vulnerability assessment. The spatial distribution of the indicators, characteristics and other related information of a watershed provide site-specific explicit information of the most vulnerable areas. It will be easier and reliable for the watershed restoration and degradation reduction processes coming in front of the policy- and decision-makers.

Watershed Sensitivity and Sustainability

The purpose of sensitivity assessment was to categorise areas (various scales of watersheds) depending on how they might behave to projected climate-induced variations in hydrological processes. Watershed sensitivity to change is influenced in part by parent geology, soils, normal climate, terrain and vegetation. Depending on the scope and site of management-related activities, human factors also have an impact on watershed resilience. Sensitivity is defined as a measure of how a watershed and its dependent human communities or ecosystems would be damaged by a given climate change condition (Glick et al. 2011). Sensitivity analysis looks at the reliability of each indicator's significance on the index output, demonstrating which input options are perhaps the most or fewest impactful (Tate 2012).

According to Furniss et al. (2013), watersheds incorporate all elements of the environment spanning upslope, floodplain and in-channel subsystems, including climate, geography, vegetation cover, anthropogenic activities, human populations and biodiversity. The three types of methodologies for determining watershed sensitivity are (i) watershed health or function indicators, (ii) biological indicators (bioindicators) and (iii) watershed models that are paired or interconnected. These divisions illustrate the various approaches of determining the sensitivity of watersheds. A few case studies often use the finest facts obtainable as indicators of watershed condition including physical, habitat conditions, geological, biological and water quality or as indicators of a watershed's biological integrity such as species diversity and habitat extent. Other example studies have utilised neither coupled models, i.e. outcomes from one model have been used as intakes to the next model, integrating climate, water, land and biodiversity, nor integrated watershed models which refer to the incorporation of all inputs into a unified modelling approach.

Climate, topography, vegetation and biological activities all play a role in watersheds. Vegetation evolution and natural disturbance, basin geomorphology, hydrologic characteristics, chemical cycles and human development and behavioural patterns are all examples of these processes. The extent of watershed hydrology damage is determined by a variety of factors including current land usage, geography, soils and vegetation (Hazbavi et al. 2018a; Hazbavi et al. 2018b). Because it is hard to ascertain environmental degradation from a single source, and because it is often difficult to prevent or anticipate, numerous indicators are needed to monitor the health or performance of a particular watershed (Liao et al. 2018). Scientists and academics are increasingly aware of the need of measuring indicators and methodologies for assessing the health of watersheds in order to achieve this goal (Ahn and Kim 2019). For example, freshwater macroinvertebrate and fish populations are widely observed to characterise the ecological health of a watershed, particularly the status of in-channel ecosystems; biological indicators are linked to indicators of watershed health (Fore 2003). Changes in biological markers, on the other hand, have been linked to vicissitudes in the ecological status of watersheds and are thus being advocated as a technique for expressing watershed sensitivity to climate change (US EPA 2008). The forms of climate change that have been observed and are anticipated across the USA have the capability to have a detrimental effect on watershed health. Relatively warm atmospheric air temperatures, prior and faster snowmelt, shorter and higher intense rainfall events and more severe droughts and floods are only a few examples (EPA n.d.). Shukla et al. (2016) compared mountainous agricultural villages in India's Uttarakhand state based simply on the strength of underlying vulnerability ranks. Uncertainty in estimating climate vulnerability can stem from a variety of factors.

As previously stated, a watershed's current structure and function are the result of a complex interaction of numerous elements, including climate, geography, land use and land cover, natural disturbance, geomorphology, physiological and biochemical dynamics, human health and development characteristics. As a replacement to using indicators, using quantitative models that specifically express the functional relationships between variables involved is an extra robust way of describing a cohesive perception of watersheds. In vulnerability analysis, two broad methodologies are utilised to quantitatively simulate linkages between climate and watershed response. One method is to use connected or coupled modelling. Watershed elements and/or mechanisms of importance are divided into sub-models using this method, with each sub-model being constructed completely independently of the others. Each sub-model, on the other hand, is built with the intention of using the estimates or yields of one sub-model as inputs to another. The soil erosion models are capable of capturing the hydrological processes. Thus, these models are data intensive and required large database for complete characterisation of the watershed. The process-based models are more capable in predicting the soil erosion, nutrient erosion, surface runoff, carbon dynamics, identifying best management practices, etc. The topography, land use/land cover, soil physicochemical and hydrological properties are the major input parameters needed for simulating the model.

These models can provide the exposure of various processes towards the climate change. For example, Gupta and Kumar (2017a) employed RUSLE model for identifying future climate change impact on soil erosion from a mid-Himalayan watershed. Also, using the CENTURY model, Gupta and Kumar (2017b) anticipated the influence of climate change on soil carbon sequestration in agro-ecosystems in the mid-Himalayan terrain. Similarly, Sooryamol et al. (2022a) predict the future soil loss using the SWAT model according to the projected rainfall amount in a lesser Himalayan watershed. Apart from this, Singh et al. (2021) used the WEPP model to model the surface runoff and sediment yield from a watershed of Himalayas. David Raj et al. (2021) used the APEX model for predicting the land use-based surface runoff and sediment loss from the highly erosion vulnerable watershed of Shiwalik Himalaya and identified the scrubland as most vulnerable to erosion. Thus, the employment of empirical- to process-based model will provide more accurate assessment of the processes and functional relationships. The exposure of each watershed value can be identified explicitly and will be beneficial for all who are in the field of policymaking. The spatial and temporal distribution of watershed processes can identify the most vulnerable area of the watershed. Monitoring and identifying the vulnerable areas are very much necessitated for attaining sustainability. The implementation of the conservation and mitigation measures is only possible after the proper understanding of vulnerability. Hence, the vulnerable watershed can be transformed into sustainable by proper vulnerability analysis and measures.

Sustainability entails preserving natural resources for future generations while also allocating them fairly among present generations (WCED 1987). Sustainable development, it was agreed, should not just encourage economic progress but also promote equitable wealth distribution, prioritise environmental concerns and empower individuals instead of just excluding them (UNDP 1994). The susceptibility of a watershed (soil erosion, hydrologic functions, species richness, habitat and so on) must be assessed for soil resources and long-term sustainability. The academicians have been extremely involved in analyses, and study on the vulnerability and robustness of natural ecosystems and

communities in specific locations is one of the seven important topics of sustainability science (Cutter 2003). Watershed sustainability indicators are separated into three categories: socio-economic development, ecosystem and biodiversity. Aditya and William (2011) used a hydrological/water quality model to build a framework for measuring watershed sustainability. They supplied water for residential use and livelihood, renewable energy from the land, water pollution, water for ecosystem support and undeveloped land as sustainability indicators. As a result, a river basin sustainability index can be produced by combining the above-mentioned socio-economic, environmental and biodiversity indicators. Some indications may be more important than others depending on the locale. As a result, these indicators must be given the appropriate weighting based on a subjective assessment of the problems in the research location. Chaves and Alipaz (2007) created a watershed sustainability index by combining indicators based on basin hydrology, environment, life and policy aspects. Ecosystem services are important components that subsidise to human health by providing security, raw material supply, health and communal consistency (Millennium Ecosystem Assessment (MEA) 2005). Because of their great significance, they have been explored and recognised at various scales such as local to worldwide and functional unit or organisation levels representing ecosystems or watersheds. Diverse ecosystem benefits have lately been evaluated at the watershed scale attributed to the prevalence of natural barriers (Pérez-Maqueo et al. 2013). Mititelu-Ionuş (2017) proposed the creation and use of a watershed sustainability index.

The climate change and anthropogenic activities are the major threat for the watershed vulnerability. The increased greenhouse gas emissions enhance the global warming tendency of the Earth's atmosphere. Accompanied to this, the land degradation is also enhancing the climate change potential. Soil erosion is a major land degradation concern which drastically removes the soil organic carbon (SOC) and leads to reduced agricultural productivity and emits this carbon to the atmosphere. This directly affects the climatological processes and makes the watersheds more vulnerable to climate change. The adaptation and mitigation measures can withstand the adverse effects of watershed vulnerability and climate change to a great extent. The climate resilience is one of the prime key solutions which can resist the adverse effect occurring in a watershed scale. Along with it, the advances in technology provide more insight to these kinds of processes. The application of geospatial approaches, erosion models, various other hydrological models and machine learning help to uncover the unfathomable understanding about climate change and watershed vulnerability. The policies adopted in the watershed level also play major role while considering in large scale. Proper policies and implementation are vital in attaining sustainability. Management practices and climate-resilient policies generated with the help of various technologies will enhance the efficiency of our responses. The sustainable development goals and climate resilience will promote the watershed sustainability in a global scale. As watershed is the natural unit of managing the natural resources, watershed sustainability will provide the overall sustainability of socio-economy and ecosystem. Figure 11.2 is a simple illustration for attaining the sustainable watershed from a vulnerable and degraded watershed which predominantly suffered with climate change and land degradation.

Climate Change Impact on Watershed Vulnerability

The anthropogenic-induced climate change is growing as a serious concern among scientists, decision-makers, administrators and common people all around the globe. The increased greenhouse gas emissions, enhancing the global atmospheric temperature and with respect to these various adverse consequences, are observed worldwide during the certain last decades. Each year, the production of coal, oil and gas releases a large amount of CO_2 into the atmosphere. Human activities are releasing a record amount of carbon dioxide into the atmosphere, with no indications of slowing down. If we do not take steps to limit global emissions, temperatures could rise above three degrees Celsius by 2100. Sea levels are rising faster than ever in polar and mountainous regions due to the melting of glaciers and glacial ice sheets in these areas. Climate-related calamities and extreme weather have long been a part of life on Earth. It's just that as the world warms, catastrophic weather phenomena like heatwaves, droughts, typhoons and hurricanes are inflicting devastation on every continent. Because the effects of climate disruptions are felt most acutely at a local level, local urban/rural bodies are at the forefront of climate action and adaptation (UN n.d.) (Table 11.4).

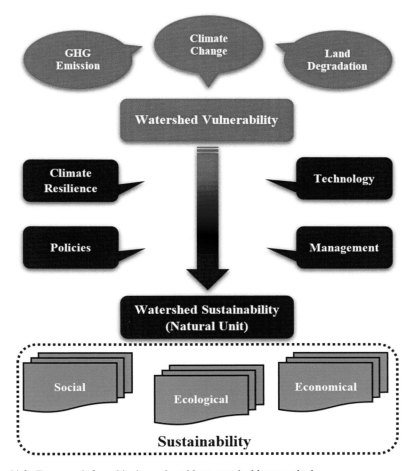

Fig. 11.2 Framework for achieving vulnerable to sustainable watershed

There is substantial year-to-year variability in precipitation, and scientists are curious whether there were any long-term changes. A global annual precipitation record (1940–2009) indicate variations in precipitation is seen in roughly 14% of the global land surface (Sun et al. 2018). The global water cycle is intensified by rising temperatures and the consequent surge in the atmosphere's water-holding capability (Trenberth et al. 2003), subsequent to increasingly intense and frequent rainfall events (Alfieri et al. 2015). The increased occurrence of extreme events is one of the key repercussions of rising global temperatures (Yaduvanshi et al. 2021). Severe weather has significant consequences to human wellbeing and economy (Robine et al. 2008) and are likely to vary consequently with climate change (Robine et al. 2008). In recent decades, a heating up tendency has been detected across South Asian (SA) regions, particularly in India, which is coherent with the global warming pattern envisaged from anthropogenic climate change (Kumar et al. 2011). Extreme

Table 11.4 Selected watershed vulnerability assessment studies around the globe

Country	Purpose	Approach	Observed/climate model	Scale of the study	References
Mexico	Vulnerability in various scales of watersheds to climate change	Indices (annual-storage to annual-surface-runoff ratio; water availability index; water use index)	GFDL-R30; CTM – EBM; CCCM	River basins/watershed	Mendoza et al. (1997)
Canada	Flood vulnerability to climate change	Mike11 NAM; hydrological modelling	CGCM2-GCM	Watershed	Huang et al. (2005)
USA (Monocacy River)	Watershed vulnerability to climate change	Integrated modelling system (BASINS-CAT)	ECHM-GCM	River basin	Imhoff et al. (2007)
Portugal	Vulnerability of water resources, vegetation productivity and soil erosion to climate change	SWAT model	Historical and GCM	Watershed	Nunes et al. (2008)
USA	Climate change on soil erosion vulnerability	Scores between 0 and 1 assigned to erodibility, slope and land cover categories for spatial analysis of vulnerable areas to mass wasting and erosion	CGCM3_GCM	Country	Segura et al. (2014)
India	Sensitivity analysis of a climate vulnerability index	Index (climate vulnerability index)	Natural disaster and impact; climate variability	Watershed	Sathyan et al. (2018)
Iran	Vulnerability of aquatic macroinvertebrates	Species distribution models (SDMs)	Observed climate data	Catchment area	Naderi (2020)
Nepal (Himalayas)	Livelihood vulnerability due to climate change	Formative composite index (livelihood vulnerability index)	Observed climate hazards and variability	Sub-watershed	Adhikari et al. (2020)
Mexico	Vulnerability of a watershed's water, environmental, economic and social resources to	Predictive vulnerability interface (MPDV 1.0) which is capable of quantifying the hydrological,	CMIP5 models-GCM	Sub-basin	Orozco et al. (2020)

(continued)

Table 11.4 (continued)

Country	Purpose	Approach	Observed/climate model	Scale of the study	References
	climate change and land use change	environmental, economic, social and global vulnerability of a hydrological basin			
Congo	Climate change and land use/land cover on domestic water vulnerability	Index (water poverty index)	Long-term observed climate data	Watershed	Chishugi et al. (2021)

BASINS-CAT U.S. Environmental Protection Agency's Better Assessment Science Integrating Point and Nonpoint Sources-Climate Assessment Tool, *ECHM-GCM* German High Performance Computing Centre for Climate and Earth System Research GCM model, *CTM-EBM* Climate thermodynamic model-Energy Balance Model, *CCCM* Canadian Climate Centre second-generation model, *GFDL-R30* Geophysical Fluid Dynamics Laboratory model, *CGCM2* Canadian Global Coupled Model Version 2, *Mike11 NAM* NAM is the abbreviation of the Danish "Nedbor-Afstromnings Model", which means precipitation-runoff model, *CGCM3* Canadian Global Coupled Model Version 3

high-temperature events have become more common in most sections of the heavily inhabited SA region (Zhou and Ren 2011). In comparison, prior to the industrialisation, a surge in severity of monsoon precipitation intensity above SA is expected to be roughly 7% as a result of 1.5 °C warming and 10% due to 2 °C warming (Kumari et al. 2019). The western coastline, innermost and Northeast India have seen the most intense precipitation occurrences (Goswami et al. 2006). According to observations, the periodicity with of severe rainfall occurrences appears to be increasing, (Gautam 2012). As a consequence of these global threats, the signatory countries to the Paris Agreement agreed to "limit global average temperature rise to well below 2 degrees Celsius above pre-industrial levels, and to pursue efforts to limit temperature rise to 1.5 degrees Celsius above pre-industrial levels, in order to reduce potential issues and threat of climate change" (UNFCCC 2015). The literature cites that as a consequence of climate change, the normal weather patterns are changing at a faster rate and may pose grave concern over various sectors. Hence, our framework for these kinds of impact-based studies is very necessary and needs to be utilised properly for the adaptation and mitigation purposes.

Climate-associated investigations, such as studies of temperature extremities, water resources, agronomy, air quality and wind power, rely heavily on fine-scale projections of the global future climate. Interpolation, statistical downscaling, dynamical downscaling and composite statistical-dynamical downscaling are all prominent methods for obtaining high-resolution projection data (Gutowski 2020; Navarro-Racines et al. 2020; Sun et al. 2015; Wilby et al. 1998). Dynamical downscaling, in contrast to approximation and statistical downscaling, may represent a broad variety of physical mechanisms in the climate system and offer a comprehensive array of continuously prolonged climate data. Conventional dynamical downscaling of the climate (future) entails combining the beginning and lateral boundary conditions from a general circulation model (GCM) with a regional climate model (RCM) (Giorgi et al. 2009). This conventional dynamical downscaling method has received a lot of attention (Tang et al. 2016). Statistical downscaling, on the other hand, cannot capture associations that do not exist in the past and therefore is not appropriate for places where there is no significant link between extensive (global) climate data and site-specific observations (local) (Maraun et al. 2010). Dynamical downscaling, on the other hand, uses a RCM associated with a global climate model (GCM) to generate fine-resolution climatic data (Giorgi and Mearns 1991). RCMs are created using physical concepts as a foundation. RCMs are evolving into regional Earth system models which are able to replicate interconnections between the atmosphere, land, water, financial sector and other factors as biogeochemical mechanisms are added (Adam et al. 2015). As a result, dynamical downscaling, as opposed to statistical downscaling, can reconcile multiple processes in climate or Earth systems, as well as their relationships. Nevertheless, dynamical downscaling is a computationally costly and laborious (time demanding) procedure, which limits its use. Mutually, the dynamical and statistical downscaling methodologies are used in the dynamical-statistical downscaling method (Han and Wei 2010). Sooryamol et al. (2022b) used MarkSim weather generator tool to predict future climate scenario and its impact on soil erosion over Lesser Himalayas.

Variations in rainfall features owing to climate change cause spatial and temporal deviation in precipitation-runoff erosivity, which has consequences for management of soil and water in undeveloped nations. In tropical places with few continuous daily rainfall records, it's critical to understand historical and future fluxes in precipitation-runoff erosivity and their implications (Amanambu et al. 2019). In many basins around the world, severe soil erosion caused by increased rainfall-runoff erosivity is a major problem (Mondal et al. 2016; Wang et al. 2017a, b). The tremendous kinetic energy of intercepting precipitation frequently detaches the topsoil elements and transports them along by way of surface runoff, and rainfall-runoff erosivity is related to the characteristic of rainfall (Yang et al. 2003). Wischmeir (1968) demonstrated this phenomenon using a rainfall-soil constituent detachment association established by Govers (1991). Because erosion is mostly dependent on precipitation, rainfall strength and duration, erosivity does not pose a straight connection with soil erosion (Salles et al. 2002). Rainfall erosivity, in combination with slope grade, represents around three by fourth of soil erosion variability, according to Doetterl et al. (2012). Due to variations s in precipitation, these factors operate as governing factors of surface runoff and soil erosion. Employing previous and predicted climatic data, the climatic component of rainfall has been used in numerous basins to discover trends in rainfall-runoff erosivity (Almagro et al. 2017; Gupta and Kumar 2017a; Panagos et al. 2017; Stefanidis et al. 2021).

Variations in rainfall patterns will affect soil erosion through numerous paths, including changes in rainfall-runoff erosivity, according to GCMs (Plangoen et al.

2013). Since erosivity rates are expected to fluctuate as a consequence of climate change, variations in erosivity have indeed been primarily linked to the impacts of soil erosion (Nearing 2001). The global escalation of land usage for agronomic reasons is connected to an increase in soil erosion over time. Hence, land use, soil management and erosion and agronomic sustainability all have a significant link (Vanwalleghem et al. 2017). Furthermore, as farmland soil is distracted, nutrient residence time is expected to decrease (Quinton et al. 2010). Over the last 8000 years, the advent of agriculture has considerably hastened soil erosion due to rainfall and runoff, mobilising roughly 783 Pg of SOC universally (Wang et al. 2017a, b). Soil erosion puts SOC stocks beneath a secure situation on eroding sites (Van Oost et al. 2007), which can result in significant CO_2 emissions in some areas (Worrall et al. 2016). Emissions of CO_2 from soil erosion can arise as a result of the eroding of soil aggregates and the possible contribution of eroded SOC by runoff and ultimately to streams and rivers (Naipal et al. 2018). Hence, it is explicit that the climate change and soil erosion processes are cyclic; one process has the capability to fuel the other processes. It indicates the proper management of soil resources which can mitigate the adverse impact of climate change.

Geospatial Modelling in Assessing Watershed Vulnerability

Geospatial modelling plays key roles in assessing watershed vulnerability due to their crucial roles in execution of various hydrological, soil erosion and water services models.

Hydrological Models

Hydrological models are simplified forms of real-world systems such as surface water, soil water, groundwater, wetland, etc., that help us to understand, predict and manage water resources much better and efficiently (Wheater et al. 2007). Hydrological models are widely used to study flow as well as quality of water, across the world. These models help in the detailed characterisation of hydrological systems and their features using statistical models, mathematical analogues, physical process-based models, computer simulations, etc. (Allaby and Allaby 1999). Majority of the models focus on water flow through various components of hydrological systems and study the influence of different parameters such as soil, climate, topography, etc., on flow intensity and discharge. The major hydrological processes (components of hydrological cycle) addressed by different models include precipitation, evaporation, transpiration, interception, infiltration, runoff, interflow, stream flow, routing, sediment/contaminant transportation, etc. Different types of classification schemes based on specific characteristics are widely being adopted for these models (Moradkhani and Sorooshian 2009). Model classification into lumped and distributed models is manly based on model input and parameters as a function of space and time, whereas classification into stochastic and deterministic categories is based on the extent of physical principles employed in the model. Models are also classified based on the time factor into static (time excluded) and dynamic (time included) models. Another most widely adopted and followed model classification is into empirical, conceptual and physical process-based models. Empirical models are mainly observation based, whereas the physical-based models provide an idealised mathematical representation of the real-world phenomenon. Numerous reviews (Sorooshian et al. 2008; Devi et al. 2015; Singh 2018) are available regarding hydrological models involving their history, classification, characteristics and their brief descriptions. Soil and water assessment tool (SWAT), MIKE SHE, HBV, TOPMODEL, variable infiltration capacity (VIC) model, etc., are some of the most widely used hydrological models (among the large number of models available) at various scales and geographical settings, nowadays.

Geospatial technology including the combined use of RS data and GIS platforms plays a key role in hydrological modelling, thus enabling us to study various stages and components of terrestrial water cycle in detail (Singh 2018). RS, the process of non-contact acquisition of data, regarding the various characteristics of Earth surface as well as atmospheric parameters nowadays provides various input data for hydrological models at different scales and high spatial as well as temporal resolutions, even in inaccessible areas. This also aids us to detect changes in land surface characteristics over a period in addition to providing a synoptic view of the area. Geospatial technology and its various components provide as well as help in the generation of spatial information regarding terrain, geology, land use/land cover, soil, geomorphology, surface water bodies, etc. Kumar (2020) presented a detailed review regarding RS satellites along with their specifications. RS-based synoptic data regarding meteorological parameters, soil properties and land parameters; information regarding various surface water bodies, snow coverage and ice fields as well as water quality parameters; etc., obtained from various space agencies have been widely used for incorporation into hydrological models (Singh 2018). Singh (2018) has also listed a number of dedicated spaceborne missions for studying various components of water cycle. A detailed review of RS applications with respect to studying terrestrial water cycle could be seen in Lakshmi et al. (2015). GIS plays an important role in storage, manipulation and analysis of various spatial as well as nonspatial data, thus enabling their integration into hydrological models and thus significantly enhancing their modelling capabilities (Mujumdar and Kumar 2012; Griffin et al. 2017).

Erosion (Including Nutrient Loss) and Watershed Services Modelling

Soil erosion, important component of hydrological processes and water cycle, represents one of the most commonly and widely occurring forms of land degradation in most of the watersheds across the world, thus playing a crucial role in defining the sustainability of watersheds. Different soil erosion processes and the associated loss of sediments/nutrients are studied in detail by making use of several erosion models, depending on the watershed characteristics as well as parameters to be estimated and their planned use. The numerous models developed and used globally could be grouped into empirical/statistical, conceptual and physical process-based models mainly based on the data requirements as well as the physical processes involved (Merritt et al. 2003). A detailed list, description and the working principles of commonly adopted erosion models have been documented by Kumar and Kalambukattu (2022). As described in relation to hydrological models in the previous paragraph, geospatial technology plays a crucial role in spatial estimation of soil erosion as well as associated nutrients too, by providing spatial data related to climatic parameters, soil, terrain, land use/land cover, vegetation, etc., which form vital inputs for most of the models. Certain models like APEX, SWAT, AGNPS, etc., could predict the associated nutrient losses also when calibrated with necessary field information as well as parameters. Integrated Valuation of Ecosystem Services and Tradeoffs (InVEST) is a model increasingly being used for mapping as well as valuation of ecosystem services provided by different terrestrial, freshwater and marine ecosystems (Sharp et al. 2018). InVEST could be run using various environmental data at diverse flexible scales (site, local, landscape, regional, national and multiscales) to study how the flow of benefits to people is being (likely to get) influenced by various ecosystem changes. The vital information generated from by the model could be used for informed decision-making regarding management and simultaneous conservation of different natural resources. RS and GS play a vital role in modelling various regulating, provisioning and cultural services using InVEST by providing vital inputs at varying spatial scales (Sharp et al. 2018).

Watershed Development and Environmental Sustainability

Soil, water and vegetation are crucial natural resources for the existence of human being. There is tremendous pressure on the natural resources to meet the need of increasing population in the world. They are causing substantial problems to agriculture sustainability, livelihood opportunity and vulnerable communities. There is utmost need to conserve soil, water and vegetation. It was realised though watershed development programmes.

FAO has initiated watershed management projects since the late 1980s. Understanding the special need and attention of watershed management programme aiming to concern fragile mountain ecosystem, FAO played a leading role in celebrating the observance of the international year of mountains (IYM) during 1998 to 2002. FAO implemented several projects throughout the world on watershed management between 1990 and 2003. The broad objectives of the projects were to collect and disseminate the information generated by implementing watershed management projects and to support and guide in development of new generation projects by integrating hydrology and ecology, human ecology and environmental economic. Now the projects mainly focus from a participatory to a collaborative approach as major shift.

India initiated wasteland/degraded land development programs in 1985. The Govt. of India (GoI) identified the development of land resources as a major instrument for rural development and poverty alleviation. Therefore, three major projects as watershed programs were launched as (i) Integrated Wasteland Development Program, (ii) Drought-Prone Areas Program and (iii) Desert Development Program. Later, these programs were brought under one umbrella and implemented as "Integrated Watershed Management Program" (IWMP) from 2008. The aim of the IWMP was "to restore the ecological balances by harnessing, conserving and developing degraded natural resources such as soil, water and vegetation cover". It primarily envisaged reduction in soil runoff, rainwater harvesting, revival of natural vegetation and restoring of the groundwater table. It helped in promoting multi-cropping, diverse agro-based activities that lead to sustainable livelihood to the people residing in the area of the watershed. In 2009, the IWMP project was merged with MGNREGS (Mahatma Gandhi National Rural Employment Guarantee Scheme) of the Ministry of Rural Development of Govt. of India to strengthen it for further support of livelihood and employment of the rural people. Since 2014, the IWMP program is being implemented as Prime Minister Krishi Sinchayee Yojana (Watershed Development Component) (WDC_PMKSY).

Climate change poses serious challenge to the livelihood of the people of Himalayan region due to its high sensitivity to the climate and fragile balance of natural resources. The Govt. of India launched the National Mission for Sustaining the Himalayan Ecosystem (NMSHE) as part of National Action Plan on Climate Change (NAPCC) towards climate change adaptation and to understand linkage between climate and Himalayan ecosystem. Watershed sustainability refers to optimal utilisation of natural resources in sustainable manner. It works to create and preserve conditions that allow humans and nature to coexist in constructive balance while also meeting the social, economic and other needs of present and future civilisations.

Watershed is the most fundamental unit of soil and water conservation. Watershed hydrology is driven by topography, soil, geology, vegetation and human activities and influenced by climatic processes. Commonly, watershed sustainability is quantified by the analysis of water resources and its quality. Health watershed supports many ecosystem services such as nutrient cycling, carbon storage, erosion/sedimentation control, flood control, water storage, water quality, biodiversity, wildlife movement, risk of invasion species, etc. Climate change may adversely affect and invite vulnerability and disruption to these services. These services are essential to social, environment and economic wellbeing. Achieving sustainability of watershed required involvement of formal and nonformal institutions and participation of people especially of local community for mutual reinforcement of the activities of watershed management.

A Case Study: Watershed Vulnerability Assessment in a Watershed of Shivalik Region, India

We carried out a case study for assessing the soil vulnerability employing geospatial approaches in the context of changing climate at watershed scale. We are approaching the watershed vulnerability assessment related to some selected soil characteristics of the area of interest. The vulnerability assessment is performed based on runoff potential of the selected watershed. A brief description of experimental site, methodology and salient findings is delivered below.

Experimental Site

The experimental watershed is located in Hamirpur district of Himachal Pradesh in India. The watershed covers an area of 500 ha which represents Northwest Shiwalik Himalayas. In Hamirpur district, two principal geological horizons are found, namely, tertiary and post-tertiary formations. The elevation of the study area ranges from approximately 500 to 800 m above mean sea level. The climate belongs to humid subtropical and receives an annual average rainfall of 1342 mm. The general drainage pattern of the streams in the study region is dendritic. The soil in Hamirpur is mainly non-calcic brown soils, and soil in most regions is neutral to acidic. The main soil texture is identified as sandy loam. The forest type is dry mixed deciduous comprising of predominantly khair (*Senegalia catechu*) and chir pine (*Pinus roxburghii*). The cropland primarily consists of maize, wheat, mustard and vegetable crops. The agriculture and livestock are the major income source of natives. Hence, the maintaining soil condition/health is of utmost importance.

Methodology

The USDA Natural Resources Conservation Service (NRCS) created the soil vulnerability index (SVI) to evaluate cropland's intrinsic vulnerability to runoff and leaching (Lohani et al. 2020). For runoff leaching potential, they used the amount of SOC, hydrologic soil group (HSG), K-factor and slope, while for runoff potential, they used HSG, K-factor and slope. Apart from this, we are incorporating rainfall factor to identify the change in soil vulnerability (runoff) according to the climate change. Experts working in the same experimental watershed identified the parameters and thresholds. Hence, we consider these five parameters as the function of runoff potential. Thus, the SVI is estimated based on organic matter present in the soil, HSG and its characteristics, soil erodibility factor (K), slope factor and rainfall factor. The soil data required for the study is obtained from the Indian Institute of Remote Sensing-Soil database. The soil landscape unit prepared by David Raj

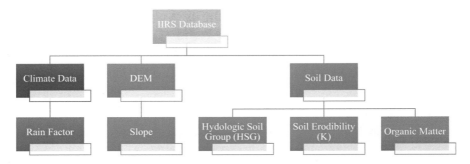

Fig. 11.3 Generation of SVI

(2020) was considered as the basic unit of watershed vulnerability assessment. The rain factor was calculated by the ratio of 30-year average rainfall for the period of 2071–2100 based on A2 scenario and baseline (1986–2015) average rainfall (Fig. 11.3). The overall methodology adopted for the study is illustrated in Fig. 11.3.

Each indicator is classified into four categorical values of low means less sensitivity and high with higher sensitivity to soil vulnerability. The value of each indicator is split into four categories based on the available indicator value distribution. This categorical classification is representing the site-specific soil quality, climate and terrain indicator. Then ranked the vulnerability category from 1 (low) to 4 (high) and average of the total indicator rank is used to determine the overall soil vulnerability. Identifying the climate change impact, we assumed the current climate condition (rainfall) as minimum vulnerability (low vulnerability) and average future (2071–2099) rainfall based on A2 scenario as maximum vulnerability (high sensitivity). We also assumed that the combined effect of soil quality indicators (HSG, K-factor, OM, slope) has 50% weightage, while future rainfall has 50% weightage for identifying climate change effect. Because we identified that the future rainfall increases up to a 1.42-fold higher, hence, it will be more vulnerable to the soil. Then, we generated the future soil vulnerability value according to the increased precipitation (climate change). The SVI is a category index with four classifications for soil vulnerability, depending on particular criteria for the soil and landscape factors given above: low, moderate, moderately high and high (Table 11.5). The final SVI map was created using a fusion of all five factors.

Discussion and Inferences

The mixed dense and open forest is the major land use/land cover of the experimental watershed (Fig. 11.4). Predominant geographical area is covering very steep slopes with more than 35° (Fig. 11.4). It indicates the complexity of the terrain. Soils of the Hamirpur experimental watershed were principally in hydrologic soil groups

11 Geospatial Approach in Watershed Vulnerability to Climate Change...

Table 11.5 Criteria for generating four classes of SVI

Sensitivity to vulnerability (rank)	HSG	K-factor	Organic matter (%)	Slope (degree)	Rain factor
Low (1)	A	0.08	>3.5	0–10	Current status is assumed as low
Moderate (2)	B	0.09	2.6–3.5	11–20	–
Moderately high (3)	C	0.10	1.1–2.5	21–35	–
High (4)	D	0.11	<1.0	>35	Future scenario is assumed as high

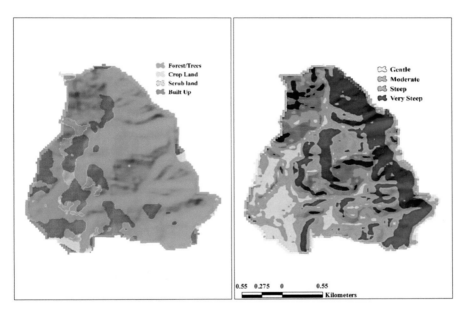

Fig. 11.4 Land use/land cover (ESRI land cover map) and slope map (Cartosat DEM)

B and C, which was categorised by moderate runoff potential and high infiltration rate (Fig. 11.5). The sandy loam soil texture decreased the soil erodibility factor.

The sandy soils are usually coarser; thus, it has low erodibility compared to other types. The organic matter of the soil ranges from 0.30% to 3.6% with higher value observed from the forest land and agricultural land while lower from the scrub land. Most of the micro-watersheds consisted mostly of areas in the higher slope categories for soil vulnerability (Fig. 11.6). Most of the soils had a soil erodibility K-factor value within 0.08–0.11. Based on SVI, the steep slope had dominant contributor of (>29.4%) highly-vulnerable areas followed by moderate (29.9%) and moderately high vulnerable area contributing 39.4% (Fig. 11.5, Table 11.5). Also, the steep slope in open forest is also included in the highly vulnerable area.

However, these high-vulnerability portions were generally in scrub land and steep slopes situated in open forest. Low-risk class for soil was in the scrub-dominated

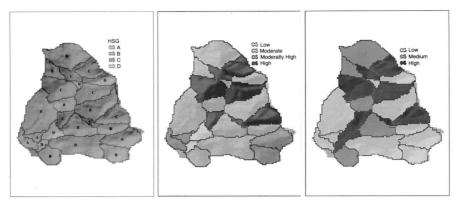

Fig. 11.5 Spatial distribution of hydrologic soil group, organic matter map and soil erodibility

Table 11.6 SVI dispersion in the experimental watershed

Vulnerability category	Current status Area (%)	Future status (2071–2099); HadCM3 A2 scenario
Low	1.3	–
Moderate	29.9	–
Moderately high	39.4	63.2
High	29.4	36.8

watershed (Fig. 11.5, Table 11.6) because of the lower slope. Hence, we can say that the slope is playing a major role in determining the soil vulnerability in the experimental watershed. The SVI was employed in this study to determine intrinsically vulnerable locations in the Shiwaliks (Hamirpur watershed). The Hamirpur watershed includes 24 micro-watersheds; most of them are characterised by sandy loam soils. The areas with the highest risk of SVI were those with a steeper slope. While the presence of dense mixed forest overcome the adverse effect erosion due to slope with a greater extent. Most of the area is covering the vulnerability class of moderate- to high-risk areas in the experimental watersheds (Fig. 11.6).

The amount of rainfall during the base period is 1342 mm, while the IPCC SRES (Special Report Emission Scenario) A2 states that it may increase to 1897 mm during 2071–2100. While assessing the climate change impact on soil vulnerability, in the future, according to the A2 scenario, the rain factor may increase to 1.41 times of the current status. As a consequence of climate change, most of the soil vulnerability classes of the watershed shifted to moderately high (63.2%) and high (36.8%). Henceforth, it is explicit that the vulnerability on soil due to climate change may increase under the projections. The vulnerability of steep slope area is increasing from base period to future period. The relatively gentle slope areas are also witnessing the vulnerability category shift from moderate to moderately high. Hence, it indicates the importance of adopting proper soil and water conservation measures at watershed scale. The watershed resilience to climate change is one vital

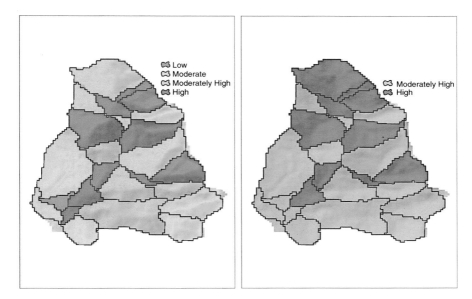

Fig. 11.6 Spatial distribution of soil vulnerability map – present (left) and future (right)

aspect which can withstand adverse effect of global warming consequences and provides sustainable societies/habitat for human

beings as well as other organisms. The limitation of this study is it doesn't consider the effect of vegetation factor in soil vulnerability and uses simple weighted and additive average of the categorical values. Thus, there may be some uncertainties in the study outcomes. Although, it is evident that climate change affects the watershed processes and functions and eventually leads to increased vulnerability. While various measures and policies can withstand it to a certain extent.

Conclusion

Watershed is considered as the natural unit for resource management, drainage line developmental activities and other hydrological processes. The watershed is the basic unit of hydrological cycle and processes occurring on the land surface. Land or soil, water and forest are the three prime constituents of the watershed. It provides various goods and services to human beings and other organisms in a watershed. The quality and quantity of goods and services provided by watershed depend on the watershed condition or watershed health. The functioning of a watershed and its components rely on the suitability of performing its functioning with respect to the environment/habitat. The perturbations in the external factors may adversely affect the optimum functioning, while at the same time the internal factors try to cope with the external disturbances which is termed as adaptive capacity. These adverse effects make the watersheds more vulnerable. The vulnerability is primarily associated with resource values (processes or functions of watershed), exposure to the threat (climate change), sensitivity to the threat and its adaptive capacity.

The warming of Earth's atmosphere creates chaos around the globe. The warming tendency has potential to make the hydrological cycle more vigorous. Hence, the watershed is one basic unit which is directly and indirectly threatened by vigorous hydrology and temperature change. Various watershed processes and functioning, including surface runoff, infiltration, saturation, natural soil erosion, carbon sequestration and habitat for organisms as well as human beings, are all affected adversely. It makes watershed more vulnerable due to climate change. The proper vulnerability assessment, monitoring and identification (indicator and modelling oriented) can help the watershed to recover by increasing adaptive capacity. Hence, the geospatial technology and watershed modelling (erosion, hydrological and carbon dynamics), with the proper amalgamation of national and international policies and frameworks (SDG, nature-based solutions, watershed management practice), can increase the resilience of watershed. The adaptation and mitigation measures acquired in this approach will help us to develop a climate resilience which can provide a healthy watershed. This will enhance the goods and services provided by the watershed.

We performed a case study in highly erosion vulnerable Shiwalik Himalayas to determine the soil vulnerability to climate change employing selected indicators. The soil quality parameters, terrain parameters and climate parameters were used as the indicator for assessment. Simple categorical method was utilised to identify the most vulnerable soil area with respect to climate change. We identified the sloping terrain along with increased rainfall as consequences of climate change that drive more vulnerability to the soils in the hilly and mountainous watershed. It may further destroy the soil quality (organic carbon and nutrients) in the watershed and eventually leads to extended land degradation from the watershed. It has the potential to reduce the productivity of highland agriculture. Hence, the timely monitoring and vulnerability assessment enable the government and policymakers to suggest proper action prior to the worsened watershed condition.

References

Abatzoglou JT, Ficklin DL (2017) Climatic and physiographic controls of spatial variability in surface water balance over the contiguous United States using the Budyko relationship. Water Resources Research 53(9):7630–7643

Adam JC, Stephens JC, Chung SH, Brady MP, Evans RD, Kruger CE et al (2015) BioEarth: envisioning and developing a new regional earth system model to inform natural and agricultural resource management. Climatic Change 129(3):555–571

Adhikari S, Dhungana N, Upadhaya S (2020) Watershed communities' livelihood vulnerability to climate change in the Himalayas. Climatic Change 162(3):1307–1321

Aditya S, William FR (2011) Developing a framework to measure watershed sustainability by using hydrological/water quality model. J Water Resour Protect

Ahn SR, Kim SJ (2019) Assessment of watershed health, vulnerability and resilience for determining protection and restoration Priorities. Environmental Modelling & Software 122:103926

Aleksandrova M, Lamers JPA, Martius C, Tischbein B (2014) Rural vulnerability to environmental change in the irrigated low lands of central asia and options for policy-makers: a review. Environ Sci. Policy. 41:77–88. https://doi.org/10.1016/j.envsci.2014.03.001

Alfieri L, Burek P, Feyen L, Forzieri G (2015) Global warming increases the frequency of river floods in Europe. Hydrol. Earth Syst. Sci. 19:2247–2260. https://doi.org/10.5194/hess-19-2247-2015

Allaby A, Allaby M (1999) Köppen climate classification. A dictionary of earth sciences.

Almagro A, Oliveira PTS, Nearing MA, Hagemann S (2017) Projected climate change impacts in rainfall erosivity over Brazil. Scientific reports 7(1):1–12

Amanambu AC, Li L, Egbinola CN, Obarein OA, Mupenzi C, Chen D (2019) Spatio-temporal variation in rainfall-runoff erosivity due to climate change in the Lower Niger Basin, West Africa. Catena 172:324–334

Baloch MA, Sahar L (2014) Development of a watershed-based geospatial groundwater specific vulnerability assessment tool. Groundwater 52(S1):137–147

Beier CM, Patterson TM, Chapin FS (2008) Ecosystem services and emergent vulnerability in managed ecosystems: a geospatial decision-support tool. Ecosystems 11(6):923–938

Besaw LE, Rizzo DM, Kline M, Underwood KL, Doris JJ, Morrissey LA, Pelletier K (2009) Stream classification using hierarchical artificial neural networks: a fluvial hazard management tool. Journal of hydrology 373(1–2):34–43

Bhattacharjee S, Kumar P, Thakur PK, Gupta K (2021) Hydrodynamic modelling and vulnerability analysis to assess flood risk in a dense Indian city using geospatial techniques. Natural Hazards 105(2):2117–2145

Brooks NK, Folliot PF, Thames JL (1991) Watershed management: a global perspective, hydrology and the management of watersheds. Iowa State University Press, Ames, pp 1–7

Cabello Villarejo V, Willaarts BA, Aguilar Alba M, Moral Ituarte LD (2015) River basins as socio-ecological systems: Linking levels of societal and ecosystem metabolism in a Mediterranean watershed. Ecology and Society 20(3)

Chatterjee S, Krishna AP, Sharma AP (2014) Geospatial assessment of soil erosion vulnerability at watershed level in some sections of the Upper Subarnarekha river basin, Jharkhand, India. Environmental earth sciences 71(1):357–374

Chaves HM, Alipaz S (2007) An integrated indicator based on basin hydrology, environment, life, and policy: the watershed sustainability index. Water Resources Management 21(5):883–895

Chishugi DU, Sonwa DJ, Kahindo JM, Itunda D, Chishugi JB, Félix FL, Sahani M (2021) How climate change and land use/land cover change affect domestic water vulnerability in Yangambi Watersheds (DR Congo). Land 10(2):165

Copeland HE, Tessman SA, Girvetz EH, Roberts L, Enquist C, Orabona A et al (2010) A geospatial assessment on the distribution, condition, and vulnerability of Wyoming's wetlands. Ecological Indicators 10(4):869–879

Creamer RE, Brennan F, Fenton O, Healy MG, Lalor STJ, Lanigan GJ, Regan JT, Griffiths BS (2010) Implications of the proposed soil framework directive on agricultural systems in Atlantic Europe-a review. Soil Use and Management 26:198–211

Cutter SL (2003) The vulnerability of science and the science of vulnerability. Ann Assoc Am Geogr 93(1):1–12

David Raj A (2020) Modelling climate change impact on surface runoff and sediment yield in a watershed of Shivalik region. B.Sc.-M.Sc. (Integrated) thesis. Kerala Agricultural University, Thrissur

David Raj A, Kumar S, Regina M, Sooryamol KR, Singh AK (2021) Calibrating APEX model for predicting surface runoff and sediment loss in a watershed – a case study in Shiva-lik region of India. International Journal of Hydrology Science and Technology. https://doi.org/10.1504/IJHST.2021.10041820

David Raj A, Kumar S, Sooryamol KR (2022) Modelling climate change impact on soil loss and erosion vulnerability in a watershed of Shiwalik Himalayas. Catena 214:106279

Devi GK, Ganasri BP, Dwarakish GS (2015) A review on hydrological models. Aquatic Proc 4: 1001–1007. ISSN 2214-241X. https://doi.org/10.1016/j.aqpro.2015.02.126

Doetterl S, Van Oost K, Six J (2012) Towards constraining the magnitude of global agricultural sediment and soil organic carbon fluxes. Earth Surface Processes and Landforms 37(6):642–655

Eakin H, Luers AL (2006) Assessing the vulnerability of social-environmental systems. Annu. Rev. Environ. Resour. 31:365–394

Elsner MM, Gangopadhyay S, Pruitt T, Brekke LD, Mizukami N, Clark MP (2014) How does the choice of distributed meteorological data affect hydrologic model calibration and streamflow simulations? Journal of Hydrometeorology 15(4):1384–1403

Engle NL (2011) Adaptive capacity and its assessment. Glob Environ Chang. 21(2):647–656. https://doi.org/10.1016/j.gloenvcha.2011.01.019

EPA (2012) Fact sheet on 'The Economic Benefits of Protecting Healthy Watersheds'. https://www.epa.gov/sites/default/files/2015-10/documents/economic_benefits_factsheet3.pdf. Accessed 27 Jan 2022

Fore LS (2003) Developing biological indicators: lessons learned from mid-atlantic streams. Prepared for: U.S. Environmental Protection Agency, Environmental Science Center, Ft. Meade, MD. EPA/903/R-03/003

Furniss MJ, Staab BP, Hazelhurst S, Clifton CF, Roby KB, Ilhadrt BL, Larry EB, Todd AH, Reid LM, Hines SJ, Bennett KA, Luce CH, Edwards PJ (2010) Water, climate change, and forests: watershed stewardship for a changing climate, General technical report PNW-GTR-812. U.-S. Department of Agriculture, Forest Service, Pacific Northwest Research Station, Portland. 75 p

Furniss MJ, Roby KB, Cenderelli D, Chatel J, Clifton CF, Clingenpeel A et al (2013) Assessing the vulnerability of watersheds to climate change: results of national forest watershed vulnerability pilot assessments, General technical report PNW-GTR-884. US Department of Agriculture, Forest Service, Pacific Northwest Research Station, Portland. 32 p. plus appendix., 884

Gautam PK (2012) Climate change and conflict in South Asia. Strategic Analysis 36(1):32–40

Giorgi F, Mearns LO (1991) Approaches to the simulation of regional climate change: a review. Reviews of geophysics 29(2):191–216

Giorgi F, Jones G, Asrar GR (2009) Addressing climate information needs at the regional level: the CORDEX framework. WMO Bull 58(3):175–183

Glick P, Stein BA, Edelson NA (eds) (2011) Scanning the conservation horizon: a guide to climate change vulnerability assessment. National Wildlife Federation, Washington, DC

Goswami BN, Venugopal V, Sengupta D, Madhusoodanan MS, Xavier PK (2006) Increasing trend of extreme rain events over India in a warming environment. Science 314(5804):1442–1445

Govers G (1991) Spatial and temporal variations in splash detachment: a field study. Loess, Geomorphological Hazards and Processes. Catena Supplement 20:15–24

Greene EA, LaMotte AE, Cullinan KA (2005) Ground-water vulnerability to nitrate contamination at multiple thresholds in the Mid-Atlantic Region using spatial probability models. US Geological Survey Scientific Investigations Report 2004-5118, 24 p

Griffin RE, Cruise JF, Ellenburg WL, Al-Hamdan M, Handyside C (2017) Geographical information systems. In: Singh VP (ed) Handbook of applied hydrology, Chap 9, pp 9-1–9-6

Gupta S, Kumar S (2017a) Simulating climate change impact on soil erosion using RUSLE model – a case study in a watershed of mid-Himalayan landscape. Journal of Earth System Science 126(3):43

Gupta S, Kumar S (2017b) Simulating climate change impact on soil carbon sequestration in agroecosystem of mid-Himalayan landscape using CENTURY model. Environmental Earth Sciences 76(11):1–15

Gutowski WJ (2020) The ongoing need for high-resolution regional climate models: process understanding and stakeholder information. Bull. Amer. Meteorol. Soc. 101(5):E664–E683. https://doi.org/10.1175/BAMS-D-19-0113.1

Hahn MB, Riederer AM, Foster SO (2009) The Livelihood Vulnerability Index: a pragmatic approach to assessing risks from climate variability and change – a case study in Mozambique. Global environmental change 19(1):74–88

Han X, Wei FY (2010) The influence of vertical atmospheric circulation pattern over East Asia on summer precipitation in the east of China and its forecasting test (in Chinese). Chin J Atmosp Sci 34:533–547

Hansen J, Kharecha P, Sato M, Masson-Delmotte V, Ackerman F et al (2013) Assessing "dangerous climate change": required reduction of carbon emissions to protect young people, future generations and nature. PLOS ONE 8(12):e81648. https://doi.org/10.1371/journal.pone.0081648

Hazbavi Z, Baartman JE, Nunes JP, Keesstra SD, Sadeghi SH (2018a) Changeability of reliability, resilience and vulnerability indicators with respect to drought patterns. Ecological Indicators 87: 196–208

Hazbavi Z, Keesstra SD, Nunes JP, Baartman JE, Gholamalifard M, Sadeghi SH (2018b) Health comparative comprehensive assessment of watersheds with different climates. Ecological Indicators 93:781–790

Hendriks MR (1993) Effects of lithology and land use on storm run-off in east Luxembourg. Hydrological processes 7(2):213–226

Hinkel J (2011) "Indicators of vulnerability and adaptive capacity": towards a clarification of the science–policy interface. Global environmental change 21(1):198–208

Huang Y, Zou Y, Huang G, Maqsood I, Chakma A (2005) Flood vulnerability to climate change through hydrological modeling: a case study of the Swift Current Creek watershed in western Canada. Water International 30(1):31–39

Hurd B, Leary N, Jones R, Smith J (1999) Relative regional vulnerability of water resources to climate change 1. JAWRA Journal of the American Water Resources Association 35(6): 1399–1409

Imhoff JC, Kittle JL Jr, Gray MR, Johnson TE (2007) Using the climate assessment tool (CAT) in US EPA BASINS integrated modeling system to assess watershed vulnerability to climate change. Water Science and Technology 56(8):49–56

Integrated Watershed Development Project, World Bank (n.d.). Available at https://projects.worldbank.org/en/projects-operations/project-detail/P041264. Accessed 3 Feb 2022

Intergovernmental Panel on Climate Change (IPCC) (2007) Parry ML, Canziani OF, Palutikof JP, van der Linden PJ, Hanson EC (eds). Climate change 2007: impacts, adaptation, and vulnerability. Contribution of Working Group II to the Fourth Assessment Report of the Intergovernmental Panel on Climate Change. Cambridge University Press, Cambridge. 976 p

Intergovernmental Panel on Climate Change (IPCC) (2014) Climate change 2013: the physical science basis. Contribution of Working Group I to the Fifth Assessment Report of the Intergovernmental Panel on Climate Change. Stocker TF et al (eds). Cambridge University Press, Cambridge

Ives JD, Messerli B (1989) The Himalayan dilemma: reconciling development and conservation. The United Nations University, Routledge/London

Ives J, Messerli B, Spiess E (2000) Mountains of the world: a global priority. Land degradation and development. Parthenon Publishing Group, Carnforth, pp 197–203

Kang SZ, Zhang L, Song XY, Zhang SH, Liu XZ, Liang YL, Zheng SQ (2001) Runoff and sediment loss response to rainfall and land use in two agricultural catchments on the Loess Plateau of China. Hydrological Processes 15:977–988

Kin-Che L, Leung YF, Yao Q (1997) Nutrient fluxes in the Shenchong Basin, Deqing County, South China. Catena 29:191–210

Knox A, Gupta S (2000) Collective action PRI technical workshop on watershed management institutions: a summary paper: collective action PRI working paper No.8. IFPRI, Washington, DC. 215p

Krishnamurthy PK, Lewis K, Choularton RJ (2014) A methodological framework for rapidly assessing the impacts of climate risk on national-level food security through a vulnerability index. Global Environmental Change 25:121–132

Kumar S (2020) Geospatial applications in modeling climate change impact on soil erosion. In: Venkatramanan V, Shah S, Prasad R (eds) Global climate change: resilient and smart agriculture. Springer, Singapore. https://doi.org/10.1007/978-981-32-9856-9_12

Kumar S, Kalambukattu JG (2022) Modeling and monitoring soil erosion by water using remote sensing satellite data and GIS. In: Bhunia GS, Chatterjee U, Lalmalsawmzauva KC, Shit PK (eds) Anthropogeomorphology: a geospatial technology based approach. Springer Nature, Cham

Kumar KK, Kamala K, Rajagopalan B, Hoerling MP, Eischeid JK, Patwardhan SK et al (2011) The once and future pulse of Indian monsoonal climate. Climate Dynamics 36(11-12):2159–2170

Kumari S, Haustein K, Javid H, Burton C, Allen MR, Paltan H et al (2019) Return period of extreme rainfall substantially decreases under 1.5 C and 2.0 C warming: a case study for Uttarakhand, India. Environmental Research Letters 14(4):044033

Kunimatsu T (1986) Management and runoff of nutrients from farming land, water management technology. 27:713–720

Lakshmi V, Alsdorf D, Anderson M, Nianmaria S, Cosh M, Entin J, Hufman GJ, Kustas W, van Oevelen P, Painter TH, Parajka J, Rodell M, Rudiger C (eds) (2015) Remote sensing of the terrestrial water cycle. Geophysical monograph 206. American Geophysical Union/Wiley, Hoboken, p 556

LaMotte AE, Greene EA (2007) Spatial analysis of land use and shallow groundwater vulnerability in the watershed adjacent to Assateague Island National Seashore, Maryland and Virginia, USA. Environmental Geology 52(7):1413–1421

Liao H, Sarver E, Krometis LAH (2018) Interactive effects of water quality, physical habitat, and watershed anthropogenic activities on stream ecosystem health. Water research 130:69–78

Lohani S, Baffaut C, Thompson AL, Sadler EJ (2020) Soil Vulnerability Index assessment as a tool to explain annual constituent loads in a nested watershed. Journal of Soil and Water Conservation 75(1):42–52

Maraun D, Wetterhall F, Ireson AM, Chandler RE, Kendon EJ, Widmann M et al (2010) Precipitation downscaling under climate change: recent developments to bridge the gap between dynamical models and the end user. Reviews of Geophysics 48(3)

Mendoza VM, Villanueva EE, Adem J (1997) Vulnerability of basins and watersheds in Mexico to global climate change. Climate Research 9(1-2):139–145

Merritt WS, Letcher RA, Jakeman AJ (2003) A review of erosion and sediment transport models. Environmental modelling & software 18(8-9):761–799. https://doi.org/10.1016/S1364-8152(03)00078-1

Millennium Ecosystem Assessment (MEA) (2005) Ecosystems and human well-being: synthesis. Island Press, Washington, DC, p 155

Mititelu-Ionuş O (2017) Watershed sustainability index development and application: case study of the Motru River in Romania. Polish Journal of Environmental Studies 26(5)

Mondal A, Khare D, Kundu S (2016) Change in rainfall erosivity in the past and future due to climate change in the central part of India. International soil and water conservation research 4(3):186–194

Moradkhani H, Sorooshian S (2009) General review of rainfall-runoff modeling: model calibration, data assimilation, and uncertainty analysis. In: Sorooshian S, Hsu KL, Coppola E, Tomassetti B, Verdecchia M, Visconti G (eds) Hydrological modelling and the water cycle. Water science and technology library, vol 63. Springer, Berlin/Heidelberg. https://doi.org/10.1007/978-3-540-77843-1_1

Mujumdar PP, Nagesh Kumar D (2012) Floods in a changing climate. Cambridge University Press, New York

Naderi M (2020) Assessment of water security under climate change for the large watershed of Doroodzan Dam in southern Iran. Hydrogeology Journal 28(5):1553–1574

Naipal V, Ciais P, Wang Y, Lauerwald R, Guenet B, Oost KV (2018) Global soil organic carbon removal by water erosion under climate change and land use change during AD 1850–2005. Biogeosciences 15(14):4459–4480

National Environmental Policy Act of 1969. Public Law 91–190. Available at http://www.epw.senate.gov/nepa69.pdf. Accessed 9 Feb 2022

National Research Council. Sustainability and The U.S. EPA (2011) Washington, DC: The National Academies Press.

Navarro-Racines C, Tarapues J, Thornton P, Jarvis A, Ramirez-Villegas J (2020) High-resolution and bias-corrected CMIP5 projections for climate change impact assessments. Sci. Data 7:7. https://doi.org/10.1038/s41597-019-0343-8

Nearing MA (2001) Potential changes in rainfall erosivity in the US with climate change during the 21st century. Journal of Soil and Water Conservation 56(3):229–232

Nelitz M, Boardley S, Smith R (2013) Tools for climate change vulnerability assessments for watersheds. Prepared by ESSA Technologies Ltd. for the Canadian Council of Ministers of the Environment

Nunes JP, Seixas J, Pacheco NR (2008) Vulnerability of water resources, vegetation productivity and soil erosion to climate change in Mediterranean watersheds. Hydrological Processes: An International Journal 22(16):3115–3134

Orozco I, Martínez A, Ortega V (2020) Assessment of the water, environmental, economic and social vulnerability of a watershed to the potential effects of climate change and land use change. Water 12(6):1682

Panagos P, Ballabio C, Meusburger K, Spinoni J, Alewell C, Borrelli P (2017) Towards estimates of future rainfall erosivity in Europe based on REDES and WorldClim datasets. Journal of Hydrology 548:251–262

Pérez-Maqueo O, Martinez ML, Vázquez G, Equihua M (2013) Using four capitals to assess watershed sustainability. Environmental management 51(3):679–693

Plangoen P, Babel MS, Clemente RS, Shrestha S, Tripathi NK (2013) Simulating the impact of future land use and climate change on soil erosion and deposition in the Mae Nam Nan sub-catchment, Thailand. Sustainability 5(8):3244–3274

Quinton JN, Govers G, Van Oost K, Bardgett RD (2010) The impact of agricultural soil erosion on biogeochemical cycling. Nature Geoscience 3(5):311–314

Raghavan Sathyan A, Funk C, Aenis T, Winker P, Breuer L (2018) Sensitivity analysis of a climate vulnerability index-a case study from Indian watershed development programmes. Climate Change Responses 5(1):1–14

Robine JM, Cheung SLK, Le Roy S, Van Oyen H, Griffiths C, Michel JP, Herrmann FR (2008) Death toll exceeded 70,000 in Europe during the summer of 2003. Comptes rendus biologies 331(2):171–178

Salles C, Poesen J, Sempere-Torres D (2002) Kinetic energy of rain and its functional relationship with intensity. Journal of Hydrology 257(1-4):256–270

Schultz MT (2001) A critique of EPA's index of watershed indicators. Journal of environmental management 62(4):429–442

Segura C, Sun G, McNulty S, Zhang Y (2014) Potential impacts of climate change on soil erosion vulnerability across the conterminous United States. Journal of Soil and Water Conservation 69(2):171–181

Sharp R, Tallis HT, Ricketts T, Guerry AD, Wood SA, Chaplin-Kramer R, Nelson E, Ennaanay D, Wolny S, Olwero N, Vigerstol K, Pennington D, Mendoza G, Aukema J, Foster J, Forrest J, Cameron D, Arkema K, Lonsdorf E, Kennedy C, Verutes G, Kim CK, Guannel G, Papenfus M, Toft J, Marsik M, Bernhardt J, Griffin R, Glowinski K, Chaumont N, Perelman A, Lacayo M, Mandle L, Hamel P, Vogl AL, Rogers L, Bierbower W, Denu D, Douglass J (2018) InVEST 3.5.0. User's Guide. 2018. The Natural Capital Project, Stanford University, University of Minnesota, The Nature Conservancy, and World Wildlife Fund

Sheng TC (1990) Watershed management field manual. Watershed survey and planning. In: FAO conservation guide. Food and Agriculture Organization of the United Nations, Rome

Shukla R, Sachdeva K, Joshi PK (2016) Inherent vulnerability of agricultural communities in himalaya: a village-level hotspot analysis in the Uttarakhand state of India. Appl Geography 74: 182–198. https://doi.org/10.1016/j.apgeog.2016.07.013

Singh VP (2018) Hydrologic modeling progress and future directions. Geosci. Lett. 5:15. https://doi.org/10.1186/s40562-018-0113-z

Singh R, Umrao RK, Singh TN (2014) Stability evaluation of road-cut slopes in the Lesser Himalaya of Uttarakhand, India: conventional and numerical approaches. B. Eng. Geol. Environ. 73:845–857. https://doi.org/10.1007/s10064-013-0532-1

Singh AK, Kumar S, Naithani S (2021) Modelling runoff and sediment yield using GeoWEPP: a study in a watershed of lesser Himalayan landscape, India. Modeling Earth Systems and Environment 7(3):2089–2100

Smit B, Pilifosova O (2001) Adaptation to climate change in the context of sustainable development and equity. Contribution of the Working Group to the Third Assessment Report of the Intergovernmental Panel on Climate Change, Cambridge University Press, Cambridge, pp 879–912

Sooryamol KR, Kumar S, Regina M, David Raj A (2022a) Modelling climate change impact on soil erosion in a watershed of north-western lesser Himalayan region. Journal of Sedimentary Environments. https://doi.org/10.1007/s43217-022-00089-4

Sooryamol KR, Kumar S, Regina M, David Raj A (2022b) Modelling climate change impact on soil erosion in a watershed of north-western lesser Himalayan region. J Sediment Environ 7(2):125–146

Sorooshian S, Hsu K-L, Coppola E, Tomassetti B, Verdecchia M, Visconti G (2008) Hydrological modelling and the water cycle – coupling the atmospheric and hydrological models. Springer, Berlin/Heidelberg, . ISBN 978-3-540-77843-1, ISSN: 0921-092X. https://doi.org/10.1007/978-3-540-77843-1

Stefanidis S, Alexandridis V, Chatzichristaki C, Stefanidis P (2021) Assessing soil loss by water erosion in a typical mediterranean ecosystem of northern Greece under current and future rainfall erosivity. Water 13(15):2002

Sun F, Walton D, Hall A (2015) A hybrid dynamical-statistical downscaling technique, part II: end-of-century warming projections predict a new climate state in the Los Angeles region. J. Climate 28:4618–4636

Sun F, Roderick ML, Farquhar GD (2018) Rainfall statistics, stationarity, and climate change. Proceedings of the National Academy of Sciences 115(10):2305–2310

Swansburg E, El-Jabi N, Caissie D (2004) Climate change in New Brunswick (Canada): statistical downscaling of local temperature, precipitation, and river discharge. Can. Tech. Rep. Fish. Aquat. Sci. 2544:42

Tang J, Li Q, Wang S, Lee DK, Hui P, Niu X et al (2016) Building Asian climate change scenario by multi-regional climate models ensemble. Part I: surface air temperature. International Journal of Climatology 36(13):4241–4252

Tapia C, Abajo B, Feliu E, Mendizabal M, Martinez JA, Fernández JG et al (2017) Profiling urban vulnerabilities to climate change: an indicator-based vulnerability assessment for European cities. Ecological indicators 78:142–155

Tate E (2012) Social vulnerability indices: A comparative assessment using uncertainty and sensitivity analysis. Nat Hazards. 63(2):325–347. https://doi.org/10.1007/s11069-012-0152-2

Tessema KB, Haile AT, Nakawuka P (2021) Vulnerability of community to climate stress: an indicator-based investigation of Upper Gana watershed in Omo Gibe basin in Ethiopia. International Journal of Disaster Risk Reduction 63:102426

The Climate Crisis – A Race We Can Win, United Nations. Available online at https://www.un.org/en/un75/climate-crisis-race-we-can-win. Accessed 29 Jan 2022

Tiburan C Jr, Saizen I, Mizuno K, Kobayashi S (2010) Development and application of a geospatial-based environmental vulnerability index for watersheds to climate change in the Philippines. In: Asia and the Pacific Symposium, vulnerability assessments to natural and anthropogenic hazards. Manila, Philippines, 30 pp

Tiburan C Jr, Saizen I, Kobayashi S (2013) Geospatial-based vulnerability assessment of an urban watershed. Procedia Environmental Sciences 17:263–269

Trenberth KE, Dai A, Rasmussen RM, Parsons DB (2003) The changing character of precipitation. Bull. Amer. Meteor. Soc. 84:1205–1217. https://doi.org/10.1175/BAMS-84-9-1205

U.S. Environmental Protection Agency (US EPA) (2008) Climate change effects on stream and river biological indicators: a preliminary analysis. EPA/600/R-07/085. Global Change Research Program, National Center for Environmental Assessment, Washington, DC. Available online: http://cfpub.epa.gov/ncea/cfm/recordisplay.cfm?deid=190304. Accessed 28 Jan 2022

U.S. Environmental Protection Agency (US EPA) (n.d.) Developing a Watershed Vulnerability Index. https://www.epa.gov/hwp/developing-watershed-vulnerability-index. Accessed 28 Jan 2022

U.S. EPA (2012) Identifying and protecting healthy watersheds: concepts, assessments, and management approaches. EPA 841-B-11-002. United States Environmental Protection Agency, Washington, DC

UNDP (1994) Informe sobre el Desarrollo Humano. Programa de la Naciones Unidas para el Desarrollo y Fondo Cultura Economica FCE, México

UNFCCC, The Paris Agreement (2015). Available at https://unfccc.int/process-and-meetings/the-paris-agreement/the-paris-agreement. Accessed 6 Feb 2022

Van Oost K, Quine TA, Govers G, De Gryze S, Six J, Harden JW, Ritchie JC, McCarty GW, Heckrath G, Kosmas C, Giraldez JV, da Silva JRM, Merckx R (2007) The impact of agricultural soil erosion on the global carbon cycle. Science 318:626–629. https://doi.org/10.1126/science.1145724

Vanwalleghem T, Gómez JA, Amate JI, de Molina MG, Vanderlinden K, Guzmán G et al (2017) Impact of historical land use and soil management change on soil erosion and agricultural sustainability during the Anthropocene. Anthropocene 17:13–29

Vieira NK, Clements WH, Guevara LS, Jacobs BF (2004) Resistance and resilience of stream insect communities to repeated hydrologic disturbances after a wildfire. Freshwater Biology 49(10): 1243–1259

Walker B, Holling CS, Carpenter SR, Kinzig A (2004) Resilience, adaptability and transformability in social-ecological systems. Ecol Soc 9(2):5. http://www.ecologyandsociety.org/vol9/iss2/art5/. Accessed 27 Jan 2022

Wang Y, Cheng C, Xie Y, Liu B, Yin S, Liu Y, Hao Y (2017a) Increasing trends in rainfall-runoff erosivity in the Source Region of the Three Rivers, 1961–2012. Science of the Total Environment 592:639–648

Wang Z, Hoffmann T, Six J, Kaplan JO, Govers G, Doetterl S, Van Oost K (2017b) Human-induced erosion has offset one-third of carbon emissions from land cover change. Nature Climate Change 7(5):345–349

WCED (World Commission on Environment and Development) (1987) Our common future. Oxford University Press, Oxford

Weil KK, Cronan CS, Lilieholm RJ, Danielson TJ, Tsomides L (2019) A statistical analysis of watershed spatial characteristics that affect stream responses to urbanization in Maine, USA. Applied Geography 105:37–46

Wheater H, Sorooshian S, Sharma K (eds) (2007) Hydrological modelling in arid and semi-arid areas, International hydrology series. Cambridge University Press, Cambridge. https://doi.org/10.1017/CBO9780511535734

Wilby RL, Wigley TML, Conway D, Jones PD, Hewitson BC, Main J, Wilks DS (1998) Statistical downscaling of general circulation model output: a comparison of methods. Water resources research 34(11):2995–3008

Williams JW, Jackson ST (2007) Novel climates, no-analog communities, and ecological surprises. Frontiers in Ecology and the Environment 5(9):475–482

Wischmeir WH (1968) Erosion rates and contributing factors in semi-arid regions. International seminar on water and soil utilisation. Brookings, South Dakota

Worrall F, Burt TP, Howden NJK (2016) The fluvial flux of particulate organic matter from the UK: the emission factor of soil erosion. Earth Surf Proc. Land. 41:61–71

Yaduvanshi A, Nkemelang T, Bendapudi R, New M (2021) Temperature and rainfall extremes change under current and future global warming levels across Indian climate zones. Weather and Climate Extremes 31:100291

Yang D, Kanae S, Oki T, Koike T, Musiake K (2003) Global potential soil erosion with reference to land use and climate changes. Hydrological Processes 17(14):2913–2928

Zhou Y, Ren G (2011) Change in extreme temperature event frequency over mainland China, 1961–2008. Climate Research 50(2–3):125–139

Part III
Agriculture and Forestry and Climate Change

Chapter 12
Agro-climatic Variability in Climate Change Scenario: Adaptive Approach and Sustainability

Trisha Roy ⓘ, Justin George Kalambukattu ⓘ, Siddhartha S. Biswas ⓘ, and Suresh Kumar ⓘ

Abstract Agriculture is the key component to support the ever-increasing population across the globe. However, the natural resources supporting agriculture, most importantly land and water resources, are shrinking at a rapid rate. Land degradation is rampant across different parts of the world, and the vagaries of climate change threaten the agricultural production. Extreme weather events in the form of increased rainfall intensity, changes in rainfall pattern, shift in the rainfall duration, increased temperature regime and more natural calamities like flood and drought threaten agricultural production. The different agro-climatic zones (ACZs) in the country face diverse challenges due to the changing climate particularly with reference to the agriculture in the area.

Under climate change scenario, the increasing temperature is likely to reduce the C gain by majority crop plants due to increased respiration rates, which will have an adverse impact on crop yield. Thus, diversification of agriculture is necessary to sustain the production. Water management and input management through sensors or precision agriculture is an important tool to combat climate change. Reduced tillage practices, crop residue retention, biochar, land management through laser levelling, micro-irrigation systems, etc., are part of mitigation and climate-proofing strategies which have been highlighted in this chapter. Also, assessing the suitability of different crops in future climate change scenarios through crop simulation models like APSIM and DSSAT is important to know how they will fare in the future. This chapter highlights the impact of climate change on Indian agriculture in particular

---

T. Roy
Human Resource Development and Social Science Division, ICAR - IISWC, Dehradun, Uttarakhand, India

J. G. Kalambukattu (✉) · S. Kumar
Agriculture and Soil Department, Indian Institute of Remote Sensing (IIRS), Indian Space Research Organization (ISRO), Dehradun, Uttarakhand, India
e-mail: justin@iirs.gov.in

S. S. Biswas
Crop Production Division, ICAR-National Research Centre for Orchids, Pakyong, Sikkim, India

and the various mitigation and climate-proofing techniques adopted at different scales and how sustainable they are in the future scenarios.

Keywords Climate Change · Agro-climatic zones · Crop productivity · Soil Health · Climate-proofing · Agriculture Sustainability

Introduction

At the present growth rate of population, the globe is likely to add 2.3 billion more people between 2009 and 2050, and it is expected to have a population of 9.7 billion by 2050. Agriculture is a key component to support this population in terms of food, nutrition and economic upliftment. To keep pace with the increasing population, agricultural production has to increase by 60–70% to meet the global food demands of 2050. Since the population growth is not evenly distributed across continents, and highest spike is predicted to occur in the poorest of the nations (the population in sub-Saharan Africa is projected to increase by 114%), the demand for agricultural commodities will also follow similar trend. The developing nations would need to increase their food production by almost double or even more to satiate the necessities and ensure quality and nutritional food for all.

Global Agriculture Scenario

Though the global agricultural sector is witnessing a growth by and large, however, agriculture sector is one of the most vulnerable sectors to various man-made and natural calamities and bears the brunt of various problems. Climate change and land degradation are the two major challenges which threaten agricultural production across the world. Currently, the contribution of this sector to the world gross domestic product (GDP) is 4%, while in some developing countries it is as high as or even more than 25%. Any loss in agricultural production and productivity will weaken the economy of many developing nations and put them at risk of extreme poverty, hunger and food and nutritional insecurity. The increased agricultural production in the coming times will be mainly driven by increased land productivity and increased cropping intensity (almost 90%), and very less will be contributed by expansion in area under agriculture. The area increase is likely to occur in sub-Saharan Africa and Latin America with an overall increase in average of about 70 million ha. On the other hand, almost 30% of the arable land is at risk to be washed away by soil erosion jeopardising the productivity and health of soil (Kendall and Pimental 1994).

Agriculture will also witness a major shift in terms of the food consumption pattern and choices followed by people across the countries. As the economies of the developing nations pace upwards significantly, the inequality in per capita income will be reduced. The inclination of people with high purchasing power and increased influence of urbanisation will shift from cereals to more animal-based diets. Though preferences for cereals will be less, cereal production has to be increased manifold putting immense pressure on cereal-based production systems, as more cereals will be needed to feed the animal population. Since animals are less efficient feed converter with an input/output ratio of feed/meat in terms of calories, substantial quantity of feed has to be diverted for livestock production. In 2016, 36% of cereals produced were diverted for animal feed worldwide. Such shift in food consumption pattern makes agricultural system more susceptible to degradation. Already the agricultural sector contributes 25% of the global greenhouse gas emission (GHG). With expansion in the livestock sector, this is expected to increase as the global livestock contributes 7.1 Gt of CO_2 equivalent per year, which is 14.5% of the total anthropogenic GHG (https://www.fao.org/news/story/en/item/197623/icode/).

Currently, with the onset of the COVID-19 pandemic, the world agriculture is facing immense challenge particularly in the poorest countries which are in deep waters. The agricultural commodity price index has increased by 17% compared to January 2021; commodity-wise, the price of wheat and maize has increased by 21% and 11%, respectively (https://www.worldbank.org/en/topic/agriculture/brief/food-security-and-covid-19). The pandemic has also exposed more population to hunger and malnutrition, and in 2020 almost 83–132 million people were added to the total number of undernourished population in the world (FAO, The State of Food Security and Nutrition in the World 2020). Thus, global agriculture has to tackle these multifaceted challenges in order to fulfil the targets of feeding quality and nutritious food to 9.7 billion people by 2050.

Indian Agricultural Scenario

Despite becoming the third largest economy in the world after China and United States of America, it is still an agrarian economy with about 58% population practising agriculture as a primary source of livelihood. Agriculture and its allied sectors are unquestionably the largest livelihood provider in India (https://www.india.gov.in/topics/agriculture) and contribute to 19% of the country's GDP (icar.org.in). Of the total geographical area of the country, 51% is under cultivation with a cropping intensity of 136%, against the world average where only 11% of area is under agriculture. Indian agriculture has been for a long time a bargain with the monsoon, as rainfed agriculture covers 65% of the total net sown area. However, since the post-independence era, India has achieved remarkable feats in the agriculture sector and transformed from a food impoverished nation to a food surplus one. Presently, India is the largest producer of milk, pulses and jute and the second largest producer of rice, wheat, sugarcane, groundnut, vegetables, fruits and cotton (fao-india.in). The economic valuation of agriculture and allied components was estimated to be around US$ 276.37 billion during financial year (FY) 2020. The total food grain production reached a record high to 296.65 million tonnes during 2020 compared to previous FY 2019 when the production was 285.21 million tonnes.

India has vast potential and poised for huge growth with respect to value addition and food processing industry. The horticulture sector is also set to achieve record high production of 331.05 million metric tonnes in 2020–2021, an increase by 10.5 million metric tonnes compared to 2019–2020. The livestock sector with the highest animal population is recording a growth of 10% almost each year (https://www.ibef.org/industry/agriculture-india.aspx). India is among the top 15 countries dominating the global market related to agricultural export with a valuation of US$ 35.09 billion in FY 2020. Various government initiatives and policy-level interventions allowed Indian agricultural sector to achieve remarkable growth over years. However, this sector and its associated human resource face multifarious challenges. Of these, climate change and land degradation are two pressing problems which not only impact the agricultural sector but all sectors as a whole. This chapter tries to highlight the problems faced by agricultural sector in different agro-climatic regions across the globe as well as in India and discusses the impacts and mitigation measures available to combat these challenges.

Agro-climatic Zones (ACZs)

ACZs as the term suggests is the zonation of land area or land parcels based on similar climatic conditions, soil types and landforms, which is suitable for cultivation of similar type of crops. The basic purpose of classifying land parcels as ACZs is to facilitate the development of the areas based on common type of potentials and constraints. The agro-ecological zonation followed by the Food and Agriculture Organization (FAO) is also in the same line as ACZ that takes into consideration soil, climate and landforms. Additionally, it also considers the length of growing period, temperature regime and soil mapping unit while delineating the agro-ecological zones (AEZs).

Global Perspective

The global agricultural scenario is classified as AEZs rather than ACZs whose concepts are overlapping but distinctly different with the LGP and temperature regime inclusion in the former classification. The FAO and the International Institute for Applied Systems Analysis (IIASA) developed the AEZ methodology for the past 30 years (1961–1990) to classify the agricultural resources and potential across the globe. The AEZ classification follows the principal of land suitability evaluation laid down by FAO from time to time (FAO 1976, 1984, 2007). The first global assessment of AEZ was done in 2000 after which the assessment has been conducted after a certain period. This assessment of AEZs at global level includes the following components (Fig. 12.1):

12 Agro-climatic Variability in Climate Change Scenario: Adaptive...

Fig. 12.1 Components under GAEZ classification at global level carried out by FAO and IIASA

The global agro-ecological zone (GAEZ) database serves as a template for diverse application including quantification of land productivity (https://www.fao.org/nr/gaez/en/#). Recently, the GAEZ version 4 database has been launched by FAO and IIASA in 2021. The GAEZv4 has been referred to as the most ambitious global assessment data to date (Fischer et al. 2021). It is an important decision-making tool to find the most suitable agricultural land utilisation options. This takes into consideration the crop factors, climatic factors and edaphic necessities of the crops to identify the resource limitations and opportunities. Accordingly, the suitability and production potential of different crops under a set of well-defined management practices is determined. This approach helps to identify the limitations and challenges in a very systematic manner and supports decision-making to deliver more sustainable and climate-resilient agricultural production systems. The GAEZv4 has also oriented itself towards addressing several Sustainable Development Goals (SDG) laid down by the United Nations and provides essential information related to crop suitability, crop production potentials and associated risks, demand for irrigation water, crop adaptation options, etc.

There are several countries across the globe that are implementing the methodology of GAEZ, for example, Kenya, Bangladesh, China, Ghana, etc. The GAEZ has been used to quantify the biofuel feedstock potential of the globe and estimate "fair value" for land parcels with the help of the World Bank to balance the investments in agriculture (https://previous.iiasa.ac.at/web/home/research/researchPrograms/water/GAEZ_v4.html#:~:text=Global%20Agro%2DEcological%20Zoning%20(GAEZ,and%20production%20potentials%20of%20land). This GAEZ methodology follows an environmental approach and finds application in various aspects like:

- Quantifying the productivity of land
- Evaluating the rainfed and irrigation production potential of food, fuel, fibre and fodder of land parcels
- Identifying the environmental constraints in agricultural production
- Identifying suitable hot spots for agricultural conversion
- Identifying geographical shifts in agricultural landforms due to climate change impact

The GAEZ globally helps to facilitate researchers and policymakers to develop strategies at regional, national or global level for making agricultural production systems more sustainable and at the same time productive keeping in view the demand from agricultural sector in the coming times.

ACZs in India

The Planning Commission in 1989 devised the concept of ACZs in India to facilitate the planning, distribution and maximum utilisation of man-made as well as natural resources for agricultural purposes. The ACZ is delineated on the basis of soil, water, temperature, rainfall, etc., and land survey, soil survey and agricultural survey provided the baseline data for delineation of the ACZs. The ACZ is defined as:

An Agro-climatic zone is a land unit in terms of major climates, suitable for a certain range of crops and cultivars. The planning, aims at scientific management of regional resources to meet the food, fibre, fodder and fuel wood without adversely affecting the status of natural resources and environment.

The prime objective of identifying ACZ was as follows:

- Optimise agricultural production
- Enhance the income of farmers
- Generating more scope of rural employment
- Prudent use and distribution of the irrigation water resources
- Phase out regional inequalities in agricultural development

The National Remote Sensing Agency delineated 15 ACZs in the country which is primarily influenced by the physical attributes and the socioeconomic condition of the region (http://jalshakti-dowr.gov.in/agro-climatic-zones):

1. Western Himalayan region
2. Eastern Himalayan region
3. Lower Gangetic Plains region
4. Middle Gangetic Plains region
5. Upper Gangetic Plains region
6. Trans-Gangetic Plains region
7. Eastern plateau and hills region
8. Central plateau and hills

9. Western plateau and hills
10. Southern plateau and hills
11. East Coast plains and hills
12. West Coast plains and Ghats region
13. Gujarat plains
14. Western dry region
15. The Islands region

The Indian Council of Agricultural Research (ICAR) also divided the country into 127 ACZs under the National Agricultural Research Project. This focussed on identifying the major bottlenecks in agricultural growth and development in the specific ACR and coming up with technological solutions to solve the same and pave a pathway for a more sustainable agriculture. The concept of AEZ in the country was introduced by the National Bureau of Soil Survey and Land Use Planning (NBSSLUP) including the LGP in different regions. There are 20 agro-ecological regions and 60 agro-ecological subregions in India as per this classification.

Brief Description of the ACZs

Each ACZ is unique in its characteristics and encompasses a wide variability in terms of soil, landform, flora, fauna, climate, water resources, geographical features and the socioeconomic conditions of the people. The Western Himalayan region, with the three hill states of Jammu and Kashmir, Himachal Pradesh and Uttarakhand and Union Territory of Ladakh, is characterised by steep slopes and very low temperatures at high altitudes. These states are highly vulnerable to the impacts of climate change. A recent study conducted by the Potsdam Institute of Climate Change and TERI has revealed that Uttarakhand (with almost 90% of the area under hills) is highly vulnerable to the changing climate. Shifting of peak rainfall from July–August to August–September and delayed winter rains continuing till February impact the crop production heavily which is mainly dependent on rainfall (Thadani et al. 2015). Incidents of cloudbursts have increased by 30% during the year 2010. The state is facing an extraordinary situation of out-migration influenced by climate change pushing the socioeconomic conditions towards acute poverty and declining the agricultural activity in the state (Upadhyay et al. 2021). The Eastern Himalayan region consists of Sikkim, Darjeeling area (West Bengal), Arunachal Pradesh, Assam hills, Nagaland, Meghalaya, Manipur, Mizoram and Tripura. It is also characterised by rugged topography, thick forest cover and subhumid climate (rainfall over 200 cm; temperature July 25–33 °C, January 11–24 °C). Shifting cultivation or *jhum* cultivation is one of the major agricultural constraints in this area which makes the already vulnerable slopes more susceptible to erosion and land degradation. Under the changing climate scenario, practising *jhum* cultivation by the tribes of the Northeast is also becoming a challenge, since shifts in monsoon pattern, decreased water availability in the hills and seasonal shifts have made the system more vulnerable (Feroze et al. 2016). The ACZ 2–6 comprises of the Indo-Gangetic plains (IGP) which bears the most fertile soil in the country. The area is highly productive and forms the bread basket of the country. The productivity of rice and wheat in the IGP is considerable higher than the national average, and since Green Revolution this zone has 70–74% share in country's wheat production and 40–45% share in total rice production (Sekar and Pal 2012). This zone is characterised by vast tracts of flood plains of moderate to low fertility status alluvial soil where intensive agriculture is practised. This is also the heart of Green Revolution in India where the technological spread in agriculture happened at a much rapid pace compared to other parts in the country (Mandal et al. 2021).

The ACZ 7–10 covers the central parts of the country. These are the table lands or plateaus which have dearth of water resources and have various land degradation problems (http://epgp.inflibnet.ac.in/epgpdata/uploads/epgp_content/S000017GE/P001789/M025466/ET/1512709815Agro-climaticzones_e-text.pdf). The region is characterised by ravine landforms which indicate extreme form of soil erosion by water and comprises about 1.468 Mha (Kumar et al. 2020). Agriculture in these areas is also rainfed with lower cropping intensity. In light of climate change, water harvesting and optimisation of water resources is the key to have good production from this part. Both the East Coast plains and the West Coast plains receive good amount of rainfall, and agricultural crops like rice, millet, maize, etc., are grown. This area has inundation by sea water and faces problem of costal salinity and alkalinity which is a major challenge. The island zone includes the Andaman and Nicobar Islands and the Lakshadweep Islands where rainfall is abundant and rice is the major crop cultivated. Zones 13 and 14 in Gujarat and Rajasthan, respectively, are in the semiarid to arid conditions, and water resources are limiting. All these zones are likely to face various challenges under the climate change scenario as the rainfall pattern changes, shift in monsoon rainfall trends is witnessed and extreme weather events become more predominant. Agriculture sector has to prepare itself very prudently to combat these challenges and device strategies to cope with the changes.

Climate Change Scenario

Humankind has become acclimated to climatic conditions that change on a day-to-day, cyclical and intraseasonal basis. A mounting number of indicators imply that, apart from the natural climate fluctuation, mean climatic states evaluated over extended time scales are changing, along with the natural fluctuation seen on various time periods. Climate change poses a significant threat to the Earth's environment. This problem can be caused by a variety of factors. Carbon release in the atmosphere is one among the primary causes. There are numerous explanations for this worldwide concern, one of which is GHG emissions. For all of the non-mitigation scenarios accounted, it was estimated that all of the models predict the global mean surface air temperature will continue to rise throughout the twenty-first century, owing primarily to intensification in anthropogenic carbon dioxide concentrations, with heating corresponding to the associated radiative forcing. Depending on a variety of different growth scenarios and model parameterisations, the IPCC found that global temperatures might probably increase around 1.4 °C and 5.8 °C during 1990 and also during 2100 if no special efforts were implemented to cut emissions. Although there are uncertainties regarding the model predictions because estimates for rainfall and wind speed were less reliable, but they do point to considerable variations.

There is new and superior evidence that the majority of the warming seen in the previous 50 years is likely due to human activity (IPCC 2001). Deforestation, land use change, aerosol and black carbon are some other key contributors to ozone depletion and climate change. The atmosphere is polluted as a result of carbon emissions, and numerous catastrophes occur on a regular basis. As a consequence, the atmosphere is becoming increasingly heated. Glaciers are melting due to the effect of unusual and rapid temperature rise, also resulting in vigorous hydrological cycle and abrupt flash floods. The agricultural sector is also suffering as a result of global warming. This will have an impact on grain productivity all across the planet. Climate change raises land and sea temperatures, as well as changing rainfall amounts and trends. As a result of rising global average sea levels, the risk of coastal erosion, etc., will occur, and it will place further strain on the fisheries and aquaculture industries. Hence, coasts and marine habitats will also be severely impacted. Extreme occurrences such as drought and flooding may occur as a result of these effects (Nema et al. 2012).

Climate predictions are often made for a variety of possible paths, scenarios or objectives that illustrate the linkages among human decisions, emissions, concentrations and temperature change. Several scenarios can be accomplished by continuing to rely on fossil fuels, whereas others could only be realised by taking conscious steps to minimise emissions. The ensuing range shows the inherent uncertainty in determining the impact of human actions including developmental change on climate. The significant proportion of the future projections included in IPCC assessment reports are based on the standard sets of time-dependent scenarios. These are utilised by the climate modelling community as feed to global climate model simulations. IPCC developed various scenarios representing future climatic conditions simultaneously. The earlier SA90 (IPCC 1990) scenarios became revised by IS92 emission scenarios (Leggett et al. 1992), which had been followed by the Special Report on Emissions Scenarios during 2000 (Nakicenovic et al. 2000) and the Representative Concentration Pathways in 2010 (RCPs) (Moss et al. 2010). Fourth and latest, a significant number of socioeconomic scenarios have been released separately from the RCPs, and a category of these has been confined by employing emission restriction policies that were reliable with their fundamental storylines, resulting in five shared socioeconomic pathways (SSPs) also with climate forcing that linked the RCP values. This combination of SSPs and RCPs is intended to convene the requirements of the impacts, adaptation and vulnerability of societies, which allow them to combine the socioeconomic scenarios with climate scenarios devised utilising RCPs to investigate socioeconomic impediments for climate mitigation and adaptation (O'Neill et al. 2014).

While discussing about the climate change scenario and consequences over Indian subcontinent, the average temperature in India increased by around 0.7 °C over 1901 and 2018. This rise in temperature is primarily caused by GHG-induced warming, which is partially offset by anthropogenic aerosol forcing and changes in land use/cover. Under the RCP 8.5 scenarios, the mean temperature over India is expected to climb by around 4.4 °C by the end of the twenty-first century, compared to the recent past average during 1976–2005. Across the last 30 years, the hottest day and coolest night of the year have surged by roughly 0.63 °C and 0.4 °C, respectively (1986–2015). In reaction to amplified atmospheric moisture content, the regularity of localised extreme precipitation has grown substantially. During the period 1950–2015, the occurrence of rainfall extremes with rainfall intensities of more than 150 mm per day has gone up by significantly 75% throughout Central India. During the 21st century, India has experienced an increment in temperature, a drop in monsoon rainfall, a surge in severe temperature and precipitation events, droughts, rise in sea levels, as well as many other disturbances in the monsoon system. Anthropogenic activities have contributed these variations in regional climate, according to technical data. India also will face the threats of climate change in various sectors which have been caused by human activities. It is critical to establish effective methods for expanding knowledge of Earth system processes, as well as to enhance observation systems and climate models and to enhance the accuracy of future climate projections, specifically in the perspective of localised forecasts (Krishnan et al. 2020). Hence, suitable widely applicable adaptation and mitigation measures are very much necessitated for checking the adverse consequence of climate change.

Climate Change Predictions and Projections

A climate prediction or forecast is the outcome of an effort to generate utmost plausible characterisation or assessment of factual development of the climate in the future, on periodic, intra-seasonal or decadal time periods. Climate projections are similar to weather forecasts, which rely on a precise depiction of the present information, primarily regarding the seas and atmosphere. A climate projection is a depiction of the climate system's reaction to various GHG developments, which largely depends on climate model simulations. Climate projections are differentiated from climate predictions to emphasise that climate projections are dependent on the radiative forcing emission and its concentration scenarios, which is based on assumptions that might or might not be achieved and thus is prone to significant ambiguity extraneous to the climate system, while climate scenario and climate projection are being used interchangeably. A scenario is a reasonable and frequently refined representation of how the future might evolve, represented by a set of logically consistent assertions about determining factors and crucial interconnections. Typically, a set of scenarios is created to cover a wide range of possibilities, for instance, what occurs to the future climate of Asia and Antarctica and also if GHG concentrations rise to 750 parts per million. There seems to be a sequence of scenarios in the realm of likely impacts, ranging from worldwide socioeconomic scenarios through climate scenarios to localised adverse effect scenarios. Various methods and their outcomes are generated around the globe based on climate prediction and projections that are depicted here.

Potential impact of future climate change on food supply is the most prime concern to policymakers (Parry et al. 2007). For quantitative impact assessment, GCM (general circulation model) scenario data is widely used to study climate change impacts on natural resources. However, major obstacles occurred with the data of coarse resolution (250–600 km). Crop yield projection requires climate inputs at higher resolution (landscape scale) typical for global climate models (Glotter et al. 2014). Fine-resolution climate data is required to assess climate change impact for crop-related phenomenon. Therefore, downscaling at finer resolution is a requisite for agriculture impact assessment than typical GCM spatial resolution. To transform low resolution to high resolution, two basic downscaling methodologies are commonly used: dynamic and statistical downscaling. Employing regional climate models, dynamic downscaling is a means of getting regional scale data (RCM). The outcome of the GCM model is used as a source within those models. The RCM model produces high-resolution data, though it is a complicated and time-consuming procedure that is entirely reliant on the GCM boundary circumstances. Statistical downscaling, on the other hand, gives a quantitative association among fine (local)-scale and coarse-scale data (Benestad et al. 2008), which is then utilised to derive future scenarios at a wide scale using GCM outcome. Statistical downscaling has a lot of benefits over dynamic downscaling, one being ease of installation and cost-effectiveness. On the other hand, a requirement of large historical meteorology station data for building a relationship with large-scale variables turns out to be the major disadvantage of statistical downscaling (Wilby 2004).

Weather typing, weather generator and regression approaches are the three kinds of statistical downscaling. Each approach has its own set of benefits and drawbacks. Weather typing encompasses the Monte Carlo approaches for producing synthetic sequences of weather patterns and resampling of measured data, as well as local and meteorological data in connection with the existing structure of air circulation. The main benefit is that it may be used in a multisite operation. The latter is a weather generator that uses a first-order Markov chain to produce rainfall. Weather generators like WGEN and LARS-WG are widely utilised. The key benefit is that it generates having similar outcome as station data and is widely utilised. The third approach is regression, which creates transfer functions among predictors and predictands (observed station data). To construct an association, regression employs a variety of methods, including linear and nonlinear regression, artificial neural networks, canonical correlation and principal component analysis. Apart from this, recently machine learning methods are also used for future climate change-related studies (O'Gorman and Dwyer 2018). The key benefit is ease of application; hence, it is widely used in climate change research (Chu et al. 2010; Goyal and Ojha 2010), although it only represents a small portion of observed climate variability (Wilby 2004). The SDSM (Statistical Downscaling Model) is a frequently employed statistical model for scenario generation. The SDSM is a hybrid model that combines multiple linear regression and a stochastic weather generator to produce estimates (Gagnon et al. 2005). While, DSSAT MarkSim weather generator is an online web application tool which provides downscaled climate projection on global basis (Jones et al. 2011). These models use IPCC emission scenarios (SRES, RCP and SSP) for future climate projections.

Climate Change Impact on Agriculture with Reference to India

Direct Impacts

Crop Yield and Productivity

Increased emission of GHGs in the atmosphere will increase the average global temperature, which can decrease yield of agricultural crops and lead to food insecurity. Agricultural activities are one of the biggest producers of GHGs. Decreased yields may push human civilisation towards more intensive cultivation practices, thereby leading to more emission of GHGs (Wang et al. 2018). Majority of the world population consumes paddy, wheat and maize as their staple food. The predicted world population by 2050 will be 9.8 billion. Thus, the demand for food grains will also increase putting immense pressure on the already shrinking land resources, which is likely to be diverted for more infrastructure building to accommodate the increased population (Godfray et al. 2010; Hawkesford et al. 2013). The increasing concentration of GHGs may raise the global temperature by 2.5–4.5 °C at the end of the twenty-first century (Solomon 2007). This phenomenon of global warming may increase the rate of plant respiration, thereby reducing the rate of C gain. This will decline the crop yields, and invasion by pathogens, pests and weeds will be on the rise (Asseng et al. 2011). For instances, a 1 °C temperature rise during wheat cultivation period can decrease its production by ~3–10% (You et al. 2009). The increased global temperature is increasing plant respiration as well as carbon metabolism which ultimately leads to decreased yield of paddy (Zhao and Fitzgerald 2013). Higher temperature also increases proportion of sterile paddy flowers and hampers yields. As per the International Food Policy Research Institute (IFPRI), climate change can reduce paddy yield by ~10–15%, which can lead to 32–37% hike in market price (Nelson et al. 2009). The ORYZA 2000 model predicted that a rise in temperature by 2 °C can cause 0.36 t ha^{-1} yield reduction of rice. Based on some highly optimistic scenarios, some studies claim that global warming can lead to 2.9–34% increase in rice yield. The probable reasons may be increase in night temperature and heating up of cooler areas (which are not suitable for rice cultivation at present for low temperature) during the 2050s and the 2080s which will become more suitable for rice cultivation (Aggarwal and Mall 2002). But the question arises: What will be the socioeconomic condition of the present rice-growing belts? Above all, in spite of getting slight yield increase, several studies agree that long-term effect of temperature rise will lead to drastic fall in crop yield at the end of this century (Welch et al. 2010). The increased concentration of CO_2 in the atmosphere can decrease the average rice yield by ~4% (Matthews et al. 1997). Millets are the most drought-resistant crops (Hadebe et al. 2017). As per the published reports, millet yield can increase with increasing temperature in the dry areas (Cao et al. 2010). The modified CSM-CERES-Pearl millet simulation model showed increased heat from 27 to 29 °C, and drought increased the millet yield by 8% and 6%, respectively (Singh et al. 2017). Combined effect of heat and drought under climate change increased millet yield by 14% (Rasul et al. 2018). Between 1970 and 2000, the temperature rise would have dropped wheat yield by 4.5%, but the increased use of resources (like fertilisers, irrigation etc.) managed this fall (You et al. 2009). Due to climate change, in some important cropping systems, the productivity either declined or became static. In North-Western India and Indo-Gangetic plain, the recent trends of stagnation or decline in the yield of rice-wheat cropping system have become a matter of concern as these regions share the major chunk of rice-wheat production of the country (Pathak et al. 2003). A report from IARI suggested that for wheat impact of climate change (i.e. temperature rise and variation in rainfall) was more severe for the areas with higher potential productivity than the areas with lower potential productivity. It was more severe for rainfed areas as compared to irrigated areas. The tropical areas were more severely impacted compared to sub-tropical areas, which indicates warmer areas of India may get more affected due to temperature rise and uncertainty of rainfall. For rice, yields may be reduced due to rise in temperature by 2–4 °C. The deceased radiation and increased temperature can impact the Eastern India most. In contrary, for Northern India, impact of climate change may be less; higher radiation may offset the yield reduction caused by increased temperature. Increased CO_2 concentration can benefit rice yield, but the negative impact of increased temperature can nullify that benefit.

Soil Productivity and Soil Health

Major changes are happening in global environment due to increasing concentration of GHGs. It results to global warming, changes in precipitation and atmospheric moisture, ozone concentration reduction in the stratosphere and increased atmospheric deposition. Those are ultimately changing numerous soil processes and affecting soil health. Soil organic matter (SOM) is considered as the main soil health indicator, because it influences many soil quality parameters. If SOM is maintained at optimum level, soil health will be maintained; at the same time, rise in CO_2 concentration in the atmosphere can also be alleviated. Climate change can influence several biological processes such as microbial and metabolic quotients, SOM decomposition and N, S and C cycling. Climate change can influence SOM dynamics directly by changing water and temperature balance as well as indirectly by changing productivity. The depositions of atmospheric N have increased significantly during the last 150 years, and by the end of twenty-first century, it may increase ~2.5 times more, which may significantly influence several soil biological processes as well as C and N cycles (Benbi 2012). The changes in climate and soil properties take place over a long term. Generally, soil biological processes, i.e. storage and decomposition of SOM, microbial and metabolic quotients and N, S and C cycling, are mainly influenced by climate change (Allen et al. 2011). Physical indicators like soil porosity and soil available water-holding capacity, chemical indicators like available nutrients and pH and electrical conductivity are used to assess the impact of climate change on soil health (Benbi 2012). Globally 1 m depth soil stores 1500–2000 Gt of soil organic carbon (SOC). It is the main component of SOM. The SOM can act as both sources and sink of atmospheric CO_2 depending on the balance between input and output of it to the soil. A soil may work as source of CO_2 if the net decomposition of SOM from it exceeds the net input. It may happen due to anthropogenic activities or due to accelerated SOM decomposing because of global warming. The Roth C model (Coleman and Jenkinson 1996) predicted that global warming-induced increased soil respiration will deplete ~54 Gt of C from SOC stock at the end of twenty-first century (Niklaus and Falloon 2006). Primary productivity is the source of input to SOC. Increased CO_2 concentration along with elevated temperature may increase primary productivity. The difference between soil SOC input through primary productivity and SOC loss through SOM decomposition decides the net change in soil C storage. Soil N and C pools are negatively correlated with temperature and positively correlated with rainfall (Post et al. 1985). The trend of SOC pool will depend on the relative temperature sensitivities of SOM decomposition and primary productivity. Atmospheric deposition of N is increasing which is attributed by biomass and fossil fuel burning. N deposited in agricultural land can give nutritional benefit and at the same time adversely influence several soil processes. It can lead to biodiversity reduction, acidification of soil, increased nitrate leaching, groundwater pollution, eutrophication, increased N_2O emissions, altered balance of nitrification and mineralisation/immobilisation. Uncertainty in rainfall can lead to flooding or draught. It can also deteriorate the quality of irrigation water. Extreme drought can raise soil salinity. Frequent flood and drought can make an arable land vulnerable to erosion. Climate change with unsustainable irrigation and management practices has increased soil salinisation, erosion, deteriorated soil health and nutrient depletion (Benbi 2012).

Indirect Impacts of Climate Change on Agriculture

Indirectly, climate change may put considerable impact on land use, spatial and temporal rainfall variability, snow melt, availability of irrigation, floods, intensity and frequency of intra- and inter-seasonal droughts, soil erosion, SOM transformations, change in pest profiles, submergence of costal lands, decline in arable areas and availability of energy. All these can put question marks on food security of any area by hampering agricultural production (Aggarwal 2003).

Change in Rainfall and Temperature

The increasing concentration of GHGs may rise the global temperature by 2.5–4.5 °C at the end of the twenty-first century (Solomon 2007). Due to climate change, in the last century, noticeable changes occurred in extreme rainfall events over India. The annual normal rainy days (ANRD) have shown large spatial variations. The ANRD varied from 130 days in Northeastern India to 10 days in extreme western parts of Rajasthan. Different data analysis predicted that the frequency of wet days decreased in most parts of India. In central and many North Indian regions, significant decreases in rain days are observed, and in Peninsular India significant increasing trends are observed. The desert areas of the country received more number of wet days in last few years (Guhathakurta et al. 2011). The extreme southern India, i.e. Tamil Nadu and Kerala, is observing more number of dry days. Similar trend of decreasing mean annual rainfall over Tamil Nadu and Kerala was also reported by Guhathakurta and Rajeevan (2008). The Intergovernmental Panel on Climate Change (IPCC 2007) projected that areas with less precipitation are going to observe more extreme drought and areas with high rainfall are going to observe an increase in precipitation.

The long-term changes in rainfall pattern across Indian states reveal that deficit rainfall is likely to occur in several pockets over parts of eastern Madhya Pradesh, Chhattisgarh and northeast region in Central and Eastern India (Rao et al. 2003). On the contrary, areas in the West Coast, northern parts of Andhra Pradesh and parts of North-West India show enhanced rainfall activity by +10 to 12% (National Action Plan on Climate Change 2008). In the southern peninsular region, shift in monthly rainfall patterns is also observed. Across all the 36 subdivisions of the country, data reveals an increased contribution in the June and August rainfall, while contribution of July rainfall has exhibited significantly declining trends (Guhathakurta 2006; Guhathakurata et al. 2015). Of the 36 subdivisions, 19 showed increased trend in June rainfall, while 17 showed a declining trend. Interestingly, the areas with reduced July rainfall activity were the ones which showed increased rainfall during August. This is prominent in the Central and West Peninsular India. Thus, in different agro-climatic regions, the rainfall patterns will change differently which will directly influence the agricultural practices like choice of crops, planning for different intercultural operations and other related activities to minimise the impact of such changes.

Increase in Natural Hazards (Flood, Drought, Fire, etc.)

In India, flash floods were observed consecutively in 2005, in the month of July in Mumbai, in the month of October and December in Chennai and in the month of October in Bangalore, all of which were resulted due to heavy precipitation events and caused severe economic damage and loss of life (Guhathakurta et al. 2011). World forest area will face significant pressure from climate change in the next century. Long-term change in rainfall pattern and temperature will have great impact on forest cover. Tropical forest extends livelihood to people of several regions of the world. But tropical forest ecosystems are most vulnerable to climate change. Combined effect of increase in temperature, prolonged dry season and industrial forestry-induced disturbance to the forest systems will make tropic forest ecosystem more prone to forest fires. Rise in temperature is likely to increase the area to be covered under forest fire and its duration and intensity, worldwide. Climate change has hampered the forest ecosystem and lives of people dependent on forest in India by increased forest fire numbers, duration and intensity and pest outbreak to the forest, woodlands, inland waters, marine ecosystem and agroecosystems. Due to increased drought and temperature, dry and moist deciduous forests can face increased risk of forest fire. In India, threat to the valuable genetic resources and biodiversity is increasing due to increase in incidence of forest fires. The forests are not only getting disturbed by the climate change, but the disturbed forest ecosystems are also influencing the climate change. Forest fires increase the emission of GHGs. Deforestation reduces the amount of CO_2 fixation in plants as well as in soils. Thus, it can increase the impact of climate changes also (Mukhopadhyay 2009). Climate change is likely to increase drought severity and frequency throughout the globe (Cook et al. 2014; Dai 2011, 2013), especially in semiarid parts of the world, which are experiencing significant water stress already (Seager et al. 2013, 2014). Climate change forecasts indicate robust declines in precipitation, thereby increasing drought in the future due to anthropogenic activity-induced global warming (Seager et al. 2014; Polade et al. 2017).

Reduction in Water Resources

Due to rise in GHGs, global temperature is rising, and it is also changing the precipitation pattern. These things lead to change in runoff patterns, rise in the sea level, change of land use and shift in population which ultimately lead to changes in water demands. Rising sea level is decreasing good-quality water for plants as well as for the animal kingdoms. The demand for the irrigation water changes depending on atmospheric CO_2 level, temperature and precipitation. Reduced precipitation and increased CO_2 level and temperature increase the demand for irrigation water. Climate change makes the weather phenomenon such as precipitation more uncertain, which is disturbing the planners to make the plans of water demand and availability at river basin and watershed levels (Frederick and Major 1997).

Increased Disease and Pest Infestation

The global warming and extreme changes in temperature and precipitation affect the incidence of insects, weeds and crop pathogens of a region. The crop and pest interaction is more influenced by the changes in precipitation pattern as compared to annual total changes in precipitation. Increased temperature can lead to faster crop development which makes them vulnerable to the pests at earlier stage of development. At high elevation and high altitude, disproportionate warming in nighttime and winter can bring repatterning of geographical distribution of production activities, affect development of crops and alter the ecological balance between the crops and its associated pests. Pest management can face some great challenges in the near future due to climate change. That will increase dependence on chemicals and increase the costs of environmental protection (Rosenzweig et al. 2001).

Biodiversity Losses

An alarming image of the global biodiversity was portrayed by the World Wildlife Fund (WWF 2020). As per the Living Planet Index 2020, the population sizes of birds, amphibians, mammals, fish and reptiles have decreased (from 1970 to 2016) on an average ~ 68% as forecasted by WWF (2016). Throughout the world, the number of threatened species of fishes, plants, amphibians, birds, insects, molluscs, mammals and reptiles is showing an increasing trend for the last two decades. From 2006 to 2015, for only 10-year period, molluscs, fishes and reptiles experienced their biodiversity loss at an annual average rate of 8.5%, 8.0% and 12.5%, respectively (Habibullah et al. 2022). WWF (2016) stated that "the decline in species is due to a variety of factors including unsustainable agriculture, fisheries, mining, habitat loss and degradation, overexploitation, climate change and pollution". Several studies are repeatedly reporting that the most important driver of biodiversity loss is climate change (MEA 2005; Guo et al. 2017). The rise of CO_2 level and temperature is the main driver of global climate change. These give rise to major changes in hydrologic cycles (i.e. evaporation and precipitation) and in frequency and magnitude of extreme weather, like cyclones, floods and droughts, which will have a profound adverse effect on biodiversity (Adler et al. 2009; Rinawati et al. 2013).

Adaptation Measures for Combating Climate Change Effects: Climate-Proofing Measures

Considering the varying impacts of climate change on different agroecosystems across the world, especially with respect to agricultural area and production, making use of adaptation strategies or climate-proofing measures is vital from a sustainability point of view. Measures aimed at improving the adaptability of agricultural sector to changing climate scenario may include different engineering, nonengineering and biophysical options (ADB 2013). The engineering options mainly include the development and maintenance of various agricultural infrastructures such as irrigation/drainage systems, storage buildings, construction of rural roads, etc. Whereas the non-engineering and biophysical options include changes with respect to land use/cropping patterns, management and conservation of different resources such as soil, water, land etc, development of climate resilient crop varieties/cultivars, and technological interventions in farming etc. (ADB 2013). Further discussions will be mainly limited to various nonengineering and biophysical options, as engineering measures are out of the scope of this chapter.

Agronomic Practices

Agronomic and soil management practices, primarily aimed at optimal utilisation and conservation of various resources and agricultural inputs, play a vital role in improving the adaptability of different agricultural systems to changing climate. Different conservation agriculture (CA)-based practices such as zero/no tillage, reduced/minimum tillage, adoption of legume-based crop rotations, cover cropping, mulching, sowing time adjustment, integrated nutrient management, etc., are known to be effective in improving the adaptability to climate change (Mrunalini et al. 2020). These practices help to check and reduce excessive use of various off-farm inputs while enhancing the production diversity and customising the intensity of production as per the capacity of the landscape. Adoption of various agronomic measurements involving efficient and optimum use of fertilisers, conservation/reduced tillage operations and proper water management is estimated to have the potential to reduce annual GHG emission from Indian region to the tune of 85.5 megatons of CO_2 equivalent (Mrunalini et al. 2020). In addition, Srinivasarao et al. (2016) have reported many other soil and crop management strategies having the potential to increase soil carbon sequestration and reduce GHG emission. These technologies mainly comprised of different (improved) cultivation methods such as System of Rice Intensification (SRI), direct seeding of rice (DSR), mechanical transplantation, direct seeding of wheat and corn, furrow irrigated raised bed planting (FIRB) for different cereal crops, precision land levelling with the aid of laser guidance or GPS, different water management approaches, precision farming, etc. Different weed management approaches, intercropping, mulching, crop diversification, etc., are also known to improve climate change adaptability by various processes.

Soil Management

Soil management has an important role in determining the climate change adaptability of different agricultural systems by mainly limiting or reducing the emission of various GHGs. Different tillage operations break soil aggregates, causing an increased oxidation of SOM, thus releasing CO_2. Around 80% of soil carbon is reported to be lost due to conventional tillage, whereas only 12% and 2% carbon loss is reported in case of reduced and minimum tillage operations (Olaya et al. 2013). Thus, the conventional tillage contributes much more (26–31% more) to the global warming in comparison to conservation tillage, which reduces C loss by means of its sequestration (Lal 2004; Mangalassery et al. 2014). Apart from this, conservation agricultural practices could reduce the global warming potential by means of reduction in fossil fuel input (Pratibha et al. 2016) as well as better management of crop residues (reduced burning), resulting in lower GHG emission and improved pools of SOC. Drainage regulation especially with respect to peat lands, the natural sink of CO_2 which can store nearly 87–350 Gt of C globally (Gumbricht et al. 2017), could aid in SOM build up and reduction in GHG emissions (thus reducing GWP). Carbon sequestration aided by various agronomic practices, crop rotations, residue management, balanced fertilisation/nutrient management approaches, land use modifications, controlled grazing, etc., is a very cost-effective approach to mitigate various ill effects of climate change (Haokip et al. 2020).

Recently, the application of biochar, a pyrogenic carbon produced by combustion of biomass under limited oxygen supply, has been widely promoted for improving soil C sequestration (Liang et al. 2014; Venkatesh et al. 2015). This reduces the GHG emission, as the recalcitrant C present in it is resistant to microbial degradation and aids in better residue management. Integrated nutrient management (INM) strategies involving the combined use of organic sources as well as chemical fertilisers could improve soil carbon pools, reduce GHG emission, increase availability of various nutrients for prolonged periods and improve microbial diversity as well as soil aggregation, all of which have profound influence on improving crop productivity (Sharma et al. 2008; Benbi 2013). Soil management with respect to optimisation of pH and E_h values reduces the GHG emission especially with respect to CH_4. Adoption of fertiliser management techniques such as altering the source or amount applied and timing of fertiliser application considering field moisture or crop stages could considerably reduce the leaching as well as volatilisation losses and thus can improve crop yield and productivity. Adoption of advanced fertilisation techniques such as the use of nitrification inhibitors, urease inhibitors, coated fertilisers, nano-formulations, use of leaf colour charts for timing fertilisation, etc., also could prolong nutrient availability and reduce losses (Haokip et al. 2020).

Water Management and Improving Water Use Efficiency (WUE)

Different water management strategies in conjunction with soil management approaches also help in adaptation of agriculture sector towards various climate change-associated extremes. Adoption of technology, namely, rubber dams, has been reported as a very effective soil and water conservation method due to its ability for flood control, regulation of water flow towards stream, etc. Pusa Hydrogel, a semisynthetic-type superadsorbent polymer developed by ICAR-Indian Agricultural Research Institute, New Delhi, has been widely used since 2012 due to its high water-holding capacity and slow-release characteristics. The product is found to have the ability to maintain soil moisture status for longer durations, reduce evaporation rates and increase crop yield by 10–25% (Roy et al. 2019; Rajput et al. 2020).

Adoption of various rainwater harvesting and farm pond technologies is also observed to improve the water availability, and extending irrigation facilities thus helps in combating the ill effects of drought. Rainwater harvesting could be adopted at varying scales ranging from micro-watersheds even to bigger catchments and is known to provide multiple benefits such as crop productivity improvement, ensuring regular food supply and income generation, improved water and fodder availability for livestock and other animals, reduction of soil erosion and sedimentation, water supply during dry spells and drought, etc. It also enhances the recharge of groundwater and aids in the hydraulic design of structures (Rejani et al. 2015). Similarly, the use of different models for hydrologic characterisation at watershed scale could aid in carrying out long-term planning for development and management of water resources in response to variations in precipitation characteristics.

Sensor technology as well as adoption of various geospatial tools is expected to play a crucial role in irrigation management and sustainable crop production under changing climate scenario. A wide array of sensors/techniques comprising of tensiometers, porous gypsum blocks, neutron probes, calcium carbide technique, time domain reflectometry (TDR), heat flux sensors, frequency domain reflectometry (FDR), etc., have increasingly been employed for scheduling irrigation activities as per the crop need and estimated soil moisture content under diverse conditions (Rajput et al. 2020). Remote sensing and GIS techniques involving sensors capable of obtaining information in near-infrared, thermal infrared and visible wavelength regions are also widely used for irrigation scheduling. These techniques provide vital information regarding water requirement of crops by means of monitoring soil moisture status, plant moisture stress and weather/climate data-based water use forecast systems. Apart from these, different climate-smart drought, flood and WUE measures, especially in arid, semiarid and rainfed farming systems, also need special mention with respect to climate change adaptation (Srinivasarao et al. 2020). Early season droughts caused by the late onset of monsoon could be managed by adjusting the planting time within the optimum planting/sowing period, community-based nursery raising, sowing of alternative varieties/crops, repeated sowing and gap filling according to the germination status and growth. Mid-season droughts may be managed mainly by adoption of in situ water conservation and moisture management measures such as mulching, weed management, land configuration, etc. Late season droughts may be managed effectively either by choosing short duration, high yielding crops/varieties and providing lifesaving irrigation by means of various rain water harvesting techniques as well as use of good-quality recycled water. Rajput et al. (2020) provided a detailed list of drought management measures for climate-smart agriculture.

Flood and sediment management technologies are also known to play a key role in ensuring sustainable crop production. Measures such as river linking, afforestation for runoff/sediment control, better designing of canal and other irrigation facilities, maintenance of high-quality hydro-met databases, installation of effective prewarning systems, etc., are found to be very effective. Different measures to improve WUE in both rainfed and irrigated agricultural systems could be achieved by the adoption of numerous agronomic, engineering, management and institutional measures. Reclamation and management of different categories of problematic soils could accelerate our efforts in combating the ill effects of climate change on agriculture. Salt-affected soils, a prominent category of problem soils, can be managed by a multifaceted approach involving hydrological, physical, chemical and biological measures. Hydrological methods involve the adoption of various water management and application methods, thus reducing the impact of salinity on crops, whereas the physical measures mainly focus on reducing the salt content within the root zone by surface management activities. Chemical measures involve the application of different amendments as per the nature of salts present, and biological measures mainly comprise of plant-based management strategies like growing of salt-tolerant crops/varieties, green manuring, etc. Groundwater management strategies aimed at increased recharge, prevention of overexploitation and pollution control are also important from a water management point of view. The adoption of watershed-based approach involving different components such as remote sensing, GIS, hydrological/erosion models, participatory management with the participation of local communities, etc., demands special attention. Making use of various protected cultivation technologies where crops are cultivated under controlled/modified environment such as poly houses, shade nets, diffident structures, etc., will help in maintaining or improving the crop productivity under the climate change scenarios.

Climate-Resilient Crop Varieties

In addition to the various management strategies, development as well as adoption of different climate-resilient crop varieties plays a significant role in our climate adaption strategy. Under the climate change scenario, the prime focus will be on developing varieties which can perform consistently across various environments as well as tolerance to different kinds of biotic and abiotic stress. Another focus will be for reducing the time involved in varietal development. Novel breeding techniques involving shuttle breeding, QTL mapping, rapid generation advance (RGA), marker-assisted recurrent selection (MARS), mutation breeding, speed breeding, genomic selection, marker-assisted backcrossing (MABC), etc., have the ability to reduce breeding cycle while efficiently screening the varieties for various biotic and abiotic stresses. These accelerated breeding approaches could be very effective in developing varieties capable of reducing the impacts of climate change, in reduced time span (Atlin et al. 2017). Gobu et al. (2020) have given a detailed description as well as notable achievements of various accelerated breeding techniques in Indian scenario. Crop simulation models/simulation genetic models, the computer-based tools having the ability to predict crop growth and development under specific environmental conditions, are known to provide vital and detailed genetic data of crops to plant breeders, which will enable them to identify the best breeding methodologies with respect to numerous genes responsible for different desirable traits. These models also provide vital data that will help the breeding community in identifying genes involved in different biochemical mechanisms as well as gene interactions in response to various micro- and macro-environmental elements/conditions (Wang 2012). These techniques will enable us to develop newer varieties with increased resistance to various pests and diseases, drought resistance, short duration, high yielding and other desirable characteristics. The major focus will be on developing multiple stress-tolerant varieties as well as crops to resist yield losses, identifying more number of resistant genes for different major and minor pests and diseases, conversion of C_3 crops to C_4 to harness the increase in CO_2 levels, detailed investigations on root architecture and systems for the ability to tolerate water stress, development of varieties and crops suitable for growth under problematic soils, flood conditions, etc. Advanced techniques such as genetic engineering and gene editing tools may be made use of for causing favourable mutations in genes playing crucial roles in different climate change adaption mechanisms. Attempts may also be extended to breeding of minor/orphan crops for crop diversification as well as identification of favourable genes for better performance under diverse climatic conditions.

Agroforestry Systems

Agroforestry systems have been identified as a good land use pattern with respect to climate change scenario owing to its high resilience towards climate change as well as its ability to perform both adaptation and adaptation roles (Suresh Ramanan et al. 2020). Agroforestry systems have been identified as a suitable and more effective alternative for various afforestation/reforestation programmes that have several limitations. Various traditional and modern agroforestry systems have their own unique characteristics and are widely practised and adopted by numerous communities globally. Rajeshwar Rao et al. (2018) reported a structure-based classification of agroforestry systems including agrisilviculture, agrihorticulture, silvipastoral, agri-silvipastoral and other specialised systems in addition to different existing classifications (Nair 1993; Nair et al. 2017). Agroforestry systems are known to improve the carbon sequestration, thus offsetting GHG emissions up to a certain extent, thus reducing the pressure on natural forests to do so. Regular agroforestry systems in India are expected to sequester C at the rate of 0.21 Mg C/ha/year (Dhyani et al. 2009). Similarly, using modelling technique, it was estimated that existing agroforestry systems in India have the ability to offset nearly 33% of GHG emissions from agricultural sector (Gupta et al. 2017). These systems could also provide communities with additional benefits such as improved food security, advanced soil health, reducing land degradation rates and providing other ecosystem services compared to tree plantations in addition to maintaining/improving the biological diversity of the ecosystems and supporting livelihoods. Among the different agroforestry systems, agri-silvipastoral system is reported to maintain the highest SOC contents as well as biomass carbon stocks compared to others (Nath et al. 2021).

Precision Farming/Digital Agriculture

Precision farming is as an important step in our efforts towards sustainability via climate-smart agriculture. It's a collective term for the adoption of various technological interventions such as the use of global positioning systems (GPS), geographic information system (GIS), different types of sensors, adoption of variable rate technologies (VRTs), yield monitoring systems, etc., at various stages of crop production. Roy and Kalambukattu (2020) have given a much-detailed description about the role of precision farming as a step towards climate-smart agriculture. Precision farming techniques are based on the concept of providing the right amount of inputs at right place and time, thereby improving the use efficiency of different inputs. This involves various technological interventions to improve precision at various steps of farming including soil preparation, seeding, crop management, irrigation management, fertilisation, harvesting along with data analysis and evaluation for further modifications and improvement. It helps in better management of on-farm variability and thus aids in the design of site-specific crop management (SSCM) practices. Adoption of mechanisation for various farm operations also helps in conservation as well as optimum utilisation of agricultural resources. Modi et al. (2020) have given a detailed account of various farm mechanisation approaches having the potential to combat the impact of climate change on crop production. These mechanisation approaches play significant roles at various stages such as seedbed preparation, land levelling, sowing/transplantation, weeding, irrigation/water management, other intercultural operations, fertiliser application/nutrient management, plant protection operations, crop harvesting/threshing, residue management, etc.

Different types of sensors including electrical/electromagnetic, optical, mechanical, electrochemical, pneumatic sensors, etc., provide quick, reliable, repetitive and real-time estimation of various soil properties including moisture as well as nutrient status. This will be helpful in devising various management strategies aimed at resource conservation and improving use efficiencies. Monitoring of grain yields/yield mapping making use of GPS-based harvesting technologies helps farmers/farm managers in assessing the spatial variability distribution of yield across the area, thus identifying the highest/lowest yielding areas (Mandal and Maity 2013). These yield maps also help in identifying and assessing various factors causing the yield variation and in devising better-input management strategies at farm level. Combining the information obtained from various sensors-based monitoring devices with various remote sensing data sources such as unmanned aerial vehicles (UAVs) will help in devising crop/field requirement-based application of different inputs like seeds, fertilisers, irrigation scheduling, plant protection chemicals, etc., as part of adopting VRTs instead of uniform application. Successful implementation of VRTs requires the availability and adoption of machineries having the ability to apply different inputs (seed drills, sprayers, irrigation systems, etc.) at varying rates within the field/farm. The increased availability of remote sensing data at much finer resolutions will aid in the preparation of various soil property maps, studying their spatial variability as well as delineation of management zones. The rate of different inputs to be applied could be adjusted as per the management zones, thus avoiding the excess application and wastage of resources as well as environmental pollution, without any compromise on yield levels.

ICT Applications

Recent technological advancements such as artificial intelligence (AI) and Internet of things (IoT) have also been increasingly adopted across the globe for improving the climate adaptability of agricultural systems (Dheeraj et al. 2020). Under the scheme of AI, various components such as machine learning/deep learning algorithms, robotics, computer vision, etc., are used for sensing as well as processing of large quantum of diverse datasets, constructing robust systems for weather data processing and interpretation. This helps to understand and predict the influence of environmental variations on crop growth/yield and thus aids in formulating effective and possible solutions for risk reduction of farmers. This also helps in easy dissemination of vital information, knowledge transfer and feedbacks across a large user network without much human interaction in a short span of time. This can also lead to improved automation of various farm operations, designing of early warning systems and prediction of climatic extremes such as flood, drought, etc., by collecting and combining multisource data. AI-based systems are known to automatically detect weeds and initiate spraying of herbicides, identify disease/pest, identify pollutants and contribute to land suitability assessment. Machine learning combined with remote sensing data is known to play vital roles in effective monitoring of emissions due to agricultural activities, automatic detection of deforestation activities, irrigation scheduling, stubble management, renewable energy production, etc. Precision agriculture is also known to mitigate GHG emissions largely by adjusting the rate and timing of fertilisation/agrochemicals, water management, adjusting soil conditions, etc.; thus, adoption of PA may have multifaceted influence on devising various climate-proofing strategies.

Agriculture Sustainability and Climate Change

Agricultural production systems are heavily dependent on the climatic parameter to maintain their productivity. Climatic parameters and weather variables amount for about 67% variation in agricultural productivity, while the rest (33%) is due to soil-related and other management factors (Yadav et al. 2020). The extremities in climatic conditions, shift in rainfall patterns, rise in temperature, variation in rainfall erosivity and increased incidences of flood and drought all are likely to impact agricultural productivity in a negative way. In recent times, drought during the year 2002 led to an estimated monetary loss of US$ 0.9 billion impacting almost

0.3 billion people in the country (Samra and Singh 2004). Thus, the adaptation of different measures and strategies and shaping of national-level policies to address these challenges at varied temporal and spatial scale are urgent. Various such adaptation measures and climate-proofing techniques have been discussed in detail in the previous section of this chapter. Here, we will fact check how far these measures are sustainable for the future.

What makes a measure/strategy climate resilient: To understand this phenomenon, we should have very thorough knowledge of how climate change impacts different activities and services in daily lives. It is stated that even if the global emission of GHGs is reduced below the current levels in the future by adaptation of the different climate goals, the impact of climate change would still be felt in various sectors. Thus, it is necessary to have an action plan for all the developmental activities carried at different levels. For example, low-lying areas under rice cultivation may be inundated permanently by rise in sea level in the coastal areas, leading to huge financial losses. Therefore, alternate land use planning like brackish water aquaculture or other suitable activities has to be planned for these areas (Hahn and Frode 2010). The bottom line to ensure the sustainability of any proposed measure is to take into consideration the impact of climate change during the planning phase to ensure its future viability. The impact analysis at a temporal scale of short-term, medium-term and long-term climate change impact on planned activities is essential for decision-making (Fig. 12.2).

To deal with the challenges emerging from the changing climate, research has been initiated at various platforms. The Indian Council of Agricultural Research initiated the network project on National Innovation on Climate-Resilient

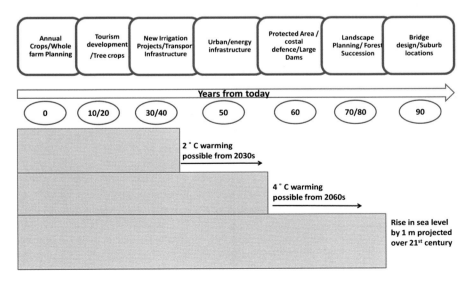

Fig. 12.2 Planning horizons: balance of options for action changes from "autonomous and incremental" to "planned and transformative". (Adapted from Stafford Smith et al. 2010)

Agriculture (NICRA, http://www.nicra.icar.in). The concept of climate-smart agriculture (CSA) and climate-smart villages (CSV) has also emerged simultaneously to help the resource-poor farmers plan and implement sustainable agricultural packages which not only acts as a climate-proofing measure but also reduces the contribution of agricultural sector to climate change (Lipper et al. 2014).

Adoption of water-saving technologies is highly sustainable particularly under Indian scenario where flooded rice is a major crop. Technologies like alternate wetting and drying can save up to 38% irrigation water without having adverse impact on crop productivity (Lampayan et al. 2015). Similarly, laser levelling of field resulted in higher rice yield, more water savings and less consumption of fuel or energy savings (Aryal et al. 2015). Adoption of conservation agricultural practices and reduced tillage is highly energy efficient and helps in increasing the SOC by favouring the formation of more micro-aggregates over macroaggregates (Six et al. 2004). Practices such as introduction of agroforestry in agricultural systems are highly robust and more sustainable and can help to sequester 0.29–15.21 Mg C ha^{-1} year^{-1} (Nair et al. 2010). Kumar et al. (2013b) reported that zero tillage in rice-wheat cropping system in the Indo-Gangetic plain resulted in 13% higher energy use efficiency compared to conventional tillage.

Various such techniques have been discussed in the previous section which is sustainable both in terms of enhancing or maintaining the production of different agricultural systems and reducing the impact of climate change. Also, these technologies are more energy efficient, thereby contributing less emission to atmosphere and making them climate-smart.

Geospatial Variability in LGP

The LGP defines the period of the year when both moisture and temperature conditions are suitable for crop production. The crop growing period provides a framework for summarising temporally variable elements of climate, which can then be compared with the requirements and estimated responses of the crop. The estimation of crop growing period is based on a water balance model which compares rainfall (P) with potential evapotranspiration (PET). GIS and simulation models are commonly employed to compute spatial LGP and crop suitability assessment for crop planning (Abou-Shleel and El-Shirbeny 2014).

Soil water balance model of BUDGET Program (Raes 2005) was used to estimate length of growing period (LGP) by determining the water storage of soils as affected by the input and withdrawal of water by a crop for a given period. The study computed general LGP using a long-term climatic database (1980–2003) to compute decadal rainfall and temperature variations of 13 IMD stations in part of Madhya Pradesh state. Crop specific LGP for soybean was also mapped based on Soil water balance model of BUDGET Program, for different soil types using soil map. Digitised soil map was employed to map spatial distribution of crop-specific LGP (Kumar et al. 2013a; Anonymous 2008).

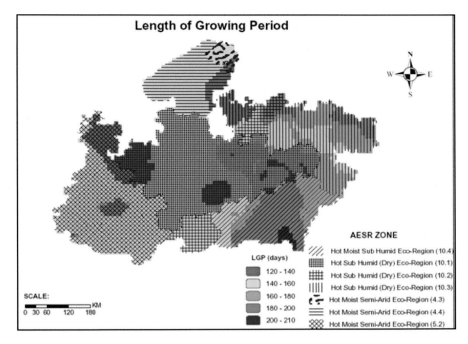

Fig. 12.3 Spatial variability in LGP in Madhya Pradesh state

Several studies of agro-ecological zoning (AEZ) computed LGP based on average monthly rainfall and PET. This approach may be acceptable for broad-scale regional studies. The estimation of growing period is based on a water balance model, which compares rainfall (P) with potential evapotranspiration (PET). The determination of the beginning of the growing period is based on the start of the rainy season. The rainfall and temperature on a decadal basis for each of the 28 stations for a period of 24 years (1980–2003) were computed. Then, the PET (mm/ month) was estimated using the Thornthwaite method. FAO New LocClim software was used to interpolate weather station data to generate spatial layer of rainfall and PET using the thin-plate splines method and to compute LGP for Madhya Pradesh state (Fig. 12.3).

The Budget Program (Raes 2005) was run by defining climatic input of daily mean 10-day evaporation and rainfall data. Soil parameters such as available water-holding capacity, effective root zone and soybean crop parameters such as crop rooting depth, sensitive crop growth stages to water stress etc were defined in the program. The programme simulates key water balance component for the crop at critical growth stages and associated relative water-limited yield potential. Crop-specific LGP was computed using decadal/monthly rainfall and PET as inputs to compute pseudo-daily estimates of actual (Eta) and potential (ETp) evapotranspiration. Crop-specific LGP for pigeon pea (Fig. 12.4a) and sorghum crops (Fig. 12.4b) of the soil map units were computed and represented spatially through GIS platform.

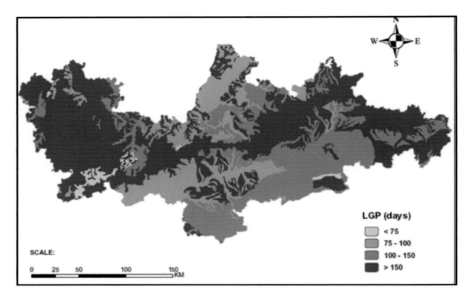

Fig. 12.4a Crop-specific LGP (days) for pigeon pea crop

Climate Change and Geospatial Analysis for Crop Suitability

Remote sensing and GIS-based assessment provide unique opportunity for the spatial distribution of crop suitability and production potential. Climate change will alter the growth period of crops (duration) and growing season. It has been predicted of high precipitation, intense precipitation and reduction in number of consecutive rainless days. This information is needed while evaluating the variation in the trend of future weather events and their impact on crop growth. Crop suitability assessment is carried out at a single time for specific crops using soil map based on requirement criteria of the crop. These crop suitability maps are diminishing their relevance over time, particularly in the face of rapid global climate change. Climate projections need to be accounted to generate crop suitability map indicating possible future suitable area under various emission scenarios. It will help in predicting crop failure and to avoid risk area and their management to maximise crop production based on geospatial climate data alone (Lal et al. 1998; Seif-Ennasr et al. 2020).

In the geospatial sciences, availability of temporal data remote sensing data, abundance of climate and biophysical data, simulation models and computational tools are capable of processing large volumes of geospatial data. Recent development in geographic information science offers considerable value for biogeographic crop suitability and enhances the impact of agricultural improvement across the world. GAEZ products distributed by the FAO of the United Nations offered the most notable crop suitability maps of various regions of the world (Fischer et al.

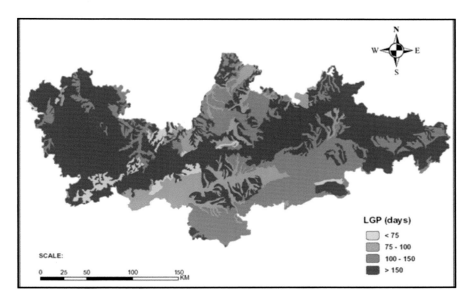

Fig. 12.4b Crop-specific LGP (days) for sorghum crop

2021). These products are immensely valuable; however, uncertainty issues are embedded in the products that utilise future climate predictions. It is necessitated to generate dynamic biogeographic crop suitability maps employing geospatial technology with integrating crop simulation models and climate change scenarios in managing crop production for achieving food security (Liu et al. 2017).

The crop growth simulation model predicts crop growth and response to variable climate, soil quality and management strategies using complex plant physiology algorithms. Various simulation models such as AquaCrop to estimate the impact of water supply on crop viability, DIVA-GIS as an open software to map the suitable geographic area of crop species with climate data extraction, DSSAT (Decision Support System for Agrotechnology Transfer) and APSIM (Agricultural Production Systems sIMulator) to simulate crop yields based on parameters defined by the user using embedded databases and functions (e.g. crop rotations and management) and meteorological parameters (e.g. solar radiation, temperature and rainfall) are available to study the impact of different climate parameters on crop growth (Daloz et al. 2021). Model simulations are important research tools to understand crop growth processes and to simulate crop yields. However, they require specialised training and expert knowledge to produce relevant results.

Web-based GIS software has the potential to transform the crop niche mapping paradigm by integrating a series of static maps to a dynamic form to derive temporal continuity with interactive geovisualisations. GEE is a web-based high-computational platform that facilitates unprecedented planetary-scale satellite imagery access and geospatial analytics38. In recent times, the emergence of GEE also supports simulation models by improving data accessibility, data processing efficiency and scalability of geographic solutions.

Conclusion

Agriculture sector is expected to be highly influenced by the changes in climate parameters. Such impacts may cause profound impact on humankind globally. The extent of influences may vary between different agro-climatic/agroecosystems due to their inherent variations in climatic parameters as well as the crops or agricultural practices adopted. Understanding various climate change scenarios as well as projections with respect to the world and Indian region as discussed will help in understanding the possible impacts. Impacts of climate change on agriculture, especially with respect to Indian regions, especially the direct impact on crop yield and productivity as well as different indirect impacts on water resources, livelihood, etc., were also discussed in detail. Further, various adaptation strategies that are suitable and could be adopted at various agro-climatic regions including agronomic, soil management, water management, precision agriculture, agroforestry interventions, etc., were elaborated from a climate-proofing point of view. Further, the role of geospatial technology in assessing the LGP making use of different crop growth simulation models was also briefly detailed.

References

Abou-Shleel SM, El-Shirbeny MA (2014) GIS assessment of climate change impacts on tomato crop in Egypt. Glob J Environ Res 8(2):26–34

ADB (2013) Guidelines for climate proofing investment in agriculture, rural development, and food security. ADB, pp 1–104

Adler PB, Leiker J, Levine JM (2009) Direct and indirect effects of climate change on a Prairie plant community. PLoS One 4(9):e6887. https://doi.org/10.1371/journal.pone.0006887

Aggarwal PK (2003) Impact of climate change on Indian agriculture. J Plant Biol New Delhi 30(2):189–198

Aggarwal PK, Mall R (2002) Climate change and rice yields in diverse agro-environments of India. II. Effect of uncertainties in scenarios and crop models on impact assessment. Clim Chang 52:331–343

Allen DE, Singh BP, Dalal RC (2011) Soil health indicators under climate change: a review of current knowledge. In: Singh BP et al (eds) Soil health and climate change. Springer, Berlin/Heidelberg, pp 25–45

Anonymous (2008) Geo-spatial approach in soil & climatic data analysis for agro-climatic suitability assessment of major crops in rainfed agro-ecosystem (A case study of Parts of Madhya Pradesh). M. Tech. thesis, IIRS. https://www.iirs.gov.in/iirs/sites/default/files/StudentThesis/aditi_finals_thesis.pdf

Aryal JP, Mehrotra MB, Jat ML, Sidhu HS (2015) Impacts of laser land levelling in rice–wheat systems of the north–western Indo-Gangetic plains of India. Food Secur 7(3):725–738. https://doi.org/10.1007/s12571-015-0460-y

Asseng S, Foster I, Turner NC (2011) The impact of temperature variability on wheat yields. Glob Chang Biol 17:997–1012

Atlin GN, Cairns JE, Das B (2017) Rapid breeding and varietal replacement are critical to adaptation of cropping systems in the developing world to climate change. Glob Food Sec 12(1):31–37

Benbi DK (2012) Impact of climate change on soil health. Ann Agric Res 33(4):204–213

Benbi DK (2013) Greenhouse gas emissions from agricultural soils: sources and mitigation potential. J Crop Improv 27(6):752–772

Benestad RE, Hanssen-Bauer I, Chen D (2008) Empirical-statistical downscaling. World Scientific, New York

Cao L, Wang Q, Deng Z, Guo X, Ma X, Ning H (2010) Effects of climate warming and drying on millet yieldin Gansu province and related countermeasures. Ying Yong Sheng Tai Xue Bao 21: 2931–2937

Chu JT, Xia J, Xu CY, Singh VP (2010) Statistical downscaling of daily mean temperature, pan evaporation and precipitation for climate change scenarios in Haihe River, China. Theor Appl Climatol 99:149–161

Coleman K, Jenkinson DS (1996) Roth C26.3 – a model for the turnover of carbon in soil. In: Datasets DS, Powlson PS, Smith JU (eds) Evaluation of soil organic matter models using existing long-term. Springer, Berlin, pp 237–246

Cook B, Smerdon JE, Seager R, Coats S (2014) Clim Dyn 43(9–10):2607. https://doi.org/10.1007/s00382-014-2075-y

Dai A (2011) Wiley Interdiscip Rev Clim Chang 2(1):45. https://doi.org/10.1002/wcc.81

Dai A (2013) Nat Clim Chang 3(1):52. https://doi.org/10.1038/nclimate1633

Daloz AS, Rydsaa JH, Hodnebrog O, Sillmann J, van Oort B, Mohr CW, Agrawal M, Emberson L, Stordal F, Zhang T (2021) Direct and indirect impact of climate change on wheat yield in the Indo-Gangetic plain in India. J Agric Food Res 4:100–132

Dheeraj A, Nigam S, Begam S, Naha S, Jayachitra Devi S, Chaurasia HS, Kumar D, Ritika Soam SK, Srinivasa Rao N, Alka A, Sreekanth PD, Sumanth Kumar VV (2020) Role of artificial intelligence (AI) and internet of things (IoT) in mitigating climate change. In: Srinivasarao C et al (eds) Climate change and Indian agriculture: challenges and adaptation strategies. ICAR-National Academy of Agricultural Research Management, Hyderabad, pp 325–358

Dhyani SK, Newaj R, Sharma AR (2009) Agroforestry: its relation with agronomy, challenges and opportunities. Indian J Agron 54(3):249–266

FAO (1976) Irrigation and drainage paper 46. Land and Water Development Division, FAO, Rome

FAO (1984) Tillage systems for soil and water conservation. FAO Soil Bulletin 54. FAO, Rome

FAO (2007) Global Administrative Unit Layers (GAUL). Available at http://www.fao.org/geonetwork/srv/en/metadata.show?id=12691&currTab=simple

FAO, IFAD, UNICEF, WFP and WHO (2020) In brief to the state of food security and nutrition in the world 2020. Transforming food systems for affordable healthy diets. FAO, Rome. https://doi.org/10.4060/ca9699en

Feroze SM, Saha B, Singh R (2016) Climate change impact on Jhum in North Eastern Himalaya: a case study of North Tripura. In: Kinhal et al (eds) Climate change combating through science and technology

Fischer G, Nachtergaele FO, van Velthuizen HT, Chiozza F, Franceschini G, Henry M, Muchoney D, Tramberend S (2021) Global Agro-Ecological Zones (GAEZ v4) model documentation. FAO & IIASA, p 303

Frederick KD, Major DC (1997) Climate change and water resources. Clim Chang 37(1):7–23

Gagnon S, Singh B, Rousselle J, Roy L (2005) An application of the statistical downscaling model (SDSM) to simulate climatic data for stream flow modelling in Québec. Can Water Resour J 30(4):297–314. https://doi.org/10.4296/cwrj3004297

Glotter M, Elliott J, McInerney D, Best N, Foster I, Moyer EJ (2014) Evaluating the utility of dynamical downscaling in agricultural impacts projections. Proc Natl Acad Sci 111(24): 8776–8781. https://doi.org/10.1073/pnas.1314787111

Gobu R, Shiv A, Anilkumar C, Basavaraj PS, Harish D, Adhikari S, Ramtekey V, Hudedamani U, Mulpuri S (2020) Climate smart soil and water management strategies for sustainable agriculture. In: Srinivasarao CH et al (eds) Climate change and Indian agriculture: challenges and adaptation strategies. ICAR-National Academy of Agricultural Research Management, Hyderabad, pp 49–69

Godfray HCJ, Beddington JR, Crute IR, Haddad L, Lawrence D, Muir JF, Pretty J, Robinson S, Thomas SM, Toulmin C (2010) Food security: the challenge of feeding 9 billion people. Science 327:812–818

Government of India (2008) National action plan on climate change. Prime Minister's Council on Climate Change
Goyal MK, Ojha CSP (2010) Robust weighted regression as a downscaling tool in temperature projections. International journal of global warming. Interscience Publishers, UK 2(3):234–251
Guhathakurta P (2006) Long-range monsoon rainfall prediction of 2005 for the districts and sub-division Kerala with artificial neural network. Curr Sci 90(6):773–779
Guhathakurta P, Rajeevan M (2008) Trends in rainfall pattern over India. Int J Climatol 28:1453–1469
Guhathakurta P, Sreejith OP, Menon PA (2011) Impact of climate change on extreme rainfall events and flood risk in India. J Earth Syst Sci 120(3):359–373
Guhathakurta P, Rajeevan M, Sikka DR, Tyagi A (2015) Observed changes in southwest monsoon rainfall over India during 1901–2011. Int J Climatol 35(8):1881–1898
Gumbricht T, Cuesta RMR, Verchot L, Herold M, Wittman F, Householder E, Herold N, Murdiyarso D (2017) An expert system model for mapping tropical wetlands and peatlands reveals South America as the largest contributor. Glob Chang Biol 2:3581–3599
Guo D, Desmet PG, Powrie LW (2017) Impact of the future changing climate on the Southern Africa biomes and the importance of geology. J Geosci Environ Protect 5:1–9. http://www.scirp.org/journal/gep
Gupta A, Dhyani S, Handa AK, Newaj R, Chavan S, Alam B, Prasad R, Ram AR, Raza J, Tipathi AU, Shakhela D, Patel R, Dalvi A, Vijay Saxena A, Parihar A, Ravalan B, Sudhagar J, Gunasekaran S (2017) Estimating carbon sequestration potential of existing agroforestry systems in India. Agrofor Syst 91. https://doi.org/10.1007/s10457-016-9986-z
Habibullah MS, Din BH, Tan SH, Zahid H (2022) Impact of climate change on biodiversity loss: global evidence. Environ Sci Pollut Res 29(1):1073–1086
Hadebe S, Modi A, Mabhaudhi T (2017) Drought tolerance and water use of cereal crops: a focus on sorghum as a food security crop in sub-Saharan Africa. J Agron Crop Sci 203:177–191
Hahn M, Fröbe A (2010) Climate proofing for development: adapting to climate change, reducing risk. In: Climate proofing for development: adapting to climate change, reducing risk. Deutsche Gessellschaft für Technische Zusammenarbeit (GTZ), Alemania
Haokip IC, Premalatha RP, Homeshwari Devi M, Chetan Jangir BH, Sunil SK, Surekha K, Srinivasa Rao C (2020) Climate change adaptation and mitigation through soil management. In: Srinivasarao C et al (eds) Climate change and Indian agriculture: challenges and adaptation strategies. ICAR-National Academy of Agricultural Research Management, Hyderabad, pp 105–130
Hawkesford MJ, Araus JL, Park R, Calderini D, Miralles D, Shen T, Zhang J, Parry MA (2013) Prospects of doubling global wheat yields. Food Energy Secur 2:34–48
http://jalshakti-dowr.gov.in/agro-climatic-zones
https://icar.org.in/files/state-specific/chapter/3.htm
https://www.adaptationcommunity.net/download/ms/mainstreaming-guides-manuals-reports/gtz-climateproofing-td-2010-en(2).pdf
https://www.canr.msu.edu/news/feeding-the-world-in-2050-and-beyond-part-1
https://www.fao.org/fileadmin/templates/wsfs/docs/Issues_papers/HLEF2050_Global_Agriculture.pdf
https://www.fao.org/india/fao-in-india/india-at-a-glance/en/
https://www.fao.org/news/story/en/item/197623/icode/
https://www.ibef.org/industry/agriculture-india.aspx
https://www.worldbank.org/en/topic/agriculture/brief/food-security-and-covid-19
https://www.worldbank.org/en/topic/agriculture/overview#1
IPCC (1990) Climate change: the IPCC scientific assessment. Cambridge University Press, 212 pp
IPCC (2001) In: Houghton JT, Ding Y, Griggs DJ, Noguer M, van der Linden PJ, Xiaosu D (eds) Climate change 2001: the scientific basis: contribution of Working Group I to the third assessment report of the Intergovernmental Panel on Climate Change. Cambridge University Press, Cambridge
IPCC (2007) Fourth Assessment Report (AR4) of the United Nations Intergovernmental Panel on Climate Change

Jones PG, Thornton PK, Giron E (2011) Web application. MarkSim GCM-A weather simulator. https://gismap.ciat.cgiar.org/MarkSimGCM

Kendall H, Pimentel D (1994) Constraints on the expansion of the global food supply. Ambio 23(3): 198–205

Krishnan R, Sanjay J, Gnanaseelan C, Mujumdar M, Kulkarni A, Chakraborty S (2020) Assessment of climate change over the Indian region: a report of the ministry of earth sciences (MOES), government of India. Springer, p 226

Kumar S, Patel NR, Sarkar A, Dadhwal VK (2013a) Geospatial approach in assessing agro-climatic suitability of soybean in rainfed agro-ecosystem. Indian Soc Remote Sensing 41(3):609–618

Kumar V, Saharawat YS, Gathala MK, Singh A (2013b) Effect of different tillage and seeding methods on energy use efficiency and productivity of wheat in the Indo-Gangetic plains. Field Crop Res 142:1–8

Kumar G, Adhikary PP, Dash C (2020) Spatial extent, formation process, reclaim ability classification system and restoration strategies of gully and ravine lands in India. In: Gully erosion studies from India and surrounding regions. Springer, Cham, pp 1–20

Lal R (2004) Soil carbon sequestration to mitigate climate change. Geoderma 123:1–22

Lal M, Singh KK, Rathore LS, Srinivasan G, Saseendran. (1998) Vulnerability of rice and wheat yields in NW India to future changes in climate. Agric For Meteorol 89(2):101–114

Lampayan RM, Rejesus RM, Singleton GR, Bouman BAM (2015) Adoption and economics of alternate wetting and drying water management for irrigated lowland rice. Field Crop Res 170: 95–108. https://doi.org/10.1016/j.fcr.2014.10.013

Leggett J, Pepper WJ, Swart RJ, Edmonds J, Filho LGM, Mintzer I, Wang MX, Watson J (1992) In: Houghton JT, Callander BA, Varney SK (eds) Emissions scenarios for the IPCC: an update. Cambridge University Press, pp 73–95

Liang C, Zhu X, Fu S, Mendez A, Gasco G, Ferreiro PJ (2014) Biochar alters the resistance and resilience to drought in a tropical soil. Environ Res Lett 9(6):1–6

Lipper L, Thornton P, Campbell BM, Baedeker T, Braimoh A, Bwalya M, Torquebiau EF (2014) Climate-smart agriculture for food security. Nat Clim Chang 4. https://doi.org/10.1038/nclimate2437

Liu DL, O'Leary GJ, Christy I, Macadam B, Wang MR, Anwar AW (2017) Effects of climate downscaling methods on the assessment of climate change impacts on wheat cropping systems. Climate Change 144(4):687–701

Mandal SK, Maity A (2013) Precision farming for small agricultural farm: Indian scenario. Am J Exp Agric 3(1):200

Mandal D, Roy T, Kumar G, Yadav D (2021) Loss of soil nutrients and financial prejudice of accelerated soil loss in India. Indian J Fertil 17(12):1286–1295

Mangalassery S, Sjogersten S, Sparkes DL, Sturrock CJ, Craigon J, Mooney SJ (2014) To what extent can zero tillage lead to a reduction in greenhouse gas emissions from temperate soils? Sci Rep 4:4586

Matthews R, Kropff M, Horie T, Bachelet D (1997) Simulating the impact of climate change on rice production in asia and evaluating options for adaptation. Agric Syst 54:399–425

MEA (Millennium Ecosystem Assessment) (2005) Ecosystems and human well-being: synthesis. Island Press, Washington, DC. https://www.cifor.org/knowledge/publication/1888/

Modi RU, Manjunatha K, Gautam PV, Nageshkumar T, Sanodiya R, Chaudhary V, Murthy GRK, Srinivas I, Srinivasa Rao CH (2020) Climate-smart technology based farm mechanization for enhanced input use efficiency. In: Srinivasarao CH et al (eds) Climate change and Indian agriculture: challenges and adaptation strategies. ICAR-National Academy of Agricultural Research Management, Hyderabad, pp 325–358

Moss RH, Edmonds JA, Hibbard KA, Manning MR, Rose SK, van Vuuren DP, Carter TR, Emori S, Kainuma M, Kram T, Meehl GA, Mitchell JFB, Nakicenovic N, Riahi K, Smith SJ, Stouffer RJ, Thomson AM, Weyant JP, Wilbanks TJ (2010) The next generation of scenarios for climate change research and assessment. Nature 463:747–756. https://doi.org/10.1038/nature08823

Mrunalini K, Rolaniya LK, Datta D, Kumar S, Behera B, Makaarana G, Singh A, Prasad JVNS, Pratibha G, Naik MR, Swamy GN, Srinivasarao (2020) Resource conservation technologies for climate change adaptation and mitigation. In: Srinivasarao C et al (eds) Climate change and Indian agriculture: challenges and adaptation strategies. ICAR-National Academy of Agricultural Research Management, Hyderabad, pp 131–156

Mukhopadhyay D (2009) Impact of climate change on forest ecosystem and forest fire in India. In: IOP conference series. Earth and environmental science, vol 6, no 38. IOP Publishing

Nair PKR (1993) An introduction to agroforestry. Springer

Nair PR, Nair VD, Kumar BM, Showalter JM (2010) Carbon sequestration in agroforestry systems. In: Advances in agronomy, vol 108. Academic, pp 237–307

Nair PK, Viswanath S, Lubina PA (2017) Cinderella agroforestry systems. Agrofor Syst 91. https://doi.org/10.1007/s10457-016-9966-3

Nakicenovic N et al (2000) IPCC special report on emissions scenarios. Cambridge University Press

Nath AJ, Sileshi GW, Laskar SY, Pathak K, Reang D, Nath A, Das AK (2021) Quantifying carbon stocks and sequestration potential in agroforestry systems under divergent management scenarios relevant to India's Nationally Determined Contribution. J Clean Prod 281:124831

Nelson GC, Rosegrant MW, Koo J, Robertson R, Sulser T, Zhu T, Ringler C, Msangi S, Palazzo A, Batka M (2009) Climate change: impact on agriculture and costs of adaptation, vol 21. IFPRI, Washington, DC

Nema P, Nema S, Roy P (2012) An overview of global climate changing in current scenario and mitigation action. Renew Sust Energ Rev 16(4):2329–2336

Niklaus PA, Falloon P (2006) Estimating soil carbon sequestration under elevated CO_2 by combining carbon isotope labelling with soil carbon cycle modelling. Glob Chang Biol 12:1909–1921

O'Neill BC, Kriegler E, Riahi K, Ebi KL, Hallegatte S, Carter TR, Mathur R, van Vuuren DP (2014) A new scenario framework for climate change research: the concept of shared socio-economic pathways. Clim Chang 122:387–400. https://doi.org/10.1007/s10584-013-0905-2

O'Gorman PA, Dwyer JG (2018) Using machine learning to parameterize moist convection: potential for modeling of climate, climate change, and extreme events. J Adv Model Earth Syst 10(10):2548–2563

Olaya AMS, Cerri CEP, Scala NL, Dias CTS, Cerri CC (2013) Carbon dioxide emissions under different soil tillage systems in mechanically harvested sugarcane. Environ Res Lett 8:1–8

Parry M, Canziani O, Palutikof J, van der Linden P, Hanson C (2007) Contribution to the fourth assessment report of the Intergovernmental Panel on Climate Change. Cambridge University Press, Cambridge

Pathak H, Ladha, Aggarwal PK, Peng S, Das S, Singh Y, Singh B, Kamra SK, Mishra B, Sastri ASRAS, Aggarwal HP, Das DK, Gupta RK (2003) Trends of climatic potential and on-farm yields of rice and wheat in the Indo-Gangetic Plains. Field Crop Res 80:223–234

Polade SD, Gershunov A, Cayan DR, Dettinger M, Pierce DW (2017) Sci Rep 7:10783. https://doi.org/10.1038/s41598-017-11285-y

Post WM, Pastor J, Zinke PJ, Strangenberger AG (1985) Global patterns of soil nitrogen. Nature 317:613–616

Pratibha G, Srinivas I, Rao KV, Shanker AK, Raju BMK, Choudhary DK, Srinivasrao K, Srinivasarao CH, Maheswari M (2016) Net global warming potential and greenhouse gas intensity of conventional and conservation agriculture system in rainfed semiarid tropics of India. Atmos Environ 145:239–250

Raes D (2005) BUDGET – a soil water and salt balance model. In Reference Manual Version 6.0. http://www.iupware.be

Rajeshwar Rao G, Prabhakar M, Venkatesh G, Srinivas I, Sammi Reddy K (2018) Agroforestry opportunities for enhancing resilience to climate change in rainfed areas. ICAR—Central Research Institute for Dryland Agriculture, Hyderabad

Rajput J, Kushwaha NL, Blessy VA, Nivesh S, Paramaguru PK, Rao KV, Kumar M, Srinivasa Rao C (2020) Climate smart soil and water management strategies for sustainable agriculture. In: Srinivasarao C et al (eds) Climate change and Indian agriculture: challenges and adaptation strategies. ICAR-National Academy of Agricultural Research Management, Hyderabad, pp 157–182

Rao AR, Hamed KH, Chen HL (2003) Non-stationarities in hydrologic and environmental time series. Springer, Dordrecht

Rasul G, Hussain A, Mahapatra B, Dangol N (2018) Food and nutrition security in the Hindu Kush Himalayanregion. J Sci Food Agric 98:429–438

Rejani R, Rao KV, Osman M, Chary GR, Pushpanjali, Sammi Reddy K, Srinivasarao CH (2015) Location specific in-situ soil and water conservation interventions for sustainable management of drylands. J Agrometeorol 17(1):55–60

Rinawati F, Stein K, Lindner A (2013) Climate change impacts on biodiversity – the setting of a lingering global crisis. Diversity 5:114–123. https://doi.org/10.3390/d5010114

Rosenzweig C, Iglesius A, Yang XB, Epstein PR, Chivian E (2001) Climate change and extreme weather events – implications for food production, plant diseases, and pests. Glob Change Human Health 2:90–104

Roy T, Kalambukattu JG (2020) Precision farming: a step towards sustainable, climate-smart agriculture. In: Venkatramanan V, Shah S, Prasad R (eds) Global climate change: resilient and smart agriculture. Springer, Singapore

Roy T, Kumar S, Chand L, Kadam DM, Bihari B, Shrimali SS, Bishnoi R, Maurya UK, Singh M, Muruganandam M, Singh L (2019) Impact of Pusa hydrogel application on yield and productivity of rainfed wheat in North West Himalayan region. Curr Sci 116(7):00113891

Samra JS, Singh G (2004) Heat wave of March 2004: impact on agriculture. Indian Council of Agricultural Research, New Delhi

Seager R, Ting M, Li C, Naik N, Cook B, Nakamura J, Liu H (2013) Nat Clim Chang 3:482. https://doi.org/10.1038/nclimate1787

Seager R, Liu H, Henderson N, Simpson I, Kelley C, Shaw T, Kushnir Y, Ting M (2014) J Clim 27(12):4655. https://doi.org/10.1175/JCLI-D-13-00446.1

Seif-Ennasr M, Bouchaou L, El Morjani Z, Hirich A, Beraaouz E, Choukr-Allah R (2020) GIS-based land suitability and crop vulnerability assessment under climate change in Chtouka Ait Baha, Morocco. Atmosfera 11:1167. https://doi.org/10.3390/atmos11111167

Sekar I, Pal S (2012) Rice and wheat crop productivity in the Indo-Gangetic Plains of India: changing pattern of growth and future strategies. Indian J Agric Econ 67:902-2016-67295

Sharma KL, Grace JK, Mandal UK, Gajbhiye PN, Srinivas K, Korwar GR, Bindu VH, Ramesh V, Ramachandran K, Yadav SK (2008) Evaluation of long-term soil management practices using key indicators and soil quality indices in a semiarid tropical Alfisol. Aust J Soil Res 46:368–377

Singh P, Boote K, Kadiyala M, Nedumaran S, Gupta S, Srinivas K, Bantilan M (2017) An assessment of yield gains under climate change due to genetic modification of pearl millet. Sci Total Environ 601:1226–1237

Six J, Ogle SM, Jay Breidt F, Conant RT, Mosier AR, Paustian K (2004) The potential to mitigate global warming with no-tillage management is only realized when practised in the long term. Glob Chang Biol 10(2):155–160. https://doi.org/10.1111/j.1529-8817.2003.00730.x

Solomon S (2007) Climate change 2007 – the physical science basis: working group I contribution to the fourth assessment report of the IPCC, vol 4. Cambridge University Press, Cambridge

Srinivasarao C, Rani YS, Veni VG, Sharma KL, Sankar GM, Prasad JVNS, Prasad YG, Sahrawat KL (2016) Assessing village-level carbon balance due to greenhouse gas mitigation interventions using EX-ACT model. Int J Environ Sci Technol 13(1):97–112

Srinivasarao C, Rao KV, Gopinath KA, Prasad YG, Arunachalam A, Ramana DBV, Ravindra Chary G, Gangaiah B, Venkateswarlu B, Mohapatra T (2020) Agriculture contingency plans for managing weather aberrations and extreme climatic events: development, implementation and impacts in India. Adv Agron 159. https://doi.org/10.1016/bs.agron.2019.08.002

Stafford Smith DM, Horrocks L, Harvey A, Hamilton C (2010) Rethinking adaptation for a four degree world. Phil Trans R Soc A 369(1934):196

Suresh Ramanan S, Soam SK, Srinivasa Rao C (2020) Can planting trees avert climate emergency? In: Srinivasarao C et al (eds) Climate change and Indian agriculture: challenges and adaptation strategies. ICAR-National Academy of Agricultural Research Management, Hyderabad, pp 183–197

Thadani R, Singh V, Chauhan DS, Dwivedi V, Pandey A (2015) Climate change in Uttarakhand. Singh V. 2015 Bishen Singh Mahendra Pal Singh, Dehradun

Upadhyay H, Vinke K, Bhardwaj S, Becker M, Irfan M, George NB, Biella R, Arumugam P, Murki SK, Paoletti E (2021) Locked houses, fallow lands: climate change and migration in Uttarakhand, India. Potsdam Institute for Climate Impact Research (PIK), Potsdam and The Energy and Resources Institute (Teri), New Delhi

Venkatesh G, Srinivasarao C, Gopinath KA, Reddy SK (2015) Low-cost portable kiln for biochar production from on-farm crop residue. Indian Farming 64(12):9–12

Wang J (2012) Modelling and simulation of plant breeding strategies. In: Abdurakhmonov I (ed) Plant breeding. InTech Press, p 19

Wang J, Vanga SK, Saxena R, Orsat V, Raghavan V (2018) Effect of climate change on the yield of cereal crops: a review. Climate 6(2):41

Welch JR, Vincent JR, Auffhammer M, Moya PF, Dobermann A, Dawe D (2010) Rice yields in tropical/subtropical Asia exhibit large but opposing sensitivities to minimum and maximum temperatures. Proc Natl Acad Sci U S A 107:14562–14567

Wilby RL (2004) Guidelines for use of climate scenarios developed from statistical downscaling methods. IPCC Task Group on Data and Scenario Support for Impact and Climate Analysis (TGICA)

WWF (World Wild life Fund) (2020) Living planet report 2020: bending the curve of biodiversity loss. LPR 2020 Full report.pdf

WWF (World Wildlife Fund) (2016) Living planet report 2016: risk and resilience in a new era. WWF International, Switzerland. https://www.google.com/url?sa=t&rct=j&q=&esrc=s&source=web&cd=&cad=rja&uact=8&ved=2ahUKEwjSr_2X-aPrAhW8IbcAHRWXDvUQFjAAegQIBRAB&url=https%3A2F%2Fawsassets.panda.org%2Fdownloads%2Flpr_2016_full_report_low_res.pdf&usg=AOvVaw2JmRg9VlAX_cKy_Svx6Mwu

Yadav RP, Panday SC, Kumar J, Bisht JK, Meena VS, Choudhary M, Nath S, Parihar M, Meena RP (2020) Climatic variation and its impacts on yield and water requirement of crops in Indian Central Himalaya. In: Agrometeorology. IntechOpen

You L, Rosegrant MW, Wood S, Sun D (2009) Impact of growing season temperature on wheat productivity in China. Agric For Meteorol 149:1009–1014

Zhao X, Fitzgerald M (2013) Climate change: implications for the yield of edible rice. PLoS One 8: e66218

Chapter 13
Peri-urban Farmers' Perception of Climate Change: Values and Perspectives – A French Case Study

Marie Asma Ben-Othmen ⓘ**, Juliette Canchel** ⓘ**, Lucie Devillers** ⓘ**, Anthony Hennart** ⓘ**, Lucie Rouyer, and Mariia Ostapchuk** ⓘ

Abstract Peri-urban agriculture often relates to sustainable agriculture and is recognized as a source of fresh and local food intended to maintain healthy populations in and around cities. It plays a significant role in dealing with pressing societal challenges such as sustainable land management, community building, economic development growth, and climate change mitigation and adaptation. Understanding how peri-urban farmers perceive climate change has important implications for a adaptation measures implementation.

This chapter aims to understand the underlying mechanisms shaping how peri-urban farmers process information about climate change and perceive this phenomenon. We argue that these mechanisms are complex and multidimensional, linking many attitudes (e.g., risk perceptions, trust in local policies and institutions *versus* skepticism and denial, etc.) and values (e.g., ecological awareness, efficient use of on-farm resources, etc.), thereby affecting farmers' willingness to implement adaptation measures. We applied a survey-based approach to collect data from 102 farmers in Rouen Normandy Metropole in France. The questionnaire included questions that sought to understand the mechanisms that shape and underlie peri-urban farmers' perception of climate change. The findings revealed five types of farmers: the resilient, the negativist, the peri-urban steward, the independent, and the regulation skeptic. The discussion put forward that identifying farmer types through regional studies helps explore possible adaptation measures to be implemented depending on the farming systems and the specific characteristics of each geographic area. Such granularity enables tackling climate change as a set of localized

---

M. A. Ben-Othmen (✉) · M. Ostapchuk
INTERACT Research Unit–Innovation, Land Management, Agriculture, Agro-Industries, Knowledge, and Technology, UnilaSalle-France, Mont-Saint-Aignan, France
e-mail: marie-asma.benothmen@unilasalle.fr; mariia.ostapchuk@unilasalle.fr

J. Canchel · L. Devillers · A. Hennart · L. Rouyer
Graduate School of Agronomy and Agroindustry, UniLaSalle-France, Mont-Saint-Aignan, France
e-mail: juliette.canchel@etu.unilasalle.fr; lucie.devillers@etu.unilasalle.fr; anthony.hennart@etu.unilasalle.fr; lucie.rouyer@etu.unilasalle.fr

© The Author(s), under exclusive license to Springer Nature Switzerland AG 2022
U. Chatterjee et al. (eds.), *Ecological Footprints of Climate Change*, Springer Climate, https://doi.org/10.1007/978-3-031-15501-7_13

environmental challenges with localized contexts worthy of specific solutions and interventions that could help spread localized urbain food systems.

Keywords Peri-urban agriculture · Attitudes · Motivations · Values · Perceptions · Environmental stewardship

Introduction

Several complex cause-and-effect relationships explain the link between agriculture and climate change (Agovino et al. 2019). Abundant literature emphasizes agriculture's contribution to climate change and global warming through the release of greenhouse gases (Smit and Skinner 2002; Brunner and Baum 2014). In particular, nitrous oxide soil emissions, fertilizer application, grazing animal dejections, and methane production by ruminants are recognized to be significant farming emission sources that trap heat in the atmosphere (Tilman et al. 2002).

Many farmers have experienced warmer and wetter winters in France. This phenomenon has increased heat concentration, raising the risks of pathogen and disease proliferation (Launay et al. 2014). However, climate change impact on agriculture is a controversial issue as several scenarios predict a favorable impact in the North compared to the South for the present farming systems (Merot et al. 2014). While drought and heat are expected to strain arable crops and vineyards in Southern France (Gammans et al. 2017), warmer temperatures leading to minimal frost risk and more extended and advantageous growing season might increase crop yields in the North (Seguin 2003). Moreover, projections put forward grassland and fodder crop potential to offer additional annual and seasonal fodder production opportunities, mainly in spring and winter (Graux et al. 2013).

Agriculture is impacted by location contexts, mainly when agricultural activities occur around cities and metropolitan areas. Indeed, most cities are characterized by unsustainable consumption and production patterns that heavily rely on fossil fuels. Transportation, heating, production, distribution of goods and services, and consumption, to name a few, are among the causes which lead to the continuous increase of carbon dioxide emission, thereby intensifying the climate change effects. Besides, the trend toward city expansion and urbanization contributes to climate change (e.g., more hot summers, increasing sea levels, and more intense rainstorms and floods). It is also associated with increased demand for food, threatening the food security of urban areas' inhabitants.

Climate change will undeniably involve socio-ecological systems adaptation across multiple spatial, temporal, and ecological scales (Adger et al. 2005). In this context, farming around cities can bring sustainability and resilience by mitigating climate change through carbon sequestration and maintaining food production (Marçal et al. 2021). Indeed, peri-urban agriculture often relates to sustainable agriculture and is recognized as a source of fresh and local food intended to maintain healthy populations in and around cities (Pribadi and Pauleit 2015). In France, peri-urban agriculture has been promoted around large cities such as Paris, Nantes, and

Marseille, to contain uncontrolled urbanization and ensure many other benefits. Indeed, peri-urban agriculture simultaneously fosters environmentally sound land-use practice, sustainable economic growth, local urban market delivery with fresh produce, and social links (Duvernoy et al. 2018). Peri-urban agriculture is also recognized for its fundamental role in implementing regional and local food systems, thereby reducing the food carbon footprint (Pradhan et al. 2020) and building urban resilience. It is therefore cornerstone to adapt cities to climate change if developed within integrated systems that account for diverse crop rotations, sustainable water management, and alternative energy sources (Lwasa et al. 2014).

There is an agreement in the literature that the positive effects of peri-urban agriculture depend on farmers' willingness and commitment to implement adaptive measures to counteract the negative effect of climate change. Nonetheless, as Smit and Skinner (2002) state, farmers' decisions and willingness to implement adaptive measures result from their personal interpretations of the short- and long-term evolution of climatic phenomena. By referring to a Canadian case study, the authors explain that farmers' decisions are made in response to climatic and non-climatic stimuli and depend on the social, political, and economic systems within which farmers operate. This generates heterogeneity in farmers' attitudes toward climate change that is acknowledged by a growing body of literature relating to both developed (Evans et al. 2011; Fleming et al. 2010; Gorton et al. 2008; Mitter et al. 2019; Running et al. 2017) and developing (Azadi et al. 2019) countries' case studies. Hence, assessing the risks posed by climate change and how farmers perceive them is crucial in motivating adaptive responses (Arbuckle et al. 2015). Many authors have shown a direct connection between farmers' environmental awareness and behavior and the heterogeneity of their risk perception and attitudes toward climate change (Davies and Hodge 2007; Mitter et al. 2019; Niles et al. 2013). Individuals are likely to protect the environment as awareness and a sense of responsibility for environmental issues intensify and risk assessment related to these issues looms up (Kaida and Kaida 2016; Story and Forsyth 2008).

While earlier literature exists on farmers' attitudes with regard to climate change and its associated issues (Hyland et al. 2016), perceptions of climate change effects related to specific geographic locations such as peri-urban areas have been largely unexplored.

Identifying common characteristics of farmers and their classification helps better understand the impact of a specific set of values and perceptions on a particular issue related to climate change. In particular, Hyland et al. (2016) point to the relevance of identifying farmer types to allow a more efficient spread of knowledge about climate change adaptation and policy initiatives.

This chapter aims to understand the underlying mechanisms shaping French peri-urban farmers' perception of climate change. We argue that these mechanisms are complex and multidimensional, linking many factors such as risk perception (Arbuckle et al. 2015), trust in local policies and institutions *versus* skepticism and denial of the regulation (Davidson et al. 2019), confidence in scientific statements (Pidgeon 2012), perception of existing adaptive measures and solutions, educational level, cultural norms (Running et al. 2017), and personal experience with severe climatic events (Lane et al. 2018). Following Hyland et al. (2016), we hypothesize that farmers' attitudes and values (e.g., farmers' focus on environmental concerns, profit maximization) influence climate change perception, which, in turn, impacts their willingness to implement adaptation measures. The methodology is based on a survey approach established to collect data from farmers operating in Rouen Metropole in France.

Located in Northern France, Rouen Metropole is the prefecture of the region Normandy, marketed by the predominance of dairy and livestock farming systems. Like many metropoles in France, Rouen opted to set up a farm-to-fork system approach by establishing a territorial food system and reducing agriculture's carbon footprint. The main goal is to countermeasure the looming threat of climate change by preserving functional agricultural territories connected to the urban centers. This could be possible through reinforcing a high-quality local food supply chain that is attractive to the metropole inhabitants and promoting adaptation measure uptake by peri-urban farmers.

Rouen metropole offers a relevant case to study the conditions of a transition from dairy and livestock systems, responsible for increasing gas emissions, to more balanced food systems intended to supply near urban area residents with local fresh food. To this end, understanding farmers' perceptions of climate change and their motivations to implement adaptation measures will help organize a win-win dialogue between farmers and territorial decision-makers by implementing a new peri-urban farming model.

Material and Method

Rouen Metropole: A Case Study

Rouen Metropole spreads over 71 municipalities with 490,000 inhabitants, while the city center has approximately 111,000 inhabitants. Major natural elements that structure this territory's landscapes include extensive forests, the Seine River, and agricultural areas covering nearly one-third of the territory (see Appendix: Supplementary Material). At least 365 farmers are spread over the entire metropole territory (Fig. 13.1). The peri-urban agricultural context is comparable to the dominant farming systems in Normandy with a dominance of dairy and livestock farming (28%), field crops (19%), and mixed crops and livestock (19%). In comparison, fruit and vegetable production only accounts for 9% of the total utilized agricultural area (Atlas Agricole et Rural de Normandie 2015). Although the high specialization in dairy and livestock – which is a common trait of the region of Normandy farming system – is a major cause of agricultural greenhouse gas emission (Ben-Othmen et al. 2020; Ben-Othmen and Ostapchuk 2019), many other constraints, such as the rising pressure of urban area expansion at the expense of farmland and the lack of support to small peri-urban farms producing local and fresh food, exacerbate climate change's adverse effects on the metropole territory.

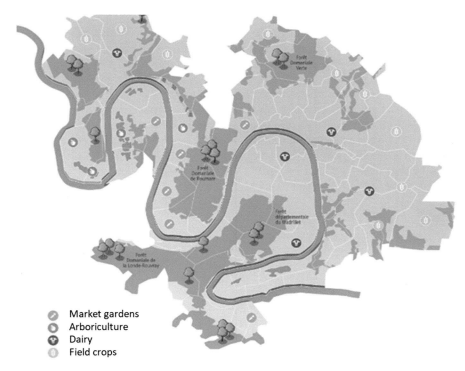

Fig. 13.1 Typology of farming practices in the territory of Rouen Metropole. (Source: Territorial Farming Chart 2018)

Rouen Metropole set out to develop incentives to encourage peri-urban farmers to adopt adaptive measures and diversify local food production in peri-urban areas. In 2018, the metropole initiated a *Territorial Farming Chart* aimed at fostering stakeholders' commitment (e.g., land-use departments, agricultural organizations, environmental preservation public agencies) for joined and coordinated actions for supporting adaptation practices in peri-urban farms aimed at containing climate change's adverse effects (Arnaudet and Mosnier 2017). In particular, diversification has been identified as a potential farm-level response to climate change, and it is believed to help operate the shift from a highly specialized agri-food system (e.g., dairy and livestock) to a more diversified system that accounts for fresh and local food production. Nevertheless, the reasons why farmers would adopt adaptation measures are not well understood to date.

Questionnaire Design and Survey

The questionnaire was designed based on the relevant literature review on farmers' perceptions of climate change. It consisted of three sections. Section one included farmers' sociodemographic information and the structural characteristics of their farms (size, number of employees in the farm) and their farming systems (produced crops, on-farm offered activities, labeled products, etc.). Section two consisted of 40 statements (20 attitudes-related statements and 20 values-related statements) for which farmers were requested to express their opinion on a 5-point Likert scale. The last section asked farmers about their viewpoints on implementing adaptive measures to countermeasure climate change's negative impact.

Before its final administration, the questionnaire was tested with ten farmers. This helped ensure the clarity of questions and the reliability of the study. In total, 324 farmers were contacted, informed about the study's goals, and asked if they would accept participating in the survey by arranging a face-to-face meeting on their farms' sites to answer the questionnaire. The survey was performed by four research assistants trained in the survey method. At the end of the survey campaign (from September to November 2021), we obtained 102 completed questionnaires representing this study's sample size.

The Analysis

The statistical analysis of the participants' responses to the statements consisted of two steps: principal component analysis (PCA) and cluster analysis (CA). PCA underscores common factors to consider; most of the variation in data has been analyzed by examining the patterns of correlations between the questionnaire statements (attitudes and values). When these statements (considered as variables) are significantly correlated, they are, in fact, *expressing the same idea* and therefore characterized as components (Field 2009). PCA was used in diverse contexts to understand how farmers perceive and conceptualize climate change by classifying farmers who share similar perceptions about climate change into groups also called "types" (Arbuckle et al. 2015). This classification of farmers has been proven effective in explaining the heterogeneity of their motivation and uncovering constraints to the use of adaptation measures in several contexts, including European (Barnes and Toma 2012), North American (Hyland et al. 2016), and African (Otitoju and Enete 2016) countries.

Hence, a PCA was applied in our analysis to define components that can be used as criteria to group the respondents into groups. Accordingly, separate sets of PCAs were conducted on attitudes and values statements. The Kaiser-Meyer-Olkin measure of sampling adequacy was higher than 0.8, thus confirming that datasets were accurate for PCA. Bartlett's test of sphericity result is significant at 5% confidence level, suggesting that PCA could be performed (Pallant 2010).

PCA is a practical tool to examine each attitude and value component independently. Nonetheless, interactions between these components are important for knowing how perceptual types could be developed (Barnes and Toma 2012). CA is used to explain these interactions based on the PCA scores by indicating how similar perceptions defined by specific attitudes and values are spread across all farmers to group them into categories.

Results

Based on the Kaiser criterion (with eigenvalues ≥1), we selected factors that explained 52.9% of the attitudes and 54.6% of the values variances. Cronbach's alpha higher than 0.5 is considered satisfactory as evidence of a common factor underpinning the responses (Nunnally 1967). The reliability of each factor was analyzed through CA by deleting inaccurate statements (Field 2009). Nine factors were attained at this level for the attitude's statements; nonetheless, five statements were excluded because of the low value of Cronbach's alpha. All the values statements had, however, significant CA scores. Finally, the analysis yielded four principal components for the attitudes (e.g., regulation skepticism, production-oriented statements). Similar results were obtained for the values statements, and four principal components were retained for post-factor analysis (Table 13.1). Ward's hierarchical and K-means clustering approaches (Burns and Burns 2008) were applied to the factor scores from the previous PCA.

The dendrogram interpretation (Köbrich et al. 2003) suggests the existence of five clusters. The elbow test result confirms the optimal number of clusters for the consecutive K-means clustering to be $n = 5$ which is in line with the dendrogram estimation.

Table 13.1 Attitudes and values principal factors and key statements

Theme	Key statements	Literature review
Attitudes		
Regulation skepticism	Environmental regulations raise farmers' workload	Haltinner (2018), Davidson et al. (2019), Running et al. (2017) and Barnes and Toma (2012)
	Environmental regulation adoption decreases farmers' income	
	Other industries cause environmental pollution more than farmers, and regulation should focus on them	
Production oriented	To make a living from a farm of this size, I must focus on production more than anything else	Morrison et al. (2017), Brobakk (2018), Barnes and Toma (2012), Gorton et al. (2008), Holloway and Ilbery (1996) and Bouzguenda (2019)
	Farmers must always target the highest production from their production	
	Focusing on supplying nearby urban centers, residents will increase my profit margins	
	Price premiums for local food will rise compared to other productions	
Positivism toward climate change	Climate change should be considered as an opportunity rather than a threat	Barnes and Toma (2012), Holloway and Ilbery (1996) and Islam et al. (2013)
	Climate change will enhance my production practices	

(continued)

Table 13.1 (continued)

Theme	Key statements	Literature review
Risk perception of climate change	I have already been a victim of climate change	Barnes et al. (2013), Fahad and Wang (2018) and Sulewski and Kłoczko-Gajewska (2014)
	Climate change is already affecting my ability to invest in my farm	
	Climate change led me to reconsider my farming practices	
	The location of my farm near urban centers exacerbates the adverse impact of climate change	
Values		
Ecological leadership	I need to adopt environmentally friendly practices	Hyland et al. (2016), Gorton et al. (2008), Holloway and Ilbery (1996) and Bouzguenda (2019)
	It is essential for me that my practices respect soil resources	
	I need to reduce the pollution produced by my farming practices	
	I need to protect local biodiversity	
	I need to reduce food miles by producing local food	
	I need to try new varieties that better adapt to climate change	
	It is essential for me to lead than to follow when it is a matter of adopting environmentally friendly practices	
Technological innovation	To farm efficiently, I need to keep up to date with innovative farming technologies	Adamides et al. (2020) and Ho and Shimada (2019)
	Using innovative farming technologies allows farming, while respecting the environment, to be profitable	
Passive behavior	Climate change will impact adversely only in the long term	Carolan (2020)
	Dairy agriculture does not contribute to climate change	
	I do not believe that climate change will require that I shift my business operations to produce more local food	
	Climate change adaptation is not a priority around cities	
Resource efficiency	I need to keep my debt as low as possible	Carolan (2020), Hyland et al. (2016) and Gorton et al. (2008)
	I need to make the best use of my farm resources	

Values and Attitudes Analysis: PCA

Our results suggest that there seems to be some recognition by farmers of climate change's adverse effects. Farmers stated that being near urban centers accentuated the negative impact of climate change. Besides, they agreed that local food production and short food chains reduced climate change's negative impact. These attitude statements are involved into the climate change risk perception factor. Few statements pertaining to farmers' profit perspective are associated with the positive perception of climate change factor, including "I think that the productivity of my farm will improve because of climate change" and "climate change has to be considered as an opportunity rather than a threat." Statements that are included in the production-oriented attitudes factor are "Farmers must always target the highest production from their production," "Focusing on supplying residents of nearby urban centers will improve my profit margins," and "Price premium for organic products will increase compared to conventional production."

Finally, the PCA performed on attitudes has revealed skeptical views toward environmental regulation. Overall, farmers agreed that climate change would increase the costs of their inputs and their workload and reduce their income.

Similarly, the PCA done on values has been classified into four main factors. Pro-environmental and leadership values factor was revealed through statements that underscore the importance attached by farmers to the adoption of environmentally friendly practices, food miles reduction by producing local food intended for supply near urban center residents, and the use of new crop varieties to better adapt to climate change

Statements such as "to farm efficiently, I need to keep up-to-date with innovative farming technologies" and "using innovative farming technologies allows farming while respecting the environment to be profitable" could be associated with the technological innovation value factor. The following two statements pertained to a logic of wise and efficient use of the farm resources: "I need to keep my debt as low as possible" and "I need to make the best use of my farm resources."

It should be noted that studies such as Gorton et al. (2008) have found a link between resource-efficient values and production-oriented attitudes. Accounting for this aspect is crucial as resource use efficiency is critical for meeting national emissions targets (Barnes and Toma 2012).

Finally, statements such as "I do not believe that climate change will require that I shift my business operations to produce more local food" and "Climate change adaptation is not a priority around cities" are included in the passive behavior factor.

Perceptions Toward Climate Change

Crops and livestock farms represented 46% of our sample (Fig. 13.2). Besides, more than 50% of farmers were members of either an agricultural union or a group of exchange of good farming practices (Fig. 13.3). While 12% of farmers sell their produce in farmers' markets, and 8% offer training, 45% do not offer other farm activities (Fig. 13.4).

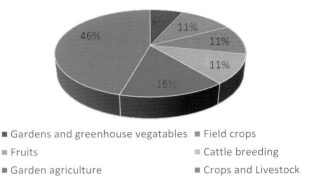

Fig. 13.2 Farmers' technical and economic orientations

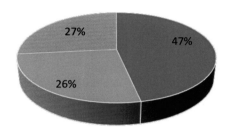

Fig. 13.3 Farmers' professional networks

Fig. 13.4 On-farm activities

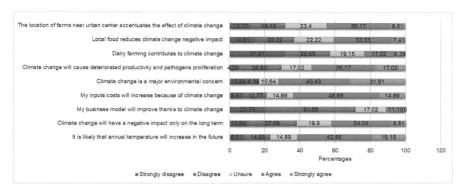

Fig. 13.5 Peri-urban farmers' perceptions toward climate change

Turning to farmers' perception of climate change (Fig. 13.5), the analysis revealed contrasting risk perceptions and concerns about climate change. Of respondents, 42.55% agreed that "it is very likely that annual temperature will rapidly rise in the future"; at the same time, 30.04% agreed to the statement "climate change is already affecting my ability to invest in my farm." While 49% of respondents think that their "on-farm operations costs will rise because of climate change impacts," only 12% of farmers believe that "their productivity will increase thanks to the climate change effect."

CA

The CA following the PCA identified five types of individual farmers: the resilient, the negativists, the regulation skeptics, the independents, and the peri-urban stewards. Radar diagrams are constructed from these cluster centers to visualize the distinctive traits of each farmer's type (e.g., perceptions of climate change) created with respect to the component obtained from PCA.

Figures 13.6 and 13.7 show the main factor loadings for each theme (e.g., values and attitudes) in the clusters presented via the radar graphs. For more intelligibility, clusters are presented in two independent graphs.

The Resilient

The resilient farmer type represents 24% of the sample. The defining qualities of farmers in this group are their high pro-environmental behavior and ecological leadership, while they also tend to use innovative farming technologies. These farmers believe that it is essential that their farming practices be as profitable and

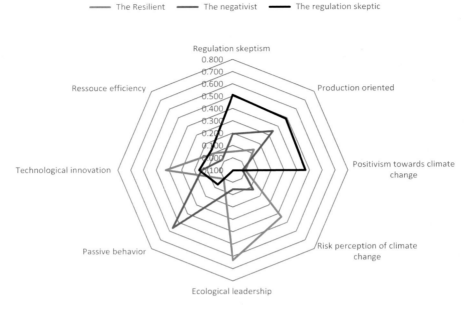

Fig. 13.6 The resilient, negativist, and regulation skeptic farmer types

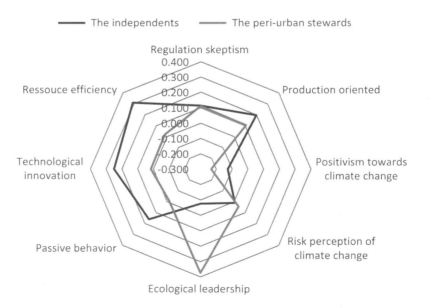

Fig. 13.7 The independent and peri-urban steward farmer types

environmentally friendly as possible. They also deem it necessary to try new crop varieties that better adapt to climate change. Besides, there is a significant agreement among farmers of this group that it is crucial to lead rather than follow when it is a matter of environmentally sound farming technique implementation.

Such farmers are also aware that climate change is occurring. They testify that they have already experienced its adverse effects on their farms (increased temperature, drought, floods, pathogen proliferation); consequently, these farmers score high in perceived risk and level of awareness about climate change.

It should be noted that the resilient farmers were the highest educated among the five clusters as 62% of the farmers of this group had a university degree. Besides, the number of years these farmers have been farming appears to define resilient farmers as a type. Indeed, while most of the other farmer types had been farming for over 28 years, the resilient farmers had been farming between 15 and 21 years. Evans also have found similar results underscoring that the longer farmers had been practicing their activities, the more inclined they were to disagree with scientific evidence about climate change and the less are their perceptions of its threats.

The Negativist

The negativist farmer type represents 22% of our sample. Farmers of this group scored high in the climate change passive behavior, which might correlate with their production-oriented attitude. Not surprisingly, they did not score high in ecological leadership values either. While they share a relative agreement on the likelihood that average annual temperature will increase in the future, farmers of this cluster did not recognize climate change as either a risk that could induce productivity loss or a cause that would lead them to reevaluate their farming and production model. Moreover, farmers of this group did not display any particular agreement with the resource efficiency values, making it difficult to identify any possible behavioral response. Finally, it is interesting to mention that despite their relative regulation skepticism, compared to the resilient type, 27% of farmers in this group display confidence and optimism toward public subsidies to support their resilience in front of climate change risks. This finding is in line with those obtained by studies on how French farmers manage production risks, such as drought (Foudi and Erdlenbruch 2012) and operating risks of agricultural enterprises (Špička et al. 2009). Overall, these studies have shown that governmental subsidies serve as a "financial pillow" increasing the level of farmers' income and extending their decision-making possibilities.

The Regulation Skeptics

The regulation skeptic farmer type represents 19% of our sample and is predominantly defined by farmers' skepticism toward environmental regulation along with their production-oriented attitudes. Farmers of this group believe that regulation will cause more workload and reduce their income and tend to recognize climate change as an opportunity to enhance their farming practices rather than a threat. Besides, values constructs do not seem to significantly contribute to their definition as a type (i.e., regulation skeptics), suggesting that production goals dictate management decisions for the farmers of this group.

Furthermore, the regulation skeptic believed that climate change could affect their farm businesses positively and criticized emissions from other industries that are causing the most harm in terms of climate change. This focus on productivity and profit maximization is a common finding in the literature (Gorton et al. 2008). Previous research by van Huik and Bock (2007), who have studied attitudes of Dutch pig farmers, and Barnes et al. (2009), who have studied farmers' attitudes toward water pollution control programs and regulations in Scotland, put forward a relationship between profit maximization behavior and regulation skepticism. Both studies underscore farmers' focus on maximizing yields as the consequence of their land management decisions (Wilson 2001), which seems to be the case in our study.

The Independent

As their name indicates, the independent farmers, who represent 18% of the sample, display the strongest propensity for independence against climate change impacts (Fig. 13.7). These farmers attach great importance to managing their farms' resources effectively, getting the best market prices for their products, and keeping debt as low as possible. Unlike the regulation skeptic and the negativist, the independent farmers did not display a pronounced production-oriented attitude. Their attitudes toward production are even substantially close to the resilient type.

In this group, the independence characteristic seems to be coupled with farmers' tendency to adopt innovative technology. Besides, independent farmers seem not to have a pronounced sense of climate change concerns, as reflected by their passive behavior and risk perception in the radar graph (Fig. 13.7). For example, 38% of farmers in this group did not believe that a short food supply chain could reduce climate change's adverse effects, while 40% did not deem that climate change adaptation should be prioritized around cities. This disconnect suggests a noticeable lack of understanding of the concept of climate change and how it relates to farming practices in peri-urban contexts. We can conclude that even if they do not expect climate change to enhance their productivity, independent farmers in our sample would more likely adopt innovative technologies that align with efficient resource use and satisfy production and profit maximization requirements.

The Peri-urban Stewards

Peri-urban stewards, who represent 10% of our sample, attach tremendous importance to preserving soil resources and local biodiversity near urban areas, which could be interpreted as an expression of their pronounced sense of responsibility and pro-environmental behavior. These farmers are, however, less prone than the resilient to trying new crop varieties as a way to adapt to climate change. Both groups differ significantly regarding their perception of climate change.

Despite a sign on the attitude expressing positivism toward climate change that is negative and a sign on the attitude expressing the risk perception of climate change that is positive compared to other groups (Table 13.2), most of the peri-urban stewards were not conscious of climate change as a phenomenon implying greenhouse emissions that could be exacerbated by the relative positions of their farms to urban centers. Besides, 41% of these farmers did not recognize dairy farming to cause climate change, and 52% agreed that climate change could only impact negatively in the long run. Interestingly, the proportion of university-educated farmers in this group was significantly lower than the resilient type. Eggers et al. (2015) have recorded similar findings in their study of grassland farmers' attitudes toward climate change in the North German Plains. The authors explain that less-educated farmers are more traditionalist and conservative and tend to explain climate change's adverse effects with natural climate variability.

Table 13.2 Farmer cluster scores as obtained from K-means estimation

	Factors	The resilient	The negativist	The independents	The regulation skeptic	The peri-urban stewards
Attitudes	Regulation skepticism	0.050	0.192	0.113	0.510	0.111
	Production oriented	0.134	0.345	0.148	0.487	0.107
	Positivism toward climate change	−0.001	−0.023	−0.128	0.469	−0.233
	Risk perception of climate change	0.437	0.126	0.009	−0.097	0.042
Values	Ecological leadership	0.632	0.056	−0.077	−0.097	0.375
	Passive behavior	0.002	0.564	0.162	0.067	−0.015
	Technological innovation	0.425	0.143	0.251	0.163	0.015
	Resource efficiency	0.096	0.01	0.311	0.134	0.021

Discussion

This chapter explored how peri-urban farmers conceptualized climate change. It determined a typology of peri-urban farmers based on their perceptions of climate change. The results can be used as a starting point to design adequate support for peri-urban farmers who would implement adaptation measures. Instead of reporting the general farmers' responses, our approach sought to understand different existing perspectives across farming communities. We, therefore, applied a descriptive approach that defines types of farmers' perceptions according to a set of values and attitudes.

Despite an expanding body of scholarships addressing farmers' belief or absence of belief in climate change in various geographic locations (Lane et al. 2018; Menapace et al. 2015; Mitter et al. 2019; Schattman et al. 2016), research on climate change perception in peri-urban and metropolitan areas remains largely absent in the literature. Our findings highlight positive and negative perceptions of climate change which aligns with previous research (Barnes and Toma 2012; Holloway and Ilbery 1996) regardless of the differences between peri-urban farms and rural farming context in terms of spatial, socioeconomic, and ecological characteristics.

Given the diversity of farmers' viewpoints and perceptions of climate change expressed within this study, our results suggest grouping them into five major types. Most of the resilient, the negativist, and the peri-urban steward farmers share a common belief that climate change will negatively impact them in the future. In particular, the resilient type composed of farmers who have already experienced production loss attributed to climate change is more likely to take a leadership role in implementing adaptation measures and trying new crop varieties. This is consistent with Menapace et al.'s (2015) findings showing that belief in climate change and personal experience with crop loss helps explain why some farmers perceive more risk than others. Similarly, while Schattman et al. (2016) acknowledge the significance of personal experience as an important motivator for adopting climate change adaptation practices, the authors reveal an apparent problem with personal experience that may bias farmers' adaptation practices as a replacement for mitigation. Indeed, both mitigation and adaptation are needed to reach national greenhouse gas reduction goals.

Despite their regulation skepticism and lack of understanding of climate change, the peri-urban stewards seem more inclined to adopt environmentally friendly farming practices than the other farmers. This finding is in line with Dietz et al.'s (2007) study, which reports no correlation between knowledge about climate change and support for environmental policy. Our results suggest that distance in time and space for the peri-urban stewards might be particularly challenging in understanding climate change as an environmental threat. For most of them, future events that could influence society are considered more theoretical and speculative. In contrast, near future and highly personal events are tangible. Trope and Liberman (2010) relate to these events as processes for which people exhibit different traversing psychological distances. Their source point is the self (e.g., in the here and now) and the various ways an object might be distanced from that point in time and space. Further explanation of the peri-urban steward behavior can relate to what is known in the literature as a judgmental heuristic which posits that an individual evaluates the probability of events by the ease with which relevant examples come to mind (e.g., availability) (Tversky and Kahneman 1973). The study by Hyland et al. (2016) echoes similar results and hypothesizes that farmers do not consider climate change as the origin of adverse weather effects.

Environmental regulation influenced with varying degrees farmers' perceptions of climate change in our sample. Despite their involvement in ecological practices, the peri-urban stewards might have difficulties aligning their environmental practices with governmental ends compared with the resilient farmers. Niles et al. (2013) provide evidence indicating that farmers' experience with local policy is a much stronger predictor of climate change attitudes than personal experience with climate change effects. Would farmers have been as skeptical of environmental regulation if they were better informed about climate change and had perceived and understood its impacts? Based on the above discussion and the differences with the resilient type: probably not.

For the independent farmers, given the dominant values and attitudes defining this type, skepticism toward environmental regulation could be directly related to an antienvironmental discourse prevalent among this group, explaining their passive behavior toward climate change. This could also relate to what Rossi and Hinrichs (2011) have defined as a rejection of ecological knowledge rooted in "deep anthropocentrism and possessive individualism." Production-oriented attitude could also explain the independent group's regulation skepticism. Evidence suggests that farmers' approaches to production, profit, and efficiency maximization are perceived to be constrained by environmental regulation (Gorton et al. 2008).

The skepticism toward environmental regulation is critical for policymakers to consider. A considerable effort has been made to communicate climate change effects through health and economic vitality lenses (Nisbet et al. 2008). If done without caution to the skeptic farmers' responses, these efforts might actually create more resistance and negatively affect motivation change (Haltinner 2018). The independent farmers could consider the climate if there are any economic incentives available to them to do so (Fleming et al. 2010). Indeed, as Islam et al. (2013) suggested, messages emphasizing low-cost win-win technologies could influence skeptic and productivist farmers' behavior toward considering the climate. In the same line of thoughts, Hyland et al. (2016) put forward that discourse framed monetarily may gain recognition among farmers with productivism tendencies and low behavioral capacity.

Furthermore, a study of the impact of government credibility on Australian farmers' attitudes highlights the importance of building better strategies to share climate change knowledge by framing the information within the local sociocultural, economic, and biophysical environment of farmers it was meant to influence (Evans et al. 2011). Hyland et al. (2016) take this reasoning a step further by emphasizing the potential and usefulness of social learning in shifting farmers' production sense toward environmental tendencies. The authors underscore the role of participatory approaches and discussion groups that could offer a platform to raise awareness and consider adopting environmentally and economically sound measures. This is remarkably accurate in peri-urban areas where climate change requires governments to pledge actions and concrete measures to contain CO_2 emissions by considering the food system inside the farming systems around urban areas. Hence, accounting for the relationship between urban foodshed, food miles, and food transport CO_2 emissions is critical (Pradhan et al. 2020). Local food systems and short food value chains could appeal to production-oriented farmers as they offer production diversification pathways rather than compromise (Gibbons et al. 2020). In this way, the metropolitan food supply chain will be an economic subsystem of a broader socio-ecological system (Paloviita and Järvelä 2015).

Conclusion

Our study findings put forward a vision of the negativist farmers of climate change as a problem for an individual farmer that could be tackled within individual farm borders despite their experience with climate change, these farmers consider that these changes did not require any immediate actions as they did not report any active behavioral response toward implementing adaptation measures as reflected by their passive behavior.

Conversely, independent farmers tend to align their production practices to the best use of the resources available on their farms which they deem necessary for survival. Focusing on the environment might seem unneeded for them. Our results suggest that the independent farmers' group behavior reflects Weber's finite pool of worries concept (Weber 1997). In his article entitled "Experience-based and description-based perceptions of long-term risks: why global warming does not scare us (yet)," Weber (2006) underscore that the rising concerns of the public about security and terrorism issues in the USA has had as a result, a decreased concern about other issues such as environmental issues and global warming. This seems to be true for the independent farmers for whom production imperatives seem to overtake any other worries about the environment and climate change.

Precise interpretations of climate change are critical determinants of pro-environmental behavior and climate change policies (Leiserowitz 2006; Lorenzoni et al. 2007). Except for the resilient farmers, analyzing farmer types reveals a divide in perceptions and understanding of agricultural emission sources and their contribution to climate change in peri-urban areas. This is particularly true for the peri-urban steward and the negativist. Although they knew that agriculture contributes to climate change, both types were doubtful about how these emissions are generated and even exacerbated in a peri-urban environment. While the low-risk awareness explains the low preparedness of the peri-urban steward (as they did neither seem to use technological innovation nor try new varieties of crops), it also justifies the negativist passive behavior toward climate change. Part of the solution can be found in the literature underscoring the role of advice and information dissemination (Hall and Wreford 2012; Wreford et al. 2017) as well as building influential outreach programs (Dessart et al. 2019) in attaining common knowledge about agriculture sources of climate change and leading to more commitment to collective and individual adaptation actions.

The French government's strategy to meet emissions goals depends on mostly voluntary farmers' adaptation measures (e.g., introducing new plant species, irrigation management, plant protection management) (Akimowicz et al. 2021) and technological uptakes (Steffen et al. 2015). This study helped identify visions of resilience and vulnerability of farmers operating in the metropolitan area of Rouen. Identifying farmer types through regional studies is helpful to explore possible adaptation measures depending on the farming systems and the specific characteristics of each geographic area. Such granularity enables tackling climate change as a set of localized environmental challenges with localized contexts worthy of specific solutions and interventions. The endless evolution of agricultural systems and food systems within which farmers permanently make decisions makes it difficult to identify the scale – urban, peri-urban, and rural – that is the most vulnerable.

Acknowledgment We thank Rouen Metropole stakeholders, in particular Léo Kazmierkzak, as well as the farmers who have accepted to take part in this study by answering the questionnaire.

Appendix: Supplementary Material (Fig. 13.8)

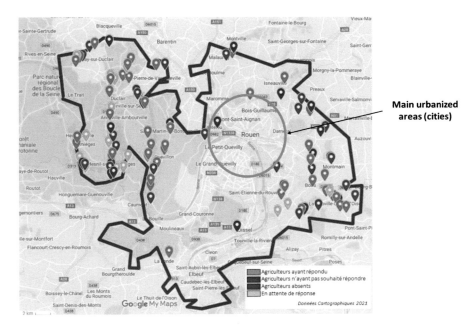

Fig. 13.8 Glocalization of peri-urban farms concerned by the survey (Google maps)

References

Adamides G, Kalatzis N, Stylianou A, Marianos N, Chatzipapadopoulos F, Giannakopoulou M, Papadavid G, Vassiliou V, Neocleous D (2020) Smart farming techniques for climate change adaptation in Cyprus. Atmosphere 11(6). https://doi.org/10.3390/atmos11060557

Adger WN, Arnell NW, Tompkins EL (2005) Successful adaptation to climate change across scales. Glob Environ Chang 15(2):77–86. https://doi.org/10.1016/j.gloenvcha.2004.12.005

Agovino M, Casaccia M, Ciommi M, Ferrara M, Marchesano K (2019) Agriculture, climate change and sustainability: the case of EU-28. Ecol Indic 105:525–543. https://doi.org/10.1016/j.ecolind.2018.04.064

Akimowicz M, Del Corso J-P, Gallai N, Képhaliacos C (2021) Adopt to adapt? Farmers' varietal innovation adoption in a context of climate change. The case of sunflower hybrids in France. J Clean Prod 279:123654. https://doi.org/10.1016/j.jclepro.2020.123654

Arbuckle JG, Morton LW, Hobbs J (2015) Understanding farmer perspectives on climate change adaptation and mitigation: the roles of trust in sources of climate information, climate change beliefs, and perceived risk. Environ Behav 47(2):205–234. https://doi.org/10.1177/0013916513503832

Arnaudet A, Mosnier A (2017) Charte Agricole de territoire. https://www.metropole-rouen-normandie.fr/sites/default/files/publication/2018/SYNTHESE_AGRICOLE_WEB.pdf

Atlas Agricole et rural de Normandie, 53 Ministère de l'agriculture de l'agroalimentaire et de la forêt 1689 (2015) https://doi.org/10.1017/CBO9781107415324.004

Azadi Y, Yazdanpanah M, Mahmoudi H (2019) Understanding smallholder farmers' adaptation behaviors through climate change beliefs, risk perception, trust, and psychological distance: evidence from wheat growers in Iran. J Environ Manag 250(September). https://doi.org/10.1016/j.jenvman.2019.109456

Barnes DKA, Griffiths HJ, Kaiser S (2009) Geographic range shift responses to climate change by Antarctic benthos: where we should look. Mar Ecol Prog Ser 393:13–26. Available at: https://www.int-res.com/abstracts/meps/v393/p13-26/7

Barnes AP, Toma L (2012) A typology of dairy farmer perceptions towards climate change. Clim Chang 112(2):507–522. https://doi.org/10.1007/s10584-011-0226-2

Barnes AP, Islam MM, Toma L (2013) Heterogeneity in climate change risk perception amongst dairy farmers: a latent class clustering analysis. Appl Geogr 41:105–115. https://doi.org/10.1016/j.apgeog.2013.03.011

Ben-Othmen MA, Ostapchuk M (2019) Farmers' preferences for grassland restoration: evidence from France. In: 172nd EAAE seminar 'Agricultural policy for the environment or environmental policy for agriculture?' May 28–29, 2019, Brussels

Ben-Othmen MA, Dubois MJ, Sauvée L (2020) Evolution Agrotechnique Contemporaine III animal & techniques (collection). Université de technologie de Belfort-Montbéliard

Bouzguenda I (2019) Towards smart sustainable cities: a review of the role digital citizen participation could play in advancing social sustainability. Sustain Cities Soc 50(May):101627. https://doi.org/10.1016/j.scs.2019.101627

Brobakk J (2018) A climate for change? Norwegian farmers' attitudes to climate change and climate policy. World Polit Sci 14(1):55–79. https://doi.org/10.1515/wps-2018-0003

Brunner S, Baum I (2014) Climate change 2014 Mitigation of climate change Working Group III contribution to the fifth assessment report of the Intergovernmental Panel on Climate Change

Burns R, Burns RP (2008) Business research methods and statistics using SPSS. Sage

Carolan M (2020) Filtering perceptions of climate change and biotechnology: values and views among Colorado farmers and ranchers. Clim Chang 159(1):121–139. https://doi.org/10.1007/s10584-019-02625-0

Davidson DJ, Rollins C, Lefsrud L, Anders S, Hamann A (2019) Just don't call it climate change: climate-skeptic farmer adoption of climate-mitigative practices. Environ Res Lett 14(3). https://doi.org/10.1088/1748-9326/aafa30

Davies BB, Hodge ID (2007) Exploring environmental perspectives in lowland agriculture: a Q methodology study in East Anglia, UK. Ecol Econ 61(2):323–333. https://doi.org/10.1016/j.ecolecon.2006.03.002

Dessart FJ, Barreiro-Hurlé J, van Bavel R (2019) Behavioural factors affecting the adoption of sustainable farming practices: a policy-oriented review. Eur Rev Agric Econ 46(3):417–471. https://doi.org/10.1093/erae/jbz019

Dietz T, Dan A, Shwom R (2007) Support for climate change policy: social psychological and social structural influences. Rural Sociol 72(2):185–214. https://doi.org/10.1526/003601107781170026

Duvernoy I, Zambon I, Sateriano A, Salvati L (2018) Pictures from the other side of the fringe: urban growth and peri-urban agriculture in a post-industrial city (Toulouse, France). J Rural Stud 57:25–35. https://doi.org/10.1016/j.jrurstud.2017.10.007

Eggers M, Kayser M, Isselstein J (2015) Grassland farmers' attitudes toward climate change in the North German Plain. Reg Environ Chang 15(4):607–617. https://doi.org/10.1007/s10113-014-0672-2

Evans C, Storer C, Wardell-Johnson A (2011) Rural farming community climate change acceptance: impact of science and government credibility. Int J Sociol Agric Food 18(3):217–235

Fahad S, Wang J (2018) Farmers' risk perception, vulnerability, and adaptation to climate change in rural Pakistan. Land Use Policy 79:301–309. https://doi.org/10.1016/j.landusepol.2018.08.018

Field A (2009) Dicovering statistics using SPSS. Sage

Fleming A, Vanclay F, Fleming A, Vanclay F, Aysha F, Frank V (2010) Farmer responses to climate change and sustainable agriculture. A review To cite this version: HAL Id: hal-00886547 Farmer responses to climate change and sustainable agriculture

Foudi S, Erdlenbruch K (2012) The role of irrigation in farmers' risk management strategies in France. Eur Rev Agric Econ 39(3):439–457. https://doi.org/10.1093/erae/jbr024

Gammans M, Mérel P, Ortiz-Bobea A (2017) Negative impacts of climate change on cereal yields: statistical evidence from France. Environ Res Lett 12(5). https://doi.org/10.1088/1748-9326/aa6b0c

Gibbons C, Morgan B, Kavouras JH, Ben-Othmen M (2020) In: Leal Filho W, Azul AM, Brandli L, Özuyar PG, Wall T (eds) Sustainability in agriculture and local food systems: a solution to a global crisis BT – zero hunger. Springer, pp 1–12. https://doi.org/10.1007/978-3-319-69626-3_116-1

Gorton M, Douarin E, Davidova S, Latruffe L (2008) Attitudes to agricultural policy and farming futures in the context of the 2003 CAP reform: a comparison of farmers in selected established and new Member States. J Rural Stud 24(3):322–336. https://doi.org/10.1016/j.jrurstud.2007.10.001

Graux A-I, Bellocchi G, Lardy R, Soussana J-F (2013) Ensemble modelling of climate change risks and opportunities for managed grasslands in France. Agric For Meteorol 170:114–131. https://doi.org/10.1016/j.agrformet.2012.06.010

Hall C, Wreford A (2012) Adaptation to climate change: the attitudes of stakeholders in the livestock industry. 207–222. https://doi.org/10.1007/s11027-011-9321-y

Haltinner K (2018) Climate change skepticism as a psychological coping strategy. December 2017: 1–11. https://doi.org/10.1111/soc4.12586

Ho TT, Shimada K (2019) The effects of climate smart agriculture and climate change adaptation on the technical efficiency of rice farming – an empirical study in the Mekong Delta of Vietnam. Agriculture 9(5). https://doi.org/10.3390/agriculture9050099

Holloway LE, Ilbery BW (1996) Farmers' attitudes towards environmental change, particularly global warming, and the adjustment of crop mix and farm management. Appl Geogr 16(2): 159–171. https://doi.org/10.1016/0143-6228(95)00034-8

Hyland JJ, Jones DL, Parkhill KA, Barnes AP, Williams AP (2016) Farmers' perceptions of climate change: identifying types. Agric Hum Values 33(2). https://doi.org/10.1007/s10460-015-9608-9

Islam M, Barnes A, Toma L (2013) An investigation into climate change scepticism among farmers. J Environ Psychol 34:137–150. https://doi.org/10.1016/j.jenvp.2013.02.002

Kaida N, Kaida K (2016) Facilitating pro-environmental behavior: the role of pessimism and anthropocentric environmental values. Soc Indic Res 126(3):1243–1260. https://doi.org/10.1007/s11205-015-0943-4

Köbrich C, Rehman T, Khan M (2003) Typification of farming systems for constructing representative farm models: two illustrations of the application of multi-variate analyses in Chile and Pakistan. Agric Syst 76(1):141–157. https://doi.org/10.1016/S0308-521X(02)00013-6

Lane D, Chatrchyan A, Tobin D, Thorn K, Allred S, Radhakrishna R (2018) Climate change and agriculture in New York and Pennsylvania: risk perceptions, vulnerability and adaptation among farmers. Renew Agric Food Syst 33(3):197–205

Launay M, Caubel J, Bourgeois G, Huard F, Garcia de Cortazar-Atauri I, Bancal M-O, Brisson N (2014) Climatic indicators for crop infection risk: application to climate change impacts on five major foliar fungal diseases in Northern France. Agric Ecosyst Environ 197:147–158. https://doi.org/10.1016/j.agee.2014.07.020

Leiserowitz A (2006) Climate change risk perception and policy: 45–72. https://doi.org/10.1007/s10584-006-9059-9

Lorenzoni I, Nicholson-cole S, Whitmarsh L (2007) Barriers perceived to engaging with climate change among the UK public and their policy implications. Glob Environ Change 17:445–459. https://doi.org/10.1016/j.gloenvcha.2007.01.004

Lwasa S, Mugagga F, Wahab B, Simon D, Connors J, Griffith C (2014) Urban and peri-urban agriculture and forestry: transcending poverty alleviation to climate change mitigation and adaptation. Urban Clim 7:92–106. https://doi.org/10.1016/j.uclim.2013.10.007

Marçal D, Mesquita G, Kallas LME, Hora KER (2021) Urban and peri-urban agriculture in Goiânia: the search for solutions to adapt cities in the context of global climate change. Urban Clim 35(October 2020). https://doi.org/10.1016/j.uclim.2020.100732

Menapace L, Colson G, Raffaelli R (2015) Climate change beliefs and perceptions of agricultural risks: an application of the exchangeability method. Glob Environ Chang 35:70–81. https://doi.org/10.1016/j.gloenvcha.2015.07.005

Merot P, Corgne S, Delahaye D, Desnos P, Dubreuil V, Gascuel C, Giteau JL, Joannon A, Quenol H, Narcy JB (2014) Évaluation, impacts et perceptions du changement climatique dans le Grand Ouest de la France métropolitaine le projet CLIMASTER. Cahiers Agricultures 23(2):96–107. https://doi.org/10.1684/agr.2014.0694

Mitter H, Larcher M, Schönhart M, Stöttinger M, Schmid E (2019) Exploring farmers' climate change perceptions and adaptation intentions: empirical evidence from Austria. Environ Manag 63(6):804–821. https://doi.org/10.1007/s00267-019-01158-7

Morrison M, Hine DW, D'Alessandro S (2017) Communicating about climate change with farmers. Oxford University Press. https://doi.org/10.1093/acrefore/9780190228620.013.415

Niles MT, Lubell M, Haden VR (2013) Perceptions and responses to climate policy risks among california farmers. Glob Environ Chang 23(6):1752–1760. https://doi.org/10.1016/j.gloenvcha.2013.08.005

Nisbet EK, Zelenski JM, Murphy SA (2008) The nature relatedness scale: linking with nature to environmental concern and behavior. Environ Behav 27(1):1–26. http://eab.sagepub.com/content/10/1/3.abstract

Nunnally J (1967) Psychometric theory, New York

Otitoju MA, Enete AA (2016) Climate change adaptation: uncovering constraints to the use of adaptation strategies among food crop farmers in South-west, Nigeria using principal component analysis (PCA). Cogent Food Agric 2(1):1178692. https://doi.org/10.1080/23311932.2016.1178692

Pallant J (2010) SPSS survival manual: a step by step guide to data analysis using SPSS. Open University Press

Paloviita A, Järvelä M (2015) Climate change adaptation and food supply chain management: an overview. In: Paloviita A, Järvelä M (eds) Climate change adaptation and food supply chain management. Routledge, pp 1–14

Pidgeon N (2012) Public understanding of, and attitudes to, climate change: UK and international perspectives and policy. Clim Pol 12(sup01):S85–S106. https://doi.org/10.1080/14693062.2012.702982

Pradhan P, Kriewald S, Costa L, Rybski D, Benton TG, Fischer G, Kropp JP (2020) Urban food systems: how regionalization can contribute to climate change mitigation. Environ Sci Technol 54(17):10551–10560. https://doi.org/10.1021/acs.est.0c02739

Pribadi DO, Pauleit S (2015) The dynamics of peri-urban agriculture during rapid urbanization of Jabodetabek Metropolitan Area. Land Use Policy 48:13–24. https://doi.org/10.1016/j.landusepol.2015.05.009

Rossi AM, Hinrichs CC (2011) Hope and skepticism: farmer and local community views on the socio-economic benefits of agricultural bioenergy. Biomass Bioenergy 35(4):1418–1428. https://doi.org/10.1016/j.biombioe.2010.08.036

Running K, Burke J, Shipley K (2017) Perceptions of environmental change and climate concern among Idaho's farmers. Soc Nat Resour 30(6):659–673. https://doi.org/10.1080/08941920.2016.1239151

Schattman RE, Conner D, Méndez VE (2016) Farmer perceptions of climate change risk and associated on-farm management strategies in Vermont, northeastern United States. Elementa 4:1–14. https://doi.org/10.12952/journal.elementa.000131

Seguin B (2003) Adaptation des systèmes de production agricole au changement climatique. Compt Rendus Geosci 335(6–7):569–575. https://doi.org/10.1016/S1631-0713(03)00098-1

Smit B, Skinner MW (2002) Adaptation options in agriculture to climate change: a typology. Mitig Adapt Strateg Glob Chang 7(1):85–114. https://doi.org/10.1023/A:1015862228270

Špička J, Boudný J, Janotová B (2009) The role of subsidies in managing the operating risk of agricultural enterprises. Agric Econ 55(4):169–179. https://doi.org/10.17221/17/2009-agricecon

Steffen W, Broadgate W, Deutsch L, Gaffney O, Ludwig C (2015) The trajectory of the anthropocene: the great acceleration. Anthropocene Rev 2(1):81–98. https://doi.org/10.1177/2053019614564785

Story PA, Forsyth DR (2008) Watershed conservation and preservation: environmental engagement as helping behavior. J Environ Psychol 28(4):305–317. https://doi.org/10.1016/j.jenvp.2008.02.005

Sulewski P, Kłoczko-Gajewska A (2014) Farmers' risk perception, risk aversion and strategies to cope with production risk: an empirical study from Poland. Stud Agric Econ 116(3):140–147. https://doi.org/10.7896/j.1414

Tilman E, Cassman KG, Matson PA, Naylor R, Polasky S (2002) Agricultural sustainability and intensive production practices. Nature 418:617–677. https://doi.org/10.1080/11263508809430602

Trope Y, Liberman N (2010) Construal-level theory of psychological distance. Psychol Rev 117(2):440–463. https://doi.org/10.1037/a0018963

Tversky A, Kahneman D (1973) Availability: a heuristic for judging frequency and probability. Cogn Psychol 5(2):207–232

van Huik MM, Bock BB (2007) Attitudes of Dutch pig farmers towards animal welfare. Br Food J 109(11):879–890. https://doi.org/10.1108/00070700710835697

Weber EU (1997) Perception and expectation of climate change: precondition for economic and technological adaptation. In: Environment, ethics, and behavior: the psychology of environmental valuation and degradation. The New Lexington Press, San Francisco

Weber EU (2006) Experience-based and description-based perceptions of long-term risk: why global warming does not scare us (yet). Clim Chang 77(1–2):103–120. https://doi.org/10.1007/s10584-006-9060-3

Wilson GA (2001) From productivism to post-productivism . . . and back again? Exploring the (un) changed natural and mental landscapes of European agriculture. Trans Inst Br Geogr 26(1):77–102. Available at: http://onlinelibrary.wiley.com/doi/10.1111/1475-5661.00007/abstract.

Wreford A, Ignaciuk A, Gruère G (2017) Overcoming barriers to the adoption of climate-friendly practices in agriculture. OECD Publishing, p 101

Chapter 14
Determinants and Spatio-Temporal Drivers of Agricultural Vulnerability to Climate Change at Block Level, Darjeeling Himalayan (Hill) Region, West Bengal, India

Deepalok Banerjee, Jyotibrata Chakraborty, Bimalesh Samanta, and Subrata B. Dutta

Abstract The Darjeeling Himalayan region is diversified with multiple cropping patterns and mostly dependent on rain-fed irrigation, hence it is very sensitive to the changes in weather parameters. The untoward impacts due to climate change are obvious on agricultural production and specially the hilly parts of the Darjeeling Himalaya report the same intimation. Therefore, it is an utmost need to identify the menace developed by such changes in weather attributes and its possible adversities on agricultural practices, food security of the large extent of populace and in this juncture, this research work attempts to figure out the determinants and spatio-temporal drivers of agricultural vulnerability in view of climate change at block level within the Darjeeling Himalayan (hill) region. The Agricultural Vulnerability Assessment (AVA) is an exclusive method as it involves numerous ideas and data factors, though initially the Principal Component Analysis has been introduced in this study to understand the magnitude of indicators contributing to the agri-vulnerability. Outcomes of the AVA come up with the Agricultural Vulnerability Index scores that range between 0.723 and 0.445 and demonstrating that high agriculturally vulnerable blocks, i.e. Kalimpong I, Rangli Rangliot, Mirik and Kurseong are facing challenges in terms of high dependency on rainfed agriculture; high percentage marginal farmers; significant yield variability. Such drivers of vulnerability help to infer the gap areas of the concerned sector and possibly pointing towards adaptive measures like integrated farm management practices to build a climate resilient agricultural society by sustaining the ecological tranquillity of the Himalayan ecosystem.

D. Banerjee (✉) · J. Chakraborty
Geoinformatics and Remote Sensing Cell, West Bengal State Council of Science and Technology, Government of West Bengal, Kolkata, West Bengal, India

B. Samanta · S. B. Dutta
Department of Science and Technology and Biotechnology, Government of West Bengal, Kolkata, West Bengal, India

© The Author(s), under exclusive license to Springer Nature Switzerland AG 2022
U. Chatterjee et al. (eds.), *Ecological Footprints of Climate Change*, Springer Climate, https://doi.org/10.1007/978-3-031-15501-7_14

Keywords Climate change · Darjeeling Himalayan (hill) region · Principal Component Analysis · Agricultural Vulnerability Index · Drivers of vulnerability · Adaptive measures

Introduction

Indian agriculture is predominantly dependent on weather and any deviations in its pattern affect the production of agriculture. Besides, some areas of the country are more vulnerable than the others depending on their adaptive capacity and the status of socio-economy (Sehgal et al. 2013). But in the recent era, the Indian Himalayan region (IHR) has been found to be very sensitive towards the negative effects of climate change and its possible vulnerabilities. A large extent of the IHR has already endured some markable long-term changes in the frequency of occurrences of extreme temperature events as observed in the few previous decades (Bhatt and Nakamura 2005; Wulf et al. 2010). Nesting upon the complex and diversified Himalayan ecosystem, the Darjeeling Himalayan region (DHR) is marked with multifarious cropping patterns and mostly dependent on rain-fed irrigation, hence it is very sensitive to the changes in rainfall pattern, intensities and varying frequencies. The overall rainfall amount and total rainy days have declined almost by 52% and 34% respectively through the last one decade in the IHR, whereas the incidence of rainfall with very high intensity has exacerbated the situation (Government of India 2016). On an average, the temperature (mean surface temperature) of the IHR is said to be projected to rise about 1.4–5.8 °C by the year 2100 as per the report of the Inter-Governmental Panel on Climate Change (IPCC) (IPCC 2001). The adverse impacts of such changes in weather parameters are magnified in the hill environment because of its sensitive nature in the ecological system and the DHR is not an exemption to this connotation. At the global level, the IPCC accounts that in the lower latitudes yields are expected to be declined most dominantly and agriculture might be found as one of the key vulnerable sectors that is going to experience such adversities of climate change (Easterling et al. 2007; Parry et al. 2007; Field et al. 2014; Stevanovic et al. 2016).

The Darjeeling Himalayan agriculture is distinguished with its complex farming systems comprising of agriculture, animal husbandry, horticulture, poultry farming and fishery etc. whereas the region is also distinctly characterised with cold winter, terraced farm plots, steep hill slopes lash tea gardens, medicinal crops and orchards. The region is also preeminent with the production of unique Mandarin orange, world famous Darjeeling tea, large cardamoms and cinchona (Bookhagen 2010; Joshi et al. 2014). Agricultural practices done so far in the Darjeeling hill region are solely dependent on monsoonal rain and they do have strong dependence on weather conditions, so agriculture is one of the most vulnerable sectors in this regard. It is the need of the hour to quantify and analyse the case-specific (sectoral) study of climate change vulnerability when it is already in concern at global level. Such studies on agricultural vulnerability are so important because the assessment of transitional impacts of climate change like how the change of climate at globe affects the regional or sectoral vulnerability and vice versa is considered to be of serious apprehension (Loi et al. 2022). The agriculture of this region also witnesses the smaller terraced farm size denoting higher marginal farmers, maximum subsistence level of production with minimal proportion towards total family income, significant yield variability, etc. In this context, the study of climatic agri-vulnerability is now a prime focus to identify the menace developed by such changes in weather attributes and its possible adversities on agricultural practices, food security and to the large extent of populace taking part in different agricultural activities (O'Brien and Leichenko 2000). Climate change and its potential repercussions have grave impact on agriculture, and therefore it may be a prime concern to evaluate and identify the vulnerabilities of this sector. It also seeks proper attention to alleviate such uncertainties developed and thereon importance should be put on probable adaptation measure and mitigation strategies (Isabel et al. 2020). Here, the successful attempts of vulnerability assessment can support to point out that, which communities are mostly exposed to the vulnerability related to climate change, where exactly they are located or their spatial extent as well and what are the thirst driving forces to such vulnerability (Bouroncle et al. 2017; Tina-Simone et al. 2018). Vulnerability assessment can help to provide proper guidance and directives, better governance and support for scientific implementation of adaptation planning and judicious justification for better accomplishment of different projects in contemplation of creating a more objective decision-making process to achieve more sustainability (Oppenheimer et al. 2014; Sherbinin et al. 2014; Sanjit et al. 2017). In this juncture, this research work focuses to identify the determinants and spatio-temporal dimensions of agricultural vulnerability in view of climate change effects at block level in the eight hilly blocks of DHR. The focus of such study is not only valid to derive the agricultural vulnerability but also to project how the vulnerability affects the food security, livelihood, etc. and it also examines to draw inferences about the relationship between climate change and responsible drivers of vulnerability (Sridevi et al. 2021). The foremost objective and prime focus of the study is to delineate and analyse the indicators driving towards vulnerability of the agricultural sectors, to get agricultural vulnerability indices of the concerned blocks and their ranking by performing statistical operations and finally to find out the pivotal drivers from the Agricultural Vulnerability Assessment (AVA).

The mountain people of the DHR are highly dependent on climate-supported agricultural practice for their survival. The main livelihood option of the DHR community is agricultural practices and thus, the worse impacts of climate change may be found extremely drastic on communities residing in the hills (Government of India 2018). Hence, such a complex agricultural setup of the DHR must be given special care and keep track of the symptoms identifying the pessimistic aftermaths of climate change on agriculture. Here, the Climate Smart Agriculture (CSA) can be considered as a unified approach for combating the conjugated challenges in terms of food security of the region and climate change consequences, and for better appreciation of likely measures, adaptation strategy through integrated farm management practice using spring water may be used as a novel technique. Under such observations, this work extends towards a comprehensive derivation of the agricultural vulnerability of the DHR and possibility of climate resilient agricultural practice adopted for a sustainable future.

Datasets and Methodology

Area of Interest

The DHR is an ecologically significant landscape of the eastern Himalaya shared by the two northernmost districts, i.e. Darjeeling and Kalimpong of West Bengal. The DHR can be referred as the land of diversity in terms of its physico-biotic resource availability and their immense environmental signification. Contemporary threats particularly to the Darjeeling Himalayan hill region induced by climate change underwent substantial enduring deviations in occurrences of acute temperature instances during the past periods of time, rainfall amount, recession of rainy days but increase in intense precipitation days, etc. Geographically, the DHR hilly part majorly incorporates five Community Development (C.D) blocks (Darjeeling Pulbazar, Rangli Rangliot, Jorebunglow-Sukiapokhri, Kurseoung and Mirik) of the Darjeeling district and three C.D blocks (Kalimpong I, Kalimpong II and Gorubathan) of the Kalimpong district and covering 2390 sq.km. in terms of total area. The latitudinal extension of the present study area is from $27°\ 13'N$ to $26°\ 44'N$ latitude whereas longitudinally the area lies in between $88°\ 53'E$ and $87°\ 59'E$ longitude as well as encircled in the east-west directions by two transboundary Rivers namely Jaldhaka and Mechi (Samanta et al. 2022). These rivers are geo-strategically important as they have demarcated international, inter-state as well as inter-district boundaries by maintaining their perennial flow across the terrain (Fig. 14.1).

Datasets and Processing Methods

The developed methodology of this work is basically outlined for a better understanding of the vulnerability concept in accordance with the general vulnerability assessment methodological framework under contemporary climate change phenomenon by following the Fifth Assessment Report of IPCC 2014. Likely method helps in determining vulnerability, produce comparable results, and a similar organised approach can assist in developing adaptation strategies on climate resilience across administrative units (IPCC 2014). The present AVA is also exclusive as it involves numerous ideas and data factors containing sensitivity or potential exposures to be adversely affected, shortage of capabilities to deal with as well adjust and can be easily enumerated by Fig. 14.2 for a transparent understanding of the method behind the AVA.

14 Determinants and Spatio-Temporal Drivers of Agricultural Vulnerability... 377

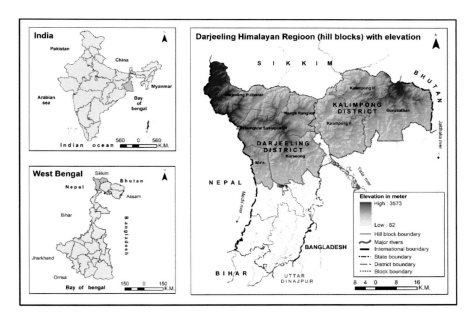

Fig. 14.1 Location map of the study area showing DHR (hill blocks) with varying elevational differences

Initially, in-house available Gaofen-2 remotely sensed satellite images fused with its panchromatic (PAN) band (spatial resolution 80 cm, path/row 1220/048, 1375/585, 1385/202, 1385/327, 1396/397, 1500/329, 1518/107, 1500/328, 1858/196, 1924/329, 1924/382, 1924/385, 1938/598, 1938/601, 2102/714, 2102/719, 2103/007, 2103/010 & 20103/030) of 2017–2018 were taken to generate requisite indicator data. However, two image processing techniques have been performed to the available images in the ERDAS Imagine v14 software involving radiometric calibration and atmospheric corrections. Multi-spectral satellite data are mostly affected by atmospheric dynamics, geometric disorders of sensor and thus, it is essential to carry out necessary image correction procedures before taking them for image-based analysis (Moran et al. 1992; Kayet et al. 2016). Aforesaid correction techniques additionally enable calculation of the apparent reflectance in satellite images by converting the DNs to radiance and then to reflectance values. Other relevant data that are necessary to compute the crop diversification within the area of interest to assess the agricultural vulnerability were collected from different sources and are demonstrated in Table 14.1.

Analysis of the Crop Diversification

Analysis of crop diversification denotes the estimation of the present crops or cropping systems of a certain region which are effectively related to the agricultural

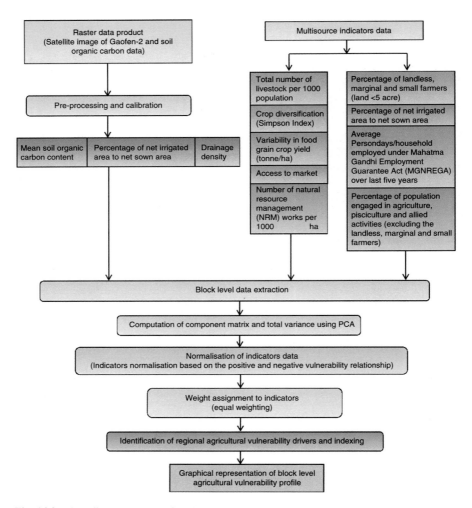

Fig. 14.2 Flow diagram enumerating the AVA methodology adopted in this study

production and linked to corresponding marketing prospects. The present study tries to evaluate the scope and dimensions of crop diversification at the block level under the distinct agro-climatic condition of the DHR hills with advent to identify the factors that influence crop diversification (Behera et al. 2007; Basavaraj et al. 2016). Usually, crop diversification in reality is an inexpensive way to get better of income uncertainties by reason of market conditions but it has huge possibilities to address climate change-induced vagaries in crop production (Joshi et al. 2004; Lin 2011). The rain-fed agriculture specifically in the concerned hills of the West Bengal is going to witness shortage of rainwater availability in the coming decades and thus making crop diversification a difficult task (Government of West Bengal 2020). In this notion, estimation crop diversification in a particular geographical area is an important measure to derive insight on the contemporary status of the agricultural

Table 14.1 Selected indicators and data sources used in the AVA study

Sl. no.	Indicators	Data source
1	Total number of livestock per 1000 population	Bureau of Applied Economics & Statistics, Government of West Bengal (BAES, GoWB), District Statistical Handbook of 2013
2	Percentage of landless, marginal and small farmers (land <5 acre)	BAES, GoWB, District Statistical Handbook of 2013
3	Mean soil organic carbon content (tonne/ha)	Soil organic carbon raster gridded data (30 arc sec) of 2015 is available from https://earthmap.org
4	Crop diversification (Simpson Index)	BAES, GoWB, District Statistical Handbook of 2013
5	Percentage of net irrigated area to net sown area	BAES, GoWB, District Statistical Handbook, 2013 and Gaofen-2 MSS + PAN merge satellite image of 2017–2018
6	Variability in food grain crop yield (tonne/ha)	BAES, GoWB, District Statistical Handbooks of 2009, 2010–2011 (Combined) & 2013
7	Drainage density	Gaofen-2 MSS + PAN merge satellite image of 2017–2018
8	Road connectivity-percentage of villages connected by pucca roads	Gaofen-2 MSS + PAN merge satellite image of 2017–2018
9	Average person-days/household employed under Mahatma Gandhi Employment Guarantee Act (MGNREGA) over the last five years	MGNREGA, online Management Information System (MIS) reports of 2015, 2016, 2017, 2018, 2019 & 2020
10	Access to market	District Census Handbook, Village and Town Release of 2011
11	Percentage of population engaged in agriculture, pisciculture and allied activities (excluding the landless, marginal and small farmers)	BAES, GoWB, District Statistical Handbook of 2013
12	Number of natural resource management (NRM) works per 1000 ha (MGNREGS) and/or other schemes	MGNREGA, MIS report of 2020

practice, dominance of crops, output invariabilities to expand in production and comparatively lesser risk than climate-sensitive single crop system (Imbs and Wacziarg 2003; Jill and Erin 2005). Among the available indices on crop diversification, the Simpson Index has been computed here and can be calculated using Eq. (14.1)

$$\text{Simpson Index (SI)} = 1 - \sum P_i^2 \qquad (14.1)$$

here $P_i = A_i/\Sigma A_i$ is the percentage of the ith activity in hectare. If the value of SI is near to 0 then it indicates that the region or the zone is close to the specialisation in the growth of a specific crop and if it is near to 1, at that time it can be said that the zone is fully diversified in terms of crops (Jill and Erin 2005; Basavaraj et al. 2016).

Construction of Agricultural Vulnerability Index

Indicators Assortment and Quantification

Agricultural practice is thriving in all the eight blocks of the DHR but due to the physiographic differentiation between the hills and valleys, the distribution of agricultural lands is not uniform and the agricultural practice is mostly labour intensive (Bhattacharya 2008). Keeping the regional physical diversity in mind, this study highlights the vulnerability scenario of the agricultural sector as the native people living in the hilly terrains of the DHR have more intense dependency on climatic factors with limited or not-so-modern technology options to use. The net sown of the hill blocks is exceptionally poor and the cropping system is predominantly rain dependent subsistence type (Government of West Bengal 2012). Therefore, it is the first step of the AVA to understand who is vulnerable, and why? and hence, at this juncture the block level AVA of the DHR has been attempted by taking a total of twelve significant region-specific indicators to weigh the influences of climate change on agricultural produce, economy and analyse the local community exposure to those who are engaged in agricultural practices (Table 14.4).

Prior to the selection of these significant twelve indicators, alongside Principal Component Analysis (PCA) has been strived using Statistical Package for the Social sciences (SPSS) software (Version 21) to identify and understand the most dominant and crucial indicators impacting towards AV (Table 14.2). PCA is regarded as a way to identify patterns in data and weighing of the most striking indicators and to highlight the similarities and difference as well (Mishra et al. 2017). PCA is usually performed to identify the amount of the variability amongst designated variables (indicators) (Raju et al. 2016). PCA also help to filter and identify smaller number of components which explain maximum variance observed among a larger datasets (Esteves et al. 2016; Raju et al. 2016). In this current study factors with eigen values having greater than one only were considered in the analysis. The rotated factor analysis resulted into three factors, with eigen values greater than 1, which account almost 80% of the total cumulative variance in the entire datasets compiled for the assessment of AVI (Table 14.3).

The first principal component (PC) recons for a larger proportion of variability in the dataset and consequently each of the succeeding PC accounts for as much as possible variability remained. Up to the third component (eigen values >1), it

Table 14.2 Component matrix extracted through PCA (3 components extracted)

Variables	Component 1	Component 2	Component 3	Index to variables Variables	Name of the indicators
X12	−0.883			X1	Total number of livestock per 1000 population
X2	0.839			X2	% of landless, marginal and small farmers (land <5 acres)
X9	0.800	0.521		X3	Mean soil organic carbon content (tonne/ha)
X11	0.762			X4	Crop diversification
X10	0.589	−0.540		X5	% of net irrigated area to net sown area
X5	−0.558			X6	Variability in food grain crop yield (tonne/ha)
X3		0.856		X7	Drainage density
X7		0.758		X8	Road connectivity-percentage of villages connected by pucca roads
X1	0.532	−0.742		X9	Average person days/household employed under MGNREGA over the last five years
X8	−0.575	0.702		X10	Access to market
X6	0.513	0.665		X11	% of population engaged in agriculture, pisciculture and allied activities (excluding the landless, marginal and small farmers)
X4	0.515		−0.679	X12	Number of NRM works per 1000 ha (MGNREGS) and/or other schemes

Table 14.3 Total variance explained by principal components for agricultural vulnerability

Component	Initial eigen values			Extraction sums of squared loadings		
	Total	% of variance	Cumulative %	Total	% of variance	Cumulative %
1	4.606	38.386	38.386	4.606	38.386	38.386
2	3.622	30.180	68.566	3.622	30.180	68.566
3	1.280	10.665	79.231	1.280	10.665	79.231
4	0.958	7.981	87.212			
5	0.791	6.589	93.801			
6	0.444	3.701	97.502			
7	0.300	2.498	100.000			
8	3.444E−16	2.870E−15	100.000			
9	2.592E−16	2.160E−15	100.000			
10	9.350E−17	7.792E−16	100.000			
11	−4.131E−19	−3.442E−18	100.000			
12	−1.757E−16	−1.464E−15	100.000			

aggregates 79.23% of cumulative variance. PC-1, i.e. factor-1 accounts for the largest share of variance (36.39%) that includes indicators according to their sequence of weight as follows NRM works per 1000 ha (MGNREGS) and/or other schemes, % of landless, marginal and small farmers (land <5 acre), average person-days/household employed under MGNREGA over the last five years, % of population engaged in agriculture, pisciculture and allied activities (excluding the landless, marginal and small farmers), access to market, % of net irrigated area to net sown area, total number of livestock per 1000 population, road connectivity-percentage of villages connected by pucca roads, variability in food grain crop yield (tonne/ha) and crop diversification. PC-2, i.e. factor-2 records for 30.18% of variation and marks with the following indicators in descending order of weights namely average person-days/household employed under MGNREGA over the last five years: access to market, mean soil organic carbon content (tonne/ha), drainage density, total number of livestock per 1000 population, road connectivity-percentage of villages connected by pucca roads and variability in food grain crop yield (tonne/ha). PC-3. i.e. factor-3 holds only 10.67% of variation and is only composed of one variable that is crop diversification.

Furthermore, this present assessment as well helps to categorise the explicitly region-based drivers of agricultural vulnerability, and these might be considered as intellectual inputs in framing out adaptation interferences of the particular area. Choice of indicators is a notable stage of vulnerability assessment, and it is depending upon various determinants viz., indicator form, their functional characteristics (sensitivity or adaptive capacity) and the identity of indicator (Brenkert and Malone 2005; Eriksen and Kelly 2007). The vulnerability cannot be measured at once, thus, the assessment chiefly demands indicators to characterise and quantify vulnerability. Meanwhile, for vulnerability assessment of specific sector, e.g. as in this work agricultural sector is the prime concern of vulnerability assessment, then it is similarly essential prerequisite condition to consider the sub-sectors criterion like irrigated agriculture, semi-arid agriculture, and rain-fed agriculture in form of indicators before carrying out the AVA (Sharma et al. 2019) (Table 14.4).

Quantification of the available data to fit them in the ultimate vulnerability assessment is also an imperative task; typically it combines mechanisms and practices for numerical valuing of indicators. This often ranges from processing of secondary data, ground survey, and modelling. Thereafter, it is quite necessary to derive selected indicators data in measurable units for vulnerability index (VI) calculation (Esteves et al. 2016). Normally, secondary data that are coming out from the dependable sources taken into consideration to quantify indicators and examine their intrinsic deviation, e.g. the indicator yield variability is quantified by using the data published by the BAES, Government of West Bengal. Even though, not all the indicators data are not measured in the same unit (e.g. crop intensity in terms of percentage, MGNREGA in terms of person-days/household/year, etc.), hence, with the intention of computing them all together in the VI equation they have to be normalised.

Table 14.4 Designated indicators, relation with vulnerability, rationale for selection and their configuration used in the block level AVA

Sl. no.	Indicators	Relationship with vulnerability	Rationale	Configuration
1	Total number of livestock per 1000 population	Adaptive capacity (Negative)	Agriculture involving crop production and animal husbandry is the main occupation for the majority of people	Proportion of total number of livestock in 2013 with 1000 population of 2011
2	Percentage of landless, marginal and small farmers (land <5 acres)	Sensitivity (Positive)	This section of agrarian people is mostly exposed to adversities of climate change	Percentage of landless, marginal and small landholders of 2011 to total land holding population of 2011
3	Mean soil organic carbon content (tonne/ha)	Adaptive capacity (Negative)	As mass of soil organic matter exists as carbon and considered to be producer of productive soil	Interpolated block level average organic carbon data in tonne/ha
4	Crop diversification	Adaptive capacity (Negative)	Crop diversification offers a variety of choice in terms of the crop production in a region	Share of individual cropping activity (major food crops) to total area under crop for individual blocks
5	% of net Irrigated area to net sown area	Adaptive capacity (Negative)	Absence of irrigation facilities in rain-fed agricultural systems poses higher degree of climate-induced threats, decline in crop production and marginal income households	The ratio between net irrigated area of 2013 and net sown area in 2016–2017
6	Variability in food grain crop yield (tonne/ha)	Sensitivity (Positive)	Increased inconsistency in crop production signifies instabilities in agro-climatic environments	Coefficient of variation (i.e. (standard deviation/arithmetic mean) × 100) of major food grains over a period 2009–2013
7	Drainage density	Adaptive capacity (Negative)	Rain-dependent agriculture is extremely susceptible to the whims of climatic phenomenon, thus, region with high drainage density can sustain their crop growing areas using river water	Sum of the length of surface drainage in km/total geographical area in sq.km of 2011–2012

(continued)

Table 14.4 (continued)

Sl. no.	Indicators	Relationship with vulnerability	Rationale	Configuration
8	Road connectivity-percentage of villages connected by pucca roads	Adaptive capacity (Negative)	It emphasises on the institutional capacities of a region that integrates the accessibility and connectivity measures between rural locations	Percentage of villages with pucca roads to total number of villages in 2011–2012
9	Average person days/household employed under MGNREGA over the last five years	Adaptive capacity (Negative)	MGNREGA scheme operated across the places act as substitute livelihood means to people and can be treated as an alternative option for asset generation	Average days of employment (equal to 100 days) provided per household under MGNREGA with period from 2015 to 2020
10	Access to market	Adaptive capacity (Negative)	It gives conceptual notions of developmental status of an area, because improved access to market comes with opportunities	Percentage of villages with access to mandi and weekly haat to total number of villages in 2011
11	% of population engaged in agriculture, pisciculture and allied activities (excluding the landless, marginal and small farmers)	Adaptive capacity (Negative)	It supports diversified income that are interrelated to the agriculture	Ratio of population engaged in agricultural sector to total working population of 2011
12	Number of NRM works per 1000 ha (MGNREGS) and/or Other Schemes	Adaptive capacity (Negative)	NRM works contribute in safeguarding natural resources and overall societal development plantations, contour trenches etc.	Average number of completed NRM work (soil conservation, groundwater recharge and irrigation) to 1000 ha area of each block in 2020

Normalisation of Indicators

The normalisation process varies based upon the indicator's nature and relationship with vulnerability (positive or negative relationship) (Ravindranath et al. 2011; Esteves et al. 2016). The normalisation process mainly looks after the comparison among indicators in a logical sense. As the indicators are in different units, that is why it cannot be simply added to imply the index value. The normalisation technique allows aggregation of indicators with different units, by eliminating the units and converting all the values into dimensionless units. The normalised values of indicators lie between 0 and 1 and thus could be aggregated (Sharma et al. 2015, 2019). Consequently, in this AVA normalisation has been adopted by following two methods based on the relationship.

Indicators Normalisation Formula of Positive Vulnerability Relationship

In positive relationship cases, higher the value of the indicator, higher will be the vulnerability. As the values of these indicators increase, greater will be the vulnerability of the agricultural sector. In such cases the variables have direct and positive functional relationship with vulnerability and the normalisation is done using the following Eq. (14.2):

$$X_{ij}^p = \frac{X_{ij} - \text{Min}_i\{X_{ij}\}}{\text{Max}_i\{X_{ij}\} - \text{Min}_i\{X_{ij}\}} \qquad (14.2)$$

where X_{ij} is the variable that is being normalised, i.e. in this case X_{ij} is the value of jth indicator for ith region and X^P_{ij} is the normalised value. Normalised value of X^P_{ij} scores will lie between 0 and 1. The value 1 will correspond to that block with maximum value and 0 will correspond to the block with minimum value (Esteves et al. 2016; Sharma et al. 2019; Samanta et al. 2022).

Indicators Normalisation Formula of Negative Vulnerability Relationship

On the other hand, the vulnerability increases with decrease in the value of the indicator in case of negetive relationship, the following normalisation Eq. (14.3) need to be adopted and the normalised score of the indicator can be computed accordingly;

$$X_{ij}^n = \frac{\text{Max}_i\{X_{ij}\} - X_{ij}}{\text{Max}_i\{X_{ij}\} - \text{Min}_i\{X_{ij}\}} \qquad (14.3)$$

here X_{ij} is the variable that is being normalised, i.e. in this case X_{ij} is the value of jth indicator for ith region and X_{ij}^n is the normalised value. Normalised value of X_{ij}^n scores will range between 0 and 1. The value 1 will resemble that block with minimum value and 0 will correspond to the block with maximum value (Ravindranath et al. 2011; Samanta et al. 2022).

Weights Assignment to Indicators

Post normalisation of indicators, attribute of indicators weight is the next step to infer responsible outcomes. Commonly, in vulnerability assessment weights assignment to indicators basically depends on their relative importance in determining vulnerability of a system and also through discussion and consultation with the stakeholders (Esteves et al. 2016). During weight assignment, it should be maintained that the weight assigned to all the indicators might add up to 100. On the other hand, aggregation of different indicators with appropriate weights is similarly essential to generate a composite index value. But here owing to the complexities of the region, acquiring better perception and to attain a comprehensive result with the decided twelve significant indicators/variables. The AVA has been carried out by giving equal weights to the indicators as all of them are of equal importance. Often, equal weights are used if a large number of indicators are present. The method for equal weighing may be described as the total weight 1, divided by the total number of indicators, i.e. $1/12 = 0.083$ which were multiplied to all indicators but while equal weights are attributed, vulnerability index is constituted by taking a simple average of all the normalised values as described by Eq. (14.4);

$$VI = \frac{\sum_j X_j^{pi} + \sum_j X_{ij}^n}{K} \qquad (14.4)$$

where VI is the vulnerability index of region i, X_{ij}^p and X_{ij}^n are normalised values of jth indicators (with positive and negative relationship with vulnerability respectively) in ith region and K is the number of indicators. Furthermore, obtained results designate the vulnerability ranking of agriculturally stressed blocks on the basis of derived Agricultural Vulnerability Index (AVI) scores and generated outputs of the vulnerability assessment have been categorised on a scale of very high to very low vulnerability and exemplified in form of block-level spatial vulnerability maps (Sharma et al. 2019).

Identification of Regional Vulnerability Drivers and Indexing

Once the assessment is carried out, different spatial units, i.e. eight blocks under the analysis of the hilly DHR have different vulnerability scores. The key objective behind blocks level representation of vulnerability profiles is to deliver information on the drivers of regional agricultural vulnerability and the accompanying uncertainties that are linked to climate change in front of decision-makers, planners and other stakeholders (Esteves et al. 2016; Sharma et al. 2019). Assessment of this type may bring crucial utility to the agricultural sector by identification of the deciding factors of vulnerability together with their comparative involvement to the specified sector and such realistic evidence about the indicators that contribute most to the sectoral vulnerability is reckoned to be valuable in prioritising aspects of regional progress and enactment of adaptation interventions (Sharma et al. 2015; Samanta et al. 2022).

Results and Discussion

Crop Diversification Evaluation and Characteristics

The present study has evaluated the degree of crop diversification at the block level within the distinct agro-climatic conditions of the DHR. Outcomes of the crop diversification analysis show that the concerned blocks are acquainted to monocropping practices with limitations to diversified multi-crop cultivation as the SI values of the blocks express a range from 0.377 to 0.671 (Table 14.5). Blocks

Table 14.5 Block level (DHR-hills) crop (major food crops) diversity calculation

Name of block	Area under paddy	Area under wheat	Area under maize	Area under potato	Total area under crops (paddy + wheat + maize + potato)	Crop diversification (SI) values using Simpson Index								
						Pi of paddy	Pi of wheat	Pi of maize	Pi of potato	Pi^2 of paddy	Pi^2 of wheat	Pi^2 of maize	Pi^2 of potato	$\sum Pi^2$ (Pi^2 paddy + Pi^2 wheat + Pi^2 maize + Pi^2 potato)	$SI = 1 - \sum Pi^2$
Darjeeling-Pulbazar	692	31	752	909	2384	0.290	0.013	0.315	0.3813	0.084	0.0002	0.099	0.145	0.329	0.671
Jorebunglow-Sukhiapokhri	0	0	711	239	950	0.000	0.000	0.748	0.2516	0.000	0.0000	0.560	0.063	0.623	0.3767
RangliRangliot	46	1	1552	526	2125	0.022	0.001	0.730	0.2475	0.001	0.0000	0.533	0.061	0.595	0.409
Kalimpong I	2531	37	5695	293	8556	0.296	0.004	0.666	0.0342	0.087	0.0000	0.443	0.001	0.532	0.468
Kalimpong II	1754	59	4071	473	6357	0.276	0.009	0.640	0.0744	0.076	0.0001	0.410	0.005	0.492	0.508
Gorubathan	643	66	1578	364	2651	0.242	0.025	0.595	0.1373	0.059	0.0006	0.354	0.019	0.433	0.567
Kurseong	29	3	806	382	1220	0.024	0.002	0.660	0.3131	0.001	0.0000	0.436	0.099	0.535	0.465
Mirik	36	0	88	59	183	0.197	0.000	0.481	0.3224	0.039	0.0000	0.231	0.104	0.374	0.626

namely Jorebunglow-Sukiapokri, Rangli Rangliot, Kurseong and Kalimpong I are going through with poor crop diversification and possibly indicate the chances of crop failure, decreases food security, deteriorating land-soil quality, negatively impacting the crop yields, reducing the adaptive measures of the farmers and thereby intensifies vulnerability.

Block level low crop diversification measures probably imply that in the DHR hills there are certain factors which dominate crop growth, yield and these are mostly associated with climate change. The spatial representation of the SI values confirms relative uniformity of their agricultural practice and cropping intensity across the studied region (Ludwig and Reynolds 1993). Here, as a whole the appraisal of block level comparative crop diversification has been used to express both the concepts of evenness and richness on agricultural produce and applied to gauge qualitative traits of crop diversity into a scalar (Jain et al. 1975; Spagnoletti Zeuli and Qualset 1987; Magurran 1991).

Explanation of the AVI

The present AVA has thoroughly explored the nature and extent of indicator's data that are contributing to the AVI. The AVI varies between 0.723 in Kalimpong I and 0.445 in Darjeeling Pulbazar blocks. The vulnerability scores of blocks are very close to each other as observed in Figs. 14.3 and 14.4 and the result of the assessment exhibits that the high agriculturally vulnerable blocks, i.e. Kalimpong I, Rangli Rangliot, Mirik and Kurseong are facing challenges in terms of high dependency on rainfed agriculture; high percentage of landless, marginal and small farmers; significant yield variability, poor access to market and inconsistent income prospects within agriculture sector. Other blocks namely Kalimpong II, Gorubathan, Jorbunglow-Sukhiapokhri and Darjeeling Pulbazar are in moderate to low vulnerability class as in these blocks high participation in MGNREGA as well as more NRM works are found completed coupled with comparatively high ratio of livestock to total population, greater access to market accompanied by well road connectivity, better crop diversity and relatively low percentage of marginal landholders (Table 14.6).

Outcome of the AVA specifies determinants of agricultural vulnerability in the Darjeeling hills that are contributing to the sector by different degrees in consonance with change of the weather parameter. Observations from the State Action Plan on Climate Change likewise imply that in the Darjeeling Himalayan belt the temperature range between maximum and minimum is projected to exaggerate at certain levels that entails the diurnal variation of temperature shall lessen and the climate shall convert to a warmer one. These alterations may affect the plant physiology and productivity. Alongside, the rainfall is likely to reduce during summer months and in the winter months doing crop rotation a difficult task and thereafter choice of crop for sowing become a complicated decision-making procedure (Government of West Bengal 2020). Similarly, with the drivers of vulnerability, climate change impacts

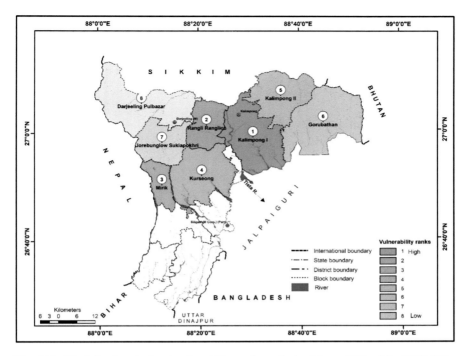

Fig. 14.3 Block level agricultural vulnerability rank as obtained from AVA of the DHR Hills, West Bengal

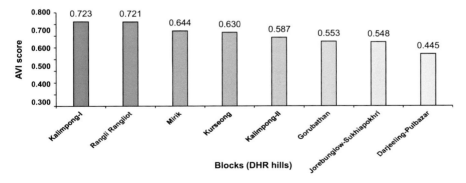

Fig. 14.4 Block level AVI values as obtained from AVA of the DHR Hills, West Bengal

focus on the pressing demand to intensify agricultural land management. Such issues must be significant in alleviating the effects of climate change on agricultural production (Meng et al. 1998; International Institute for Applied Systems Analysis 2002; Parker et al. 2019). Only then, with the advantages of sufficient rainfall, generous surface water resources in conjunction with abundance of human population engaged in agricultural sector will play an elementary role in meeting the food

Table 14.6 Block level (DHR-hills) AVA indicators data-Actual Values (AV), Normalised Values (NV), AVI and rank of blocks

Block	Indicator												
	Total number of livestock per 1000 population		% of landless, marginal land, small farmers (Land < 5 Acre)		Average person days/household employed under MGNREGA over last five years		Mean soil organic carbon content (tonne/ha)		Crop diversification (Simpson Index)		% of net irrigated area to net sown area		Variability in food grain crop yield (tonne/ha)
	AV	NV	AV	NV	AV	NV	AV	NV	AV	NV	AV	NV	AV	NV
Darjeeling Pulbazar	1509	1.000	6.223	0.201	270.340	0.000	45.236	0.000	0.671	0.000	2.151	0.886	18.568	0.834
Jorebunglow-Sukiapokhri	2036	0.667	4.358	0.082	31.720	1.000	37.011	0.745	0.377	1.000	9.930	0.075	5.558	0.000
Rangli Rangliot	2011	0.682	11.498	0.539			38.578	0.603	0.405	0.904	1.058	1.000	13.480	0.508
Kalimpong I	1892	0.757	13.186	0.646			35.931	0.843	0.468	0.688	2.782	0.820	19.378	0.886
Kalimpong II	2523	0.358	17.545	0.925			40.454	0.433	0.508	0.553	7.638	0.314	21.158	1.000
Gorubathan	3090	0.000	18.720	1.000			35.511	0.881	0.567	0.351	2.543	0.845	8.783	0.207
Kurseong	1915	0.743	3.070	0.000			34.275	0.993	0.465	0.700	10.655	0.000	8.304	0.176
Mirik	2132	0.606	5.388	0.148			34.196	1.000	0.626	0.152	1.998	0.902	12.809	0.465

Block	Indicator												
	Drainage density		Road connectivity -percentage of villages connected by pucca roads				Access to market		% of population engaged in agriculture, pisciculture and allied activities		Number of NRM works per 1000 ha (MGNREGS) and/or other schemes			Vulnerability rank
	AV	NV	AV	NV	AV	NV	AV	NV	AV	NV	AV	NV	AVI	
DarjeelingPulbazar	6.571	0.000	100.000	0.000	113.660	0.657	13.953	0.852	4.286	0.615	0.505	0.950	0.445	8
Jorebunglow-Sukiapokhri	5.852	0.266	100.000	0.000	255.980	0.060	11.905	0.921	3.166	0.820	3.936	0.000	0.548	7
Rangli Rangliot	5.193	0.510	100.000	0.000			20.000	0.650	2.184	1.000	1.430	0.694	0.721	2
Kalimpong I	5.571	0.370	84.000	0.553			16.000	0.784	3.474	0.763	0.969	0.821	0.723	1

Kalimpong II	5.279	0.478	94.872	0.177	201,600	0.288	10.256	0.976	6.853	0.144	0.876	0.847	0.587	5
Gorubathan	4.794	0.658	71.053	1.000	189,600	0.338	39.474	0.000	7.639	0.000	0.324	1.000	0.553	6
Kurseong	4.160	0.892	91.304	0.300	56,900	0.894	15.942	0.786	3.135	0.826	1.930	0.555	0.630	4
Mirik	3.869	1.000	85.714	0.494	57,955	0.890	9.524	1.000	4.347	0.603	2.785	0.319	0.644	3

Note: AVs are derived by calculation or sorting the raw data, NVs are derived using normalisation methods. Here, with varied indicators NV values and the AVI has been computed by aggregating all the obtained NV as described in the methodology section for the equal weight method security of the DHR as till date, the region sustains its populace by producing a large quantity of rice, potato besides being a large grower of pineapple, flowers and mandarin orange (Government of India 2018).

Categorisation of Regional Agricultural Vulnerability Drivers

The AVA not only highlights the block level sectoral vulnerability characteristics of the hilly DHR as well it exhibits the regional drivers of vulnerability of the agricultural sector. The most contributing in addition to governing agricultural vulnerability factors of the region are, e.g. lack of access to market, depleted soil condition, fewer number of NRM works per 1000 ha, lesser area under irrigation, variable distribution of livestock to population and poor crop diversification accompanied by inconsistent yield have contributed towards vulnerability. In contrast, satisfactory person-days generated under MGNREGA over the last five years, high drainage density around the region and improved percentage of villages connected by paved roads are showing nearly the adaptive capacity measures (Fig. 14.5).

The derived agricultural vulnerability score of each block helps to prioritise targeted adaptation planning to minimise vulnerability. Categorisation of drivers of vulnerability insights a profile of agricultural vulnerability of the DHR (hills) for

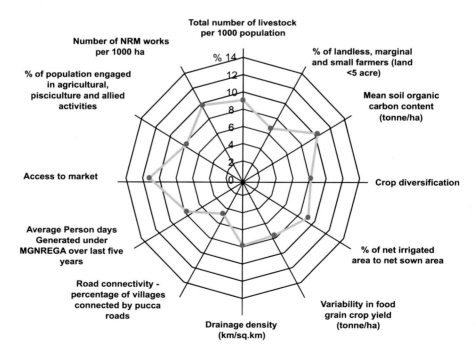

Fig. 14.5 District level contribution of indicators to vulnerability as observed from the modified agricultural vulnerability study

framing out adaptation strategies and preparedness. In view of this, impending studies on major crops and their contemporary climate-sensitive behaviour are essential to recognise options for adaptation to future climate change (Parker et al. 2019). The challenges of climate change and irregularity for agricultural research now relate to interpret the differences in the flow and storage of materials, the ecology of pests and diseases and not only limited to the dynamics of rainfall-temperature regimes (Bouroncle et al. 2017). Thus, the adaptation of food production systems to weather calamities, among other important issues that need to be tackled urgently, may use these outputs of AVA index as a necessary rational input. Consulting the obtained regional agricultural vulnerability driver's data in crop models, sensitivity of the key crops to climate change, cohesive exposure from natural hazards which affect agricultural systems can be addressed and it harmonises towards a clearer understanding of the agricultural vulnerability to climate change (Vos et al. 2016; Thiault et al. 2018).

Conclusion

Assessment of agricultural vulnerability correspondingly provides opportunities to the policymakers, development administrators engaged in decision-making process, Civil Society Organisations (CSOs) and scientific planners to prioritise the blocks-gram panchayats-villages and enhance to improve different unique adaptation planning, with the overall objective of minimising agricultural vulnerability to climate change issues in such fragile region. This AVA can aid in substantiate decision-making when it comes to procure adaptation measures, based on the derived drivers of agri-vulnerability with their index values. Indications from the drivers of vulnerability show the gap in the current scenario and focus on sector-specific demand to enable precise planning in terms of capacity building for adaptation strategies like crop diversification at the farm level under such distinct agro-climatic conditions of the region may be adopted in near future, which will also reflect farmers concern for food security, setting up of commercial market, cold chains for preservation, better infrastructural facilities concerning communication and power supply, etc.

Likewise, natural resource management also demands much importance for the sustenance of the livelihood of the mountain people. Inserting a contemporary extension to integrated farm management practices, the CSA can be used comprehensively using spring water management system and to generate its all possible produces involving traditional knowledge to attain a sustainable goal and give certain strength to adaptation planning addressing climate change resilient agriculture (Fig. 14.6). The DHR hill agriculture already witnesses of practising such integrated farm management using natural spring water. Despite limitation and hindrances, there are huge scopes and prospect for cultivation of fruits, flowers, medicinal plants-herbarium and is supposed that in approaching years it will adequately back the transitional crop production by providing necessary stability options.

Fig. 14.6 Integrated farm management (**a**) vegetable production (**b**) pisciculture in concrete tanks using spring water using spring water at Upper Burmiak village, Sangsey gram panchayat, Kalimpong II block, Kalimpong district, West Bengal, India

Acknowledgements The authors are grateful to the Principal Secretary, Department of Science and Technology and Biotechnology, Govt. of West Bengal (DSTBT, GoWB) for according all necessary support in the attainment of this study. The authors are indebted to the Department of Science and Technology, Govt. of India (DST, GoI) for aiding this research work under the project National Mission for Sustaining the Himalayan Ecosystem (NMSHE). The authors are thankful to Dr. P.B. Hazra, Principal Scientist, DSTBT, GoWB for his continuous support and encouragement for carrying out this work. Sincere thanks to concerned line State/Central Government Departments for providing all necessary co-operation and assistance during the collection of ancillary data.

Declaration

- *Compliance with Ethical Standards*

   This research work has been done in accordance with transparency, moral values, research ethics, honesty and hard work. In doing this work, no human or animal experiments were involved.
- *Funding*

   This research work has been carried out under the NMSHE project, Climate Change Programme Division funded by the Department of Science and Technology, Government of India.
- *Conflict of Interest*

   Any co-authors have no conflict of interest directly or indirectly in this work.
- *Ethical Approval*

   According to our knowledge and belief the used methodology, findings and conclusions made here belong to the original research work.
- *Informed Consent*

   Everything has been coordinated with the collective decision and proper consent.

References

Basavaraj ND, Gajanana TM, Satish kumar M (2016) Crop diversification in Gadag District of Karnataka. Agric Econ Res Rev 29(1):151–158. https://doi.org/10.5958/0974-0279.2016.00027.6

Behera UK, Sharma AR, Mahapatra IC (2007) Crop diversification for efficient resource management in India: problems, prospects, and policy. J Sustain Agric 30(3):97–127

Bhatt BC, Nakamura K (2005) Characteristics of monsoon rainfall around the Himalayas revealed by TRMM precipitation radar. Mon Weather Rev 133(1):149–165

Bhattacharya R (2008) Crop diversification: a search for an alternative income of the farmers in the state of West Bengal in India. In: International conference on applied economics, pp 83–93

Bookhagen B (2010) Appearance of extreme monsoonal rainfall events and their impact on erosion in the Himalaya. Geomat Nat Haz Risk 1(1):37–50

Bouroncle C, Imbach P, Rodríguez-Sánchez B, Medellín C, Martinez-Valle A, Läderach P (2017) Mapping climate change adaptive capacity and vulnerability of smallholder agricultural livelihoods in Central America: ranking and descriptive approaches to support adaptation strategies. Clim Chang 141:123–137. https://doi.org/10.1007/s10584-016-1792-0

Brenkert AL, Malone EL (2005) Modelling vulnerability and resilience to climate change: a study of India and Indian states. Climate Change 71(1–2):57–102

de Sherbinin A, Caffrey P, Farmer A (2014) Spatial climate change vulnerability assessments: a review of data, methods, and issues. African and Latin American resilience to climate change project. USAID. Prosperity, Livelihoods, and Conserving Ecosystems (PLACE), Tetra Tech ARD

Easterling WE, Aggarwal PK, Batima P, Brander KM, Erda L, Howden SM, Kirilenko A, Morton J, Soussana JF, Schmidhuber J, Tubiello FN (2007) Food, fibre and forest products. Climate change 2007: impacts, adaptation and vulnerability. In: Parry ML, Canziani OF, Palutikof JP, van der Linden PJ, Hanson CE (eds) Contribution of working group II to the fourth assessment report of the Intergovernmental Panel on Climate Change. Cambridge University Press, Cambridge, pp 273–313

Eriksen SH, Kelly PM (2007) Developing credible vulnerability indicators for climate adaptation policy assessment. Mitig Adapt Strateg Glob Chang 12:495–524. https://doi.org/10.1007/s11027-006-3460-6

Esteves T, Ravindranath D, Bedamatta S, Raju K, Sharma J, Bala G, Murthy I (2016) Multi-scale vulnerability assessment for adaptation planning. Curr Sci 110(7):1225–1239

Field CB, Barros VR, Mach KJ, Mastrandrea MD, van Aalst M, Adger WN, Arent DJ, Barnett J, Betts R, Bilir TE, Birkmann J, Carmin J, Chadee DD, Challinor AJ, Chatterjee M, Cramer W, Davidson DJ, Estrada YO, Gattuso J-P, Hijioka Y, Hoegh-Guldberg O, Huang HQ, Insarov GE, Jones RN, Kovats RS, Romero-Lankao P, Larsen JN, Losada IJ, Marengo JA, McLean RF, Mearns LO, Mechler R, Morton JF, Niang I, Oki T, Olwoch JM, Opondo M, Poloczanska ES, Pörtner H-O, Redsteer MH, Reisinger A, Revi A, Schmidt DN, Shaw MR, Solecki W, Stone DA, Stone JMR, Strzepek KM, Suarez AG, Tschakert P, Valentini R, Vicuña S, Villamizar A, Vincent KE, Warren R, White LL, Wilbanks TJ, Wong PP, Yohe GW (2014) Technical summary. In: Field CB, Barros VR, Dokken DJ, Mach KJ, Mastrandrea MD, Bilir TE, Chatterjee M, Ebi KL, Estrada YO, Genova RC, Girma B, Kissel ES, Levy AN, MacCracken S, Mastrandrea PR, White LL (eds) Climate Change 2014: Impacts, adaptation, and vulnerability. Part A: Global and sectoral aspects. Contribution of Working Group II to the fifth assessment report of the Intergovernmental Panel on Climate Change. Cambridge University Press, Cambridge/New York, pp 35–94

Government of India, Ministry of Agriculture and Farmers Welfare (2016) Handbook of agriculture. Indian Council of Agricultural Research

Government of India, National Institution for Transforming India Ayog (2018) Sustainable tourism in the Indian Himalayan Region report

Government of West Bengal, Department of Environment (2012) West Bengal state action plan climate change report

Government of West Bengal, Department of Environment (2020) West Bengal state action plan climate change report

Imbs J, Wacziarg R (2003) Stages of diversification. Am Econ Rev 93(1):63–86

International Institute for Applied Systems Analysis (2002) United Nations Institutional Contract Agreement No. 1113 on "Climate Change and Agricultural Vulnerability as a contribution to the World Summit on Sustainable Development", Johannesburg

IPCC (2001) The regional impacts of climate change: an assessment of vulnerability: special report of IPCC working group II. In: Watson RT, Zinyowera MC, Moss RH (eds) Intergovernmental panel on climate change. Cambridge University Press, Cambridge, p 517

IPCC (2014) Summary for policymakers. In: Field CB, Barros VR, Dokken DJ, Mach KJ, Mastrandrea MD, Bilir TE, Chatterjee M, Ebi KL, Estrada YO, Genova RC, Girma B, Kissel ES, Levy AN, MacCracken S, Mastrandrea PR, White LL (eds) Climate change 2014: impacts, adaptation, and vulnerability. Part A: global and sectoral aspects contribution of working group II to the fifth assessment report of the Intergovernmental Panel on Climate Change. Cambridge University Press, Cambridge/New York, pp 1–32

Isabel PP, Reboredo FH, Ramalho JC, Pessoa MF, Lidon FC, Silva MM (2020) Potential impacts of climate change on agriculture – a review. Emirates J Food Agric 32(6):397–407. https://doi.org/10.9755/ejfa.2020.v32.i6.2111

Jain SK, Qualset CO, Bhatt GM, Wu KK (1975) Geographical patterns of phenotypic diversity in a world collection of durum wheat. Crop Sci 15:700–704

Jill LC, Erin OS (2005) Land use and income diversification: comparing traditional and colonist population in the Brazilian Amazon. Agric Econ 32(3):221–237

Joshi PK, Gulati A, Birthal PS, Tewari L (2004) Agriculture diversification in South Asia: patterns, determinants and policy implications. Econ Polit Wkly 39(18):2457–2467

Joshi S, Kumar K, Joshi V, Pande B (2014) Rainfall variability and indices of extreme rainfall-analysis and perception study for two stations over Central Himalaya, India. Nat Hazards 72(2):361–374

Kayet N, Pathak K, Chakrabarty A (2016) Urban heat island explored by co-relationship between land surface temperature vs multiple vegetation indices. Spat Inf Res 24:515–529

Lin BB (2011) Resilience in agriculture through crop diversification: adaptive management for environmental change. BioScience 61(3):183–193

Loi DT, Huong LV, Tuan PA, Hong Nhung NT, Quynh Huong TT, Hoa Man BT (2022) An assessment of agricultural vulnerability in the context of global climate change: a case study in HaTinh Province, Vietnam. Sustainability 14:1282. https://doi.org/10.3390/su14031282

Ludwig JA, Reynolds JF (1993) Statistical ecology: a primer on methods and computing. Wiley, New York

Magurran A (1991) Ecological diversity and its measurement. Princeton University Press, New Jersey, United States.

Meng ECH, Smale M, Bellon M, Grimanelli D (1998) Definition and measurement of crop diversity for economic analysis. In: Smale M (ed) Farmers gene banks and crop breeding: economic analyses of diversity in wheat maize and rice. Springer, Dordrecht. https://doi.org/10.1007/978-94-009-0011-0_2

Mishra S, Sarkar U, Taraphder S, Datta S, Swain D, Saikhom R et al (2017) Multivariate statistical data analysis- Principal Component Analysis (PCA). Int J Livest Res 7(5):60–78. https://doi.org/10.5455/ijlr.20170415115235

Moran MS, Jackson RD, Slater PN, Teillet PM (1992) Evaluation of simplified procedures for retrieval of land surface reflectance factors from satellite sensor output. Remote Sens Environ 41:169–184

O'Brien KL, Leichenko RM (2000) Double exposure: assessing the impacts of climate change within the context of economic globalization. Glob Environ Chang 10(3):221–232

Oppenheimer M, Campos M, Warren R, Birkman J, Luber B, O'Niel B (2014) Emergent risks and key vulnerabilities. In: Field CB, Barros VR, Dokken DJ, Mach KJ, Mastrandrea MD, Bilir TE, Chatterjee M, Ebi KL, Estrada YO, Genova RC, Girma B, Kissel ES, Levy AN, Mac-Cracken S, Mastrandrea PR, White LL (eds) Climate change 2014: impacts, adaptation, and vulnerability. Part A: global and sectoral aspects. Contribution of working group II to the fifth assessment report of the Intergovernmental Panel on Climate Change. Cambridge University Press, Cambridge/New York, pp 1039–1099

Parker L, Bourgoin C, Martinez-Valle A, Läderach P (2019) Vulnerability of the agricultural sector to climate change: the development of apan-tropical Climate Risk Vulnerability Assessment to inform sub-national decision making. PLoS ONE 14(3):e0213641. https://doi.org/10.1371/journal

Parry ML, Canziani OF, Palutikof JP, van der Linden PJ, Hanson CE (eds) (2007) Climate Change 2007: impacts, adaptation and vulnerability. Contribution of Working Group II to the fourth assessment report of the Intergovernmental Panel on Climate Change. Cambridge University Press, Cambridge, p 982

Raju KV, Deshpande RS, Satyasiba B et al (2016) Socio-economic and Agricultural Vulnerability Across Districts of Karnataka. Climate Change Challenge (3C) and Social-Economic-Ecological Interface-Building, Environmental Science. https://doi.org/10.1007/978-3-319-31014-5_11

Ravindranath N, Rao S, Sharma N, Nair M, Gopalakrishnan R, Rao A, Malaviya S, Tiwari R, Sagadevan A, Munsi M, Krishna N, Govindasamy B (2011) Climate change vulnerability profiles for north East India. Curr Sci 101:384–394

Samanta S, Chakraborty J, Dutta SB (2022) Village level landslide probability analysis based on weighted sum method of multi-criteria decision-making process of Darjeeling Himalaya, West Bengal, India. In: Shit PK, Pourghasemi HR, Bhunia GS, Das P, Narsimha A (eds) Geospatial technology for environmental hazards, Advances in geographic information science. Springer, Cham. https://doi.org/10.1007/978-3-030-75197-5_17

Sanjit M, Sujeet KJ, Sanchita G, Arindam N, Bera AK, Vijay Paul RC, Upadhaya SMD (2017) An assessment of social vulnerability to climate change among the districts of Arunachal Pradesh, India. Ecol Indic 77:105–113

Sehgal VK, Singh MR, Chaudhary A, Jain N, Pathak H (2013) Vulnerability of agriculture to climate change: district level assessment in the Indo-Gangetic Plains. Indian Agricultural Research Institute, New Delhi

Sharma J, Chaturvedi RK, Bala G, Ravindranath NH (2015) Assessing "inherent vulnerability" of forests: a methodological approach and a case study from Western Ghats, India. Mitig Adapt Strateg Glob Chang 20(4):573–590. https://doi.org/10.1007/s11027-013-9508-5

Sharma J, Murthy KI, Esteves T, Negi P, Sushma S, Dasgupta S, Barua A, Bala G, Ravindranath NH (2019) Climate vulnerability and risk assessment: framework, methods and guidelines for the Indian Himalayan region, report

Spagnoletti Zeuli PL, Qualset CO (1987) Geographical diversity for quantitative spike characters in a world collection of durum wheat. Crop Sci 27:235–241

Sridevi G, Jyotishi A, Mahapatra S, Jagadeesh G, Bedamatta S (2021) Climate change vulnerability in agriculture sector: indexing and mapping of four southern Indian states. Quaderni—working paper DSE N 966. https://ssrn.com/abstract=2503834

Stevanovic M, Popp A, Lotze-Campen H, Dietrich JP, Muller C, Bonsch M (2016) The impact of high-end climate change on agricultural welfare. Sci Adv 2:e1501452–e1501452. https://doi.org/10.1126/sciadv.1501452

Thiault L, Marshall P, Gelcich S, Collin A, Chlous F, Claudet J (2018) Mapping social-ecological vulnerability to inform local decision making: mapping social-ecological vulnerability. Conserv Biol 32:447–456. https://doi.org/10.1111/cobi.12989

Tina-Simone N, Lotten W, Tomasz O, Erik G, Björn-Ola L (2018) Evaluation of indicators for agricultural vulnerability to climate change: the case of Swedish agriculture. Ecol Indic 105:571–580

Vos R, Cattaneo A, Stamoulis K, Semedo MH, Salazar RC, Frick M (2016) The State of Food and Agriculture. Climate change, agriculture and food security. Food and Agriculture Organization of the United Nations. Office for Corporate Communication, Rome

Wulf H, Bookhagen B, Scherler D (2010) Seasonal precipitation gradients and their impact on fluvial sediment flux in the Northwest Himalaya. Geomorphology 118(1–2):13–21

Chapter 15
Forest Landscape Dynamic and People's Livelihood Dependency on Forest: A Study on Bankura District, West Bengal

Abira Dutta Roy and **Santanu Mandal**

Abstract Forest is the key element of the environment and ecosystem. The forests of Bankura district are of deciduous type and the dominant species here is Sal (*Shorea robusta*). Due to the development processes and increasing demand of agricultural land, the forests have been destroyed. In this study, we have analysed the spatio-temporal change of land use land cover from the year 1990 to 2020 with 10 years of interval aided by Landsat 4-5TM and Landsat 8OLI images. The study also looked into the degree of fragmentation through various matrices such as Percentage of Landscape (PLAND), Patch Number (NP), Patch Density (PD), Total Edge (TE), Edge Density (ED), Landscape Shape Index (LSI), Patch Cohesion Index (COHESION), Perimeter-Area Fractal Dimension (PAFRAC), and Aggregation Index (AI) using Fragstat 4.2 software. It was evident that agricultural land and built-up area had increased at the expense of forest. Moreover, the matrices highlight the forest areas have become more degraded and fragmented during the study period. Considerable portions of the population in the district earn their livelihood from various forest products and dwell in and around the forest areas. A household survey was conducted in different blocks of the district to understand the people's extent of dependency on the forest and their perception on the forest fragmentation. The levels of awareness of forest degradation were though evident but keen interests in participation towards mitigating the problem were not observed among the respondents.

Keyword Forest landscape · Forest patch · People's livelihood

---

A. D. Roy · S. Mandal (✉)
Department of Geography, Bankura Zilla Saradamani Mahila Mahavidyapith, Bankura University, Purandarpur, West Bengal, India
e-mail: abira6520@bzsmcollege.org

© The Author(s), under exclusive license to Springer Nature Switzerland AG 2022
U. Chatterjee et al. (eds.), *Ecological Footprints of Climate Change*, Springer Climate, https://doi.org/10.1007/978-3-031-15501-7_15

Introduction

FAO, 2000 defines forest as a land area of more than 0.5 ha, with a tree canopy cover of more than 10%, which is not primarily under agricultural or other specific non-forest land use. Forests are an important part of the environment and their importance is immense, and it is extremely important. They aid as biodiversity repositories (Li et al. 2009), restrain soil erosion (Nandy et al. 2011), prevent landslides given that tree roots bind soil, regulate air humidity, temperature and mitigate global warming (Cabral et al. 2010) by absorbing 30% of fossil fuel CO_2 emissions (Pan et al. 2011). There has been a general consensus that forests play a vital role in maintaining the ecological balance, provide economic resources, and perform social functions such as water and nutrient cycling, maintaining species and genetic diversity, regulation of greenhouse gases, controlling soil erosion, balancing the global carbon cycle, etc. (Devendra 2011; Rao and Pant 2001). Loss of forest area is a global, national and regional problem mainly due to industrial and agricultural revolution (Reddy et al. 2013; Willims 2003; Crist and Cicone 1984), and the forests of Bankura district too have not been able to flee from this problem either (Dutta 2021). Henceforth scientific land use planning is indispensable for forest resource conservation at local, regional and global levels with all the stakeholders involved.

Complete obliteration is also accompanied by the process of forest degradation. In recent years, forest habitats are being fragmented due to multiple reasons in different parts of the world (Bera et al. 2020b). Forest fragmentation is the process or mechanism where continuous or grandiose forest land is converted into small segments or isolated forest regions (Bera et al. 2020b). The process of forest fragmentation mainly occurs due to human activities such as logging or conversion of forests into agricultural areas and suburbanization. Expanding human pressure over nature and anthropogenic stress over natural resources degrades the health of the forest cover which is the principal reason of forest fragmentation and it expands the research area of landscape ecologists (Whitmore 1997; Liu and Taylor 2002; Roy et al. 2013). Forest cover changes can be recorded through land use land cover change detection mechanisms. Forest degradation on the other hand is quantified by measuring the degree of forest fragmentation. The term "Fragmentation" has been defined as a simultaneous reduction of forest area, increase in forest edge, and subdivision of large forest areas into smaller non-contagious fragments (Laurance et al. 2011). Fragmentation is a dynamic process in which the habitat of organisms is progressively reduced into smaller patches that become more isolated and affected by edge effects (Forman and Godron 1986; Reed et al. 1996; Franklin 2001). The degree of fragmentation has been described as a function of the varying size, shape, spatial distribution, and density of patches (Mladenoff et al. 1993). Forman (1995) has identified fragmentation as the most important factor contributing to the decline and loss of species diversity worldwide (Noss and Cooperrider 1994). Taking into consideration the global trend in forest degradation it also becomes imperative to carry out a forest fragmentation analysis for Bankura where according to the State Forest Report, 2007, the *Sal Jungle* occupies 21.5% of the total geographical area.

Study Area

Bankura district is located in the western part of the state of West Bengal. It is placed between 22° 38′ and 23° 38′ north latitude and between 86° 36′ and 87°46′ east longitude. On the north and north-east, the Damodar River separates Bankura from Bardhaman. On the south-east, it is bounded by Hooghly district and on the south by Paschim Medinipur. On the western flank of Bankura lies Purulia. Bankura district has been described as the connecting link between the plains of Bengal on the east and the Chotanagpur plateau on the west.

It has an area of 6882 square kilometres (sq km). Almost 21.5% area of the district is covered by forest (state forest report 2007). There is 0.046 ha of forest land per capita in the district, whereas 0.02 ha per capita for the whole of West Bengal (Bankura district forest Report). The forest comprises 44.48 km^2 of reserved forest, 1391.95 km^2 of protected forest, and 27.13 km^2 of unclassified state forest (state forest report 2007). Bankura district forest is predominantly Sal and its allied species and plantation forest of Eucalyptus and Akashmoni. The district has the best quality of Sal forest in West Bengal particularly at Radhanagar, Sonamukhi, and Patrasayer blocks and the entire Bishnpur subdivisional jurisdiction. Figure 15.1 shows the location of Bankura district. The district is drained by the Darakeshwar River, which is a tributary of the Damodar. The economy is most agrarian-based. Agriculture accounts for almost 70% of the district's income whereas 80% of the farmers are small and marginal. Erratic and sporadic rainfall in the region hinders agricultural practices to a great extent. However, due to protective irrigation system, land reforms, and use of high fertile & hybrid crops, the district is now not so poor as it was. Also, cottage and small-scale industries, e.g., stone-crushing, weaving, oilseed-crushing, handicraft units like Dokra, Terra-cotta, Baluchari Sari, etc. play a key economic role of the district. Bagli (2015) writes about the multi-dimensional poverty in Bankura district. In order to improve the socio-economic standards the tribal and rural population restore to forest degradation and encroachment. Majumdar (2018) states how for biomass fuel the forest areas are exploited and sustainable forest management practices are being hampered. Several studies by Mandal and Chatterjee (2018, 2019, 2020, 2021) have researched upon the forest degradation patterns of Bankura district, but their works were restricted to only Panchet and Radhanagar forest divisions.

On the contrary, this paper attempts to look into the forest cover dynamics of the entire district and looks into the various fragmentation matrices to understand both the reduction in cover as well as thinning in tree density that is forest degradation. It also looks into the perception of the people towards the depletion of this natural resource on which their existence depends.

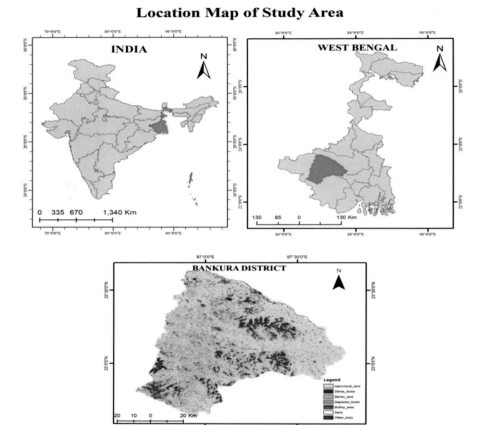

Fig. 15.1 The location map of the study area

Material and Methodology (Fig. 15.2)

The state forest report, 2017 has emphasized that about 21% of the land area of Bankura district belongs to forest land but the area of forest cover has been steadily declining. To determine the extent of deforestation or afforestation, conversion and modification of land use land cover (LULC) multispectral satellite images enlisted in Table 15.1 have been downloaded from https://earthexplorer.usgs.gov/. All images have been taken in December because of cloudless sky and full canopy conditions. For the image classification and analysis, QGIS 3.10, ERDAS IMAGINE 15, ArcGIS 10.4, and Fragstat4.2 software were used.

After downloading the Landsat data, it has been clipped by Bankura district shapefile. All clipped bands of satellite images are merged or stacked into a single image for generation of false colour composite image and supervised classification

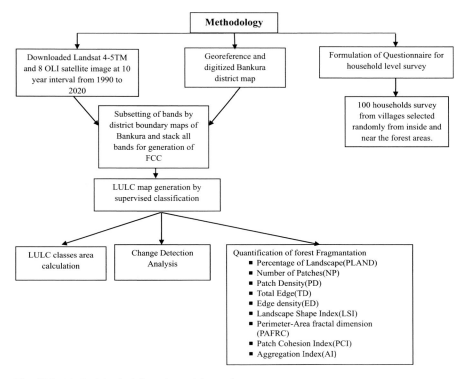

Fig. 15.2 Methodological flow chart of the study

Table 15.1 Details of Landsat images used

Sensor	Path number	Row	Date of acquisition	Spatial resolution
Landsat 4-5 TM	139	44	23/12/1990	30 m
Landsat 4-5 TM	139	44	18/12/2000	30 m
Landsat 4-5 TM	139	44	14/12/2010	30 m
Landsat 8 OLI	139	44	25/12/2020	30 m

in ERDAS IMAGINE 15 was carried out. The images were classified into seven LULC classes as can be seen in Fig. 15.1. Accuracy assessment was carried out post classification with the help of Google Earth Pro and field survey as seen in Fig. 15.3. The recoded supervised image was run in Fragstat 4.2 for forest fragmentation analysis. The following indicators are used for forest fragmentation.

Percentage of Landscape (PLAND)

PLAND quantified the proportional ratio of each patch type in the landscape. It equals the sum of the area (m^2) of all patches of the corresponding patch type, divided by the total landscape area (m^2), and multiplied by 100 (to convert to percentage) (Herold et al. 2005).

Fig. 15.3 Field photographs for LULC verification

$$\mathrm{PLAND} = \mathrm{P}i = \frac{\sum_{j=1}^{n} \mathrm{a}ij}{A}(100) \qquad (15.1)$$

$\mathrm{P}i$ = proportion of the landscape occupied by patch type (class) i.
$\mathrm{a}ij$ = area (m^2) of patch ij,
A = total landscape area (m^2).

The Number of Patches (NP)
It is the measure of discontinuous urban areas or individual units in the landscape (Gezahegn Awake Abebe, 2013). Due to the rapid core development, the number of patches is expected to increase due to the emergence of new fragmented patches around the cores. The number of patches indicates the diversity or richness of the landscape. In other words, it gives a simple measure of the extent of subdivision or

fragmentation of the patch type (Herold et al. 2005). NP = 1 when the landscape contains only one patch of the corresponding patch type.

$$NP = ni \tag{15.2}$$

where ni = number of patches in the landscape of patch type.

Patch Density (PD)

It is one more measure of landscape fragmentation of the patches of a land cover class which specifies the density of the fragmented urban units within a quantified area. Values of this indicator are affected by the size of the pixel and also the minimum mapping unit since this is the significant factor for describing individual patches. This usually expresses the number of patches on a per unit area basis that facilitates comparisons among landscapes of the varying size.

$$PD = \frac{ni}{A}(10,000)(100) \tag{15.3}$$

where ni = number of patches in the landscape of patch type (class) i.
A = total landscape area (m^2).

Total Edge (TE)

Total edge is an absolute measure of the total edge length of a particular patch type. In applications that involve comparing landscapes of varying size, no edge in the landscape; that is, when the entire landscape and landscape border, if present, consists of a single patch and the user specifies that none of the landscape boundary and background edge be treated as edge. Value of total edge ranges greater than zero.

$$TE = E \tag{15.4}$$

where E = total length (m) of edge in the landscape.

Edge Density (ED)

ED equals the sum of the lengths (m) of all edge segments in the landscape. Its value ranges from zero to greater than zero.

$$ED = \frac{E}{A}(10,000) \tag{15.5}$$

E = total length (m) of edge in the landscape.
A = total landscape area (m^2).

Landscape Shape Index (LSI)

Landscape shape index provides a standardized measure of total edge or edge density that adjusts for the size of the landscape:

$$\text{LSI} = \frac{E}{\min E} \quad (15.6)$$

E = total length of edge in landscape in terms of number of cell surfaces; includes all landscape boundary and background edge segments.

min E = minimum total length of edge in landscape in terms of number of cell surfaces (see below).

Perimeter-Area Fractal Dimension (PAFRAC)

PAFRAC is the slope of regression line obtained by regressing the logarithm of patch area (m²) against the logarithm of patch perimeter. $1 \leq \text{PAFRAC} \leq 2$.

$$\text{PAFRAC} = \frac{\left[N \sum_{i=1}^{m} \sum_{j=1}^{n} (\text{Inp}_{ij} - \text{Inp}_{ij}) - \left[\left(\sum_{i=1}^{m} \sum_{j=1}^{n} \text{Inp}_{ij} \right) \left(\sum_{i=1}^{m} \sum_{j=1}^{n} \text{Ina}_{ij} \right) \right] \right]^{2}}{\left(N \sum_{i=1}^{m} \sum_{j=1}^{n} \text{Inp}_{ij}^{2} \right) - \left(\sum_{i=1}^{m} \sum_{j=1}^{n} \text{Inp}_{ij} \right)^{2}} \quad (15.7)$$

a_{ij} = area (m²) of patch ij.
p_{ij} = perimeter (m) of patch ij.
n_i = number of patches in the landscape of patch type (class) i.

Patch Cohesion Index (COHESION)

Patch cohesion index measures the physical connectedness of the corresponding patch type below the percolation threshold. Patch cohesion is sensitive to the aggregation of the focal class. Patch cohesion increases as the patch type become more clumped or aggregated in its distribution; hence, more physically connected. Above the percolation threshold, patch cohesion does not appear to be sensitive to patch configuration (Gustafson and Parker 1994).

$$\text{PCI} = \left[1 - \frac{\sum_{j=1}^{n} p_{ij}^{4}}{\sum_{j=1}^{n} p_{ij}^{4} \sqrt{a_{ij}^{4}}} [1 - \frac{1}{\sqrt{z}}]^{-1} \right] * 100 \quad (15.8)$$

p_{ij} = perimeter of patch ij in terms of number of cell surfaces.
a_{ij} = area of patch ij in terms of number of cells.
A = total number of cells in the landscape.

Aggregation Index (AI)

Aggregation index is calculated from an adjacency matrix, which shows the frequency with which different pairs of patch types (including like adjacencies between the same patch type) appear side-by-side on the map. The aggregation index is scaled to account for the maximum possible number of like adjacencies given any Pi. The maximum aggregation is achieved when the patch type consists of a single, compact patch, which is not necessarily a square patch. The value of AI index ranges between 0 and 100.

$$AI = \left[\frac{g_{ii}}{\max - g_{ii}} \right] (100) \qquad (15.9)$$

g_{ii} = number of like adjacencies (joins) between pixels of patch type (class) i based on the *single-count* method.

max-g_{ii} = maximum number of like adjacencies (joins) between pixels of patch type (class) i (see below) based on the *single-count* method.

Household survey at ten villages as shown in Fig. 15.4 was done using a standard questionnaire and personal interview techniques to assess the dependency of people on the forest area and it was attempted to associate their response to forest fragmentation. In order to select the villages for the questionnaire-based survey, the forest division map prepared by the Forest department of Bankura district was used. The

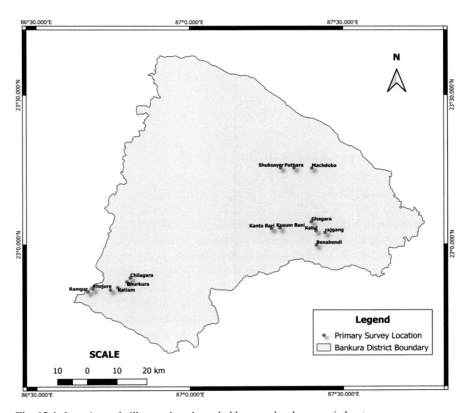

Fig. 15.4 Locations of villages where household survey has been carried out

Fig. 15.5 Field photographs of household survey

village boundary map was overlaid on the forest division map. The village boundaries that were either inside the forest patches or nearest to them were sorted out. Random sampling technique was later employed to select the villages under each forest division and ten villages were selected for the survey. Hundred households were surveyed with a structured questionnaire. Figure 15.5 shows some of the photographs of the survey conducted.

Result and Discussion

Spatio-Temporal Changes in Forest Cover

LULC are the two main variables of landscape. Land cover is the physical process and land use is the subject of how people are using the land. Due to the rapid rate of population explosion along with the execution of different developmental projects, the land use and land cover patterns have been tremendously changed over time (Bera et al. 2020a, b). Land cover of Bankura district in general and forest cover in specific have undergone a rapid change to cater the growing demand of food (Dutta 2021).

Satellite images (1990, 2000, 2010 and 2020) of Bankura district have been classified into seven land use classes, namely agricultural land, barren land, built-up area, degraded forest, dense forest, sand bars, and water body. Agricultural lands are shown to have occupied almost 72.43% area to the total area of the district. The area under the agricultural land has increased from the past. In 1990 the percentage of agricultural land occupied was 68.73% (4720 sq km) as has been calculated. Most

Table 15.2 LULC change statistic of Bankura district

Land use class	1990 Area (sq km)	Area (%)	2000 Area (sq km)	Area (%)	2010 Area (sq km)	Area (%)	2020 Area (sq km)	Area (%)
Agricultural land	4720	68.73	4533	66	4944	71.99	4974	72.43
Barren land	95	1.39	77	1.12	46	0.68	71	1.03
Built-up area	146	2.12	196	2.85	204	2.96	212	3.08
Degraded forest	766	11.15	560	8.15	571	8.31	351	5.12
Dense forest	936	13.63	1162	16.93	908	13.22	1044	15.2
Sand deposition	95	1.38	61	0.89	64	0.93	82	1.19
Water body	110	1.6	279	4.06	131	1.91	134	1.95
Total	6868	100	6868	100	6868	100	6868	100

Source: Computed and calculated by researcher from Landsat 4-5TM & 8 satellite images

of the agricultural land is seen to be located near the four main rivers Damodar, Darakeswer, Shilabati, and Kangsabati for having access to irrigation facilities. The percentage share of barren land has seen to have decreased in different periods (in 1999–1.39%, 2000–1.12, 2010–0.68, and 2020–1.03), which means barren lands have been converted to other land uses as can be seen from Table 15.2 and Fig. 15.6. Increase in the built-up area has been significant over the years. In 1990, 2000, 2010 and 2020, the percentage share of built-up area has been 2.12%, 2.85%, 2.96% and 3.08% respectively showing a rapid rise. The change of forest cover in this district too has been noteworthy. In 1990, 2000, 2010 and 2020, forest area to the total area of the district has been calculated to about 24.78% (1702 sq km), 25.08% (1722 sq km), 21.53% (1479 sq km), and 20.14% (1395 sq km) respectively, which definitely hints optimistic one. Between 2000 to 2010 and 2010 to 2020 forest areas have decreased by 243 sq km and 84 sq km respectively. The results comply to a considerable extent to the state forest report of 2019. In 1990, 1.38% and 1.6% area has under the sand deposition and water bodies respectively. The area under the sand deposition and water body has occupied 0.89% and 4.06% to the total area of the district in 2000. Later in 2010, 0.93% area was covered under the sand deposition and 1.91% was covered by the water body. The areas under the sand deposition and water body have occupied 1.19% and 1.95% of the total area respectively in 2020. The detailed yearly LULC data and their respective maps can be seen in Table 15.2 and Fig. 15.6.

From Table 15.3 it is quite evident that the areas under dense forest have been progressively converted to degraded forest and also agricultural land during the study period. Some efforts of afforestation can also be observed from the LULC change detection matrix.

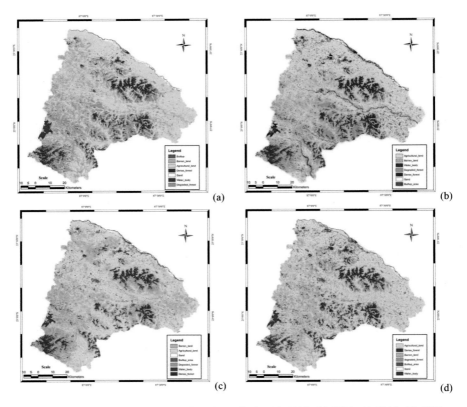

Fig. 15.6 Land use land cover maps of Bankura district (**a**) 1990, (**b**) 2000, (**c**) 2010, (**d**) 2020

Forest Fragmentation

From a global point of view, societal development has a negative impact on environment as well as forest regions. Forests are being fragmented due to execution of some developmental projects such as road construction, building and multiplex construction, construction of railway, and extension agricultural land (Southworth et al. 2004; Bera et al. 2020b). Exponential population growth along with various anthropogenic activities mainly in tropical and sub-tropical regions have created conflicts related to forest land and resource allocation (Dutta et al. 2017; Bera et al. 2020b). Forest fragmentation has thus led to habited loss and hence biodiversity loss in a region (Fahrig 2003).

Fragstat metrics generated to assess the extent of fragmentation in the forests of Bankura yielded the following results.

Bankura district is known as Janghalmahal district but the forest is continuously being cleaned up from the region. Percentage of forest landscape is 25% in 1990, 25.3% in 2000, 21.97% in 2010, and 20.5% in 2020. Total forest cover is decreasing

Table 15.3 Change detection matrix of LULCC

1990-2000							
	Agricultural Land	Dense Forest	Degraded Forest	Built-up	Sand	Water Body	Barren Land
Agricultural Land	3935.11	254.83	219.38	100.32	11.38	147.36	55.87
Dense Forest	124.89	760.17	35.64	13.23	0.0009	1.73	0.4
Degraded Forest	330.34	131.4	283.99	0.35	0.13	6.44	13.44
Built-up	51.83	8.69	3.01	17.48	0.0027	3.1	0.07
Sand	9.25	0.03	0.22	0.02	40.09	44.87	0.26
Water Body	24.07	2.34	0.86	4.61	4.24	72.97	0.88
Barren Land	63.53	5.15	17.64	0.03	0.13	1.8	6.71
2010-2020							
	Agricultural Land	Dense Forest	Degraded Forest	Built-up	Sand	Water Body	Barren Land
Agricultural Land	4130.03	299.25	272.68	134.79	15.1	38.9	55.03
Dense Forest	180.05	682.32	32.22	11.86	0.06	1.44	0.35
Degraded Forest	467.13	46.9	38.11	2.58	1.04	1.92	13.16
Built-up	69.28	8.42	4.6	40.24	0.3	4.72	0.08
Sand	29.95	0.19	0.09	1.14	41.51	7.46	0.01
Water Body	29.42	7.7	0.76	2.5	13.28	75.87	0.01
Barren Land	38.26	2.1	3.33	0.26	0.49	0.18	1.11
2000-2010							
	Agricultural Land	Dense Forest	Degraded Forest	Built-up	Sand	Water Body	Barren Land
Agricultural Land	3904.69	120.46	380.11	84.3	4.29	16.75	28.43
Dense Forest	371.77	735.39	45.8	5.85	0.01	1.46	2.32
Degraded Forest	384.5	36.81	126.47	0.62	0.02	0.31	11.01
Built-up	87.4	12.88	1.04	27.28	0.01	7.35	0.6
Sand	7.68	0.02	0.15	0.07	40.72	7.33	0.02
Water Body	127.62	2.08	5.41	9.37	35.26	96.22	2.3
Barren Land	62.22	0.67	11.86	0.014	0.02	0.11	1.59

In this table, the yellow colour highlights are to show that in the change detection matrix, Dense Forest and Degraded Forest have been converted to agricultural land. The cells highlighted in saffron colour show that agriculture and degraded forests have changed into dense forest as a result of agroforestry, afforestation and joint forest management schemes. The portions of these LULC occupy the major proportion of the LULC classes and hence have been highlighted to draw the reader's attention to the major players in the LULC dynamics of the area.

continuously in Bankura district. The area of dense forest of the district has increased 3.33% to the total geographical area between 1990 and 2000. In 2010 percentage of dense forest cover rather decreased to 13.48% and in 2020 it showed an increase to 15.33%. Percentage of the degraded forest has shown a steep decline in areal coverage from 1900 to 2020 as can be shown from the PLAND statics in Fig. 15.7a.

Forest pixels comprising small forested areas surrounded by non-forested land cover is known as forest patch (Ramachandra et al. 2016). Forest fragmentation occurs when a large region of forest is broken down, or fragmented into a collection of smaller patches of forest habitat (Wilcove et al. 1986; Collingham and Huntley 2000; Fahrig 2003). The number of patches is also calculated for forest landscape analysis which is another important indicator of forest fragmentation analysis.

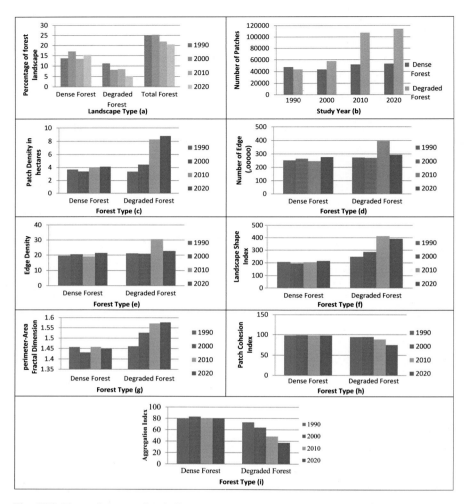

Fig. 15.7 Forest fragmentation indices (**a**) PLAND, (**b**) NP, (**c**) PD, (**d**) TE, (**e**) ED, (**f**) LSI, (**g**) PAFRAC, (**h**) COHESION, (**i**) AI

Increased numbers of forest patches signify fragmentation. Excluding 2000, the number of forest patches has increased at every 10-year interval during the study period. The number of forest patches in dense forest has rapidly increased between 2000 and 2010 which is 43,851 and 52,065, but the scenario is worse in the case of degraded forest as can be seen from Fig. 15.7b.

Increased patch density highlights increased forest fragmentation. Over the years from 1990 to 2020 increased patch density has been observed in both dense and degraded forest. The rate of expansion of patch density is higher in degraded forests than the dense forest. Figure 15.7c shows the steeper rise in patch density in degraded forests of Bankura leading to more thinning of the forest densities.

Fragstat results show that the total number of edges (TE) in dense forest increased between 1990 and 2000 which is little bit slower and the total number is near about 235 lakh and 265 lakhs (Fig. 15.7d). In 2000 to 2010 total edge was decreased to 247 lakh but in last decade it increased in very faster rate and reached to 277 lakh. The scenario is different in the degraded forest. Highest total edge is found in the year 2010 (395 lakhs). The increase in edge means forest is more fragmented. In this observation, we can see that the fragmentation rate is higher in the degraded forest than in the dense forest.

Like patch density, edge density is very sensitive to the spatial resolution of images used (McGarigal and Marks 1995). In higher resolution imagery where the smallest patches can be identified and distinguished, it is reasonable that edge density will be higher as more edges are identified and quantified (Iris and James 2019). Edge density is another parameter to assess the forest fragmentation. In this observation, we can see the edge density is increasing in both types of forest. In the dense forests of Bankura the edge density is increasing in slower rate than the degraded forest area. The rate of increase of edge density of dense forest in 1990–2000 is 0.093 per year. The rate is negative in the next decade (2000–2010) which is −0.146 per year. Between 2010 and 2020 the edge density has continued to increase 0.236 m/ha per year. Edge density of degraded forest has declined between 1990 and 2000 but the rate of decline has increased in the last decade. The rate of decline in ED means forests are tending to concentrate (Fig. 15.7e).

In this paper it has been observed that in Bankura the value LSI throughout the study years are very high. If the value of LSI is close to zero, it indicates that the landscape is a simple shape with high aggregation. LSI value of the dense forest is a little bit lower than the degraded forest of all consecutive study years. High LSI values indicate higher forest fragmentation. The LSI value from 1990 to 2000 clearly shows that the degraded forest has been more fragmented than the dense forest.

Perimeter-area fractal dimension is appealing because it reflects shape complexity across a range of spatial scales (patch sizes). Patch shape complexity is another measure of forest fragmentation. PAFRAC of the dense forest has not significantly changed in the period of study, which means the fragmentation rate is very slow. But if we look into the value of PAFRAC of the degraded forest it is higher which means forests are fragmented to a large extent.

Patch cohesion index measures the physical connectedness of the corresponding patch type. Below the percolation threshold, patch cohesion is sensitive to the aggregation of the focal class. Patch cohesion increases as the patch type becomes more clumped or aggregated in its distribution; hence, more physically connected. Above the percolation threshold, patch cohesion does not appear to be sensitive to patch configuration (Gustafson and Parker, 1994). In Bankura district, the patch cohesion index is continuously decreasing for all forest classes. Patch cohesion index of the dense forest has been fluctuating over the years in the study period with a declining trend over the last decade (Fig. 15.7h). Patch cohesion index is rapidly decreasing in degraded forests. This means the degraded forest is rapidly deforested.

Aggregation index increases as the landscape is increasingly aggregated and equals 100 when the landscape consists of a single patch. AI of dense forest is almost the same for all the years and the value of AI is close to 80% from which it can be deciphered that the dense forest is not highly fragmented. On the other hand, the degraded forest experiences a steady decline in the value of AI throughout the study period. The AI value of the degraded forest is 73.22 in 1990 and 37.30 in 2020 which is an indicator of the rapid fragmentation of the degraded forest.

People's Perception on Forest Fragmentation

The survey of farmers, villagers and tribal people staying in the vicinity of the forest patches were surveyed through structured and unstructured questionnaire and it was found that many people were conscious about forest fragmentation. In order to comprehend and interpret the responses provided by the forest dwellers it becomes imperative to know their socio-economic background (Fig. 15.8a, b). Lower educational qualification suggest poorer economic condition and joblessness, thus higher dependency on forest, which includes timber and non timber products.

It can be observed from Fig. 15.8a that around 30% of the respondents have their primary occupation as agriculture and 10% have it as secondary occupation. A major portion of the surveyed population are unemployed, but most of them have the basic education as only 2% are illiterate (Fig. 15.8b). Thus their awareness levels are high enough to comprehend about the problem of forest degradation. But unemployment is highly responsible in increasing their dependency on forest for fuel wood, agricultural land and non timber products which leads to forest fragmentation.

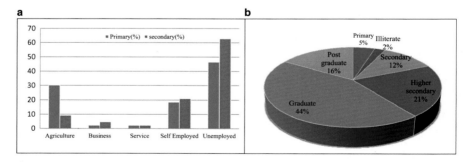

Fig. 15.8 The diagrams in Figure 15.8 shows the basic socio-economic status of the respondents. The survey was done among tribals residing near and within the forest area. It is essential to understand their literacy level and their occupation to comprehend the nature of their response. Lower educational qualification suggest poorer economic condition and joblessness, thus higher dependency on forest, which includes timber and non timber products. But unemployment is highly responsible in increasing their dependency on forest for fuel wood, agricultural land and non timber products which leads to forest fragmentation. Thus the label (**a**) shows the primary and secondary occupations and label (**b**) shows the educational level

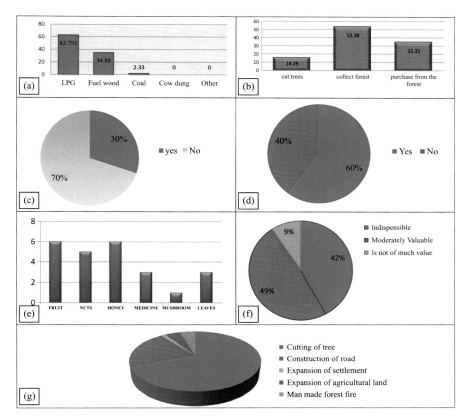

Fig. 15.9 (**a**) Types of fuel used for cooking by respondents, (**b**) process of collection of fuel wood, (**c**) socio-cultural dependency on forest, (**d**) dependency on timber products from forest, (**e**) dependency on non-timber forest products, (**f**) value or importance of forest among respondents, (**g**) people's perception on causes of forest fragmentation

When enquired about the various aspects of their dependency on forest, it was found that 60% of the respondents replied to a yes for being dependent on the forest for timber products, mainly fuel wood (Fig. 15.9a, d). Besides fuel wood timber products mainly sought for were sal wood, eucalyptus, neem, teak wood, bamboo wood, teak, mahogany, Indian rosewood, etc., which were used as building material, used in special wood products such as furniture (chairs, tables, chairs, beds).

The main method of procurement of fuel wood was said to be through collecting of twigs and branches (52.38%), followed by purchasing from forest depots and then through illegal felling (Fig. 15.9b). The majority of responds revealed that 79% regional people were dependent on the non-timber products and 21% people on non-timber products such as fruits, nuts, honey, medicinal herbs, mushrooms, twigs and honey (Fig. 15.9e) and their monthly income ranged from around Rs 4000/– to Rs 15,000/– per month. 16% people reported an income of Rs 4000/month, 25% (Rs 5000/–), 22% (Rs 6000/–), 6% earn (Rs 7000/–), 3% earn (Rs 8000/–), 13%

(Rs.10,000/−), 9% (Rs 12,000/−), and 6% of respondents reported they earn Rs 15,000/− per month from non-timber products. When enquired about their socio-cultural dependency on the forest 70% replied that their socio-economic and cultural behaviour was not dependent on the forest patches they were dwelling in, whereas 30% reported that festivals Ban Devi Puja, Bantustan, Poush, Banmela, Shaloy Puja, etc. were celebrated as a result of their worship of the forest resources (Fig. 15.9c).

Almost 68% of respondents have said that the forest cover has increased and 32% of respondents reported that the forest cover has decreased. But in the case of forest density almost all the respondents say that it has decreased. 67% of respondents have said forests are fragmented rapidly. 46% of respondents said forests are fragmented due to deforestation, 14% said due to road construction, 1% said due to expansion of rural and urban settlement, 3% said expansion of agricultural land was responsible for forest degradation cum fragmentation and 3% reported man-made forest fire were responsible (Fig. 15.9g). 33% of respondents are not aware about forest fragmentation. The perception of people of this region too has indicated that the forest is continuously being fragmented.

All the indicators of Fragstat and people perception about forest fragmentation are indicating that the forest area of the district has fragmented into small patches of forest. Another point comes out that the degraded forest is rapidly fragmented than the dense forest of the district.

People's Perception of Forest Changes and Their Livelihood

Bankura is a backward rural district where villagers live within the peripheral areas and depend directly on forests in search of alternative livelihood. Forest areas of the district provide both major and minor non-timber forest products to the native villagers to sustain their daily life (Dutta, 2021). People depend on forest but there is also enormous uncertainty about how many forest-dependent people are there in the world (Chao, 2012). Dependency of people on forest and the nature of forest is different in different part of the district. People of Southern part of the district is more dependent on the forest for their livelihood. Almost 81% of the respondents are directly or indirectly depend on the forest in the Southern division. But 21% despondent of North division and Penchet division depends upon forest. Khatra subdivision is the dominant area of ST population. People of this area have no alternative employment. Most of the families are involved in primary activity. Bankura district is known as the backward district of West Bengal. Comparably forest surrounded people of the district are educationally and economically backward (Fig. 15.8). In this area 31%, 20% and 49% respondents are general, schedule caste (SC) and schedule tribe (ST) respectively.

Forests are a source of food, fuel, fibre, and income for millions of people globally (Newton et al. 2016). In our study, we have seen that the 61% of the respondents are dependent on the forest for their livelihood and respondents of the South division experienced most of the families are attached with forest for their livelihood. Small number of families are involved in forest dependent economy in north and Panchet division. Only 6% of respondents have their agricultural plots inside the forest and they grow different vegetables in this plot.

It was observed that 52% families of respondents collect wood and leaf from the forest and 16% families of the respondents collect only wood and 32% families of the respondents are not involved in any kind of forest-related non-agricultural activity. Timbers are used only for domestic uses and leaves are used for commercial purposes. Almost 44% of the respondents sell the stitched leaf in the local market at the rate of Rs-35 per hundred stitched leaf. Half of the families of the respondents use timber for their cooking, 6% use both timber and LPG, 32% use LPG, and 12% use other sources of fuels. Timbers are collected from local forest and the average collection of timber is 178 kg/month/family.

Almost all the respondents experience human-animal conflict. Most of the time conflict occurred with the elephant. Penchet and north division experienced more human-animal conflict than the south division. Almost all the respondents have experienced the human-elephant conflict.

68% of respondents say that the forest cover has increased but forest density has decreased and 32% of respondents said that the forest cover and density both have decreased. Sal and Eucalyptus are planted in afforested areas and also Sal is the leading species that is cut in deforested areas. All respondents have experienced that forest.

Local peoples were not happy about the role of forest department. They say forest department has not taken any strong steps to Shut off the woody shipment. Co-management or Joint Forest management or Community management has evolved as one of the most viable options for both poverty reduction, local level economic development, and biodiversity conservation, and also recognizes the importance of the inclusion of local communities along with the government to ensure sustainability (Bowler et al. 2012; Carlsson and Berkes 2005; Nath et al. 2016; Shackleton et al. 2002). But in some areas, Forest Protection Committee is very active to protect the forest. The forest department has planted new trees and conserved that by JFM scheme.

Conclusion

This geospatial study has paid attention to the changing behaviour of forest landscape, people's perception and livelihood of the forest dwellers in Bankura district. It has been assessed that the percentage of forest landscape has decreased over time, number of patch and patch density has increased means forests are increasingly getting fragmented. Other indicators of forest fragmentation have also positively indicated the increase of fragmentation. Even the forest dwellers whose livelihood is heavily dependent on the forest have admitted forest degradation and fragmentation. People have reported that some of the causes of fragmentation are illegal cutting of trees, road and building construction inside the forest, expansion of agricultural land,

and manmade forest fire. To decrease deforestation and forest fragmentation, the government has to take some necessary steps. People's awareness is the most important to conserve and increase the forest cover. So the government's initiative to improve the people's awareness and forest conservation through programs like Joint Forest Management (JFM), community forest management and social forestry should be boosted in this district.

References

Abebe GA (2013) Quantifying urban growth pattern in developing countries using remote sensing and spatial metrics: a case study in Kampala, Uganda. M.Sc thesis, Geo-Information Science and Earth Observation of the University

Bagli S (2015) Multi-dimensional poverty: an empirical study in Bankura District, West Bengal. J Rural Dev 34:327–342

Bera, B., Saha, S. & Bhattacharjee, S. (2020a) Estimation of forest canopy cover and forest fragmentation mapping using Landsat satellite data of Silabati River Basin (India). KN J Cartogr Geogr Inf 70:181–197. https://doi.org/10.1007/s42489-020-00060-1

Bera B, Saha S, Bhattacharjee S (2020b) Forest cover dynamics (1998 to 2019) and prediction of deforestation probability using binary logistic regression (BLR) model of Silabati watershed, India. Trees For People 2(2020):100034. https://doi.org/10.1016/j.tfp.2020.100034

Bowler DE, Buyung-Ali LM, Healey JR, Jones JPG, Knight TM, Pullin AS (2012) Does community forest management provide global environmental benefits and improve local welfare? Front Ecol Environ 10:29–36. https://doi.org/10.1890/110040

Cabral AIR, Vasconcelos MJ, Oom D, Sardinha R (2010) Spatial dynamics and quantification of deforestation in the central-plateau woodlands of Angola (1990–2009). Appl Geogr 31:1185–1193

Carlsson L, Berkes F (2005) Co-management: concepts and methodological implications. J Environ Manag 75:65–76. https://doi.org/10.1016/j.jenvman.2004.11.008

Chao S (2012) Forest peoples: numbers across the world. Forest Peoples Programme, Moreton-on-Marsh

Collingham YC, Huntley B (2000) Impacts of habitat fragmentation and patch size upon migration rates. Ecol Appl 10(1):131–144. https://doi.org/10.2307/2640991

Crist EP, Cicone RC (1984) Application of the tasseled cap concept to simulated thematic mapper data. Photogramm Eng Remote Sens 50(3):43–352. https://www.asprs.org/wp-content/uploads/pers/1984journal/mar/1984_mar_343-352.pdf

Devendra K (2011) Monitoring forest cover changes using remote sensing and GIS: a global prospective. Res J Environ Sci 5:105–123. https://doi.org/10.3923/rjes.2011.105.123

Dutta S, Chatterjee S, Dey F (2021) Quantification of decadal deforestation and afforestation scenario in Bankura District of West Bengal using geospatial techniques. In: Bhunia GS, Chatterjee U, Kashyap A, Shit PK (eds) Land reclamation and restoration strategies for sustainable development, modern cartography series, vol 10. Academic Press, pp 627–639. https://doi.org/10.1016/B978-0-12-823895-0.00016-6

Dutta S, Sahana M, Guchhait S (2017) Assessing anthropogenic disturbance on forest health based on fragment grading in Durgapur Forest range, West Bengal, India. Spatial Information Research 25:501–512. https://doi.org/10.1007/s41324-017-0117-3

Fahrig L (2003) Effects of habitat fragmentation on biodiversity. Annu Rev Ecol Evol Syst 34:487–515. https://doi.org/10.1146/annurev.ecolsys.34.011802.132419

Food and Agricultural Organization of the United Nation (2000) Global forest resources assessments

Forest Survey of India (2007) India state forest report
Forest Survey of India (2017) India state forest report
Forest Survey of India (2019) India state forest report
Forman TTR (1995) Some general principles of landscape and regional ecology. Landsc Ecol 10: 133–142
Forman RTT, Godron M (1986) Landscape ecology. Wiley, Brisbane, p 619
Franklin S (2001) Remote sensing for sustainable forest management. Lewis Publishers, Boca Raton, p 407
Gustafson EJ, Parker GR (1994) Using an index of habitat patch proximity for landscape design. Landsc Urban Plan 29:117–130
Herold M, Couclelis H, Clarke KC (2005) The role of spatial metrics in the analysis and modeling of urban land use change. Comput Environ Urban Syst 29:369–399. http://bit.ly/2Q2hxLB
Iris EMF, James C (2019) Forest fragmentation analysis from multiple imaging formats. J Landsc Ecol 12(1):1–15. https://doi.org/10.2478/Jlecol-2019-0001
Laurance WF, Camargo JLC, Luizão RCC, Laurance SG, Pimm SL et al (2011) The fate of Amazonian forest fragments: a 32-year investigation. Biol Conserv 144:56–67. https://goo.gl/qLUw7p
Li M, Huang C, Zhu Z, Wen W, Xu D, Liu A (2009) Use of remote sensing coupled with a vegetation change tracker model to assess rates of forest change and fragmentation in Mississippi, USA. Int J Remote Sens 30:6559–6574
Liu J, Taylor W (2002) Integrating landscape ecology into natural resources management. Cambridge University Press, Cambridge. https://doi.org/10.1017/CBO9780511613654
Majumdar A (2018) A critical evaluation on forest dependence and forest out comes in West Bengal in the context of joint forest management programme – a case study of Bankura District. Indian J Econ Dev 6:1–11
Mandal M, Chattarjee ND (2018) Quantification of habitat (forest) shape complexity through geo-spatial analysis: an ecological approach in Panchet forest division in Bankura, West Bengal. Asian J Environ Ecol 6:1–18
Mandal M, Chattarjee ND (2019) Forest core demarcation using geo-spatial techniques: a habitat management approach in Panchet Forest division, Bankura, West Bengal, India. Asian J Geogr Res 2:1–8
Mandal M, Chattarjee ND (2020) Land use alteration strategy to improve forest landscape structural quality in Radhanagar forest range under Bankura District. Eur J For Sci 8:1–10
Mandal M, Chattarjee ND (2021) Spatial alteration of fragmented forest landscape for improving structural quality of habitat: a case study from Radhanagar Forest Range, Bankura District, West Bengal, India. Geol Ecol Landsc 5:252–259
McGarigal K, Marks BJ (1995) Fragstats: spatial pattern analysis program for quantifying landscape structure, USDA forest service general technical report PNW-GTR-351. Corvallis, Oregon
Mladenoff DJ, White MA, Pastor J, Crow TR (1993) Comparing spatial pattern in unaltered old growth and disturbed forest landscapes. Ecol Appl 3:294–306. https://bit.ly/30U2rtG
Nandy S, Kushwaha SPS, Dadhwal VK (2011) Forest degradation assessment in the upper catchment of the river tons using remote sensing and GIS. Ecol Indic 11:509–513
Nath TK, Jashimuddin M, Inoue M (2016) Community-Based Forest Management (CBFM) in Bangladesh. Springer, Berlin/Heidelberg. https://doi.org/10.1007/978-3-319-42387-6
Newton P, Miller CD, Augustine M, Bynkya A, Agrwal A (2016) Who are forest-dependent people? A taxonomy to aid livelihood and land use decision-making in forested regions. Land Use Policy 57:388–395. Elsevier. https://doi.org/10.1016/j.landusepol.2016.05.0320264-8377
Noss RE, Cooperrider AY (1994) Saving nature's legacy. Protecting and restoring biodiversity. Econ Bot 50:317
Pan Y, Birdsey RA, Fang J, Houghton R, Kauppi PE, Kurz WA, Phillips OL, Shvidenko A, Lewis SL (2011) A large and persistent carbon sink in the world's forests. Science 333(6045):988–993

Ramachandra TV, Setturu B, Chandran S (2016) Geospatial analysis of forest fragmentation in Uttara Kannada District, India. For Ecosyst 3:10. https://doi.org/10.1186/s40663-016-0069-4

Rao KS, Pant R (2001) Land use dynamics and landscape change pattern in a typical micro watershed in the mid elevation zone of central Himalaya, India. Agric Ecosyst Environ 86(2):113–124. https://doi.org/10.1016/S0167-8809(00)00274-7

Reddy CS, Jha CS, Dadhwal VK (2013) Assessment and monitoring of long-term forest cover changes in Odisha, India using remote sensing and GIS. Environ Monit Assess 185(5):4399–4415. https://doi.org/10.1007/s10661-012-2877-5

Reed R, Johnson-Barnard J, Baker W (1996) Fragmentation of a forested rocky mountain landscape, 1950–1993. Biol Conserv 75:267–277

Roy P, Murthy M, Roy A, Kushwaha S, Singh S (2013) Forest fragmentation in India. Curr Sci 00113891(105):774–780

Shackleton S, Campbell B, Wollenberg E, Edmunds D (2002) Devolution and community-based natural resource management: creating space for local people to participate and benefit? Nat Resour Perspect 76:1–6. https://doi.org/10.1016/j.ecolecon.2012.09.001

Southworth J, Munroe D, Nagendra H (2004) Land cover change and landscape fragmentation—comparing the utility of continuous and discrete analyses for a western Honduras region. Agric Ecosyst Environ 101:185–205. https://doi.org/10.1016/j.agee.2003.09.011

Whitmore TC (1997) Tropical forest disturbance, disappearance, and species loss. In: Laurance WF, Bierregaard RO (eds) Tropical forest remnants: ecology, management and conservation of fragmented communities. The University of Chicago Press, Chicago, pp 3–12

Wilcove D. S., McLellan C. H., Dobson A. P., 1986 - Habitat fragmentation in the temperate zones. In: M. E. Soulé (ed.), Conservation biology. Sinauer Ass., Sunderland (Mass.), pp. 237–256

Willims M (2003) Deforesting the earth: from prehistory to global crisis. The University of Chicago Press, Chicago

Chapter 16
Forest Fire Risk Modeling Using GIS and Remote Sensing in Major Landscapes of Himachal Pradesh

Shreyasee Dutta, Akanchha Vaishali, Sadaf Khan, and Sandipan Das

Abstract Forest fires increase at an alarming rate destroying forest landscapes and ecosystems. Thousands of forest fire incidents occur in the foothills of the Himalayas. This study utilized remote sensing, geospatial technology, and mathematical tools to develop a "forest fire risk model" in landscapes of Himachal Pradesh, namely the Chamba and Kangra districts. This work aims to design a prototype to delineate the risk of forest fire hazards by using geospatial and geostatistics methods to evaluate forest fire impacts. Methods like multicriteria analysis and a knowledge-based AHP approach were adopted to develop a forest fire risk index map using independent variables ("slope, elevation, aspect, LULC, LST, drainage, roadways, and settlements"). Thus, the developed risk index map classifies the total area into five zones. Therefore, the created index maps were overlapped with the dependent variable (MODIS fire point data). The created hazard map finally gets validated using a statistical method like the Kernel density estimation method. About 85.62% and 4.24% of the total study area fall under the category of high and very high-risk zones, respectively. Planners study the hazardous impact of forest fire risk maps, and it helps develop fire management strategies to protect future forest fires and destruction rates.

Keywords Forest · Fire · Risk · AHP · Multicriteria · Kernel density

Introduction

India is rich in biodiversity, both in terms of fauna and flora. About 21.67% of India's total geographical area falls under forest cover (ENVIS RP on Forestry and Forest Related Livelihoods, 2021). Moreover, it experiences several climatic zones, ranging from northwestern hot deserts and Himalayan cold deserts to tropical in the south (Roy and Dun 2005). The forest cover of India has proved to be both

---

S. Dutta · A. Vaishali · S. Khan · S. Das (✉)
Symbiosis Institute of Geo-Informatics (SIG), Symbiosis International (Deemed University), Pune, Maharashtra, India

economically and environmentally valuable. However, it is decreasing at an alarming rate due to increasing population density and changing climatic conditions. Forest fires have been shaping landscape structure, pattern, and ecosystem. It has also been a source of disturbance for over a long period. Due to anthropogenic and natural determinants, the forest catches fire. Lightning, cigarette buds, campfires, and solar energy are significant heat sources that spark forest fires (Kanga et al. 2017). Due to the increase in temperature, wind velocity, and relative humidity, the annual forest fire incidents and approximate burnt areas are also increasing slowly. The consequences of forest fires are loss of flora and fauna, growing landslide hazards, rate of desertification and soil erosion, and changes in the climate of the surrounding areas. These fires also affect the human population in several ways, thus causing economic damage for the community (Mangiameli et al. 2021). The average perineal destruction of forest fire worldwide is around six to fourteen million hectares. The worst scenarios are caused in the forest due to wildfires, triggering industrial accidents due to ongoing climatic changes and rapid industrialization (Naderpour et al. 2019). Fire plays a vital role in ecological processes by altering flora and fauna composition, conserving water, enhancing soil quality, and thus promoting biodiversity. The statistical study of forest conditions and its analysis conducted by the Forest Survey of India (FSI) research tells that 95% of the country's forest fires are due to anthropogenic reasons. In comparison, 50% of the total forest cover is vulnerable to burning, according to Indian Forest Service.

The total number of forest fires is increasing and has forced individuals to design models for effective coping mechanisms. To decrease the rate of destruction created by forest fires, fire control managers need methods to assess fire hazards, vulnerability, and susceptibility. Forest fire hazards refer to "potential fire behavior that has caused damage to human beings and environments." Forest fire vulnerability refers to "the location of properties, housing, inhabitants, infrastructures, and other physical human resources located in forest fire susceptible areas." Thus, mapping of hazard vulnerability is done to calculate the amount of loss caused by forest fires and identify the vulnerable areas (Naderpour et al. 2019). Modeling of Forest Fire involves hazard vulnerability assessment and its evaluation. It is a challenging job to analyze the behavior of forest fires. The collection of data of the required parameters has helped achieve the desired solutions. Models, maps, and databases were used with geospatial technology to simulate forest fire occurrence and its highest vulnerability factor (Naderpour et al. 2019).

Remote sensing analysis and satellite imagery are widely used to detect spectral characteristics between forest types and fire incidents, especially in the electromagnetic spectrum that indicates infra-red (Kanga et al. 2017). Moreover, they help monitor fire behavior in operational activities (Qayum et al. 2020). This study considers dependent and independent variables to design a forest fire hazard zonal map to model the potential vulnerable forest fire zones (Parajuli et al. 2020). Weights were assigned to all the independent variables through a knowledge-based approach following their fire influencing capacity, and the fire hazard zone was delineated. Decision-making techniques like the AHP process are helpful and widely used in the multicriteria analysis. The forest fire risk index map, created by a spatial tool, is analyzed with the help of hazardous risk factors (Kanga et al. 2017). The prepared fire risk index map is overlapped with MODIS fire spots data collected from the Forest Survey of India portal to develop fire incidents risk map. The validation of the created fire risk index map is done with the help of Kernel density estimation. When done precisely, the mapping of forest fire risk zones helps know the future impacts of forest fires (Jaiswal et al. 2002).

The Chamba and Kangra districts of Himachal Pradesh is chosen as a study area. The stated district has a predominant forest cover spread, with more than 70% of the total area coming under the forest category. And they have a history of forest fire incidents occurrence. Time series analysis is drawn here, showing the fire incidents occurrence from 2000 to 2019, hence helping develop a predictive risk model. GIS and remote sensing techniques are appropriate for mapping forest fire risk index and completing the stated study. The designed GIS-based model is the best approach for the Indian condition, where most of the land under forest cover gets exploited by human interference. Thus, such maps help fire managers to create a fire management plan and help reduce hazard activities within forests caused due to forest and wildfires (Jaiswal et al. 2002).

Materials and Methods

In this study, independent and dependent variables are considered the most critical factors that have influenced forest fire in many ways. The details of the data used in this study are shown in Table 16.1 (Parajuli et al. 2020).

The collected data are procured from different sources and processed in ArcGIS software. Figure 16.1 shows the methodological framework followed in this study.

Table 16.1 Details of the collected data

Datasets	Type of data	Details	Spatial resolution	Sources
Fire point data	Vector	Archive fire points data of 2000–2018	1 km	MODIS and FIRMS from FSI Portal
Cartosat 1 (DEM)	Raster	Topographical "elevation, slope, aspect"	30 m	BHUVAN Portal
Satellite imagery	Raster	Data of 2020 collected, classified, and processed in ArcGIS PRO to obtain LULC and forest types	30 m	Sentinel 2A
Land surface temperature	Rasters	Surface temperatures of daily and night (May–August 2018)	1 km	NASA, MODIS, MOD11C3
Study area	Vector	Administrative boundaries of all districts	1:25000	DIVA GIS
Built-up/ roads/ drainage	Vector	Roads and the associated cluster of settlements and drainage network	1:25000	BHUKOSH Portal

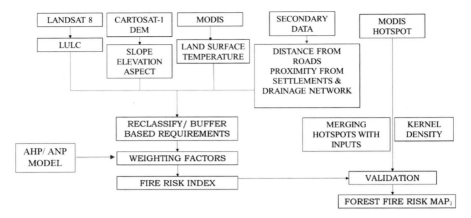

Fig. 16.1 Methodology workflow

Independent Variable

Independent variables are positively related to the occurrence of forest fires (Parajuli et al. 2020). LULC, slope, elevation, aspect, LST, roadways proximity, drainage network proximity, and proximity from settlements are the independent variables considered in this study.

Slope

The topography and terrain values of a region play a significant role in defining the survival percentage of forest and wildfire. Hence, variation in topographic data, which includes slope, aspect, and elevation, was extracted from Cartosat 1 DEM (30 m) as mentioned in Table 16.1. Slope (Fig. 16.2) and elevation (Fig. 16.3) data points get extracted by GIS software like ArcGIS Pro (Parajuli et al. 2020). Slope ranges from 0.00 to 82.76. The higher slope gets assigned a weight of 5 as the upper limit. As the gradient goes downward, less water makes it more effective at convective preheating. Steep slopes have more importance and are given more weightage in the statistical model (Appendix).

Elevation

Elevation of the area ranges between 195.0 and 5920 m (Fig. 16.3). An elevated area has less risk compared to an area with a lower elevation. Thus, the low elevated area gets more weightage in the statistical model.

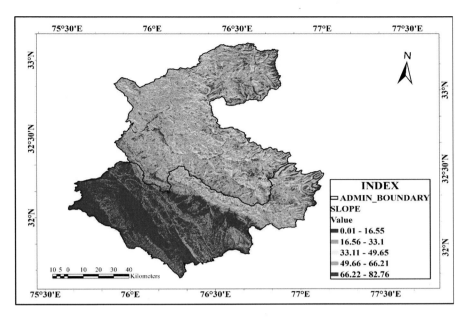

Fig. 16.2 Slope map of study area

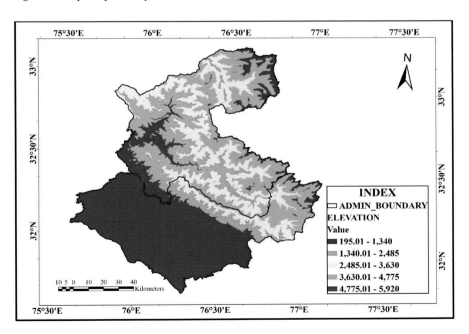

Fig. 16.3 Elevation map of study area

Aspect

As aspect is a non-numerical variable, proportional values are further classified into nine categories. Southeast and southwest being the most fire-prone areas due to the amount of insolation received from the sun are highest in this zone compared to others. Thus, these regions get higher weightage. Moreover, fire erupts rapidly in those regions where it receives direct sunlight (Fig. 16.4).

Land Use and Land Cover

One of the essential variables responsible for the widespread forest fire is land cover land use. The LULC map analysis of the study area was done using ArcGIS Pro software's help using the supervised classification technique (Fig. 16.5). Kappa accuracy was also done on the stated classified map to validate the classification process. And it is seen that coniferous forest and the deciduous forest get highest weightage, and lowest weightage is assigned to snow and water. Because the dry leaves encourage forest fires, snow and water discourage the occurrence of a forest fire.

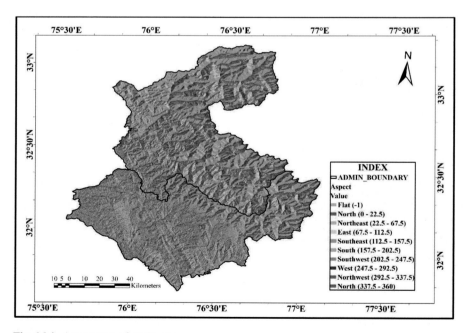

Fig. 16.4 Aspect map of study area

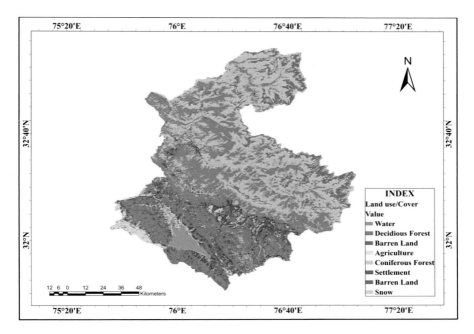

Fig. 16.5 Land use and land cover map

Land Surface Temperature

Temperature with high value shares a direct proportional relationship with moisture content and relative humidity of the fuels. The LST algorithm uses temperature with higher brightness values in the 31 and 32 MODIS bands to cause night and day outputs based on LST at a spatial resolution of 1 km. The average of all the monthly temperate years is taken to generate an LST layer to create an effective model (Fig. 16.6). The temperature ranges from −19 °C to 42 °C.

Proximity to Drainage, Roads, and Settlement

Distance of roads, settlement, and drainage are considered as valuable parameters to identify the risk zone, mainly those influenced by human activities. Forest fires are high in numbers in areas closer to settlements and roads due to more human movements. The phenomenon is opposite in the case of the drainage network. Areas closer to roads and settlements get higher weightage than those closer to roads and settlements, whereas the situation is opposite in the drainage network (Appendix). Proximity to the settlements (Fig. 16.7), proximity to the drainage network (Fig. 16.8), and proximity to the road (Fig. 16.9), and with resolution 1:

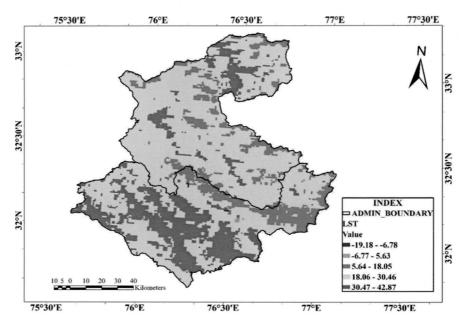

Fig. 16.6 Land surface temperature Map

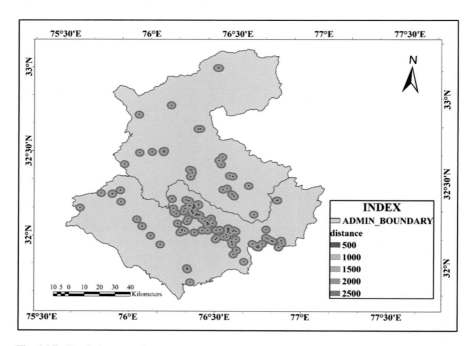

Fig. 16.7 Proximity to settlements map

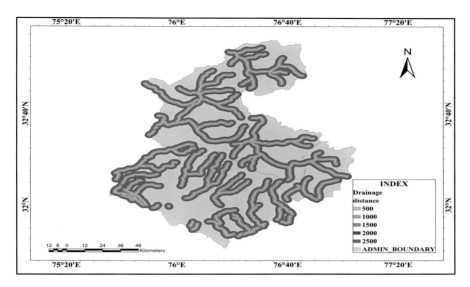

Fig. 16.8 Proximity to drainage network map

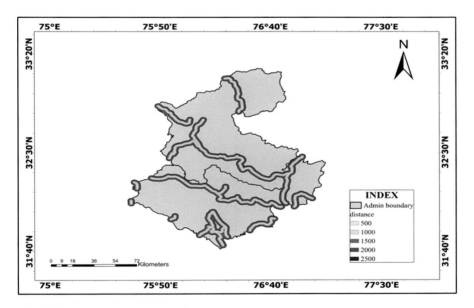

Fig. 16.9 Proximity to roads network map

250000, are used to group the buffers into five classes: 500, 1000, 1500, 2000, and 2500 m. Spatial tools like multiple ring buffers are used in ArcGIS PRO to create a buffer zone around the stated parameters with an interval of 500 m.

Dependent Variables

Fire Incidents and Burned Area

Forest fire incidents and their burn area or size are the only results of the driven force when all of the aforementioned independent variables are in a favorable environment for ignition. As a result, this study used burnt incidents and their size as dependent variables. MODIS Archive Data Tool was used to extract archive fire points, and a polygon covering the study area from 2000 to 2018 was created (Fig. 16.10). FSI Portal provides information on active fires detected by the MODIS instrument (1 km resolution) onboard NASA's Aqua and Terra satellites for burn area detection (NASA 2002). The detection confidence is estimated and ranges from 0% to 100%, with higher confidence indicating greater accuracy. As a result, more than 30% of confidence data were used in this study.

Time Series Analysis

Time series analysis is a graphical representation of data points collected for a span of time. It is used for future predictions of any historical occurrences. Here, a time series analysis of forest fire incidents over a period of time from 2000 to 2020 is done

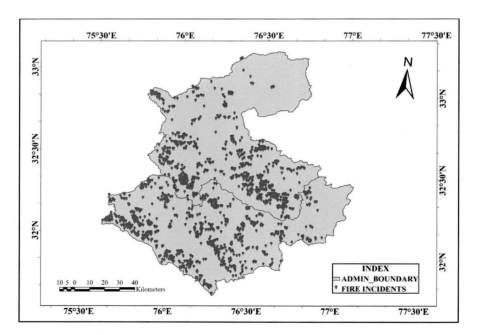

Fig. 16.10 MODIS fire hot spot data map

in R studio with the help of programming language (Fig. 16.11). Different packages and libraries were imported into the R studio for plotting.

Analytical Hierarchy Process (AHP)

Analytical hierarchy process (AHP) is an exemplary technique used in suitability analysis, and it helps in giving a methodical approach in selecting the site. It is also considered one of the most important statistical models indicating forest fire spots. Studies have proved the effectiveness and usage of forest fires in using these models. Even numbers (2,4 and 6) refer to similar characteristics, but odd ones stand out (Table 16.2).

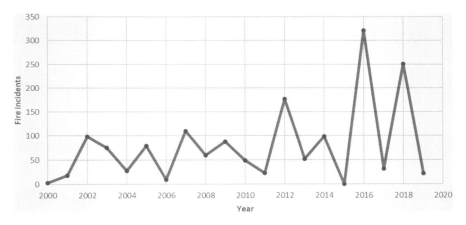

Fig. 16.11 Time series analysis map

Table 16.2 Ratio scale and AHP explanation of each class

Importance	Definition	Explanations
1	Equally preferred	Equally vital to the target
3	Moderately preferred	Compared to the overall profit or damage
5	Strongly preferred	A strong preference for one factor over another
7	Very strongly preferred	Considered above all others, has gained prominence, and is superior
9	Extremely preferred	One element is preferred in comparison to another when proven firmly with evidence.
2,4,6,8	Inter values	

Source: Saaty (1987)

Assigning Weights to the Variables Using the AHP Model

AHP is how different parameters are evaluated by compiling and prioritizing other variables (Saaty 1987). Integration of these variables with AHP has gained wide usage in vulnerability studies. AHP framework provides importance to indicators on a ratio scale, which several evaluators review for various possibilities (Al-harbi 1990). ANP model calculates the relationship of these nine variables with different classes. The steps followed in AHP and ANP for assigning weights to thematic layers are as follows:

(a) Model Construction: Forest reserve mapping achieved by different models. It is essential to identify the problem statement at both thematic and abstract levels before selecting which model to use and thus putting the layers in the model together.
(b) Pairwise Comparison Matrices: Synthesized matrix is obtained by dividing each cell's values with sum values with their respective rows. The relative values assigned in pairwise comparison matrix are as follows: Score 9 determines extreme preferred importance with one of the themes, whereas one means equally preferred of two themes. Table 16.4 indicates classes are written in a particular order and are prioritized accordingly. The method suggested by Saaty opted to do a consistency check. While generating a pairwise matrix, consistency ratio (CR) is used in AHP to check the consistency. CR expresses as the ratio between consistency index (CI) and random index (RI), whereas the value of RI has taken from random consistency with matrix size 6x6. Therefore, CI was calculated by the Satya method using Eq. 16.1.

$$CI = (\lambda max - n)/(n - 1) \qquad (16.1)$$

Thus, λmax is "the largest eigenvalue in the n size matrix." The equation expresses the consistency measure in pairwise consistency (MC) as follows:

$$CR = \frac{CI}{RI} \qquad (16.2)$$

Tables for ranking and empirical analysis, are derived from RIs. Thus, ranking tables do include empirical tables and ratios in countries that use and consider those indexes used. Table 16.3 shows RI values of various n values. If variability does not increase by 0.1, it is good. When CR is greater than 10%, our judgments clarify (Al-harbi 1990). The nine pairwise consistencies are represented in Table 16.4. Count value: 9.00; CR: 0.079; λmax: 10.179; CI =0.147; CR=0.098; constant=1.49.

Table 16.3 Random consistency index

n	1	2	3	4	5	6	7	8	9	10	11	12
R.I	0.0	0.0	0.58	0.90	1.12	1.24	1.32	1.41	1.45	1.49	1.51	1.48

Source: Saaty (1987)

Table 16.4 Pairwise comparison matrix showing parameters priority factor

Item number	LULC	Aspect	Slope	Proximity to roads	Elevation	Proximity to settlements	LST	Proximity to drainage network
LULC	4	5	3	4	4	3	4	4
Aspect	1	1	2	1	0.5	4	0.5	4
Slope	1	0.5	1	1	1	3	2	5
Proximity to roads	1	1	1	1	1	2	2	2
Elevation	1	2	1	1	1	1	0.75	2
Proximity to settlements	0.5	0.25	0.33	0.55	0.5	1	1	4
LST	1	1	0.6	0.75	0.5	0.5	0.7	1
Proximity to drainage network	0.5	0.2	0.33	0.5	0.5	0.25	0.33	3
	10	10.95	9.26	9.8	9	14.75	11.28	25

The multicriteria model includes independent variables, and they were as follows: topography (slope, elevation, and aspect), LULC, proximity to road, drainage, settlements; land surface temperature. These variables were assigned weighted values and shown in Table 16.5.

Determining Fire Risk Model

To avoid errors and get the practical conclusion of the designed model, weights were assigned to each variable based on their influencing capacity using the AHP approach. Each independent variable categorizes into different classes based on their forest fire influencing capability. After determining the weights of all variables by the AHP approach, they were reclassified and overlaid in ArcGIS Pro with the help of the weighted overlay analysis tool. Finally, a risk index model was designed based on multicriteria analysis, including all the stated independent variables. The equation of the risk model is stated below:

$$\text{FRI} = 27.98\%\text{LC} + 15.88\%\text{LT} + 12.75\%\text{S} + 8.89\%\text{E} + 10.47\%\text{A} \\ + 7.63\%\text{PS} + 11.64\%\text{PR} + 4.76\%\text{PD} \quad (16.3)$$

where FRI stands for forest fire risk index, LC stands for land use and land cover, LT stands for land surface temperature, S stands for Slope, E stands for elevation, A stands for aspect, PS stands proximity for settlement, PR stands for proximity for roadways, and PD stands for proximity for drainage. Finally, a risk model map was developed based on the above-stated analyses.

Model Validation

Validating the model will expand the model's precision percentage and prognostic power. Firstly, it is done by using the dependent variables (the actual world MODIS fire spots data downloaded from the FSI portal) for validation. They created a risk index map that is merged with the MODIS fire incident point data. It is found that each risk zone is overlaid with the archive fire count data, assuming forest fire occurrences are high in areas with higher index values and are riskier to forest fire occurrences. Secondly, Kernel density estimation (KDE) was used to compare the result of the fire risk model. The goal of KDE is to generate a smooth density surface of point events over space by using event intensity as an estimation. MODIS hot spots were used as an input for both methods.

Table 16.5 Weighted values assigned to each class

Variables	LULC	Aspect	Slope	Proximity to roads	Elevation	Proximity to settlements	LST	Proximity to drainage network	Weightage (%)
LULC	0.3	0.4	0.3	0.45	0.36	0.15	0.16	0.15	27.98
Aspect	0.05	0.08	0.18	0.09	0.05	0.2	0.05	0.16	10.47
Slope	0.08	0.04	0.09	0.09	0.09	0.15	0.2	0.16	12.75
Proximity to roads	0.05	0.08	0.09	0.09	0.09	0.1	0.21	0.07	11.64
Elevation	0.06	0.16	0.09	0.09	0.09	0.05	0.1	0.07	8.89
Proximity to settlements	0.08	0.02	0.03	0.04	0.09	0.05	0.02	0.13	7.63
LST	0.08	0.08	0.03	0.02	0.15	0.15	0.05	0.01	15.88
Proximity to drainage network	0.05	0.04	0.02	0.06	0.01	0.01	0.16	0.02	4.76

Results and Discussion

Several factors influence forest fire. These factors are discussed and analyzed separately. Moreover thus, they help in developing a fire risk model. The appendix shows the index value according to the capability of influencing forest fire. No-risk and low-risk classes indicate zones not risky to forest fires, whereas high-risk, moderately high-risk, and very high-risk classes indicate riskier zones to forest fires. The most significant variable influencing forest fires is land use and land cover (Fig. 16.5). Coniferous and deciduous forests pose the most significant threat, followed by plantation cover and cropland. The region has a maximum area covered by coniferous forest, with pine and fir being the predominant species. Dry deciduous forest covers are more susceptible to fire incidents, and they mainly occur in the Chamba District. AHP model is adopted, and the highest weightage is given to LULC and forest types variable (Table 16.5). The appendix shows the assigned index value to different classes to LULC and forest types keeping in mind its influencing capacity to create a forest fire. 82% of forest fire incidents occur in the coniferous and deciduous forest and 18% in the plantation area.

Table 16.5 shows different weightage given to other variables. Land surface temperature is the second most important variable in influencing wildfire. It is followed by slope, proximity to roadways, aspect, elevation, proximity to settlement, and proximity to drainage. The land surface temperature of the area does play a crucial role in determining the forest fire hazards. Table 16.5 shows the weightage given to LST, and Fig. 16.6 shows the region and its temperature variation over a period of time. The appendix shows that the highest index value indicates higher temperatures because the land surface temperature increases due to forest fire. The situation is the opposite when the temperature is less.

The slope varies from 0 degrees to 82.76 degrees (Fig. 16.2). 70% of the region have a slope of fewer than 47.65 degrees. The risk factor of forest fire rises with an increase in the degree of the slope. Thus, the fire spread easily and quickly in areas with steeper slopes, as shown in Fig. 16.12. Figure 16.9 depicts the road network proximity. 500m is the lowest buffer limit, and 2500m is the highest. Thus, roads with lower buffer limits are riskier to fires than the highest ones. As stated, areas within lower buffer limits are more susceptible to forest fires. The elevation direction is displayed in the created aspect map (Fig. 16.4). The appendix shows the assigned index value, and the southern part is most vulnerable to fire hazards, whereas the northern part is less vulnerable. Figure 16.3 shows that the area has an elevation ranging from 195.01 to 5920 m. Most of the areas under Kangra District have elevation ranging between 195.01 and 3630 m, whereas Chamba elevation has a wide variation, ranging from 195.01 to 5920 m. Appendix shows that low elevated areas are riskier to forest fires, whereas high elevated areas possess less risk, so index values are assigned accordingly. And the value starts decreasing as elevation increases.

Settlement areas also play a crucial role in influencing and igniting forest fires, mainly those caused due to anthropogenic reasons. Thus, they get higher weights.

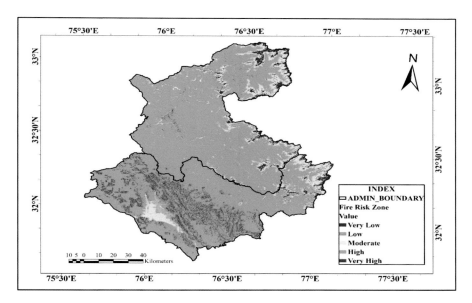

Fig. 16.12 Forest fire index map

Figure 16.7 shows proximity to settlements and calculations associated with them. The buffer rings surrounding the drainage network get lower weights, whereas areas around the periphery of the drainage network get higher weights. Figure 16.8 shows the buffer index zone for the drainage network with five classes having a 500 m interval. The risk index map is obtained by overlaying all the stated independent variables using the weighted overlay spatial tool in ArcGIS Pro (Fig. 16.12). Thus, it helps to classify the whole study area into six classes ranging from no-risk zone to very high-risk zone. Table 16.5 shows that 3.82% falls under the category of very high, followed by 84.38% in high, 9.36% in moderate, 0.18% in low, and 2.25% in very low. Different shades of green represent high-risk areas to forest fires, as shown in Fig. 16.13. It clearly shows that most of the study region is in the alert zone, and these areas are designated as very high risk and higher risk. These also imply that the area is significantly more vulnerable. The high-risk zones are indicated by the colors in dark green and green.

MODIS forest fire hot spots data are overlaid to validate the model or check the accuracy of the created forest fire index map and finally generate the forest fire risk zone map. Figure 16.13 shows that fire incidents lie completely on the risk zones, which states the correctness of the index and risk mapping. Finally, a statistical model like the Kernel density estimation method was adopted to validate the study. According to Zhang et al. (2017), KDE works as a proper development process for the decision support system. This method has been used as a validation tool and technique for forest fire risk maps. Integrating the fire risk map and fire events results in a consistent and complete final risk map. It allows the map to be integrated into a

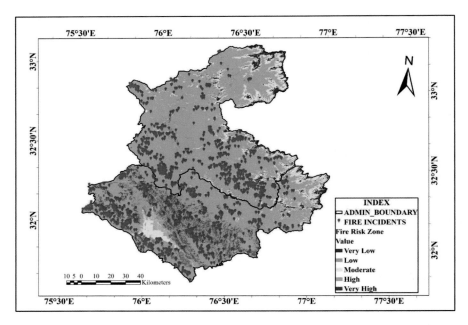

Fig. 16.13 Forest fire risk map with fire incidents data

cohesive risk assessment process. The majority of fire incidents fall into the very high-risk and high-risk categories of fire counts.

Figure 16.14 shows the map generated by the KDE method. The comparison of risk areas identified by fire point validation and KDE in table shows that high- and very high-risk areas are nearly identical to the fire risk index map generated. In KDE, on the other hand, the low-risk zone has shrunk, and the very low-risk zone has expanded. It could be due to the lower concentration of fire points in low and very low-risk areas (Table 16.6).

Challenges and Limitations

Many false alarms get recorded by the MODIS instruments, and thus, the collected fire point data consist of a number of errors and ambiguity. The anthropogenic independent variables such as roadways and settlements are subjected to circumstantial changes over time due to developmental purposes. And natural independent variable changes due to climate change. Thus, the accurate predictive fire risk modeling for the future may vary from the one created at present.

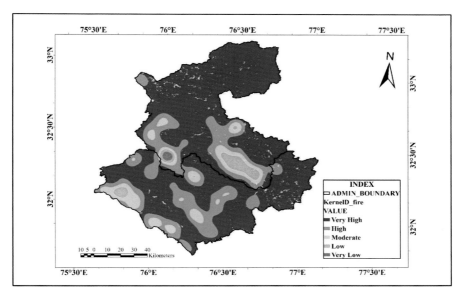

Fig. 16.14 Forest fire risk map with KDE method

Table 16.6 Comparison of risk area % calculated from FRI and KDE method

Risk zone	Percentage of risk area from FRI	Percentage of risk area from KDE method
Very high	3.82	4.24
High	84.38	85.62
Moderate	9.36	7.58
Low	0.18	0.56
Very low	2.25	2

Conclusion

From the results found in this study, we can easily conclude that firefighters often use GIS techniques to scale the destruction rate caused due to forest fires toward both flora and fauna species. Forest fire hazard assessment, done with the help of collected data, assists firefighters in protecting wildfires. The developed forest fire risk model acts as an introductory level and helps administrators predict the risk of fire incidents and the trend pattern of its occurrence. It will also assist in researching the socioeconomic effect of forest fires and climatic changes. Socioeconomic analysis of the area and fire hazard analysis is done from the collected CENSUS data.

Forest fires affect lives in both environmental and socioeconomic ways. Fire incidents are increasing due to increased climatic changes and the global warming rate, thus sharing a direct correlation with the land surface temperature of the region. In contrast, wildfires share an indirect correlation with the physiographic feature of the area, i.e., slope and aspect. Sites with a large concentration of human settlements and road networks are more likely to forest fires, whereas areas with a good drainage network discourage forest fires.

Assessing the MODIS area of interest and all impacting factors from 2000 to 2019 shows that the region is amazingly powerless against forest fires in contrast with different regions. The accompanying variables, specifically, are liable for the high danger of timberland fire in these two scenes: deciduous backwoods, normal temperature of 25–30 °C incline 82.76, south and southwestern viewpoint, the elevational scope of 5920 m inside 500 m, 1000 m, and 1500 m of settlements, and distance of 1000 m from the street. A high fire hazard zone covers 84.3% of the complete space of Chamba and Kangra. Notwithstanding, just 3.82% of the populace in the locale is probably going to be in the high-hazard class. In Himachal, forests are generally defenseless against fierce blazes during the long stretch of April. During the dry season, Himachal encounters vast forest fires, requiring the improvement of an effective backwoods fire hazard appraisal, cautioning, and checking framework. These factors ought to be thought of, and regions affected by these elements should be firmly observed by local area based firefighting bunches consistently.

This study supports community development by integrating wildfires data in the decision-making process and helping develop fire protection strategies. Thus, the created GIS model was adopted in Chamba and Kangra districts, Himachal Pradesh, to identify fire risk zones. The model helps local administrators and forest officials to design a management plan that would reduce the destruction rate caused by forest fires. Thus, citizens and animals can migrate from more risky areas to less dangerous areas.

Appendix (Table 16.7)

Table 16.7 Assigned and normalized weight of the different classes of each theme

Variables	Classes	Values assigned	Fire risk zone
LULC	Settlement	2	Low
	Water	1	No risk
	Snow	1	No risk
	Agriculture	3	Moderate
	Coniferous forest	5	Very high
	Barren land	3	Moderate
	Deciduous forest	4	High
Slope	0.0–16.5	1	No risk
	16.5–33.1	2	Low
	33.11–49.65	3	Moderate
	49.66–66.21	4	High
	66.22–82.76	5	Very high

(continued)

Table 16.7 (continued)

Variables	Classes	Values assigned	Fire risk zone
Elevation	195.0–1340 m	5	Very high
	1340.01–2485 m	4	High
	2485.01–3630 m	3	Moderate
	3630.01–4775 m	2	Low
	4775.01–5920 m	1	No risk
Aspect	Flat	1	No risk
	North (0–22.5); North (337.5–360); Northeast (22.5–67.5)	1	No risk
	East (67.5–112.5)	3	Moderate
	Southeast (112.5–157.5)	4	High
	South (157.5–202.5), Southwest (202.5–247.5)	5	Very high
	West (247.5–292.5); Northwest (292.5–337.5)	2	Low
Proximity to roads	500 m	5	Very high
	1000 m	4	High
	1500 m	3	Moderate
	2000 m	2	Low
Proximity to settlements	500 m	5	Very high
	1000 m	4	High
	1500 m	3	Moderate
	2000 m	2	Low
	2500 m	1	No risk
Proximity to drainage network	500 m	1	No risk
	1000 m	2	Low
	1500 m	3	Moderate
	2000 m	4	High
	2500 m	6	Very high
Land surface temperature	(−)19.18–(−)6.78 °C	1	No risk
	(−)6.77–5.63 °C	2	Low
	5.64–18.05 °C	3	Moderate
	18.06–30.46 °C	4	High
	30.47–42.87 °C	5	Very high

References

Al-harbi KMA (1990) The analytic hierarchy process. Eur J Oper Res 45(2–3):378. https://doi.org/10.1016/0377-2217(90)90209-t

Bhukosh-GIS (2021, July 27). Retrieved from https://bhukosh.gsi.gov.in/Bhukosh/Public

Bhuvan-NRSC (2021, July 27). Retrieved from https://bhuvan-app3.nrsc.gov.in/data/download/index.php

Copernicus Open Access Hub (2021, July 27). Retrieved from https://scihub.copernicus.eu/

DIVA-GIS (2021, July 29). Retrieved from https://www.diva-gis.org/datadown

Earth Explorer-USGS (2021, July 29). Retrieved from https://earthexplorer.usgs.gov/

FSI Fire Alert Systems (2021, July 28). Retrieved from http://117.239.115.41/smsalerts/index.php

Jaiswal RK, Mukherjee S, Raju KD, Saxena R (2002) Forest fire risk zone mapping from satellite imagery and GIS. Int J Appl Earth Obs Geoinf 4(1):1–10. https://doi.org/10.1016/S0303-2434(02)00006-5

Kanga S, Tripathi G, Singh SK (2017) Forest fire hazards vulnerability and risk assessment in Bhajji Forest Range of Himachal Pradesh (India): a geospatial approach. J Remote Sens GIS 8(1):25–40. https://www.researchgate.net/publication/325848317

Mangiameli M, Mussumeci G, Cappello A (2021) Forest Fire Spreading using free and open-source GIS technologies. Geomatics 1(1):50–64. https://doi.org/10.3390/geomatics1010005

Naderpour M, Rizeei HM, Khakzad N, Pradhan B (2019) Forest fire induced Natech risk assessment: a survey of geospatial technologies. Reliab Eng Syst Saf 191(September 2018):106558. https://doi.org/10.1016/j.ress.2019.106558

Parajuli A, Gautam AP, Sharma SP, Bhujel KB, Sharma G, Thapa PB, Bist BS, Poudel S (2020) Forest fire risk mapping using GIS and remote sensing in two major landscapes of Nepal. Geomat Nat Hazard Risk 11(1):2569–2586. https://doi.org/10.1080/19475705.2020.1853251

Qayum A, Ahmad F, Arya R, Singh RK (2020) Predictive modeling of forest fire using geospatial tools and strategic allocation of resources: eForestFire. Stoch Env Res Risk A 34(12):2259–2275. https://doi.org/10.1007/s00477-020-01872-3

Roy PS, Dun D (2005) Forest fire and degradation assessment using satellite remote sensing and geographic information system. 361–400

Saaty RW (1987) The analytic hierarchy process-what it is and how it is used. Math Modell 9(3–5):161–176. https://doi.org/10.1016/0270-0255(87)90473-8

Zhang Z, Feng Z, Zhang H, Zhao J, Yu S, Du W (2017) Spatial distribution of grassland fires at the regional scale based on the MODIS active fire products. Int J Wildland Fire 26(3):209

Part IV
Food Security and Livelihoods

Chapter 17
Climate-Smart Agriculture Interventions for Food and Nutritional Security

Manpreet Kaur, D. P. Malik, Gurdeep Singh Malhi, Muhammad Ishaq Asif Rehmani, and Amandeep Singh Brar

Abstract Human health is mainly dependent on nutrition. In this era of modern civilization, many parts of the world (countries) are still struggling with the availability of sufficient food grains. Agriculture production is largely dependent on climatic conditions of the region. Owing to large dependence of agriculture on climatic conditions, it is adversely affected by the weather vagaries, rising temperature, and fluctuations in precipitation pattern. Moreover, the growing population has placed a huge burden on agriculture production for food security. Low food demand due to low per capita household income also adds to lower consumption, consequently leading to food insecurity. Climate-smart agriculture (CSA) is a blend of various approaches to transform and re-orient agriculture to ensure food security under changing climatic conditions. The main target of CSA is adaptability to climate change, alleviation of greenhouse gases, energy efficiency, and ultimately food security. Various smart technologies are incorporated in CSA to increase water, nutrient, and energy use efficiencies using advance knowledge and climate weather services. The adoption of CSA technologies has proven to efficiently raise agriculture production and consequently the household consumption of food items because of improved and better availability of quality food. The increased income of farming households also adds to the food security of the people and improves their food consumption pattern.

Keywords Agricultural productivity · Climate-smart agriculture · Climate-smart technologies · Food security

---

M. Kaur · D. P. Malik
Department of Agricultural Economics, CCS Haryana Agricultural University, Hisar, Haryana, India
e-mail: mkaur2515@gmail.com; dpmalik@hau.ac.in

G. S. Malhi · A. S. Brar
Punjab Agricultural University (PAU), Krishi Vigyan Kendra (KVK), Moga, Punjab, India
e-mail: 89gurdeep.malhi@gmail.com; amanbrar@pau.edu

M. I. A. Rehmani (✉)
Department of Agronomy, Ghazi University, Dera Ghazi Khan, Pakistan
e-mail: mrehmani@gudgk.edu.pk

Introduction

Food insecurity and malnourishment are the biggest challenges of the modern world. Worldwide, the hunger is at a moderate level and the highest hunger and undernutrition levels were observed in south Asia and Africa. India is facing a serious level of hunger issue, as evident through the score of 27.5 and a rank of 101 (in 116 countries) in the global hunger index (GHI 2022). The most serious consequence of food insecurity is malnutrition causing poor growth, especially in children. According to state of food insecurity, India is estimated to have 217 million undernourished inhabitants. The main reasons for food insecurity are the inaccessibility (economic or social) to sufficient nutritious food, and inadequate food utilization, mainly because of the shrinking income of the agriculture sector (Chakravarty and Dand 2010).

High economic growth achieved by India has not shown considerable impression on the food security of the Indian population. It was reported that cereal consumption of the bottom 30% of the population is much lesser than the top 20% of population, despite having other options like fruits, vegetables, meat, etc. available with the upper strata. The calorie consumption of bottom half of the population is consistently declining since 1987. Around 39% of the Indian children were undernourished in 2014 while half of the Indian women and three-quarters children were anemic (Saxena 2018). The present condition of undernourishment in India is the result of imbalanced supply and demand. The main issue of the increasing food insecurity in Indian people is due to their decreasing purchasing power, as around half of the Indian population is dependent on agriculture. The achievement of food security is not possible without raising the level of agricultural production which can ensure food security and promote the reduction of poverty (Pandey 2015).

As evident through Fig. 17.1, the production and productivity are showing an increasing trend since independence, however, the rate of growth has declined over the years. But, the annual per capita food grain consumption of the Indians has shown an increasing trend up to 1991, after that subsequently declined over a decade and further remained constant over the past decade. Despite the increased production, the consumption of food grains has been persistent and not showing any significant improvement mainly because of the increasing population.

Although the agriculture sector in India is predominantly cereal-oriented and India ranks second in the world's rice and wheat production, the stagnation in cereal consumption is evident. Other commodities, such as pulses and oilseeds, also have equal importance in the diet of Indian people. The gap in demand and supply of edible oils necessitates importing 60% of our requirement which accounted for 14.01 million tons worth ₹73,048 crores in 2016–2017. Despite 3.89% growth, the country needed to import oilseeds worth billions because of the huge growth rate of demand due to the increasing population (NFSM 2018). India produces, consumes, and imports the maximum in context of pulses, accounting for 25, 27, and 14% of the total pulses in the world, respectively. The production of pulses has recorded a significant increase from 8.41 million tons (1950–1951) to 23.40 million tons

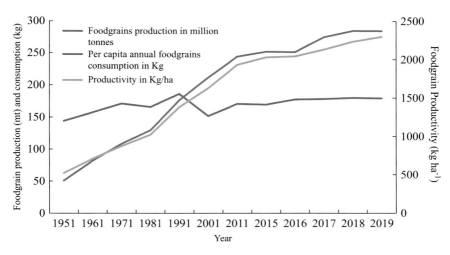

Fig. 17.1 Food grains production, productivity, and per capita consumption over the years in India

(2018–2019) (Agricultural Statistics at a Glance 2020) which is mainly attributed to increased acreage and productivity enhancement. However, according to IIPR, the pulses requirement of India is estimated to be 32 MT, and to fulfill this requirement, an additional area of 3–5 million ha is to be brought under pulses cultivation and the productivity also needs to be augmented to 1361 kg ha^{-1} (Mohanty and Satyasai 2015). Vegetable consumption in India has been gradually increasing, however, per capita average vegetables' consumption in India is less than the WHO recommendation of 300 grams per day (Venkatanarayana 2020). The availability and consumption of fruits and vegetables are significant determinants, which can only be raised with enhanced agricultural production and improved supply chains (Mukhrejee et al. 2016).

Disparity between the demand and supply of food commodities is the major reason for food insecurity. Apart from the low supply, the low household income and resource base of the Indian population also lead to lower demand. The improved technologies have the potential to increase both production and consumption by raising agricultural productivity and consequently income, which needs huge efforts in the era of climate change. The changing climate has a huge influence on agricultural productivity that make it more difficult to fulfill the food and nutritional requirement of the 7.50 billion world population (World population review 2019). The world's agricultural production is required to be raised by 60% from 2005/2007 to 2050 constituting a rise in agricultural production by 24% and 77% in developed and developing nations, respectively to feed the population by 2050 (Alexandratos and Bruinsma 2012). Global agricultural production is predicted to be decreased by climatic anomalies, and is projected to reduce global maize production by 3.8% and global wheat production by 5.5% (Lobell et al. 2011). This is mainly because of increased human activities which have significantly altered the atmospheric structure, especially during the last two decades (IPCC 2007). The carbon dioxide emission has a considerable share in greenhouse gases (63%) (Sathaye et al. 2006) that has increased to 36.14 Gt in 2014 (Abeydeera et al. 2019). Since 1975, the global average temperature has increased at a rate of 0.15–0.20 °C decade^{-1} (NASA 2020) and is likely to be risen by 1.4–5.8 °C by 2021 (Arora et al. 2005; Rehmani et al. 2011). As per another estimate, the average temperature is predicted to further increase by 1.5 °C till 2050 (Arora et al. 2019). There is a substantial spatio-temporal disparity in increasing temperature (Rehmani et al. 2021), which will have serious implications for crop production and food security, both at global and regional scales (Meng et al. 2020; Mirzaei et al. 2021; Rehmani et al. 2014).

The increased atmospheric CO_2 concentration has a positive impact on the growth and productivity of C_3 crops because of a higher rate of photosynthesis. However, C_3 crops perform photorespiration instead of photosynthesis at a very high temperature which leads to energy wastage. So, the increased growth is offset by raised temperature which leads to raised crop respiration rate, higher pest infestation, higher evapotranspiration, reduced crop duration, and a shift in weed flora (Mahato 2014; Malhi et al. 2021). The annual economic losses instigated by the hostile influence of climate change on agriculture account for 0.3% of future GDP by the year 2100. The burning of crop residue releases around 7300 kg CO_2-equivalent ha^{-1} of greenhouse gases in northwestern India (Bhattacharyya et al. 2021). The detrimental effect of climate change on global agricultural welfare especially after 2050 will be mainly due to losses in consumer surplus which would outweigh the gains in producer surplus (Stevanovic et al. 2016). However, the negative inferences of climate change on food production in South Asia may not be significant. But in the long term (beyond the 2050s), the productivity of crops is most likely to be declined owing to huge variations in climate, pest incidences, and virulence. At least 4–10% losses in net cereal production in South Asia are most likely to be witnessed under most conservative climate change estimates (regional warming of 3 °C) (Lal 2010). Comparatively, 1.5 °C warming has lesser influence because of carbon fertilization; however, the warming of 3 °C is predicted to cause significant damage (US$84 billion) in South Asia (Mendelsohn 2014).

Climate Change Vulnerability and Indian Agriculture

India is much susceptible to changing climatic conditions owing to its dependence on climate-sensitive sectors (agriculture & allied activities) and varied agro-climatic conditions. In India, the annual mean temperature has been increasing at a rate of 0.42 °C per 100 years with a rise of 0.92 °C/100 years and 0.09 °C/100 years in mean maximum and minimum temperature, respectively (Arora et al. 2005). In India, the temperature during winter and post-monsoon months is enhanced by 1 °C and 1.1 °C, respectively since the last century. The minimum temperature during the summer monsoon is also reduced and it shows a considerable increase in post-monsoon months thereby creating a variation of 0.8 °C in the minimum temperature of two seasons (Dash et al. 2007). Moreover, India faced a negative trend for rainfall during the last 30 years (Radhakrishnan 2017). The rise in average annual temperature in the subtropical and tropical regions of India is also predicted by 1.5–2.5 °C, and 1.2–2.2 °C, respectively in 2050 (USAID 2020).

The changing climate projected during 2010–2039 is estimated to reduce crop yields by 4.5–9% and in absence of long-term adaptations; it can lead to a yield loss of around 25% (Guiteras 2009). The yield of irrigated rice is predicted to reduce by 7% and 10% in 2050 and 2080, respectively while that of rainfed rice has already shrunk by 6% and only a marginal reduction is likely to be observed in 2050 and 2080 scenarios. The yield of the wheat crop is also projected to decrease by 6–23% by 2050 and 15–25% by 2080 in India. Moreover, these negative impacts are more likely to be severe in higher GHG emission scenarios (Kumar et al. 2014). The maize yield in the monsoon season is likely to have adverse effects but the increased rainfall will counterbalance those losses. However, maize yield in winter is likely to be reduced in the mid Indo-Gangetic plains and southern plateau (Byjesh et al. 2010). The negative influence of changing climatic conditions on sorghum will be more evident by the 2030s in the Indian context mainly due to increased vulnerabilities as predicted through data used from the FAO-EcoCrop database and EcoCrop model (Ramirez-Villegas et al. 2013).

Considering changes in climate over years, CSA is a tactic guiding the actions to transform and reorient the system of agriculture to ensure food security and development in the changing climate scenarios (FAO 2011). The main objectives are to increase agricultural productivity through sustainability, climate change adaptation, and reduction of greenhouse gases emission. CSA works within a framework with its narrow focus on technical efficacy at production (Taylor 2017). CSA aims to raise the adaptive capacity of farmers and endorse climate-resilient ways by building evidence, increasing local institutions' efficiency, raising rationality between agricultural policies with climate change, and linking agricultural funding with climate change (Lipper et al. 2014). An agricultural practice or production system is said to be climate-smart only when it is suitable to local climatic and socio-economic conditions, a biophysical context determining the efficiency of the agricultural practices in terms of increasing productivity, ensuring resilience as well as mitigation benefits (Williams et al. 2015).

Climate-Smart Agriculture and Enhanced Agricultural Productivity

The practices and technologies merged in CSA are reasonably effective in achieving higher productivity for ensuring food security, climate change adaptation, and GHG mitigation with the cost-effective generation of additional benefits (Dinesh et al. 2015). CSA has huge potential for carbon sequestration, although, the extent varies depending on types of intercropping, specific farming practices, etc. (Raj et al. 2018). The components of CSA comprise smart technologies of weather, carbon, water, nutrient, knowledge, and energy management.

Water-Smart Technologies

The water-smart technologies include laser land leveling (LLL), drainage management, rainwater harvesting (RWH), drip irrigation, furrow irrigated bed planting, cover crops methods, etc. Rainwater harvesting (RWH) ensures site-specific control of water resources, primarily used to satisfy local water requirements. RWH has various ecosystem services (Sarwar et al. 2021) and benefits viz. indoor use, irrigation of small areas, and overflow to dry wells for groundwater recharge in India. It also yielded maximum net present value (₹21,764–38,851) as analyzed for 1999–2011. Moreover, it provides the highest net present value (NPV) of ₹21,764–38,851, has the least payback period of 0.30–0.98 years, and recharged groundwater to the extent of 40% of onsite rainfall (Stout et al. 2017). Drip irrigation also yielded significant social and private profits. The value of social benefit-cost ratio (BCR) lies between 4.33 and 5.19 at 2% discount rate under various scenarios across southern India (Kumar and Palanisami 2011). Laser land leveling technology lowers the water use for irrigation and energy use by 21% and 31% as evident through a study conducted in 2009–2011 in western Uttar Pradesh province of India. The total irrigation depth was reduced in rice, wheat, and sugarcane by 14.7, 13.3, and 20.3% when compared with traditional levelled fields. The water use efficiency has increased by 48 and 22% in rice, 47 and 19% in wheat, and 49 and 20% in sugarcane, in precisely leveled fields and traditionally leveled fields, respectively over unlevelled fields. The average annual returns also increased by 14, 13.5, and 23.8% in rice, wheat, and sugarcane, respectively in precisely leveled fields through LLL over traditionally leveled fields (Naresh et al. 2014).

Furrow irrigated raised bed system (FIRBS) has led to enhanced irrigation water productivity by 76, 44, and 54% in pigeon pea, cluster bean, and green gram, respectively in comparison with flat sown crops. The use of irrigation water is also reduced by 25% compared with flat sown crops. Apart from agronomic benefits, the additional net returns of ₹2820, ₹830, and ₹1465 were obtained in pigeon pea, cluster bean, and green gram, respectively with FIRBS (Dhindwal et al. 2005). Layering of two technologies namely precision land leveling and furrow irrigated raised bed planting (FIRBP) had shown higher crop yield and nutrient use efficiency along with less irrigation. The wheat yield was augmented by 16.6% with nearly half amount of irrigation water when compared with traditional land leveling with flat planting. The agronomic productivity and uptake efficiency of nutrients (N, P, and K) was also significantly improved (Jat et al. 2011).

Energy-Smart Technologies

The energy-smart technologies include zero tillage and minimum tillage. Use of resource conservation technologies (RCTs) such as minimum tillage and zero tillage (ZT) has shown immense potential for enhancing the productivity of crops with less water and energy use. ZT practices even without full retention of residues had shown a yield gain of 498 kg ha^{-1} (19%) in wheat over conventional tillage (CT) in different agro-ecological zones. The economic benefits obtained because of raised yield and reduced cultivation cost amounted to 6% of total annual income of sampled households in Bihar state of India (Keil et al. 2015). The combined benefits of yield gains and reduced cost had made adoption worthwhile in South Asia where the most prominent cropping system is rice-wheat (Erenstein et al. 2008). The cumulative emission of CO_2 had been considerably reduced in minimum tillage by around 30% over 3 months after autumn plowing and 28% along spring-summer cultivation for similar maize yields in comparison to conventional tillage. The emission of N_2O was also reduced by 40% for fallow and 18% for maize. Minimum tillage also had the potential to reduce greenhouse gas (GHGs) emissions from the cropped soils in both short and long term (Forte et al. 2017). The minimum tillage along with residue management improved the soil water content and soil organic matter, and consequently increased the yield and oil content of canola particularly in rainfed conditions (Abdullah 2014). The intercropping system along with minimum tillage also enhanced the soil organic matter (SOM) and soil organic carbon (SOC) and thus, played a significant role in soil carbon sequestration (Maia et al. 2019).

Nutrient-Smart Technologies

Understanding the processes regulating nutrient fluxes, uptake, accumulation, and distribution in plants is vital for nutrient efficiency, crop yield and quality, and environmental concerns (Gastal and Lemaire 2002). Site-specific nutrient management (SSNM), green manuring, leaf color chart (LCC), intercropping with legumes, etc. are nutrient-smart technologies. The nutrient dose in SSNM is calculated based on yield targets, indigenous nutrient supply, and nutrient demand. In Punjab state of India, the agronomic use efficiency of N in transplanted rice was enhanced by 83% in SSNM than farmers' fertilizer practices (FFP). It also increased the N recovery efficiency to 0.30 kg kg^{-1} in SSNM plots. Moreover, judicious use of fertilizers in SSNM also raised the gross return over fertilizer cost (GRF) by 14% than FFP (Khurana et al. 2007). Long-term data analysis of various locations of India revealed that SSNM enhanced the yield of rice and wheat crops by around 12 and 17% which consequently increased the profitability by 14% in rice and 13% in wheat (Sarkar et al. 2017). SSNM application in no tillage-based system also improved the nutrient use efficiency, yield, and economic profitability in wheat crop. Apart from this, it also decreased the global warming potential of wheat crop compared with the conventional tillage system in northwestern India (Sapkota et al. 2014; Malhi et al. 2022).

The green manuring (*Sesbania aculeate*) in wetland rice enhanced the water stable aggregates, reduced the soil bulk density, and raised the infiltration rate. The results, however, are limited in maize planted soils. The green manuring improved the root density and wheat grain yield cultivated after rice which shows the potential of green manure in improving the soil physical properties (Boparai et al. 1992). Green manuring with legume crops has the potential to improve soil physico-chemical and biological properties, SOC, nutrient availability, and consequently crop productivity. As the legume crops fix atmospheric nitrogen and improved soil nitrogen level is positively correlated with soil organic carbon. The legume crops, therefore, help in optimization of biological nitrogen fixation and ensure soil sustainability (Meena et al. 2018). Green manuring (GM) with nitrogen fertilizer also significantly improved the morphological traits viz. chlorophyll content (SPAD value), light interception (% LI), leaf area index (LAI), and net assimilation rate (NAR) in the succeeding crop which consequently led to improved yield. The green manuring also improved the macro- and micro-nutrient content of post-harvest soil (Islam et al. 2018). Maize-legume intercropping facilitates biological fixation of atmospheric nitrogen and helps in restoring soil fertility (Dwivedi et al. 2015). Thus, the intercropping of cereals with legumes improves land-use efficiency and enhances fixation of atmospheric nitrogen. The productivity of legumes was reduced but the overall productivity of baby corn yield equivalent was increased in baby corn-groundnut intercropping (Banik and Sharma 2009).

Estimation of crop nutrient requirement using leaf color chart (LCC) has also helped in real-time nitrogen (N) management by farmers and led to enhanced fertilizer N use efficiency in both transplanted and direct-seeded rice (DSR) crops (Singh et al. 2006; Virk et al. 2021; Yadvinder-Singh et al. 2007). The use of urea, based on LCC, reduced the nitrous oxide emission in rice and wheat crops. At LCC ≤ 4, the application of 120 kg N ha^{-1} reduced the nitrous oxide and methane emission by 16 and 11%, respectively over split application of urea in rice. Whereas the application of urea in wheat at LCC ≤ 4 reduced the emission of nitrous oxide by 18% over the conventional method. Thus, the global warming potential (GWP) of rice and wheat could be reduced with the judicious application of nitrogenous fertilizer with the help of LCC (Bhatia et al. 2011).

Plant nutrient dilution curves are stable in-season diagnostic indicators and helpful for improving the accuracy and efficiency of fertilizer management (Ata-Ul-Karim et al. 2017; Zhang et al. 2020; Zhao et al. 2021). Nutrient dilution curves are widely used (more than 40 countries) for different nutrients including nitrogen, potassium, phosphorous, and sulfur (Chen et al. 2021; Makowski et al. 2020; Fontana et al. 2021; Nyiraneza et al. 2021), and applied to more than 30 plant species including cereals (wheat, maize rice) (Ata-Ul-Karim et al. 2016; Makowski et al. 2020), legumes (Salvagiotti et al. 2021), etc. (Chen et al. 2021).

Carbon-Smart Technologies

Agro-forestry, fodder management, concentrate feeding for livestock, and integrated pest management (IPM) etc. are categorized as carbon-smart technologies. Agroforestry works as a carbon sink that removes atmospheric carbon dioxide. It also reduces land clearing while maintaining carbon in existing vegetation. Carbon storage to the extent of 9 t C ha^{-1} in semi-arid, 50 t C ha^{-1} in humid, 21 t C ha^{-1} in sub-humid, and 63 t C ha^{-1} in temperate eco-zones had been reported with agro-forestry (Schroeder 1994). Apart from carbon sequestration, agro-forestry has the potential to conserve biodiversity, enrichment of soil, and improve air and water quality (Jose 2009). The agroforestry systems improved the economic benefits of the farmers as evident through the paulownia (*Paulownia elongta*) intercropping system in northern China and the tea (*Camellia sinensis* O. Ktze) intercropping system in southern China. The energy output/input ratio and economic output/input ratio were significantly improved in paulownia intercropping by 9.45 and 7.56%, respectively over non-paulownia intercropping. The energy output/input ratio and economic output/inputs also showed significant improvement by 18.7 times and 64.29%, respectively in tea intercropping over non-tea intercropping (Jianbo 2006). The carbon footprint (CF) in the livestock sector is positively related to concentrate use, herd structure, and milk yield. The promotion of balanced feed and concentrate according to the requirement of cow-based on its livestock cycle is to be promoted to reduce CF of milk production (Wilkes et al. 2020).

Weather-Smart Technologies

The weather-smart technologies include climate-smart housing for livestock, weather-based crop advisory, and crop insurance. Weather-based agro-met advisory services (AAS) are quite beneficial to farmers as they provide timely information of weather and suggest suitable agronomic practices accordingly. The users of AAS make judicious use of inputs and consequently reduce their input costs and improve the net benefits in wheat, carrot, and rice cultivation in the NCR region of Delhi (Vashisth et al. 2013). The farmers managing the operations according to AAS can earn additional benefits in crop growth (80.91%), plant protection (77.27%), irrigation (79.09%), income (79.09%), and adverse weather issues related to advisory services (66.36%) (Manjusha et al. 2019). AAS has the potential to simplify the lab-to-land transmission of technologies with ICT tools to help farmers in avoiding weather risks (Dheebakaran et al. 2020). The AAS provided through mobile phones has reduced the information asymmetry among the farmers; particularly between males and females. The farmers reported that precise and timely weather-based AAS had helped in the reduction of the cost of inputs and saved irrigations (Mittal 2018). Apart from AAS, crop insurance plays a pivotal role in reducing weather risks. About 60% of yield variations in India are attributed to weather-related issues and around 70% of crop production is exposed to weather vagaries in India which makes crop insurance very important in the Indian context (Raju et al. 2016). The CSA adoption by farmers for residue management practices could be improved with crop insurance. Thus, the provision of insurance services with climate-smart technologies and practices has the potential to manage the risk by farmers more efficiently (Kramer and Ceballos 2018).

Knowledge-Smart Technologies

Contingent crop planning (CCP), improved crop varieties, seed, and fodder banks, etc. are considered as knowledge-smart technologies. CCP is the cultivation of crop suited to a region sown despite the normally sown crop fails because of aberrant weather conditions. Various kinds of weather vagaries viz. insufficient and irregular rainfall, long gap in rainfall, early and late onset of monsoon, early cessation of rains, prolonged dry spells, etc., occur which could be detrimental to agriculture and need careful crop planning. Change of crops, short duration varieties, increased seed rate, mulching, etc., are some of the methods to attain maximum possible farm returns. The fodder maize cultivation or cluster bean for vegetable or cowpea for grain or groundnut for dry pod or pigeon pea for grain were reported to have higher yield and monetary returns if they are sown in the first fortnight of July under the scarcity zone of Maharashtra, which face climate vagaries leading to compromised crop productivity (Satpute and Dhadge 2016). The sowing of crops in the second fortnight of August was having higher productivity in rainfed alfisols of Hyderabad and the highest pod yield (16,113 kg ha^{-1}) and net returns (₹203,688 ha^{-1}) were observed in cluster bean (Reddy et al. 2017).

The adoption of improved rice varieties has a consistently positive influence on domestic food security as evidenced through two parameters of domestic dietary diversity score and subjective food security (Lu et al. 2021). Improved rice varieties also yielded higher output per unit of land and their adoption was influenced by educational attainment, the experience of farming, credit access, extension contact, and access to quality seed of improved rice varieties (Bello et al. 2020). In eastern India, the acceptance of modern rice varieties has increased the per capita household expenditure. The adoption of these varieties was positively influenced by age, education, land size, risk aversion, yield, disease resistance, perception, and availability of modern high-yielding varieties. On the other hand, household size, farmer's experience, amount of credit received, off-farm jobs, cost of seeds, insecticides, and fertilizers were negatively affecting the adoption of modern high-yielding varieties. The intensity of adoption was, however, positively influenced by the size of the household, risk aversion, cost of insecticides, land size, perceptions of modern high-yielding varieties as high yielding and their availability and negatively influenced by farmer's experience, marketability of modern high-yielding varieties, and cost of fertilizer. The incidence of poverty was higher among the non-adopters (Bannor et al. 2020).

The production of genetically improved wheat varieties increased the wheat yield steadily during the past half-century and has a significant influence on food security (Dixon et al. 2006). The adoption of high-yielding wheat varieties raised the average wheat yield (1 to 1.1 t ha^{-1}) and consequently raised the income of the farmers (35–50%) in Ethiopia. The adoption of the varieties was significantly and positively influenced by the gender of the family leader, livestock ownership (Tesfaye et al. 2016). Climate change is making agriculture more vulnerable to pest infestation by affecting their biology, outbreak potential, and distribution among various crops across landscapes. Climate-smart pest management is an approach that targets to reduce the pest-related yield losses, enhance ecosystem services and strengthen the agricultural resilience to climate change (Heeb et al. 2019). The climate-smart push-pull technology was assessed, and a huge proportion of farmers were willing to adopt it. The adoption of the technology was positively influenced by gender, level of awareness about relevant technology, perceptions, and market access (Murage et al. 2015).

Evidence on Climate-Smart Agriculture on Increased Welfare

Food insecurity is considerably linked with family income, living standard, and expenses on food items (Chatterjee et al. 2012). A study conducted in an urban slum of Delhi revealed 51% of households to be food insecure (Agarwal et al. 2009). The marginalized sector of urban society possesses high food insecurity as observed in the urban resettlement colony (Chinnakali et al. 2014). In rural households, the issue of food insecurity is equally severe as 25% of the household had access to food less than 2400 calories and only 10% of households had safe drinking water. The monthly income of most households was found less than ₹3000. The low income of the farmers with small landholdings was the major factor that leads to food insecurity in Uttar Pradesh (Mushir et al. 2012).

The adoption of CSA has caused a significant change in the consumption pattern of people. A positive relation was reported between the improved technology adoption and welfare outcomes of the households such as consumption expenditure among smallholder farmers of Ethiopia and Tanzania. The non-adopters were having systematically different characteristics from adopters. When the non-adopters adopted the cultivation of improved pigeon pea and chickpea varieties, their consumption expenditure per adult equivalent was enhanced by 99.4 and 20.9%, respectively (Asfaw et al. 2012). Another study investigated the impact of yield-increasing crop technologies on the socioeconomic aspects of Ethiopian farmers living in the marginality hotspot districts. It was stated that around 90% of farm households were food insecure and poor in the sampled area of Ethiopia. There was low adoption of yield-increasing technologies, but those technologies had the potential to increase yield and incomes and thereby reducing the poverty. The net benefits accrued from the adoption of improved seeds with appropriate agronomic packages per household were 17,000 birrs in Halaba and 18,000 birrs in Baso Loben. Net benefits of the districts varied from 71 to 182 million birrs per year while the net additional benefit of a person in a day ranged from 5 to 13 birr. These additional benefits had the potential to lift the poorer out of poverty, however, these earnings were not sufficient to raise the ultra-poor out of poverty (Kotu and Admassie 2016). The cumulative effect of CSA technologies varied within and between various practices in East Africa. Components of CSA, e.g., nutrient management (synthetic fertilizers and organic sources), a component of CSA, had a positive impact on crop productivity in Tanzania and Uganda while other practices such as agroforestry had an inconsistent impact on productivity within and between countries. In Tanzania, post-harvest improvements had the largest effect on productivity followed by soil and water management practices while in Uganda, the largest effect on productivity was witnessed in livestock diet management followed by nutrient and soil management practices (Lamanna et al. 2016).

Lan et al. (2018) studied the rice-based agricultural system of Vietnam in the Mekong delta region (sample of 170), the cocoa-based agriculture system of Nicaragua north-central region (sample of 180) and diversified system of Uganda northern region (sample of 453) in 2015. The entire population is divided into three clusters depending on their income and farm size; households having the lowest income and smallest farm size in all cases (cluster 1), medium level of income and farm size (cluster 2), and better off households having large farm size, better education, and less dependence on agriculture (cluster 3). The results revealed that in Vietnam, the adoption rate of lower-income and middle-income clusters was determined by the cost and profitability of CSA practices. The adoption rate was found to be the same among all the clusters in the case of low cost and low returns practices while the practices having low cost and higher benefits were more readily adopted by lower income groups while high cost and high benefits were more affordable to cluster 3. In Nicaragua, the aggregate impact of CSA adoption was higher in groups constituting a larger number of households while in Uganda, both initial investment and benefit were high in comparison to the income of all the clusters so the conservation agriculture (CA) having high cost and returns was expected to be adopted by 90% of clusters.

The adoption of CSA technologies was influenced by extension facilities and knowledge of farmers. Regression analysis showed that CSA adoption is negatively associated with the number of years in farmers' group, last time extension services were received, frequent meetings, age, household size, and size of land. The results of HDDS (Household Diversity Dietary Score) regression and HFCS (Household Food Consumption Score) pairwise correlation showed that use of organic fertilizers, intercropping, crop rotation, and soil conservation adds to food security and there was a positive relationship with CSA technologies and food security of the area (Jelagat 2019).

In northern Uganda, the adoption of drought tolerant maize varieties (DTMVs) was positively and significantly affected by households perceiving higher temperature, education level of household head, ownership of assets, and agricultural knowledge while being negatively influenced by household asset index. On the other hand, the adoption of maize legume intercropping (MLI) was significantly influenced by increasing temperature, education of household head, number of kinship links, and non-agricultural asset index. It was also reported that the adoption of DTMVs and MLI raised the mean yield and reduced the yield variance suggesting a higher stable yield under heat stress. Moreover, the period of food shortage was reduced for MIL and DTMV adopters by 8–12 days and 7–8 days, respectively. The adoption of these CSA practices also enhanced the household income per adult equivalent with 20–25% increase in income of MLI adopters and 25–34% increase in income of DTMV adopters. The food consumption score of the adopters was also increased and was reported to be 2.3–3.1 points more for DTMV adopters and 2.4 points more for MLI adopters (Shikuku and Mwongera 2019).

The impact of adoption of CSA technologies reduced the overall deprivation (2–3%) and increased the income and consumption of households due to gain in farm production. The severely deprived households have the higher impact of CSAs adoption with a reduction in multi-dimensional poverty due to increased income, expenditure, and consumption. The impact of these technologies was mainly through the non-food expenditure pathway (Habtewold 2021).

In Ethiopia, total factor productivity (TFC) analysis, of major crops by performing a dis-aggregated analysis of growth components, highlighted the significant effect of inputs on aggregate output. Further, technical efficiency (TE) of around 59% of inefficient farming households can be improved consequently leading to increased production. Technical change (TC) value of 22% showed that anticipation of improved rate of responsiveness for technical change. These TC and TE were the major factors influencing TFP growth of crops. The gradual increase in TE indicated that the outputs could be increased with the same level of inputs with improved technology adoption and efficiency (Meja et al. 2021). A compilation of studies showing the impact of agricultural technologies on enhanced food security is shown in Table 17.1.

Conclusion

The agriculture sector is primarily weather dependent, so the changing climate conditions are causing significant impact on crop productivity. The rising levels of carbon dioxide because of anthropogenic activities are the major reason reasons for increase in raising the earth's temperature. The increased CO_2 raises the rate of photosynthesis; however, warming caused by elevated CO_2 also results in increased crop respiration rate. Moreover, higher temperatures caused higher incidents of insects, pests, and diseases. This necessitates the development of an approach which makes agriculture production viable and efficient in these conditions. The CSA is the approach of re-orienting and transforming the agriculture system toward better production, higher energy efficiencies, and higher food security. It helps in increasing agricultural productivity while adapting to climate change and mitigating greenhouse gases. CSA technologies include water-smart technologies, nutrient-smart technologies, carbon-smart technologies, weather-smart technologies, and knowledge-smart technologies. Water-smart technologies either reduce water use, water application, or increase water use efficiency while maintaining or increasing agricultural production. The carbon and energy-smart technologies help in the reduction of carbon dioxide emissions and ensure better utilization of waste resources to increase the energy use efficiency and carbon footprints of crop production. The nutrient-smart technologies also help in the judicious use of

Table 17.1 Impact of climate-smart agricultural technologies on enhanced consumption vis-à-vis food security

Location	Crop	Climate-smart technology	Enhanced efficiency	Incremental economic benefit	References
Tanzania and Ethiopia	Pigeon pea and chickpea	Improved varieties	Welfare of household, consumption expenditure	99.4% and 20.9% increase in consumption expenditure per adult eq. in pigeon pea and chickpea	Asfaw et al. (2012)
Ethiopia	Maize, Tef, and wheat	Crop technologies	Improved income and reduction of poverty	Net benefits of 17,000 bir in Halaba and 18,000 bir in Baso Loben per household	Kotu and Admassie (2016)
East Africa	Various crops	Organic and inorganic fertilizers, green manure, improved varieties, reduced tillage, water harvesting, mulching	Increased productivity	–	Lamanna et al. (2016)
Vietnam, Nicaragua, and Uganda	Various crops	Number of CSA practices	Enhanced income	–	Lan et al. (2018)
Kenya	Various crops	CSA practices	Enhanced food security	–	Jelagat (2019)
Uganda	Various crops	Drought tolerant maize varieties, maize legume intercropping	Increased yield and reduced yield variance, reduction of food shortage period, increased income, increased food consumption score	20–25% increase in income (MLI) and 25–34% increased income (DTMV)	Shikuku and Mwongera (2019)
Ethiopia	Various crops	CSA practices	Increased income and household consumption	Reduced multidimensional poverty	Habtewold (2021)
Ethiopia	Various crops	Technologies	Improved technical efficiencies	TC value of 22%	Meja et al. (2021)

nutrients. Apart from the injudicious use of resources; the dependence of agriculture on weather conditions also adversely affects the production. The weather-smart technologies come to rescue in that case and help in policymaking for agricultural activities based on weather forecast. The knowledge-smart technologies also help in ensuring the availability of modern agricultural technologies to the farmers and raise their productivity under changing climate. The adoption of climate-smart agriculture technologies has proven to be beneficial for the farmers that could enhance the agriculture production under changing climatic conditions. The increased agricultural production has a dual impact of increasing food availability to the consumers influencing the supply side and raising the income of farmers influencing the demand side. The improvement in supply and demand conditions with the adoption of climate-smart agriculture technologies has proven to have the potential of increasing food and nutritional security.

References

Abdullah AS (2014) Minimum tillage and residue management increase soil water content, soil organic matter and canola seed yield and oil content in the semiarid areas of Northern Iraq. Soil Tillage Res 144:150–155

Abeydeera LHUW, Mesthrige JW, Samarasinghalage TI (2019) Global research on carbon emissions: a scientometric review. Sustainability 11(3972):1–24

Agarwal S, Sethi V, Gupta P, Jha M, Agnihotri A, Nord M (2009) Experiential household food insecurity in an urban underserved slum of North India. Food Secur 1:239–250

Alexandratos N, Bruinsma (2012) World Agriculture towards 2030/50. ESA Working Paper No. 12-03. FAO

Arora NK (2019) Impact of climate change on agriculture production and its sustainable solutions. Environ Sustain 2:95–96

Arora M, Goel NK, Singh P (2005) Evaluation of temperature trends over India/evaluation de tendances de température en Inde. Hydrol Sci J 50(1):81–93

Asfaw S, Shiferaw B, Simtowe F, Lipper L (2012) Impact of modern agricultural technologies on smallholder welfare: evidence from Tanzania and Ethiopia. Food Policy 37:283–295

Ata-Ul-Karim ST, Cao Q, Zhu Y, Tang L, Rehmani MIA, Cao W (2016) Non-destructive assessment of plant nitrogen parameters using leaf chlorophyll measurements in rice. Front Plant Sci 7. https://doi.org/10.3389/fpls.2016.01829

Ata-Ul-Karim ST, Zhu Y, Cao Q, Rehmani MIA, Cao W, Tang L (2017) In-season assessment of grain protein and amylose content in rice using critical nitrogen dilution curve. Eur J Agron 90: 139–151

Banik P, Sharma RC (2009) Yield and resource utilization efficiency in baby corn-legume-intercropping system in the eastern plateau of India. J Sustain Agric 33(4):379–395

Bannor RK, Kumar GAK, Oppong-Kyeremeh H, Wongnaa CA (2020) Adoption and impact of modern rice varieties on poverty in eastern India. Rice Sci 27(1):56–66

Bello LO, Baiyegunhi LJS, Danso-Abbeam G (2020) Productivity impact of improved rice varieties' adoption: case of smallholder rice farmers in Nigeria. Econ Innov New Technol. https://doi.org/10.1080/10438599.2020.1776488

Bhatia A, Pathak H, Jain N, Singh PK, Tomer R (2011) Greenhouse gas mitigation in rice-wheat system with leaf color chart-based urea application. Environ Monit Assess 184:3095–3107

Bhattacharyya P, Bisen J, Bhaduri D, Priyadarsini S, Munda S, Chakraborti M, Adak T, Panneerselvam P, Mukherjee AK, Swain SL, Dash PK, Padhy SR, Nayak AK, Pathak H,

Kumar S, Nimbrayan P (2021) Turning the wheel from waste to wealth: economic and environmental gain of sustainable rice straw management practices over field burning in reference to India. Sci Total Environ 775(145896):1–17

Boparai BS, Singh Y, Sharma BD (1992) Effect of green manuring with sesbania aculeata on physical properties of soil and on growth of wheat in rice-wheat and maize-wheat cropping systems in a semiarid region of India. Arid Land Res Manag 6(2):135–143

Byjesh K, Kumar SN, Aggarwal PK (2010) Stimulating impacts, potential adaptation and vulnerability of maize to climate change in India. Mitig Adapt Strateg Glob Chang 15:413–431

Chakravarty S, Dand SA (2010) Food insecurity in India: causes and dimensions. Working paper. IIMA institutional repository

Chatterjee N, Fernandes G, Hernandez M (2012) Food insecurity in urban poor households in Mumbai. Food insecurity 4:619–632

Chen R, Zhu Y, Cao W, Tang L (2021) A bibliometric analysis of research on plant critical dilution curve conducted between 1985 and 2019. Eur J Agron 123:126199

Chinnakali P, Upadhyay RP, Shokeen D, Singh K, Kaur M, Singh AK, Goswami A, Yadav K, Pandav CS (2014) Prevalence of household-level food insecurity and its determinants in an urban resettlement colony in North India. J Health Popul Nutr 32(2):227–236

Dash SK, Jenamani RK, Kalsi SR, Panda SK (2007) Some evidence of climate change in twentieth-century India. Clim Chang 85:299–321

Dheebakaran G, Panneerselvam S, Geethalakshmi V, Kokilavani S (2020) Weather based automated agro advisories: an option to improve sustainability in farming under climate and weather vagaries. In: Venkatramanan V, Shah S, Prasad R (eds) Global climate change and environmental policy. Springer, Singapore. https://doi.org/10.1007/978-981-13-9570-3_11

Dhindwal AS, Hooda IS, Malik RK, Kumar S (2005) Water productivity of furoow-irrigated rainy-season pulses planted on raised beds. Indian J Agron 51(1):49–53

Dinesh D, Frid-Nielsen S, Norman J, Mutamba M, Loboguerrero, RAM, Campbell B (2015) Is climate-smart agriculture effective? A review of selected cases. CCAFS working paper no. 129. Copenhagen, Denmark: CGIAR Research Program on Climate Change, Agriculture and Food Security (CCAFS). Available online at: www.ccafs.cgiar.org

Dixon J, Nalley L, Kosina P, Rovere RL, Hellin J, Aquino P (2006) Adoption and economic impact of improved wheat varieties in the developing world. J Agric Sci 144(6):489–502

Dwivedi A, Dev I, Kumar V, Yadav RS, Yadav M, Gupta D, Singh A, Tomar SS (2015) Potential role of maize-legume intercropping systems to improve soil fertility status under smallholder farming systems for sustainable agriculture in India. IJLBPR 4(3):145–157

Erenstein O, Farooq U, Malik RK, Sharif M (2008) On-farm impacts of zero tillage wheat in South Asia's rice-wheat systems. Field Crop Res 105(3):240–252

FAO (2011) Climate smart agriculture managing ecosystems for sustainable livelihoods. Food and Agriculture Organization of United Nations

Fontana M, Bélanger G, Hirte J, Ziadi N, Elfouki S, Bragazza L, Liebisch F, Sinaj S (2021) Critical plant phosphorus for winter wheat assessed from long-term field experiments. Eur J Agron 126: 126263

Forte A, Fiorentino N, Fagnano M, Fierro A (2017) Mitigation impact of minimum tillage on CO_2 and N_2O emissions from a Mediterranean maize cropped soil under low-water input management. Soil Tillage Res 166:167–178

Gastal F, Lemaire G (2002) N uptake and distribution in crops: an agronomical and ecophysiological perspective. J Exp Bot 53:789–799

Global hunger index @ gobalhungerindex.org. Accessed on 8 Sept 2022

Guiteras R (2009) The impact of climate change on Indian Agriculture. PhD thesis. University of Maryland

Habtewold TM (2021) Impact of climate-smart agricultural technology on multidimensional poverty in rural Ethiopia. J Integr Agric 20(4):1021–1041

Heeb L, Jenner E, Cock MJW (2019) Climate-smart pest management: building resilience of farms and landscapes to changing pest threats. J Pest Sci 92:951–969

IPCC (2007) Climate change 2007: impacts, adaptation and vulnerability. Working group II contribution to the Fourth Assessment Report of the Intergovernmental Panel on Climate Change. Cambridge University Press. 978 0521 88010-7 Hardback; 978 0521 70597-4 Paperback

Islam MM, Urmi TA, Rana MS, Alam MS, Haque MM (2018) Green manuring effects on crop morpho-physiological characters, rice yield and soil properties. Physiol Mol Biol Plants 25:303–312

Jat ML, Gupta R, Saharawat YS, Khosla R (2011) Layering precision land levelling and furrow irrigated raised bed planting: productivity and input use efficiency of irrigated bread wheat in Indo-Gangetic plains. Am J Plant Sci 2:578–588

Jelagat J (2019) Effects of climate-smart agricultural awareness on food security among smallholder farmers: the case of Kaptumo-Kaboi ward, Nandi county. M.A. thesis. Faculty of Arts, University of Nairobi, Kenya

Jianbo L (2006) Energy balance and economic benefits of two agroforestry systems in northern and southern China. Agric Ecosyst Environ 116(3–4):255–262

Jose S (2009) Agroforestry for ecosystem services and environmental benefits: an overview. Agrofor Syst 76:1–10

Keil A, D'souza A, McDonald A (2015) Zero-tillage as a pathway for sustainable wheat intensification in the Eastern Indo-Gangetic plains: does it work in farmers' fields? Food Secur 7:983–1001

Khurana HS, Phillips SB, Bijay-Singh DA, Sidhu AS, Yadvinder-Singh PS (2007) Performance of site-specific nutrient management for irrigated transplanted rice in Northwest India. Agron J 99(6):1436–1447

Kotu BH, Admassie A (2016) Potential impacts of yield increasing crop technologies on productivity and poverty in two districts of Ethiopia. In: Gatzweiler FW, Braun JV (eds) Technological and institutional innovations for marginalized smallholders in agricultural development. Springer, Cham. https://doi.org/10.1007/978-3-319-25718-1_20

Kramer B, Ceballos F (2018) Enhancing adaptive capacity through climate-smart insurance: theory and evidence from India. 30th international conference of agricultural economist. July 28–Aug 2, 2018, Vancouver

Kumar S, Palanisami K (2011) Can drip irrigation technology be socially beneficial? Evidence from Southern India. Water Policy 13(4):571–587

Kumar SN, Aggarwal PK, Rani DNS, Saxena R, Chauhan N, Jain S (2014) Vulnerability of wheat production to climate change in India. Clim Res 59:173–187

Lal M (2010) Implications of climate change in sustained agricultural productivity in South Asia. Reg Environ Change 11:79–94

Lamanna C, Namoi N, Kimaro A, Mpanda M, Egeru A, Okia C, Ramirez-Villegas J, Mwongera C, Ampaire E, Van Asten P, Winowiecki L, Laderach P, Rosenstock TS (2016) Evidence-based opportunities for out-scaling climate-smart agriculture in East Africa. CCAFS Working Paper no. 172. CGIAR Research Program on Climate Change, Agriculture and Food Security (CCAFS). Copenhagen, Denmark

Lan L, Sain G, Czaplicki S, Guerten N, Shikuku M, Grosjean G, Laderach P (2018) Farm-level and community aggregate economic impacts of adopting climate smart agricultural practices in three mega environments. PLoS One:1–21

Lipper L, Thornton P, Campbell BM, Baedeker T, Braimoh A, Bwalya M, Caron P, Cattaneo A, Garrity D, Henry K, Hottle J, Ackso L, Jarvis A, Kossam F, Mann W, McCarthy N, Meybeck A, Neufeldt H, Remington T, Sen PT, Sessa R, Shula R, Tibu A, Torquebiau EF (2014) Climate-smart agriculture for food security. Nat Clim Chang 4:1068–1072

Lobell DB, Schlenker W, Costa-Roberts J (2011) Climate trends and global crop production since 1980. Science 333:616–620

Lu W, Addai KN, Ng'ombe JN (2021) Impact of improved rice varieties on household food security in Northern Ghana: a doubly robust analysis. J Int Dev 33(2):342–359

Mahato A (2014) Climate change and its impact on agriculture. Int J Sci Res Publ 4(4):1–6

Maia SMF, Otutumi AT, Mendonca EDS, Neves JCL, Oliveira TSD (2019) Combined effect of intercropping and minimum tillage on soil carbon sequestration and organic matter pools in the semiarid region of Brazil. Soil Res 57(3):266–275

Makowski D, Zhao B, Ata-Ul-Karim ST, Lemaire G (2020) Analyzing uncertainty in critical nitrogen dilution curves. Eur J Agron 118:126076

Malhi GS, Kaur M, Kaushik P (2021) Impact of climate change on agriculture and its mitigation strategies: a review. Sustainability 13:1318. https://doi.org/10.3390/su13031318

Malhi GS, Kaur M, Singh A, Singh VK, Saini SP, Jatav HS (2022) Agronomic and economic assessment of site-specific nutrient management in crop production. In: Jatav HS, Rajput VD (eds) Ecosystem services: types, management and benefits. Nova Science Publishers, USA

Manjusha K, Nitin P, Suvarna D, Vinaykumar HM (2019) Exposure, perception and advantages about weather based agro-advisory services by selected farmers of Anand district, India. Int J Curr Microbiol App Sci 8(5):1934–1944

Meena BL, Fagodiya RK, Prajapat K, Dotaniya ML, Kaledhonkar MJ, Sharma PC, Meena RS, Mitran T, Kumar S (2018) Legume green manuring: an option for soil sustainability. In: Meena R, Das A, Yadav G, Lal R (eds) Legumes for soil health and sustainable management. Springer, Singapore. https://doi.org/10.1007/978-981-13-0253-4_12

Meja MF, Alemu BA, Shete M (2021) Total factor productivity of major crops in southern Ethiopia: a dis-aggregated analysis of the growth components. Sustainability 13:3388. https://doi.org/10.3390/su13063388

Mendelsohn R (2014) The impact of climate change on agriculture in South Asia. J Integr Agric 13(4):660–665

Meng L, Zhou Y, Li X, Asrar GR, Mao J, Wanamaker AD, Wang Y (2020) Divergent responses of spring phenology to daytime and nighttime warming. Agric For Meteorol 281:107832

Mirzaei S, Kahrizi D, Hassan SS (2021) Climate change impacts on agriculture and food security; a global overview. Cent Asian J Environ Sci Technol Innov 2:184–197

Mittal S (2018) Role of mobile phone-enables climate information services in gender-inclusive agriculture. Gend Technol Dev 20(2):200–217

Mohanty S, Satyasai KJ (2015) Feeling the pulse: Indian pulse sector. NABARD Rural Pulse. 10. Department of Economic Analysis and Research

Mukherjee A, Dutta S, Goyal TM (2016) India's phytonutrient report: a snapshot of fruits and vegetables consumption, availability and implications for phytonutrient uptake. Academic Foundation, New Delhi

Murage AW, Midega CAO, Pittchar JO, Pickett JA, Khan ZR (2015) Determinants of adoption of climate-smart push-pull technology for enhanced food security through integrated pest management in eastern Africa. Food Security 7:709–724

Mushir A, Hifzur R, Murshid HS (2012) Status of food insecurity at household level in rural India: A case study of Uttar Pradesh. Int J Physic Soc Sci 2(8):227–244

Naresh RK, Singh SP, Misra AK, Tomar SS, Kumar P, Kumar V, Kumar S (2014) Evaluation of the laser levelled land levelling technology on crop yield and water use productivity in western Uttar Pradesh. Afr J Agric Res 9(4):473–478

NASA earth observatory (2020) Goddard Space Flight Centre, United states (www.earthobservatory.nasa.gov). Assessed on 15 May 2020

NFSM (2018) National Food Security Mission, https://www.nfsm.gov.in/StatusPaper/NMOOP2018.pdf

Nyiraneza J, Bélanger G, Benjannet R, Ziadi N, Cambouris A, Fuller K, Hann S (2021) Critical phosphorus dilution curve and the phosphorus-nitrogen relationship in potato. Eur J Agron 123:126205

Pandey A (2015) Food security in India and states: key challenges and policy option. J Agric Economic Rural Develop 2(1):12–21

Radhakrishnan K, Sivaraman I, Jena SK, Sarkar S, Adhikari S (2017) A climate trend analysis of temperature and rainfall in India. Clim Change Environ Sustain 5(2):146–153

Raj A, Jhariya MK, Bargali SS (2018) Climate smart agriculture and carbon sequestration. In: Pandey CB, Gaur MK, Goyal RK (eds) Climate change and agroforestry: adaptation mitigation and livelihood security. New India Publishing Agency (NIPA), New Delhi, pp 1–19

Raju KV, Naik G, Ramseshan R, Pandey T, Joshi P, Anantha KH, Rao AVRK, Moses, Shyam D, Kumara, Charyulu D (2016) Transforming weather index-based crop Insurance in India: protecting small farmers from distress. Status and a way forward. Research Report IDC-8. Technical Report. ICRISAT, Patancheru

Ramirez-Villegas J, Jarvis A, Laderach P (2013) Empirical approaches for assessing impacts of climate change on agriculture: the EcoCrop model and a case study with grain sorghum. Agric For Meteorol 170:67–78

Reddy GK, Reddy ST, Reddy PM (2017) Studies on contingent crop planning for rainfed alfisols. J Res ANGRAU 45(4):44–49

Rehmani MIA, Zhang J, Li G, Ata-Ul-Karim ST, Wang S, Kimball BA, Yan C, Liu Z, Ding Y (2011) Simulation of future global warming scenarios in rice paddies with an open-field warming facility. Plant Methods 7:41

Rehmani MIA, Wei G, Hussain N, Ding C, Li G, Liu Z, Wang S, Ding Y (2014) Yield and quality responses of two indica rice hybrids to post-anthesis asymmetric day and night open-field warming in lower reaches of Yangtze River delta. Field Crop Res 156:231–241

Rehmani MIA, Ding C, Li G, Ata-Ul-Karim ST, Hadifa A, Bashir MA, Hashem M, Alamri S, Al-Zubair F, Ding Y (2021) Vulnerability of rice production to temperature extremes during rice reproductive stage in Yangtze River Valley, China. J King Saud Univ – Science 33:101599

Salvagiotti F, Magnano L, Ortez O, Enrico J, Barraco M, Barbagelata P, Condori A, Di Mauro G, Manlla A, Rotundo J, Garcia FO, Ferrari M, Gudelj V, Ciampitti I (2021) Estimating nitrogen, phosphorus, potassium, and sulfur uptake and requirement in soybean. Eur J Agron 127:126289

Sapkota TB, Majumdar K, Jat ML, Kumar A, Bishnoi DK, McDonald AJ, Pampolino M (2014) Precision nutrient management in conservation agriculture based wheat production of Northwest India: profitability, nutrient use efficiency and environmental footprint. Field Crop Res 155:233–244

Sarkar D, Meena VS, Haldar A, Rakshit A (2017) Site-specific nutrient management (SSNM): a unique approach towards maintaining soil health. In: Rakshit A, Abhilash P, Singh H, Ghosh S (eds) Adaptive soil management: from theory to practices. Springer, Singapore. https://doi.org/10.1007/978-981-10-3638-5_3

Sarwar A, Ahmad SR, Rehmani MIA, Asif Javid M, Gulzar S, Shehzad MA, Shabbir Dar J, Baazeem A, Iqbal MA, Rahman MHU, Skalicky M, Brestic M, Sabagh AEL (2021) Mapping groundwater potential for irrigation, by geographical information system and remote sensing techniques: a case study of district lower Dir, Pakistan. Atmosphere 12:669

Sathaye J, Shukla PR, Ravindranath (2006) Climate change, sustainable development and India: global and national concerns. Curr Sci 90(3):314–325

Satpute NR, Dhadge SM (2016) Crop contingent planning for aberrant weather conditions under scarcity zone of Maharahtra. Adv Life Sci 5(7):2866–2870

Saxena NC (2018) Hunger, under-nutrition and food security in India. In: Mehta A, Bhide S, Kumar A, Shah A (eds) Poverty, chronic poverty and poverty dynamics. Springer, Singapore. https://doi.org/10.1007/978-981-13-0677-8_4

Schroeder P (1994) Carbon storage benefits of agroforestry systems. Agrofor Syst 27:89–97

Shikuku KM, Mwongera C (2019) Food security, downside risk, and resilience effects of agricultural technologies in northern Uganda. Invited paper presented at the 6th African conference of agricultural economists, September 23–26, 2019, Abuja, Nigeria

Singh B, Gupta RK, Singh Y, Gupta SK, Singh J, Bains JS, Vashishta M (2006) Need-based nitrogen management using leaf color chart in wet direct-seeded rice in northwestern India. J New Seeds 8(1):35–47

Stevanovic M, Popp A, Lotze-Campen H, Dietrich JP, Muller C, Bonsch M, Schmitz C, Bondirsky BL, Humpenoder S, Weindl I (2016) The impact of high-end climate change on agricultural welfare. Sci Adv 2(8):e1501452. https://doi.org/10.1126/sciadv.1501452

Stout DT, Walsh TC, Burian SJ (2017) Ecosystem services from rainwater harvesting in India. Urban Water J 14(6):561–573

Taylor M (2017) Climate-smart agriculture: what is it good for? J Peasant Stud 45(1):89–107. https://doi.org/10.1080/03066150.2017.1312355

Tesfaye S, Bedada B, Mesay Y (2016) Impact of improved wheat technology adoption on productivity and income in Ethiopia. Afr Crop Sci J 24(1):127–135

USAID (2020) Climate risk profile India: country profile. Fact sheet, United States

Vashisth A, Singh R, Das DK, Baloda R (2013) Weather based agromet advisories for enhancing the production and income of the farmers under changing climate scenario. Int J Agric Food Sci Technol 4(9):847–850

Venkatanarayana M (2020) Vegetable consumption in India: supply chain and prices. Munich Personal RePEc Archive. Paper no. 101979

Virk AL, Farooq MS, Ahmad A, Khaliq T, Rehmani MIA, Haider FU, Ejaz I (2021) Effect of seedling age on growth and yield of fine rice cultivars under alternate wetting and drying system. J Plant Nutr 44:1–15

Wilkes A, Wassie S, Odhong C, Fraval S, Dijk SV (2020) Variation in the carbon footprint of milk production on smallholder dairy farms in central Kenya. J Clean Prod 265:121780

Williams TO, Mul M, Cofie O, Kinyangi J, Zougmore R, Wamukoya G, Nyasimi M, Mapfumo P, Speranza CI, Amwata D, Frid-Nielsen S, Partey S, Girvetz E, Rosenstock T, Campbell BM (2015) Climate smart agriculture in the African context. Background Paper. Feeding Africa Conference 21–23 October 2015

World population review (2019) Walnut, United states (www.worldpopulationreview.com). Assessed on 12 May 2020

Yadvinder-Singh B-S, Ladha JK, Bains JS, Gupta RK, Jagmohan-Singh BV (2007) On-farm evaluation of leaf color chart for need-based nitrogen management in irrigation transplanted rice in northwestern India. Nutr Cycl Agroecosyst 78:167–176

Zhang K, Wang X, Wang X, Tahir A-U-KS, Tian Y, Zhu Y, Cao W, Liu X (2020) Does the organ-based N dilution curve improve the predictions of N status in winter wheat? Agriculture 10:500

Zhao B, Ata-Ul-Karim ST, Lemaire G, Duan A, Liu Z, Guo Y, Qin A, Ning D, Liu Z (2021) Exploring the nitrogen source-sink ratio to quantify ear nitrogen accumulation in maize and wheat using critical nitrogen dilution curve. Field Crop Res 274:108332. https://doi.org/10.1016/j.fcr.2021.108332

Chapter 18
Critical Appraisal and Evaluation of India's First Carbon Neutral Community Project – A Case of Meenangadi Panchayat, Kerala, India

Arunima KT and Mohammed Firoz C

Abstract Along with the sixth anniversary of the Paris Agreement on global climate change, a promising movement for carbon neutrality is taking form with the country's enormous efforts to realize its objectives. India is a significant emitter of greenhouse gases (GHG), generating 2.46 billion tons (6.8%) of carbon. The country needs to adopt a bottom-up approach involving each village, city, and town for a green economy. Hence, other than concentrating only on the country's major urban centers, there is a need to focus on the emerging urban centers and villages. Hence, the purpose of the study is to critically value India's first carbon-neutral community project that targeted developing the Meenangadi grama panchayat (Village) of Wayanad district within the southern state of Kerala, India as a model carbon-neutral panchayat. With an in-depth passive and participatory observation, this research analyzed every initiative taken by the core team to achieve the target within the planned timeframe. The overall performance level of the project was evaluated on the premise of appropriateness, affordability, and implementation rate. The analysis reveals that the project lacked coherence and effectiveness and could not achieve its goal within the timeframe. The result concludes that the project was satisfactory even though it fell short of expectations, but the positive results were dominating. Although Meenangadi did not succeed within the planned timeframe, it points the way ahead and would undoubtedly serve as a model to inspire heavily polluted Indian cities and emerging urban centers to achieve carbon-neutral status.

Keywords Critical appraisal · Carbon neutrality · Carbon-neutral community · Climate change

A. KT (✉) · M. Firoz C
Department of Architecture and Planning, National Institute of Technology Calicut, Kozhikode, India
e-mail: arunima_m200355ar@nitc.ac.in; firoz@nitc.ac.in

Introduction

Human activities have been the leading cause of climate change since the nineteenth century, mainly due to burning fossil fuels such as coal, oil, and gas (UN, Climate Change 2021). Burning fossil fuels creates greenhouse gas emissions that act like a blanket wrapped around the earth, trapping solar heat and increasing temperatures (UN, Climate Change 2021). The connection between greenhouse gas (GHG) emissions and climate change is a scientific certainty (Pat Ganase 2015) and the concept of carbon neutrality combats this phenomenon. Every aspect of our lives is related to the environment, from food, transportation, and energy to waste management. The way we produce and consume makes a negative footprint on our planet. "Carbon neutrality reduces the negative impacts on the environment to achieve a net result of zero emissions. Carbon neutrality describes a state in which the GHG emissions released to the atmosphere by a stakeholder (individual, organization, company, country, etc.) have been reduced or avoided, and the remaining ones are compensated with carbon credits" (CNN Guidelines 1995). Removing and storing carbon dioxide from the atmosphere is known as carbon sequestration (CLEAR Center 2019). For achieving net-zero emissions, all global greenhouse gas (GHG) emissions must be balanced by carbon sequestration (europarl 2021).

Paris Agreement and Carbon-Neutral Missions

A global assembly of scientists on climate change in Villach, Austria, held in October 1985, concluded that growing concentrations of greenhouse gases would lead to a historical upward push in worldwide temperature (Oppenheimer Michael 2007). The meeting also raised questions on directing these concerns to influence government action worldwide. As a result, with the help of US government support, the Intergovernmental Panel for Climate Change (IPCC) got established by the United Nations Environment Programme (UNEP) and World Meteorological Organization (WMO) in the year in 1988. IPCC issued its first assessment in 1990, stating that human activity would produce unparalleled warming (Oppenheimer Michael 2007). Since then, four such evaluations have been released by IPCC; and these reports led to international policy developments such as the UN Framework Convention on Climate Change (UNFCCC), Kyoto Protocol, and the Paris Agreement. Paris Agreement is a historic deal to fight global warming adopted by 196 countries at COP 21 in Paris on December 12, 2015, and entered into force on November 4, 2016. The Paris Agreement had three main objectives: restricting the average global temperature rise below 2 °C, enhancing resilience, and aligning economic flows towards climate-resilient developments (Biniaz Sue 2020).

The Paris Agreement is a milestone in the various climate change processes because, for the first time in history, it brought all nations into a common cause to undertake ambitious efforts to combat climate change and adapt to its effects (longer

Articles 2021). The Paris accord talks about reducing greenhouse gases emitted to the same level of carbon sequestration. In this context, carbon neutrality was proposed as a condition in which CO_2 removals balance CO_2 emissions (Dhanda and Hartman 2011). The agreement mentions the need to review each country's efforts to reduce GHG emissions every five years. The Paris Agreement adopted a bottom-up structure compared to most international environmental law treaties, which have a top-down approach. The national pledges by countries to reduce emissions are voluntary, but the agreement needs all parties to put optimum efforts through Nationally Determined Contributions (NDC).

Since the Paris agreement, bold visions for cutting global warming have emerged globally. With several governments pledging to become carbon neutral by mid-century, a solid carbon-neutral strategy is required. India is the third largest contributor of GHG emissions and aims to become carbon neutral by 2070. China, the second most significant contributor after the US, targets carbon neutral by 2060. Both China and India are focusing on reducing the share of fossil fuel-based energy production and increasing the share of renewable energy. Increasing the area of carbon sink within the country by afforestation programs is also being considered. Wanting to become carbon neutral by the middle of the century, South Korea has planned to invest $7 billion in green projects. They also plan to launch a carbon tax on companies to reduce emissions and discourage funding of overseas coal plants (Bazilian Morgan 2020).

Indian States and Decarburization[1] Pathway

According to the Press Statement given on 2 October 2015, by the Ministry of Environment, Forest, and Climate Change, Government of India, the NDC of India is comprised of eight goals: (a) Sustainable Lifestyles, (b) Cleaner Economic Development, (c) Reduced Emission intensity, (d) Increased renewable energy, (e) Enhancing Carbon Sink (Forests), (f) Adaptation, (g) Mobilizing Finance, (h) Capacity Building & Technology Transfer. India has committed to reducing its GDP's greenhouse gas emission intensity by 2% by 2030, doubling its renewable energy capacity by 2015, and creating an additional 2.53 billion tons of carbon sinks (MOEF 2015). India has several ambitious energy efficiency measures taken in various industries to achieve lower emissions intensity in the automotive and transport sectors. Thrust is on renewable energy, climate-resilient cities, and sustainable green mobility networks. Solar energy in India has grown remarkably as solar missions are significant initiatives taken by the Indian government. Urban transport policies promote the mobility of people instead of vehicles with a

---

[1] "Decarbonization means zero net emissions of CO_2 – as well as the stabilization of emissions of short-lived greenhouse gases such as methane that dissipate in the atmosphere in days, weeks, or decades" – World bank group.

substantial focus on Mass Rapid Transport Systems (MRTS). In addition to the 236 km of installed rail metro, around 1150 km of metro projects are planned for Pune, Ahmedabad, and Lucknow (MOEF 2015). Delhi Metro was India's first MRTS project to earn carbon credits and reduce about 0.57 million tons of CO_2 equivalent annually (MOEF 2015). India also plans to use renewable energy sources on the way to a cleaner environment, energy independence, and a more resilient economy. Forests provide regulatory services such as carbon sequestration and storage. India also has the long-term goal of increasing its forest area through a planned afforestation campaign, various programs and initiatives such as the India Green Mission, Green Road Policy, financial incentives for forests, plantations along rivers, REDDPlus, and other measures, and the management of compensatory afforestation funds and planning authority (CAMPA) (MOEF 2015).

Under decentralized planning,[2] states were supposed to prepare the adaptation plan, but the central government decided on renewable energy, energy efficiency, and carbon intensity targets. Several Indian states have taken up renewable energy initiatives, enhanced energy efficiency, and commitments to emissions reduction. Climate action momentum in Indian states includes state action plans for climate change, LED village campaigns, solar and wind policies, state-level Energy Conservation Building Code (ECBC) policies, electric vehicle policies, and afforestation drive (Singh 2019). According to a study conducted by Klynveld Peat Marwick Goerdeler (KPMG) International Limited for The Climate Group, Kerala is one of the top 10 high-performing states in terms of the climate action plan. Gujarat and Tamil Nadu are in the spotlight with the successful implementation of proactive climate policies. Karnataka has the highest Energy Efficiency Index, followed by Gujarat (Kumar et al. 2020). Examples of top-down and bottom-up approaches can be witnessed in the states of India. In 2017, the Bihar government announced the installation of 3433 megawatts (MW) of renewable energy by 2022.

However, the state has only installed 194 MW of renewable capacity in the past four years, which is only 5% of the target, and the state still has to reach the remaining 95% by next year (Pandey Kundan 2011). Pioneering initiatives and voluntary commitments are being explored in different parts of the country to achieve carbon neutrality. For example, infrastructure developments in Ladakh focused on solar parks, geothermal energy from hot water springs, and micro-irrigation using solar power set examples of ecosystem service-based solutions (Verma Ayush 2020). Sikkim, India's first fully organic farming state, is also keen on retaining a carbon-neutral form by focusing on climate inventory and monitoring systems (Avni Agarwal 2021). The Metropolitan city of Bangalore is another sub-region aiming to achieve carbon neutrality by 2050 in an inclusive and participatory manner (Malvika Kaushik 2021). With towns and sub-regions coming forward to cut carbon emission, cross-sectoral coordination mechanisms, financial models, and inclusive planning framework are the need of the hour.

---

[2]Local organization responsible for formulating, adopting, and executing actions and plans for betterment of the locals without intefernce from central government.

Rationale of the Study

India lacks a strategic transferal method to translate the national-level target to a state-level target; hence, the gap widens between the announcement of big goals in the central government and the implementation (Pandey Kundan 2011). Kerala provides an example of how the participatory model, a bottom-up approach, can make a big difference at lower levels of governance. Helping India achieve its goals Meenangadi grama panchayat[3] of Wayanad District in Kerala has set a target to become India's first "carbon-neutral panchayat." The panchayat has integrated climate-resilient strategies into its development plans in various sectors and realigned development projects to achieve carbon neutrality. The concept of carbon-neutral panchayat presents developing zero-carbon and self-preserved food and energy at the local government level. Detailed studies are yet to be conducted to critically evaluate and elucidate the need to translate the national-level goals into regional goals. In this context, this chapter assesses the first carbon-neutral panchayat project in India and formulates a framework for transferring national level targets to state and local self-government levels. The objectives of the study were as follows:

1. To explore the carbon neutrality concept, significance, and relevance in the planning and management of cities and settlements.
2. To systematically assess the carbon-neutral Meenangadi project initiatives and activities of Meenangadi panchayat.
3. To propose a futuristic vision for translating India's carbon-neutral commitments to sub-regional levels.

Background Study

This section narrates how the project was conceptualized and describes the site's ecological and socio-economic features.

Conceptualization of the Project

Climate change is everyone's problem, and global *warming* impacts *everyone's* food and water security. In 2015–2016 locals of Meenangadi were victimized by climate change. The major crops cultivated in Meenangadi include tea, pepper, and spices, all of which are climate-sensitive (Santhoshkumar 2010). With the rise in temperatures, they had a significant dip in the harvest (Prasad 2014). Along with the government, locals wanted to tackle the situation but did not have the know-how.

---
[3]Functional institutions of grassroots governance in villages or small level towns in India.

Fig. 18.1 Site location. (Source: Author generated)

At the same time, Dr. TM Thomas Isaac, then finance minister of the government of Kerala, returned from the United Nations Climate Change Conference (COP 21) held in Paris with an ambitious vision. Soon after his return, letters were sent to all the panchayat in Kerala asking them to express interest in a state-sponsored carbon-neutral pilot project. The Meenangadi panchayat authority came forward and decided to take up this ambitious project. On June 5, 2016, the world environmental day, Meenangadi panchayat (Fig. 18.1) in Wayanad district of Kerala, India, pledged to become carbon neutral by 2020. Organizations like Kannur University, Swaminathan Research Foundation, and a local Non-Governmental Organization (NGO) named "Thanal" joined hands with the panchayat to flag the project. With the leadership of the NGO Thanal and the help of locals, detailed study and assessment of the site were carried out. Multiple surveys were carried out to calculate carbon footprint, energy audit, waste audit, and survey of household tree (The climate group 2020).

The Setting of the Study

Meenangadi panchayat is a part of the Western Ghats (Tropical Evergreen Forest Regions) – a biodiversity hotspot and world heritage site. It is a land of locals and tribal communities at a relatively lower altitude away from the tourist crowds. It covers an area of 53.52 km^2, with agricultural area dominating the region with 49.2 km^2. Meenangadi panchayat houses a population of 34,601 with a population density of 646 per km^2 (Census 2011). As a grama panchayat, it has less local autonomy but has a solid organizational structure with three standing committees for Development, Health & Education, and Welfare (Fig. 18.2). They also have a strong group of implementing officers led by the panchayat secretary (J.B.Rajan 2017). The team includes an Agricultural Officer (AO), Medical Officer (MO), Assistant Engineer (AE), Headmaster (HM) of the local government school, Village extension officer (VEO), and Integrated Child Development Services (ICDS) coordinator. The panchayat has 19 wards, and elected ward members from these 19 wards select the committee members. 70% of the panchayat area is a plateau and 20% fertile. The forest covers almost 1.45 km^2 (3%) of the total area in mountain valley patterns (Thanal 2018). The primary ecosystem types that predominate in Meenangadi grama

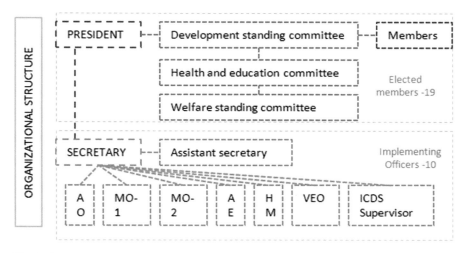

Fig. 18.2 Organizational structure of Meenangadi panchayat. (Source: Authors generated)

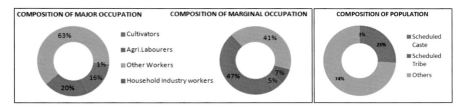

Fig. 18.3 (**a**) Occupational composition. (**b**) Population composition. (Source: J. B. Rajan 2017)

panchayat are: wetlands, forests, plantations, and farms. panchayat is located in two revenue villages, Krishnagiri and Purakkadi (Thanal 2018). Meenangadi panchayat has four rivers and 23 streams and a good aqueduct connection to the Karappuzha Dam (Thanal 2018). The panchayat also has an elevation of 773 m and has a relatively flatter topography compared to the surrounding area. The panchayat has a welcoming and affirming environment.

The occupational composition of Meenangadi panchayat clearly shows that 47% of the marginal occupation comes under the agricultural sector (Fig. 18.3). The actual wages in coffee plantations and paddy cultivation attract locals to prefer agricultural labor. The tribal population accounts for about a quarter of the total population of the panchayat (Fig. 18.3). Paniyar, Kurichyar, Kurumar, Kaattunaykkar, and Vettakkurumar (Urali) are the tribal communities of Meenangadi (J.B.Rajan 2017). These tribal communities are close to nature and work to preserve it as they depend on nature for their livelihood. The literacy rate of the panchayat is 90%, which is at par with the literacy rate of Wayanad District (89.2%) and Kerala State (93.91%) (Census 2011).

Meenangadi has a tropical climate with significant rainfall around the year. The average annual temperature is 23.2 °C (73.8 °F), and precipitation is 2944 mm (115.9 inches) per year. The warmest month is April, with an average temperature of 25.6 °C (78.1 °F), whereas December is the coldest month, with temperatures averaging 21.7 °C (71.1 °F). January is the driest month, with 4 mm (0.2 inches) of rainfall. In July, the most precipitation falls, averaging 1011 mm (39.8 inches). State Action Plan on Climate Change categories Wayanad district under hotspot region and warns of a rise by 2 °C– 4.5 °C by 2050. Temperature rise would affect the production of various staple climate-sensitive crops cultivated in Wayanad, such as coffee, tea, pepper, and cardamom.

Land Use Details

The local government, Meenagadi panchayat, and the local NGO Thanal have carried out a detailed assessment report to assess the settlement's carbon neutrality (Thanal 2018). The assessment has undertaken a detailed spatial analysis to understand the land use pattern of the area (Fig. 18.4). Accordingly, 93% of the area was agricultural land and less than 1% fallow land, and the remaining 6% was dedicated to non-agricultural activities. Hence agricultural activities dominate the landscape of the panchayat. The cultivation of perennial crops takes up approximately four thousand six hundred seventy-four hectares of land, 60% of which is coffee on farms and other mixed crops. Large coffee farms occupy six percent of the perennial crops (dark green). Around 1005 hectares of land have been used for seasonal,

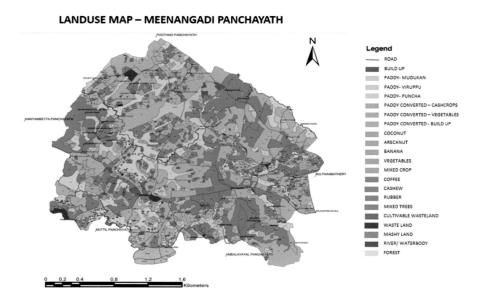

Fig. 18.4 Land use map. (Source: Thanal 2018)

climate-sensitive crops, mainly rice and bananas (Thanal 2018). Ginger is another critical thermo-sensitive crop that takes up about five percent of cultivated land. For economical and climatic reasons, cash crop cultivation slowly expands from farmlands to rice fields (Archana 2021). The reduction in paddy fields is negatively affecting the groundwater table of the region. A minimal area is commercialized as a town area and has less built-up area throughout the settlement, and rural lifestyle characteristics dominate existing built-up.

Methodology

The section investigates and systematically describes the conceptualization, assessments, and programs initiated as a bottom-up approach to assessing India's first carbon-neutral project (Fig. 18.5). An explanatory system is adapted to identify the project's performance level, which has never been studied before. With an in-depth passive and participatory observation, this chapter has analyzed every initiative taken by the Meenangadi panchayat and the core team with the support of locals to achieve the target within the planned timeframe. Remote studies were based on documents published by the panchayat authority, assessment and recommendation reports by the core team, and articles from local and national newspapers. The content-based analysis helped study the site's relevance, scope, and limitation of the project and critical issues faced during the implementation stage. Each program initiative was evaluated based on appropriateness, affordability, and implementation rate to conclude the performance level and success rate.

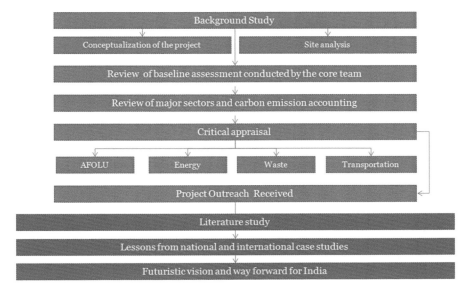

Fig. 18.5 Methodology

Critical Appraisal

Baseline Assessment and Recommendations

In 2016–2017, a baseline assessment study was conducted in Meenangadi Grama Panchayat to assess the greenhouse gas emissions and sequestration profile (Thanal 2018). The national inventory of GHG emissions lacked research at the regional or district level. Therefore, there was no established method for assessing "carbon status." The team, led by Thanal, followed a bespoke research methodology to estimate greenhouse gas emissions at the panchayat level by strictly following the IPCC guidelines for each sector identified (Thanal 2018).

The assessment estimated total carbon emission for sectors like Agriculture, Forestry, and Other Land Use (AFOLU), energy, transportation, and waste (Table 18.1: Total GHG emission from Meenangadi grama panchayat (Source: Thanal 2018)). The study also estimated carbon sequestration for different ecosystem types (Table 18.2). Emission inventory was limited to methane, carbon dioxide, and nitrous oxide. Similarly, carbon sequestration estimations only included soil carbon and carbon stored in forests, homesteads, and plantations.

The study used simple average-based projections and concluded that Meenangadi must offset fifteen thousand tons of CO_2 equivalent to become carbon neutral by 2020. Based on the baseline assessment study, the carbon balance was estimated to be 11,412.57 tons of CO_2 equivalent (Thanal 2018). Future emissions were also calculated to plan and achieve net zero emission.

The whole project was outlined into two phases. Phase 1 was baseline assessment, which studied Meenangadi panchayat in detail to account for carbon emissions

Table 18.1 Total GHG emission from Meenangadi grama panchayat

Sector	Emission in CO_2Eq tons
Transportation	14910.700
Energy	13082.560
Waste	948.100
AFOLU	4433.739
Total	**33375.099**

Source: Thanal (2018)

Table 18.2 Total carbon sequestrated in Meenangadi grama panchayat

Sector	Sequestrated CO_2 Eq (tons)
Homestead coffee	4975.57
Large coffee plantation	445.28
Forests	8370.08
Homestead trees	7425.22
Soil	746.38
Total	**21962.53**

Source: Thanal (2018)

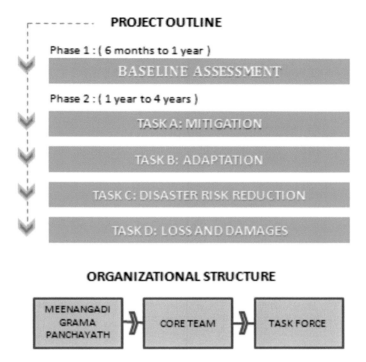

Fig. 18.6 Proposed project Outline and organizational structure carbon-neutral Meenangadi. (Source: Author generated based on (Concept 2016))

and identifies significant emission sectors and existing carbon stock. The project's second phase, with strategies and recommendations under four tasks listed below (Fig. 18.6), was formulated for the next four years.

Mitigation measures focused on industries and technology, energy and transport, integrated resource and waste management, soil and water. Strategies were to create green jobs by shifting to eco-friendly technologies (Concept 2016). These measures contemplated gender-equitable livelihoods and empowering informal sectors. At the same time, adaptation measures aligned along with industries like agriculture and food, responsible tourism, forests, and biodiversity. Disaster risk reduction measures included capacity building, awareness creation, and making people resilient to extreme climate events. Task-D focused on developing a framework to monitor the loss and destruction by climate change in the region. Major projects were solar lighting, energy-efficient stoves for households, biogas plants, Energy Park, source-level aerobic composting, material recovery and recycling, tree-banking, and Climate Emergency Response Units (CERU) (Thanal 2018).

Sector-Wise Critical Appraisal

Meenangadi panchayat implemented the carbon-neutral project as a comprehensive project that encompasses all sectors. With financial support from the government of Kerala, the "Tree Bank Scheme" became the main project of the Meenangadi Carbon Neutral Initiative (Thanal 2020). The panchayat had earmarked Rs 110 million for this ambitious project, with Rs 100 million allotted from the state government budget (Vishnu Verma 2020). Meenangadi panchayat has an average income of Rs 30 million per year, out of which 5.3 million are own funds generated through various taxes. Panchayat smartly integrated other government-funded projects into carbon-neutral proposals making it economical to spend on the project initiatives. However, several of the projects advocated by the Thanal committee, such as a solar park (about four crores), LED light production unit, electric car distribution, and so on, are not feasible and were beyond the scope of the panchayat fund. Panchayat authority and the technical support from Thanal undertook programs and project initiatives in all the sectors identified during the baseline assessment (Thanal 2018). Through the resource center and its intended activities, the core team aimed to enhance public participation, expand knowledge sharing and technology transfer, create a platform to share success stories and learnings, and also information dissemination, outreach, and awareness programs.

The following section critically details every initiative taken by the panchayat under the categorized sectors and evaluates the same based on appropriateness, affordability, and implementation rate. Accordingly, four sectors were critically reviewed.

Sector 1: AFOLU (Agriculture, Forest, and Other Land Use)
Under this sector, four projects were identified, namely the tree banking project, Punyavanam project, Rainwater harvesting initiatives, and the Attakolli' Jaiva park (Sakshi Krish 2017).

Tree Banking Project
The tree banking project, under the sector of forest and plantation, was considered to be the backbone project of the carbon-neutral Meenangadi Initiative. If implemented well, the project was estimated to have the capacity to offset 15,000 tons of CO_2 eq. The tree banking project offers interest-free loans to farmers that they do not need to repay if the trees are uncut (cris 2021). According to the assessment, around 600 thousand trees were required to create additional sequestration capacity (Thanal 2018). Panchayat planted more than three lakh trees from 2017 to 2020. The team used an app called Tree Banking to map the coordinates on Google Maps. Photos, species, and ID numbers were digitally documented and geo-tagged to keep track of the trees (Fig. 18.7). However, due to unforeseen interruptions tree survey and loan process were not delivered as planned. Only 157 farmers from two wards (Choothupara and Appad) were identified as beneficiaries of the first payment installment by October 2020 (cris 2021). The tree banking project of Meenangadi was also considered an experimental project that would mark the beginning of thinking of trees as agricultural products. Eventually, National Bank for Agriculture

Fig. 18.7 Tree survey. (Credits: Ajith Tomy, Thanal)

and Rural Development (NABARD) and other financial institutions may start offering loans by considering tree security.

This project was regarded as one of the most appropriate nature-based solutions to climate change. This ecological experiment encouraged people to plant more trees and safeguard them. After three years of tree planting, farmers could avail the loan, which induced them not to cut the plants short term (Rajendran 2022). The plant sampling was carefully selected based on their environmental suitability, and many public and private nurseries joined hands with the panchayat for this project. The project gave incentives to let the trees survive and eventually increase the carbon sequestration and storage in trees and soil.

Regarding affordability, the project did not face any financial issues as the state government deposited 100 million in the cooperative bank to guarantee the incentives (The times of India 2020). Intergenerational equity in private land was one of the significant challenges during implementation (Isaac 2020). Also, the samplings supplied by the panchayat were of long-gestation trees, which did not benefit the current generation. Only 300 thousand samples were distributed, representing 50% of the total required trees (Neha Bhatt 2020). Due to unforeseen issues, the tedious tree survey for initiating incentives got delayed beyond the planned timeframe. The project formulated was entirely in line with the action plans, but the overall performance level of the project was insufficient.

Punyavanam Project

As part of redeveloping forest land, the Punyavanam (holy forest) project was initiated by the panchayat authorities in collaboration with the village temple devasom board.[4] Under the project, around 38 acres of land adjoining the Manikav temple was restricted as a sacred forest (Fig. 18.8) (M S Swaminathan Research

---

[4] Socio-religious trusts in India. Members of the trutst are nominated by the government and the neighbourhood.

Fig. 18.8 View of Punyavanam adjoining Manikav temple. (Credits: Authors)

Foundation & Community 2018). The team planted various plant sampling in a phased manner with the support of the Mahatma Gandhi National Rural Employment Guarantee Act[5] (MNREGA) team and the forest department. Today the Punyavanam has turned into a significant carbon sink in the area and symbolizes the togetherness of Meenagadi's people toward the carbon-neutral project.

Punyavanam project got completed as per the planned timeframe, and today it is one of the greenest localities in the whole panchayat. Under the protection of locals and the temple authorities, the 38 acres of land turned into a lush green zone. This initiative had outstanding results and exceeded expectations.

Rainwater Harvesting Initiatives
Panchayat undertook various nature-based water harvesting solutions an d increased the region's groundwater levels. Panchayat dug 456 individual ponds with the help of the MNREGA team in different parts of the panchayat. Out of the 456,423 ponds were used for fish cultivation (Basil Poulose 2020) (Fig. 18.9). The public immensely supported this initiative and also helped in successfully implementing it.

Attakolli' Jaiva Park
Attakolli Jaiva park for organic farming was one of the initial programs under the carbon-neutral project. This project brought self-help groups like kudumbashree[6] workers under one umbrella, attempting to reverse the use of pesticides to cultivate fruits and vegetables. The project also aimed at making Meenangadi panchayat self-sufficient in food production. Around 4.7 million (Basil Poulose 2020) was spent on the program, and it was the most successful program under the carbon-neutral Meenangadi mission.

---

[5] A labour law in India which guarantees "right to work".

[6] 'Kudumbashree means "prosperity of the family" – it's a Kerala state government mission to eradicate poverty and empower women.

Fig. 18.9 Natural pond dug for water harvesting. (Credits: Authors)

The core team also wanted to enhance the livelihood of the farmers by branding and certifying the products. Certification of the products of Meenangadi under the carbon-neutral tag would fetch a better income for the farmers. The initiative is in progress and would succeed if more state and national institutions joined hands with the panchayat. Another extended project under the list is a responsible tourism plan. The team has already conducted various awareness campaigns and has had positive results.

Sector 2: Waste Management (Zero Waste Project)
Waste production in Meenangadi panchayat contributes 3% of the GHG emission (Thanal 2018). Zero waste projects were proposed to offset 100% of GHG emissions from the waste sector. Aligned with India's solid waste management guidelines, source segregation of waste was followed by the collection of biowaste and composting. Under the initiative of the Meenangadi panchayat, two units (4 bins) of the aerobic composting unit were installed in the town's center, near Krishi Bhavan (Mission 2018). The powdered waste from these composite units was then used as manure.

Haritha karma sena,[7] a kudumbashree team, is responsible for collecting household waste and waste from commercial areas of Meenangadi. They installed plastic collection bins in every neighborhood and monitored them weekly (Fig. 18.10). Panchayat also trained 38 women in the Haritha Karma Sena, two from each ward, as a part of capacity building (cris 2021). Collected plastic from houses and organizations was divided into two. The non-recyclable ones were shredded in the panchayat's shredding unit.

In contrast, the recyclable ones were sold to Clean Kerala Company,[8] which became a small source of income for the panchayat. They organized various awareness campaigns on scientific waste management, reducing plastic usage, and sustainable resource consumption. The awareness campaigns helped change the

---

[7] The green task force of every local body, an integral part of the waste management system in the State of Kerala.

[8] A company formed under LSG department to ensure hygiene management of Kerala state.

Fig. 18.10 Plastic waste collection bins placed under zero waste initiative. (Credits: Authors)

behavioral economics of the people and reduce the amount of waste generated. Currently, Haritha karma sena has 32 workers; according to them, people are more responsible nowadays, and plastic debris collected from the household has decreased commendably (Eco India 2020). Panchayat also promoted household bio-composite, and composite bins were distributed in every ward under subsidy schemes.

The zero waste project was also successful and generated green jobs and finance for the panchayat. The project increased the hygiene of the market areas with scientific waste management measures. Incorporating LSG-level regulation on waste management would help the authorities continue this initiative for an extended period. Overall this project had positive results and fewer shortcomings.

Energy Management

Panchayat focused on reducing wastage of electricity, encouraging electric autos and solar energy programs. Two electric vehicles were distributed to the kudumbashree by the municipality. The village crematorium, which once used wood, was promoted into machinery which runs on LPG, a much cleaner fuel. Panchayat also conducted awareness programs with stakeholders to promote the use of electric auto.

A design for a commercial hub (shopping complex and panchayat buildings) with solar panels installed on the terrace was made with the help of engineering students but was not implemented due to funding insufficiency. The core team also proposed solar street lighting and replacing incandescent bulbs with led bulbs to reduce electricity consumption. It was estimated that LED bulbs could reduce 11% of CO_2 emissions from the Energy Sector (Thanal 2018). A LED bulb manufacturing unit was set up in the locality, and women were trained for green jobs (Ameerudheen 2017). Energy-efficient LED bulbs were distributed to all households in the region in collaboration with the Kerala state electricity board. The core team also focused on converting waste into energy. The cattle population of Meenangadi is more than 2000, hence bio-gas plant was installed in selected households.

The energy sector is an important sector contributing 39% of carbon emissions. Projects and programs under this sector were beyond a local government's funding capacity. Most of the projects remained just in papers and did not get implemented. Overall, the initiative in this sector was inadequate and lacked clarity on implementation mechanisms.

Transportation (Promotion and Conversion to EVs)

According to the baseline assessment, the transportation sector is the most significant contributor to greenhouse gas emissions in Meenangadi. Emission inventory in the transportation sector was limited to the internal traffic of Meenangadi. Panchayat initiatives focused on reducing fossil fuel dependency by introducing electric vehicles and promoting shared and mass transportation. Awareness programs promoting the use of EVs were initiated. Auto drivers were encouraged to shift into electric autos, but the high ownership cost and inadequate no. of loans for electric vehicles were the main barriers to this initiative. The projects formulated under this sector are innovative and modern, and panchayat requires support from both public and private sectors to implement them successfully.

A summary of the critical assessment outlined above is provided below (Table 18.3). Targets achieved were analyzed based on passive and participatory observations. The study compared the goals set and attained to determine the overall performance of each project (Fig. 18.11). The study revealed that the project lacked coherence and effectiveness and could not achieve its goal within the timeframe. The result concludes that the project was satisfactory even though it fell short of expectations.

Project Outreach

Knowledge sharing and technology transfer at all levels of governance was one primary goal of the carbon-neutral project. The district administration greatly appreciated the project and scaled it up to the Wayanad district level with an extended target to sequester more than 600 thousand tons of CO_2 eq. The district-level mission focused on solar power generation, forestry, and agricultural interventions in all

Table 18.3 Summary table of critical appraisal

Targets formulated	Targets achieved	Critical appraisal
Feasibility and baseline assessments		
Carbon accounting and assessments.	Social preparation, engagement, and outreach with the help of Thanal (resource center). Developed a carbon estimation framework for LSGs.	Carbon accounting only took CO_2, N_2O, and CH_4 into account; and ignored other greenhouse gases. Carbon accounting for the transportation sector was limited to vehicles within the Meenangadi panchayat. Ecological data has to be documented at the LSG level for better assessment and accounting of carbon stocks.
Inventory of data and records to monitor the impact of the project.	Inventory of birds and trees grown in homesteads were documented.	Unforeseen interruptions delayed the assessment process, and the tree survey was only conducted for 2 out of 19 wards in Meenangadi panchayat.
Projects/schemes/initiatives		
SECTOR 1: AFOLU		
Tree banking project.	Three hundred thousand trees were distributed. A tree banking app was developed. The tree survey and loan process was not delivered as planned.	Natural disasters, pandemics, and state government elections were the many interruptions during the project implementation. A tree survey is a mammoth task that involves field surveys, mapping, and documentation. A fully developed tree banking app with automated data collection and evaluation would reduce the task in future. This app can also connect farmers and financial institutions more efficiently.
Punyavanam project.	Four hundred thousand trees were planted in a prime location within the panchayat and converted into a thick forest area.	Conservation of forests is essential for the carbon-neutral trajectory of development. Concerned authorities must draft laws and regulations to protect the generated carbon sink.
Rainwater harvesting.	More than 450 ponds were dug in the panchayat for water conservation.	The ponds were dug with the support of the MNREGA team and maintained by kudumbashree workers. The community participation and support make this project a successful one.

(continued)

Table 18.3 (continued)

Targets formulated	Targets achieved	Critical appraisal
Promotion of agro-ecological farming (Attakolli' Jaiva park).	Seventy acres of land are used for organic cultivation. Self-sufficiency in vegetable production.	Massive support from the locals of the region was the main reason for the successful completion of the project. Innovative farming options and skill development for farmers can boost the livelihood of farmers in the region.
Branding, certification, and marketing of organic products.	Not achieved yet.	The project has been pushed forward and will be implemented soon. *Certified organic products are generally more expensive than their conventional counterparts;* hence, this project would increase income generation from the agriculture sector.
Responsible tourism.	Awareness drives conducted. Restricted use of plastics in tourist spots.	Not a tourism-rich panchayat.
SECTOR 2: WASTE		
Zero waste project.	Two units of the aerobic composting unit were installed. Generated green jobs and finance for the panchayat.	Joined efforts Haritha karma Sena and the core team emerged fruitful in the successful implementation of the zero waste project.
SECTOR 3: ENERGY		
Energy park.	The architectural design for a commercial hub with solar panels installed on the terrace was made with the help of engineering students. Cost estimation for the project was carried.	The cost estimated for the project (40 million) was beyond the scope of the panchayat fund. Meenangadi panchayat need to focus on innovative funding mechanism to implement such projects. Public and private partnerships can be encouraged for such projects.
Biogas plant.	Small biogas plants were built in selected households to turn organic waste into energy.	The panchayat has more than 5000 cattle and has a high potential for generating energy from waste. The project can be better implemented and managed through self-help groups like the kudumbashree unit.
Led lamp manufacturing unit.	Set up led manufacturing unit. Trained women in green jobs.	The new manufacturing unit has created jobs for the locals.

(continued)

Table 18.3 (continued)

Targets formulated	Targets achieved	Critical appraisal
Renewable energy projects.	The village crematorium, which once used wood, now runs on liquefied petroleum gas. Solar streetlight not implemented. Awareness programs were conducted.	Geographical restrictions for the installation of wind turbines and solar panels. Lack of awareness as well as lack of financial options for house-owners.
SECTOR 4: TRANSPORTATION		
Green transport initiatives.	Two electric vehicles were distributed to the kudumbashree by the municipality.	The primary challenges for the project are High ownership cost. Very few banks provide loans for electric vehicles. Limited support from state and national levels.
Awareness campaigns.	Awareness programs were conducted with stakeholders to promote the use of electric auto. Awareness campaigns for reducing the use of plastics.	Awareness campaigns, inception workshops, and participatory discussions positively impacted the locals. Usage of plastics has reduced considerably.

Source: Authors generated

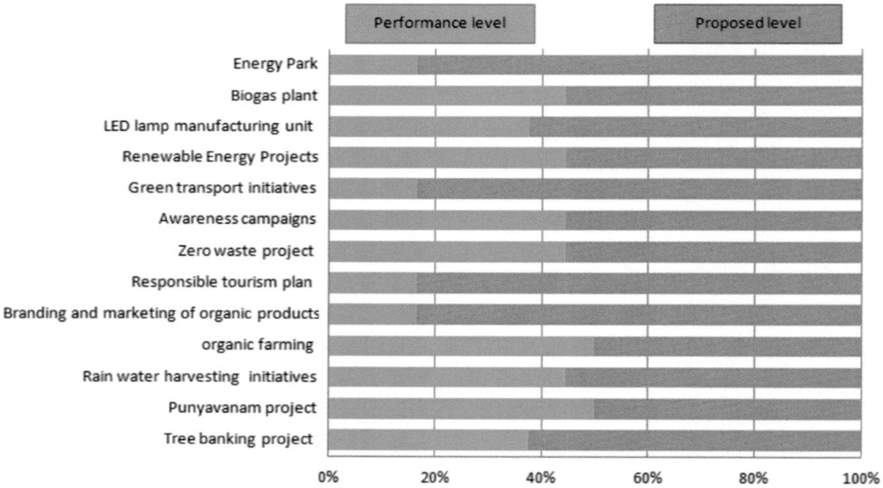

Fig. 18.11 Performance level of carbon-neutral project initiative. (Source: Authors generated based on passive and participatory observations)

local bodies. The Meenangadi Grama panchayat was selected in Green Kerala express reality show for its initiative and is regarded as an example for other panchayats in Kerala. The project also inspired students of various educational and research institutions to refer to Meenangadi's initiative as a modal project for their thesis and research. Students of "The Academy of Climate Change Education and Research" (ACCER) took up a similar project called "carbon-neutral Kunnamkulam[9]" as part of their academic referring to Meenangadi's project. The project did not receive much outreach at the national level, even though articles were often published in national newspapers and magazines. The project did catch the eyes of a few NGOs and authorities working in line with Climate sensitive planning and was appreciated for the visionary approach of the panchayat. Meenangadi has been receiving awards of "best panchayat" for MNREGA implementation for the past three consecutive years. Meenangadi's carbon-neutral program was the main focus of the webinar series on Climate Technologies for Carbon-Neutral Agriculture organized by TU Delft University. In conclusion, the project received good outreach at district and State level governance but failed to impress National and international audiences.

Lessons from National and International Case Studies

Key takeaways from case studies and research papers were analyzed in the context of India, and adaptable strategies are discussed below. Preserving and enhancing nature's ability to shield climate variations cost less than restoring lost ecosystem functions with technology (WRI 2021). Hence nature-based solutions are the best forward for sensitive ecological regions like Meenangadi, whereas more technologically advanced solutions can be adopted in the polluted cities of India.

"Muhammodayam" is a pilot project taken up by the Muhamma panchayat in Kerala associated with ATREE,[10] aiming to conserve the Vembanad Lake ecosystem by developing sustainable development models and livelihood options in various sectors (CERC ATREE 2020a). The panchayat focused on responsible tourism and waste management, solid waste management using Black Soldier Fly Larvae (BSFL) or Hermetia illucens adopted here is an ingenious technique that can be quickly adopted in a developing country like India. The low-cost BSFL Biopod for household organic waste management can be adopted at the LSG level with the help of kudumbashree workers. Participatory wetland conservation and awareness programs are other initiatives that make the panchayat stand out. Canal protection forums engaging all stakeholders have acted as a bridge between the locals of the panchayat and the State Wetland Authority, which led to the successful rejuvenation of 10 canals in Muhamma (CERC ATREE 2020b).

Forests are significant carbon sinks. Conservation of forests prevents carbon emissions from the felling of trees and enables carbon sequestration. In Indonesia,

---

[9]Municipal town situated in the Thrissur District of Kerala in India.
[10]Ashoka Trust for Research in Ecology and the Environment is a research institution based in Bangalore, India focusing in the areas of biodiversity conservation and sustainable development.

Bukit Barisan Selatan National Park (BBSNP) set an example of a joint partnership between the government, companies, and communities to improve livelihoods and conserve a vital forest landscape (Dominik Schwad et al. 2021). This innovative partnership helped preserve National Park and enhance the profitability and resilience of coffee farming groups (Dominik Schwad et al. 2021). The Indian government needs to address the drivers of deforestation and focus on Urban Forestry. Spatial planning done at all levels of government must begin with incorporating green corridors for biodiversity conservation.

Agricultural practices must be aligned to enhance soil carbon storage and gain momentum. Livestock is another primary source of carbon emission. Emission reductions in this sector are possible through technological innovations like feed compounds that improve cattle digestion and manure management improvements (Waite and Rudee 2020). Reducing groundwater tables is another crisis for India. ABC water program in Singapore, launched in 2016, adopted a blue-green infrastructure approach and helped restore concrete channels into a soft river edge using techniques like riparian buffers, rain gardens, soil bioengineering, green roof tanks, etc. (Khoo Teng Chye 2017). This program brought people close to natural resources and taught them to value its services. Integrating socio-ecological initiatives like blue-grey infrastructure, ecosystem approach, and landscape approach in spatial planning can be a way out for Indian cities.

Reducing carbon dioxide emissions from the electricity sector is currently the primary focus of the Indian government. India has already set targets to reduce fossil fuel consumption, and a significant shift to renewable energy has been visualized for the coming decades. Besides, adding smartness or intelligence to the power system, from generation through transmission, would help efficient distribution and reduce energy losses. The low-carbon economy model of Germany and the Netherlands focuses on improving energy efficiency and creating a green economy (Dzikuć et al. 2021). The use of renewable energy was promoted by subsidizing green energy by local government authorities. They also focus on research and eco-innovations for a low-carbon economy. Along with renewables, biogas plants, biofuel, heat pumps, and solar collectors can be considered for Indian cities (Dzikuć et al. 2021).

Futuristic Vision and Way Forward for India

India has already set its goal of becoming carbon neutral by 2070. But the policymakers and government are still brainstorming on working out a robust pathway to achieve this same. The climate change assessment of the Indian region published by the Indian government recently stated an increase in the frequency and intensity of droughts (Kulkarni 2020). India is also facing high water stress, and the country is running out of ground and surface water. These issues can only be acknowledged with the broadest possible cooperation by all LSGS and their participation effectively and appropriately. The current action plans set by India have to be successfully translated in a decentralized manner to states and local governments. A top-bottom approach alongside a bottom-up has to be planned to implement the set targets successfully. Creating "Green villages/towns" for future expansion will help distribute the national level targets to sub-regional targets. These Green villages/ Towns can later form clusters and establish a common management framework. These frameworks can be dynamic for continuous improvements based on the performance of the green clusters. In this section, an implementation framework has been worked out for India. It follows part (LSG) to the whole(country) and whole to part approach in successfully translating the NDCs to different states of India.

The part of the whole approach starts with LSGs taking up initiatives to account for and assess carbon footprint within their locality with the support of educational institutions and organizations. The assessment should be community-focused, involve stakeholders' participation, and address any quarries of the locals without any delay. Local self-governments can modify and adapt the carbon assessment methodology developed by the Meenangadi panchayat. Once the carbon footprint statistics of all the LSGs are prepared, it is the state government's responsibility to review them with the support of an expert panel. Evaluation and selection of LSGs to take up pilot projects can be based on indicators and indexes formulated at State level. A few examples of indicators are: percentage area of forest cover, vehicular density, and nitrogen level of the soil. After the State government announces the list of potential LSGs, three months should be given for each LSG to prepare their best proposals to be submitted to the central government. LSGs can reach out to educational institutions, NGOs, environment think tanks, and the Government department to prepare the proposals. The central government evaluated the LSGs based on criteria set to match the promised NDCs. Two or more LSGs can be selected from each state to take up pilot projects. The central government has the final word on the number of LSGs to be chosen. The export panel at the central government level decides on the definitive list of LSGs to take up the pilot project within 30 days (Fig. 18.12).

The whole-to-part approach is worked out once the list of LSGs is finalized to implement pilot projects. An agreement is drafted between the LSG, State government, and central government stating the responsibility of each level of governance. National-level institutions and organizations like the National Institute of Urban Affairs (NIUA), The Energy and Resources Institute (TERI), and Environmental think tanks can pitch in to support the LSGs in successfully implementing the project. These institutions will act as the technical core system for the project. LSGs are responsible for setting up the Special purpose vehicle (SPV) as the implementing authority for the project. The management will include officials from central and state governments. LSGs are free to choose the best suitable approach for implementing the project. Studies and analyses should be carried out (Fig. 18.13). State and central government should provide financial support as a transparent channel. Every LSG selected has to complete the project within five

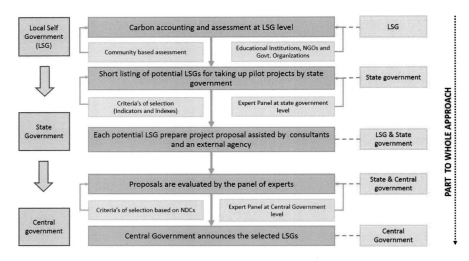

Fig. 18.12 Implementation framework of achieving carbon neutrality by 2070: PART TO WHOLE. (Source: Authors generated)

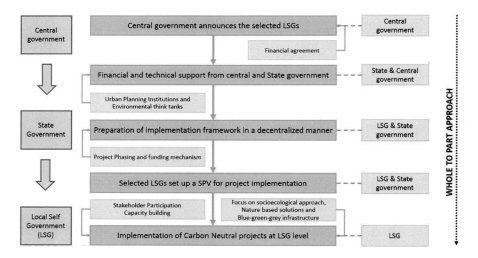

Fig. 18.13 Implementation framework of achieving carbon neutrality by 2070: WHOLE TO PART. (Source: Authors generated)

years from the selection date. After five years, the process will be repeated with a new list of LSGs. The central government can also introduce national awards for best implementation and innovations and encourage the LSGs to put their best foot forward.

Conclusion

The critical appraisal of Meenangadi's carbon-neutral project helped me understand the lacunas and drawbacks of implementing a bottom-up approach to achieving carbon neutrality in India. Even though Meenangadi did not entirely succeed within the planned timeframe, it set an example for Indian cities. The journey was inspiring, drawing light on the importance of public participation, a bottom-up approach, and having a clear vision. The courage of the authorities and the core team has to be appreciated. Encouraging electric vehicles, changing mobility patterns, and initiatives to use bio-fuels can be worked out in Meenagadi's context. Food & travel footprint can be significantly reduced with conscious effort, and a push for cleaner stoves can help cut the emission at the household level. Meenangadi panchayat plans to continue the project, and a new target of 2027 has been set.

The vision of Meenangadi can be taken back to all the gram panchayat in India. With comprehensive support from the state and central government, an integrated bottom-up and top-down approach can be the key to achieving India's targets. India has a prodigious opportunity to enforce the climate action momentum at local governance levels. Local bodies need to have proper access to information and policy actions across all governance levels. Government should focus more on peer learning and engagement, capacity building, and long–term climate strategies. Meenangadi has set an example for other regions and proved that they have the power, influence, and aspiration to break the climate and energy solutions barriers. India should focus on creating a platform to share the understanding of climate action for local bodies, which would help identify the benefits and lacunas in the initial stage itself. Climate governance and engagement with external partners need to be strengthened state-wide.

The study had its limitations as an explorative approach where most data collected were from secondary sources with limited field investigations. Critical appraisal was inclined to passive and participatory observations made by the authors and relied more on focus group interviews with experts, government officials, and stakeholders. The project's complexity was also amplified due to the pandemic. As a pilot project, the carbon-neutral Meenangadi project and its methodology are yet to be recognized at the national level. However, this study is only an initial attempt to evaluate the project critically from a resident's perspective. Planning think tanks and institutions have to work out a solid implementation framework to successfully translate the NDCs target set by the central government to state government and local government. As a developing country, India does have immense potential to turn the tables and emerge as a green economy.

Acknowledgments The authors wish to thank the Department of Architecture and Planning faculty members, National Institute of Technology Calicut, whose input has greatly helped develop the manuscript, which was initially done as part of an academic seminar presentation. The authors place their highest accolades on the NGO "Thanal," with all its limitations of the project did an exhaustive study (first of its kind in India) to come up with a report, using which this critical appraisal of the village panchayat could be done. We are also grateful to the government officials and the residents of the panchayat, without whose cooperation, the manuscript would not have seen the light.

References

Ameerudheen T (2017) Meenangadi in Kerala is well on its way to being India's first carbon-neutral panchayat. https://scroll.in/article/825260/meenangadi-in-kerala-is-well-on-its-way-to-being-indias-first-carbon-neutral-panchayat

Archana TVAKN (2021) Resilient community planning: meenangadi panchayat. https://issuu.com/nimasatheesh/docs/major_project_report_group_2

Avni Agarwal RM, Saransh B (2021, March 30) Exploring Carbon Neutral Development for India's Subnational Regions I WRI INDIA. https://wri-india.org/blog/exploring-carbon-neutral-development-india's-subnational-regions

Basil Poulose. (2020). Kerala village first to become carbon neutral village in India. https://www.thebetterindia.com/103551/kerala-village-meenangadi-first-carbon-neutral-panchayat/

Bazilian Morgan GD (2020, December 10) Five years after Paris: How countries' climate policies match up to their promises, and who's aiming for net zero emissions. https://www.downtoearth.org.in/blog/climate-change/five-years-after-paris-how-countries-climate-policies-match-up-to-their-promises-and-who-s-aiming-for-net-zero-emissions-74602

Biniaz Sue IM (2020, August 21) Yale Experts Explain The Paris Climate Agreement. https://sustainability.yale.edu/explainers/yale-experts-explain-paris-climate-agreement

Census (2011) District census handbook wayanad, 590

CERC ATREE (2020a) Model wetland village. https://www.vembanad.org/modelwetlandvillage

CERC ATREE (2020b, October 23) Black Soldier Fly Larvae. https://www.vembanad.org/bsflarvae

CLEAR Center (2019, September 20) What is Carbon Sequestration and How Does it Work? I CLEAR Center. https://clear.sf.ucdavis.edu/explainers/what-carbon-sequestration

CNN Guidelines (1995) PS: political science & politics. 28(3):582. https://doi.org/10.1017/s1049096500058042

Concept D (2016) Creating a carbon-neutral meenangadi grama panchayat. 34601(June):1–7

cris (2021, January 2) This Kerala panchayat "rewards" farmers for not cutting down trees I The News Minute. The News Minute. https://www.thenewsminute.com/article/kerala-panchayat-rewards-farmers-not-cutting-down-trees-140724

Dhanda KK, Hartman LP (2011) The ethics of carbon neutrality: a critical examination of voluntary carbon offset providers. J Bus Ethics 100(1):119–149. http://www.jstor.org/stable/41475831

Dominik S, Matt L, Leonie L, Arnaud G, & Janice WS. (2021) WCS CASE STUDIES AND RECOMMENDATIONS. Wildlife Conservation Society. Retrieved September 9, 2022, from https://c532f75abb9c1c021b8ce46e473f8aadb72cf2a8ea564b4e6a76.ssl.cf5.rackcdn.com/2021/02/10/rcpdnuyof_WCS_Forest_First_2021_screen.pdf

Dzikuć M, Gorączkowska J, Piwowar A, Dzikuć M, Smoleński R, Kułyk P (2021) The analysis of the innovative potential of the energy sector and low-carbon development: a case study for Poland. Energ Strat Rev 38(November). https://doi.org/10.1016/j.esr.2021.100769

Eco India (2020) This town in Kerala has set its sights on becoming India's first carbon-neutral panchayat – Vikalp Sangam. https://vikalpsangam.org/article/this-town-in-kerala-has-set-its-sights-on-becoming-indias-first-carbon-neutral-panchayat/

europarl (2021, June 24) What is carbon neutrality and how can it be achieved by 2050? I News I European Parliament. https://www.europarl.europa.eu/news/en/headlines/society/20190926STO62270/what-is-carbon-neutrality-and-how-can-it-be-achieved-by-2050

Isaac TT (2020, July 26) Towards a carbon neutral Wayanad – The New Indian Express. Newindianexpress. https://www.newindianexpress.com/opinions/2020/aug/26/towards-a-carbon-neutral-wayanad-2188319.html

Khoo TC (2017) The active, beautiful, clean waters programme: water as an environmental asset. In Centre for Liveable Cities (CLC), Singapore. https://www.clc.gov.sg/docs/default-source/urban-systems-studies/rb172978-mnd-abc-water.pdf

Kulkarni RKSGM (2020) Assessment of climate change over the Indian region a report of the Ministry of Earth Sciences (MoES), Government of India

Kumar S, Sangeeta MPD (2020) State Energy Efficiency Index:2020

Longer Articles (2021, January 10). The paris agreement explained – planetary concerns. https://www.planetaryconcerns.com.au/the-paris-agreement-explained-climate-change/

M S Swaminathan Research Foundation, & Community (2018) Sacred Grove Augmentation – MSSBG. https://mssbg.mssrf.org/know-the-garden/action-framework/conservation/sacred-grove/

Malvika K (2021, March 4) Decentralize for net zero Bengaluru | Deccan Herald. Deccan Herald. https://www.deccanherald.com/opinion/main-article/decentralise-for-net-zero-bengaluru-957847.html

Mission S (2018) Carbon Neutral Meenangadi. In Local Self Governmental Department Govt: of Kerala

MOEF (2015) India's Intended Nationally Determined Contribution is Balanced and Comprehensive: Environment Minister. https://pib.gov.in/newsite/PrintRelease.aspx?relid=128403

Neha B (2020, December 28) How a "tree mortgage" scheme could turn an Indian town carbon-neutral | Global development | The Guardian. https://www.theguardian.com/global-development/2020/dec/28/how-a-tree-mortgage-scheme-could-turn-an-indian-town-carbon-neutral

Oppenheimmer M (2007, November) How the IPCC got started. https://blogs.edf.org/climate411/2007/11/01/ipcc_beginnings/

Pandey K (2011, May 11) Net-zero debate: Where do Indian states stand on the decarbonization pathway? https://india.mongabay.com/2021/05/where-are-indian-states-participating-in-net-zero-debate/

Pat G (2015) The Greenhouse gas accountant. http://www2.sta.uwi.edu/uwiToday/archive/may_2015/article13.asp

Prasad KVD (2014) Examining the influence of climate change impacts on agrobiodiversity, land use change, and communities: an empirical experience from Wayanad-Kerala. Green India: strategic knowledge for combating climatic change prospects & challenges, November, 203–210. https://doi.org/10.13140/RG.2.1.1050.0326

Rajan JB (2017) Twelfth five year plan 2012–2017 tribal sub-plan (TSP) Meenangadi village panchayat

Rajendran K (2022, February 2) Indian village banks on tree mortgages in a bid to go carbon-neutral | Reuters. https://www.reuters.com/markets/commodities/indian-village-banks-tree-mortgages-bid-go-carbon-neutral-2022-02-04/

Sakshi K (2017, June 7) This tiny Keralan village is the first to achieve zero carbon emissions – Homegrown. https://homegrown.co.in/article/801241/this-tiny-keralan-village-is-the-first-to-achieve-zero-carbon-emissions

Santhoshkumar AV (2010, July) The home gardens of Wayanad. www.infochangeindia.org

Singh J (2019) Driving climate action: state leadership in India. May, 9. https://www.theclimategroup.org/sites/default/files/india_report_web_singles.pdf

Thanal (2018) Carbon Neutral Meenangadi – Assessment and Recommendations. http://thanal.co.in/uploads/resource/document/carbon-neutral-meenangadi-assessment-recomendations-87546380.pdf

Thanal (2020, October 22) Tree Banking Scheme-Online Inauguration – Thanal. https://thanaltrust.org/events/tree-banking-scheme-online-inauguration/

The Climate Group (2020) Carbon neutral Meenangadi: a bottom-up model for integrating climate action into development planning. https://www.theclimategroup.org/under2-coalition

The Times of India (2020, October 21) Kerala farmers get loans by pledging live trees | Kozhikode News – Times of India. https://timesofindia.indiatimes.com/city/kozhikode/farmers-get-loans-by-pledging-live-trees/articleshow/78776339.cms

UN, Climate Change (2021). https://www.un.org/en/climatechange/what-is-climate-change

Verma A (2020, August 17) PM Modi brings 7.5 GW Ladakh solar project back into focus. Saurenergy.Com. https://www.saurenergy.com/solar-energy-news/pm-modi-brings-7-5-gw-ladakh-solar-project-back-into-focus

Vishnu V (2020, October 22) In Kerala village, farmers can mortgage their trees for interest-free bank loans | India News, The Indian Express. The Indian Express. https://indianexpress.com/article/india/kerala/in-kerala-village-farmers-can-mortgage-their-trees-for-interest-free-bank-loans-6836458/

Waite R, Rudee A (2020) 6 ways the US can curb climate change and grow more food. https://doi.org/10.1088/1748-9326/AAB908/PDF

WRI (2021) Reckoning and recovery. www.wri.org

Part V
Infrastructure and Resilient Cities and Settlements

Chapter 19
Land Use and Land Cover Change Dynamics and Modeling Future Urban Growth Using Cellular Automata Model Over Isfahan Metropolitan Area of Iran

Bonin Mahdavi Estalkhsari, Pir Mohammad, and Alireza Karimi

Abstract Exploring changes in land use land cover (LULC) is essential to know the dynamics of urban sprawl, as rapid urbanization affects a city's ecology and thermal environment. The present study aims to understand the spatial-temporal variability of LULC over the Isfahan metropolitan area of Iran of four different periods, every ten years, 1900, 2000, 2010, and 2020, using multispectral images of Landsat satellite. The results of LULC reveal that the built-up area witnessed a significant increase of approximately 72.9% of the study area during 1990–2020, at the expense of decreasing vegetation and bare land areas throughout the study region. The present research was also applied to simulate urban land use changes and to predict spatial patterns using an artificial neural network-based cellular automata model. The model was validated using the Kappa index statistics, resulting in 85% overall accuracy. Finally, based on the transition rules produced during the calibration process, the future land cover map for 2030 was predicted. The predicted maps suggest a significant increase in an urban built-up area of about 24.72% in a near-future scenario compared to the present year. The result of the present chapter will be helpful for urban policymakers and stakeholders in the future sustainable development plan of the city.

Keywords Land use land cover · Cellular automata · Artificial neural network · Urban growth · Remote sensing · GIS · Modelling

---

B. Mahdavi Estalkhsari
Department of Landscape Architecture, Shahid Beheshti University, Tehran, Iran
e-mail: b.mahdaviestalkhsari@Mail.sbu.ac.ir

P. Mohammad (✉)
Department of Earth Sciences, Indian Institute of Technology, Roorkee, Uttarakhand, India
e-mail: pirmohammad291@gmail.com

A. Karimi
Instituto Universitario de Arquitectura y Ciencias de la Construcción, Escuela Técnica Superior de Arquitectura, Universidad de Sevilla, Sevilla, Spain
e-mail: alikar1@alum.us.es

© The Author(s), under exclusive license to Springer Nature Switzerland AG 2022
U. Chatterjee et al. (eds.), *Ecological Footprints of Climate Change*, Springer Climate, https://doi.org/10.1007/978-3-031-15501-7_19

Introduction

Approximately two centuries ago, only 3% of the world's population lived in cities (Wang et al. 2012), increasing to 10% in 1900 by increasing the human population. Today, more than 53% of the inhabitants of the planet live in cities (Sampson et al. 2020), and it is predicted that by 2030 and 2050, this amount will reach 66% and 72%, respectively (Bongaarts 2006; Gerten et al. 2019). By increasing urban sprawl, numerous international, national, regional, and urban issues have caused social, economic, and cultural problems in these settlements (Gomes and Hermans 2018; Sadowski and Bendor 2019). Problems such as housing (Li et al. 2018), rural immigrants' rights (Lu Dadao and Hui 1997), illegal behaviors (Tang et al. 2021), environmental pollution (Wu et al. 2020), urban heat islands (Mohammad and Goswami 2021b, 2021c), increased energy consumption (Karimi et al. 2020), and economic vulnerability (Canh and Thanh 2020) are only parts of the problems in these spaces, which will become more complex if not addressed.

Iran's rural and urban population from 1950 until now has shown an increasing trend. Its forecast until 2050 shows that over time, the share of the rural population of Iran is decreasing, while the urban population has increased significantly, and approximately 86% of the population of Iran will live in urban areas (Qelichi et al. 2017). Before 1980, a more significant proportion of Iran's population lived in rural areas. However, in 2018, these ratios reversed, and more than 75% migrated to cities. In the previous three decades, as a hub of industrial and tourism, Isfahan province has not been an exception to this rule and has increased its urban population and development (Mohammad and Goswami 2021a; Shirani-Bidabadi et al. 2019). According to the results of the last general population and housing census of 2006, Isfahan, with about 1266,000 habitats, ranks after Tehran and Mashhad, which is considered the third most populous city in Iran.

The rate of change of an urban population, urban growth (Afsharzadeh et al. 2021), uses the land for the construction of buildings and impermeable surfaces (Black and Henderson 1999), while urbanization is mainly due to urban growth, which can be because of the indexes such as natural population growth, reclassification of cities/rural systems, and rural-urban migration (Khanna 2020). In Iran, the increasing rate of urbanization is attributed to factors such as natural population growth based on mortality and birth rate, reassignment of urban and rural boundaries due to political laws, and changing the definition of rural areas. Moreover, due to general public welfare and job opportunities in urban areas, rural migrations to urban areas are another influential factor (Assari and Mahesh 2011a, b). One of the outward characteristics of urban growth that affects the land use/cover of the area is called urban sprawl, which is the emergence of residential areas in suburban districts. According to Parsi and Farmihani Farahani (2014), urban sprawl is mainly caused by industrial developments and political decisions in Isfahan. In different urban studies, it is essential to monitor land use/cover and urban sprawl, using remote sensing, one of the helpful observation methods.

Remote sensing can help map (Wu et al. 2018), monitor (Vihervaara et al. 2017), and evaluate current and past urban growth trends (Bhatta 2010; Vanden Borre et al. 2011; Kantakumar et al. 2019; Mohammad et al. 2019). Current remote sensing technology provides data analysis from ground-based, atmospheric, and Earth-orbiting platforms with GPS data layers, GIS data layers, and new modeling capabilities (Gohain et al. 2020; Mohammad and Goswami 2019; Treitz and Rogan 2004). LULC has two separate terms, often used interchangeably (Liping et al. 2018). Land cover refers to a set of biophysical characteristics of the earth's surface, including the distribution of soil, water, vegetation, and other physical characteristics, while land use refers to how humans use it (McConnell 2015). LULC results from complex interactions between humans and the physical environment (Pielke Sr et al. 2011) that is the primary driver of global change and is involved in ecosystem processes, biological cycles, and biodiversity (Jewitt et al. 2015).

Based on the research, the process of monitoring and assessing change detection, such as urban growth, is being conducted using various tools and software, such as Google Earth Engine (GEE) ArcGIS (Stevens et al. 2007; Tayyebi et al. 2011), Python-based model (Tripathy and Kumar 2019), and numerical models (Clarke et al. 1996). These tools use a variety of techniques to predict the future urban growth of an area, such as the cellular automata model (CA) (Falah et al. 2020), Markov chain (Aburas et al. 2021; Kushwaha et al. 2021), artificial neural network (Lee et al. 2020; Yatoo et al. 2020), decision trees (Jin and Mountrakis 2013), and agent-based models (Jokar Arsanjani and Kainz 2011). Based on Musa et al. (2017), these models have limitations and strengthen points that can influence their selection in research; however, the cellular automata model is the preferable model among all. Due to C.A.'s simplicity in performing complex calculations using only spatial information, this model is more satisfactory than differential equations. In addition to its dynamism and compatibility with remote sensing and GIS data, the C.A. model is still more flexible than other models.

A collection of research on LULC and urban growth prediction over the past decade has caught the attention of most Asian researchers, including Saxena and Jat (2020) in Ajmer, Han et al. (2015) in Beijing, Abdalla et al. (2020) in Ningxia, Mumtaz et al. (2020) in Lahore, Baqa et al. (2021) in Karachi, and Khan et al. (2020) in Quetta. These results indicate the need for appropriate policies and regulations to maintain sustainable development in different regions. In addition, the impact of LULC and urban growth prediction in the future have been recorded as one of the popular activities of researchers in Iran (Ansari and Golabi 2019; Karimi et al. 2018; Mazloum et al. 2021; Mosammam et al. 2017; Parsa and Salehi 2016). Isfahan city is no exception, for example, in a research conducted by Asghari Sarasekanrood and Asadi (2021), LULC changes on the land surface temperature (LST) were investigated between 2000 and 2018. It was concluded that there is a relationship between LULC and LST, mainly concerning vegetation and humidity changes. According to estimates made by Kalkhajeh and Jamali (2019), between 1986 and 2016, more than 10% of agricultural land was destroyed in Isfahan, while urban growth increased by 23.4%, and the southern parts of the river and the north of the city played a more significant role. Furthermore, another research (Soffianian and Madanian 2015) investigated changes related to the conversion of agricultural lands to impenetrable surfaces, undeveloped lands to agricultural lands, agricultural and undeveloped lands to impenetrable surfaces. The growth of population, traffic conditions, and the industrialization of today's cities have been proposed as the most important factors affecting them. Therefore, planning to conserve and restore green cover should be an essential program for decision makers in the study areas.

In this research, the integration of time-series satellite images and models based on artificial neuron networks is used to identify and analyze urban expansion patterns in the fast-growing city of the Isfahan metropolitan area of Iran. This chapter aims to assess the amount and pattern of urban growth in the past and present using multispectral remote sensing data between 1990 and 2020, at each successive ten-year interval period. The LULC change pattern in time series is then used to predict the future urban growth scenario. The main objective of this research is: (1) LULC classification of different change years (1990, 2000, 2010, and 2020) using Google Earth Engine; (2) analysis and comparison of the spatio-temporal patterns of LULC changes concerning the urban growth; (3) determining of transition probabilities of other LULC types change to the urban area and simulating the change using ANN-CA integrated modeling approach; and (4) based on time series a transition potential matrix, determining the future urban expansion for the year 2030. The result of the present study will provide information on the past and current trends of LULC. It will also help planners and decision-makers assess the changing pattern and prepare and manage the expected future urban expansion to achieve sustainability of the Isfahan metropolitan area.

Study Area

This study was carried out in Isfahan province, which consisted of 23 districts, and is located in central Iran, in addition to the lush ZayandehRud plain and the foothills of the Zagros mountain range (31° 43′ N to 34° 22′ N and 49° 38′ E to 55° 31′ E). According to the latest census (2016), more than 4,629,312 people live in this province (Abbasnia et al. 2016), which is known as one of the most important cities in Iran with the fastest population growth rate and migration due to the significant reduction in agricultural areas (Ramezankhani et al. 2018). According to the Koppen-Geiger climate classification, Isfahan has a characteristic dry climate (Kottek et al. 2006) with a maximum temperature of 40 °C and cold winters with a minimum of −7 °C. The average temperature in the coldest month of January is 4.8 °C and in the warmest month of July is 30.6 °C. Furthermore, it experiences low annual average rainfall (116.9 mm) due to low humidity. The selected area is shown in Fig. 19.1.

19 Land Use and Land Cover Change Dynamics and Modeling Future Urban...

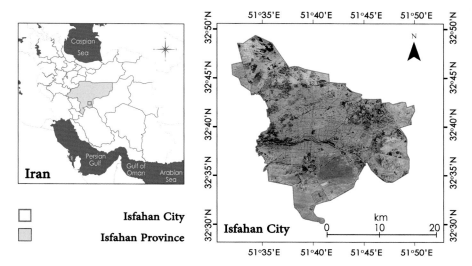

Fig. 19.1 Location map of the Isfahan metropolitan area, Iran

Data and Methods

Data and Image Pre-processing

In this research, land use and land cover (LULC) mapping was prepared from Landsat 5 (1990, 2000, and 2010) and Landsat 8 (2020) surface reflectance datasets in the Google Earth Engine (GEE) platform. The initial stage of extracting the LULC information is image synthesis to process the Landsat images of frost-free seasons of 1990, 2000, 2010, and 2020. First, after defining the study area, in all Landsat images of this research, the "CLOUD COVER" filter has been used to obtain the best possible quality and a lower than 5% existence of cloud masks. In the next step, the bands of "NDVI" (Normalized Difference Vegetation Index), "NDBI" (Normalized Difference Built-up Index), and "NDWI" (Normalized Difference Water Index) were added to the image collection to detect the existence of live green vegetables, built-up areas and liquid water, respectively. The other variables that have been taken into bands were "Slope" and "Elevation" to obtain the effects of topography on the image process, especially in the east and west parts of Isfahan, where there are differences in topographical levels. Subsequently, a "reduce region" function was applied as a part of assessing the evaluation of the "Principal Component Analysis (PCA)" to identify the transformation of the original variables into other variables in image processing with the probability of essential information capture, as Estornell et al. (2013) and Loughlin (1991). After all the above steps, we obtained a composite image containing all the calculated bands.

Supervised Classification

Langer et al. (2020) denote that in supervised classification, some classes should be defined with some characteristics established to categorize the data in them. In LULC characterization, training samples represent each type of LULC class. Therefore, in this investigation, based on visual investigations of the high-resolution composite image derived by overlapping the bands mentioned in the previous stages, we introduced four classes of LULC in the study area: Built area, vegetation, barren land, and water bodies. After that, we selected some points in each class as training samples. Each point has a specific name, color, and properties that reflect its class. In all years 1990, 2000, 2010, and 2020, the selected points of each defined class were more than 100 points, a total of 4412 points, with a homologous distribution throughout the study area, to ensure the precision of the assessments. These sample points have been used as the ingredients of the accuracy assessment stage.

Processing the Training Points

To process the training data and ensure their accuracy, we used the Random Forest (R.F.) algorithm in this investigation. Based on Breiman (2001), in R.F., different training data are processed by multiple decision trees to identify the most popular class. In other words, the trees vote on the classes of the training points, and the most popular type will be selected. R.F. can deal with large and full input noise data without being overfitted, so the uncertainty of accuracy assessment would be reduced in this algorithm rather than in others. In this investigation, to assess the precision, the selected point samples of each class have been split into two groups, training samples (70%) and validation samples (30%). The number of decision trees that used 70% of the points was 500, so a total of 3088 samples were used to train the R.F. model. The outputs of the R.F. assessments have been used in the validation assessment by constructing a confusion matrix.

Cellular Automata (C.A.)-Based Urban Growth Modeling

In this study, the C.A. model was used to simulate the future urban growth of the year using the MOLUSCE plugin in open-source QGIS software. The CA model is the most widely used in recent times, as it considers both static and dynamic aspects of land cover changes. The prediction of land cover change required both dependent and independent variables in this plugin. In this study, five dependent variables were considered: elevation, slope, distance to major roads, distance to water bodies, and distance to the central building. The details of these variables are presented in the next section and Fig. 19.2.

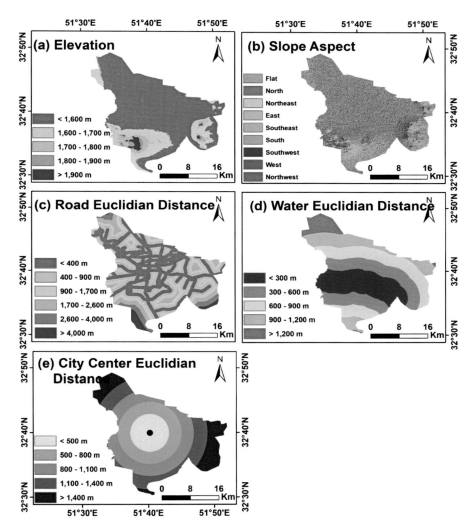

Fig. 19.2 Different driving layers for urban growth prediction

Parameters Used for Urban Growth Modeling

The study has used multiple parameters to accurately collect data from different sources to represent the urban expansion of Isfahan city. This study has considered five other parameters for modeling urban growth.

Elevation In this study, we have used the SRTM digital elevation model to get the elevation information of the area. It is seen that a higher elevation is observed in the south-west part of the city compared to the northern region (Fig. 19.2a).

Aspect The aspect layer of the study area is prepared from SRTM DEM. The area is generally a flat topography area with slight undulation. The figures show that most of the area is in flat topography in the northern part (Fig. 19.2b). However, much more variation in aspect is observed in the southern and eastern regions.

Road Euclidian Distance The road network is extracted from the open street map (OSM). The road is always the main parameter responsible for urbanization, as shown in Fig. 19.2c for Isfahan city. There is general agreement that everyone wants better road connectivity for their house, which would provide many benefits to their livelihood.

Water Euclidian Distance The water body layer is obtained from the Open Street Map (OSM). Since ancient history, it can be seen that the most ancient civilization was developed along the river coast. Water is a vital part of the sustainability of the earth. Therefore, this study uses the Euclidian distance from the water bodies as one of the driving parameters of the LULC (Fig. 19.2d).

Central Building Distance (CBD) Proximity to the central business district has been considered an essential parameter in the study to represent multi-urban functional areas and connectivity. In this study, the city center is regarded as the central building point, and from there, a buffer of equal size was drawn to the city's outer boundary (Fig. 19.2e). The inner part of CBD is associated with rapid urban growth, while the outer part is associated with lower urbanization.

Results

Validation Assessment

Validation of classifications in this research has been carried out using the confusion matrix with the parameters of overall accuracy (O.A.), producer accuracy (P.A.), consumer accuracy (C.A.), and Kappa coefficient (K.C.). As declared by Abijith and Saravanan (2021), the O.A. is the total number of correct training samples divided by the total number of training samples. While the P.A. in each class indicates the precision of the samples, the C.A. is the users' beliefs about their precision. The Kappa coefficient acts as a reliability statistic of the classification. Table 19.1 shows all the parameters in all LULC classes from 1990 to 2020.

The percentage of overall classification accuracy in the whole period is more than 95, and that of Kappa accuracy is more than 93. Based on the data in Table 19.1, during the last 30 years, the lowest P.A. percentage is dedicated to bare land in both 1990 and 2000, and the highest P.A. relied on water bodies in 2000. On the other hand, considering the C.A. percentages, the class with the lowest accuracy percentages is the built area in 1990. In addition, vegetation in 1990 and the water bodies in 2000 have the highest rates. The average precision of vegetation is higher than in other classes, and bare land is lower during the whole period.

Table 19.1 Validation assessments of the LULC change detection of Isfahan, from 1990 to 2020

LULC	in % 1990		2000		2010		2020	
	PA	CA	PA	CA	PA	CA	PA	CA
Built area	96	89	94	96	97	96	91	95
Vegetation	97	100	98	96	93	98	100	97
Bare land	89	95	89	91	96	96	95	90
Water bodies	98	97	100	100	98	91	94	98
Overall accuracy	95		95		96		95	
Kappa accuracy	93		94		94		93	

P.A. Producer accuracy, C.A. Consumers' accuracy

LULC Change Detection

This part demonstrates the changes in land use/cover (LULCC) in the study area from 1990 to 2020 at intervals of every ten years. Four main land cover classes have been categorized using the high-resolution satellite imageries and field surveys; built area, vegetation, water bodies, and barren land. The area of each of the LULC classes has been calculated to be studied as a change detection measure. Figure 19.3 shows that in all years, the trends of LULCC are the same throughout all classes, with an increase in built area and a decrease in all other types of land cover.

Table 19.2 defines that the category of built areas experienced growth throughout the period, while there was a decrease in vegetation, bare land, and water bodies. In 1990, bare land with 306.34 km^2 (46.76%) of the total area was the largest LULC category. Vegetation and built area were 183.14 km^2 (27.95%) and 163.52 km^2 (24.96%), respectively. The water bodies category had the lowest area percentage and ranged from 2.09 km^2 (0.31%) to 1.09 km^2 (0.16%) throughout the period. Furthermore, from 2000, the built area started to grow gradually, and with a 36.91% increase, it reached 282.78 km^2 (43.16%) in 2020. However, the bare land, which had the highest area percentage until 2010, was second, after the built area, with 249.19 km^2 (38.03%) of the total area. Between 2000 and 2020, the total size of the vegetation, with a decreasing pattern, changed from 61.11 km^2 (9.33%) to 122.03 km^2 (18.62%). The category most changed was the built area; this emphasizes the urban expansion of the study area through vegetation and bare land over the last 30 years.

Gains and Losses of Different Land Cover Types

Analyzing the shift rates between different LULC categories can help identify the factors influencing the changes of each category during the study time.

Table 19.3 shows only ascending or descending trends in each decade, and there are no fluctuations in each land cover class. The built area experienced growth, while

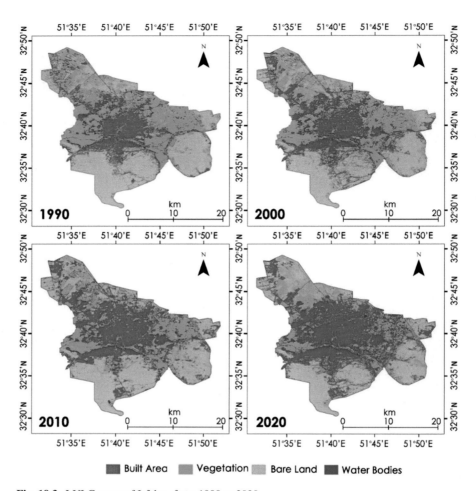

Fig. 19.3 LULC maps of Isfahan from 1990 to 2020

Table 19.2 The percentages of four LULC classes in Isfahan, from 1990 to 2020

	Area							
	1990		2000		2010		2020	
LULC	km²	%	km²	%	km²	%	km²	%
Built area	163.52	24.96	206.54	31.52	244.72	37.35	282.78	43.16
Vegetation	183.14	27.95	167.68	25.59	137.10	20.92	122.03	18.62
Bare land	306.34	46.76	279.15	42.61	272.14	41.54	249.19	38.03
Water bodies	2.09	0.31	1.72	0.26	1.13	0.17	1.09	0.16
Total	655.09	100	655.09	100	655.09	100	655.09	100

Table 19.3 Increase and decrease trends between LULC classes

LULC	Area							
	1990–2000		2000–2010		2010–2020		1990–2020	
	km²	%	km²	%	km²	%	km²	%
Built area	+43.02	+26.3	+38.18	+18.48	+38.06	+15.55	+119.26	+72.93
Vegetation	−15.46	−8.44	−30.58	−20.06	−15.07	−18.23	−61.11	−33.36
Bare land	−27.19	−8.87	−7.01	−2.51	−22.95	−8.43	−57.15	−18.65
Water bodies	−0.37	−17.7	−0.59	−34.3	−0.04	−3.53	−1	−47.84
Total	655.09	100	655.09	100	655.09	100	655.09	100

there was a decline in the other categories. With the highest LULC transformation rate, the built area increases by more than 15% every decade. On the other hand, every ten years, the reduction trend in vegetation and bare land and water bodies is not more than 18.23%, 8.87%, and 34.3%, respectively. Based on Fig. 19.4 and Table 19.3, the highest increase amount is dedicated to the built area from 1990 to 2000, and the highest decrease amount is for vegetation between 2000 and 2010.

Decadal Change of Each Land Use/Cover Class

From 1990 to 2020, the LULC conversion matrix has been calculated: the conversion from 1990 to 2000, 2000 to 2010, 2010 to 2020, and 1990 to 2020. This calculation was to investigate how much the area of each LULC class had been converted to other types. For instance, between 1990 and 2000, how much of the water bodies class changed to vegetation category. The results have been depicted in Fig. 19.5 and Table 19.4. It is essential to mention that a conversion of a type to itself is equivalent to no changes in that category. Although there are changes between one class to another and vice versa, which causes a mere balance in the area percentage, changing the landscape's characters should not be considered of low importance.

From 1990 to 2000, most of the area of each class remained unchanged. However, the total changes to the built area were more than in other categories (70.58 km²). There was a transition of bare land, vegetation, and water bodies to a built area, with 41.22 km², 38.73 km², and 0.4 km² changes. The transformation of bare land to the built area, 46.64 km², was the most significant change in this period, which had the most amount among all other decades, and after that, it was the vegetation to bare land, 28.37 km², the vegetation to the built area, 23.78 km², and the built area to vegetation (21.16 km²). In conversion to water bodies, the change of bare land was the most among others with a 0.24 km² switch, and after that, there is vegetation and built area classes with 0.09 and 0.07 km² transformations.

From 2000 to 2010: the total conversion was dedicated to the changes into the built area category, 75.74 km². The amount of bare land change in the urban area, 40.41 km², was the most significant change in this decade, and then there was vegetation to the built area, 35.27 km², and the built area to bare land, 31.31 km². It is essential to mention that the transformation of the built area to vegetation in this

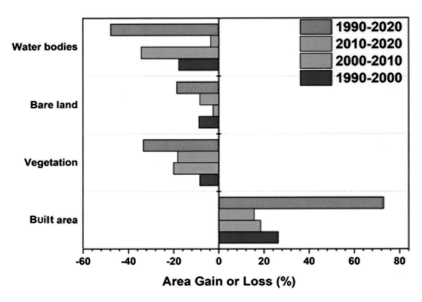

Fig. 19.4 Distribution of gains and losses of different land cover categories in the study area

period was half of the last decade, 11.57 km^2, and that of the bare land to vegetation was almost the same as in the previous decade 16.75 km^2. Most of the conversion of the water bodies was to vegetation, 0.11 km^2, and its change to the built area and bare land was almost the same, 0.06 km^2. In this period, the transition of vegetation and the built area to water bodies was the most notable change.

Between 2010 and 2020, the total change to bare land was less than the past decade, 38.88 km^2, while conversions to water bodies were significantly higher than the past two decades, 0.8 km^2. The transformation of the bare land to the built area of 46.41 km^2 was the most significant change. After that, there was vegetation to the built area 27.17 km^2, the built area to the bare land 23.41 km^2, the built area to the vegetation 17.13 km^2, and the vegetation to the bare land 15.39 km^2, respectively.

Totally, from 1990 to 2020, the most notable change was the conversion of bare land into the built area, 65.7 km^2, and vegetation into the built area, 61.7 km^2. The number of LULC categories converted to the built area was more significant than other categories. Subsequently, there were changes in bare land, 38.17 km^2, vegetation 35.49 km^2, and water bodies 1.14 km^2, respectively. In the vegetation changes, the bare land 23.12 km^2 was a pioneer, and in the transformation to bare land, the vegetation class had the majority, 25.95 km^2. It is noteworthy that, among all types, vegetation had the enormous change into water bodies of 0.41 km^2; there was bare land of 0.39 km^2 and built area of 0.34 km^2, respectively.

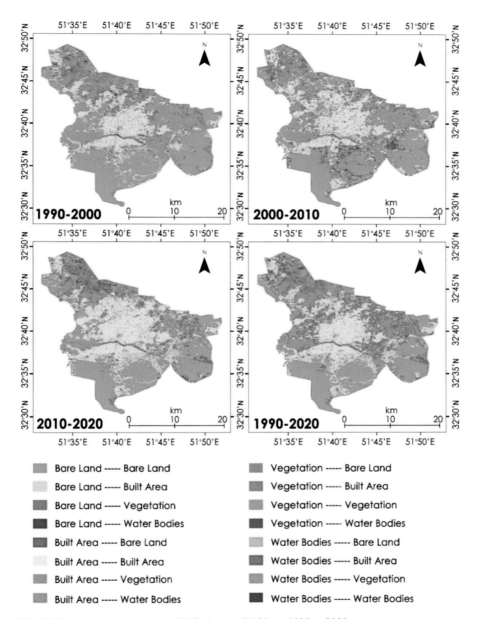

Fig. 19.5 Transformation in the LULC classes of Isfahan, 1990 to 2020

Table 19.4 Change matrix of different LULC classes in specific periods

LULC change matrix (km²)					
Period	LULC classes	Built area	Vegetation	Bare land	Water bodies
1990–2000	Built area	102.85	21.16	12.81	0.07
	Vegetation	23.78	102.14	28.37	0.09
	Bare land	46.64	17.37	192.85	0.24
	Water bodies	0.16	0.2	0.04	0.49
	Total change	70.58	38.73	41.22	0.4
2000–2010	Built area	130.3	11.57	31.31	0.23
	Vegetation	35.27	86.32	18.94	0.34
	Bare land	40.41	16.75	176.76	0.13
	Water bodies	0.06	0.11	0.06	0.65
	Total change	75.74	28.43	50.31	0.7
2010–2020	Built area	165.34	17.13	23.41	0.17
	Vegetation	27.17	71.57	15.39	0.61
	Bare land	46.41	12.9	167.75	0.04
	Water bodies	0.28	0.19	0.08	0.8
	Total change	73.86	30.22	38.88	0.82
1990–2020	Built area	111.63	12.17	12.75	0.34
	Vegetation	61.7	66.3	25.95	0.41
	Bare land	65.7	23.12	167.89	0.39
	Water bodies	0.19	0.2	0.01	0.48
	Total change	127.59	35.49	38.17	1.14

Analysis of LULC in Each of the Classes

Changes in Built Area

Due to its unique culture, handicrafts, arts, and architecture, the historic city of Isfahan has always been fascinating to rulers, tourists, and business people throughout its history. People from other towns and rural areas would like to live in Isfahan to have a better life and employment opportunities. This population growth causes the need for more urban facilities, infrastructures, and houses. In a religious city like Isfahan, the neighborhoods are shaped primarily by the demands of different social and religious classes that prefer to live in separate communities apart from each other. For example, Muslims live in different neighborhoods from Christians. Therefore, the expansion of the city in Isfahan is inevitable and progressive. In this research, the built area refers to the routes and buildings. Table 19.3 declares that from 1990 to 2020, the increase in the area of this class is approximately 73%. Based on Fig. 19.3, the city expansion is from the center, which is dense due to the ZayandehRud River, to the north and northwest, and gradually it is expanding to the southern parts of the city.

Changes in Vegetation

In this research, agricultural land and urban greeneries have been classified as the vegetation class. According to Table 19.3, more than 33% of the vegetation class has declined from 1990 to 2020. Furthermore, as Fig. 19.3 shows, in the 30 years, in the northern part of the ZayandehRud River, the area of cultivation zones was more significant than in the south, and over time with urban growth, the size of agricultural land in the east and northwest of Isfahan has been reduced. Although urban growth is the most compelling reason for the change in land cover in this class, considering that Isfahan is one of the arid regions of Iran, the issue of water supply is a critical problem in its agricultural field. Due to climate change, a decrease in the level of the ZayandehRud River, and progress in industries, Isfahan experienced drought, so farming has become a problematic activity. Therefore, to supply water to keep factories' machines cool and to make cultivation possible, a wastewater treatment plant effluent has been constructed in Isfahan. The effects of the Isfahan wastewater collection system on cultivation soils have been investigated in recent years. Eslamian et al. (2013) and Esmaeili et al. (2014) declare that in Isfahan, reclaimed water used for agricultural purposes has contaminated groundwater resources and the cultivation soils polluted nitrate. Therefore, agriculture in Isfahan has become a challenge that can reduce its prosperity and reduce the area under cultivation.

Changes in Bare Land

The bare land in Isfahan is located mainly in the southern part of the ZayandehRud, and urban growth has exceeded this land cover category over the last 30 years and caused more than 33% loss in the area. The mountainous topography of the southern Isfahan lands and the existence of the industrial zones of Mobarakeh Steel and the Bama Lead-Zinc Mine affect the usability of these lands in terms of agricultural or settlement usage. Field surveys have shown that the bare southern lands of Isfahan have become a place for garbage accumulation, which reduced the life quality of the surrounding citizens, made their protests to the municipality due to bad smells and demanded a change in the use of these barren lands.

Changes in Water Bodies

Figure 19.3 shows a reduction in the width of the ZayandehRud river through studying time. Between 1990 and 2020, more than 47% of the water bodies class experienced a drop. This river has the primary water resource that has made Isfahan more thriving than other arid cities in Iran. Although the Qanat effects of the underground water management system should not be neglected in the vitality of this city, it is considered a hidden water source for us, so it would not be considered in this investigation. ZayandehRud originates from the Zagros Mountains, and after passing through hundreds of urban and rural settlements, it reaches the Gavkhoni wetland. As Ziaei (2020) declares, the river has faced a water shortage crisis due to environmental and human factors in the last two decades. Human factors such as the construction of the ZayandehRud dam and the Koohrang tunnel and the establishment of the enormous steel industry are considered to lack political and environmental management, which can change the class of water bodies over history.

Future Prediction

Validation of the Model

The spatial accuracy of the result has been estimated about the spatial location of the land built up in satellite-based LULC. The transition probability matrix has been computed using the temporal LULC maps. The validation process measures the kappa coefficient using the MOLUSCE Plugin in QGIS. Different input parameters are required to validate the CA-ANN-based LULC prediction, like the number of samples taken 1000 pixels of each land cover class randomly, the learning rate of 0.001, hidden layers 10, maximum iteration of 100, and momentum of 0.001. In this chapter, three different kappas were obtained from the model, as Kappa local, Kappa histogram, and Kappa overall, which vary between 0.82, 0.91, and 0.85, respectively. CA-ANN validation has also been obtained by measuring the percentage of correctness, which was 85.23%. Considering whether the present change in land use would continue in the same manner, we anticipate the 2030 LULC, in this case, using the 2010 and 2020 LULC data, which will be discussed in the next section.

Future Urban Growth

In this research, the future prediction analysis for the year 2030 has been conducted using the ANN-based C.A. method. The simulated LULC map, area percentages, and change patterns of 2030 are shown in Fig. 19.6 and Table 19.5.

The analysis shows that between 2020 and 2030, the vegetation and bare land classes area experienced loss, while that of the built area and water bodies gained. The scope of built areas had a tremendous change with a 69.92 km^2 (24.72%) expansion; after that, there was vegetation with 23.93 km^2 (19.6%) and bare land with a 46.05 km^2 (18.47%) decrease. Furthermore, the class of water bodies gained up to 0.05 km^2 (4.58%); this increasing trend was the opposite of the past years. Isfahan will experience urban growth during the next decade due to vegetation and bare land loss, and increased water bodies.

Considering the whole process from 1990 to 2030, tremendous area changes will occur in all classes. The built area will have the dominant area amount from 163.52 km^2 (24.96%) in 1990 to 352.7 km^2 (53.83%) in 2030, a gain of 115.69%

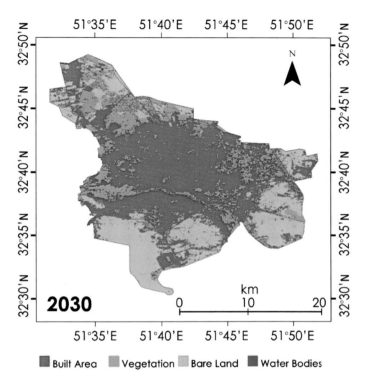

Fig. 19.6 The LULC prediction map of Isfahan, in 2030

Table 19.5 The percentages of four LULC classes in Isfahan, 2030

LULC	Built area	Vegetation	Bare land	Water bodies	Total
Km²	352.7	98.1	203.14	1.14	655.09
%	53.83	14.97	31	0.17	100

throughout the period; This growth will be due to the expansion of the steel and iron industries (Table 19.6). Subsequently, while the built area class increases, the other categories experience area droppings. The decrease in vegetation and water body percentages was somewhat similar, −46.43% and −45.45%, respectively. The reduction in the vegetation class can be considered due to the drought of the leading water bodies of Isfahan, ZayandehRud, and the industrial pollution of the soil. Furthermore, the results show that over 40 years, urban growth in Isfahan has been less inclined toward bare land than the vegetation class because the changing trend in the bare land is less than others, 33.68%, in the whole period.

Table 19.6 Increase and decrease trends between LULC classes

LULC	2020–2030 km²	%	1990–2030 km²	%
Built area	69.92	24.72	189.18	115.69
Vegetation	−23.93	−19.6	−85.04	−46.43
Bare land	−46.05	−18.47	−103.2	−33.68
Water bodies	0.05	4.58	−0.95	−45.45
Total	655.09	100	655.09	100

Conclusion

In the present chapter, the LULC change patterns of Isfahan city have been calculated from 1990 to 2020, which consisted of four classes: built area, vegetation, bare land, and water bodies. Moreover, its urban growth has been predicted up to 2030. The change detection processes were carried out using remote sensing satellite data in GEE, and the future prediction part was used with the cellular automata model in QGIS. This research exposed the expansion of the built area from 1990 to 2020. Consequently, the bare land and vegetation classes were converted into built spaces. Urban growth was mainly due to the advanced steel industries in Isfahan, which encouraged immigrants from other urban or rural districts to live in this city. The 2030 prediction maps demonstrate a considerable increase in the built area, and this progressive increase in urban growth causes further decreases in all other LULC types. Urban and industrial expansion can be considered an alarm for the continued existence of natural and social resources in Isfahan because soil and water pollution and the lack of agricultural land can cause food shortages and health problems in the future. Furthermore, to minimize the issues of city expansion in a city like Isfahan, it is essential to properly manage how to expand the city and its industries using specialized solutions such as architecture, urban planning, agriculture, environment, geography, and sociology. Using the information obtained in this chapter that shows the past to future LULC situations of Isfahan may lead to better decision-making to achieve sustainability of the Isfahan metropolitan area.

References

Abbasnia M, Tavousi T, Khosravi M, Toros H (2016) Investigation of interactive effects between temperature trend and urban climate during the last decades: a case study of Isfahan-Iran. Eur J Sci Technol 4(7):74–81

Abdalla EM, Fang D, Abd El-Hamid H (2020) Simulation and prediction of LULC change detection using Markov Chain and geo-spatial analysis, a case study in Ningxia North China. Glob Res Dev J Eng:20–32

Abijith D, Saravanan S (2021) Assessment of land use and land cover change detection and prediction using remote sensing and CA Markov In the Northern Coastal districts of Tamil Nadu, India. Research Square, pp 1–26

Aburas MM, Ho YM, Pradhan B, Salleh AH, Alazaiza MYD (2021) Spatio-temporal simulation of future urban growth trends using an integrated CA-Markov model. Arab J Geosci 14(2):1–12

Afsharzadeh M, Khorasanizadeh M, Norouzian-Maleki S, Karimi A (2021) Identifying and prioritizing the design attributes to improve the use of Besat Park of Tehran, Iran. Iran Univ Sci Technol 31(3):1–24

Ansari A, Golabi MH (2019) Prediction of spatial land use changes based on LCM in a GIS environment for Desert Wetlands–a case study: Meighan Wetland, Iran. Int Soil Water Conserv Res 7(1):64–70

Asghari Sarasekanrood S, Asadi B (2021) Analysis of land use changes and their effects on the creation of thermal islands in Isfahan City. J Geogr Res Desert Areas:217–246

Assari A, Mahesh TM (2011a) Demographic comparative in heritage texture of Isfahan City. J Geogr Reg Plan:463–470

Assari A, Mahesh TM (2011b) Urbanization process in Iranian cities. Asian J Dev Matter:151–154

Baqa M, Chen F, Lu L, Qureshi S, Tariq A, Wang S et al (2021) Monitoring and modeling the patterns and trends of urban growth using urban Sprawl matrix and CA-Markov Model: a case study of Karachi, Pakistan, Land

Bhatta B (2010) Analysis of urban growth and sprawl from remote sensing data. Springer, Berlin

Black D, Henderson V (1999) A theory of urban growth. J Polit Econ 107(2):252–284

Bongaarts J (2006) United nations department of economic and social affairs, population division world mortality report 2005. Popul Dev Rev 32(3):594–596

Breiman L (2001) Random forests. Mach Learn:5–32

Canh NP, Thanh SD (2020) Domestic tourism spending and economic vulnerability. Ann Tour Res 85:103063

Clarke KC, Hoppen S, Gaydos L (1996) Methods and techniques for rigorous calibration of a cellular automaton model of urban growth. In: Third international conference/workshop on integrating GIS and environmental modeling, 21–25.

Eslamian S, Tarkesh Esfahany S, Nasri M, Safamehr M (2013) Evaluating the potential of urban reclaimed water in area of north Isfahan, Iran, for industrial reuses. Int J Hydrol Sci Technol:258–269

Esmaeili A, Moore F, Keshavarzi B, Jaafarzadeh N, Jafarkhani Kermani M (2014) A geochemical survey of heavy metals in agricultural and background soils of the Isfahan industrial zone, Iran. Catena:88–98

Estornell J, Gavilá JM, Sebastiá MT, Mengual J (2013) Principal component analysis applied to remote sensing. Modell Sci Educ Learn 6:83–89

Falah N, Karimi A, Harandi AT (2020) Urban growth modeling using cellular automata model and AHP (case study: Qazvin city). Model Earth Syst Environ 6(1). https://doi.org/10.1007/s40808-019-00674-z

Gerten C, Fina S, Rusche K (2019) The sprawling planet: simplifying the measurement of global urbanization trends. Front Environ Sci 7:140

Gohain KJ, Mohammad P, Goswami A (2020) Assessing the impact of land use land cover changes on land surface temperature over Pune city, India. Quat Int, March, 0–1. https://doi.org/10.1016/j.quaint.2020.04.052

Gomes SL, Hermans LM (2018) Institutional function and urbanization in Bangladesh: how peri-urban communities respond to changing environments. Land Use Policy 79:932–941

Han H, Yang C, Song J (2015) Scenario simulation and the prediction of land use and land cover change in Beijing, China. Sustain For 7:4260–4279

Jewitt D, Goodman PS, Erasmus BFN, O'Connor TG, Witkowski ETF (2015) Systematic land-cover change in KwaZulu-Natal, South Africa: implications for biodiversity. S Afr J Sci 111(9–10):1–9

Jin H, Mountrakis G (2013) Integration of urban growth modelling products with image-based urban change analysis. Int J Remote Sens:5468–5486

Jokar Arsanjani J, Kainz W (2011) Integration of spatial agents and Markov Chain Model in simulation of urban Sprawl. In: Proceedings of the 4th AGILE international conference on geographic information science. AGILE, Utrecht, pp 18–22

Kalkhajeh RG, Jamali AA (2019) Analysis and predicting the trend of land use/cover changes using neural network and systematic points statistical analysis (SPSA). J Indian Soc Remote Sens 47(9):1471–1485

Kantakumar LN, Kumar S, Schneider K (2019) SUSM: a scenario-based urban growth simulation model using remote sensing data. Eur J Remote Sens 52(sup2):26–41

Karimi H, Jafarnezhad J, Khaledi J, Ahmadi P (2018) Monitoring and prediction of land use/land cover changes using CA-Markov model: a case study of Ravansar County in Iran. Arab J Geosci 11(19):1–9

Karimi A, Sanaieian H, Farhadi H, Norouzian-Maleki S (2020) Evaluation of the thermal indices and thermal comfort improvement by different vegetation species and materials in a medium-sized urban park. Energy Rep 6. https://doi.org/10.1016/j.egyr.2020.06.015

Khan Z, Saeed A, Bazai M (2020) Land use/land cover change detection and prediction using the CA-Markov model: a case study of Quetta city, Pakistan. J Geogr Soc Sci:164–182

Khanna NP (2020) Urbanization and urban growth: sustainable cities for safeguarding our future. Sustain Cities Commun:953–965

Kottek M, Grieser J, Beck C, Rudolf B, Rubel F (2006) World map of the Köppen-Geiger climate classification updated. Meteorol Z 15(3):259–263

Kushwaha K, Singh MM, Singh SK, Patel A (2021) Urban growth modeling using earth observation datasets, Cellular Automata-Markov Chain model and urban metrics to measure urban footprints. Remote Sens Appl Soc Environ 22:100479

Langer H, Susanna F, Hammer C (2020) Chapter 2 – Supervised learning. In: Langer IH, Susanna F, Hammer C (eds) Advantages and pitfalls of pattern recognition. Elsevier, Amsterdam, pp 33–85

Lee H-Y, Jang KM, Kim Y (2020) Energy consumption prediction in vietnam with an artificial neural network-based urban growth model. Energies 13(17):4282

Li Y, Jia L, Wu W, Yan J, Liu Y (2018) Urbanization for rural sustainability–Rethinking China's urbanization strategy. J Clean Prod 178:580–586

Liping C, Yujun S, Saeed S (2018) Monitoring and predicting land use and land cover changes using remote sensing and GIS techniques—a case study of a hilly area, Jiangle, China. PLoS One 13(7):e0200493

Loughlin WP (1991) Principal component analysis for alteration mapping. Photogramm Eng Remote Sens:1163–1169

Lu Dadao YS, Hui L (1997) China regional development report: urbanization and Spatial Sprawl. The Commercial Press, Beijing

Mazloum B, Pourmanafi S, Soffianian A, Salmanmahiny A, Prishchepov AV (2021) The fate of rangelands: revealing past and predicting future land-cover transitions from 1985 to 2036 in the drylands of Central Iran. Land Degrad Dev 32:4004–4017

McConnell WJ (2015) Land change: the merger of land cover and land use dynamics A2—Wright, James D. In: International Encyclopedia of the Social & Behavioral Sciences. Elsevier, Oxford

Mohammad P, Goswami A (2019) Temperature and precipitation trend over 139 major Indian cities: an assessment over a century. Model Earth Syst Environ 5(4):1481–1493. https://doi.org/10.1007/s40808-019-00642-7

Mohammad P, Goswami A (2021a) A Spatio-Temporal Assessment and Prediction of Surface Urban Heat Island Intensity Using Multiple Linear Regression Techniques Over Ahmedabad City, Gujarat. J Indian Soc Remote Sens 49(5):1091–1108. https://doi.org/10.1007/s12524-020-01299-x

Mohammad P, Goswami A (2021b) Quantifying diurnal and seasonal variation of surface urban heat island intensity and its associated determinants across different climatic zones over Indian cities. GISci Remote Sens 00(00):1–27. https://doi.org/10.1080/15481603.2021.1940739

Mohammad P, Goswami A (2021c) Spatial variation of surface urban heat island magnitude along the urban-rural gradient of four rapidly growing Indian cities. Geocarto Int 1–23. https://doi.org/10.1080/10106049.2021.1886338

Mohammad P, Goswami A, Bonafoni S (2019) The impact of the land cover dynamics on surface urban heat island variations in semi-arid cities: a case study in Ahmedabad City, India, Using Multi-Sensor/Source Data. Sensors 19(17):3701. https://doi.org/10.3390/s19173701

Mosammam HM, Nia JT, Khani H, Teymouri A, Kazemi M (2017) Monitoring land use change and measuring urban sprawl based on its spatial forms: the case of Qom city. Egypt J Remote Sens Space Sci 20(1):103–116

Mumtaz F, Tao Y, Bashir W, Kareem M, Gengke W, Li L, Bashir B (2020) Transition of Lulc and future predictions by using Ca-Markov chain model (a case study of metropolitan city Lahore, Pakistan). ESMY 4:146–151

Musa SI, Hashim M, Md Reba MN (2017) A review of geospatial-based urban growth models and modelling initiatives. Geocarto Int:813–833

Parsa VA, Salehi E (2016) Spatio-temporal analysis and simulation pattern of land use/cover changes, case study: Naghadeh, Iran. J Urban Manag 5(2):43–51

Parsi H, Farmihani Farahani B (2014) Analysis the urban Sprawl in the Peripheral Metropolitan Areas (case study: Northern Peripheral Areas of Isfahan, Iran). Urban Stud:49–62

Pielke RA Sr, Pitman A, Niyogi D, Mahmood R, McAlpine C, Hossain F, Goldewijk KK, Nair U, Betts R, Fall S (2011) Land use/land cover changes and climate: modeling analysis and observational evidence. Wiley Interdiscip Rev Clim Chang 2(6):828–850

Qelichi MM, Murgante B, Feshki MY, Zarghamfard M (2017) Urbanization patterns in Iran visualized through spatial auto-correlation analysis. Spat Inf Res 25(5):627–633

Ramezankhani R, Sajjadi N, Jozi SA, Shirzadi MR (2018) Climate and environmental factors affecting the incidence of cutaneous leishmaniasis in Isfahan, Iran. Environ Sci Pollut Res 25(12):11516–11526

Sadowski J, Bendor R (2019) Selling smartness: corporate narratives and the smart city as a sociotechnical imaginary. Sci Technol Hum Values 44(3):540–563

Sampson L, Ettman CK, Galea S (2020) Urbanization, urbanicity, and depression: a review of the recent global literature. Curr Opin Psychiatry 33(3):233–244

Saxena A, Jat MK (2020) Land suitability and urban growth modeling: development of SLEUTH-Suitability. Comput Environ Urban Syst 81:101475

Shirani-Bidabadi N, Nasrabadi T, Faryadi S, Larijani A, Roodposhti MS (2019) Evaluating the spatial distribution and the intensity of urban heat island using remote sensing, case study of Isfahan city in Iran. Sustain Cities Soc 45:686–692

Soffianian A, Madanian M (2015) Monitoring land cover changes in Isfahan Province, Iran using Landsat satellite data. Environ Monit Assess 187(8):1–15

Stevens D, Dragicevic S, Rothley K (2007) iCity: a GIS–CA modelling tool for urban planning and decision making. Environ Model Softw 22(6):761–773

Tang P, Feng Y, Li M, Zhang Y (2021) Can the performance evaluation change from central government suppress illegal land use in local governments? A new interpretation of Chinese decentralisation. Land Use Policy 108:105578

Tayyebi A, Pijanowski BC, Tayyebi AH (2011) An urban growth boundary model using neural networks, GIS and radial parameterization: an application to Tehran, Iran. Lands Urban Plan 100(1–2):35–44

Treitz P, Rogan J (2004) Remote sensing for mapping and monitoring land-cover and land-use change-an introduction. Prog Plan 61(4):269–279

Tripathy P, Kumar A (2019) Monitoring and modelling spatio-temporal urban growth of Delhi using cellular Automata and geoinformatics. Cities 90:52–63

Vanden Borre J, Paelinckx D, Mücher CA, Kooistra L, Haest B, De Blust G, Schmidt AM (2011) Integrating remote sensing in Natura 2000 habitat monitoring: prospects on the way forward. J Nat Conserv 19(2):116–125

Vihervaara P, Auvinen A-P, Mononen L, Törmä M, Ahlroth P, Anttila S, Böttcher K, Forsius M, Heino J, Heliölä J (2017) How essential biodiversity variables and remote sensing can help national biodiversity monitoring. Glob Ecol Conserv 10:43–59

Wang H, He Q, Liu X, Zhuang Y, Hong S (2012) Global urbanization research from 1991 to 2009: a systematic research review. Landsc Urban Plan 104(3–4):299–309

Wu J, Li Y, Li N, Shi P (2018) Development of an asset value map for disaster risk assessment in China by spatial disaggregation using ancillary remote sensing data. Risk Anal 38(1):17–30

Wu H, Gai Z, Guo Y, Li Y, Hao Y, Lu Z-N (2020) Does environmental pollution inhibit urbanization in China? A new perspective through residents' medical and health costs. Environ Res 182:109128

Yatoo SA, Sahu P, Kalubarme MH, Kansara BB (2020) Monitoring land use changes and its future prospects using cellular automata simulation and artificial neural network for Ahmedabad city, India. GeoJournal:1–22

Ziaei L (2020) Zayandeh Rud River Basin: a region of economic and social relevance in the Central Plateau of Iran. In: Mohajeri S, Horlemann L, Besalatpour AA, Raber W (eds) Standing up to climate change. Springer, Cham, pp 91–105

Chapter 20
Analysing Spatio-temporal Changes in Land Surface Temperature of Coastal Goa Using LANDSAT Satellite Data

Venkatesh G. Prabhu Gaonkar, F. M. Nadaf,
Vikas BalajiraoKapale, Siddhi Gaonkar, Sumata Shetkar,
and Merel D'Silva

Abstract Rising temperatures, ice cap melting, sea level rising, heatwaves, droughts, flooding, extreme cold and snow, tropical cyclones and extratropical storms are some of the key environmental concerns confronting the planet today due to climate change and anthropogenic actions. To detect climate change, and estimate the temperature, land surface temperature (LST) is a very vital parameter. Evaluation of LST helps in understanding temperature differences, which in turn, is affected by Normalized Difference Vegetation Index (NDVI), altitude, and land cover. At a given place and time both natural and anthropogenic activities affect LST.

The studies conducted in different parts of the world including Goa and India indicate the impacts of climate change on different ecosystems. Hence, to quantify the spatio-temporal variability of the LST in Coastal Goa, Landsat series for 1991–2021 were used. A six-fold process using geospatial tools is employed to determine the land surface temperature. The present study shows that the processed mean land surface temperatures, data obtained from data access viewer as well as meteorological data exhibit a similar trend.

Keywords Land Surface Temperature (LST) · Earth surface · Satellite Data · Normalized Difference Vegetation Index (NDVI) · Land use land cover

---

V. G. Prabhu Gaonkar · V. BalajiraoKapale
Department of Geoinformatics, S. P. Chowgule College, Margao, Goa, India
e-mail: vgp005@chowgules.ac.in

F. M. Nadaf (✉)
Department of Geography, DPM's Shree Mallikarjun ShriChetan Manju Desai College, Canacona, Goa, India

S. Gaonkar · S. Shetkar · M. D'Silva
Department of Geography, Government College, Khandola, Goa, India
e-mail: merel.dsilva@khandolacollege.edu.in

Introduction

Land surface temperature (LST) and air temperature are two different parameters. NASA defines land surface temperature as 'a condition of the Earth at a given geographic location that indicates the heat of the surface'. The sensors fitted to the satellite, observe the surface which includes snow cover, grass, the roof of a structure, or the greenery which is perceived by the sensors as the earth's 'surface' (Li and Duan 2017). LST is also well-defined as the emitted temperature of the skin of the earth's surface received by satellite-based distant sensors (Copernicus 2021). On the other hand, air temperature is the condition just above the surface which is normally recorded using weather instruments.

Land surface temperature is an important parameter in land surface processes, not only because it serves as an indicator of climate change, but also because it controls upward terrestrial radiation (Sun 2008). Determination of LST requires thorough attention towards the intensity of solar radiation (energy received per unit time per unit surface area of the Earth). During the winter solstice, the angle of depression of the Sun is lowest vis-a-vis the angle of incidence of solar radiation in the northern hemisphere is highest which in turn increases the distance solar radiation has to travel through the atmosphere which directly affects the amount of energy received and emitted by the Earth's surface. The solar energy intercepted by the earth fluctuates by $\pm 3.3\%$ about the mean value due to variations in the sun-earth distance, with the maximum at the start of January and the lowest at the start of July (Landsberg 1961).

Land surface temperature is a precise estimation tool for signifying the energy exchange balance between the Earth and the Atmosphere (Youneszadeh et al. 2015). It is also a vital variable in estimating the temperature of the surface of the earth which helps in understanding global climate change (Devi et al. 2020). The level of LST is fundamentally affected by the altitude, gradient and aspect. Notwithstanding, physiography is one of the variables that regulates distribution of soil moisture and applies an extra impact on the LST (Youneszadeh et al. 2015).

Varied land use categories have different surface reflectance and roughness, resulting in variances in land surface temperature (Deng et al. 2018). Specular surfaces have uniform reflection whereas Lambertian surfaces have diffused reflection (Griffiths and De Haseth 2007). Non-evaporative and non-porous urban materials tend to have high heat capacity and low solar reflectivity compared to organic surfaces. Thermal inertia is very high for materials such as concrete masses, asphalt roads and metal surfaces (Farina 2012).

The underlying premise behind using satellite imagery to estimate soil moisture is that soil moisture has an impact on surface properties that can be measured through remote sensing techniques. Biophysical elements like vegetation cover can be assessed by vegetation indices, and the surface energy balance can be measured by surface temperature (Petropoulos et al. 2009; Wang and Qu 2009; Shafian and Maas 2015). Soil moisture and vegetation cover are two de-facto features that have complex interdependency and the highest weighted value in the determination of

LST (Carlson et al. 1994; Entezari et al. 2019). We must take into account the fact that barren land and moisture-rich soil have different thermal signatures (Moukalled et al. 2006).

Vegetation may be recognized from most other materials using remote sensing data due to its noticeable absorption in the red and blue regions of the visible spectrum. The spatial and spectral resolution helps to understand to determine NDVI. This index has been used to correct land surface emissivity and also serves as a pivot point in the calculation of LST.

Many studies have been conducted by scientists to study LST using Landsat data (Ravanelli et al. 2018). To comprehend the physical processes, Landsat data has opened the floodgates through remote sensing (Yuvaraj 2020). Surface temperature is a vital parameter to map and monitor environmental complications (Becker and Li 1990).

For changes in land use and land cover, human and natural actions are principally blameworthy (Brovkin et al. 2013). Alterations in land cover have impacted the environment framework through the release of Green House Gases, for example, CO_2 and CH_4 and adjustment of land surface albedo, evapotranspiration, and surface roughness. Changes in the land cover affect both regional and global climates (Brovkin et al. 2006). Due to a blend of physical and anthropogenic reasons, there are modifications in the climatic patterns of a geographical area over a long period (Lambeck 2010).

Over the last century, the Earth's climate has warmed by about 0.6 °C, with two major periods of warming occurring between 1910 and 1945, and from 1976 onwards (Walther et al. 2002). Climate change has a direct impact on land surface temperature (LST) and accelerates permafrost thawing, which affects land degradation, sea level rise, etc. (Lawrence and Slater 2005; Zimov et al. 2006; Petropoulos et al. 2012; Haigh and Cargill 2015).

Urbanization brings key changes in land use thereby affecting LST by destabilizing the surface energy balance (Imran et al. 2021). Urban hydrology, urban heat islands (UHI), temperature regimes, and others are some of the major changes inflicted by the land use/cover changes which have led to unsustainable environments (Grover and Singh 2016). Further UHI affects local and global climate change due to city power consumption (Chotchaiwong and Wijitkosum 2019). Climate change is perhaps the most serious problem planet is facing. Past research indicates that climate change significantly affects the land surface temperature and other parameters (Mustafa et al. 2020).

Many research studies have investigated the various aspects of rainfall, temperature and climate change in Goa. However, only a few studies have examined land surface temperature and land use land cover change in Goa.

Dhorde and Korade have studied trends in surface temperature variability over Panaji City of Goa and indicated that Panaji has shown signs of warming with change in seasons (Dhorde 2015). T. V. Ramachandra, Uttam Kumar and Anindita Dasgupta in their technical report "Analysis of land surface temperature and rainfall with landscape dynamics in Western Ghats" have observed that area under dense forest has declined and area under agricultural/grassland has amplified in the

Western Ghats. The rainfall showed a decreasing trend in the pattern over forest and agricultural/grassland. Further, the study specifies various responses to changing LST and precipitation with reference to NDVI of dense vegetation (Ramachandra et al. 2016).

The study of Ramaiah, Avtar and Rahman endeavours to look at Panaji and Tumkur cities of India with the help of landscape sensitivity analysis and its effect on LST. The study uncovers that water bodies and vegetation are effectively responsible in reducing LST (Ramaiah et al. 2020). The IMD, Pune report "Climate of Goa State", makes an attempt to study the Climate of Goa with the help of meteorological data (India Meteorological Department 2019).

As per the "State Action Plan on Climate Change for the State of Goa for Period of 2020–2030", between 1901 and 2018, the mean annual temperature of Goa has increased more than 1 °C with noticeable rise during 1990–2018. Further, the report predicts that under the high emission circumstances, the mean annual temperatures may rise by 2 °C in 2030s and 4 °C by 2080s in relation to 1901–1950. The state of Goa will begin encountering hot spells (>40 °C) beyond 2040s due to rise in temperature by 5 °C towards the culmination of this century under high emission conditions. Under high emission conditions, the minimum temperatures are relied upon to upsurge more than to 8 °C by the end of this era (NABARD 2012). Hence, in the light of the above observations, this chapter makes an attempt to explore the spatio-temporal changes in land surface temperature using Landsat data.

Objectives

The key objectives of this study include

1. To examine spatio-temporal changes in land surface temperature from 1991 to 2021 in coastal Goa using Landsat series data.
2. To validate land surface temperature with the help of Data Access Viewer and Meteorological data.
3. To identify and examine the spatial pockets revealing the rise in land surface temperature.

Study Area

Goa located along the coast of the Arabian Sea is a geographically tiny yet highly diversified state. Administratively state is divided into 12 talukas (administrative units) of which 7 talukas namely Pernem, Bardez, Tiswadi, Mormugao, Salcete, Quepem and Canacona are coastal talukas. The coast of Goa is 105 km long with varying widths (Nadaf 2019). The coast of Goa is dotted with some of the breathtaking beaches of the world Arambol, Anjuna, Baga, **Vagator, Calangute,** Miramar, Colva, Agonda, Palolem, Patnem and Galgibag and incredible estuarine rivers and mangrove ecosystem. Arambol and Galgibag beaches are identified for Olive Ridley Turtle nesting sites (Nadaf 2020).

For this study, the costal Goa is explicitly divided into three regions, i.e. Northern Coast, Mid Coast and Southern Coast. The Northern Coast consists of the coast of Pernem, Bardez and Tiswadi. Mormugao and Salcete make the mid-coast, whereas Quepem and Canacona encompass the Southern Coast (Fig. 20.1).

Goa is a tourist paradise for both domestic and international tourists which is also the treasure of mineral wealth. It is known for some of its spectacular beaches and wildlife sanctuaries. Tourism in Goa began immediately after its liberation in 1961, since 1975 growth has been alarming (Table 20.1). Just before Covid-19 lockdowns, 80,03,795 tourists visited the fascinating beaches of Goa. Such a huge tourist footfall requires strong infrastructure such as stared hotels, resorts, pubs, shacks, housing apartments, transport network connectivity, etc. (Table 20.2).

The coastal Goa is dotted with 3900 hotels and a large number of beach shacks with 48,534 rooms and 83,706 beds. Such a huge infrastructure in a tiny state of Goa which is sandwiched between the Western Ghats and the Arabian Sea is possible only by changing land use land cover.

The study conducted by the Government of India suggests that many beaches have exceeded the carrying capacity. This has led to the adverse impact on the coastal ecosystem ((NCSCM, GOI) 2017).

The Western Ghats of Goa are known to have rich reserves of iron ore and manganese. Of the 600 square kilometres of the total area of the Western Ghats, mineral wealth is found in 350 square kilometres of area. For a very long time, mining remained as the backbone of Goan economy along with tourism. The mining that is carried out in the Western Ghats is responsible for cascading effects of flooding in low-lying areas during the monsoon season (Alvares 2002).

Of late, the Arabian Sea and the coast of Goa have been experiencing severe cyclones. According to a study conducted by the Indian Institute of Tropical Meteorology, severe cyclones have increased by 150% in the Arabian Sea with 260% rise in their duration (Deshpande et al. 2021). Climate change is a major issue in coastal areas. Rise in sea level and occurrence and intensity of cyclonic storms are the major impacts of global climate change on the coast and people (TERI 2015). The recent cyclones such as Ockhi, Maha, Kyarr, Vayu, Nisarga and Tauktae have caused irreparable damage to the coast of Goa.

Materials and Methods

This study is an outcome of both primary and secondary data. Primary data is obtained from field observations and ground-truthing. Secondary data includes satellite images from the LANDSAT Series of TM, ETM+ and OLI (Table 20.3).

Out of 30 years, data for only 21 years was available for analysis due to technical errors. In addition, several sources were considered for data validation, such as NASA's Data Access Viewer and data from India Meteorological Department (Fig. 20.2).

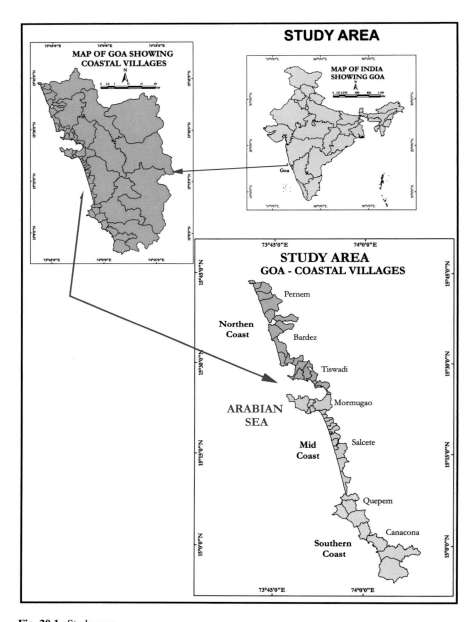

Fig. 20.1 Study area

Table 20.1 Domestic and foreign tourist arrivals in Coastal Goa

Year	Domestic tourists	Foreign tourists	Total tourists
1965	61,252	1939	63,191
1975	1,98,110	14,521	2,12,631
1985	6,64,692	87,599	7,52,291
1995	8,78,487	2,29,218	11,07,705
2010	19,94,711	3,66,940	23,61,651
2015	48,36,711	5,84,032	54,20,743
2018	69,51,467	9,22,766	78,74,233
2019	71,05,587	8,98,208	80,03,795

Source: Department of Tourism, Government of Goa & Statistical Handbooks of Goa

Table 20.2 Tourism infrastructure

Total number of Hotels/Paying Guest House, Rooms and Beds as on 31.03.2019 (including Star Category and Heritage Hotels)			
Category	No. of hotels	No. of rooms	No. of beds
A	84	9034	15,189
B	252	10,156	17,467
C	715	11,450	20,495
D	2784	12,508	20,516
Star category hotels	63	5362	10,001
Heritage hotels	02	24	38
TOTAL	**3900**	**48,534**	**83,706**

Source: Department of Tourism, Government of Goa

Table 20.3 Satellite images used in the study

Sr. No	Satellite	Sensor	Years	Resolution	Path	Row	No of bands	Cloud cover
1	Landsat	TM	1991, 1992, 1993, 1994,1995,1997, 1999	30 m Thermal band (120 m)	146, 147	49, 50	7	Nil
2	Landsat	ETM+	2001, 2002, 2005, 2008, 2009, 2010, 2011	30 m Thermal band (60 m)	146, 147	49, 50	8	Nil
3	Landsat	OLI	2014, 2015, 2016, 2017, 2019, 2020, 2021	30 m Thermal band (100 m)	146, 147	49, 50	11	Nil

DIVA-GIS was used for obtaining shapefiles for the area under investigation. For processing, Arc Map 10.7.1 version was utilized. Raster datasets were superimposed with shapefiles of the study area and then extracted. Mosaic operations were used to combine the clipped raster datasets.

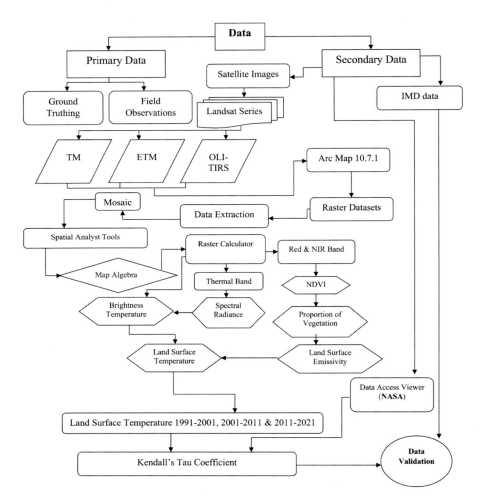

Fig. 20.2 Flowchart for LST retrieval

Spatial analyst tools were employed to conduct the analysis. To calculate various algorithms Raster Calculator was used. To derive near accurate results formulas (U.S. Geological Survey 2016) have been used to calculate surface radiance, brightness temperature, NDVI, proportion of vegetation, land surface emissivity, and land surface temperature (Carlson and Ripley 1997).

The output of spectral radiance is used to calculate brightness temperature using the thermal infrared band. On the other hand, red band and near infrared (NIR) band are applied for the Normalized Difference Vegetation Index (NDVI) and the proportion of vegetation. The proportion of vegetation is required for the determination of land surface emissivity. Finally, data validation is carried out using NASA's Data Access Viewer and meteorological data by applying Kendall's Tau coefficient.

Top of Atmospheric Radiance

The first step in the calculations of LST is a conversion of DN value to the top of atmospheric radiance. LST, as well as emissivity, have to be derived from active or passive electromagnetic radiation leaving the Earth's surface. These measurements are done by satellite sensors integrating radiating effect of various surface features (Dash et al. 2002). DN is the Digital Number value assigned to each pixel in a given raster. These values represent the intensity of electromagnetic radiation captured by the sensors onboard the satellite.

For Landsat 5 (TM) and Landsat 7(ETM+)

The following formula is taken from Landsat 7 data:
   Band_6 for Landsat 5(TM) andBand_6_VCID 1 for Landsat 7 (ETM+) is selected for the calculations (Ihlen and Zanter 2019).

$$L_\Delta = \left(\frac{L_{Max\Delta} - L_{Min\Delta}}{Q_{cal(max)} - Q_{cal(min)}} \right) \times \left(Q_{cal} - Q_{cal(min)} \right) + L_{min\,\Delta}$$

where

L_Δ = Spectral Radiance
$L_{Max\Delta}$ = Spectral Radiance of band 6 (Read from MetaData as RADIANCE_MAXIMUM_BAND_X)
$L_{Min\Delta}$ = Spectral radiance of band 6 scaled to $Q_{cal(min)}$ (From Meta Data read RADIANCE_MINIMUM_BAND_X)
Q_{cal} = Quantized calibrated pixel value (While performing calculations Insert Raster file of BAND_X in the raster calculator)
$Q_{cal(max)}$ = Maximum quantized calibrated pixel value (Form MetaData QUANTIZE_CAL_MAX_BAND_X)
$Q_{cal(min)}$ = Minimum quantized Pixel value (From MetaData QUANTIZE_CAL_MIN_BAND_X)
X represents Band Number

For Landsat 8 (OLI and TIRS)

For sensor spectral radiance, the following formula is used:
   Band_10 is used for the calculations (U.S. Geological Survey 2016).

$$L_\Delta = M_L \times Q_{cal} + A_l$$

where

L_Δ = Spectral radiance.
M_L = Radiance multiplicative scaling factor for band 10 (read from MetaData RADIANCE_MULT_BAND_10)
Q_{cal} = Pixel value in DN (Insert raster of Band 10 in the calculator)
A_I = Radiance additive scaling factor for band 10 (read from MetaData RADIANCE_ADD_BAND_10)

For this study of LST, two sets of formulas are used as the data acquired is from Landsat 5 (TM), Landsat 7 (ETM+) and Landsat 8 (OLI and TIRS) is used.

Top of Atmospheric Brightness Temperature

The next step in the determination of LST is, finding Top of Atmospheric Brightness Temperature. For this, we have to convert spectral radiance values into more meaningful physical parameters such as temperature. Values obtained from the following calculations give us temperature viewed by a satellite at the top of the atmosphere considering Earth as a perfect blackbody. An ideal blackbody is a perfect absorber as well as a perfect emitter of radiation with an emissivity value of 1 (Strojnik et al. 2016; Ihlen and Zanter 2019)

$$T_B = \frac{K_2}{\ln\left(\frac{K_1}{L_\Delta} + 1\right)}$$

where

T_B = At Satellite Temperature measured in Kelvin

TIRS Thermal Constants;

K_1 = Band-specific thermal calibration constant 1.
(Read from Meta Data for Landsat 8 (TIRS) K1_CONSTANT_BAND_X)
K_2 = Band-specific thermal calibration constant 2.
(Read from Meta Data for Landsat 8 (TIRS) K2_CONSTANT_BAND_X)

Rectification of at Sensor Temperature Through Emissivity Correction

Considering Earth as a perfect blackbody is a crude assumption but the results of T_B obtained through the above formula can be rectified with emissivity correction by using NDVI values.

Determination of NDVI

NDVI is the most commonly used index to infer vegetation health; however, this index is useful for this study as emissivity estimation is based on NDVI values of a pixel (Avdan and Jovanovska 2016). For the calculations of NDVI for Landsat 5 (TM) and Landsat 7(ETM+), Red bad 3 and NIR Band 4 are used, whereas, for Landsat 8(TIR), Red Band 4 and NIR Band 5 are used. The formula for NDVI is

$$\text{NDVI} = \frac{\text{NIR} - \text{Red}}{\text{NIR} + \text{Red}}$$

Here NIR and Red Band represent surface reflectance values of wavelengths in the near-infrared region averaged over ($\lambda \sim 0.77 - 0.90$ μm) and in the visible region averaged over ($\lambda \sim 0.63 - 0.69$ μm) (Carlson and Ripley 1997).

The NDVI value of a pixel can range from -1 to $+1$.

The next step is to calculate Vegetation proportion (Carlson and Ripley 1997):

$$P_v = \left[\frac{\text{NDVI} - \text{NDVI}_{\min}}{\text{NDVI}_{\max} - \text{NDVI}_{\min}} \right]^2$$

where

$$\text{NDVI}_{\max} = 0.5$$
$$\text{NDVI}_{\min} = 0.2$$

NDVI= Insert NDVI raster image.

Above values are considered for the global conditions (Sobrino et al. 2004a, b). For the current study, we have segregated Earth's surface into three broad categories water cover, bare soil, vegetation cover and one subcategory of mixed pixels of bare soil and vegetation. NDVI threshold method provides a particular value for different surface features. The obtained values of NDVI are further used for emissivity estimation. NDVI < 0 is attributed to water bodies (Sobrino et al. 2008).

NDVI pixel values $0 \leq \text{NDVI} < 0.2$ ranging between 0 and 0.2 are considered as bare soil whereas $0.2 \leq \text{NDVI} \leq 0.5$ is considered as a composition of a mixture of bare soil and vegetation and NDVI > 0.5 pixels are considered as dense vegetation cover (Sobrino et al. 2004a, b). Corresponding emissivity values for Landsat 5 (TM) and Landsat (ETM+) emissivity of soil is 0.97 and for dense vegetation cover it is 0.99 and for the mixture of bare soil and vegetation emissivity is calculated by (Sobrino et al. 2004)

$$\xi_6 = 0.986 + 0.004 P_v$$

where ξ_6= emissivity calculated for thermal band 6 in TM and ETM+ sensor (Landsat 5 and Landsat7)

P_v = vegetation proportion

For Landsat 8(TIR) corresponding emissivity values for soil, vegetation is 0.996 and 0.973 (Wang et al. 2015)

For mixed pixels of soil and vegetation, emissivity can be calculated as (Sobrino et al. 2008);

$$\xi_{10} = \xi_s + (\xi_v - \xi_s)P_v$$

For the given values of emissivity of soil and vegetation formula becomes:

$$\xi_{10} = 0.966 + 0.007 P_v$$

where ξ_{10} = emissivity calculated for thermal band 10 in TIRS sensor (Landsat 8)

ξ_s = emissivity of soil
ξ_v = emissivity of vegetation.

Values shown in Table 20.4 are used for emissivity estimation of land surface

Determination of Land Surface Temperature with Emissivity Correction (Weng et al. 2004)

$$\mathrm{LST} = \frac{T_B}{1 + \left[\left(\frac{\lambda \times T_B}{\rho}\right) \times \ln \xi\right]}$$

where

T_B = brightness or at satellite sensor temperature.
λ = wavelength of emitted radiance.
ξ = insert emissivity image in raster calculator.

Table 20.4 Emissivity estimation values of land surface

Surface feature	NDVI	Emissivity (ξ) TM sensor	Emissivity (ξ) TIR sensor
Water	NDVI < 0	0.991	0.991
Soil	$0 \leq$ NDVI < 0.2	0.970	0.966
Mixture of bare soil and vegetation	$0.2 \leq$ NDVI ≤ 0.5	$\xi_6 = 0.986 + 0.004 P_v$	$\xi_{10} = 0.966 + 0.007 P_v$
Vegetation	NDVI > 0.5	0.99	0.973

$$\rho = \frac{h \times c}{\sigma}$$

Plank's constant $h = 6.626 \times 10^{-34}$ Js, speed of light $c = 2.998 \times 10^8$ m/s
σ is Blotzmann constant its value is 1.38×10^{-23} J/K
For Landsat 5 (TM) and Landsat 7(ETM+) $\lambda = 11.457$ μm (Jiménez-Munoz and Sobrino 2003)
And for Landsat 8(TIR) Band10 $\lambda = 10.869$ μm (Yu et al. 2014)

Results and Discussion

Following the above-specified methodology, the LST maps of the northern coast, mid coast and southern coast of Goa are prepared for three decades, i.e., 1991–2000, 2001–2010 and 2011–2021 (Figs. 20.3, 20.4 and 20.5).

It is observed that on the Northern Coast during 1991–2000 the minimum land surface temperature was 24.02 °C, which increased to 24.19 °C during 2001–2010. In the subsequent decade, 2011–2021 there was a drop in the minimum land surface temperature that is 23.94 °C. Whereas the maximum land surface temperature during the above three decades was 32.95 °C, 34.66 °C and 33.90 °C respectively. Though there was a drop in the maximum land surface temperature in 2011–21 decade however, it was more than 1991–2000 decade (Fig. 20.3 and Table 20.5).

During 1991–2021, the Mid Coastal Region has experienced minimum land surface temperature between 24.22 °C and 24.27 °C. Similarly, the maximum land surface temperature has shown a substantial increase from 32.98 °C to 36.70 °C during 1991–2021 (Fig. 20.4 &Table 20.5).

In the Southern Coastal Region, the minimum land surface temperature during 1991–2000 was 24.79 °C. In the subsequent decades 2001–2010 and 2011–2021, the minimum land surface temperature was 23.79 °C and 24.28 °C respectively, showing a decrease with reference two the last two decades. The maximum land surface temperature has increased from 36.08 °C to 37.90 °C (Fig. 20.5, Table 20.5).

It is evident from the above figures (Figs. 20.3, 20.4, and 20.5) that the mean land surface temperature in the entire coastal Goa has increased from 28.42 °C during the decade 1991–2002 to 28.90 °C during 2001–2010. The subsequently, mean land surface temperature has further increased to 29.11 °C during the current decade 2011–2021.

It is interesting to note that in all three regions of coastal Goa, the minimum land surface temperatures have decreased, maximum land surface temperatures have increased and mean land surface temperature has shown a rise.

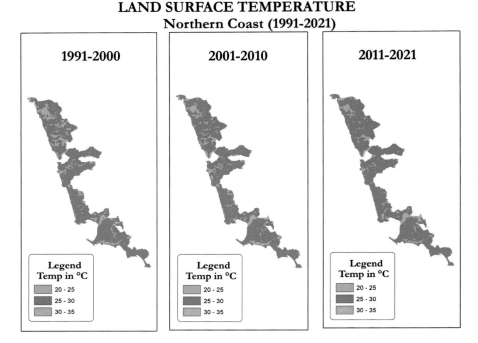

Fig. 20.3 Land surface temperatures 1991–2021 for Northern Coast of Goa

To further investigate into the impact of climate change on land surface temperature over the coastal Goa, 15 spatial pockets representing varied land surface features were identified; further mean land surface temperature was examined temporally and tabulated below (Figs. 20.6, 20.7 and 20.8).

It is observed from the above figures (Figs. 20.7, and 20.8) the rise in land surface temperature has largely occurred in those spatial pockets where much physical and anthropogenic transformations have not taken place. Further, it is important to note that the Southern Coast, which is a part of the Western Ghats has mostly witnessed rise in both minimum and maximum land surface temperatures in the past three decades in comparison with Northern and Mid Coast. Similarly, situations are also found in Northern and Mid coastal region. Hence, the rise in land surface temperature can be attributed to global climatic changes.

A change in course of the Earth's climatic trend can be identified using a non-parametric Mann-Kendall test (Mohorji et al. 2017); this trend can be either monotonically increasing or decreasing. Mann-Kendall test only identifies the existence of trend while Kendall's tau coefficient quantifies the inter-correlation and the degree of agreement between two variables, this test is immune to skewness in the data distribution (Hamed 2011). It also confirms the existence of a monotonic trend

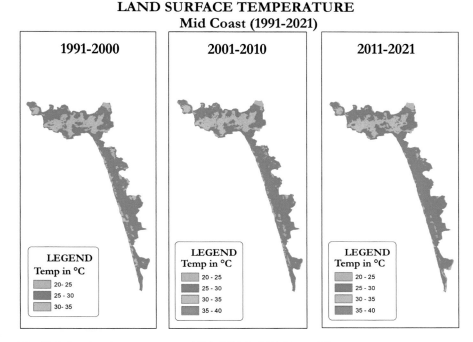

Fig. 20.4 Land surface temperatures 1991–2021 for Mid Coast of Goa

in time series data. It acts as a test of trend and gives the direction of the trend as well (El-Shaarawi and Niculescu 1992). In this study, the trend significance is studied for mean land surface temperature over the past three decades from 1991 to 2021.

A statistical trend test is necessary for this scenario to determine univariate trend significance with ordered time series (Mann 1945). The trend of decadal mean land surface temperatures was assessed. The mean land surface temperatures were calculated using a raster calculator in ArcMap software. A local operation 'average' by using multiple rasters (Dixon 2016) allows us to take an average of the same pixel having identical spatial extent over different rasters. A map is generated with each pixel being loaded with mean land surface temperature values for a given spatial extent. Each raster should have identical pixel size and map extension while taking an average of pixel values of multiple rasters (Dixon 2016).

A Kendall's tau correlation coefficient value is found to be 1 at the significance level of 0.01; this indicates a perfect relationship between mean decadal land surface temperature and the time order.

A positive trend for mean temperature with time is verified using the Data Access Viewer portal provided by NASA. 30 years of annual mean temperature data from 1991 to 2020 was obtained. This portal provides Earth surface temperatures at 2 meters. The data is utilized to find Kendall's tau. Data access viewer uses remotely sensed data and meteorological data to estimate the temperature (Stackhouse 2020).

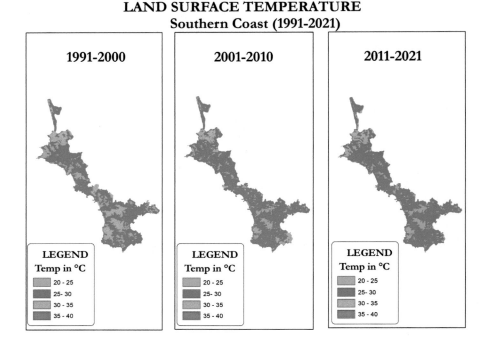

Fig. 20.5 Land surface temperatures 1991–2021 for Southern Coast of Goa

Tau value for the average temperature of three decades was found to be 1 at a significance level of 0.05; this value agree with observations made in this study (Table 20.6).

For a very long time Indian Meteorological department has been monitoring weather conditions at Panjim and Mormugao weather stations, these two stations are considered for aggregation of weather parameters for the entire state. Hence to identify the trend with reference to meteorological data, same stations are used (Table 20.7).

The mean temperature of Panjim has consistently increased from 27.50 °C to 27.72 °C from 1990 to 2019, showing an increase by 0.22 °C. Similarly, Mormugao has shown a slight increase in temperature that is 27.8 °C to 28 °C during the same period, indicating an increase of 0.20 °C. The mean temperature for the entire state has increased from 27.65 °C to 27.86 °C during 1990–2019. It is evident from the above table that the mean temperature for all the three decades for both Panjim and Mormugao stations is showing a positive trend.

Table 20.5 Minimum and maximum land surface temperature for coastal Goa

Region	Decades					
	1990–1999		2000–2010		2011–2021	
	Minimum temperature in °C	Maximum temperature in °C	Minimum temperature in °C	Maximum temperature in °C	Minimum temperature in °C	Maximum temperature in °C
Northern coast	24.02	32.95	24.19	34.66	23.94	33.90
Mid coast	24.22	32.98	24.27	35.09	24.22	36.70
Southern coast	**24.79**	**36.08**	23.79	**35.45**	**24.28**	**37.90**

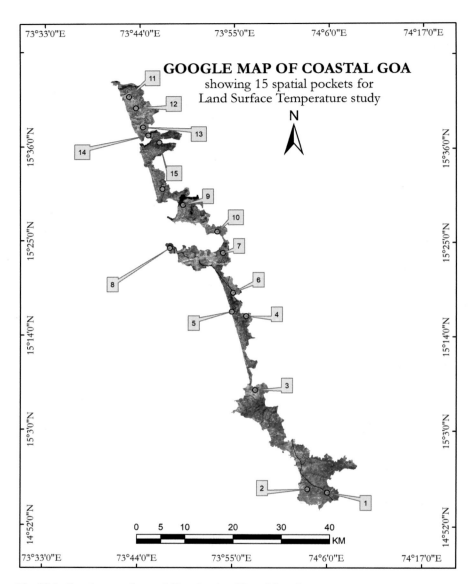

Fig. 20.6 Google map of coastal Goa showing 15 spatial pockets

Spatial pockets	Google Image 2003	Google Image 2021	1991-2000	2001-2010	2011-2021
			Temperature in °C		
1	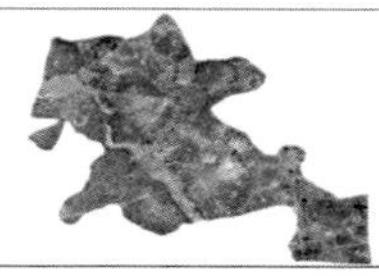	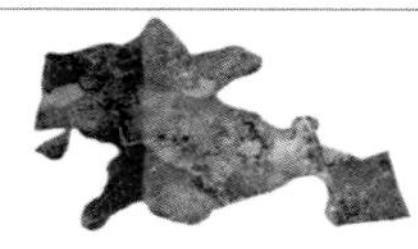	33.16	31.68	32.73
2	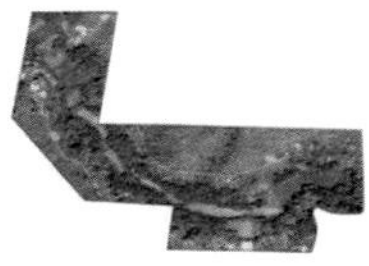	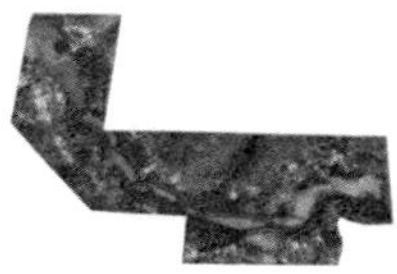	33.67	32.57	33.96
3	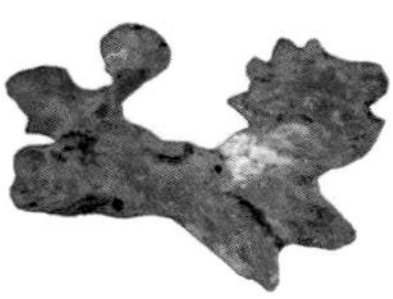	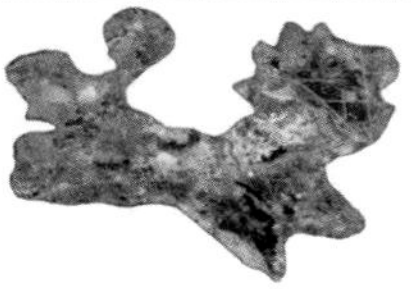	33.75	32.17	34.18
4		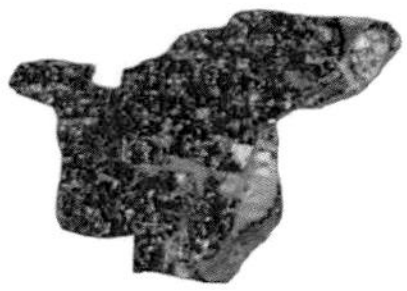	26.72	27.07	27.02
5			28.69	29.52	29.08
6			29.12	29.32	29.58

Zoomed Spatial Pockets of Google Map indicating rise in LST due to Climate Change (Areas least affected by Land Use Land Cover Change)

Fig. 20.7 Identified Google locations indicating the rise in LST

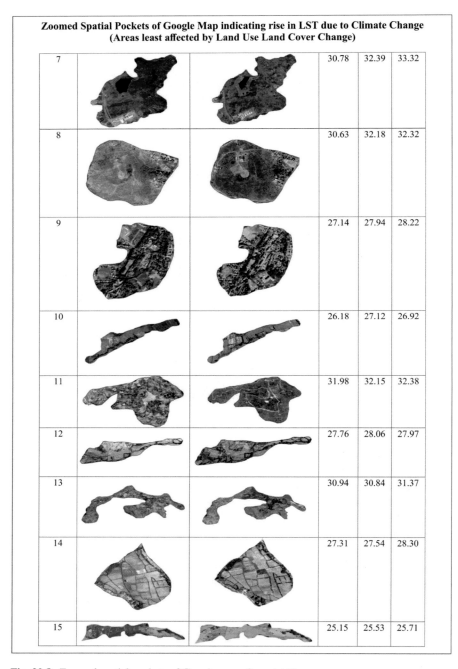

Fig. 20.8 Zoomed spatial pockets of Google map of coastal Goa

Table 20.6 Kendall's tau test result

Kendall's tau correlation coefficient for decadal mean land surface temperature: 1991–2021				Decades	Mean_Temperature
Kendall's tau_b	Decades	Correlation coefficient		1.000	1.000
		Sig. (2-tailed)		.	.
		N		3	3
	Mean_Temperature	Correlation coefficient		1.000[a]	1.000
		Sig. (2-tailed)		.	.
		N		3	3

[a]Correlation is significant at the 0.01 level (2-tailed)

Table 20.7 Mean temperature data of Panjim and Mormugao stations

Station	Decades		
	1990–1999	2000–2009	2010–20,219
	Temperature in °C	Temperature in °C	Temperature in °C
Panjim	27.50	27.6	27.72
Mormugao	27.80	28.00	28.00
Mean temperature of Goa state	27.65	27.8	27.86

Source: Statistical Handbooks of Goa and India Meteorological Data

Conclusion

Goa is a coastal geographic entity. Coastal areas are the most sensitive and highly productive ecosystems on Planet. These areas are the tourist heavens on Earth. They are known to experience unique atmospheric conditions leading to special climates. The coastal areas are attaining a catastrophe due to global climatic changes and anthropogenic meddling, which is also true in the case of Goa.

The oceans have a great impact on the weather and climate. The effects of the Arabian Sea are felt on the Bay of Bengal vice versa. The geographic expanse of about 100 km offshore to 100 km inland is greatly influenced by Coastal Meteorology. Hence, comprehending coastal meteorology needs the proper understanding of the interaction layers between the hydrosphere, lithosphere and atmosphere, interplay between air and sea, large-scale dynamics of the atmosphere, and the oceanic circulation (Hsu 1988).

Among all the problems faced by human society, climate change is by far the most complicated one since it is threatening the livelihood of the people irrespective of place, country and region. This has promoted us to assess variations in land surface temperature as it serves as an indicator of global climate change (World Meteorological Organization 2021).

It is apparent from the present study that the processed mean land surface temperatures, data obtained from data access viewer as well as meteorological data exhibit a similar trend. Hence, it can be concluded that the rising temperature is primarily because of global climatic changes.

Various reports of expert committees on climate change suggest that the climate change is going to stay here for long. Hence, adapting to climate change is going to be a new norm. People need to be sensitized on the ways and methods of adapting to climate change.

Further research needs to be conducted on the impact of climate change on land surface temperature particularly in the Goan context.

- Farmers in the affected areas need to modify agricultural practices such as seeds, and irrigation methods that will suit the change in climatic conditions.
- We must encourage citizens to use nature-based mitigations.
- Mangroves afforestation program must be carried out on a large scale because mangroves are believed to provide a defence system against coastal erosion.
- The areas that have undergone deforestation due to mining and other activities are required to undergo a reforestation program to control flooding in the low-lying coastal areas. While reforestation, care must be taken to plant indigenous trees to avoid ecological disasters.

References

(NCSCM, GOI), N. C. F. S. C. M (2017) Carrying capacity of beaches of for providing shacks & other temporary seasonal structures in private areas Goa

Alvares C (2002) Fish, curry & rice, 2nd edn. THe Goan Foundation, Mapusa

Avdan U, Jovanovska G (2016) Algorithm for automated mapping of land surface temperature using LANDSAT 8 satellite data. J Sens 2016. https://doi.org/10.1155/2016/1480307

Becker F, Li ZL (1990) Towards a local split window method over land surfaces. Int J Remote Sens 11(3):369–393. https://doi.org/10.1080/01431169008955028

Brovkin V et al (2006) Biogeophysical effects of historical land cover changes simulated by six earth system models of intermediate complexity. Clim Dyn 26(6):587–600. https://doi.org/10.1007/s00382-005-0092-6

Brovkin V et al (2013) Effect of anthropogenic land-use and land-cover changes on climate and land carbon storage in CMIP5 projections for the twenty-first century. J Clim 26(18): 6859–6881. https://doi.org/10.1175/JCLI-D-12-00623.1

Carlson TN, Gillies RR, Perry EM (1994) A method to make use of thermal infrared temperature and NDVI measurements to infer surface soil water content and fractional vegetation cover. Remote Sens Rev 9(1–2):161–173. https://doi.org/10.1080/02757259409532220

Carlson TN, Ripley DA (1997) On the relation between NDVI, fractional vegetation cover, and leaf area index. Remote Sens Environ 62(3):241–252. https://doi.org/10.1016/S0034-4257(97)00104-1

Chotchaiwong P, Wijitkosum S (2019) Relationship between land surface temperature and land use in Nakhon Ratchasima city, Thailand. Eng J 23(4):1–14. https://doi.org/10.4186/ej.2019.23.4.1

Copernicus (2021) Home | Copernicus Global Land Service. VITO NV en nombre del Centro Común de Investigación de la Comisión Europea (JRC), p 1. Available at: https://land.copernicus.eu/global/index.html

Dash P et al (2002) Land surface temperature and emissivity estimation from passive sensor data: theory and practice-current trends. Int J Remote Sens 23(13):2563–2594. https://doi.org/10.1080/01431160110115041

Deng Y et al (2018) Relationship among land surface temperature and LUCC, NDVI in typical karst area. Sci Rep 8(1):1–12. https://doi.org/10.1038/s41598-017-19088-x

Deshpande M et al (2021) Changing status of tropical cyclones over the North Indian Ocean. Clim Dyn 57(11):3545–3567. https://doi.org/10.1007/s00382-021-05880-z

Devi RM et al (2020) Spatial and temporal analysis of land surface 17(1):45–56

Dhorde K (2015) Trends in Surface Temperature variability over Panaji City of Goa, India. Conference: Bharatiya Vigyan Sammelan & Expo, 2015At: Goa. Available at: https://www.researchgate.net/publication/305059734_Trends_in_Surface_Temperature_variability_over_Panaji_City_of_Goa_India

Dixon B (2016) Chapter goals. https://doi.org/10.1002/9781118826171.ch11

El-Shaarawi AH, Niculescu SP (1992) On kendall's tau as a test of trend in time series data. Environmetrics 3(4):385–411. https://doi.org/10.1002/env.3170030403

Entezari M, Esmaeily A, Niazmardi S (2019) Estimation of soil moisture and earth's surface temperature using landsat-8 satellite data. International Archives of the Photogrammetry, Remote Sensing and Spatial Information Sciences – ISPRS Archives 42(4/W18):327–330. https://doi.org/10.5194/isprs-archives-XLII-4-W18-327-2019

Farina A (2012) Exploring the relationship between land surface temperature and vegetation abundance for urban heat island mitigation in Seville, Spain. LUMA-GIS Thesis nr (15):50. Available at: http://lup.lub.lu.se/luur/download?func=downloadFile&recordOId=3460284&fileOId=3460402

Griffiths PR, De Haseth JA (2007) Diffuse reflection. Chem Anal 171:349–362. https://doi.org/10.1002/9780470106310.ch16

Grover A, Singh RB (2016) Monitoring Spatial patterns of land surface temperature and urban heat island for sustainable megacity: a case study of Mumbai, India, using landsat TM data. Environ Urban ASIA 7(1):38–54. https://doi.org/10.1177/0975425315619722

Haigh JD, Cargill P (2015) Solar Radiation at the Earth. The Sun's Influence on Climate. https://doi.org/10.23943/princeton/9780691153834.003.0004

Hamed KH (2011) La distribution du tau de Kendall pour tester la significativité de la corrélation croisée dans des données persistantes. Hydrol Sci J 56(5):841–853. https://doi.org/10.1080/02626667.2011.586948

Hsu SA (1988) Coastal meteorology. Coast Meteorol (July):889–895. https://doi.org/10.1201/9780203756980-9

Ihlen V, Zanter K (2019) Landsat 7 (L7) Data Users Handbook. USGS Landsat User Services 7 (November):151

Imran HM et al (2021) Impact of land cover changes on land surface temperature and human thermal comfort in Dhaka City of Bangladesh. Earth Syst Environ 5(3):667–693. https://doi.org/10.1007/s41748-021-00243-4

India Meteorological Department (2019) Climate of Goa State, (25). Available at: https://www.imdpune.gov.in/library/public/CLIMATEOFGOA_EBOOK.pdf

Jiménez-Munoz JC, Sobrino JA (2003) A generalized single-channel method for retrieving land surface temperature from remote sensing data. J Geophys Res Atmos 108(22). https://doi.org/10.1029/2003jd003480

Lambeck K (2010) The science of climate change – questions and answers. Australian Acad Sci (August):1–24

Landsberg HE (1961) Solar radiation at the earth's surface. Sol Energy 5(3):95–98. https://doi.org/10.1016/0038-092X(61)90051-2

Lawrence DM, Slater AG (2005) A projection of severe near-surface permafrost degradation during the 21st century. Geophys Res Lett 32(24):1–5. https://doi.org/10.1029/2005GL025080

Li ZL, Duan SB (2017) Land surface temperature. Compr Remote Sens 1–9(February 2000): 264–283. https://doi.org/10.1016/B978-0-12-409548-9.10375-6

Mann HB (1945) Non-parametric test against trend. Econometrica 13(3):245–259

Mohorji AM, Şen Z, Almazroui M (2017) Trend analyses revision and global monthly temperature innovative multi-duration analysis. Earth Syst Environ 1(1):1–13. https://doi.org/10.1007/s41748-017-0014-x

Moukalled F et al (2006) Heat and mass transfer in moist soil, part II. application to predicting thermal signatures of buried landmines. Numer Heat Transfer Part B Fundamentals 49(5):487–512. https://doi.org/10.1080/10407790500510965

Mustafa EK et al (2020) Study for predicting land surface temperature (LST) using landsat data: a comparison of four algorithms. Adv Civil Eng 2020. https://doi.org/10.1155/2020/7363546

NABARD (2012) State action plan on climate change for the state of Goa period from 2020–2030. (91)

Nadaf FM (2019) Geographical analysis of the Coastal Landforms of Canacona, Goa. Res Rev Int J Multidiscip 4(2):655–661

Nadaf FM (2020) Mukt Shabd Journal ISSN NO : 2347-3150. Mukt Shabd J IX(Iv):1953–1959

Petropoulos G et al (2009) A review of Ts/VI remote sensing based methods for the retrieval of land surface energy fluxes and soil surface moisture. Prog Phys Geogr 33(2):224–250. https://doi.org/10.1177/0309133309338997

Petropoulos G et al (2012) Assessment of land surface temperature variation due to change in elevation of area surrounding Jaipur, India. Procedia Technol 6(24):224–250. https://doi.org/10.1029/2005GL025080

Ramachandra TV, Kumar U, others (2016) Analysis of land surface temperature and rainfall with landscape dynamics in Western Ghats, India. Journal of the Indian Institute of Science. Available at: https://www.researchgate.net/profile/Sahyadri_Environmental_Information_System/publication/318786685_Analysis_of_Land_Surface_Temperature_and_Rainfall_with_Landscape_Dynamics_in_Western_Ghats_India/links/597ec28aa6fdcc1a9accb94e/Analysis-of-Land-Surface-T

Ramaiah M, Avtar R, Rahman MM (2020) Land cover influences on LST in two proposed smart cities of India: comparative analysis using spectral indices. Land 9(9). https://doi.org/10.3390/LAND9090292

Ravanelli R et al (2018) Monitoring the impact of land cover change on surface urban heat island through Google Earth Engine: proposal of a global methodology, first applications and problems. Remote Sens 10(9):1–21. https://doi.org/10.3390/rs10091488

Shafian S, Maas SJ (2015) Index of soil moisture using raw Landsat image digital count data in Texas High Plains. Remote Sens 7(3):2352–2372. https://doi.org/10.3390/rs70302352

Sobrino JA et al (2004a) Land surface temperature and emissivity estimation from passive sensor data: theory and practice-current trends. Remote Sens Environ 23(4):434–440. https://doi.org/10.1016/j.rse.2004.02.003

Sobrino JA et al (2008) Land surface emissivity retrieval from different VNIR and TIR sensors. In: IEEE transactions on geoscience and remote sensing, pp 316–327. https://doi.org/10.1109/TGRS.2007.904834

Sobrino JA, Jiménez-Muñoz JC, Paolini L (2004b) Land surface temperature retrieval from LANDSAT TM 5. Remote Sens Environ 90(4):434–440. https://doi.org/10.1016/j.rse.2004.02.003

Stackhouse P (2020) Methodology. POWER Datale, NASA. Available at: https://power.larc.nasa.gov/docs/methodology/. Accessed: 5 Mar 2022

Strojnik M, Scholl MK, Garcia-Torales G (2016) Black-body radiation, emissivity, and absorptivity. In: Infrared Remote Sensing and Instrumentation XXIV, p 997310. https://doi.org/10.1117/12.2238569

Sun Y (2008) Retrieval and application of land surface temperature. Geo Utexas Edu 1(1):1–27. Available at: http://www.geo.utexas.edu/courses/387H/PAPERS/Termpaper-Sun.pdf

TERI (2015) Climate Resilient infrastructure services, Case study brief: Panaji. Available at: http://www.teriin.org/eventdocs/files/Case-Study-Vishakhapatnam.pdf

U.S. Geological Survey (2016) Landsat 8 Data Users Handbook. Nasa 8(June):97

Walther GR et al (2002) Ecological responses to recent climate change. Nature:389–395. https://doi.org/10.1038/416389a

Wang F et al (2015) An improved mono-window algorithm for land surface temperature retrieval from landsat 8 thermal infrared sensor data, pp 4268–4289. https://doi.org/10.3390/rs70404268

Wang L, Qu JJ (2009) Satellite remote sensing applications for surface soil moisture monitoring: a review. Front Earth Sci China 3(2):237–247. https://doi.org/10.1007/s11707-009-0023-7

Weng Q, Lu D, Schubring J (2004) Estimation of land surface temperature-vegetation abundance relationship for urban heat island studies. Remote Sens Environ 89(4):467–483. https://doi.org/10.1016/j.rse.2003.11.005

World Meteorological Organization (2021) State of the Global Climate 2020 (WMO-No. 1264). Available at: https://library.wmo.int/index.php?lvl=notice_display&id=21880#.YHg0ABMzZR0

Youneszadeh S, Amiri N, Pilesjo P (2015) The effect of land use change on land surface temperature in the Netherlands. International Archives of the Photogrammetry, Remote Sensing and Spatial Information Sciences – ISPRS Archives 40(1W5):745–748. https://doi.org/10.5194/isprsarchives-XL-1-W5-745-2015

Yu X, Guo X, Wu Z (2014) Land surface temperature retrieval from landsat 8 TIRS-comparison between radiative transfer equation-based method, split window algorithm and single channel method. Remote Sens 6(10):9829–9852. https://doi.org/10.3390/rs6109829

Yuvaraj RM (2020) Extents of predictors for land surface temperature using multiple regression model. Sci World J 2020. https://doi.org/10.1155/2020/3958589

Zimov SA, Schuur EAG, Stuart Chapin F (2006) Permafrost and the global carbon budget. Science 312(5780):1612–1613. https://doi.org/10.1126/science.1128908

Chapter 21
Analysing the Relationship Between Rising Urban Heat Islands and Climate Change of Howrah Sadar Subdivision in the Past Two Decades Using Geospatial Indicators

Parama Bannerji and Radhika Bhanja

Abstract Rapid urbanization, concretization and emission of heat from human activities have transformed the land use/land cover (LULC) dynamics of city peripheries, which in turn have plummeted the formation of urban heat islands in small pockets of the urban hinterlands. This study identifies the new urban zones that have emerged in Howrah Sadar subdivision in the last two decades and investigates the change in land surface temperature (LST), formation of urban heat islands and its impact on the microclimate of the surrounding areas. The spatial and temporal change in the cityscape was mapped using multispectral and thermal satellite images from Landsat imagery. Further, spectral indices like NDVI, NDBI and NDWI were used to examine the impact of rise in LST on the rates of urbanization and vegetation degradation. The results indicate a rise in urban heat islands and microclimates across the Howrah Sadar subdivision, which is further severely influenced by the global climate change events.

Keywords Anthropogeomorphology · Landform · Geomorphology · Remote sensing · GIS · Urban Heat Island

Introduction

Climate change is evident and its impact on the planet is immeasurable. The global climate patterns are influenced by local changes in the weather and vice versa. Among the different players that contribute to climate changes, the role of

---

P. Bannerji (✉)
Department of Geography, Nababarrackpore Prafulla Chandra Mahavidyalaya, Kolkata, West Bengal, India
e-mail: paramabannerji3@gmail.com

R. Bhanja
Department of Geography, Presidency University, Kolkata, West Bengal, India
e-mail: radhika.bhanja@gmail.com

© The Author(s), under exclusive license to Springer Nature Switzerland AG 2022
U. Chatterjee et al. (eds.), *Ecological Footprints of Climate Change*, Springer Climate, https://doi.org/10.1007/978-3-031-15501-7_21

urbanization cannot be ignored. Urbanization and the consequent growth of infrastructure in cities and peripheries contribute to the changing thermal dynamics of cities. Thermal properties of urban areas have changed resulting in significantly higher air temperature in cities than in the surrounding rural areas, leading to the urban heat island (UHI) effect. With the introduction of remote sensing technology, works exhibiting the relationship between urban heat island and land use distribution have increased. Generally, there is a bountiful of these types of studies for the important cities of the developed and developing world. However, there is a dearth of research with respect to this direction in the major urban pockets of India, particularly the non-metropolitan cities. These cities are characterized by high population rates, a scattered yet unplanned urban morphology, expansion of urban infrastructure from city core to the peripheries, rapid transformation of vegetated or vacant land to concrete structures, horizontal development in comparison to vertical development, and a need for policy regulation to control the abrupt change in built-up morphology. Understanding the complex thermal structure of the spatial region gives a clear picture of the need to monitor and regulate the rising microclimatic disturbances in small pockets. Such a measure will help policy makers to cope with the extreme microclimatic changes from an earlier stage.

Microclimate of a region can be identified as a zone which is different from the usual climatic condition that prevails in an area and such localized events vary spatially. The behaviour of concentrated weather patterns in cities is different from that of a rurban or rural area. In previous studies, the variation of localized climatic characteristics in cities is studied thoroughly, however there exists a small gap of understanding the impact of such events in a rurban setting. Keeping this research gap in mind, the study selects the Howrah Sadar subdivision of Howrah district of West Bengal to investigate the change in land surface temperature (LST), formation of urban heat islands and its impact on the microclimate of the surrounding areas. A total of 20 years is considered to track and map the thermal and built-up characteristics of the region. Given the lack of information about the thermal nature of urban-periurban regions of Howrah district and their potential in affecting the microclimate of the region, this research provides a meaningful insight into the spatial variation of the changing climatic scenario of the study area, thus, providing a base for a long-term research platform to the policy makers.

Review of Literature

Lexically, microclimate refers to the climate of a small region or that of confined space which can be buildings, fauna or urban region whose climate is different from the general climate of the region. While anthropological activities have induced climate change, urban areas with dense population and human activities are facing even more challenges. An interesting observation is that although the urban areas cover only 2% of the global area, more than 50% of the world population reside in cities (United Nations 2019). Allen et al. (2018) estimate that 20–40% world population have experienced a temperature increase greater than 1.5 °C.

Urbanization is one such causative factor that brings about biophysical changes in the ecological composition of the landscape (Ali et al. 2017). The differences are reflected in the changing green and blue spaces of the urban fabrics, and over a long period of time, they influence the rise in surface temperature in such pockets of rapid concretization (Ahmed et al. 2013).

Construction activities have changed the thermal properties of underlying surfaces resulting in significantly higher air temperature in cities than in the surrounding rural areas. This is commonly known as the urban heat island (UHI) effect (Oke 1989). It is generally observed that the land surface temperature (LST) is inversely proportional to the vegetation cover of the area (Weng and Lu 2008). Urbanization has brought changes to land cover patterns, which has in turn brought environmental changes. Several studies have shown the relationship between urban heat island and land use particularly with the introduction of remote sensing technology. Such analyses were conducted by extracting the LST using satellite-derived data having a range of sensors, like Landsat 4 and 5 (TM), 7 (ETM+), 8 (TIRS 1 and 2), Advanced Spaceborne Thermal Emission and Reflection (ASTER), and Moderate Resolution Imaging Spectroradiometer (MODIS) (Weng and Lu 2008). To understand the influence of land use and land cover (LULC) change on the thermal environment of the urban areas, Ding and Shi (2013) utilized the Landsat images of Beijing and related LULC with LST. Similarly, Jusuf et al. (2007) studied the effect of LULC change on the UHI in Singapore with the help of LST data. Focussing on India, a comparative analysis of two metropolitan cities of Delhi and Mumbai shows that due to low vegetation and urban area expansion, the UHI in Mumbai was significant as compared to Delhi (Grover and Singh 2015).

Vegetation index is often a preferred indicator of the extent of LULC change. Lo and Quattrochi (2003) pointed out that NDVI values point to the degree of greenness of vegetation. It has been observed that though a number of studies in this direction have been made globally, proportion in this regard is low for India (Grover and Singh 2015). Mallick et al. (2008) did a noteworthy study for Delhi. Generally, there is a bountiful of these types of studies for the major cities of the developed world, with a paucity of research in this direction for the urban areas of India, particularly the non-metropolitan cities.

An increase in the LST of Kolkata and its surroundings has been reported by numerous studies (Sharma et al. 2015; Biswas and Ghosh 2021; Halder et al. 2021). However, due to the compact spatial structure of the city, the spatial variance in the temperature changes is overshadowed by the importance in the temporal change of the surface temperature pattern. An investigation of the spatial variation over temperature changes over a large area, considering a significant time period, is necessary to identify if small pockets of microclimatic changes can be observed. This study is novel in such case, where a significant proportion of urbanized district is considered to map whether the urbanscape has the potential to influence the formation of microclimatic zones, different from the regular weather pattern experienced by a region. Microclimate of regions is significantly influenced by the temperature, humidity and wind conditions. With the rising amount of urban sprawl and spatial expansion in cities, the morphology of cities is everchanging. The cityscape has an impact on the temperature of the area. Therefore, the growing pace of urbanization highlights the spatial trajectory of heat island formation in cities. The potential microclimatic zones can be identified from these hotspots and provisions to lessen the impact of adverse temperatures can be taken beforehand to avoid consequent rise in temperature over time. This study identifies the new urban zones that have emerged in Howrah Sadar subdivision in the last two decades and investigates the change in land surface temperature (LST), formation of urban heat islands and its impact on the microclimate of the surrounding areas.

Datasets and Method

Howrah Sadar Subdivision – Study Area

The location of Howrah district lies between 22°48′ N to 22°12′ N latitudes and 88°23′ E to 87°50′ E longitudes. The district is a plain land bounded by the Hooghly and Rupnarayan River on the east and southeast of the district, respectively. Howrah district is further split into two subdivisions: Howrah Sadar and Uluberia. The smaller portion of the district belongs to the Howrah Sadar, constituted by Howrah Municipal Corporation and five community development blocks (CD blocks): Jagatballavpur, PanchlaBally–Jagacha, Domjur, and Sankrail. The five blocks are comprised of 44 census towns and 67 Gram Panchayats. The subdivision has its headquarters at Howrah. The Howrah Sadar subdivision has shown considerable growth in the past few decades and therefore its changes in the microclimate of the region need to be assessed to deduce the implications of changes in the land cover of the region on the weather imbalances of the region (Fig. 21.1).

Datasets

To assess the temporal change in LST and land cover of Howrah Sadar subdivision, Landsat images of 5-year interval from 1900 to 2020 were downloaded from the USGS website as shown in Table 21.1. The images were collected for the month of January to March, as the cloud cover over the study area was prevalent in other months. Thus, the selected scenes were cloud free in nature. Thermal bands of different Landsat sensors were essential to determine the LST of the area of interest and the method used to determine it is mentioned in the next section.

Determination of LST

After the satellite images were downloaded, they were taken up to atmospheric correction. The spectral radiance of the obtained images was transformed to digital number using the formula:

21 Analysing the Relationship Between Rising Urban Heat Islands and...

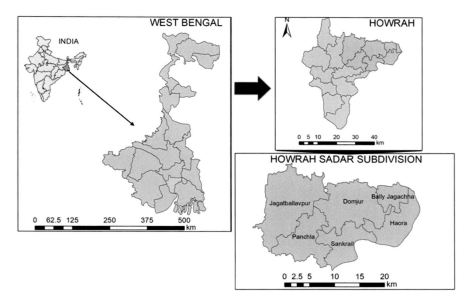

Fig. 21.1 Study area

$$LT_\lambda = \frac{LT_{max} - LT_{min}}{QCAL_{max} - QCAL_{min}} * (QCAL - QCAL_{min}) + LT_{min}$$

where LT_λ is the spectral radiance of the sensor in watts/(sqm*sr*μm); LT_{max} is the maximum radiance which is scaled to $QCAL_{max}$; LT_{min} is the minimum radiance; QCAL is the DN value of the pixel; while $QCAL_{max}$ and $QCAL_{min}$ represent the maximum and minimum DN value of pixels, respectively. The spectral reflectance data was converted to top of atmosphere (TOA) brightness values assuming the presence of uniform emissivity values, through the formula:

$$T_B = \frac{K_2}{l_n\left(\frac{K_1}{LT_\lambda+1}\right)} - 273.15$$

where T_B is the brightness temperature (K) of the satellite image, LT_λ is the spectral radiance (W/m^2*sr*μm), K_2 and K_1 are the calibration constant (W/m^2*sr*μm). NDVI is then calculated. Soil and vegetation are considered as main surface cover for the terrestrial components (Table 21.2). After surface emissivity is calculated from the NDVI threshold method, fractional vegetation, F_v, of each pixel in the image, was further calculated from the NDVI values using the following equation:

Table 21.1 Description of satellite images used in the study

Year	1900	1995	2000	2006	2010	2015	2020
Sensor	Landsat 5 TM	Landsat 5 TM	Landsat 5 TM	Landsat 5 TM	Landsat 5 TM	Landsat 8 OLS/TIRS	Landsat 8 OLS/TIRS
Date of acquisition	1990/01/28	1995/01/30	2000/02/11	2006/02/27	2010/04/11	2015/03/08	2020/02/02
Path/row	138/044	138/044	138/044	138/044	138/044	138/044	138/044
Thermal bands	6	6	6	6	6	10–11	10–11

Table 21.2 Indices used in the study

Index	Formula	Source
NDVI	(NIR − RED)/(NIR + RED)	Rouse et al. (1974)
NDBI	(SWIR − NIR)/(SWIR + NIR)	Zha et al. (2003)
NDWI	(NIR − SWIR)/(NIR + SWIR)	McFeeters (1996)

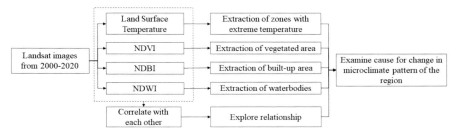

Fig. 21.2 Method followed in the study

$$F_v = \left(\frac{(\text{NDVI} - \text{NDVI}_{\min})}{(\text{NDVI}_{\max} - \text{NDVI}_{\min})} \right)^2$$

The minimum and maximum values are designated as $\text{NDVI}_{\min}$ and $\text{NDVI}_{\max}$, respectively.

The final expression for the emissivity ε is calculated as:

$$\varepsilon = 0.004 * F_v + 0.986$$

Finally, the LST was derived using the following equation:

$$\text{LST} = \frac{T_B}{\left(1 + \left(\lambda * \frac{T_B}{c2}\right) * l_n(\varepsilon)\right)}$$

where λ is the effective wavelength, σ is Boltzmann constant ($1.38 \times 10-23$ J/K), h is Plank's constant ($6.626 \times 10-34$ Js), c is the velocity of light in a vacuum (2.998×10^{-8} m/s), $c2 = h*c/\sigma$ and ε is emissivity (Fig. 21.2).

Determination of Land Cover Indices

The land cover composition in the region of interest and the change in their temporal evolution, mapped using Normalized Difference Vegetation Index (NDVI), Normalized Difference Built-up Index (NDBI), and Normalized Difference Water Index (NDWI), were used to identify the presence of green zones, concrete spaces and water bodies, respectively in the study area. The formulae used are given in Table 21.2.

The bands differ for Landsat TM 5 and Landsat OLI and TIRS bands, thus while calculating LST and the indices, the wavelength distribution should be noted. After the LST and indices were estimated, a spatio-temporal analysis was conducted to understand the change in temperature and land cover over the last two decades. In order to map the regions with high temperatures, the LST values were subjected to quartile distribution, such as values falling in the fourth quartile range describe extreme temperature conditions and therefore, regions with extreme temperature were mapped and delineated to understand the expansion of areas with extreme temperature in the region in the past two decades. Further, 1000 random points were generated and the LST, NDVI and NDBI. NDWI and MNDWI values were extracted and correlated to determine the relationship between these variables.

Findings

Temperature Variation in Howrah Sadar Subdivision

A significant rise in the temperature distribution pattern can be seen in the Howrah Sadar subdivision in the past three decades, with the lowest temperature ranging from 20° to 25° in the warmer months of 1900 to 40° to 45° in the warmer months of 2020. In 1990, the extreme temperature zone is limited to the Howrah and Bally Municipality, while the other districts display a pleasant summer season. With every passing decade, a 10° rise in temperature is visible. From 2010, a rise in temperature can be seen in small pockets of the district. The growth of Census Towns and the change in their urban structure has given rise to temperature in these small pockets. However, in the next five years, a rapid growth of urban areas from the transformation of cultivable land to urban areas has led to a rise in temperature in these CD blocks. Due to the pandemic situation, a distinct contrast in the temperature in the urban, peri-urban and rural regions can be seen in 2020, where the municipalities and corporations have higher LST values than their neighbouring peripheral areas and rural areas. Such an observation can be attributed to the low occupation of rurban areas by inhabitants, lesser functioning of factories, lesser electricity consumption, and lesser mobility. Figure 21.3 shows the LST distribution in the Howrah Sadar subdivision from 1990 to 2020.

The zones of extreme temperature were mapped. The urban heat islands are concentrated in these regions. Table 21.3 shows the spatial and temporal change in the extreme LST distribution zones of the Howrah Sadar subdivision. The range of quartile values selected for determining extreme temperature distribution is more the 27° in 2000 and 2006, more than 33° in 2010, and more than 37° in 2015 and 2020. Howrah Sadar subdivision was majorly a rural zone with few built-up spaces near the bank of River Ganga, before 2000. The rise in concrete spaces was visible from the 2000s. Initially, only 10% of the region displayed extreme temperature

21 Analysing the Relationship Between Rising Urban Heat Islands and... 551

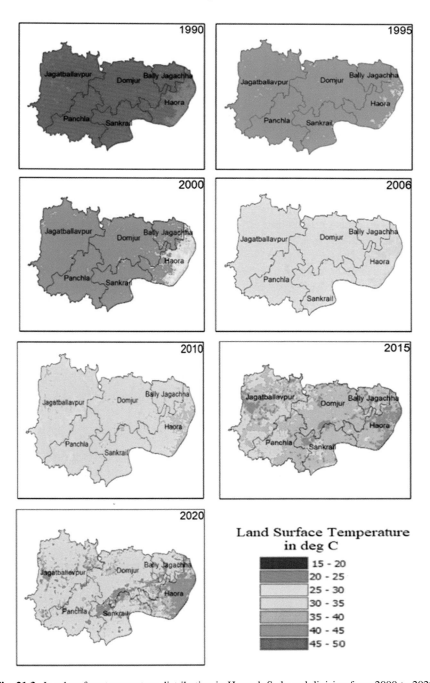

Fig. 21.3 Land surface temperature distribution in Howrah Sadar subdivision from 2000 to 2020

Table 21.3 Area occupied by extreme temperature conditions in the summer season of the Howrah Sadar subdivision

Year	Area covered	Percentage	Change in area (%)
1990	16.00	3.54	–
1995	8.03	1.78	−1.76
2000	42.26	9.35	7.57
2006	50.18	11.10	1.75
2010	70.87	15.68	4.57
2015	155.96	34.50	18.82
2020	176.38	39.02	4.51

conditions, however, by the end of two decades, more than one-third of the region Howrah Sadar subdivision experienced massive increase in LST between 2010 and 2015. This also correlated with the rate of urban expansion in the region. Such patterns influence the microclimate of the study area.

Regions which previously exhibited bearable warm summers are turning into minor urban heat islands, which further simulate the microclimate of the region. The rise and fall in temperature, influenced by the thermal imbalance and urban geometries in such regions lower the urban comfort. Census Towns like Domjur, Alampur, Dankuni, Panchla and Jagatballavpur have expanded over these years, their built-up structure and loss in vegetation cover have also led to the formation of microclimates over these regions, influenced by the urban heat island phenomenon. Figure 21.3 illustrates the expanding extreme temperature zone in the study area between 1990 and 2020.

LST vs. Land Cover Indices

Three metrics of landcover were selected to analyse the pattern of development that has taken place in the Howrah Sadar subdivision over the past few decades and their contribution in the variation in the microclimate of the region has been studied. NDVI acted as an important tool to distinguish regions with concentrated vegetation cover in the study areas, such as values with more than 0.3 exhibited high concentration of vegetation. Vegetation cover acts as an urban element which lessens the impact of high temperatures and heat islands in cities and their peripheries. They reflect the solar radiation to keep the region cool, evapotranspiration reduces the air temperature and photosynthesis reduces the concentration of carbon dioxide in the surrounding areas. Similarly, NDBI, which was used as an indicator to determine the built-up area zones, separated the impermeable structures from the bare and barren lands. Positive NDBI values of more than 0.2 indicated built-up zones. The concrete fabric and the urban canyon structure, and the anthropogenic heat from buildings and transport increase the temperature of the region. The microclimate of a region is simulated by the presence of vertical buildings and their morphologies, coupled by pollution from automobiles. Finally, NDWI with values more than 0.5 was used to delineate the waterbody present in the study area. Water bodies act as a carbon sink and help in reducing the overall thermal imbalances produced by built-up spaces over a region. The vegetation cover, impervious surface and water body are mapped and displayed in Fig. 21.4.

Discussion

A westward expansion of urban areas has taken place in the Howrah Sadar subdivision. Earlier the trade and commerce, transport development was more frequent along the bank of river Ganga, but with time, land crunch has led to more rural to urban conversion of landscape towards the interior in this case. Accordingly, the vegetated areas with long stretched forest covers and agricultural land have converted to built-up areas to accommodate the growing population of the Howrah Sadar subdivision. In the last two decades, new Census Towns have emerged, and infrastructural development has taken place. Howrah city serves as a hub to its neighbouring CD blocks and its market base has led to the development of better transport routes and more job opportunities for the inhabitants of its surrounding rural areas. Such parameters have accentuated the expansion of urban areas towards the rural areas of the district. With the development, the microclimate of the region gets stimulated. Rural areas once covered with lush vegetation are now diminishing in area (Fig. 21.5).

The temperatures in these regions were regulated as long vegetation stretch keeps extreme weather under control. During summers the evenings would be colder than its surrounding urban areas, which are devoid of long stretches of trees and grasses. The winters of rural areas are much colder than that of urban areas because the warmth is trapped in the concrete structures of urbanscape, a phenomenon not usually visible in rural areas. The summers of rural areas are moderately hot due to more evapotranspiration and presence of water bodies acts as a carbon sink and temperature regulator at a microscale. With the expansion of urban areas, microclimate of the region gets affected, the regions which previously exhibited pleasant to moderate temperatures changes to extreme temperature conditions. High temperature makes people vulnerable to unpredictable climate, exposes them to heat waves, and makes it worse for communities with poor access to air conditioning, ventilation and urban green space. Extreme temperatures can overheat the body, making them lose the ability to regularize body temperature, thus exacerbating body ailments, pertinent disease, and often leading to death. Thus, sudden changes in the microclimate of a localized region need to be monitored to prepare communities for sudden changes in the environment that can be harmful for their existence.

Table 21.4 explains the relationship between the increasing temperature of the region and different land cover indices. The comparison is performed to understand how the rise in temperature is caused by changing land dynamics of the Howrah Sadar subdivisions. Numerous works have confirmed the relationship between LST and the mentioned indices. With the diminution of vegetation coverage in the subdivision, it can be observed that the LST has increased tremendously. The strong

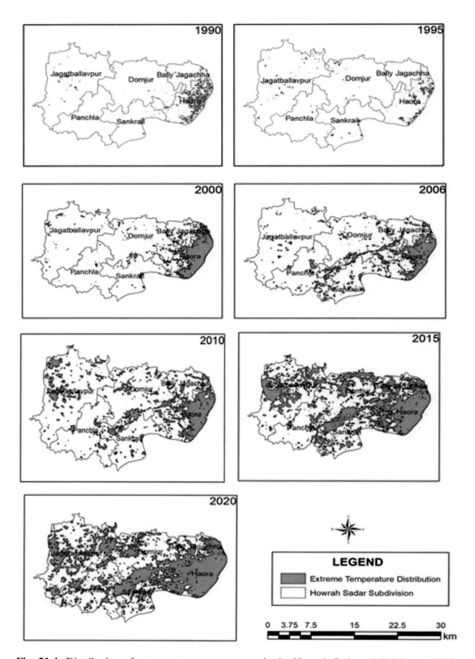

Fig. 21.4 Distribution of extreme temperature zones in the Howrah Sadar subdivision and their changes over time

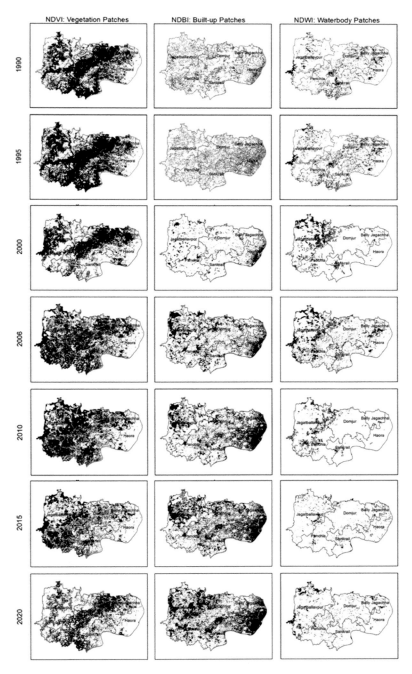

Fig. 21.5 Vegetation, built-up area and water body distribution in the Howrah Sadar subdivision and their changes over time

Table 21.4 Relation between land surface temperature and land cover indices

		LST 1990	NDVI 1990	NDBI 1990	NDWI 1990
LST 1990	Pearson correlation	1	−.346**	.279**	−.279**
	Sig. (2-tailed)		0.001	0.002	0.002
	N	1200	1200	1200	1200
		LST 1995	NDVI 1995	NDBI 1995	NDWI 1995
LST 1995	Pearson correlation	1	−.372**	.391**	−.390**
	Sig. (2-tailed)		0.001	0.001	0.000
	N	1200	1200	1200	1200
		LST 2000	NDVI 2000	NDBI 2000	NDWI 2000
LST 2000	Pearson correlation	1	−.216**	.488**	−.488**
	Sig. (2-tailed)		0.001	0.001	0.000
	N	1200	1200	1200	1200
		LST 2006	NDVI 2006	NDBI 2006	NDWI 2006
LST 2006	Pearson correlation	1	−.612**	.548**	−.548**
	Sig. (2-tailed)		0.000	0.000	0.000
	N	1200	1200	1200	1200
		LST 2010	NDVI 2010	NDBI 2010	NDWI 2010
LST 2010	Pearson correlation	1	−.753**	.795**	−.795**
	Sig. (2-tailed)		0.000	0.000	0.000
	N	1200	1200	1200	1200
		LST 2015	NDVI 2015	NDBI 2015	NDWI 2015
LST 2015	Pearson correlation	1	−.671**	.719**	.671**
	Sig. (2-tailed)		0.000	0.000	0.000
	N	1200	1200	1200	1200
		LST 2020	NDVI 2020	NDBI 2020	NDWI 2020
LST 2020	Pearson correlation	1	−.593**	.569**	−.569**
	Sig. (2-tailed)		0.000	0.000	0.000
	N	1200	1200	1200	1200

negative correlation between the variables explains how the vegetation cover is important to minimize the impact of LST and promote a pleasant living climate condition in the subdivision. Similarly, the presence of water bodies has a moderate to strong negative correlation with the increasing surface temperature conditions of the subdivision. This can be due to the lack of presence of large rivers in the area. Small ponds, lakes, wetlands and dry riverbeds are not sufficient to regularize extreme weather conditions, further the filling up of wetlands and converting them to concrete landscape has further reduced their temperature regulation capacity in the region.

Urban areas, on the other hand, are increasing at a greater pace and a strong correlation between NDBI and surface temperature of the region portrays a pattern of temporal change in the microclimate, influenced by the changing urban morphology of the region. The impervious concrete landscape prevents the heat to escape between the tall congested urban structures, thus altering the wind path and radiation in its boundaries. The microclimate of the region is significantly affected due to the imbalance in temperature, humidity and wind distribution in the urban scape, leading to higher LST and thus, stimulating the temperature conditions in the urban fabrics of a region. In the case of Howrah Sadar subdivision, the LST has risen significantly in areas with growing NDBI scores, thus the strong positive correlation established between these two regions is highly justified. To avoid the effect of microclimate in these regions, it is necessary to prevent more land use conversion to urban areas and constrict migration and population growth in the cities. Further investing in green infrastructures, planting trees along roads, in vacant spaces, and developing parks is also beneficial to the cities. Awareness about green roofs and green infrastructure is a necessity to invest in green technologies in cities and move towards sustainable development in the future.

Conclusion

The change in LST and land cover characteristics were mapped at a 5-year interval for 30 years since 1990, and it was observed that with the growth of the impervious layer in the district subdivision, the thermal nature has altered. The range of temperature has increased by 10° per decade, and numerous pockets of heat zones have emerged, exhibiting greater temperature variations. The study rings an alarm bell for policy makers to design the city in a planned way to avoid localized concentration of extreme temperature, as can be seen in cities with high rises. Census Towns like Domjur, Alampur, Dankuni, Panchla and Jagatballavpur have expanded over these decades; their built-up structure, conversion of dried-up river beds to practise farming and loss in vegetation cover have also led to the formation of microclimates over these regions, influenced by urban heat island phenomenon. Howrah Sadar subdivision is gradually exhibiting a shift in horizontal built-up expansion to the vertical rise, therefore the policy regulations concerning green buildings, promotion of renewable energy usage and green corridors and pavements need to be introduced and mandated to avoid future imbalance in existing climatic patterns of the region.

References

Ahmed B, Kamruzzaman MD, Zhu X, Rahman M, Choi K (2013) Simulating land cover changes and their impacts on land surface temperature in Dhaka, Bangladesh. Remote Sens 5(11): 5969–5998

Ali SB, Patnaik S, Madguni O (2017) Microclimate land surface temperatures across urban land use/land cover forms

Allen MR, Dube OP, Solecki W, Aragón-Durand F, Cramer W, Humphreys S et al (2018) Framing and context. Global Warming 1(5)

Biswas S, Ghosh S (2021) Estimation of land surface temperature in response to land use/land cover transformation in Kolkata city and its suburban area, India. Int J Urban Sci:1–28

Ding H, Shi W (2013) Land-use/land-cover change and its influence on surface temperature: a case study in Beijing City. Int J Remote Sens 34(15):5503–5517

Grover A, Singh RB (2015) Analysis of urban heat island (UHI) in relation to normalized difference vegetation index (NDVI): a comparative study of Delhi and Mumbai. Environments 2(2):125–138

Halder B, Bandyopadhyay J, Banik P (2021) Monitoring the effect of urban development on urban heat island based on remote sensing and geo-spatial approach in Kolkata and adjacent areas, India. Sustain Cities Soc 74:103186

Jusuf SK, Wong NH, Hagen E, Anggoro R, Hong Y (2007) The influence of land use on the urban heat island in Singapore. Habitat Int 31(2):232–242

Lo CP, Quattrochi DA (2003) Land-use and land-cover change, urban heat island phenomenon, and health implications. Photogramm Eng Remote Sens 69(9):1053–1063

Mallick J, Kant Y, Bharath BD (2008) Estimation of land surface temperature over Delhi using Landsat-7 ETM+. J Ind Geophys Union 12(3):131–140

McFeeters SK (1996) The use of the Normalized Difference Water Index (NDWI) in the delineation of open water features. Int J Remote Sens 17(7):1425–1432

Oke TR (1989) The micrometeorology of the urban forest. Philos Trans R Soc Lond B Biol Sci 324(1223):335–349

Rouse JW, Haas RH, Schell JA, Deering DW (1974) Monitoring vegetation systems in the Great Plains with ERTS. NASA Spec Publ 351(1974):309

Sharma R, Chakraborty A, Joshi PK (2015) Geospatial quantification and analysis of environmental changes in urbanizing city of Kolkata (India). Environ Monit Assess 187(1):1–12

United Nations, Department of Economic and Social Affairs, Population Division (2019) World urbanization prospects: the 2018 revision. United Nations, New York

Weng Q, Lu D (2008) A sub-pixel analysis of urbanization effect on land surface temperature and its interplay with impervious surface and vegetation coverage in Indianapolis, United States. Int J Appl Earth Obs Geoinf 10(1):68–83

Zha Y, Gao J, Ni S (2003) Use of normalized difference built-up index in automatically mapping urban areas from TM imagery. Int J Remote Sens 24(3):583–594

# Chapter 22
Assessment of Site Suitability of Wastelands for Solar Power Plants Installation in Rangareddy District, Telangana, India

Dhiroj Kumar Behera, Aman Kumari , Rajiv Kumar, Mohit Modi, and Sudhir Kumar Singh

Abstract Renewable energy in today's growing global economies has become essential and an integral part in answering the global major problem such as mitigating climate change, promoting sustainable development and conserving natural resources. Decentralized generation of solar power with photovoltaic (PV) panel installation in the wastelands, accompanied by the setting up of grid-connected systems emerges to be the befitting solution. In this work, an effort is being made to ascertain the solar potential for Rangareddy District, Telangana, India for utilization of vast wasteland patches utmost from satellite-derived insolation data. The research contains optimum utilization of vast wasteland patches and to identify potential sites for installing solar power plants which include generating global solar radiation, global insolation (direct and diffuse), and direct duration radiation maps using Cartodem. The LISS III satellite imagery (23 m) and Cartosat-2 DEM (10 m) data are the major data sources for extracting thematic layers. A multi-criteria decision analysis (MCDA) model was designed to rank and locate the potential sites by taking various socio-economic factors such as near to town headquarters, transmission lines, built-up land, and road connectivity based on their overall performance. They are normalized and integrated using a raster calculator created from Saaty's AHP approach, based on the appropriate weights of selected parameters. The intersection of the constraint and potential factor layers in a GIS platform was also used to extract acceptable places for solar power unit

---

D. K. Behera (✉) · R. Kumar · M. Modi
Land Use and Cover Monitoring Division, National Remote Sensing Centre, ISRO, Hyderabad, India
e-mail: dhirojkumar_behera@nrsc.gov.in; rajiv_kumar@nrsc.gov.in; mohit_m@nrsc.gov.in

A. Kumari
Centre of Advanced Study in Geography, Panjab University, Chandigarh, India
e-mail: aman.nehra@pu.ac.in

S. K. Singh
K. Banerjee Centre of Atmospheric and Ocean Studies, IIDS, Nehru Science Centre, University of Allahabad, Prayagraj, UP, India
e-mail: sudhirinjnu@gmail.com

© The Author(s), under exclusive license to Springer Nature Switzerland AG 2022
U. Chatterjee et al. (eds.), *Ecological Footprints of Climate Change*, Springer Climate, https://doi.org/10.1007/978-3-031-15501-7_22

installation. The result reveals that the study region has suitable insolation conditions for establishment of photovoltaic plant (602.34–773.14 KWh/m^2/day) along with vast wasteland area (981.23 km^2) which is approximately 13% to the total geographical area of the district.

Keywords Solar energy · Potential sites · Cartosat · Multi-criteria decision analysis · Sustainable development

Introduction

Solar power plant installation units in wasteland areas are not only helpful in job opportunity creation but also help in water resources conservation, infrastructural development, energy revenue generation, and above all land resource management. The continuous usage of non-renewable resources of limited stocks for consumption of energy results in negative environmental impacts (Mishra 2020). These concerns have forced many nations globally to go for energy combustion resources which are renewable in nature. With the scarcity of natural resources and an incessant demand for electric power, there is an increasing need for renewable energy for sustainable development. The burning of fossil fuel and non-renewable energy resources has the adverse impact on the environment like an increase in the concentration of pollutants in the lower atmosphere which results in the increase of global temperature, climate change, unusual pattern of rainfall, etc. (Ahire et al. 2022). This unprecedented changes in the environmental conditions forced the world to adopt renewable and more environment-friendly resources for day-to-day energy consumption in order to stabilize the environment. Solar energy, which is more than 2500 Terawatts (TW) of energy, is a cheap and environmentally acceptable energy source, and solar power plants may be utilized to store this vast quantity of energy (Georgiou and Skarlatos 2016). A grid-connected photovoltaic system including solar panels, inverters, power conditioning units, and grid connection equipment makes up a solar power plant (Kumar and Sudhakar 2015). Photovoltaic panels are made up of semiconductors that convert sunlight directly into electricity. Photovoltaic cells were utilized to create electricity, which was then used to electrify rural settlements in the region. It is a renewable energy source that is safe, efficient, dependable, and environmentally beneficial, and may be utilized for a long time (Makrides et al. 2010). Thus, in the years to come solar power finds much more scope and is more affordable.

According to spatial data analysis, over 58% of the country's geographical areas might be solar hotspots, with yearly average global insolation of greater than 5 kWh/m2/day (Ramachandra et al. 2011). It is an easy and flexible multiple criteria technique and has been employed by several researchers in a variety of applications, including land suitability assessments (Kontos et al. 2005; Masera et al. 2006; Tegou et al. 2010; Georgiou et al. 2012). Mishra (2020) used the AHP approach in conjunction with the cost distance function to analyze the site suitability of solar energy facilities in the Trans-Yamuna Upland Region of Allahabad District, India. It was discovered that the research region had adequate insolation for the establishment of a solar power plant, with an insolation of 703.73 kWh/m^2/day. About 17% of the study area was identified to be most suitable for solar power installation plant. Andreas Georgiou and Dimitrios Skarlatos (2016) examined the best-suitability index for solar energy utilization using a decision analysis methodological framework in the Limassol district in Cyprus. The finding reveals that 3.0% of the study region lies in the best-suitability index for solar resource exploitation. Effat (2013) examined Egypt optimal location for solar energy production in the world using multi-criteria analysis and SRTM for the selection of ideal sites for solar photovoltaic plant in Ismailia Governorate. In the Shodirwan Region of Iran, Asakereh et al. (2014) applied the Fuzzy Analytical Hierarchy Process (Fuzzy AHP) and geographical mapping models using GIS to pick solar energy sites. Annual sun insolation in Shodirwan is quite good, according to GIS interpolation, and can be used to locate prospective solar farm locations. Carrión et al. (2008) have discussed the need of environmental decision-support system (EDSS) for selecting optimal sites for grid-connected photovoltaic power plants in European Union countries. Khan and Rathi (2014) selected an optimal site for solar PV plant in Rajasthan using GIS. Many factors have been used for criteria and classified as analysis and exclusion criteria. Availability of solar radiation, availability of vacant land, distance from highways and existing transmission lines, etc., were considered as analysis criteria. Variations of local climate, module soiling, the topography of the site, etc. are exclusion criteria. The resulting maps revealed that a large part of the area is highly suitable. Munkhbat U and Choi Y. (2021) identified sites that could be appropriate for the installation of large-scale solar power facilities. Annual worldwide horizontal radiation, temperature, elevation, slope, aspect, distance from a significant road, and distance from a major power line were all taken into account, and raster datasets were created using GIS. The AHP was used to rate the GIS layers and calculate their weightings. A PV system installation suitability map of Mongolia was created by integrating the seven rated GIS layers with their weightings. Guaita-Pradas et al. (2015) investigated renewable energy for sustainable development in India and Pakistan and suggested that both these countries should take advantage of their location with maximum solar radiation. This step will lead to sustainable economic growth in both of the countries. Sadhu et al. (2015) studied the role of solar power in the sustainable development of India. Due to the position of India in the tropical belt, it has a giant potential of solar energy. Their study examined the existing state of solar power, challenges and solution to these challenges in solar industry, and also future potential of solar power in the sustainable development of India. Yousefi et al. (2018) identified a spatial site for solar power plants in the Markazi province of Iran using the Boolean-Fuzzy Logic Model. This study takes into account technical, economic, and environmental considerations. By combining above mentioned two methods, it becomes a powerful analytical tool for optimal site selection studies with the help of GIS.

Energy Scenario in India

India which accounts for 17% of the world population (1.21 billion people, Census of India 2011) is the world's second most populous country after China. This population explosion made India a significant energy consumer. India's demand and consumption of energy continued to increase also during the global financial crises. India, being located in the tropical region, has an immense prospective for the exploitation of solar energy (Muneer et al. 2005). It has an estimated insolation that varies between 4 to 7 kWh/m^2 per day and with an exposure of nearly 300 sunny days in years (Sudhakar et al. 2013). The maximum amount of country's energy is used to meet the demand in commercial, residential, and agricultural activities compared to Japan, China, United States, and European Union. The major concern of the Ministry of New and Renewable Energy (MNRE), Government of India is to make the country " Energy self-sufficient" (Lalwani and Singh 2010).

In the non-renewable energy resources, coal is the major source of energy production which accounts about 69.5% of total energy production of the country (India). Around 33% of India's energy needs are fulfilled by the imported sources which include oil as well. India faces a major challenge to ensure the safe and clean energy to all its citizens particularly in the rural areas. About 85% of its rural population is dependent on solid fuel to meet the household needs which is unsafe for the nearby people as well as for the environment and only about 55% of the rural population has access to electricity. India has an estimated renewable energy potential of about 1095GW of which wind potential is of more than 300GW, solar potential of 750GW assuming 3% wasteland is made available, the small hydro potential is about 20GW and the bioenergy potential has been estimated at 25GW (MNRE Annual Report, 2017–2018).

The utility electricity sector had an installed capacity of 344GW as of 31st March 2018 (Central Electricity Authority, March 2018). Renewable power plants constituted 33.23% of total installed capacity and non-renewable power plants constituted the remaining 66.77%. In India per capita, electricity consumption is 1122 kWh as of 31st March 2018 (Central Electricity Authority 2017). The total installed capacity power generation capacity as on 31st March 2018 with type-wise break up is given in Table 22.1. The total solar power capacity installed is 17052.37 MW in India as on 31st December (Ministry of New and Renewable Energy, Report 2017–2018). The anticipated a cumulative capacity of around 40,000 MW is proposed to be installed under different Solar Programmes by end of Financial Year 2020–21, in India (Ministry of New and Renewable Energy, Report 2020–2021).

Table 22.1 Total installed utility power generation capacity (in MW)

Thermal				Renewable		
Coal	Gas	Diesel	Nuclear	Hydel	Other renewable	Total
197171.50	24897.46	837.63	6780	45293.42	69022.39	344002.39
57.31%	7.23%	0.24%	1.9%	13.17%	20.06%	100%

Source: Installed Capacity Monthly Report, Central Electricity Authority, 31 March (2018)

Table 22.2 RES total installed capacity (in MW) of India

Small hydro power	Wind power	Bio-power		Solar power	Total capacity
		Biomass power	Waste to energy		
4485.81	34046.00	8700.80	138.30	21651.48	69022.39
6.49%	49.32%	12.6%	0.2%	31.36%	100%

As on 31st March 2018; Source: Installed Capacity Monthly Report, Central Electricity Authority, March (2018)

India is one of the world's most active participants in renewable energy sources (RES) other than hydel, particularly in solar and wind generation. Renewable energy sources installed capacity of India as on 31st March 2018 are shown in Table 22.2.

Solar power plants constituted 31.36% of total installed capacity and other renewable energy power plants constituted the remaining 68.64% (Installed Capacity Monthly Report, March 2018). Government of India initiated so many missions to increase the solar power potential in India because India is endowed with vast solar energy potential.

For the current study, GIS interface method is developed to calculate solar potential in the study area. Cartosat-2 DEM (10 m) is used to generate the solar potential maps by area solar analyst tool. The present study aims to develop and present a framework using multi-criteria decision analysis (MCDA) using GIS in locating suitable land for installation of solar power plant at Rangareddy district of Telangana State by taking various socio-economic factors such as near to town headquarters, transmission lines, built-up land and road connectivity based on their overall performance. Telangana was a power-deficit state when it was separated from Andhra Pradesh in 2014. Over the previous seven years, however, the government has made a number of steps to improve the situation and increase energy production. In 2015, the state administration unveiled its Solar Policy (Kumar 2021). Rising demand and unavailability of conventional energy sources have prompted scientists to pursue study in the field of renewable energy sources, particularly solar energy (Goura 2015). So, the present study aims to analyze the following objectives in the context of Rangareddy district: (a) to delineate the wastelands patches and (b) to utilize wasteland patches for installing solar power plants using multi-criteria decision analysis toward sustainable development.

Study Area Description

The Rangareddy district of Telangana State, India is located in the Deccan Plateau. Its latitudinal extent is from 16° 19′ and 18° 20′ North and longitudinal extent is from 77° and 80° 47 East (Fig. 22.1). It is surrounded by Medak district in the North, Nalgonda district in the East, Mahbubnagar district in the South, and Gulbarga district of Karnataka state in the West. It covers an area of 7493 km² and is inhabited

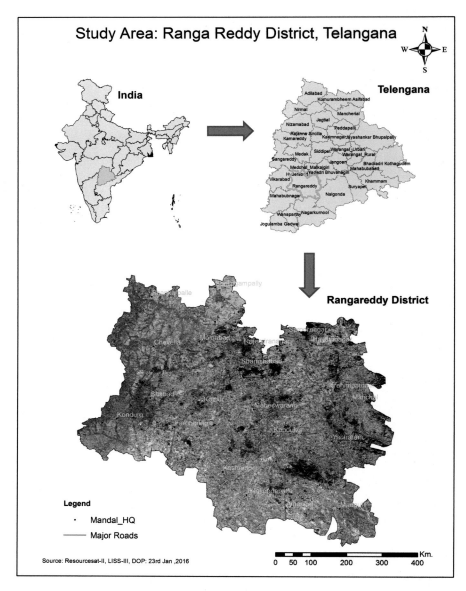

Fig. 22.1 Study area location map

by 52, 96,741 persons as per Census (2011). The population density of the district is 707 persons per km^2. There are 37 Mandals and 870 villages in the district. The climate of the district is generally dry except during the southwest monsoon season and characterized by hot summer. The months of April and May are very hot while the winter season is from January to February. The district receives around 781 mm of rainfall mostly by southwest monsoon. The Anantagiri hills are composed of laterite and traverse the district from Mahbubnagar in the south to Vikarabadmandal. The most dominant soil in the district is red soil followed by black cotton soil. The main rivers of the district are Musi and Kagna.

Materials and Methods

(a) LISS III multispectral data (23.5 m, Year 2016).
(b) Cartosat-2 DEM data (Year-2016, Resolution- 10 m) downloaded from nrsc.gov.in.
(c) ArcGISv 10.5 and ERDAS IMAGINE v 2014 were used for image processing and analysis.

Global solar radiance, slope, aspect, transport network, transmission grids, and wasteland layers were used as input into AHP environment to estimate the suitable sites as shown in Fig 22.2.

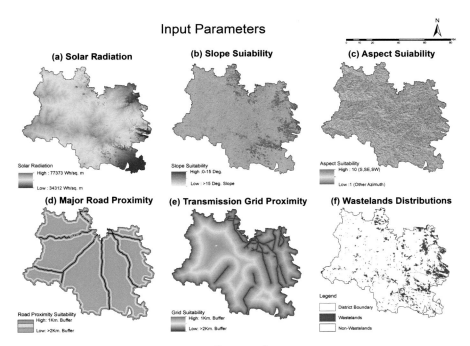

Fig. 22.2 The maps show the input layers (from **a** to **f**)

Methods

The methodology followed in the study is discussed in the below section:

Preparation of Database

Cartosat-2 DEM data is used to create a solar radiation layer that shows the total quantity of incoming solar radiation over the course of a year in kWh/m^2. The DEM is also used to create slope and azimuth (aspect) layers. Wastelands map is prepared by visual interpretation method to delineate the different type of wasteland patches using LISS III image and for ground truth verification high resolution satellite Google imagery is used by taking a sufficient amount of ground check points spreading over all classes. Wastelands classified map has been prepared by following the 23-fold classification (National Wastelands Change Analysis 2015, NRSC) and the summary of it is shown in Table 22.3. Transmission lines and road networks are digitized using Google Earth images and the SOI toposheet. All input layers are reclassified for standardization and considered for further analysis as shown in Fig. 22.2.

Multi-criteria Decision Analysis

A multi-criteria decision analysis model is developed to select the optimum sites for utilization of available wastelands by taking other socio-economic layer rankings. It is important to give weightage to each of the participating elements according to their relative importance in the study, for this purpose Analytical Hierarchy Process (Saaty 1977) is used for the analysis. The method employs pairwise comparisons within a reciprocal matrix, with the number of rows and columns determined by the number of criteria (Murmu et al. 2019; Singh et al. 2021; Popov et al. 2021; Pandey et al. 2022).

Table 22.3 Summary of area statistics for wastelands classes

WL2016	Category	Area (ha)
1	Gullied and/or ravinous land (medium)	24.4383
3	Land with dense scrub	20065.0564
4	Land with open scrub	12112.5872
7	Land affected by salinity/alkalinity (medium)	1668.5539
8	Land affected by salinity/alkalinity (strong)	368.0464
11	Under-utilized/degraded forest (scrub domain)	10096.9028
12	Under-utilized/degraded forest (agriculture)	1744.2962
13	Degraded pastures/grazing land	920.9347
20	Mining wastelands	1281.7137
21	Industrial wastelands	143.9597
22	Barren rocky/stony waste	5939.9750

Weight Assignments Through Saaty's AHP Method

For the installation of solar power plants five layers were prepared and standardized i.e., slope, aspect, transmission line, road network, and solar radiation. The pairwise rank comparison matrix is constructed in the Microsoft Excel sheet by assigning rankings to all 5 characteristics of the solar radiation layer according to their relative relevance using the 9-point rating system of Saaty. A range of 1–9 was employed in this rating scale, with 1 indicating equal significance between two factors and 9 indicating high importance of one parameter over the other (Saaty 1980; Malczewski 1999; Feizizadeh et al. 2014; Patel et al. 2022; Kumar et al. 2022).

As we know that for any developmental activity to be done in any area, understanding the land cover pattern plays a vital role in decision-making. Land use and land cover pattern by classifying them into 54-fold classes (Behera et al. 2017) of the study site is shown as in Fig. 22.3.

The reciprocity factors in the comparison matrix are represented mathematically as $n(n-1)/2$ for n different components in a pairwise comparison matrix (Saaty 1980; Akinci et al. 2013). After computing the pairwise matrix, Saaty's methodology (Saaty 1980) is used to measure relative weights/eigenvectors. Furthermore, the analytic hierarchy approach also detects and calculates the anomalies of call manufacturers (Saaty 1980; Feizizadeh et al. 2014):

$$\text{CR} = \frac{\text{CI}}{\text{RI}} \tag{22.1}$$

Equation 22.1 shows CR i. e. consistency ratio where CI is the consistency index and RI is the random index. The consistency index and random index are the most important factors in determining the CR (Singh et al. 2021).

The consistency relationship assists in the determination of future events and tests the decision maker's rational contradictions (Cengiz and Akbulak 2009; Chen and Wang 2010a, b). It indicates the probability of the matrix decisions being formed at random (Park et al. 2011; Saaty 1977; Murmu et al. 2019).

$$\text{CI} = (\lambda \max - n)/(n-1) \tag{22.2}$$

Equation 22.2 depicts that the CI when λmax is the highest eigenvector of the computed matrix and n is the order of the matrix, Eq. 22.2 denotes consistency index (CI). It denotes the probability that the matrix decisions were created at random (Park et al. 2011; Saaty 1977).

The random index (RI) is the mean value of the consistency index based on the computed matrix order as by Saaty (1977). If CR>0.10, the weight values of the matrix suggest contradictions, and the procedure (AHP) does not provide useful results (Saaty 1980). The measured CR in this analysis was 0. 0.04, which is within reasonable limits.

Using two alternative methodologies, the consistency index (CI) was 0.04 or 0.016. To continue the AHP analysis and examine the consistency of the judgments

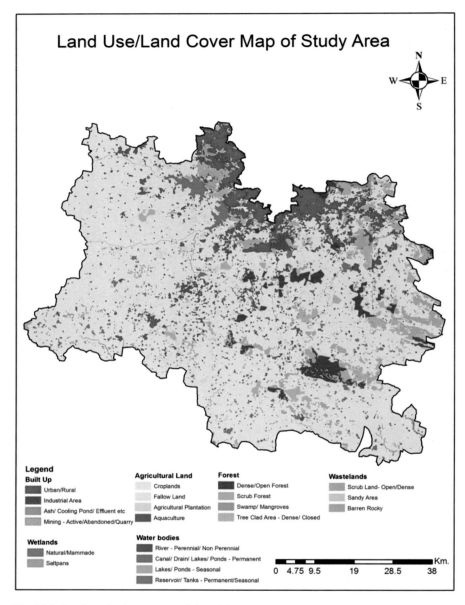

Fig. 22.3 Land use/land cover map of the study area

(i.e., assigned weight), a CI of 0.10 or less is satisfactory. Raster calculator is used to integrate all of the standardized layers according to their relevant parameters. By merging five parameter layers, such as global solar radiation, slope, aspect, road proximity, and transmission line proximity, the prospective location for the solar power plant may be determined. Table 22.4 describes the AHP environment developed for getting the weights for different parameters which was used as a multiplication factor for getting the suitability.

Suitable Site Selection

For the suitable region identification, it involves the overlying of different GIS data. To find out the suitable region for the installation of solar power plant, the intersection method was used. As a result, a GIS layer depicting the potential solar energy production region was created, which was merged with a limitation factor layer, removing the areas of water bodies, reserve forests, agriculture, built-up, and stone quarry regions and including only wasteland areas. In the study, two alternate ways have been adopted to obtain the suitable sites according to the user requirement, based on the economical parameters including or according to the availability of the wastelands which have been depicted from the research.

Results and Discussion

Suitable Areas

The study depicts that because of its geographical location and climatic environments, the entire study area has outstanding conditions for harvesting solar energy and is thus expected to meet national and state solar energy policy commitments. Many regions fall into the appropriate region for the construction of a solar PV power plant with an insolation value of (602.34–773.14 KWh/m^2) year of solar radiation. According to literature, a solar PV power plant with a capacity of nearly 49 MW would need 1 km^2 of ground. Thus, if 1 km^2 area has installed capacity of 49.42 MW (Kumar and Sudhakar 2015), the overall capacity of the study zone is 6263.71 MW, which is further confirmed by Jalaun Solar Power Plant (Uttar Pradesh) and Bhadla Power Plant (Rajasthan), which have 50 MW and 1365 MW solar capacity with an area of 1.012 km^2 and 40 km^2, respectively. The suitable areas are marked by a low population density, with sparse settlement in remote, isolated areas majorly found in Yacharam, Amangal, Kandukur, and Hayathnagar.

There is a direct relationship between environment, energy and sustainable development. Any country seeking sustainable development must utilize renewable energy resources which have minimum environmental impact. Renewable energy is replenished naturally on human timescale, and utilization of solar energy for the production of electricity in a tropical country like India is remarkable progress in sustainable development practices.

The wastelands of different categories have been given different weights depending on the scope of reclaiming them by installing solar panels in the scale of 1 to 10 where 10 is given to the most suitable category and 1 is for least suitable class shown in Fig. 22.4.

Table 22.4 AHP environment for weight calculation

Pairwise comparison matrix

	C2 (solar radiation)	C3 (aspect)	C4 (slope)	C5 (road proximity)	C6 (transmission lines proximity)
C2 (solar radiation)	1	9	7	5	3
C3 (aspect)	0.11	1	0.5	0.25	0.166
C4 (slope)	0.142	2	1	0.5	0.25
C5 (road proximity)	0.2	4	2	1	0.5
C6 (transmission lines proximity)	0.33	6	4	2	1
SUM	1.78	22.00	14.50	8.75	4.92

Normalized matrix

	C2 (solar radiation)	C3 (aspect)	C4 (slope)	C5 (road proximity)	C6 (transmission lines proximity)	Average
C2 (solar radiation)	0.561	0.409	0.483	0.571	0.610	0.527
C3 (aspect)	0.062	0.045	0.034	0.029	0.034	0.041
C4 (slope)	0.080	0.091	0.069	0.057	0.051	0.070
C5 (road proximity)	0.112	0.182	0.138	0.114	0.102	0.130
C6 (transmission lines proximity)	0.185	0.273	0.276	0.229	0.203	0.233
						1

					Consistency measure	
Consistency index	0.0448826				MMULT (pairwise matrix, priority vector)/priority vector	
Approach 1:		Approach 2:			C2 (solar radiation)	5.180
λ_{MAX}	5.127	Consistency index	0.019		C3 (aspect)	5.017
		Consistency ratio	0.017		C4 (slope)	5.025
		Consistency ratio			C5 (road proximity)	5.049
Consistency index	0.032	Consistency index/random index			C6 (transmission lines proximity)	5.103
Consistency ratio	0.028		0.04007			

22 Assessment of Site Suitability of Wastelands for Solar Power...

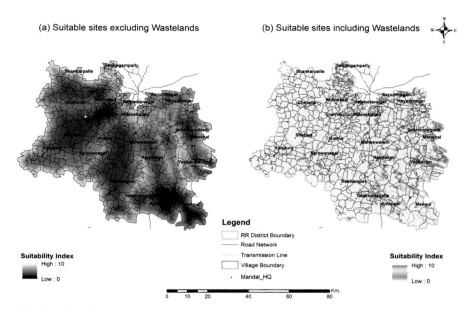

Fig. 22.4 Suitable site selection

Expected Impact on Economic Landscape Toward Strengthening the Sustainable Development

Landowners who participate in the construction of solar power plants might earn a portion of the revenue generated from their unused or uncultivated property. Furthermore, the tax base of the region may be enhanced by the building of solar power plants, which can then be utilized for the territory's general growth. From an environmental impact assessment point of view, the electricity generated by solar power plant reduces emission of carbon and other greenhouse gases, therefore it helps in cleaning the environment and helps to prevent climate change. Similarly, energy produced by solar power plants can be used in cold storage, warehouse and other industries. It can enable rural areas energy sufficient, thereby leading a way toward the development and self-resilient economy. With the installation of solar power plants in the region road connectivity may be promoted which further helps in the establishment and increase of other business activities, which can lead ways for the local people to get their means of livelihood in their areas of residence. The energy produced by solar power plants can be transmitted to the surrounding city area which helps in creating revenue and such can be used to develop the infrastructure of the area like schools, hospitals, and training institutes etc. Like other thermal power plants the installation of solar power plants can be revolutionary in terms of developmental activities, for example, it creates a lot of job opportunity in the region, can supply 24×7 energy, wasteland will be optimally used, roads would be constructed, and small-scale & agro-based industries will be established. The coal and nuclear power plants based electricity production affects the air and water quality of the surrounding environment due to emergence of large amount of trace gases, particulate matter and discharge of heated water. The discharge of hot water affects the pelagic communities (Protasov et al. 2022). However, the solar power plants rely on the renewable source of energy i.e. from the sun and hence does not have much negative impacts on the surrounding environment. Thus, establishment of solar power plants in any area of scope helps in preventing poverty in the region by creating job opportunities and other business activities and also keep the environment fit for sustenance.

Key Challenges

Despite the fact that the Rangareddy area has adequate solar radiation and land, plant construction is expected to be difficult. The research region's wasteland and extended fallow-land areas provide the greatest land for solar power plant construction. A solar power plant having 10 MW electricity production capacities requires at least more than 0.21 km^2 of land with even terrain and the total wasteland area of Rangareddy district is 13 % (981.23 km^2) of total geographical area. Hence, this area can attract a large part of investment related to solar energy deployment. To establish a 750 MW enabled solar power plant it will require about an investment of Rs. 45,000 million and spread in an area of nearly 16 km^2. Cleaning solar panels placed at the solar power plant twice a month is a requirement for long-term outcomes in order to acquire optimum insolation for a greater power yield production, which would generate pressure in the pre-existing water shortage situation, particularly during the peak summer season. Another key challenge to the government or any other agency who wishes to establish it is the lack of smooth well-connected roads for the proper functioning of solar power plants because they require timely movement of workers working in the particular plant. Apart from the villages connected to NH7 and NH9, the region's road conditions are the worst, limiting its accessibility.

Scope and Recommendations

- Use of high-resolution DTM will help in increasing the accuracy of results.
- Solar mapping can be a great tool for planners and municipalities who are trying to encourage communities to adopt solar technology.
- Solar potential of rooftops, being available to users and decision makers, gives insight into the amount of solar energy available and the amount that can be utilized.

- The results of this project can be used to understand the power generation capabilities and further be used for installation of solar PV system in conjunction with other necessary information.
- Use of continuous cloud cover data will help in calculating the accurate solar potential.
- Different approaches may be adopted for a comparative study in judging the precision of the results.

A solar PV installation and handling awareness and training program, as well as an Environmental Impact Assessment/Socio-Economic Assessment analysis for potential sites, are all needed. As a result, an appropriate location will be identified in accordance with the rules and regulations, and the final decision will be based on the stakeholders and relevant concerns.

Conclusions

Foreseeing the potential of solar power and its intensifications, reconsideration of the conventional energy structure is the current state of concern. The befitting solution is decentralized power generation by optimally utilizing the wastelands.

It is not possible to always measure the solar potential of an area with help of field measurements, in situ observations and site analysis as it is a time-consuming, tedious and highly expensive phenomena. This study focuses mainly on satellite-based approach to undertake the analysis and assessment of the solar potential of an area using remote sensing data.

The essential parameters utilized in this study are in raster format, which are integrated and normalized in a raster calculator based on their given weights, which are derived using Saaty's AHP approach. In the AHP comparative matrix, five different thematic layers along with socio-economic parameters were used with their corresponding calculated weights; assist in making correct decisions which generated the most scientifically sound and impactful results. The accuracy ratio measured for the AHP analysis was 0.0407, which was less than 0.10, indicating that the chosen method was precise. For the extraction of appropriate areas, the intersection between the potential and constraint factor layers is employed. The weight and standardization criteria for each criterion are easily adaptable, allowing them to be altered and adapted in response to economic and technical viability, as well as social and environmental concerns, giving them higher priority in the government's collaborative policymaking decisions, as well as local villagers and investors. Therefore, the result of the study shows that the given region gets sufficient insolation (602.34–773.14 KWh/m^2/day) and also it has vast wasteland area (981.23 km^2) which accounts about 13% of the total geographical area of the concern district which is enough for the installation of solar power plant.

References

Ahire V, Behera DK et al (2022) Potential landfill site suitability study for environmental sustainability using GIS-based multi-criteria techniques for nashik and environs, Environ Earth Science, Springer. Environ Earth Sci 81(178). https://doi.org/10.1007/s12665-022-10295-y

Akıncı H, Özalp AY, Turgut B (2013) Agricultural land use suitability analysis using GIS and AHP technique. Comput Electron Agric 97:71–82, S0168169913001567. https://doi.org/10.1016/j.compag.2013.07.006

Asakereh A, Omid M, Alimardani R, Sarmadian F (2014) Developing a GIS-based fuzzy AHP model for selecting solar energy sites in Shodirwan region in Iran. Int J Adv Sci Technol 68:37–48

Behera DK, Saxena MR, Ravi Shankar G (2017) Decadal landuse and landcover change dynamics in east coast of India –case study on Chilika Lake. Indian Geogr J 92(1):73–82

Carrion AJ, Estrella EA, Dols AF, Toro ZM, Rodriquez M, Ridao RA (2008) Environmental decision-support systems for evaluating the carrying capacity of land areas: optimal site selection for grid-connected photovoltaic power plants. Renew Sustain Energy Rev Elsevier 12:2358–2380

Cengiz T, Akbulak C (2009) Application of analytical hierarchy process and geographic information systems in land-use suitability evaluation: a case study of Dümrek village (Çanakkale Turkey). Int J Sust Dev World 16(4):286–294. https://doi.org/10.1080/13504500903106634

Central Electricity Authority (2017) Growth of electricity sector in India from 1947–2017. Government of India, Delhi

Central Electricity Authority (2018) Installed capacity monthly report. Government of India, Delhi

Chen MK, Wang S-C (2010a) The critical factors of success for information service industry in developing international market: Using analytic hierarchy process (AHP) approach. Expert Syst Appl 37(1):694–704, S0957417409005570. https://doi.org/10.1016/j.eswa.2009.06.012

Chen MK, Wang S-C (2010b) The use of a hybrid fuzzy-Delphi-AHP approach to develop global business intelligence for information service firms. Expert Syst Appl 37(11):7394–7407, S0957417410003040. https://doi.org/10.1016/j.eswa.2010.04.033

Effat HA (2013) Selection of potential sites for solar energy farms in Ismailia Governorate, Egypt using SRTM and multicriteria analysis. Int J Adv Remote Sens GIS 2:205–220

Feizizadeh B, Roodposhti MS, Jankowski P, Blaschke T (2014) A GIS-based extended fuzzy multi-criteria evaluation for landslide susceptibility mapping. Comput Geosci 73:208–221, S0098300414001873. https://doi.org/10.1016/j.cageo.2014.08.001

Georgiou A, Skarlatos DS (2016) Optimal site selection for sitting a solar park using multi-criteria decision analysis and geographical information systems. Geosci Instrum Method Data Syst 5:321–332

Georgiou A, Polatidis H, Haralambopoulos D (2012) Wind energy resource assessment and development: decision analysis for site evaluation and application. Energy Sources Part A Recovery Util Environ Eff 34(19):1759–1767

Goura R (2015) Analyzing the on-field performance of a 1-641 megawatt-grid -tied PV system in South India. Int J Sustain Energy 34:1–9

Guaita-Pradas I, Ullah S, Soucase BM (2015) Sustainable development with renewable energy in India and Pakistan. Int J Renew Energy Res 5:575–580

Khan G, Rathi S (2014) Optimal site selection for solar PV power plant in an Indian state using geographical information system (GIS). Int J Emerg Eng Res Technol 2:260–266

Kontos TD, Komilis DP, Halvadakis CP (2005) Siting MSW landfills with a spatial multiple criteria analysis methodology. Waste Manag 25(8):818–832

Kumar M (2021) Telangana focuses on clean energy to enhance its power supply. https://india.mongabay.com/2021/10/telangana-focuses-on-clean-energy-to-enhance-its-power-supply/

Kumar BS, Sudhakar K (2015) Performance evaluation of a 10 MW grid-connected solar photovoltaic power plant in India. Energy Rep 1:184–192

Kumar M, Singh SK, Kundu A, Tyagi K, Menon J, Frederick A, Raj A, Lal D (2022) GIS-based multi-criteria approach to delineate groundwater prospect zone and its sensitivity analysis. Appl Water Sci 12(4):71. https://doi.org/10.1007/s13201-022-01585-8

Lalwani M, Singh M (2010) Conventional and renewable energy scenario of India: present and future. Can J Electr Electron Eng 6:122–140

Makrides G, Zinsser B, Norton M, Georghiou GE, Schubert M, Werner JH (2010) Potential of photovoltaic systems in Countries with high solar irradiation. Renew Sust Energ Rev 14: 754–762

Malczewski J (1999) GIS and multicriteria decision analysis. Wiley, Canada, USA

Masera O, Ghilardi A, Drigo R, Trossero MA (2006) WISDOM: a GIS-based supply demand mapping tool for woodfuel management. Biomass Bioenergy 30(7):618–637

Ministry of New and Renewable energy (2018) Annual Report 2017–18. Government of India, Delhi. https://mnre.gov.in/img/documents/uploads/d6982ee8cce147288e7bf9434eebff55.pdf. (Accessed on date: 07.09.2022)

Ministry of New and Renewable Energy (2020) Annual Report 2020–21. Government of India, Delhi. https://mnre.gov.in/img/documents/uploads/file_f-1618564141288.pdf

Mishra D (2020) Site suitability analysis of solar energy plants in stony wasteland area: a case study of trans-yamuna upland region, Allahabad district, India. J Indian Soc Remote Sens 48(4): 659–673

Muneer T, Asif M, Munnawar M (2005) Sustainable production of solar electricity with particular reference to Indian economy. Renew Sust Energ Rev 9(5):444–473

Munkhbat U, Choi Y (2021) GIS-based site suitability analysis for solar power systems in Mongolia. Appl Sci 11:1–13

Murmu P, Kumar K, Lal D, Sonker I, Singh SK (2019) Delineation of groundwater potential zones using geospatial techniques and analytical hierarchy process in Dumka district Jharkhand India. Groundw Sustain Dev 9:100239, S2352801X18302339100239. https://doi.org/10.1016/j.gsd.2019.100239

Orgi. Census of India Website: Office of the Registrar General & Census Commissioner, India [Online]. Census of India Website: Office of the Registrar General & Census Commissioner, India. http://censusindia.gov.in/ [5 May 2019]

Pandey HK, Singh VK, Singh SK (2022) Multi-criteria decision making and Dempster-Shafer model–based delineation of groundwater prospect zones from a semi-arid environment. Environ Sci Pollut Res 29(31):47740–47758. https://doi.org/10.1007/s11356-022-19211-0

Patel A, Singh MM, Singh SK, Kushwaha K, Singh R (2022) AHP and TOPSIS based sub-watershed prioritization and tectonic analysis of Ami River Basin Uttar Pradesh. J Geol Soc India 98(3):423–430. https://doi.org/10.1007/s12594-022-1995-0

Popov M et al (2021) Long-term satellite data time series analysis for land degradation mapping to support sustainable land management in Ukraine. In: Singh TP, Singh D, Singh RB (eds) Geo-intelligence for Sustainable Development. Advances in Geographical and Environmental Sciences. Springer, Singapore. https://doi.org/10.1007/978-981-16-4768-0_11

Protasov A, Tomchenko O, Novoselova T, Barinova S, Singh SK, Gromova Y, Curtean-Bănăduc A (2022) Remote sensing and in-situ approach for investigation of pelagic communities in the reservoirs of the electrical power complex. Front Biosci (Landmark) 27(7):221. https://doi.org/10.31083/j.fbl2707221

Ramachandra TV, Jain R, Krishnadas G (2011) Hotspots of solar potential in India. Renew Sustain Rev 15:3178–3186

Saaty TL (1977) A scaling method for priorities in hierarchical structures. J Math Psychol 15(3):234–281, 0022249677900335. https://doi.org/10.1016/0022-2496(77)90033-5

Saaty TL (1980) The analytic hierarchy process. McGraw-Hill, New York

Sadhu M, Chakraborty S, Das N, Sadhu PK (2015) Role of solar power in sustainable development of India. Indones J Electr Eng 14:34–41

Singh VK, Kumar D, Singh SK, Pham QB, Linh NTT, Mohammed S, Anh DT (2021) Development of fuzzy analytic hierarchy process based water quality model of Upper Ganga river basin India. J Environ Manag 284:111985, S0301479721000475111985. https://doi.org/10.1016/j.jenvman.2021.111985

Soyoung P, Seongwoo J, Shinyup K, Choi C (2011) Prediction and comparison of urban growth by land suitability index mapping using GIS and RS in South Korea. Landsc Urban Plan 99(2):104–114, S0169204610002367. https://doi.org/10.1016/j.landurbplan.2010.09.001

Sudhakar K, Srivastava T, Satpathy G, Premalatha M (2013) Modelling and estimation of photosynthetically active incident radiation based on global irradiance in Indian latitudes. Int J Energy Environ Eng 4(21):2–8

Tegou LI, Polatidis H, Haralambopoulos D (2010) Environmental management framework for wind farm sitting: methodology and case study. J Environ Manag 91:2134–2147

Yousefi H, Hafeznia H, Sahzabi AY (2018) Spatial site selection for solar power plants using a GIS-based Boolean-Fuzzy Logic Model: a case study of Markazi province, Iran. Energies 11:1–18

Chapter 23
Integrated Study on Tsunami Impact Assessment in Cilacap, Indonesia: Method, Approach, and Practice

Ranie Dwi Anugrah and Martiwi Diah Setiawati

Abstract Cilacap was one of the areas hit by the tsunami that occurred on July 17, 2006, along a subduction zone off Central and West Java, Indonesia, due to an earthquake (Mw = 7.7). The tsunami risk was assessed in this study by combining the tsunami hazard (i.e., the tsunami inundation map) with the vulnerability index based on human and economic factors. We used a numerical method to calculate tsunami inundation, considering the tsunami's wave height, slope, and surface roughness. Meanwhile, we examined the vulnerability index using a multicriteria approach based on a weighted linear combination in a geographic information system (GIS). According to the findings, 72.8 ha of coastal regions were classified as high or very high risk, whereas 380.2 ha were moderate to very low risk. In addition, we estimated that the tsunami affected 20,850 people, with a total economic loss of USD 645,893. The commercial area was the most severely affected in this case study, followed by the agricultural sector. This risk assessment, which includes evacuation measures such as temporary evacuation shelters and evacuation routes, was developed as a countermeasure to decrease the tsunami's impact in the future. This result could also be utilized as input and considered by stakeholders in the tsunami disaster prevention and mitigation framework, such as minimizing the damage and losses that will occur in tsunami-affected coastal areas.

Keywords Risk assessment · Hazard · Vulnerability · GIS · Remote sensing · Disaster prevention and mitigation

---

Authors Ranie Dwi Anugrah and Martiwi Diah Setiawati have equally contributed to this chapter.

R. D. Anugrah (✉)
Detailed Spatial Planning of Environmental Carrying Capacity Area – Region 1, Ministry of Agrarian Affairs and Spatial Planning, National Land Agency, Jakarta, Indonesia

M. D. Setiawati
Research Center for Oceanography, The National Research and Innovation Agency (BRIN), Jakarta, Indonesia

Introduction

There are three major tectonic plates located in Indonesia, namely Eurasian, Pacific, and Australian-Indian plate. All of them are actively moving and establishing a complicated network of plate boundaries (Bock et al. 2003). There were 173 large and small tsunami events in Indonesia between 1629 and 2016 (WorldData.info 2018). The speed of plate movement in the western part was between 5 and 6 cm per year, while in the southern part was 7 cm per year (Karig et al. 1979). As in the east northern region, the speed of plate movement reaches 12 cm per year. Thus, the degree of seismicity is quiet high which caused the area vulnerable to tsunami.

The tsunami in the Indian Ocean (December 2004) and the Tohoku region (March 2011) highlights the importance of conducting holistic analysis on hazards and vulnerability of tsunami along various coastal areas worldwide. The southern coast of Java is one of most vulnerable coastal area to the impact of tsunami in the southeast Indian region. This coast was hit by tsunami on June 3, 1994, and July 17, 2006, which happened in Banyuwangi (magnitude 7.8), East Java; Pangandaran (magnitude 7.7), West Java; and Cilacap (magnitude 7.7), Central Java. The 2006 tsunami death toll was estimated to be over 600, with more than 75,000 people displaced along a 200-km stretch of coastline (Lavigne et al. 2007). Meanwhile, in 1994, tsunami killed more than 200 people in Banyuwangi, East Java.

Despite recent efforts to compile tsunami catalogs, identify the deposits, map the inundation area, and assess the building vulnerability (Medina et al. 2011; Omira et al. 2009, 2013; Atillah et al. 2011), integrated risk assessment and its evacuation routes along the south Java coasts remain poorly investigated. The data limitation of high spatial resolution topographic map is a common barrier to perform inundation simulations for detailed tsunami hazard assessment. While validated building fragility curves for this study area are lacking, the adoption of geographic information system (GIS) (Atillah et al. 2011; Omira et al. 2009; Silva et al. 2003) stands a viable option for this study.

A tsunami can cause a number of deaths, damage the infrastructure, and cause significant economic losses (Papathoma et al. 2003). Therefore, an integrated tsunami risk assessment is crucial to conduct in the study area. In this paper, we consider risk equal to the likelihood of loss from the exposed essential feature as a result of the complex sequence of hazard and vulnerability (Blaikie et al. 1994; Jelinek et al. 2012; Rynn and Davidson 1999; Tinti et al. 2008).

We aim to assess a tsunami risk defined as the integration of the tsunami hazard (i.e., the tsunami inundation) and the degree of vulnerability as per sociodemographic and monetary aspects. This research could be used as input and consideration in decision support systems by stakeholders within the framework of the tsunami disaster prevention and mitigation, in particular reducing the damage and losses that will arise in coastal areas hit by the tsunami.

Data and Method

Pilot Sites

Cilacap City is the capital city of Cilacap Regency in Central Java Province, located on the south coast of Java and directly adjacent to the Indian Ocean, as shown in Fig. 23.1. This area consists of 3 subdistricts: Cilacap Utara, Cilacap Tengah, and Cilacap Selatan, with 50.1 km^2. This city has flat topography (i.e., 6 m above the sea level) and vulnerable to tsunami where the surface altitude in the north and northeast region is higher than in the west and southwest. There were 8 villages located in this study area, namely Cilacap Village, Sidakaya Village, Sidanegara Village, Gunungsimping Village, Tegalkamulyan Village, Kebonmanis Village, Gumilir Village, and Mertasinga Village.

Data

In this study, we utilized the statistical data, raster satellite images, and vector maps. Furthermore, all supported information for evacuation routes was compiled through

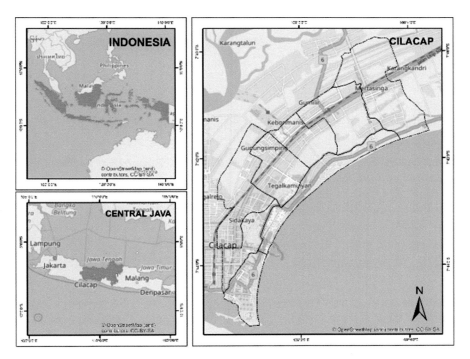

Fig. 23.1 Study area. (Source: authors)

Table 23.1 Dataset used for risk assessment

No	Data	Format	Source
1	Digital elevation model (DEM)—TerraSAR (9 m)	Raster	Geospatial Information Agency (BIG)
2	Land cover	Vector	Bing Imagery and Geospatial Information Agency (BIG)
3	Historical tsunami event (wave height in m)	Tabular	U.S. Geological Survey (USGS), the Integrated Tsunami Data Base (ITDB), National Oceanic and Atmospheric Administration (NOAA)
4	Human aspect—population density (people/km^2)	Raster	WorldPop
5	Human aspect (a) People with disability (number of people) (b) Children population (number of people) (c) Elderly population (number of people) (d) Fisherman (number of people) (e) Existence of evacuation routes (f) Existence of temporary shelters (g) Development of evacuation drill in communities (number of people)	Tabular	Cilacap Statistics Bureau (BPS Cilacap)
6	Economic aspect (a) Industrial area (b) Commercial area (c) Agricultural area (d) Residential area (e) Road network	Vector	Geospatial Information Agency (BIG) and Open Street Map

Source: authors

literature studies, institutional data, and field surveys. Table 23.1 described a list of data used in this research.

Most of the data used are spatial data, both in vector and raster formats. Then, for tabular, data format was converted into spatial dataset.

Research Framework

Figure 23.2 shows the overall research framework of the tsunami risk index. We consider the risk as a complex system. Thus, integrating physical, social, and economic aspects was applied. The analysis was computed into three steps: First was the calculation of tsunami inundation as a hazard map, second was the calculation of tsunami vulnerability index by considering both human and economic

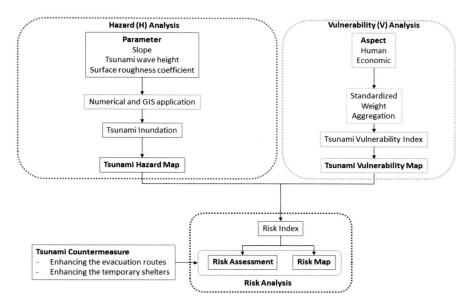

Fig. 23.2 Research framework for risk analysis. (Source: authors)

aspects, and third was the integration among them, which resulted in the risk map and its risk assessment. The risk assessment includes the human and economic impact analysis, enhancement of evacuation routes, and temporary shelter.

Tsunami Inundation

The tsunami inundation map was referred to the calculation of run-up height and inundation distances (i.e., the furthest range inland which is exposed by a tsunami). The method used to determine the tsunami inundation using the following equation. Equation 23.1 for the inundation distance developed by Pignatelli et al. (2009):

$$X_{\max} = 0.06(H_t)^{1.33} n^{-2} \cos S \qquad (23.1)$$

Equation 23.2 for run-up height developed by McSavaney and Rattenburry (2000):

$$Y_{\text{loss}} = \left(\frac{167 n^2}{H_t^{\frac{1}{3}}}\right) + 5 \sin S \qquad (23.2)$$

where $X_{\max}$ is the maximum distance that the wave height can penetrate inland (meter), H_t is the wave height which reached the coast (meter), S is the slope of the

Table 23.2 Value of surface roughness coefficient

Land cover type	Roughness coefficient	Reference
Water	0.007	Amri et al. (2016)
Scrubs/bush	0.04	
Forest	0.07	
Farmland	0.035	
Open field	0.015	
Agricultural land	0.025	
Settlement/built-up area	0.045	
Mangroves	0.025	
Embankment	0.001	

Source: authors

land surface (degree), n is the surface roughness coefficient, and Y_{loss} is wave height loss per meter of inundation distance (meter).

The slope was calculated from the digital elevation model (DEM) of TerraSAR satellite images, with the spatial resolution of 9 m. Meanwhile, the roughness data were computed by extracting land cover and then classified by the value of surface roughness coefficient using the Manning coefficient's standard by the Indonesian National Board for Disaster Management (BNPB), as shown in Table 23.2 (Amri et al. 2016).

Vulnerability Analysis

Vulnerability is defined as a strong tendency to suffer the negative impact of the hazard and depicts a personal or community capacity for preparation, response, and recovery from disaster (Twigg 2004). This analysis is essential to know what aspects are potentially for loss due to a hazardous event in the study area, and it was conducted into four stages as shown in Fig. 23.3: selection of tsunami vulnerability parameter, standardization of vulnerability parameter, estimation of weights of each parameter, and aggregation of tsunami vulnerability. Vulnerability analysis depends on the distance between the source of the hazard to the communities and their socioeconomic aspects (Cutter et al. 2000).

Selection of Vulnerability Parameter

The following are the parameters that determine the human vulnerability to the tsunami.

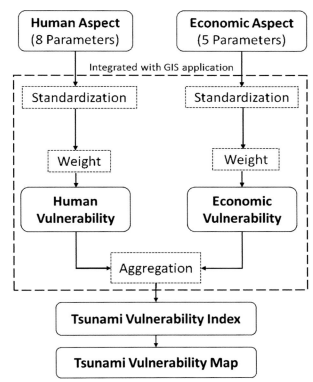

Fig. 23.3 Framework of the vulnerability analysis. (Source: authors)

(a) Population Density

Population density typically describes the distribution number of people in a certain area. The more people in an area, the more difficult evacuation will be, and the greater the potential number of causalities if a tsunami strikes (Papathoma et al. 2003).

(b) People with Disability

One of the vulnerable populations against tsunamis is people with disabilities. They have difficulties in response to the tsunami event, and they also cannot recover fast after the disaster due to their reliance (Cutter et al. 2003). Types of people with disability that require the disaster protection and evacuation related to physical and intellectual are deaf, blind, speech impaired, quadriplegic, mentally illness, and mentally disabled.

(c) Children Population

Children are defined as a population under 14 years old. This population has a lower capacity to recover fast after the disaster due to their reliance on families and communities.

(d) Elderly Population

The elderly are referred to the aspects associated to movement limitation, higher reliance on the social insurance, and minor unofficial assistance and networks which potentially affected the recovery process after the disaster (Cutter et al. 2003). The definition of the elderly population is the number of persons over 65 years old.

(e) Fisherman

Agung 2006 reported that fishing activities are the most impacted job due to the tsunami which endures a significant loss and damage of vessels and other fishing equipment that might be long term to get back. In this study, we only consider fisherman as the primary profession to fulfill their daily life necessities.

(f) Existence of Evacuation Routes

Evacuation planning required major and additional routes within tsunami-prone areas with evacuation signs to safe areas. The evacuation route planning will save the time to move out the exposed communities or individuals to a safer place (Kim et al. 2007). The definition of this parameter is the number of current evacuation routes along the Cilacap coastal area.

(g) Existence of Temporary Shelters

Temporary shelter is defined as the appropriate structures in the safe area that could provide short-term hosting of many evacuees. Because evacuees cannot quickly reach higher ground, they are advised to seek shelter in tall structures (built constructions) in an emergency. Thus, to perform the vulnerability analysis, the number of temporary shelters for a tsunami must be included.

(h) Development of Evacuation Drill in Communities

This variable defines a simulation of how people would be evacuated in the incidence of a tsunami or other disaster. People can receive coastal danger notifications and alerts and notify at-risk populations and individuals who take action in response to the alarm. As a result, coastal communities are educated and socialized on disaster preparedness and mitigation, lowering the chance of a tsunami.

Individuals, groups, and nations' levels of vulnerability are all influenced by their economic circumstances. The impoverished are more susceptible to disasters as they shortage of funds to construct solid structures and implement other engineering steps to protect themselves from disasters. The following are the parameter that determines the economic vulnerability:

(a) Industrial Area

An industrial area is a section of land zoned and developed specifically for the growth of the industry. The density of the structures gives a measure of the region/nation of the financial circumstances and potential losses in a community's business, as well as longer-term problems with postdisaster recovery (Cutter et al. 2003). The denser industrial area will increase the value of the loss caused by the tsunami.

(b) Commercial Area

A commercial building is an area zoned and planned for commercial activities such as trade or business profit. The denser commercial area will increase the value of the loss caused by the tsunami.

(c) Agricultural Area

The agricultural area is defined as land used explicitly for agricultural purposes to raise crops or livestock. Agriculture is relatively important, and the disaster will affect the coastal economy. If this area is located along a coastal area, it is highly vulnerable to tsunami flow because the area got inundated every high tide. Coastal communities dependent on subsistence agriculture for their survival are the most affected by the tsunami.

(d) Residential Area

Residential area refers to a land-use type mainly used for residential purposes where parks, internal roads, and other public lands are not included. The denser residential area in the coastal region will increase the value of the damage and loss caused by the tsunami.

(e) Road Network

Tsunami vulnerability can be reduced if adequate road infrastructure is in place. Residents fled to safer places using the access road as an evacuation route—the less accessible the road network, the greater the vulnerability to tsunamis.

Standardization of Vulnerability Parameter

The standardization is necessary to set the value of a parameter to a standard scale for comparison since each parameter has a different measurement unit. This study used a linear fuzzy membership function as a standardization method where the minimum and maximum values ranged from 0.0 to 1.0. The advantages of fuzzy membership work well with continuous variables and data uncertainties and thus are very suitable to predict the vulnerability (Baučić 2020). The data format used in the processing of fuzzy membership is more objective with an area of coverage in pixels. The type of data used can be quantitative and qualitative.

Before standardizing the parameter, each parameter must be determined in advance of its linear membership function between the parameters and vulnerability. There are two types of linear, a linear increase and a linear decrease based on the directional differences in vulnerability in a provided parameter. In this study, the membership function was determined from the previous studies and as per the theoretical assumptions as described in Table 23.3.

The linear increase represents a positive functional relationship with vulnerability. The following Eq. 23.3 was used for standardization (Nyimbili and Erden 2020).

Table 23.3 Membership function of tsunami vulnerability parameter

Aspects	Parameter	Membership function	References
Human	Population density	Increase	Cutter et al. (1997)
	Number of children population	Increase	Cutter et al. (1997, 2003)
	Number of elderly population	Increase	Cutter et al. (1997, 2003)
	Number of disabled person	Increase	Cutter et al. (1997, 2003)
	Number of fisherman	Increase	Eddy (2006)
	Existence of evacuation routes	Decrease	Kim et al. (2007) and Dewi (2010)
	Existence of temporary shelters	Decrease	Dewi (2010)
	Existence of evacuation drills	Decrease	Dewi (2010)
Economic	Industrial area	Increase	Cutter et al. (2003)
	Commercial area	Increase	Cutter et al. (2003)
	Agricultural area	Increase	Luther (2010)
	Residential area	Increase	Ruther et al. (2002)
	Road network	Increase	Cutter et al. (2003)

Source: authors

$$X_i = \frac{x_i - x_i^{\min}}{x_i^{\max} - x_i^{\min}} \quad (23.3)$$

where X_i is the standardized parameter value (i), x_i is the actual parameter value, $x_i^{\max}$ is the maximum value of parameter, and $x_i^{\min}$ is the minimum value of the parameter.

The linear decrease represents a negative functional relationship with vulnerability, where standardization was computed based on Eq. 23.4 (Nyimbili and Erden 2020).

$$X_i = \frac{x_i^{\max} - x_i}{x_i^{\max} - x_i^{\min}} \quad (23.4)$$

Estimating Weights

Parameter weighting aims to measure and distinguish the relative priority of each parameter amid a list of parameters. The principal eigenvector method was applied to determine the significance of weights. The calculation of eigenvector values was processed using a Principal Component Analysis (PCA) tool in ArcGIS software.

The eigenvalues of the PCA represent the variance associated with them, while the loadings give the participants of the set variables.

Aggregation of Vulnerability Parameter

The vulnerability parameter is aggregated using a weighted linear combination for determining the decision rule. The method is based on the concept of a weighted average as shown in Eqs. 23.5 and 23.6:

$$\text{TVI} = \sum w_i s_i \qquad (23.5)$$

$$\text{TVI} = V_{\text{human}} + V_{\text{economic}} \qquad (23.6)$$

where TVI is the suitability for tsunami vulnerability index, w_i is a weighting of parameter i, and s_i is the standardized value of parameter i. The TVI is calculated by multiplying the significant weight attributed for each parameter by the standardized value provided to the parameter and summing the products over all parameters. Figure 23.4 shows the overall illustration from steps from standardization until the aggregation process.

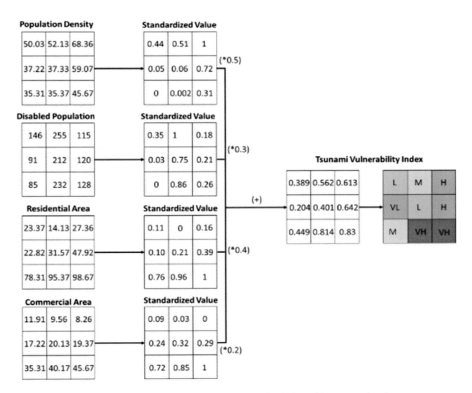

Fig. 23.4 Example for generating a tsunami vulnerability index. (Source: authors)

Risk Analysis

Risk is the likelihood of an adverse event with a specific force at a particular location and time. There are two elements essential in risk analysis: hazard as the likelihood of incidence of a harmful natural disaster and vulnerability as the sensitivity and threat of a person, a society, resources, or systems to the effects of hazard as shown in Eq. 23.7 (Kohler et al. 2004).

$$\text{Risk} = \text{Hazard} \times \text{Vulnerability} \qquad (23.7)$$

The concept of the formula is the more significant risk resulting from a higher possibility of a severe hazard incident and the more exposed vulnerable population.

Human and Economic Impact Analysis

Calculations for analyzing human impacts on tsunamis are indicated by the number of people affected by the tsunami inundation per one grid or one hectare. The economic impact analysis of the tsunami was calculated based on the type of affected land cover. There were four types of land cover in the Cilacap coastal area, namely residential, agricultural, commercial, and industrial areas. The detailed calculation criteria were based on BAPPENAS 2008 as shown in Table 23.4.

Table 23.4 Criteria of economic impact

Land cover type	Low damage	Moderate damage	High damage
Residential	Area (ha) × damage to house building, furniture, and home appliances is estimated at $360 × 20%	Area (ha) × damage to houses, furniture, and home appliances is estimated at $720 × 40%	Area (ha) × damage to house building, furniture, and home appliances is estimated at $1440 × 60%
Agricultural	Area (ha) × production cost of $504 per hectare and yield of 5 tons per hectare, assuming the price of the harvest is $0.1728 per kg × 20%	Area (ha) × production cost of $504 per hectare and yield of 5 tons per hectare, assuming the price of the harvest is $0.1728 per kg × 40%	Area (ha) × production cost of $504 per hectare and yield of 5 tons per hectare, assuming the price of the harvest is $0.1728 per kg × 60%
Commercial	Area (ha) × cleaning and repainting fee of $360 × 20%	Area (ha) × repair, cleaning and repainting costs of $3600 × 40%	Area (ha) × the cost of dismantling, rebuilding, and cleaning costs is $21,600 × 60%
Industrial	Area (ha) × cleaning and repainting fee of $360 × 20%	Area (ha) × repair, cleaning, and repainting costs of $3600 × 40%	Area (ha) × the cost of dismantling, rebuilding, and cleaning costs is $21,600 × 60%

Source: authors

Tsunami Countermeasure

Temporary Evacuation Shelters

A tsunami evacuation shelter is an emergent and temporary evacuation facility. An evacuation shelter is crucial, especially if residents cannot reach the higher surface because of the great distance and time constraints issues. Therefore, the constructions are settled within accessible span by populous areas and along the evacuation routes. We investigate the possibility for both the horizontal and vertical evacuation shelter. A horizontal evacuation shelter is a method that moves out the communities from dangerous areas to secure areas on higher elevation or away from the coastline. A vertical tsunami evacuation shelter is a structure of a sufficient height to uplift evacuees above the point of tsunami flooding and designed and constructed with the power and resiliency required to deal with the pressure of tsunami energy (FEMA 2008). These characteristics for temporary evacuation shelter are as follows:

(a) Structure

The structure/construction of a temporary shelter should be earthquake and tsunami-resistant. Also, it must be designed to be unharmed or damaged only to the degree that it does not endanger lives due to the disaster. The following are the requirements for good structural attributes. These include robust designs with backup ability to accommodate intense power, aspects that allow water current to pass through with little resistance, flexure systems that withstand powerful forces with no breakdown, and the duplicative mechanism that can resist with the half of force loss without structural failure.

(b) Evacuation Floor

The second component is the evacuation floor, where the tsunami should not flood this area. Thus, the base of the evacuation area must be designed higher than the worst case of tsunami wave height. For most circumstances, the construction is multistory, with citizens having the option of evacuating to the upper floor. Moreover, the principal task of the evacuation center is to create enough capacity to house many more evacuees during or after a temporary evacuation.

(c) Design Function

Due to the long return period of the tsunami disaster, there is no particular construction that has been intended or designated solely for evacuation purposes. Thus, an evacuation shelter is mainly set at the intended structure with supplementary use (i.e., multipurpose system). It is primarily a public facility such as schools, mosques, government offices, hotels, restaurants, convention buildings, parking lot, and shopping center.

(d) Capacity

Evacuation shelters must have enough capacity to host more evacuees, and according to BAPPENAS 2005, the ideal building space for evacuation shelters was at least 1 m² per person. They also could use empty spaces in evacuation shelters that are periodically or infrequently utilized.

(e) Horizontal Accessibility

Evacuation shelters must be within a reasonable walking or running span from residential areas in the risk zones. Furthermore, the attainable distances of 500, 1000, 1500, and 2000 m correspond to journey times of 5, 10, 15, and 20 min, respectively, according to the Republic of Indonesia's national master plan for tsunami recovery and reconstruction (BAPPENAS 2005).

(f) Vertical Accessibility

Evacuation shelters must have enough stairs or ramps to fulfill the building safety criteria and regulations. Therefore, it is crucial to take into account the width of the stairs to accommodate the mobilization of at least two evacuees.

Evacuation Routes

Spatial data in the form of road network data, tsunami inundation areas, and land use were used to determine tsunami evacuation routes. Numerous factors must be considered while selecting a tsunami evacuation route, including the following: selected routes are arterial roads, collector roads, local roads, and environmental roads. Arterial roads are defined as state roads which have the main function for transportation with long-distance features and fast average pace. The access roads are efficiently limited by road width >8 m. Collector road is referred as a state road which collects or splits carriers with medium distance features and medium average pace, and the entrance is restricted by road width >5 m. Local roads are state roads with characteristic of short-distance feature and low average pace, and entrance number is not limited by road width >3 m. While the environmental road defined as road which intended for environmental transport with the characteristics of traveling at close range, and low average speed with the width of the road does not exceed 2 m. The route is outside the tsunami inundation area.

In this research, we used ArcGIS network analyst to model evacuation routes. This assessment will assist planners in determining the optimal path to evacuate societies and choosing the most convenient area for supplementary shelters.

Result

Tsunami Wave Height and Magnitude in Indonesia Throughout the History

To understand the worst-case scenario of wave height scenario for tsunami inundation, we conducted literature review to summary the historical tsunami in the study area and surrounding area (see Appendix 23.1). Nineteen tsunamis caused by the earthquake occurred in the study area from 1818 to 2007 (189 years). The highest tsunami wave height in the study area was about 8 m, which occurred on July 17, 2006, due to an earthquake with the moment magnitude of 7.7 Mw. This incident claimed 668 lives, 65 missing (assumed dead), and 9299 others injured.

Tsunami Inundation

Figure 23.5 presents the tsunami inundation simulation as the representative of the hazard map in the study area. The result shows that the range of inundation depth was between 0 and 8 m with the total area of 4.53 km^2, respectively. Meanwhile, the maximum inundation distance in the study area was 778.26 m, respectively, from the coastline. It also shows that the spatial distribution of tsunami inundation area decreased from the eastern to the western region of the village as shown through the blue gradation on the map; the darker the color, the higher the depth level of the tsunami. Since 75% of Cilacap coastal areas have a flat slope and a small roughness coefficient, the tsunami runoff flowed easily to the inland.

We classified the inundation depth into five classes, as shown in Table 23.5. The result stated that about 68% and 28.7% of the area was inundated below 1 m and above 2 m. The table also described that six villages were affected by tsunami, namely Cilacap, Gumilir, Gunungsimping, Mertasinga, Sidakaya, and Tegalkamulyan. The most affected village by the tsunami was Tegalkamulyan, followed by Cilacap and Mertasinga. The high affected areas were mainly located along the coastline.

Tsunami Vulnerability Analysis

Human Vulnerability

This element was defined as the tendency of individuals to be harmed or killed, which includes problems pertaining to movement inadequacies and weak disparity points caused by age or disability. Figure 23.6 shows the human vulnerability map where the red color indicates very high vulnerability, while dark green shows very low vulnerability. Based on the figure, the vulnerability index was higher in the western and southern parts than in the northern part of the Cilacap coastal area. It was because the population density was higher in the region compared with the north of the area. In addition, the western and southern areas were categorized as the urban areas where most of the land covers were residential and commercial regions. Meanwhile, in the northern area, agricultural land was dominant.

Figure 23.7 described the comparison of human vulnerability areas by the village. The result stated that Sidanegara is one of the villages that had an extensive area of high human vulnerability index which reaches 275 hectares, followed by Sidakaya and Cilacap villages.

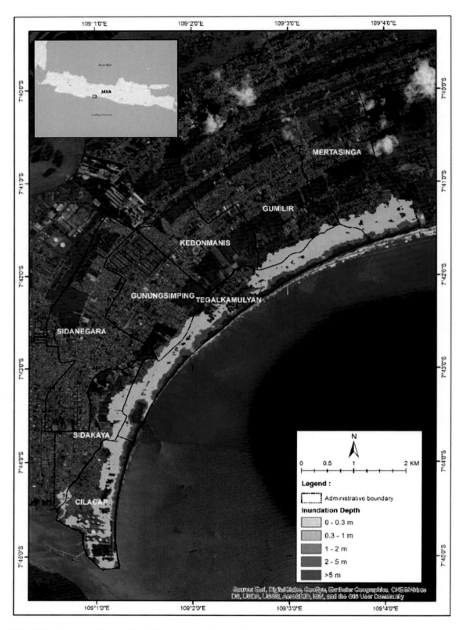

Fig. 23.5 Tsunami hazard in Cilacap coastal area. (Source: authors)

Table 23.5 Villages affected by tsunami inundation

Villages	The classification of tsunami inundation area (ha)					Total
	Very low	Low	Moderate	High	Very high	
Cilacap	61	1	5	10	17	94
Gumilir	21	0	0	0	0	21
Gunungsimping	11	0	0	0	0	11
Mertasinga	49	0	2	8	7	66
Sidakaya	26	0	0	0	1	27
Tegalkamulyan	127	12	9	38	48	234
Total	**295**	**13**	**16**	**56**	**73**	**453**

Source: authors
Note: The inundation depth classification was defined as follow: very low (0–0.3 m), low (0.3–1 m), moderate (1–2 m), high (2–5 m), and very high (>5 m)

Economic Vulnerability

The following figure showed the economic vulnerability of the pilot sites to the tsunami (Fig. 23.8). It revealed that about 49% of the region was categorized as a very high vulnerability index, where the residential area was the dominant land cover type. Moreover, among eight villages, Sidanegara has the highest economic vulnerability index with a total area of 265 ha.

Tsunami Risk Analysis

Figure 23.9 shows the tsunami risk map in the study area. The result reported that the very high-risk area was located along the coastline, covering 15% of the site. Moreover, most of the region was categorized as low-risk level. Out of eight villages, six were exposed to the risk of a tsunami: Cilacap, Sidakaya, Gunungsimping, Tegalkamulyan, Gumilir, and Mertasinga. More details on the percentage of risk areas in each village were shown in Table 23.6.

Discussion

This chapter aims to investigate the tsunami risk assessment in Cilacap City by integrating the physical, social, and economic variables through numerical models, remote sensing, and GIS approaches. In the current study, we used the worst-case scenario of wave height for tsunami hazard map where the spatial pattern of inundation was varied on the location, indicating the highest water level (i.e., >2 m) along the coastline with the maximum distance 778.26 m (Fig. 23.5). Higher degrees of hazard map along the coastline indicated that the distance from the coastline was the most vulnerable area. In addition, the flat slope and its small

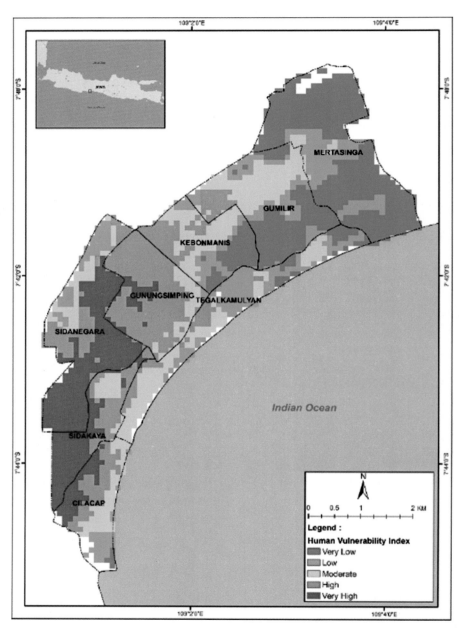

Fig. 23.6 Human vulnerability map in Cilacap coastal area. (Source: authors)

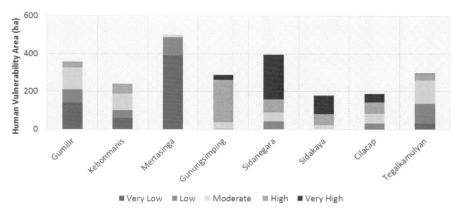

Fig. 23.7 Comparison of human vulnerability area by village. (Source: authors)

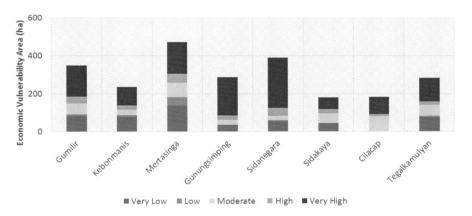

Fig. 23.8 Comparison of economic vulnerability area by village. (Source: authors)

roughness coefficient caused runoff could flow easily without even being significantly disrupted by geomorphologic varieties. Furthermore, among eight villages in Cilacap City, six were hit by the tsunami, where Tegalkamulyan was the most affected village, with 234 ha inundated (Table 23.5).

Vulnerable communities not only are at risk due to hazard exposure, but also relate to the complex socioeconomic condition. Thus, we computed the vulnerability index from two main categories: humanity and economic aspect. As shown in Figs. 23.7 and 23.8, Sidanegara village is the most vulnerable area due to higher population density, a higher number of vulnerable people, less evacuation area, and higher economic value. Within this village, 75% was categorized as a very high vulnerability index, respectively. Sidakaya and Gunungsimping were the other villages with a higher vulnerability index.

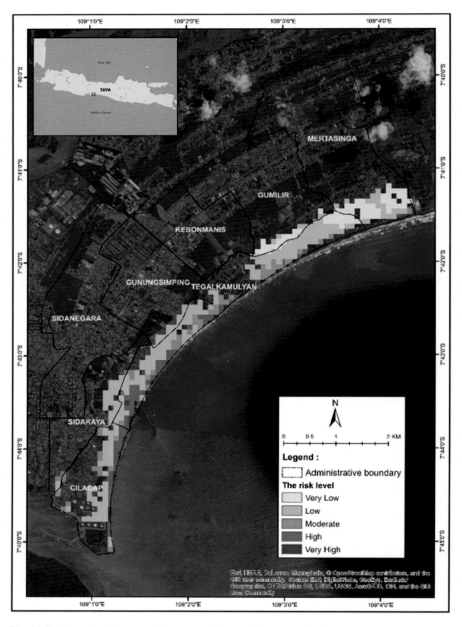

Fig. 23.9 Tsunami risk map in Cilacap coastal area. (Source: authors)

Overall, the combination between hazard and vulnerability index produced the risk map. As shown in Fig. 23.9, a higher degree of risk is located along the coastline where Tegalkamulyan village was the largest affected area (49%), followed by

Table 23.6 Percentage of risk area per each village in Cilacap coastal area

No	Village	Sub District	Percentage of risk area (%)
1	Cilacap	Cilacap Selatan	19.70
2	Sidakaya	Cilacap Selatan	6.98
3	Gunungsimping	Cilacap Tengah	2.99
4	**Tegalkamulyan**	Cilacap Selatan	**48.88**
5	Gumilir	Cilacap Utara	5.24
6	Mertasinga	Cilacap Utara	16.21

Source: authors

Table 23.7 Estimated number of people affected by village

No	Village	Sub District	Total
1	Cilacap	Cilacap Selatan	5453
2	Sidakaya	Cilacap Selatan	2213
3	Tegalkamulyan	Cilacap Selatan	10,352
4	Mertasinga	Cilacap Utara	1441
5	Gumilir	Cilacap Utara	611
6	Gunungsimping	Cilacap Tengah	780

Source: authors

Cilacap and Mertasinga. By this analysis, we also understand that under the 8 m tsunami wave height scenario, the vulnerable area concerning socioeconomic aspects was not affected by the tsunami. However, if a bigger tsunami comes in this area, a highly vulnerable area may be affected.

The tsunami caused a loss in human casualties and property belonging to both private and public property. It can disrupt the social and economic activities of the population. Humans will die, disappear, wound, and evacuate, while social and economic infrastructure will be damaged or destroyed for public facilities. Based on the tsunami risk analysis results in the study area, an estimated 20,850 people will be affected by the tsunami under an 8 m wave height scenario.

As shown in Table 23.7, Tegalkamulyan village has the highest number of affected people, as many as 10,352 people. From these estimates, how many people were affected and how the impacts can be generated for the community can be seen. Thus, this type of analysis can help the local government/policymaker estimate how many evacuation posts are needed if a tsunami hits the area and how many goods should be supplied to comply with the requirements of the affected communities.

This paper also considers the economic impact analysis due to tsunami with the 8 m wave height scenario. The land cover-based approach was used to estimate the financial loss. We classified the land cover into four classes: residential, industrial, commercial, and agricultural. Among them, the farm area was the most damaged with the extent of 213 ha (Table 23.8). Furthermore, Tegalkamulyan village has the largest damaged area, covering about 215 ha, where agricultural area covers almost half of it.

Table 23.8 Estimated area of land cover affected per each village in Cilacap coastal area

No	Villages	The classification of land cover area (ha) by villages			
		Residential	Agricultural	Commercial	Industrial
1	Cilacap	29	24	23	14
2	Sidakaya	9	11	4	0
3	Gunungsimping	10	0	0	1
4	Tegalkamulyan	76	108	30	0
5	Gumilir	3	18	0	0
6	Mertasinga	11	52	0	0

Source: authors

Table 23.9 Estimated level of damage by tsunami based on commercial and industrial area

Villages	Commercial			Industrial		
	Low	Moderate	High	Low	Moderate	High
Cilacap	$576.00	$5760.00	$142,560.00	$864.00	$2880.00	$25,920.00
Tegalkamulyan	$144.00	$10,080.00	$272,160.00	$0.00	$0.00	$0.00
Sidakaya	$288.00	$0.00	$0.00	$0.00	$0.00	$0.00
Gunungsimping	$0.00	$0.00	$0.00	$72.00	$0.00	$0.00
Gumilir	$0.00	$0.00	$0.00	$0.00	$0.00	$0.00
Mertasinga	$0.00	$0.00	$0.00	$0.00	$0.00	$0.00
Villages	**Agricultural**			**Residential**		
	Low	Moderate	High	Low	Moderate	High
Cilacap	$3,348.30	$1030.25	$8499.53	$9000.00	$720.00	$4320.00
Tegalkamulyan	$10,560.02	$4636.11	$45,588.37	$12,600.00	$16,560.00	$25,920.00
Sidakaya	$2833.18	$0.00	$0.00	$2880.00	$0.00	$1440.00
Gunungsimping	$0.00	$0.00	$0.00	$3600.00	$0.00	$0.00
Gumilir	$4636.11	$0.00	$0.00	$1080.00	$0.00	$0.00
Mertasinga	$9529.77	$1030.25	$10,817.58	$3960.00	$0.00	$0.00

Source: authors

Based on the land cover type, we calculated the damage value into the monetary value unit as shown in Table 23.9. Overall, the total cost of damage reached $645,893.4, with the highest percentage damage cost was the commercial area (i.e., $431,568). Then, Tegalkamulyan was the village that had the highest economic loss (i.e., $398,248.5) where the commercial area was the most affected (i.e., $431,568).

Evacuation Plan

Evacuation systems in tsunami-prone areas need to be developed and improved to anticipate the worst possibilities when a tsunami strikes (Shibayama et al. 2013). Our field observations revealed that the topography of the study area is very flat with no plateau or hills, which creates difficulty in finding horizontal evacuation shelters. Moreover, most of the roads in the coastal area of Cilacap are parallel and not away from the beach, so it makes it difficult during the evacuation process. Therefore, vertical evacuation is one of the alternatives. Figure 23.10 shows our evacuation map, which describes the distribution of evacuation shelters and their routes. The travel time required by residents to evacuate from settlements on the coast to temporary shelters is about 5–15 min on foot. In addition, each evacuation route is marked with evacuation directions to make it easier for residents to find the nearest temporary shelter.

We also found 23 buildings as temporary evacuation shelters (i.e., triangle yellow) scattered in the research area with a total building capacity of 37,740 people (Fig. 23.10). In addition, our study revealed that that the capacity of the buildings provided is enough to accommodate people who will be evacuated if a tsunami occurs in the future (Table 23.7). Most of the buildings used as temporary shelters have a number of building levels of more than two levels, and its structure is confined masonry (Fig. 23.11). The building used consists of schools, mosques, hospitals, town halls, stadiums, and offices (see Appendix 23.2).

Recommendations

The assessment demonstrated that a higher risk area of the tsunami was located along the coastline. Thus, areas with dense activity in potential tsunami hazard areas such as the settlements area and commercial areas need to relocate in safe areas and away from the tsunami disaster risk. Improving the evacuation system is also crucial, such as creating a protection wall (tsunami dyke) along the coastline of Cilacap as the tsunami breakwater. Modeling of tsunami hazards with different tsunami wave height scenarios for risk assessment is also essential. Furthermore, socialization for early prevention in the community is essential, as well as completing comprehensive disaster management training from the scenario prior to the disaster to the aftermath. So that people are aware of what they need to do in order to mitigate the risk. The use of village-level data in this study provides greater information regarding local disaster risk reduction needs. Regardless of the geographical location, this method can be used with a variety of local data. As a result, it has the potential to be a universal risk assessment tool.

Conclusions

This research is proposed to define the tsunami impact assessment in Cilacap City generated by a GIS and numerical approach, specifically the determination of tsunami risk areas based on tsunami hazard and vulnerability analysis. There are six villages at risk by the tsunami: Cilacap Village, Tegalkamulyan Village,

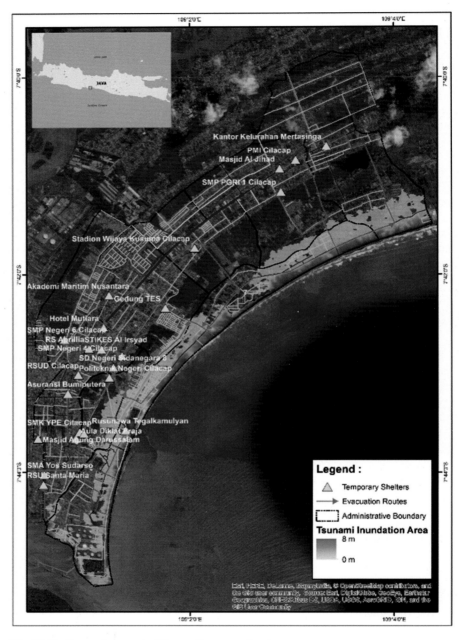

Fig. 23.10 Evacuation map in Cilacap. (Source: authors)

Fig. 23.11 One of school as a temporary shelter in Cilacap. (Source: authors)

Sidakaya Village, Gunungsimping Village, Gumilir Village, and Mertasinga Village. Among them, Tegalkamulyan Village is the most susceptible, where 48.88% of its area was at risk, with the estimated population affected by 10,352 people with an economic impact of $398,248.5. Thus, it is necessary to decrease vulnerability by relocating populations and their property as far away from the hazard in an evacuation system. Therefore, we also developed evacuation plans such as temporary evacuation shelters and evacuation routes to reduce disaster risk in the future. This research result could be used as input and consideration in a decision support system by stakeholders within the tsunami disaster prevention and mitigation framework, such as reducing the damage and losses that will arise in coastal areas hit by the tsunami.

Acknowledgement We would like to thank Asian Development Bank-Japan Scholarship Program (JSP) for funding this research. We want to thank Prof. Kensuke Fukushi and Prof. Hiroaki Furumai for the wonderful suggestion. We also gratefully acknowledge the USGS, the ITDB, the NOAA for providing the historical data record, BIG for providing DEM-TERRA SAR and land cover vector data, BPS Cilacap for local statistical data, WorldPop homepage for population grid data, and open street map for detailed vector data.

Author Contribution Ranie Dwi Anugrah as the first author did conceptualization, conducted field survey, performed data analysis, and wrote the draft paper, while Martiwi Diah Setiawati did conceptualization, wrote the draft paper, and reviewed and edited the manuscript. Ranie Dwi Anugrah and Martiwi Diah Setiawati contributed equally.

Appendices

Appendix 23.1 List of tsunami events in Cilacap and surrounding

No	Year	Earthquake (Mw)	Wave height (m)	Location of epicentrum	Distance of epicentrum to Cilacap (km)	Focal depth (km)	Category	Data source
1	1818	8.5	2.00	Java—Flores	889.405	600.0	Sea	ITDB
2	1840	7.0	4.00	South Java	171.374	150.0	Inland	ITDB
3	1843	6.0	2.00	Genteng	558.892	70.0	Sea	ITDB, Rynn (2002)
4	1852	6.5	0.50	Sunda Strait, Java	416.024	150.0	Sea	Nakamura (1979)
5	1852	6.8	0.50	Java	1600.848	100.0	Sea	ITDB
6	1859	7.0	1.00	Pacitan	264.128	33.0	Sea	Rynn (2002)
7	1867	8.0	7.70	Central Java, Yogyakarta	164.488	-	Inland	Latief et al. (2002)
8	1889	6.0	2.80	Java	389.671	70.0	Inland	ITDB
9	1921	7.5	0.10	Cilacap	426.477	-	Sea	Rynn (2002)
10	1926	6.8	1.20	Java	1055.813	-	Sea	ITDB
11	1930	7.0	0.70	South Java	439.436	100.0	Sea	ITDB
12	1930	6.2	1.50	South Java	476.944	33.0	Sea	ITDB, Rynn (2002)
13	1930	6.5	0.10	South Java	609.627	33.0	Sea	ITDB
14	1957	5.5	0.70	Java	202.905	33.0	Sea	Rynn (2002)
15	1963	6.5	2.00	Java	433.876	53.0	Sea	Rynn (2002)
16	1994	7.8	3.70	East of Java	525.142	34.0	Sea	Rynn (2002)
17	2004	9.1	0.51	Off west coast of Sumatera	1904.931	87.0	Sea	NOAA
18	2006	7.7	8.00	Pangandaran	251.861	25.3	Sea	NOAA
19	2007	8.5	5.00	Bengkulu	975.188	30.0	Sea	NOAA

Appendix 23.2 List of temporary evacuation shelters in Cilacap

Type of object	Object name	Address	Building level	Capacity	Building structure	Building wall	Building roof	Access roof
Flats	Rusunawa Tegalkamulyan	Jl. Lingkar Selatan	4	2000	Confined masonry	Brick	Tile	No
Hospital	RS Aprilia	Jl. Gatot Subroto no. 95	3	1000	Confined masonry	Brick	Tile	No
Hospital	RSUD Cilacap	Jl. Gatot Subroto no. 28	3	2000	Confined masonry	Brick	Tile	No
Hospital	RSU Santa Maria	Jl. Jenderal Ahmad Yani No. 38	2	200	Confined masonry	Brick	Tile	No
Hotel	Hotel Mutiara	Jl. Gatot Subroto no. 136	2	300	Confined masonry	Brick	Tile	No
Mosque	Masjid Al Jihad	Jl. Urip Sumoharjo no. 202	2	70	Confined masonry	Brick	Tile	No
Mosque	Masjid Agung Darussalam	Jl. Jendral Sudirman no. 34	2	500	Confined masonry	Brick	Tile	No
Office	Asuransi Bumiputera	Jl. S. Parman no. 78	2	500	Confined masonry	Brick	Tile	No
Office	Kantor Kelurahan Mertasinga	Jl. Urip Sumoharjo no. 89	2	300	Confined masonry	Brick	Tile	No
Office	PMI Cilacap	Jl. Urip Sumoharjo no. 174	2	400	Confined masonry	Brick	Tile	No
Office	Gedung TES	Jl. Kalimantan no. 49A	2	300	Confined masonry	Brick	Tile	No
School	SMP Negeri 6 Cilacap	Jl. Rinjani no. 43	2	270	Confined masonry	Brick	Tile	No
School	SMP PGRI 1 Cilacap	Jl. Rama no. 22	2	500	Confined masonry	Brick	Tile	No
School	SMP Negeri 4 Cilacap	Jl. Dr. Sutomo no. 13	2	500	Confined masonry	Brick	Tile	No
School	SD Negeri Sidanegara 8	Jl. Dr. Sutomo	2	300	Confined masonry	Brick	Tile	No
School	SMA Yos Sudarso	Jl. Ahmad Yani no. 54	3	1000	Confined masonry	Brick	Concrete	No
School	Akademi Maritim Nusantara	Jl. Kendeng no. 307	8	5000	Confined masonry	Brick	Concrete	Yes
School	Politeknik Negeri Cilacap	Jl. Dr. Sutomo no. 1	3	1000	Confined masonry	Brick	Tile	No
School	STIKES Al Irsyad	Jl. Cerme no. 24	2	400	Confined masonry	Brick	Tile	No
School	SMK YPE Cilacap	Jl. Dr. Sutomo no. 8	2	400	Confined masonry	Brick	Tile	No
School	Akademi Kebidanan Graha Mandiri	Jl. Dr. Sutomo no. 4B	2	500	Confined masonry	Brick	Tile	No
Stadium	Stadion Wijaya Kusuma Cilacap	Jl. Dr. Rajiman no. 65	2	20000	Confined masonry	Brick	None	No
Town hall	Aula Diklat Praja	Jl. Jendral Sudirman no. 12	2	300	Confined masonry	Brick	Concrete	No

References

Agung F (2006) Tsunami risk assessment of Indonesian coastal city: case study of Padang City, West Sumatera. Aarhus University, Aarhus

Amri MR, Yulianti G, Wiguna S, Adi AW, Ichwana AR, Radongkir RE, Septian RT (2016) Risiko Bencana Indonesia. BNPB, Jakarta, p 39

Atillah A, El Hadani D, Moudni H, Lesne O, Renou C, Mangin A, Rouffi F (2011) Tsunami vulnerability and damage assessment in the coastal area of Rabat and Sal_e, Morocco. Nat Hazards Earth Syst Sci 11:3397–3414

BAPPENAS (2005) Master plan for the rehabilitation and reconstruction of the regions and communities of the province of Nanggroe Aceh Darussalam and the island of Nias, Province of North Sumatera. BAPPENAS, 126, Jakarta

BAPPENAS (2008) Penilaian Kerusakan dan Kerugian-Tim Koordinasi Perencanaan dan Pengendalian Penanganan Bencana (P3B). BAPPENAS, Jakarta, p 43

Baučić M (2020) Household level vulnerability analysis—index and fuzzy based methods. ISPRS Int J Geo Inf 9(4):263. https://doi.org/10.3390/ijgi9040263

Blaikie PM, Cannon T, Davis I, Wisner B (1994) At risk: people's vulnerability and disasters. Routledge, London/New York

Bock Y, Prawirodirdjo L, Genrich JF, Stevens CW, McCaffrey R, Subarya C, Puntodewo SSO, Calais E (2003) Crustal motion in Indonesia from global positioning system measurements. J Geophys Res 108(BB):2367. https://doi.org/10.1029/2001JB000324

Cutter SL, Boruff BJ, Shirley WL (1997) Handbook for conducting a GIS-based hazards assessment at country level. South Carolina Emergency Preparedness Division Office of the Adjutant General.

Cutter S, Mitchell J, Scott M (2000) Revealing the vulnerability of people and places: a case study of Georgetown County, South Carolina. Ann Assoc Am Geogr 90:713–737

Cutter S, Boruff B, Shirley W (2003) Social vulnerability to environmental hazards. Soc Sci Q 84: 242–261

Dewi R (2010) A GIS based approach to the selection of evacuation shelter buildings and routes for tsunami risk reduction: a case study of Cilacap coastal area, Indonesia. Gajah Mada University and University of Twente

Eddy (2006) GIS in disaster management: A case study of tsunami risk mapping in Bali, Indonesia. James Cook University, Townsville

FEMA (2008) Guidelines for design of structures for vertical evacuation from tsunami. https://www.fema.gov/media-library/assets/documents/14708. Retrieved 16 June 2018

Jelinek et al (2012) Approaches for tsunami risk assessment and application to the city of Cadiz, Spain. Nat Hazards 60:23–37

Karig DE, Moore GF, Hehanussa PE (1979) Structure and Cenozoic evolution of the Sunda Arc in the Central Sumatra region. Assoc Am Petrol Geol Memoir 29:223–237

Kim S et al (2007) Evacuation route planning: Scalable Heuristics. ACM GIS 07

Kohler A, Jülich S, Bloemertz L (2004) Risk analysis – a basis for disaster risk management. Deutsche Gesellschaft für Technische Zusammenarbeit (GTZ), Berlin

Latief H, Puspito NT, Imamura F (2002) Tsunami catalog and zones in Indonesia. J Natural Disaster Sci 22:25–43

Lavigne et al (2007) Field observation of the 17 July 2006 Tsunami in Java. Nat Hazards Earth Syst Sci 7:177–183

Luther GC (2010) Integrated soil and crop management for rehabilitation of vegetable production in the tsunami-affected areas of Nanggroe Aceh Darussalam province, Indonesia. ACIAR final report SMCN/2005/075. Australian Centre for International Agricultural Research, Canberra

McSaveney M, Rattenbury M (2000) Tsunami impact in Hawke's Bay. Client Report 2000/77. Institute of Geological and Nuclear Sciences

Medina F, Mhammdi N, Chiguer A, Akil M, Jaaidi EB (2011) The Rabat and Larache boulder fields; new examples of high-energy deposits related to storms and tsunami waves in north western Morocco. Nat Hazards 59:725–747

Nakamura S (1979) On statistics of tsunamis in Indonesia. South East Asian Stud 16(4)

Nyimbili PH, Erden TA (2020) Combined model of GIS and fuzzy logic evaluation for locating emergency facilities: a case study of Istanbul. In: Bandrova T, Konečný M, Marinova S (eds) Proceedings of the 8th international conference on cartography and GIS, Nessebar, Bulgaria, 15–20 June 2020, pp 191–203

Omira R, Baptista MA, Matias L, Miranda JM, Catita C, Carrilho F, Toto E (2009) Design of a sea-level tsunami detection network for the Gulf of Cadiz. Nat Hazards Earth Syst Sci 9: 1327–1338

Omira R, Baptista MA, Leone F, Matias L, Mellas S, Zourarah B, Miranda JM, Carrilho F, Cherel J-P (2013) Performance of coastal sea-defense infrastructure at El-Jadida (Morocco) against Tsunami threat – lessons learned from the Japanese March 11, 2011 Tsunami. Nat Hazards Earth Syst Sci 13:1779–1794

Papathoma M, Dominey-Howes D, Zong Y, Smith D (2003) Assessing tsunami vulnerability, an example from Herakleio, Crete. Nat Hazard Earth Syst Sci 3:377–389

Pignatelli C, Sanso G, Mastronuzzi G (2009) Evaluation of tsunami flooding using geomorphologic evidence. Mar Geol 260:6–1

Ruther H, Hagai MM, Mtalo EG (2002) Application of snakes and dynamic programming optimization in modeling of buildings in informal settlement areas, ISPRS. J Photogramm 56:269–282

Rynn J (2002) A preliminary assessment of tsunami hazard and risk in the Indonesian region. Sci Tsunami Hazards 20:193–215

Rynn J, Davidson J (1999) Contemporary assessment of tsunami risk and implication for early warnings for Australia and its island territories. Sci Tsunami Hazard 17(2):107–125

Shibayama T, Esteban M, Nistor I, Takagi H, Thao ND, Matsumaru R et al (2013) Classification of tsunami and evacuation areas. Nat Hazards 67(2):365–386. https://doi.org/10.1007/s11069-013-0567-4

Silva FND et al (2003) Chapter XXI: evacuation planning and spatial decision making: designing effective spatial decision support system through integration of technologies. In: Mora M, Forgionne G, Gupta JND (eds) Decision making support system; achievement and challenge for the new decade. IDEA Group Publishing, London, 358 p

Tinti S, Tonini R, Pontrelli P, Pagnoni G, Santoro L (2008) Tsunami risk assessment in the Messina straits, Italy with application to the urban area of Messina. Geophys Res Abstr 10:EG2008-A-08045

Twigg (2004) Disaster risk reduction: mitigation and preparedness in development and emergency programming

WorldData.info. Tsunamis in Indonesia. https://www.worlddata.info/asia/indonesia/tsunamis.php. Retrieved 30 May 2018

Chapter 24
The Public Health Risks of Waterborne Pathogen Exposure Under a Climate Change Scenario in Indonesia

Martiwi Diah Setiawati, Marcin Pawel Jarzebski, Fuminari Miura, Binaya Kumar Mishra, and Kensuke Fukushi

Abstract Indonesia, particularly in the urban area, is prone to extreme events, especially floods. In addition, the duration, frequency, and intensity of such disasters are projected to increase due to the climate change. However, in Indonesia, little is known about the effects of those climate disasters on human health. Given the population's extremely limited adaptive capacity, this is a critical information gap. We estimated the risk probability of waterborne diseases caused by flood utilizing the quantitative microbiological risk assessment (QMRA) approach. We used the RCP 4.5 climate scenario of daily precipitation data of MRI-CGCM 3, MIROC5, HadGEM to determine the future projection of waterborne disease under the flood 50-year return period scenario in 2030. In two of the main cities in Indonesia, Medan

M. D. Setiawati (✉)
Research Center for Oceanography, National Research and Innovation Agency (BRIN), Jakarta, Indonesia
e-mail: mart009@brin.go.id; martiwi1802@gmail.com

M. P. Jarzebski
Tokyo College, The University of Tokyo, Tokyo, Japan
e-mail: jarzebski@tc.u-tokyo.ac.jp

F. Miura
Centre for Infectious Disease Control, National Institute for Public Health and the Environment, Bilthoven, The Netherlands

Center for Marine Environmental Studies (CMES), Ehime University, Matsuyama, Ehime, Japan
e-mail: miura.fuminari.bt@ehime-u.ac.jp

B. K. Mishra
School of Engineering, Faculty of Science and Technology, Pokhara University, Pokhara, Kaski, Nepal
e-mail: bkmishra@pu.edu.np

K. Fukushi
Institute of Future Initiatives, The University of Tokyo, Tokyo, Japan

United Nations University Institute for the Advanced Study of Sustainability, Tokyo, Japan
e-mail: fukushi@ifi.u-tokyo.ac.jp

© The Author(s), under exclusive license to Springer Nature Switzerland AG 2022
U. Chatterjee et al. (eds.), *Ecological Footprints of Climate Change*, Springer Climate, https://doi.org/10.1007/978-3-031-15501-7_24

and Surabaya, we looked at norovirus as a worst-case scenario waterborne disease. We found that the mean probability of the health risk during the extreme event in the future is 0.079 (increased by 22%) and 0.12 (increased by 35%) in Medan and Surabaya, respectively. The total number of infected people will have at least a twofold increase in Medan and Surabaya. It is critical to address the growing risk of norovirus in the face of climate change and to find local context-relevant adaptation measures by conducting a workshop with local stakeholders in each city. The most relevant and urgent actions listed during the workshop were raising awareness in hygiene and behavior during the flood, increasing green space, and providing leak-proof communal septic tanks.

Keywords Climate change · Waterborne diseases · QMRA · Climate scenario · Climate adaptation

Introduction

Diarrheal diseases caused by waterborne pathogens are still a major public health concern in many countries. Previous research reported that nearly 1.7 million people died from diarrheal diseases in 2016, including 0.5 million children under five (GBD 2016). Notably, low- and middle-income countries (LMICs), including Indonesia, bear the brunt of this burden (Julian 2016), owing to insufficient water and sanitation infrastructure. It has become more prevalent due to changes in the climate system, including the shift in rainfall and extreme occurrences (Curriero et al. 2001). Moreover, this phenomenon is expected to worsen this burden in the future decades as well as aggravate the control efforts (Lo Iacono et al. 2017).

Many urban areas in Indonesia frequently suffer tropical monsoon-related flood and inundation damage as an inadequate infrastructure of storm drain and less accurate weather information. Exposure to contaminated floodwater can be dangerous to one's health. People living in flooded areas may have health problems due to exposure to pathogens in dirty water and limited access to clean drinkable water, resulting in decreased sanitation and hygiene. Furthermore, the frequency and intensity of extreme precipitation will likely increase in Indonesia in 2100 under the RCP 8.5 scenario (Stocker et al. 2013). Also, Yamamoto et al. (2021) estimated that the volume of floodwaters would rise by threefold in the Indonesia river. Another study conducted by Budiyono et al. (2016) stated that Jakarta, as the capital city, will experience a significant rise of the flood risk, almost double in the near future. A number of studies have also revealed a remarkable linkage between extreme rainfalls and the incidence of waterborne diseases (gastroenteritis and leptospirosis) (Su et al. 2011; Huang et al. 2016). Given this condition, a deeper understanding of health risks associated with flooding is necessary and quantitative information of infection routes and the fate of microbial pathogens in the face of climate change is required.

According to the Indonesian Ministry of Health 2020, diarrhea was the leading cause of mortality rate for kids in Indonesia. In addition, about seven million people in this country were infected by diarrhea where nearly half of them are children (Indonesian Ministry of Health 2020). Also, Garg et al. (2018) revealed that neither heavy rainfall nor flooding plays a significant role in village health outcomes in Indonesia. Despite the high risk of waterborne diseases and growing flooding problems in Indonesia, we have not found any attempt to conduct the risk assessment quantitatively for waterborne diseases in this country.

The quantitative microbial risk assessment (QMRA) is an important tool for estimating microorganism-related health hazards (Haas et al. 1999; WHO 2016) and for developing water safety plans (Petterson and Ashbolt 2016). However, information on specific pathogen content in a particular location is required in order to employ this approach. Furthermore, because the essential data for many pathogens are unavailable, conducting a QMRA at each type of illness will take a lot of time. Thus, WHO (2004) suggested utilizing reference pathogens of a similar nature, but in emergent nations where relevant data are rare and expensive, the reference pathogens have not yet been fully utilized.

This study aims to employ the QMRA approach for estimating the likelihood of infection risk such as norovirus pathogen with the flood risk association in two metropolitan areas in Indonesia (i.e., Surabaya and Medan) under climate scenario RCP 4.5. The study will provide science-based risk estimations to policymakers and other stakeholders for identifying intervention methods and locations for reducing the health risks and related economic damages. We demonstrated that the simplified QMRA is a useful tool for assessing public health risk considering the data limitation in the LMICs.

Data and Methodology

Study Area

Two major metropolitan centers in Indonesia, Surabaya and Medan, were selected as study sites. Surabaya and Medan are part of 15 pilot project sites for the Indonesia National Adaptation Plan. The decision to conduct the study was made collaboratively, between the Indonesian Government and research team in order to create a model for replicating climate change adaptation strategies in other growing cities around the country. Surabaya and Medan are the second and third most prominent cities in Indonesia, with around three million people living in 2019. Furthermore, the urban population is expected to increase in 2035 by 71% since 2010 (Jones 2010). Along with the population increase, the built-up area has been also expanding, only between 2002 and 2015; it has increased by 10% in Surabaya and 40% in Medan

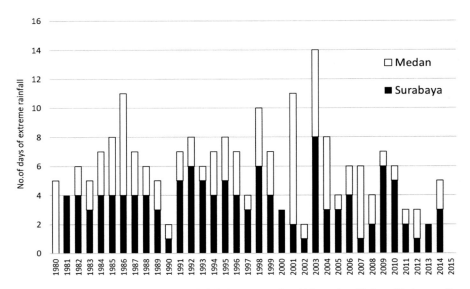

Fig. 24.1 Number of days of extreme rainfall (i.e., percentile of 99 equal to 71.6 mm/day) annually in Medan and Surabaya. (This graphic was plotted based on the rain gauge data from meteorological station in Medan and Surabaya)

(Setiawati et al. 2021). The annual rainfall average from 1980 to 2015 in Surabaya and Medan were 6900 mm and 7800 mm, respectively. Furthermore, both cities also experienced extreme precipitation (i.e., a percentile of 99 equal to 71.6 mm/day), as shown in Fig. 24.1.

Research Framework

In this study, the QMRA framework was used to develop numerical simulation models for assessing the health risk by pathogens in the flooded area. We expressed the health risk as the probability of gastroenteritis cases and then quantified using flood inundation, pathogen concentration, and other input data. We selected norovirus as the reference pathogen as a worst-case scenario; (1) infectivity of an infectious virus particle is high, (2) high persistence in the environment. This pathogen also was found in river water, domestic wastewater, and flooded water in the tropical countries. We assumed that the residents were exposed to pathogens via unintended ingestion of floodwater during outdoor activities such as wading and swimming. Finally, the risk calculation steps combined with the flood model (Section "Flood simulation") simulate the spatial distribution of expected infection

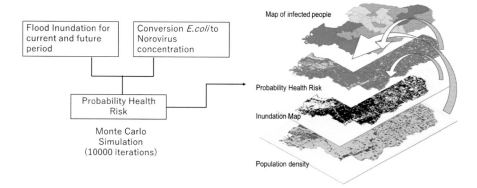

Fig. 24.2 Research framework

Table 24.1 Detailed data list for health characterization

Parameter group	Variable	Source
Flood simulation	Daily rainfall data 1980–2014	Data online BMKG 2016 (Indonesia meteorological agency)
	Digital elevation model	SRTM (Shuttle Radar Topography Mission)
	Land-use map	Landsat images (https://earthexplorer.usgs.gov/)
	Basin information	Ministry of Environment and Forestry (KLHK) (https://nfms.menlhk.go.id/peta)
	Daily precipitation data of general circulation models (GCMs) 1985–2004	Coupled Model Intercomparison Project Phase 5 (CMIP5)
	Daily precipitation data of GCMs 2020–2039	CMIP 5
QMRA	*E. coli* concentration	Masago et al. (2018)
	Current population	Surabaya and Medan Statistical Agency (BPS Medan and BPS Surabaya) https://surabayakota.bps.go.id/ https://medankota.bps.go.id/
	Future population	World Pop homepage (https://www.worldpop.org/)

risks under climate change scenarios for current and future conditions. Thus, this assessment is necessary to understand which area needs to be prioritized for adaptation planning in the city. Figure 24.2 depicts the entire research structure, whereas Table 24.1 provides the comprehensive data list.

Flood Simulation

The primary input for flood inundation modeling was the value of extreme rainfall, digital elevation model (DEM), land cover, and flood control measures. At the beginning, we performed bias correction to generate a robust analysis to the impact of climate change. Then, we involved flood modeling, where the flood hydrographs were simulated using HEC-HMS hydrologic modeling initially at the inlet point, followed by creating flood inundation using these hydrographs for the lower regions which become inundated. The inundation areas were mapped using FLO-2D. This routing model can simulate a flood hydrograph as well as foreseeing inundation areas. Furthermore, in order to create the inundation map, the following parameter is required for FLO-2D simulation such as land cover map, soil map, elevation map, daily rainfall data, and flood hydrograph in inlet. In addition, the model output must be performed into a grid in an FLO-2D project dataset.

This study projected future climatic variables using general circulation models (GCMs) to forecast the likelihood of increased flood risk. To scale the GCM climate model's precipitation output data to account for their systematic errors, we conducted a bias correction with in situ precipitation data for the frequency and intensity of the precipitation by using quantile mapping (QM) method. In this study, we conducted bias correction for the following GCMs such as HadGEM, MRI-CGCM3, and MIROC5. To assess the entire study area's current and future extreme rainfall occurrences, we created intensity–duration–frequency (IDF) curves. For further detail, the quantile mapping technique utilized in the study was clearly described in Mishra et al. (2018).

The hydrological response to precipitation in a catchment area is frequently estimated using hydrological modeling. There are various models available in our research; the model choice mainly aimed for generating flood hydrographs in the inlet site with specific statistical return periods. Then, we simulate the discharge from the upstream area through employing the HEC-HMS model, which can analyze the runoff from hourly to daily rainfall data. The aforementioned model is not specific hydrological model, but it is an open-source software which has various popular mathematical model for hydrology.

This study compared the flood damage of present and future conditions using specific climate scenarios. In addition, we utilized ArcGIS 10.4 to visualize the flood inundation area both for present and future conditions. This simulation result would raise policymakers' awareness and help them develop effective flood risk prevention and reduction strategies.

Quantitative Microbiological Risk Assessment (QMRA)

At first, we assumed the concentrations of E. coli from previous studies on flood water and river water in Jakarta (Kido et al. 2009; Phanuwan et al. 2006; Masago et al. 2018). The mean and standard deviation distribution parameters were estimated using the reference paper's E. coli concentrations monitoring point. Subsequently, we converted the concentrations to natural logarithms and transformed them to the concentration of noroviruses using the slope regression-based previous research by Masago et al. (2018).

Ingestion of Floodwater

The exposure assessment was conducted for two groups of people: non-adults (less than 18 years old) and adults (equal or higher than 18 years old). The quantity of total floodwater consumed by people was determined by multiplying the amount of time spent outside (i.e., 1.7 hours/day) (Masago et al. 2018) and the amount of water consumed per hour. In this paper, unintentional water ingestion was proportional at 75% of the total height, with the normal log distribution where $\mu = 2.92$ and $\sigma = 1.43$ (Schoen and Ashbolt 2010). Also, we assumed that residents are not to evacuate from their homes in a flood.

Risk Calculation

According to the flood modeling result, we then calculated the likelihood of norovirus infection and illness using the dose–response model for the Norovirus GI.1 8fIIb strain (Teunis et al. 2008) as shown in Eqs. 24.1 and 24.2.

The daily dose, D, was calculated as follows:

$$D = C \times I \times \left(\frac{T}{60}\right) \times F \qquad (24.1)$$

where C = pathogen concentration (CFU/ml); I = ingestion rate (ml/hour); T = time spent outdoors (mins/day), and F = fraction of time spent in the water. The value of I, T, and F was explained in the ingestion rate subsection.

The infection risk was calculated by utilizing the beta-Poisson dose–response model as shown in Eq. 24.2.

$$\text{Risk} = 1 - \left[1 + \left(\frac{D}{\beta}\right)\right]^{-\alpha} \qquad (24.2)$$

where risk = probability of risk of infection; D = daily dose (amount of agent or pathogen); β and α are dose–response parameters with the value of 65,000 and 0.631, respectively.

Based on a previously established method, we also evaluated the risk of secondary infection within a household (Miura et al. 2016). Given the primary infection risk derived above, we multiplied it by the risk ratio that measures the total risk of infection (i.e., the sum of primary and secondary infections) depending on the household size. Since the inundation events force individuals to stay inside their households, both infections within a household were defined as the exposure route in this analysis. A single family was defined as the household (i.e., father, mother, child, and baby), and all participants were classified in the simulation as susceptible or infected. The exposure assessment method of secondary infection was based on Miura et al. (2016). Then, the Monte Carlo technique was used to estimate the risk using a Python-based application with 10,000 iterations. Although the likelihood of infection risk is associated with the magnitude of flooding, population density also plays a big part in the number of infected people.

Workshop

We conducted workshops to disseminate the case study findings and formulate the recommendation as an effort for mainstreaming adaptation into policy. In Medan and Surabaya, the workshops were conducted on February 10 and 12, 2018. This workshop is part of the Climate Change Adaptation Initiative Project in Indonesia, in which the health sector is one of the priority issues. Regional governmental employees represented the majority of participants in the health sector group. Their expertise covered public health, regional administration, agriculture, forestry, natural ecosystems, natural resources, coastal areas, water environment, industry, and urban planning.

First, we asked all participants about their opinions about climate change as well as their experiences related to the effect of the changing climate on public health and possible actions that could be taken to tackle these effects. Secondly, invited experts explained climate change impact assessment on local public health and adaptation science-based options approach. Finally, we integrated the adaptation option from both parties based on the feasibility and priority. The overall workshop design is presented in Fig. 24.3.

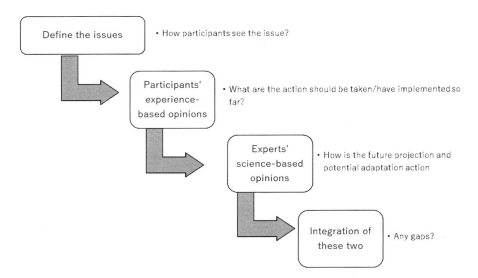

Fig. 24.3 Workshop design

Table 24.2 Daily maximum precipitation for 50-year return periods under current and future climate scenarios

Pilot study	Daily maximum rainfall (mm)	
	Current (1985–2004)	Future climate GCM of RCP 4.5 (2020–2039)
Medan	177.7	208.8
Surabaya	197	417.38

Note: Medan only used one GCM model MRI, while in Surabaya used average value of three GCMs (MRI, MIROC5, and HadGEM)

Result

In this study, precipitation was used as a factor in estimating the impact of climate change on flooding. This variability was measured for the flood analysis, focusing on the maximum daily and averaged monthly precipitation. The daily precipitation data of 8699 point from Sampali station–Medan and 9132 data from Meteorologi Perak 1–Surabaya were used to assess precipitation change over the study area. We chose those stations because they have long-term daily precipitation dataset available from the year 1985 to 2015. However, we only used the dataset range from 1985 to 2004 to assess precipitation change and it's pattern.

We utilized the bias-corrected GCM data of current (1985–2004) and future climate condition (2020–2039) to evaluate the effects of climate change on precipitation at the pilot site river basins. To analyze future climate conditions, we extracted the daily precipitation data resulted from bias correction (i.e., QM technique) of selected GCMs model; MIROC5, HadGEM2-ES, and MRI-CGCM3. Moreover, we then measured rainfall intensity by utilizing the IDF curves for specified return periods and durations. Among three GCMs, the MRI-CGCM3 was the most accurate in Medan, while in Surabaya, we used all three models.

Table 24.2 summarizes the daily maximum rainfall for the 50-year return periods and its future projection. These data were used to conduct the simulation of flood inundation in the pilot studies. We found that under the RCP 4.5 scenario, an increase of a 17.5% and 111.66% was emerged in Medan and Surabaya city, respectively. This substantial surge suggests that climate change should be taken into account in future flood mitigation initiatives to ensure sustainable cities waterways.

Urban Flooding

Despite the installation of river improvement and drainage projects in the pilot areas, the flooding still continues. In this study, we assumed that the peak flood discharge in Medan was 314 m^3/s (Masago et al. 2018) and in Surabaya was 1500 m^3/s (JICA 1998). Table 24.3 shows the inundation depth and its area in the pilot sites. The

Table 24.3 Inundation depth under current and future climate scenarios

	Area (km²)			
	Surabaya		Medan	
Inundation depth (m)	Current (1985–2004)	Future (2020–2039)	Current (1985–2004)	Future (2020–2039)
<0.2	260.57	216.2	221.09	208.13
0.2–0.5	32.58	33.08	29.51	33.33
0.5–1.5	30.653	62	31.79	37.41
>=1.5	2.99	15.53	6.63	10.16

Table 24.4 2.5th percentile, median, 97th percentile and mean daily probabilities of infection risk for each city both for current and future scenario

Pilot area	Time period	2.5 Percentile P infection risk ($\times 10^{-2}$)	Median P infection risk ($\times 10^{-2}$)	97 Percentile P infection risk ($\times 10^{-2}$)	Mean P infection risk ($\times 10^{-2}$)
Medan	Current (1985–2004)	0.39	1.05	6.48	1.66
	Future (2020–2039)	0.62	1.59	7.85	2.27
Surabaya	Current (1985–2004)	0.83	2.05	8.96	2.78
	Future (2020–2039)	1.34	3.21	12.12	4.11

result indicates that flood inundation will increase significantly in the future. The flood inundation is forecasted to rise along the Deli River, in the center and northern parts of Medan, in the north part of Surabaya, and along the Brantas River in 2030. In particular, the inundation higher than 1.5 m will increase significantly in both cities. As a result, the local management plan for flood protection must emphasize tackling the challenge of increased flooding in Medan.

Waterborne Infectious Diseases

Table 24.4 summarized the 2.5 percentile, median, 97 percentile, and mean probability of infection risk of norovirus. The result found that the likelihood infection risk range between 0.39% and 12.12%, respectively. The lower bounds of the infection probability (i.e., 2.5 percentile) for the baseline/current period were less than 1% in both cities. However, the increasing trend of 0.62% and 1.34% occurs in the future for Medan and Surabaya cities. Furthermore, the upper bounds of the infection probability (i.e., 97th percentile) of the baseline/current period in Medan and Surabaya cities were 6.48% and 8.96%, respectively. Meanwhile, in the future

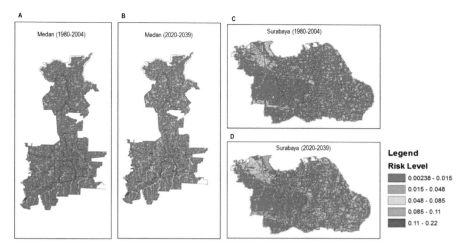

Fig. 24.4 Spatial distribution of mean probability of infection risk in Medan and Surabaya under climate change scenarios. (**a**) and (**c**) refer to current conditions under a 50-year return period, and (**b**) and (**d**) are future periods

period, the upper bounds of infection probability are 7.85% in Medan and 12.21% in Surabaya, respectively.

Figure 24.4 demonstrates the spatial distribution of the simulated risk of floodwater-borne under current and future climate change scenarios. The value is expressed as the mean probability of infection risk per grid (i.e., 100 × 100 m). The red grid defines high-risk areas (the probability is higher than 11%), while the blue grid represents the low risk (less than 1.5% cases per grid). As shown in Fig. 24.4a, c, the risk was higher at the central of Medan and the northern part of Surabaya city under the current scenario. However, under the climate scenario, the risk level will increase one or two levels higher in the future for both cities.

In Medan cities, the high-risk areas (red) will have a fourfold increase in four subdistricts in the future, namely Medan Maimun, Medan Perjuangan, Medan Area, and Medan Denai. Meanwhile, in Surabaya, the high-risk areas will experience a twofold increase at four subdistricts in the future, namely Semampir, Kenjeran, Tambaksari, and Wonokromo.

Table 24.5 shows the total number of infected people under current and future scenarios. Under current conditions, it ranged from 8000 people to 260,000 in both cities. In the future, the affected people range from 19,000 to 460,000 people. On average, the number of affected people will double in both cities in the future. The elevated risk was calculated due to the change of precipitation and population. Therefore, we strongly encourage all parties to take action to lessen the risk of infectious diseases caused by flooding.

Table 24.5 Total cases

Pilot area	Time period	2.5 P infection risk (thousand people)	Median P infection risk (thousand people)	97 P infection risk (thousand people)	Mean P infection risk (thousand people)
Medan	Current	8	23	142	37
	Future	19	48	236	68
Surabaya	Current	24	59	260	81
	Future	51	122	460	156

Table 24.6 Adaptation option (norovirus infections) based on scientific assessment

No.	Adaptation options	Effort	Cost	Effect
A. Providing information in public				
1.	Rising awareness (behavior during the flood and hygiene)	Int.	Int.	High
B. Urban planning				
2.	Improve drainage system	High	High	High
3.	Increasing green space to reduce runoff	High	High	Int.
4.	Improving drainage system and stormwater detention pond	High	High	Int.
5.	Multipurpose facilities for stormwater detention	High	High	High
6.	Leak-proof septic tank and sewage system	High	High	High
C. Technology				
7.	Improving wastewater treatment	High	High	High

Note: Int. defined as intermediate

Adaptation Option

Table 24.6 shows the example of adaptation options that need to be addressed to reduce the number of infected people due to waterborne disease. It covers information provision for the public, urban planning, and technology. This adaptation option was also included as the material for workshops with policymakers since adaptation measures should be incorporated with the national or city master plan.

We used a quadrant table to gather the information from both perspectives, as shown in Fig. 24.5. The quadrant consists of two main factors: priority and feasibility. We asked the participants to cluster the adaptation action into four quadrants (i.e., high priority with high feasibility, low priority with high feasibility, high priority with low feasibility, and low priority with low feasibility). The action can be from their perspective (knowledge/current program in their agency) or based on the list provided by the researchers.

The result stated that eight adaptations need to be tackled (i.e., high priority with high feasibility) such as rising awareness, increasing green space to reduce runoff, providing leak-proof communal septic tanks, government policies and commitment, cross-sectoral cooperation, improving water detention ponds, improving drainage system, and wastewater treatment. In general, both cities had similar sets of priority actions. Still, the feasibility scale was perceived differently, such as improving the

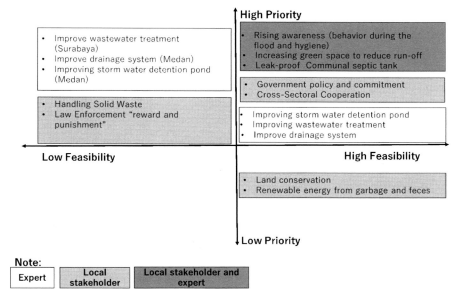

Fig. 24.5 Adaptation option and its priority based on scientific and local stakeholder input

water detention pond and drainage system was marked as high feasibility in Surabaya but low feasibility in Medan. On the other hand, improving wastewater treatment had high feasibility in Medan, but low feasibility in Surabaya.

Discussion

Under climate change scenarios, the extreme precipitation and flooded areas in Medan and Surabaya will increase in the future (Tables 24.2 and 24.3). Previous studies reported that heavy rainfall during the early phases of floods might induce the environmental change, resulting in a faster growth and regeneration of pathogens (Schwartz et al. 2006; Funari et al. 2012). High rainfalls and floods cause overflow of sewage facilities, animal waste runoff, redistribution of contaminated sediments, and mobilization of pathogens into the streams, wells, and estuaries (Ivers and Ryan 2006). Furthermore, there is a possibility cross-contamination of drinking water and sewerage pipe occurred, as it is passing through municipalities' waterways (Xu et al. 2017). Moreover, following floods, water will be contaminated as consequence of a breakdown in the purifying and wastewater disposal, which results in waterborne outbreaks of diarrheal diseases (Xu et al. 2017). Given the lack of safe drinking water and sanitation, steadily increasing fecal–oral transmission is indeed a threat. Also, there have been several outbreaks of high precipitation-related waterborne infection risks. For example, a previous study found a strong association between excessive precipitation and waterborne infection risk outbreaks on a national scale in the United States (Curriero et al. 2001). The research was based on 548 outbreaks recorded between 1948 and 1994, where there is an association between rainfall occurrences over the 90th percentile with 51% of waterborne infection risk outbreaks in the country. In France, excessive precipitation and wastewater plan facility disruption were the leading cause of the international gastroenteritis epidemics (Le Guyader et al. 2008).

We utilized a QMRA to estimate the gastroenteritis infection risk to the exposed residents from floodwater. The results confirm that floods have a significant influence on the increasing threat of disease, with the mean of risk probability ranging from 1.66% to 2.27% for Medan and 2.78% to 4.11% for Surabaya (Table 24.4). However, the upper bound of calculated risk (Table 24.4) is substantially greater than the appropriate level of the U.S. bathing water guidelines, ranging from 3% to 6%, respectively (USEPA 2012). The QMRA is dependent on assumptions, and some constraints could affect the risk assessment's accuracy. In this paper, we did not derived the dose–response element from clinical trials. Rather, the information was gathered from earlier investigations by (Masago et al. 2018) involving young and healthy adults. On the other hand, individual vulnerability to waterborne viral diseases may vary based on a variety of characteristics, including an individual's immunological condition and age, also from the virus's virulence, serotype, and infection pathways (Van Abel and Taylor 2018). As a result, specific demographic communities, such as kids, the aged people, and people with certain medical conditions, were not taken into account when collecting the data.

However, we need to be aware of uncertainties as an element of risk assessment in order to measure the level of confidence and assess the impacts and constraints of the risk assessment. QMRA modeling involves various input parameters, which could be a possible cause of uncertainty in the model result, where the calculation process depends on these input parameters. This paper found that the total water ingestion by the residents (i.e., associated with inundation depth) and the parameter value of dose–response have such a greater effect on the risk of waterborne disease, and it is consistent with the research outcome by Eregno et al. (2016).

In this study, we investigated the health risk of unintentionally ingesting water from flood inundation. However, we could not quantify the risk associated with other water uses because of the data limitation on the amount of contaminated water consumed. On the other hand, such unquantified hazards may be severe; therefore, even a small number of virus particles unintentionally absorbed in our bodies are able to infect exposed residents. Furthermore, people who drink the water in the flooded river might be subjected to an unacceptably high risk of infection since the water has not been treated in any way. According to recent findings, there is a higher concentration of indicator bacteria and the societies with high prevalence gastrointestinal illness following the flooding occurrences in the Betna River Basin, Bangladesh (Islam et al. 2017; Islam and Islam 2020).

Finally, this assessment can be used as a tool for strategic decision making on climate change adaptation policy. It quantifies the health risks for selected waterborne pathogens at specific locations under the ensemble mean of selected climate change scenarios. This assessment also can help policymakers to set the priority area to tackle the waterborne. In the conducted workshops for each city, we found that listed actions and priorities were not just the responsibility of one agency but a range of them (Fig. 24.5). Cross-sectoral cooperation appeared as one of the highest priority actions. In fact, all priority actions listed in Fig. 24.5 exist in the midterm strategic planning for related regional agencies. For instance, local public health agencies have an existing program related to raising awareness, called PHBS, which promotes a clean and healthy lifestyle. However, this program is only part of the business as a usual development program, and it does not consider the climate change scenarios. Another example of such a program coming from public agencies is a program for waste management and sanitation improvement, but this program is neither a cross-sectoral development program nor considering the climate change scenario. We found that despite various regional development programs existing, they do not yet consider adaptation strategies.

There are a few reasons why adaptation has not been a part of the current regional development planning. Unlike climate change mitigation which has been already included in the presidential regulation on The National Action Plan for Reducing Greenhouse Gas Emissions (i.e., Perpres no. 612011), however, adaptation did not receive strong attention when current regulations and programs have been designed to support the action for the regional level. Secondly, we believe there was a challenge in translating adaptation into regional regulations in Indonesia due to a lack of clear indications of how to approach it and measure it. On the other hand, mitigation has had a clear and straightforward indicator: reduction of greenhouse gas emission stated in the guideline of the Medium-Term Regional Development Plan arrangement (i.e., Home affair regulation No-86-Year 2017). However, no specific indicator related to the adaptation program was in the guidelines. The lack of measures against national standards was hindering regional actions. Thirdly, there is ubiquity in differencing between business as usual and adaptation development programs (e.g., the difference between the regular dam maintenance and improvement as a part of adaptation measure). Lastly, meteorological data access is only provided at a high expense, even for governmental agencies. Thus, large sectors such as agriculture or industry install their own weather stations instead of purchasing state-collected data.

Due to the listed above challenges, a collaboration between policy and research is necessary. Through this study, we were able to share our results with policymakers on norovirus outbreak risk. On the other hand, researchers were able to learn the local experts and government agencies' perception of climate change impacts on the health and status of an ongoing development program and proposed program and discuss the most urgent climate change adaptations for the local development program.

The parameter we used at the city level provides information regarding local susceptibility to waterborne disease. The methodology could be replicated for other places regardless the geographic location, making it a universal tool for assessing health risks related to waterborne diseases. The method might also be adapted to various local conditions by using data on river water quality and data from household surveys.

Conclusion

This study quantified the linkage between floods and gastroenteritis infection risk in Medan and Surabaya cities in Indonesia, drawing on longitudinal data from 1985 to 2004 and 2020 to 2039. The performed health risk-based QRMA assessment demonstrated that Surabaya had a higher probability of infection risk under climate change future projections. Furthermore, the highest level and median of health risk identified in both cities ranged from 6.48% to 12.12% and 1.05% to 3.2% in the future, respectively, as the contribution of the climate change factor. And yet, it may differ on the area, climate, and land-use planning. Visualizing the spatial distribution of QMRA analysis could facilitate planning the distribution of the adaptive measures, especially areas with higher risk. After learning the QMRA analysis results, local experts and government agency employees listed the urban greening, leak-proof communal septic tank, cross-sectoral cooperation, government policy, and commitment and raised awareness of hygienic behavior as the most important, effective, and the highest priority adaptation measures that should be implemented. We learned that delivering science-based evidence on current and projected health risks to policymakers and regional and local governments can be a vital help for national, regional, and local governments to plan necessary adaptation strategies when such science-supported results nor clear guidelines for adaptation are available. Applying locally appropriate and sustainable adaptation measures has a higher potential to safeguard health while minimizing economic damage caused by climate change. The assessment comes with certain limitations and uncertainty in results, but water quality sampling during the flood for determining E. coli concentration and using a higher spatial resolution of DEM data could improve accuracy.

Acknowledgement We would like to thank the MOEJ (Ministry of the Environment Japan) and BAPPENAS (the Ministry of National Development Planning of the Republic of Indonesia) for funding and facilitating the project of Climate Change Adaptation in Indonesia. We also want to thank data online BMKG (Indonesia Meteorological Agency) and BPS Medan and Surabaya (Medan dan Surabaya Statistical Bureau) for providing local climatic and demographic data.

Author Contribution All authors did conceptualization, reviewed, and edited the manuscript. Martiwi Diah Setiawati carried out the data collection, workshop arrangement, and GIS visualization and wrote the draft manuscript. Fuminari Miura conducted QMRA data analysis. B K Mishra conducted bias correction and flood simulation analysis. Martiwi Diah Setiawati is the main contributor to this manuscript.

References

Budiyono Y, Aerts JCJH, Tollenaar D, Ward PJ (2016) River flood risk in Jakarta under scenarios of future change. Nat Hazards Earth Syst Sci 16:757–774. https://doi.org/10.5194/nhess-16-757-2016

Curriero FC, Patz JA, Rose JB, Lele S (2001) The association between extreme precipitation and waterborne disease outbreaks in the United States, 1948–1994. Am J Public Health 91(8): 1194–1199. https://doi.org/10.2105/AJPH.91.8.1194

Eregno FE, Tryland I, Tjomsland T, Myrmel M, Robertson L, Heistad A (2016) Quantitative microbial risk assessment combined with hydrodynamic modelling to estimate the public health risk associated with bathing after rainfall events. Sci Total Environ 548–549:270–279

Funari E, Manganelli M, Sinisi L (2012) Impact of climate change on waterborne diseases. Ann Ist Super Sanita 48(4):473–487. https://doi.org/10.4415/ANN_12_04_13

Garg T, Hamilton SE, Hochard JP, Kresch EP, Talbot J (2018) (Not so) gently down the stream: river pollution and health in Indonesia. J Environ Econ Manag 9235–9253, S0095069618305333. https://doi.org/10.1016/j.jeem.2018.08.011

GBD (2016) Diarrhoeal disease collaborators. Estimates of the global, regional, and national morbidity, mortality, and aetiologies of diarrhoea in 195 countries: a systematic analysis for the Global Burden of Disease Study 2016 [Published online 19 Sept 2018]. Lancet Infect Dis. https://doi.org/10.1016/S1473-3099(18)30362-1

Haas CN, Rose JB, Gerba CP (1999) Quantitative microbial risk assessment. Wiley, New York

Huang L-Y, Wang Y-C, Wu C-C, Chen Y-C, Huang Y-L (2016) Risk of flood-related diseases of eyes, skin and gastrointestinal tract in Taiwan: a retrospective cohort study. PLoS One 11(5): e0155166. https://doi.org/10.1371/journal.pone.0155166

Indonesian Ministry of Health (2020) Indonesian health profile- year 2019. Indonesian Ministry of Health, Jakarta. https://pusdatin.kemkes.go.id/resources/download/pusdatin/profil-kesehatan-indonesia/Profil-Kesehatan-indonesia-2019.pdf. Accessed 2 Oct 2021

Islam MMM, Islam MA (2020) Quantifying public health risks from exposure to waterborne pathogens during river bathing as a basis for reduction of disease burden. J Water Health 18(03):292–305. https://doi.org/10.2166/wh.2020.045

Islam MMM, Hofstra N, Islam MA (2017) The impact of environmental variables on faecal indicator bacteria in the Betna river basin, Bangladesh. Environ Process 4:319–313

Ivers LC, Ryan ET (2006) Infectious diseases of severe weather related and flood-related natural disasters. Curr Opin Infect Dis 19:408–414

JICA (1998) Development of the Brantas River basin: cooperation of Japan and Indonesia. Japan International Cooperation Agency. https://openjicareport.jica.go.jp/617/617/617_108_11968989.html. Accessed 2 Oct 2021

Jones GW (2010) The 2010–2035 Indonesian population projection: understanding the causes, consequences and policy options for population and development. Available at: https://indonesia.unfpa.org/sites/default/files/pub-pdf/Policy_brief_on_The_2010_%E2%80%93_2035_Indonesian_Population_Projection.pdf. Accessed 29 Aug 2021

Julian TR (2016) Environmental transmission of diarrheal pathogens in low and middle income countries. Environ Sci: Processes Impacts 18(8):944–955. [PubMed: 27384220]

Kido M, Yustiawati, Syawal MS et al (2009) Comparison of general water quality of rivers in Indonesia and Japan. Environ Monit Assess 156:317. https://doi.org/10.1007/s10661-008-0487-z

Le Guyader FS, Le Saux J-C, Ambert-Balay K, Krol J, Serais O, Parnaudeau S et al (2008) Aichi virus, norovirus, astrovirus, enterovirus, and rotavirus involved in clinical cases from a French oyster-related gastroenteritis outbreak. J Clin Microbiol 46(12):4011–4017. https://doi.org/10.1128/jcm.01044-08

Lo Iacono G, Armstrong B, Fleming LE, Elson R, Kovats S, Vardoulakis S et al (2017) Challenges in developing methods for quantifying the effects of weather and climate on water-associated diseases: a systematic review. PLOS Negl Trop Dis 11(6):e0005659. [PubMed: 28604791]

Masago Y, Mishra BK, Jalilov SM, Kefi M, Kumar P, Dilley M, Fukushi K (2018) Future outlook of urban water environment in Asian cities. United Nations University – Institute for the Advanced Study of Sustainability, Tokyo

Mishra B, Rafiei Emam A, Masago Y, Kumar P, Regmi R, Fukushi K (2018) Assessment of future flood inundations under climate and land use change scenarios in the Ciliwung River Basin, Jakarta. J Flood Risk Manag 11:S1105–S1115. https://doi.org/10.1111/jfr3.12311

Miura F, Watanabe T, Watanabe K, Takemoto K, Fukushi K (2016) Comparative assessment of primary and secondary infection risks in a norovirus outbreak using a household model simulation. J Environ Sci 50:13–20. https://doi.org/10.1016/j.jes.2016.05.041

Petterson SR, Ashbolt NJ (2016) QMRA and water safety management: review of application in drinking water systems. J Water Health 14(4):571–589. https://doi.org/10.2166/wh.2016.262

Phanuwan C, Takizawa S, Oguma K, Katayama H, Yunika A, Ohgaki S (2006) Monitoring of human enteric viruses and coliform bacteria in waters after urban flood in Jakarta, Indonesia. Water Sci Technol 54(3):203–210. https://doi.org/10.2166/wst.2006.470

Schoen ME, Ashbolt NJ (2010) Assessing pathogen risk to swimmers at non-sewage impacted recreational beaches. Environ Sci Technol 44(7):2286–2291. https://doi.org/10.1021/es903523q

Schwartz BS, Harris JB, Khan AI et al (2006) Diarrheal epidemics in Dhaka, Bangladesh, during three consecutive floods: 1988, 1998, and 2004. Am J Trop Med Hyg 74:1067–1073

Setiawati MD, Jarzebski MP, Gomez-Garcia M, Fukushi K (2021) Accelerating urban heating under land-cover and climate change scenarios in Indonesia: application of the universal thermal climate index. Front Built Environ 7:65. https://doi.org/10.3389/fbuil.2021.622382

Stocker TF, Qin D, Plattner GK, Alexander LV, Allen SK, Bindoff NL, Bréon FM, Church JA, Cubasch U, Emori S, Forster P, Friedlingstein P, Gillett N, Gregory JM, Hartmann DL, Jansen E, Kirtman B, Knutti R, Krishna KK, Lemke P, Marotzke J, Masson-Delmotte V, Meehl GA, Mokhov II, Piao S, Ramaswamy V, Randall D, Rhein M, Rojas M, Sabine C, Shindell D, Talley LD, Vaughan DG, Xie S-P (2013) Technical summary. In: Stocker TF, Qin D, Plattner G-K, Tignor MMB, Allen SK, Boschung J, Nauels A, Xia Y, Bex V, Midgley PM (eds) Climate change 2013: the physical science basis. Contribution of working group I to the fifth assessment report of the Intergovernmental Panel on Climate Change. Cambridge University Press, Cambridge/New York

Su HP, Chan TC, Chang CC (2011) Typhoon-related leptospirosis and melioidosis, Taiwan, 2009. Emerg Infect Dis 17(7):1322–1324. [pmid: 21762606]

Teunis PFM, Moe CL, Liu P, Miller SE, Lindesmith L, Baric RS, Le Pendu J, Calderon RL (2008) Norwalk virus: how infectious is it? J Infect Dis 1476:1468–1476

USEPA (2012) Recreational water quality criteria. Available at: http://water.epa.gov/scitech/swguidance/standards/criteria/health/recreation/index.cfm. Accessed 14 Mar 2019

Van Abel N, Taylor MB (2018) The use of quantitative microbial risk assessment to estimate the health risk from viral water exposures in sub-Saharan Africa: a review. Microb Risk Anal 8:32–49

WHO (2004) Guidelines for drinking-water quality, 3rd edn. Incorporating First and Second Addenda. World Health Organization, Geneva

World Health Organization (2016) Quantitative microbial risk assessment: application for water safety management. World Health Organization. https://apps.who.int/iris/handle/10665/246195

Xu X, Ding G, Zhang Y, Liu Z, Liu Q, Jiang B (2017) Quantifying the impact of floods on bacillary dysentery in Dalian City, China, from 2004 to 2010. Disaster Med Public Health Prep 11(2):190–195. https://doi.org/10.1017/dmp.2016.90

Yamamoto K, Sayama T, Apip A (2021) Impact of climate change on flood inundation in a tropical river basin in Indonesia. Prog Earth Planet Sci 8(1):5. https://doi.org/10.1186/s40645-020-00386-4

Chapter 25
Perceived Impact of Climate Change on Health: Reflections from Kolkata and Its Suburbs

Sudarshana Sinha and **Anindya Basu**

Abstract Climate change is one of the greatest challenges, and it has a varied range of health implications on different sections of the society which varies according to the geographical region, but there is a lack of clarity about its impact on the health and well-being of people staying in of developing countries like India. The major objective of this paper is to study the impact of climate change on the mental and physical health of people staying in Kolkata and its suburbs. A random stratified sampling method has been used (age group of 20–50 years); the data were collected through online surveys. Twenty-two subindicators which were a part of either of the two domains were used for the analysis. These subindicators indicators were divided into three broad domains which deal with the kinds of loss faced by the respondents and how their physical health and mental well-being were affected. Likert scale was used, and the data were processed using SPSS software by correlation analysis, PCA, ANOVA. For physical health analysis, parameters such as allergies and respiratory disorders were considered, and for mental health experiences centering on posttraumatic stress disorder, stress due to livelihood uncertainty and fluctuations in productivity levels and alteration in their endurance level was considered. It has been observed that inter- and intra-variations in the responses lie between different age groups, sex ratio, and place of residence. However, the way in which climate change has affected the mental and physical health and well-being of the respondents is extremely unique and hard to homogeneously categorize. This is dependent on the perception of the respondents based on their memories and past experiences.

Keywords Climate change · Mental health · Physical health · Mitigation

---

S. Sinha (✉)
Department of Humanities and Social Sciences, Indian Institute of Technology, Kharagpur, West Bengal, India

A. Basu
Department of Geography, Diamond Harbour Women's University, Sarisha, West Bengal, India

© The Author(s), under exclusive license to Springer Nature Switzerland AG 2022
U. Chatterjee et al. (eds.), *Ecological Footprints of Climate Change*, Springer Climate, https://doi.org/10.1007/978-3-031-15501-7_25

Introduction

IPCC (2014) has stated that climate change involves an increase in temperatures and humidity, reduction in the number of frosty days, increase in the frequency and intensity of rainfall, rise in the sea level, reduction in snow cover, glaciers, increase in heat waves, hurricanes, cyclones, and thunderstorms, and these changes have left behind the severe impact on the human lives (Church et al. 2013, Trekenberth 2012). Changes in the mean temperature and precipitation pattern have a direct correlation to the increase in the intensity of hurricanes, heat waves, and crop failure (Jentsch and Beierkuhnlein 2008). Despite an increase in preventive measures, greenhouse gas emissions have continued to increase persistently over the last decade (IPCC 2014), and as per Patz and Thomson (2018), climate change has become the invincible cause that has directly and indirectly affected the lives of millions of people. However, there has been an ongoing debate among various academicians who have tried to assess the intensity of the impact which climate change has wrecked upon human lives. Watts et al. (2015) have stated that there would be a drastic increase in the number of deaths per year due to climate change which could also lead to the emergence of a large number of medical emergencies in the near future (Diffenbaugh and Scherer 2011).

Extremities in weather conditions tend to wreak havoc on human health (Hayes et al. 2018; Costello et al. 2009). Bourque and Willox (2014) have observed that climate change has increased the occurrence of various waterborne and vector-borne diseases in various countries like Canada, Asia, and Australia. Climate change has resulted in an increase in harmful pollutants, allergens, and particulate matter that has resulted in the deterioration of the air quality (Manisalids et al. 2020; Hong et al. 2019), especially in the case of Japan, Philippines, Germany, Madagascar, India, China, Sri Lanka, and Kenya (Kaur and Pandey 2021), and has increased the extremities in the temperature owing to these changes in the climatic conditions such as heat-induced stress (Cianconi et al. 2020), heat-induced violence (Levy et al. 2017; Novelo and Anderson 2019), respiratory illness, and cardiovascular illness which have become an increasingly common phenomenon (Costello et al. 2009; Myers and Patz 2009). Hsiang and Burke (2014) based on global data analyses have opinionated that heat-related morbidity tends to increase sociopolitical conflicts. Doherty and Clayton (2011) have stated climate change to have profound psychological and mental health impacts, among which posttraumatic stress disorder (PTSD) due to migration and loss of lives and property are expected to take the worst toll on the health and the well-being of the people.

Scholars have defined marginalized communities in various ways; they comprise of people from select age groups (such as children, senior citizens, and adults), sex ratio (such as women), people coming from lower socioeconomic status, and people experiencing serious health conditions (Berry et al. 2014; Costello et al. 2009). Climate change has rendered the marginalized communities to be extremely vulnerable due to the limited opportunities for them to mitigate the impacts (Watts et al. 2017; Berry et al. 2010). Climate change has indeed reinforced the need for societies to bring out more efficient strategies to reduce its harmful consequences. The least developed countries have ranked high in the climate change sensitivity index and devising proper strategies to mitigate the harmful consequences is a much-needed requirement (Kabir et al. 2016).

Despite the number of studies focusing on people's perception of climate change-induced health risk, most of such studies have been conducted in the context of developed countries (Akompab et al. 2013; Bai et al. 2013; Yu et al. 2013) and a few studies have been conducted in the context of developing countries like Bangladesh by Kabir et al. (2016). There is a dearth of literature that explores the impact of climate change on human health in the context of developing countries like India (Kabir et al. 2016). Thus, the main aim of this chapter is to analyze the impact of climate change on physical and mental health in the context of Kolkata and its suburbs.

Rationale Behind the Study

The intensity of the impact of climate change is heterogeneous in nature, and a vast categorization can be made even between the marginalized sections depending upon their perception and vulnerability to climate change (Patz and Thompson 2018). The definition of extremities of climate change is a unique case study in itself which is solely dependent upon the personal losses and experiences of the people staying in a particular area and those associated with particular livelihood practices (Trenberth 2012). In general, children and people belonging to the older generation tend to face more difficulty while adjusting themselves to climate change (Stanberry et al. 2018). Women also tend to face a greater level of difficulty while adjusting to climate change and on top of that socioeconomic problems, dependency on agriculture, and chronic health problems tend to aggravate the vulnerabilities of most people (Sorensen et al. 2018). In addition to affecting the economic status of people, climate change has also affected the health of the people. Cianconi et al. (2020) have expected that in the future climate change will demand more social readjustments.

The mental health of people includes an incubus of terms that deal with mental disorders, psychological well-being, wellness, and emotional resilience, and these factors are intrinsically related to each other (Butler et al. 2014). Cianconi et al. (2020) based on meta-analysis and previously published literature that is contextualized on different regions have observed that the impact of climate change on mental health tends to range from minimal distress symptoms to suicidal thoughts, which are the result of their perceptions, and the efforts made by them to adopt the mitigation strategies are unique to every community. Cianconi et al. (2020) have expected that in the future climate change will demand more social readjustments. Among the direct ways in which climate change has affected the mental health and well-being of people, the frequent occurrence of mood swings, behavioral disorders, and neurotic disorders has increased over the course of the years. Wang and Horton (2015) based on the observations and trajectories of climate change as presented in the Lancet Commission have observed that the mental illness and well-being of individuals across the globe have intensified owing to mood fluctuations, behavioral disorders, and neurotic disorders. Occurrences of PSTD, grief, survivor guilt, despair, and trauma are some of the ways in which the mental health of the people was affected by climate change (Berry et al. 2010). Ciaconi et al. (2020) have also observed that forced migration of people due to climate change would also result in the loss of their sense of place, and such feelings would give momentum to a feeling of powerlessness and helplessness among the people. Bei et al. (2013) have observed that an increase in flooding due to climate change would affect the mental health of the people due to an increase in PTSD due to the loss of their family members and belongings.

Climate change has also affected the physical health of people in both direct and indirect ways. An increase in temperature due to climate change has led to heat-related deaths. A strong correlation exists between the number of hot days and mortality rates among people (Honda et al. 2014). Among the several direct implications of climate change on human health, McMichael et al. (2006) have observed an increase in food and waterborne diseases leading to increased morbidity and mortality. Climate change has also led to alterations in temperature, humidity, and precipitation patterns which have heightened the occurrence of several vector-borne diseases such as malaria (Babaie et al. 2018). Other than this, climate change has also left behind several indirect implications on human health. It has given rise to food insecurity, melanoma cancer due to increased exposure to UV radiation, and malfunctioning of the kidneys (Watts et al. 2017). Padhay et al. (2015) have observed that increased exposure to heat UV radiation and prolonged exposure to heat waves are also associated with symptoms of heat stress and mood fluctuations which at times result in death. Cedeño et al. 2018 based on a study located in the Greater Boston region of Massachusetts have observed that increased UV radiation and heat waves have also affected the proper cognitive development and functioning of the people. Vergunst and Berry (2021) based on previously observed literature observed such imbalances have also resulted in an increase in temperament and increase in aggressiveness among individuals. Willox et al. (2015) based on the case study of circumpolar north observed that extreme variations in temperature conditions due to climate change also result in an increase in restlessness, fatigue leading to substance abuse, and an increase in self-harm among the people. Climate change has given rise to droughts, changes in natural landscapes, disruption in natural resources, and changes in the discourse of agro-economy these factors have increased the violence arising out of stress, aggression, and displacement of various agricultural groups from their place of residence (Watts et al. 2017; Nurse et al. 2010). Among the people who had to face the brunt of forced migration due to violent struggle over the ownership of scant and limited resources is expected to increase in the future. An increase in flooding especially inundation of coastal lands due to climate change is also expected to have harmful effects on the physical health of the marginalized community due to overexposure to toxins and contamination in the groundwater (Fahy et al. 2019; Nahar et al. 2014), and increased instance of flooding activities is associated with an increase in morbidity and mortality among people.

It can be observed from the literature review that, previously systematic review of literature about the impact of climate change on health has been observed by Rocque et al. (2021) based on research papers that had been previously published in the context of Western countries, a similar line of research has also been conducted by Cianconi et al. (2020); Hayes et al. (2018) based on empirical observations in the context of western countries has observed the impact of climate change on mental health, Grasso et al. (2012) using case studies from US had observed the health implications of climate change using quantitative techniques and time- series modeling, based on meta-analysis had observed the impact of rise in temperature on cognitive development of children, impact of climate change based on data obtained from previously published literary work contextualized on case study situated all over the world has been observed by Stanberry et al. (2018) in case of children and a similar observation has been conducted by Sorensen et al. (2018) in the context of women especially residing in low and middle income countries. However, Cianconi et al. (2020) have observed that there is a dearth of studies which assess the impact of climate change on the mental health of the people owing to the complexity of the issue. Other than this, impact of climate change on a marginalized section of people coming from lower socioeconomic status is lacking (Hayes et al. 2018).

As stated by Kabir et al. (2016), contextualization of the problem of climate change in the context of developing countries is extremely important as these regions are extremely sensitive to even the most minor of changes. It has been widely agreed by several scholars like Akompab et al. (2013), Bai et al. (2013), and Yu et al. (2013) that people's attitude toward climate change and an assessment of the risk involved is heavily dependent on the adaptive behavior of the people to the climate change. Moreover, as stated in COP24 Special Report: Health and Climate Change, the importance and analysis of health stands at the peak where all cost–benefit analyses of climate change mitigation need to be addressed (World Health Organization 2018). However, as pointed out by Bierbaum et al. (2013), Bouzid (2013), and Hess et al. (2014), a scant number of such studies have focused on this problem in the context of developing countries. Previously in the case of India, impact of climate change on agricultural productivity has been observed by Balasubhramanium and Birundha (2012), the effect of climate change on forestry, marine ecosystems, and glaciers have been observed by Dubey (2019), and the effect of climate change on economic growth has been observed by Sandhvani et al. (2020), the vulnerability on food grain production, economic growth, and alteration in GDP has been observed by Mauskar and Modak (2021), and the trajectory of climate change on and its possible effects has been stated by Panagarya (2009). It can be observed that in the case of India, scant literature exists that explores the perceived impact of climate change on the mental and physical health of the people health. Contextualization of this problem in the context of the urban agglomeration of Kolkata is extremely important because firstly it falls within the spectre of the developing nations and secondly as per the records of Census 2011, and it has been ranked as the third largest urban agglomeration of India. Other than this as per the report stated in the Down To Earth (2020), West Bengal can be categorized as one of the most vulnerable regions to climate change because of several reasons such as

(1) it has experienced some of the most intense thunderstorm-high casualty incidents over the past decade, (from 2000 to 2018, the Bay of Bengal region has experienced sixteen out of twenty-two category 4+ levels of cyclones that can be categorized under very severe cyclonic storms and extremely severe cyclonic storms), and (2) this region has experienced an eastward shift in tropical cyclone genesis locations in the Bay of Bengal region, especially during post-monsoon seasons. Hence, the main aim of this chapter is to analyze the perceived impact of climate change on the physical and mental health of the respondents situated in Kolkata and its suburbs using a cross-sectional study.

Materials and Methods

The method and methodology adopted for the chapter are stated below (Fig. 25.1).

Sample Population

The data presented in this chapter have been obtained from the primary survey; the questionnaire for which was circulated online. Previously 3000 questionnaires were circulated online; among all the responses, only the responses of those respondents who were aware of climate change and were part of the age group of 20–50 years were initially considered for the survey; secondly the age group of 20–50 years was selected as respondents belonging to this age group tend to have better access to the online mediums. Following this, only 560 questionnaires which were completely filled were only considered for the study. The respondents are residents of Kolkata and its suburbs (Baruipur, Maheshtala, and Pujali municipalities which are part of the Kolkata Metropolitan Authority) (see Fig. 25.2).

Data

The data were obtained via a primary survey. The data were collected from the respondents belonging to the age groups of 20–50 years who currently reside in Kolkata and its suburbs. The above-stated age group was considered for the study because people belonging to this age group are generally more aware of the recent trends in climate change. The questionnaires were circulated via Google Forms. The questionnaires were circulated via Google Forms. Initially, in the introductory e-mail, a question was put forth where the respondents were asked whether they had any prior knowledge about the effects of climate change on mental and physical health and are willing to partake in this academic survey, following which only the respondents who had given their consent and had a notion about the impact of

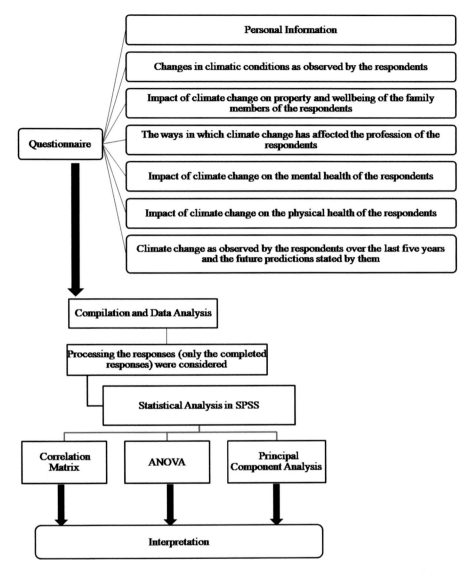

Fig. 25.1 Methodological framework involving questionnaire, data computation, and statistical analysis

climate change on mental and physical health were initially selected. The questionnaire schedule was divided into seven broad categories; the first subsection enquired the personal information about the respondents such as their place of residence, gender, family income, educational profile, and occupational status; the second subsection recorded the views expressed by the respondents with regards to the

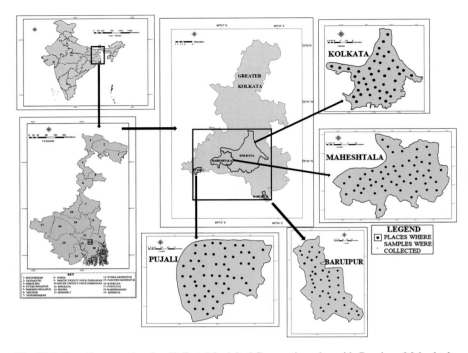

Fig. 25.2 Location map showing Kolkata Municipal Corporation, alongside Baruipur, Maheshtala, and Pujali municipalities under Kolkata Metropolitan Authority

changes in the climatic conditions; third subsection explored the impact of climate change on property and well-being of the family members of the respondents; fourth subsection asked about the ways in which climate change has affected the profession of the respondents; fifth and sixth subsection probed into the impact of climate change on the physical health of the respondents and mental health of the respondents, respectively, and seventh subsection probed into the climate change as observed by the respondents over the last 5 years and the future predictions stated by them. The questions were designed in a structured and semi-structured way. Five-point Likert scale was used for the analysis, where 1, 2, 3, 4, and 5 refer to "strongly disagree," "disagree," "neutral," "agree," and "strongly agree," respectively.

Statistical Analysis

Following this, the data were subjected to various statistical analyses.

Correlation Matrix

Pearson correlation coefficient is a multivariate analysis that was calculated among the subindicators by using the formula:

$$r = \frac{[(N\sum xy - (\sum x) * (\sum y)]}{\left\{[N\sum x^2 - (\sum x)x^2]\left[N\sum y^2 - (\sum y)^2\right]\right\}}, \text{ where}$$

r = Pearson correlation, N = number of pairs of scores, $\sum xy$ = sum of the products of paired scores, $\sum x$ = sum of x scores, $\sum y$ = sum of y scores, $\sum x2$ = sum of squared x scores, and $\sum y2$ = sum of squared y scores.

ANOVA

ANOVA analysis is commonly used to examine the significant difference between and significant differences within the various subindicators. It was performed to assess whether there is a significant difference in the mental and physical health conditions among the respondents. It was calculated using the formula as follows:

$$F = \frac{\text{MST}}{\text{MSE}}, \text{ where}$$

F = variance ratio for the overall test, MST = mean square due to treatments/groups (between groups), and MSE = mean square due to error (within groups, residual mean square)

$$\text{MST} = \frac{\left[\sum_{i=1}^{k} = 1(t_i/n_i) - G2/n\right]}{(k-1)}$$

$$\text{MSE} = \frac{\left[\sum_{i=1}^{k}\sum_{j=1}^{k} Y2ij - \sum_{i=1}^{k}(t_i/n_i)\right]}{(k-1)}, \text{ where}$$

Y_{ij} = an observation, T_i = group total, G = grand total of all observations, n_i = the number in group i, and n = total number of observations

Principal Component Analysis (PCA)

PCA is a variable reduction technique; it aims at reducing a large set of variables to smaller sets by selecting the most important variable which accounts for most of the variance. It takes into its account the variation between the eigenvalue.

Software

The data were collected via Google Forms which was circulated online. Statistical analysis was performed using SPSS (Statistical Package for Social Science Version 27).

Results and Discussions

Descriptive Analysis of the Respondents

A total of 560 participants were a part of the study, these respondents were aged between 20 and 50 years, and the majority of these respondents stayed in urban areas (64.3%) and belonged to various socioeconomic groups (Tables 25.1 and 25.2). 200 respondents had stated that they were students, and among those 75% of the respondents had stated that they also have a part-time occupation as well. 60 respondents had stated that they were directly involved in the agrarian sector, whereas 30 other respondents had stated that they were indirectly involved in the agrarian sector. 45 respondents stated that they were housewives. 54 respondents had stated that they had taken voluntary retirement, 54 respondents had stated that they were involved in the teaching profession, and 117 respondents stated that they were involved in services.

Table 25.1 Showing the descriptive analysis

Age	Gender		Total
	Male	Female	
20–30	80	280	360
30–40	40	120	160
40–50	0	40	40
Total	120	440	560
Place of residence	**Frequency**	**Percent (%)**	
Urban	360	64.3	
Suburban	80	14.3	
Rural	120	21.4	
Family income	**Frequency**	**Percent (%)**	
Less than Rs 10,000	40	7.1	
Rs 10,001–Rs 20,000	160	28.6	
Rs 20,001–Rs 30,000	80	14.3	
Rs 30,001–Rs 40,000	120	21.4	
Rs 40,001–Rs 50,000	160	28.6	

Source: computed by the authors

Table 25.2 Showing the details about the various subindicators

Subindicators ID	Details about the variables	References
AYPH	Has climate change affected your physical health conditions (*conditions pertaining to changes in respiratory tract ailments, feeling of breathlessness, changes in cardiac problems, changes in finesses levels, constant feeling of fatigue, changes in skin ailments, malaria, diarrhea, and dysentery were considered*)	Fahy et al. (2019) and Nahar et al. (2014)
AYMH	Has climate change affected your mental health conditions(*like feeling of restlessness, paranoid fear, stress, tension, sleeplessness, feeling of perceived helplessness, reduction in work productivity, deterioration of physical health conditions, depression, frequent mood swings, inability to cope with the current lifestyle*)	Berry et al. (2010)
CCNDO	Has climate change led to the loss of near and dear ones	Bei et al. (2013)
CCB	Has climate change led to the loss of your belongings (*such as land, ancestral property, loss of residential houses, farmland, other invaluable pieces of possessions*)	Gronlund et al. (2019)
CCFIO	Has climate change affected your family income and occupation	Abel et al. (2019)
CCCL	Climate change has resulted in crop loss	Trenberth (2012)
CCPAL	Climate change affected the productivity of agricultural land	Sorensen et al. (2018)
CCF	Climate change affected the fisheries	Trenberth (2012)
CCFL	Has climate change affected the fertility of lands (*as perceived by the respondents based on their direct connection with ancestral rural connection or shared experiences*)	Sorensen et al. (2018)
CCQWS	Climate change has affected the quality of water and made it more saline	Stanberry et al. (2018)
CCCF	Climate change resulted in crop failure (*as perceived by the respondents based on their perception or shared experiences*)	Stanberry et al. (2018).
CCLP	Has climate change has resulted in the loss of your property	Wang and Horton (2015)
CCD	Occurrence of diarrhea and dysentery	Hsiang et al. (2013)
CCM	Occurrence of malaria	Fahy et al. (2019) and Nahar et al. (2014)
CCSA	Occurrence of skin ailments	Rocque et al. (2021)
CCHS	Occurrence of heat stroke	Grasso et al. (2012)
CCRTD	Occurrence of respiratory tract disorder	Cedeño et al. (2018)
CCCP	Occurrence of cardiac problems	
CCA	Occurrence of various kinds of allergies	Rocque et al. (2021)
CCF	Experience fatigue due to climate change	Hayes et al. (2018)

(continued)

Table 25.2 (continued)

Subindicators ID	Details about the variables	References
CCPTSD	Experience posttraumatic stress disorder due to climate change (*symptoms such as eco anxiety, inability to cope with the loss of belongings, inability to let go of past memories or past experiences, psychiatric disorder arising out of loss of family members/friends/personal property, perceived feeling of everything going wrong, inability to adapt to new lifestyle pattern, constant flashbacks of the past trauma, increased feeling of vulnerability, frequent occurrence of nightmares*)	Cianconi et al. (2020)
CCLU	Experience stress due to livelihood uncertainty due to climate change (*such as staggered heavy rainfall, increasing cyclones, heavy waterlogging*)	Gronlund et al. (2019)
CCRWP	Experience reduction in work productivity due to climate change (such as staggered heavy rainfall, increasing cyclones, heavy waterlogging)	Gronlund et al. (2019)
CCRF	Experience reduction in fitness due to climate change	Willox et al. (2015)

Table 25.3 Showing the ways in which the respondents were affected

Climate change has affected their (%)					
Category	1	2	3	4	5
Physical health conditions	0	28.6	14.3	42.9	14.3
Mental health conditions	7.1	21.4	42.9	14.3	14.3
Loss of near and dear ones	57.1	7.1	21.4	7.1	7.1
Loss of your belongings	28.6	28.6	14.3	7.1	21.4

Source: computed by the authors

Changes in Climatic Conditions as Perceived by the Respondents

When the respondents (based on their perception/shared experiences) were asked about the specific changes, in the climatic conditions that they could attribute to the changing climate (Table 25.3). As per the observation of most of the respondents, they have stated that in Kolkata and its suburbs there have been vast changes in climatic conditions owing to the increase in flooding activities (92.9%) and increase in extreme weather conditions (82.3%); in addition to this, most of the respondents have opinionated that there has been a frequent occurrence of cyclones in this area (91%). It can be observed from the survey that most of the respondents had agreed (42.9%) that climate change has affected their physical health conditions. It can be observed from the survey that most of the respondents stated that climate change had not affected or caused any loss of their family members (57.1%) or resulted in the loss of their family's income and occupation (64.3%). A mixed response could be observed among those who had stated that climate change has resulted in the loss of their belongings where although a greater percentage share disagreed (57.2%) with the statement, a significant share of respondents also agreed to the statement (28.5%) and they reported incidents like cyclone-related household property damage and urban waterlogging-related issues. Thus, it can be observed that the intensity of the impact of climate change on the mental or physical health of the respondents is hard to generalize or delimit into a set category owing to the uniqueness and intricacies enshrouding their opinion.

Other urban residents have used their shared knowledge when they heard about crop losses from their close circles. Some of the respondents had stated that back at their maternal grandfather or ancestral home they had heard and overseen that some of the crops were completely destroyed by the rains before the harvesting season. In addition to the previous statement, 50% of the respondents additionally had stated that they had also experienced a substantial percentage share of crop loss over the years. 64.3% of the respondents stated that climate change has also resulted in the loss of the productivity of agricultural land and has also resulted in the depletion of the fertility of the lands (42.9%). 46.2% of respondents had stated that climate change has affected the quality of water and increased its salinity.

Most of the respondents (Tables 25.3 and 25.4) have strongly agreed that climate change has resulted in the loss of crops (71.4%), and it has also resulted in crop failure (50%). Other than this, climate change has resulted in the loss of the productivity of agricultural land (64.3%) and has also affected the fertility of lands (42.9%). Climate change has also wreaked havoc on the marine environment, and it has affected the quality of water and made it more saline over the course of the years (46.2%). Some of the respondents who have their ancestral roots in south twenty-four parganas district in places like Diamond Harbour and Bakkhali, near to the

Table 25.4 Showing the affect climate change has on land and water bodies

Climate change has resulted in %					
Category	1	2	3	4	5
Crop loss (*refers to complete damage of crops for the just before the harvesting season*)	0	0	7.1	21.4	71.4
Loss in the productivity of agricultural land (*refers to decrease in the yield rate due to excessive rainwater/ change in soil composition/saltwater intrusion*)	0	0	7.1	28.6	64.3
Affected the productivity of fisheries (*refers to the yield rate decrease*)	0	9.0	0	58.0	33.0
Decreased the fertility of land (*refers to decreased nutrient content which hinders growth of various crops*)	0	7.1	7.1	42.9	42.9
Decreased water fertility (*refers to the rising salinity*)	0	0	15.4	38.5	46.2
Crop failure (*refers deterred growth of crops and eventual low yield*)	0	3.0	6.0	50.0	41.0

Source: computed by the authors

coastal area have first-hand experience of crop loss due to increasing cyclonic events, flooding, water stagnation, and resultant salinization. 58% of respondents, including the ones having parallel rural holdings in south twenty-four parganas and East Midnapore Districts, have agreed that climate change has also affected the productivity of fisheries (58%) as well through increasing temperature and high cyclonic frequencies.

Impact of Climate Change on Physical Health

When the respondents were questioned about the intensity of physical discomforts experienced by them due to climate change (Table 25.5), most of the respondents stated that they were experiencing issues with skin ailments such as patchy skin, increase in spots and redness (42.9%), heat stroke (42.9%), and respiratory tract disorder (50%) and tended to experience different sorts of skin allergies and occurrence of blister-like formation in the skin (50%). Among the respondents who had stated that there was an increase in allergic reactions due to climate change, they mentioned that sun allergy and dust allergy are the most predominant ones. During the survey, most of the respondents stated that due to climate change the respondents mostly experienced a feeling of increase in breathlessness and increase in the number of people who had asthma. It can be observed from the survey that mostly the respondents experienced respiratory tract ailments and physical discomforts due to allergenic reactions; however, the chances of occurrence of vector-borne diseases such as malaria (43.9%) as opinionated by the respondents are pretty low. Other than this, most of the respondents also reported that they had observed an increase in the frequency of heat stroke among their family members as well.

Impact of Climate Change on Mental Health

Most of the respondents mentioned that they tend to suffer from various mental discomforts due to climate change as well (Table 25.6). Most of the respondents

Table 25.5 Showing the intensity of physical discomforts due to climate change as perceived by the respondents

Types of physical discomforts	In (%)		
	1	2	3
Diarrhea and dysentery	57.1	14.3	28.6
Malaria	42.9	21.4	35.7
Skin ailments	28.6	28.6	42.9
Heat stroke	35.7	21.4	42.9
Respiratory tract disorders	28.6	21.4	50.0
Cardiac problems	42.9	21.4	35.7
Kinds of allergies	28.6	21.4	50.0

Source: computed by the authors

Table 25.6 Showing the intensity of mental discomforts due to climate change as perceived by the respondents

Mental discomforts	In (%)				
	1	2	3	4	5
Fatigue	0	21.4	28.6	42.9	7.1
Posttraumatic stress disorder	35.7	7.1	14.3	42.9	0
Livelihood uncertainty	14.3	14.3	28.6	35.7	7.1
Reduction in work productivity	14.3	14.3	28.6	28.6	14.3
Reduction in fitness	7.1	21.4	21.4	28.6	21.4

Source: computed by the authors

stated that they experienced fatigue (42.9%) and posttraumatic stress disorder (42.9%). As stated by the respondents, they agreed that these discomforts primarily arose due to various reasons such as livelihood uncertainty (35.7%), reduction in work productivity (28.6%), and reduction in their fitness levels (28.6%). The feeling of posttraumatic disorders in the case of men had taken a turn for the worst among those who primarily depended upon agriculture. This feeling also became overwhelming for those male members who had faced the loss of their agricultural land due to inundation or were forcefully evicted due to climate change. Some of the respondents stated that due to climate change there is constant flooding in the urban areas, i.e., in the municipalities themselves, due to water stagnation especially along canals and creeks and inundation of acres of land causing damage to their ancestral home and property in the coastal areas of south twenty-four parganas and East Midnapore districts. As per their responses, most of these had occurred especially in the case of those regions that are lying adjacent to the rivers, creeks, and in close vicinity to the Bay of Bengal. Among those respondents who stated that they experienced posttraumatic stress disorder, they also mentioned that climate change had affected the fertility of their agricultural areas and resulted in crop failure due to the increase in monsoons, prolonged duration of the rainy season, and increase in the quantity of rainfall; these situations had worsened their financial status as they were not able to repay their loans to the money lenders; they had mentioned that thought of repaying their loans as well as bringing food to the table constantly haunted them, and this worsened both their mental health and physical health. Most of the respondents stated that they tend to feel fatigued ever so often when working outdoors, and the intensity of this tiredness has increased over the years. Most of them experience traumatic disorders, which have mostly stemmed from their perceived thoughts of helplessness due to the abnormality of the seasonal variation, inability to cope with the loss of belongings, inability to let go of their past memories/past experiences, personal property, and inability to adapt to new lifestyle pattern. They are mostly haunted by the thoughts that the frequency of flooding and water stagnation owing to the rapid occurrence of cyclones and thunderstorms would affect their personal lives and lead to the loss of their belongings in the urban as well as the rural areas. The respondents had stated that they could previously accomplish more tasks and are currently unable to continue working at the same pace.

The subindicators were tested on a 0.05 and 0.01 significance level (Table 25.7). It was observed from the correlation matrix that mental health and physical health of the respondents are intrinsically related to each other and deterioration in one would also proportionately affect the other. The loss of relatives or family members tends to have a significant impact on the mental and physical health of the respondents. Crop failure, deterioration in soil fertility, increase in the salinity of water, and loss of personal belongings tend to affect the mental and physical health of the respondents. A high correlation between the occurrence of heat stroke and respiratory tract disorder tends to affect the well-being of the respondents. A high correlation exists between reduction in work productivity, increase in uncertainty of livelihood opportunities, increase in fatigue, decrease in fitness levels, and the mental and physical well-being of the respondents. A high correlation exists between the increase in the salinity levels and the impact it has on the fisheries. Even the most minuscule of changes is likely to leave behind a huge impact on the productivity of the fisheries. A high correlation exists between reduction in work productivity, increase in uncertainty of livelihood opportunities, increase in fatigue, decrease in fitness levels, and the mental and physical well-being of the respondents. A strong correlation also exists between alteration in the salinity of the water and the occurrence of crop failure. This shows that economic loss also tends to affect the well-being of the respondents.

ANOVA analysis was conducted for the subindicators among the respondents belonging to different age groups, gender, and places of residence that is urban, suburban, and rural (Table 25.8). Among the subindicators categorized under mental and physical health conditions, a statistically significant difference exists in case of most of the categories between the means of respondents belonging to a different gender, place of residence, and age groups. It can be observed that generally among the male members decrease in agricultural productivity, crop failure, and crop loss tend to affect the mental health of the respondents to a greater extent, as an alteration in any one of these categories is likely to have a direct impact on their income as well. Among the different age groups, respondents who are between 40 and 50 years of age tend to face a greater level of difficulty while adjusting to climate change. This age group tends to face a greater number of mental health issues and PTSD, and they have a harder time facing their past experiences and letting go of their bad memories. This shows the existence of significant variation between different groups owing to the unique way in which climate change has affected each one of the respondents and their family members. This variation is also a result of their past experiences and current challenges faced by them and their family members due to climate change.

PCA was computed (Table 25.9), among the subindicators which are a part of physical health. Occurrence of diarrhea/dysentery (3.506), malaria (1.342), and skin ailments (1.046) due to climate change are considered the primary factors, whereas in case of the mental health, the occurrence of fatigue (2.941) and posttraumatic stress disorder (1.022) are considered as the primary factors affecting the lives of the respondents.

Table 25.7 Correlation matrix among the subindicators

	AYPH	AYMH	CCNDO	CCB	CCFIO	CCCL	CCPAL	CCF	CCFL	CCQWS	CCCF	CCLP	CCD	CCM	CCSA	CCHS	CCRTD	CCCP	CCA	CCF	CCPTSD	CCLU	CCRWP	CCRF
AYPH	1																							
AYMH	0.654	1																						
CCNDO	0.312	0.099	1																					
CCB	0.189	0.146	0.146	1																				
CCFIO	-0.066	0.117	-0.184	-0.432	1																			
CCCL	0.573	0.464	0.447	0.487	0.127	1																		
CCPAL	0.5	0.462	0.526	0.604	-0.083	0.913	1																	
CCF	0.415	0.739	0.353	0.066	0.189	0.465	0.488	1																
CCFL	0.373	0.588	0.317	0.115	0.17	0.69	0.705	0.683	1															
CCQWS	0.343	0.719	0.457	-0.078	0.123	0.433	0.433	0.81	0.608	1														
CCCF	0.373	0.512	0.507	0.282	-0.11	0.69	0.838	0.468	0.807	0.619	1													
CCLP	0.02	0.048	0	-0.278	0.413	0.052	-0.034	0.014	0.359	-0.101	0.186	1												
CCD	0.411	0.218	-0.085	-0.053	-0.027	-0.131	0.026	0.062	0.056	-0.12	0.056	0.067	1											
CCM	0.17	-0.425	0.073	-0.11	-0.185	0.045	-0.132	-0.32	-0.288	-0.165	-0.288	-0.344	-0.043	1										
CCSA	-0.012	-0.535	0.065	-0.156	-0.165	-0.1	-0.256	-0.396	-0.256	-0.277	-0.356	-0.128	-0.038	0.891	1									
CCHS	-0.063	-0.318	-0.282	0.243	0.122	0.292	0.106	-0.282	-0.038	-0.409	-0.253	-0.207	-0.062	0.445	0.396	1								
CCRTD	0.061	-0.548	0.274	0.215	-0.497	-0.063	0.041	-0.365	-0.328	-0.41	-0.119	-0.268	0.201	0.657	0.693	0.133	1							
CCCP	-0.288	-0.425	0	0.341	-0.348	0.045	0.177	-0.196	-0.176	-0.247	-0.064	-0.545	0.108	0.352	0.314	0.569	0.536	1						
CCA	-0.281	-0.63	-0.206	-0.086	-0.195	-0.357	-0.247	-0.482	-0.224	-0.544	-0.224	-0.08	0.482	0.415	0.477	0.365	0.548	0.536	1					
CCF	0.294	0.588	-0.063	0.393	-0.11	0.418	0.305	0.146	0.228	0.418	0.324	0.012	-0.334	-0.176	-0.256	-0.038	-0.433	-0.288	-0.641	1				
CCPTSD	0.421	0.213	0.148	0.206	-0.504	0.188	0.118	-0.089	0.07	0.166	0.219	-0.029	-0.029	0.236	0.055	0.006	-0.046	-0.198	-0.046	0.519	1			
CCLU	0.385	-0.004	0.047	-0.068	-0.2	0.137	0.141	-0.256	-0.087	-0.116	0.27	0.174	0.206	0.26	0.011	-0.142	0.254	-0.071	0.099	0.128	0.479	1		
CCRWP	0.663	0.514	0.394	-0.28	0.014	0.443	0.355	0.265	0.438	0.562	0.572	0.205	0.115	0.177	0.02	-0.265	-0.103	-0.364	-0.319	0.372	0.48	0.585	1	
CCRF	0.655	0.825	0.266	0.108	0.035	0.645	0.572	0.52	0.602	0.74	0.67	0.095	-0.026	-0.101	-0.229	-0.22	-0.333	-0.335	-0.625	0.737	0.378	0.282	0.805	1

Source: computed by the authors

Table 25.8 ANOVA analysis for the subindicators

		Gender					Place of residence					Age groups				
		Sum of squares	df	Mean square	F	Sig.	Sum of squares	df	Mean square	F	Sig.	Sum of squares	df	Mean square	F	Sig.
Mental health																
AYMH	Between groups	0.779	1	0.779	0.643	>0.05	141.587	2	70.794	73.682	<0.05	407.143	2	203.571	419.96	<0.05
	Within groups	676.364	558	1.212			535.556	557	0.962			270	557	0.485		
	Total	677.143	559				677.143	559				677.143	559			
CCF	Between groups	31.255	1	31.255	45.533	<0.05	65.397	2	32.698	52.203	<0.05	215.397	2	107.698	301.616	<0.05
	Within groups	383.03	558	0.686			348.889	557	0.626			198.889	557	0.357		
	Total	414.286	559				414.286	559				414.286	559			
CCPTSD	Between groups	145.541	1	145.541	149.55	<0.05	73.016	2	36.508	33.035	<0.05	168.571	2	84.286	90.283	<0.05
	Within groups	543.03	558	0.973			615.556	557	1.105			520	557	0.934		
	Total	688.571	559				688.571	559				688.571	559			
CCLU	Between groups	54.113	1	54.113	42.95	<0.05	108.254	2	54.127	46.462	<0.05	1.587	2	0.794	0.585	>0.05
	Within groups	703.03	558	1.26			648.889	557	1.165			755.556	557	1.356		
	Total	757.143	559				757.143	559				757.143	559			
CCRWP	Between groups	3.117	1	3.117	2.01	>0.05	113.016	2	56.508	41.658	<0.05	299.683	2	149.841	146.71	<0.05
	Within groups	865.455	558	1.551			755.556	557	1.356			568.889	557	1.021		
	Total	868.571	559				868.571	559				868.571	559			

25 Perceived Impact of Climate Change on Health: Reflections from Kolkata…

		SS	df	MS	F	p	SS	df	MS	F	p	SS	df	MS	F	p
CCRF	Between groups	0.087	1	0.087	0.057	>0.05	66.349	2	33.175	23.623	<0.05	578.571	2	289.286	596.786	<0.05
	Within groups	848.485	558	1.521			782.222	557	1.404			270	557	0.485		
	Total	848.571	559				848.571	559				848.71	559			
Physical health																
AYPH	Between groups	8.658	1	8.658	7.94	0.05	250.471	2	125.258	190.248	0	194.921	2	97.46	128.571	<0.05
	Within groups	608.485	558	1.09			366.667	557	0.658			422.222	557	0.758		
	Total	617.143	559				617.143	559				617.143	559			
CCD	Between groups	34.632	1	34.632	99.643	<0.05	113.016	2	56.508	272.379	<0.05	33.016	2	16.508	47.019	<0.05
	Within groups	193.939	558	0.348			115.556	557	0.207			195.556	557	0.351		
	Total	228.571	559				228.571	559				228.571	559			
CCM	Between groups	3.117	1	3.117	5.694	<0.05	6.349	2	3.175	5.851	<0.05	68.571	2	34.286	79.571	<0.05
	Within groups	305.455	558	0.547			302.222	557	0.543			240	557	0.431		
	Total	308.571	559				308.571	559				308.571	559			
CCSA	Between groups	3.117	1	3.117	4.512	<0.05	6.349	2	3.175	4.626	<0.05	109.683	2	54.841	109.53	<0.05
	Within groups	385.455	558	0.691			382.222	557	0.686			278.889	557	0.501		
	Total	388.571	559				388.571	559				388.571	559			
CCHS	Between groups	7.013	1	7.013	11.957	<0.05	14.286	2	7.143	12.433	<0.05	64.286	2	32.143	66.31	<0.05
	Within groups	327.273	558	0.587			320	557	0.575			270	557	0.485		
	Total	334.286	559				334.286	559				334.286	559			

(continued)

Table 25.8 (continued)

		Gender					Place of residence					Age groups				
		Sum of squares	df	Mean square	F	Sig.	Sum of squares	df	Mean square	F	Sig.	Sum of squares	df	Mean square	F	Sig.
CCRTD	Between groups	12.468	1	12.468	20.353	<0.05	58.73	2	29.365	55.341	<0.05	75.397	2	37.698	75.292	<0.05
	Within groups	341.818	558	0.613			295.55	557	0.531			278.889	557	0.501		
	Total	354.286	559				354.286	559				354.286	559			
CCCP	Between groups	3.117	1	3.117	5.694	<0.05	48.571	2	24.286	52.027	<0.05	33.016	2	16.508	33.369	<0.05
	Within groups	305.455	558	0.547			260	557	0.467			275.556	557	0.495		
	Total	308.571	559				308.571	559				308.571	559			
CCA	Between groups	58.528	1	58.528	110.42	<0.05	25.357	2	12.698	21.506	<0.05	145.397	2	72.698	193.85	<0.05
	Within groups	295.758	558	0.53			328.889	557	0.59			208.889	557	0.375		
	Total	354.286	559				354.286	559				354.286	559			

Source: computed by the authors

Table 25.9 Calculating PCA involving the subindicators

Component	Initial eigenvalues			Extraction sums of squared loadings			Rotation sums of squared loadings
	Total	% of Variance	Cumulative %	Total	% of Variance	Cumulative %	
Physical health							
CCD	3.506	50.081	50.081	3.506	50.081	50.081	3.103
CCM	1.342	19.179	69.256	1.342	19.175	69.256	1.638
CCSA	1.046	14.945	84.202	1.046	14.945	84.202	2.333
CCHS	0.589	8.419	92.62				
CCRTD	0.303	4.324	96.945				
CCCP	0.125	1.781	98.726				
CCA	0.089	1.274	100				
Mental health							
CCF	2.941	58.825	58.825	2.941	58.825	58.825	2.438
CCPTSD	1.022	20.44	79.265	1.022	20.44	79.265	2.215
CCLU	0.699	13.99	93.254				
CCRWP	0.306	6.114	99.369				
CCRF	0.032	0.631	100				

Extraction method: principal component analysis

Source: computed by the authors

Assumed Impact of Climate Change on Future

Over the last 5 years, the respondents had noticed a drastic change in the climatic conditions. Most of the respondents stated that over the last few years the heat has increased during the summer season and is expected to continue in the near future, and 71.4% of respondents mentioned that the intensity and quantity of rainfall have increased over the last 5 years; the respondents had stated that rainfall, cyclones, and thunderstorms are expected to become more frequent and intense over the years. 92.9% of the respondents had stated that climate change-induced ailments, mental health, and physical health problems are expected to intensify in near future. The respondents stated that due to climate change they had observed rapid inundation of areas adjacent to the rivers and Bay of Bengal and rapid alteration in coastlines in the near future.

Conclusion

It can be observed from the survey that indeed climate change has affected both the mental and physical health of the respondents. Respondents facing the loss of their family members or those facing livelihood uncertainties are the ones who are the worst affected. It can be observed from the survey that the responses obtained are pretty unique in themselves owing to the spatial heterogeneity in their location, shared experiences, and memories. Most of the respondents face a hard time adapting themselves to climate change because of their lack of knowledge about the perfect adaptive strategy which would help them to minimize their losses and help them find a bit of stability. After conducting the survey, it can be observed that the respondents who are a part of the lowest socioeconomic status and dependent on agro-economy for their livelihood can be classified among the most vulnerable ones because even a slightest change in the climatic conditions tends to immensely impact them as they have to face a huge financial setback because of crop failure and inability on their part to repay the loans. A similar observation was made by Abel et al. (2019) in their study. It can be observed from the previous literature as stated by Babaie et al. (2018) and from the survey that both men and women are vulnerable to climate changes in their unique ways, although it has been observed from the survey, and as per the perception, occurrence of heatstroke has rapidly increased and a drastic deterioration in health and well-being can be traced. Some of the respondents have reported a drastic decline in their work productivity, especially during the summer and the monsoon season. The respondents have mentioned that they tend to feel very tired after a short span of work in the sun. The respondents have mentioned that owing to the increase in heat waves during the summer months their cardiovascular problems and breathlessness have multiplied over the course of the years. Extremities of weather and an increase in the intensity of the monsoons over the last couple of years have made the respondents acutely vulnerable due to various issues such as water stagnation, an increase in the occurrence of waterborne diseases, and contamination of groundwater. Most of the respondents have stated that they have developed symptoms of dust allergies, heat allergies, skin rashes, and pigmentation of their skin color to such an extent that in certain cases medications fail to work. Most of the respondents stated that their mental health well-being and inner peace have completely gone for a toss and they tend to find it very hard to adapt themselves to the changing climate. Respondents who have faced inundation of their property, crop loss, forced eviction, and reduction in their livelihood opportunities are the ones who are highly affected by PTSD. Respondents belonging to the age group above 40 years have stated that livelihood uncertainty has become a major stress in their daily lives, and for them, the chances of finding a profitable employment opportunity are bleak. Facing the reality of the alteration of the physical health levels has become overwhelming for respondents. Respondents working in the food delivery sector have mentioned that they were unable to pursue their job during day time due to their deterioration of health conditions. Respondents, who are involved in agrarian business, had stated that increased occurrences of floods and increase in heat waves have also affected their livelihood opportunities to a huge extent and they find it hard to store their produce. The respondents who have witnessed the loss of family members due to heat stroke and cardiac ailments are haunted by the thought of the rapidly changing climate, and they feel after a point of time the medications do not work and their lifestyle has been affected. Respondents experiencing severe health conditions such as chronic illness and ailments, skin allergies, and asthma tend to feel helpless due to climate change. Most of the respondents mentioned that the lack of knowledge and ability to adapt mitigation strategies has affected their well-being.

References

Abel GJ et al (2019) Climate, conflict and forced migration. Glob Environ Chang 54:239–249. https://doi.org/10.1016/j.gloenvcha.2018.12.003

Akompab DA et al (2013) Awareness of and attitudes towards heatwaves within the context of climate change among a cohort of residents in Adelaide, Australia. Int J Environ Res Public Health 10:1–17

Babaie J et al (2018) A systematic evidence review of the effect of climate change on malaria in Iran. J Parasit Dis 42:331–340

Bai L et al (2013) Rapid warming in Tibet, China: public perception, response and coping resources in urban Lhasa. Environ Health 12(71):1–10

Balasubhramanium M, Birundaha DV (2012) Climate change and its impact on India. IUP J Environ Sci:1–17

Bei B et al (2013) A prospective study of the impact of floods on the mental and physical health of older adults. Aging Ment Health 17:992–1002. https://doi.org/10.1080/13607863.2013.799119

Berry HL et al (2010) Climate change and mental health: a causal pathways framework. Int J Public Health 55(2):123–132

Berry P et al (2014) Human Health. Canada in a changing climate: sector perspectives on impacts and adaptation. Government of Canada, Ottawa, pp 191–232

Bierbaum R et al (2013) A comprehensive review of climate adaptation in the United States: more than before, but less than needed. Mitig Adapt Strat Gl 18(3):361–406

Bourque F, Willox CA (2014) Climate change: the next challenge for public mental health? Int Rev Psychiatry 26(4):415–422. https://doi.org/10.3109/09540261.2014.92585

Bouzid M, et. al., (2013) . The effectiveness of public health interventions to reduce the health impact of climate change: a systematic review of systematic reviews. PLoS One 8(4), e62041.

Butler CD et al (2014) Mental health, cognition and the challenge of climate change. Clim Change Glob Health 26:251

Cedeño LJG, Williams A, Oulhote, Y, Zanobetti A, Allen JG, Spengler JD (2018) Reduced cognitive function during a heat wave among residents of non-air-conditioned buildings: an observational study of young adults in the summer of 2016. PLoS Med 15(7):e1002605. https://doi.org/10.1371/journal.pmed.1002605

Church JA et al (2013) Sea level change. In: Climate change 2013: the physical science basis. Contribution of Working Group I to the Fifth Assessment Report of the Intergovernmental Panel on Climate Change. Cambridge University Press, Cambridge/New York

Cianconi P et al (2020) The impact of climate change on mental health: a systematic descriptive review. Front Psychiatry 11(74):1–3. https://doi.org/10.3389/fpsyt.2020.00074

Costello A et al (2009) Managing the health effects of climate change. Lancet 373(9676): 1693–1733

Diffenbaugh NS, Scherer M (2011) Observational and model evidence of global emergence of permanent, unprecedented heat in the 20(th) and 21(st) centuries. Climate Change 107:615–624. https://doi.org/10.1007/s10584-011-0112-y

Doherty T, Clayton S (2011) The psychological impacts of global climate change. Am Psychol 66(4):265–276

Dubey CL (2019) Observed impacts of climate change in India. J Govern:272–285

Fahy B et al (2019) Spatial analysis of urban flooding and extreme heat hazard potential in Portland. Int J Disaster Risk Reduction 39:101117. https://doi.org/10.1016/j.ijdrr.2019.101117

Grasso M et al (2012) The health effects of climate change: a survey of recent quantitative research. Int J Environ Res Public Health 9:1523–1547. https://doi.org/10.3390/ijerph9051523

Gronlund CJ et al (2019) Assessing the magnitude and uncertainties of the burden of selected diseases attributable to extreme heat and extreme precipitation under a climate change scenario in Michigan for the period 2041–2070. Environ Health 18:1–17. https://doi.org/10.1186/s12940-019-0483-5

Hayes K et al (2018) Climate change and mental health: risks, impacts and priority actions. Int J Ment Heal Syst 12(28):1–11. https://doi.org/10.1186/s13033-018-0210-6

Hess JJ et al (2014) An evidence-based public health approach to climate change adaptation. Environ Health Perspect 122(11):1177–1186

Honda Y et al (2014) Heat-related mortality risk model for climate change impact projection. Environ Health Prev Med 19:56–63

Hong C et al (2019) Impacts of climate change on future air quality and human health in China. PNAS 116(35):17193–17200

Hsiang SM, Burke M (2014) Climate, conflict, and social stability: what does the evidence say? Climate Change 123:39–55

Hsiang SM et al (2013) Quantifying the influence of climate on human conflict. Science 341:1212–1226. https://doi.org/10.1126/science.1235367

IPCC (2014) Climate change 2014: Synthesis report, Summary for Policymakers. Contribution of Working Groups I, II and III to the Fifth Assessment Report of the Intergovernmental Panel on Climate Change. IPCC, Geneva [cited 2021 Dec 11]. Available from: https://www.ipcc.ch/pdf/assessment-report/ar5/syr/AR5_SYR_FINAL_SPM.pdf

Jentsch A, Beierkuhnlein C (2008) Research frontiers in climate change: effects of extreme meteorological events on ecosystems. Compt Rendus Geosci 340:624–628. https://doi.org/10.1016/j.crte.2008.07.002

Kabir MI et al (2016) Knowledge and perception about climate change and human health: findings from a baseline survey among vulnerable communities in Bangladesh. BMC Public Health 16(266):1–10

Kaur R, Pandey P (2021) Air pollution, climate change, and human health in Indian cities: a brief review. Front Sustain Cities:1–10

Levy SV et al (2017) Climate change and collective violence. Annu Rev Public Health 38:241–257

Manisalidis I et al (2020) Environmental and health impacts of air pollution: a review. Front Public Health:1–39

Mauskar JM, Modak S (2021) The imperatives of India's climate response. Observ Res Found 335:4–40

McMichael AJ et al (2006) Climate change and human health: present and future risks. Lancet 367(9513):859–869

Myers SS, Patz JA (2009) Emerging threats to human health from global environmental change. Annu Rev Environ Resour 34:223–252

Nahar N et al (2014) Increasing the provision of mental health care for vulnerable, disaster-affected people in Bangladesh. BMC Public Health 14:1–9. http://www.biomedcentral.com/1471-2458/14/708. https://doi.org/10.1186/1471-2458-14-708

Novelo MA, Anderson AC (2019) Climate change and psychology: effects of rapid global warming on violence and aggression. Curr Clim Change Rep:1–12

Nurse J et al (2010) An ecological approach to promoting population mental health and well-being – a response to the challenge of climate change. Perspect Public Health 130(1):27–33

Padhy SK et al (2015) Mental health effects of climate change. Indian J Occup Environ Med 19:3–7. https://doi.org/10.4103/0019-5278.156997

Panagarya A (2009) Climate change and India: implications and policy options. India Policy Forum:73–147

Patz JA, Thomson MC (2018) Climate change and health: moving from theory to practice. PLoS Med 15(7):e1002628. https://doi.org/10.1371/journal.pmed.1002628

Rocque RJ et al (2021) Health effects of climate change: an overview of systematic reviews. BMJ Open 11:1–14.e046333. https://doi.org/10.1136/bmjopen-2020-046333

Sandhani M et al (2020) Impact of climate change on economic growth: a case study of India, Working paper 204/2020. Madras School of Economics, pp 5–45

Sorensen C, Murray V, Lemery J, Balbus J (2018) Climate change and women's health: impacts and policy directions. PLoS Med 15(7):e1002603. https://doi.org/10.1371/journal.pmed.1002603

Stanberry LR, Thomson M, James W (2018) Prioritizing the needs of children in a changing climate. PLoS Med 15(7):e1002627. https://doi.org/10.1371/journal.pmed.1002627

Trenberth KE (2012) Framing the way to relate climate extremes to climate change. Climate Change 115:283–290. https://doi.org/10.1007/s10584-012-0441-5

Vergunst F, Berry LH (2021) Climate change and children's mental health: a developmental perspective. Clin Psychol Sci:1–19. https://doi.org/10.1177/21677026211040787

Wang H, Horton R (2015) Tackling climate change: the greatest opportunity for global health. Lancet 386(10006):1798–1799

Watts N et al (2015) Health and climate change: policy responses to protect public health. Lancet:60854–60856. https://doi.org/10.1016/s0140-6736

Watts N et al (2017) The Lancet countdown on health and climate change: from 25 years of inaction to a global transformation for public health. Lancet 17:32464–32469. https://doi.org/10.1016/s0140-6736

Willox AC et al (2015) Examining relationships between climate change and mental health in the Circumpolar North. Reg Environ Chang. https://doi.org/10.1007/s10113-014-0630-z

World Health Organization (2018) COP24 special report health and climate change. World Health Organization, Geneva

Yu Hao et al (2013) Public perception of climate change in China: results from the questionnaire survey. Nat Hazards 69:459–472

Part VI
Global Health, Sustainable and Adaptive Approaches and Sustainability

Chapter 26
Health Implications, Leaders Societies, and Climate Change: A Global Review

Ansar Abbas, Dian Ekowati, Fendy Suhariadi, and Rakotoarisoa Maminirina Fenitra

Abstract Perhaps the most recent difficulty for human beings across the globe is to comprehend climate change effects. It substantially impacts human rights, public health, and socio-economic benevolence. There are responsible leaders worldwide who are cognizant of its wide-ranging effects. They are undertaking appropriate initiatives (using their personality charisma/influence) to disseminate accurate information and increase awareness of global negative environmental effects. The ability of a community to anticipate, cope with, resist, and recover from the effects of weather events determines its vulnerability to climate change. The impact of climate change will be seen in the industrial and agricultural sectors, health, and quality of life of nations. Some argue that climate change would exacerbate disparities between rich and poor, minority, and politically marginalized groups. In contrast, others argue that these disparities will be worsened by shifting transportation, health, and energy infrastructures. For the sake of humanity, it is vital to take action to raise awareness of climate change. During the discussion of climate change, this chapter focuses on the health implications and the role of leadership in waking up the world's cultures. The literature review conclusions are open to various researchers and practitioners alike.

A. Abbas (✉)
Department of Economics and Business, Jawa Timur, Surabaya, Indonesia
e-mail: ansar.abbas-2018@feb.unair.ac.id

D. Ekowati
Department of Economics and Business, & Planning, Jawa Timur, Surabaya, Indonesia
e-mail: d.ekowati@feb.unair.ac.id

F. Suhariadi
Department of Postgraduate Program, Jawa Timur, Surabaya, Indonesia
e-mail: fendy.suhariadi@psikologi.unair.ac.id

R. M. Fenitra
ADS Scholar, Faculty of Economics and Business, Universitas Airlangga, Surabaya Indonesia, Jawa Timur, Surabaya, Indonesia
e-mail: maminirina.fenitra.r-2018@feb.unair.ac.id

© The Author(s), under exclusive license to Springer Nature Switzerland AG 2022
U. Chatterjee et al. (eds.), *Ecological Footprints of Climate Change*, Springer Climate, https://doi.org/10.1007/978-3-031-15501-7_26

Keywords Social psychology · Individuals · Leaders · Organizations · Well-being · Human health and change · Climate change · Global perspective

Introduction

Researchers concerned about climate change think that social psychological theories and findings are essential to solving some of the most pressing issues (Fielding et al. 2014). They stressed the relevance of the psychological process in dealing with climate change. Their research aims to highlight how modern social psychological research may benefit everyone. Since climate change social psychology is still in its infancy, social psychology models might give numerous theoretical and empirical techniques to combat climate change. Several notable scientists have shown that attitude can be affected concurrently by changing human cognition by aggravating awareness (Barnes et al. 2020). One may argue that if information processing is appropriately channeled, attitudes and beliefs can control people's reactions to believing that climate change facts are indisputable. If it does, it will help make the vast array of climate change findings theoretically coherent and orderly. Although this is a recent development, it is encouraging to see the breadth and depth of social psychology research on important issues related to climate change (Swim et al. 2011). It is critical to get the public's attention to promote public understanding of the importance of climate change. Leaders from throughout the world are making an enormous difference in this field. For example, Kjeldsen (2013) underlined the importance of presenting Former US Vice President Al Gore's climate change theory. Following his resignation as Vice President, he has dedicated himself to environmental conservation. The speech inspired the Oscar-winning film "An Inconvenient Truth." His stance explores that those efforts should be made to increase public awareness of man-made global warming. The world is looking to lay the groundwork for policies that prevent such rapid change. His work helps us understand the importance of social psychology and how climate change information is presented to the public and understood and reminds readers of climate change discourse. It encourages them to engage with governments and share their mandate. This chapter will develop social psychology leadership to understand the significance of this topic, which will help generate many approaches for further research.

This work contains four sections that make up this chapter; in the first section, we will discuss the connection linking human health and climate change. The second section discusses the emerging climate challenges. It will follow this section to highlight the importance of attention to climate change. Third, human sustainability and adoptive approach literature will be combined in the next section to grasp the human strategic perspective better. The final section will outline the importance of leadership in climate change social psychology.

Human Health, Awareness, and Climate Change

Several scientists agree that human activities are warming the Earth's surface and causing other climatic changes by increasing the number of human emissions (Chen et al. 2020; Khare et al. 2020; Mikhaylov et al. 2020). Based on top modeling organizations' published data, the Intergovernmental Panel on Climate Change (IPCC) forecasts an increasing average temperature globally. Climate change can impact higher and lower latitudes; higher latitudes and land will grow tremendously (Ahmed 2020; Anderson et al. 2020). However, the land will become drier in many regions with mid- and lower latitudes. As temperatures rise in specific locations and the world's average annual rainfall increases, precipitation events such as floods can be more severe (Tabari 2020; Woolway et al. 2020). Climate scientists have predicted that a warming planet will enhance climate variability (Griggs and Noguer 2002). Anthropogenic increases in greenhouse gases are primarily to blame for the exceptionally rapid rise in global temperatures since the mid-1970s. The current warming has had a variety of repercussions on non-human systems, including increased mortality and psychological effects. Figure 26.1 reveals the World Health Organization's mortality ratio, which has been linked to climate change.

Experts discussed psychological difficulties in human health in two ways (Swim et al. 2011). Primarily, the effects of global climate change pose significant hazards to human health and well-being. Moreover, climate change is a multi-disciplinary challenge involving human, technical, and natural systems. A collaborative, interdisciplinary approach is required, and human behavior can content social-cognitive methodologies. In this way, human activity affects climate change and its prevention. Thus, raising social psychology awareness is crucial to combating change and sustaining culture (Ayanlade et al. 2020; Hrabok et al. 2020). For example, psychologists participate in numerous essential concerns that could assist prevent this impact; hence, their efforts on climate change are more vital than other activities.

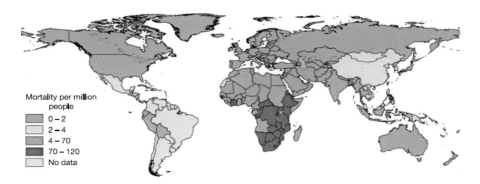

Fig. 26.1 WHO estimated mortality (per million people) attributable to climate change by 2000. (Source: Patz et al. 2005)

In Latin America and Europe, climate change is seen as a human-caused drive consequence (Cushman 2011; Nelson 2021). While in many African and Asian nations, people's views of local temperature change are more influential (Douville et al. 2001). Other important aspects, including public awareness and risk perceptions, point to the necessity of developing nation-specific climate communication strategies. Besides increasing public support and involvement in climate change policymaking, Lee et al. (2015) require better access to foundational education, climate literacy, and knowledge of local climate consequences. Climate change is thus not a regional issue. However, long-term climate changes are challenging to detect in a local setting. However, because the outcome varies per country, it is necessary to have a global approach to grasp the more comprehensive picture. Droughts and floods, for example, are ignored in personal experience-based judgments at the local level unless they are recent and overblown. Therefore, the public also largely ignores the hazards and benefits of climate change.

An emotional response to climate change affects how people perceive the risk of climate change, according to Ojala's research (Ojala 2021). When it comes to this problem, the emotional responses of the public are likely to be mixed or nonexistent. According to this school, even humans and artificially created organisms cannot affect global environmental systems. As a result, cultural values and beliefs significantly influence people's attitudes on climate change. An example of how people's opinions shape their perspectives on climate change is illustrated in this discussion. Spreading knowledge about global climate change requires that people be well-versed in the subject matter.

Emerging Climate Change Challenges

Climate depicts a region's temperature and precipitation distributions across time. Climate is less important to the average person. However, climate information is used for planning and decision-making by farmers and business people alike (Harvey et al. 2021). A human adaptation is more significant because it aims to understand that tremendous attempts to prevent global warming have a vast history of over 650,000 years. The difficulty of accepting the threat of climate change varies by individual abilities to comprehend these issues (Guile and Pandya 2018). Thus, ignorance is the most fundamental impediment to human awareness of climate change challenges and strategies to tackle them. In recent years, human civilization has come to appreciate the need to address this issue together with climate change (Rooney-Varga et al. 2018). Changes in the orbit of the Earth have affected the amount of solar energy absorbed by the planet. Supporting the findings of a recent study, data from NASA satellites show that human activity was 95% responsible for climate change in the early twentieth Century (Verma 2021). According to a study, the Earth's average temperature has risen by roughly 0.9°, owing primarily to the atmospheric emissions of carbon dioxide and other harmful gases (Levitus et al. 2012). Also, since 1880, ocean levels have risen to 20.32 cm higher, and they are anticipated to grow by another 10 cm by 2100. This increase would happen due to glacial ice melting and sea-level rise caused by warming. According to NASA data, since 1993, this issue has lost billions of tons annually in Antarctica and Greenland (Ohring et al. 2005). Global warming, according to the majority of experts, is caused by the greenhouse effect, which is something we humans are responsible for.

Climate change altered temperature and weather patterns. Although extreme weather is not always a sign of climate change, it is a gradual change in average and extreme temperatures, precipitation, and other features. Individual instances are commonly misattributed to climate change. Pattern recognition in uncertain circumstances is affected by people's expectations of change or stability. Hanlon et al. (2021) argued that climate change had worsened the impact of extreme weather events that caused a usual weather condition, including flood and drought. Torres et al. (2021) noted that variations in water temperature, ocean acidity, and sea-level rise are all part of the impact of climate change. From both physical, psychological health perspectives, many researchers investigate how climate change contributes to hydro-meteorological and climate-related disasters and the resulting health effects (Hwong et al. 2021; Lewandowsky 2021). They also look at ways to reduce climate change at a particular time.

Generally, scientists agree that climate change is a severe challenge nowadays (Gross 2018). Climate shifts have been observed on our planet, and we have learned to live with them. Plantinga and Scholtens (2021) found that the Industrial Revolution was one of the critical causes of this climatic shift. It is contributing a significant amount of Greenhouse Gases emissions to the atmosphere. Consequently, climate change is "a change in global or regional climate patterns," according to Worldwide Fund for Nature. De La Fuente et al. (2017) asserted that the primary source that causes climate change is the consumption pattern of human's daily needs. Global warming is a critical issue, but it is also a source of concern. However, there is often a misassumption on confusing climate change, and global warming, distinctly global warming is one of the results of climate change.

This global problem can disrupt the global socio-economic and environmental. The impact of global climate change on Asian economies has been particularly severe, posing numerous challenges to long-term livelihoods in these regions (Lata and Nunn 2012). Also, it affects Pacific countries' tiny low-lying islands (Campbell and Barnett 2010). These current and projected impacts will have severe consequences for Pacific communities, including dwindling freshwater supplies, deteriorating agricultural areas, and risks to houses and other community infrastructure.

The existing literature and knowledge on climate change have been increased. Nevertheless, the reality remains that we are currently in an abnormal climate phase. Tackling these issues also required changes. A shift in an individual's character is essential since it impacts economic decisions like what people buy and spend their money. Changing one's daily routine and appreciating the importance of simple actions can reduce the problem (Reisch et al. 2021). As the first generation to experience climate change, we must eliminate pollution to make the world a better place for future generations. Considerably, climate change and its consequences have become the major problem that human society is experiencing (Torres et al.

2021). Climate change and the natural disasters it caused are probably the product of human activity. People are familiar with a global warning which they often confuse with climate change. However, global warming is merely one of the consequences of climate change.

Furthermore, past evidence exhibits that climate change causes an increasing frequency of extreme weather. As a consequence of the long-term tendency, each location will experience a change in its particular climate. Indeed, many of the world's extreme weather events occurred recently, including certain astronomical events, abrupt weather changes, cloud bursts as well as ocean circulation weather events. In addition, several incidents occurred worldwide recently, such as a strong breeze or an unexpected significant snowfall. The most basic preparations are needed to adapt to climate change and reduce the risk of disasters. Hence, we must minimize the damage caused by human activity in the process. Additionally, when a specific natural disaster cannot be prevented or mitigated by human conduct, we must learn to respond to the emergency and minimize damage.

Human Sustainability and Adaptive Approaches

If the focus is a methodological innovation in resource management, then adaptive management to execute policies as experiments is recommended (Walters 1986), known as strategic adaptive management (Holling 1978). In addition to new means, the adaptive approach implies altered aims. As the name suggests, this approach prioritized learning in management. In contrast, traditional methods emphasize the opposite (Lee 1999). The effectiveness of strategic planning as a public policy technique is evaluated by assessing its intellectual, technological, equity, and practical strengths and limits, as well as its limitations. Lee (1999) discovered three conclusions after conducting extensive research successfully. In the first instance, adaptive management is more significant as an idea rather than a practical method of getting insight into ecosystems' behavior. It is related to humans' use and occupies, at least in the short term. The second point is that adaptive management should only be utilized once the opposing parties acknowledge the adaptive approach. However, this technique was never used in this way in the past. Instead, it has been misused due to a lack of understanding of the concept. The last conclusion of Lee (1999) was that efficiency and effectiveness in social learning and adaptive management make them possible, which can be considered strategically important. According to Dawson et al. (2021), in the context of governance of the ecological system, adaptive management is highly regarded to be employed. These challenges highlight leadership's critical role in ensuring that adaptive management is successful. The top management's main character is an apparent disposition. Leaders typically control the flow of benefits resulting from harvesting from or protecting the ecosystem. They often play a critical role in motivating those whose cooperation is required to gather information, analyze, and diagnose surprises.

In socio-ecological research, spanning five-time intervals over 16 years in Papua New Guinea determine the fundamental characteristics of a long-lasting traditional adaptive reef management. In their assessments, resource users identified high levels of compliance, strong leadership and social cohesion, and participatory decision-making among community members as key features of the rotating fishery closure system (Cinner et al. 2019). Cinner et al. (2019) struggled to understand the community's perception toward traditional management and how this approach impacts and improves their lives. The researchers conducted household surveys. They followed a mixed-method technique to determine and investigate local people's beliefs. The three most important characteristics of this long-lasting adaptive management system were revealed. These terms were named compliance, leader, and societal integration, and participative decision among community members was a notable three pillars of community development. This quest was an outcome of early research that showed that there are still significant uncertainties despite the widespread application of adaptive management.

Moreover, it was unclear how it could contribute to a long-term positive result for humans and natural systems (Keith et al. 2011). Numerous research in the context of adaptive management found that adaptive management outcomes have found that they are associated with low costs, flexible, and polycentric institutions. Research suggests these relationships allow solid social networks to foster participation and experimentation, debate, learning, information sharing, resource monitoring, and user support for implementing rules (Bardsley and Sweeney 2010; Folke et al. 2005). In contrast, these essential characteristics have been developed from theoretical or analogy research, not from the viewpoint of the actual resource users. Insights into local views can shed light on critical socio-cultural-political-economic elements vital to the effectiveness of management efforts in a specific setting (Adams and Sandbrook 2013). However, this is challenging because the first step ensures that leadership is not culturally relevant or inequitable. Political leadership legitimacy and public appeal are the foremost critical in social development and cohesion. It is necessary to assess the legality and acceptability of governance (Bennett 2016). We do this because it is essential to ensure that governance is neither culturally inappropriate, inequitable, nor distant from the public interest at the end of the day (Lecuyer et al. 2018). Therefore, a better understanding of critical traits that are greater efficiencies derived would significantly impact what is derived from the residents' point of view. Leadership strategies may be beneficial in shedding light on the angel of practices and processes in which outsiders have no clear opinion about or conflict with local perspectives. Consequently, this argument is especially valid for adaptive management systems built on long-term customary practices (Cinner et al. 2019).

Individuals worldwide have retrieved information and other intellectual benefits in this digital technology era. As digital technology continues to develop, one of its most essential characteristics is its multimodality: the capacity to incorporate script, pictures, music, and even haptic input, which enable humans to continue to get an overflowing and engaging experience rising availability of digital systems (Jiang et al. 2021). Making the previously inaccessible apparent is a significant possibility provided by digital advances. In addition, digital technology's role in public awareness is impactful. Technologies are possible and valuable information tools for individuals to connect with environmental impacts in more precise and engaging ways. It enables them to visualize something that would otherwise be invisible to them, such as their environmental impact (Fauville et al. 2016).

In today's ever highly dynamic situations, as the corporate environments and business practices frequently and quickly change, leaders must think creatively to challenge the unforeseen problem. They should be able to adapt quickly to those changes. The ability of leaders to quickly modify information technology strategies is essential (Heifetz and Laurie 1997). The modern world of labor has been shaped by a slew of technological, social, and organizational developments (Brem and Voigt 2009). The digitalization of employment, as well as the gig economy, are examples of this trend.

Moreover, organizations can now manage data in advanced and innovative techniques thanks to a wide range of new technologies, such as sight and monitoring devices (Brynjolfsson and McAfee 2014). Nowadays, it has become relevant in a tumultuous environment, including erratic competition and technology conditions. As a result, Trough digital transformation, leaders can improve their ability to transition from their current state to the desired future state. It is often about uncovering and digitalizing items in a new critical technique that can only be done with digital technologies (Dunn 2017). Such essential intelligence has already been generated everywhere and requires consistent new tools and approaches in human sustainability development. A crucial difference from the previous practices is that this collection of procedures must successfully accommodate the temporal gap between author/creator and digitizer/analyst of the material depicting the place.

Furthermore, literature on digital learning and skills needed for enterprises' digital transformation is scarce. Sousa and Rocha (2019) revealed that effective digital transformation of businesses based on the most recent developments in skills becomes the key objective of the research. The perspectives of individuals on issues encountered by organizations and prospects for emerging disruptive businesses are also becoming relevant. An empirical study was conducted to determine the necessity of digital skills for a successful transformation such a range of advanced digital technology. Digitalization was identified as an essential skill. Mobile technologies, tablets, and smartphone applications becoming increasingly popular among employees were identified as the most critical digital learning contexts. Sousa and Rochas' analysis suggests that firms should reassess their strategy in light of the challenges posed by digital transformation. Because of this, we may conclude that the obstacles and opportunities for new disruptive firms and individual perception and recent trends inabilities are directly associated. Trenerry et al. (2021) assessed the positive correlation between opportunities for new disruptive businesses and new trends in skills. Therefore, the organizational digital transformation's unfavorable impressions require responding to new opportunities and new trends in skills development.

Indeed, digital learning is typically haphazard and unplanned, regardless of the context. It can teach learners critical thinking to solve complicated problems, collaborate and work in a team, interact with others effectively, and have independence in the learning process (Sousa et al. 2019). According to another study's findings, taking advantage of the potential and embracing the latest trend in skill development can assist firms to thrive and expand in the future (Sousa and Rocha 2019). Concerning human sustainability and adaptive approaches, these two arguments are helpful. Johnson and Wetmore (2021) asserted that changes in the way people think about technology's impact on society are needed. In a qualitative study focusing on the social robot. Khaksar et al. (2021) outlined that human perception of technology use is the greatest obstacle to technology acceptance. This hesitance of technology acceptance may delay the development of adaptive solutions to ensure that the general public reaps the benefits of technology. Human life could benefit significantly from technology if people were more accepting of the idea of incorporating it into their daily routines. Adaptive and sustainable human development can then be achieved by applying technology. Technologies can play a vital role in changing human perception, bringing new and rising client categories and diverse cultures into the global marketplace. The uncertain economic and increased customer aspirations and the pace change of market demand also become common, triggering higher service quality demand. Since digital transformation is becoming an integral process, many organizations and company are struggling to manage it. Increasing numbers of high-level employment have emerged in the job market that demand flexibility and problem-solving skills (Markowitsch et al. 2002). Despite increasing research, it is necessary to produce work that provides a different description of "perceived problems." Nevertheless, the research agenda in the context of digital transformation in companies should consider this matter (Sousa and Rocha 2019). Furthermore, it could help remove barriers to human perception (Lata and Nunn 2012) because it hinders environmental awareness (Bennett 2016).

The Role of Leadership Is Social Psychology during Climate Change

Lawrence et al. (2010) raised a question about choosing a particular topic while conflict between conscious act and participation for general public interest among politicians can be understood from history. However, the understanding of the dispute is still limited, and researchers do not yet know how common or rare it is. Lawrence pointed out that an observer should give a general answer to a theory's public inquiry. Further, the theorist should be prompted to ask more particular queries on these conflicting ideas' significance. Besides, the theorist needs to generate a broader understanding of the best ways to deal with it. The future of social psychology, particularly concerning climate change awareness, should be considered in this field and the role of leadership. As part of this discussion, it is essential to consider how political leaders can raise public awareness and what tools they have at their disposal.

Sousa (2017) examined talents and their importance to organizations as strategic areas of human resources education. However, internally within corporations or academic institutions, determining and honing one's skills can be difficult Abbas et al. (2022a). Therefore, it was efforted to look at human resources management abilities and how humans should be educated throughout this view. Sousa's point was that university courses in human resources management need to concentrate on these abilities. Sousa's study investigated whether or not they are delivered in the learning environments of institutions of higher learning. Leaders need to have the ability to employ social cognition to knit their organizations together. They can use their social identity and social exchange to inform, build, and choose social progress as a leader (Abbas et al. 2021c).

In recent decades, authorities, policymakers, and governments worldwide have emphasized democracy, the Sustainable Development Goals (SDGs), and their sub-goals. Democracy has been broadly embraced worldwide; society's leaders have played a more prominent role in directing state affairs (Harvey and Novicevic 2004). Moreover, leaders worldwide are well aware of the need for sustainable development and actively participate in achieving these goals (Grover et al. 2021). Global leaders are active on digital media (Grover et al. 2019). While social media is a worldwide phenomenon, it can be used for information awareness because it is widely used (Aswani et al. 2018). Sousa and Rocha (2019) found that literature on digital learning is littered with conflicting interpretations of the phenomenon. Sousa and Rocha begin their investigation with the definition provided by academicians. Many people believe that all forms of digital learning, including smartphones to computers, are unstructured and undefined. This argument means every day learning from digital platforms is related to their work, skills, or learning interest (Abbas et al. 2020).

In contradiction, the learner's perspective is never planned is also believed, as Tan and Andriessen (2021) discussed that social interaction is thought to promote personal growth. This learning happens because in-group and out-group affiliations can be strengthened through communication amongst like-minded users. However, communication between people with divergent views can weaken group identification (Yardi and Boyd 2010). Connections are within a user's group, whereas out-of-group connections are made to welcome outsiders to come under one umbrella platform of the organization (Iyengar and Westwood 2015). When two people or more are debating, interacting, and exchanging ideas, one gives their self-perception a higher rating when the other person is opposed to it (Lee 2007). As a result, each person challenges themselves to make an excellent idea in the framed discussion. As a result, they grow personally and intellectually as they challenge their brain to think beyond. Groups of people who have the same political values congregate (Kim 2015; Lee et al. 2014). Voters who have little interest in politics are ideologically moderate, making them easy targets for political polarization (Lawrence et al. 2010). Whether digital learning is beneficial for personal or professional development, both arguments are there. However, we can rest assured that those who share our views will band together to establish a supportive social network.

To believe that leadership psychology has the most significant impact on perception, trust, and societal cohesion at this point. Leadership is the most important topic of diversity management (Abbas et al. 2021a). This topic has generated enormous research submitted to the most significant societies globally (Abbas et al. 2021c). Many of these topics were brought into the organizational sphere, where individual differences are almost manageable (Abbas et al. 2021b). In organizational leadership psychology, this debate is critical. Hence, this study relies on relevant research and aims to keep the existing literature up to date. Many scholars have attempted to understand better the relationship between flow and effective leadership and employees' attitudes (Abbas et al. 2021d; Smith et al. 2012). These scholars have raised a debate over how positive psychology manifests itself in political leadership is needed.

The world has witnessed how political leadership, the media, and other digital channels were leveraged to the fullest extent to raise public awareness (Saud et al. 2020). As a global leader, Al Gore is likely the most committed to increasing the general understanding of environmental concerns through public appearances (Pielke and Sarewitz 2002). Al Gore, who was 28 years old when he was elected to Congress in 1976, has been an outspoken proponent of environmental issues, including hazardous waste and climate change. He uses his social identity as one of the Democrats' Greens to promote social cognition for climate change issues, a critical success for his party since the 1980s (Gore 2017). In addition to Al Gore, Pakistani Prime Minister Imran Khan is one of the world's most recognized leaders actively participating in public awareness campaigns about the dire repercussions of climate change (Rubbani et al. 2021).

Rabbani and colleagues conducted a Critical Discourse Analysis of Pakistani Prime Minister Imran Khan's political speech on 27 September 2019, as part of a content qualitative approach. Humans construct civilization and, when gathered together, become nations; they said in their concluding remarks. Role of leadership in social constructions in this scenario becomes more pertinent since climate change substantially impacts Pakistan's ability to secure its supply of electricity, water, and food (Hussain et al. 2020; Qureshi et al. 2016). Climate change is a significant problem for Asian developing countries, as dreadful as its effects are. However, nothing has been said (Hossain et al. 2020). Climatology change Indian scholars have felt the heat as climate change, and its inevitable symptoms have become a part of people's daily lives (Das and Ghosh 2020). China's widespread worry about global warming and climate change is relatively lower. According to a recent cross-nation study, the level of concern varies widely among Chinese residents, between provinces, and across coastal and inland locations (Liu et al. 2020). Arguments suggesting social psychology is significantly dependent on what is spread were built using the literature cited above. Social media is a widely used technique for influencing people's beliefs and behaviors. That kind of straining tool is designed for people who have the power to sway the thinking of others.

The Rationale of the Study

This chapter, which fits the book's overall theme, aims to raise awareness of global health concerns, leadership societies, and climate change. Modern organizations make it possible to raise awareness about social psychology's enormous and never-ending subject matter. Moreover, politicians should strive for greater public welfare to make this achievable. It is also critical to understand that political leadership and digital media significantly impact social psychology. Therefore, this chapter's goal is to introduce pertinent information and arguments into the debate so that researchers can begin to think about it in the context of their research. The researchers can use the best literature to combine ideas that are not unfamiliar, attainable, and simple to comprehend into a cohesive entity. Human progress is inevitable if political, organizational, and digital media leaders work together. A public awareness and involvement process for more outstanding social goodwill begins due to this method.

Materials and Methods

This chapter used literature from throughout the world to support a point (Snyder 2019). Understandable and straightforward ideas can be generated by synthesizing world literature (Baron et al. 2017; Barron et al. 2012; Wyborn et al. 2018). Literature synthesizing means combining two or more points in their most basic form. Instead of describing each source's main points in detail, it is more beneficial to blend the ideas and conclusions of multiple sources. Researchers can do this by comparing and contrasting the facts gathered. Thus, the reader will discern when information from different sources converges and diverges. Researchers could benefit from conducting a literature review and a synthesis to form this idea (Harvey 2014).

Results and Discussion

There are five major components to this study, summarizing several reviewed studies. For the beginning, this study presents the outcomes of scholarly investigations that looked at knowledge dispositions. Then, we include behavior components of environmental literacy—human psychology of individuals and prosocial behavior about climate change and its effect on human health. Social perception is paramount in group cohesiveness, social psychology, and public awareness domains. It becomes more complex and time-consuming when countries engage in the process, which necessitates greater attention and optimization of approach. The issue of how climate change education and awareness should be implemented has become increasingly difficult. Climate change occurs in this context, yet individuals' perceptions are hazy. When confronted with a crisis, people's conduct might be challenging to predict because of their complicated nature. Public education has long been expected to help the general public better understand environmental science and physical systems and processes.

This context is outlined that the climate change issues are inseparable from individual psychological and external determinants. The focus should target how individuals or groups interact with the environment in any organization or community. This dimension encompasses both leader and follower behaviors that involve an inevitable psychological, affective, and cognitive process. To deal with climate change, humans need to raise our level of consciousness, and there is no other way to do it. As an individual, we have our way of tackling this matter. As a group, we can combine our ability, competencies, and virtue to make a powerful instrument to mitigate climate change. Prerequisites such as recognizing and analyzing environmental issues and evaluating them are part of this dimension. For micromanagement of individual perceptions to be possible, political leaders must encourage the participation of organizational leaders for worthwhile purposes. All essential questions, arguments, and strategies formation and assessment to address various environmental challenges must be discussed at the micro-level to realize public accountability. Environmentally responsible behavior should be promoted by the individual community and business (Fenitra et al. 2021a, b). Adapting this particular behavior, our new way of living would minimize climate change-related issues.

Besides, climate change has indirect health implications, such as reduced food and water supplies and reduced agricultural yields caused by social, economic, and political disturbances. Regional winners and losers will be due to climate change on cereal grain yields, with a 5–10% increase worldwide undernutrition. As a result of climate change, the world has become more unstable. It is thought to be the root cause of various humanitarian and global health crisis concerns such as catastrophes, conflicts, migration, and refugee flows. When discussing climate change, there must be a focus on the microaspects of human health, such as food shortages and basic sanitation. As long as they are presented correctly, public awareness initiatives can have a long-lasting impact. Climate change-related health issues may necessitate research into rapidly disseminating information about them through mass media. Thus, if everyone participates personally, social media can take on a central role in our society. For example, social media platforms to spread awareness about climate change are essential to make humans more resilient and adapt to this problem. Suppose these efforts are recognizable to the public to raise awareness about societal concerns and inspire people to take responsibility for their actions. The substantial impact of climate change on human health, life quality, social justice, and political landscape is enormous and should be counted. The table below presents significant findings from current and prior literature for a critical examination (Table 26.1).

Overall, social and climate change-related health issues affect the entire community. Thus, building more understanding about its severity in public could bring responsibility to address them. More comprehensive public education will improve people's lives, and governments will devote more resources to various adaptation

Table 26.1 Impact of climate change and its dispositions

Primary area		Key findings	Source
Social	Shortage	Food insecurities	(cf. Qureshi et al. 2013)
	Crisis	Increasing health inequalities	(cf. Mazhin et al. 2020)
	Scarcity	Water shortage	(cf. Omer et al. 2020)
Environmental	Catastrophes	Political discussion	(cf. Pierrehumbert 2005)
	Air Pollution	Multifactorial stress intensity	(cf. Zandalinas et al. 2021)
	Drought	Weather-related uncertainties	(cf. Mukherjee et al. 2018)
Economic	Global	Global climate change	(cf. Gough 2010)
	Regional	Increased poverty	(cf. Hallegatte and Rozenberg 2017)
Political	Prevention	Regional Rifts and Confits	(cf. Koubi 2019)
	Emergency	Lowering Relocations Trends	(cf. Hugo 2013)

strategies. Scholars play an essential role in creating intelligence and knowledge to grasp this problem from the ground up. Instead of constantly disseminating information, academics should focus on formulating new approaches to problem-solving, including modeling. Often practitioners experience difficulty implementing strategies based on the knowledge produced by scholars only. Scholars, practitioners, and politicians should work together to develop realistic solutions to this problem. For example, GIS modeling can be used to examine how plant disease affects rainfall and vegetation, for example, by major research institutions. Techniques like these can be used across the board to defend against natural disasters and improve the detection of infectious disease sentinel cases. Modeling can be used to integrate information effectiveness.

According to a recent study, limiting water supply in densely populated places on the African and Asian continents has another fatal effect. Table 26.2 shows data on per capita accessible water in these two continents' most notable countries, which are severely impacted and where a water shortage could be catastrophic in 2025. This water shortage might jeopardize these countries' health, economic, political, and social lives, which is just the beginning. The gravity of the challenges raising awareness about these concerns could have far-reaching consequences beyond human comprehension.

The authors of this chapter feel that a procedure that requires community involvement and participation using advanced technology and a large-scale public awareness campaign is urgently needed. Because they have to experience the effects of climate change quickly, the community should have vital supplementary knowledge. This approach will also make it easier for people to comprehend, apprehend, accept, and cope with implicit measures. The world's leaders are currently aware of rising hazards to human health due to the recent discovery of COVID-19. It is a startling occurrence that has sparked widespread awareness debates in informal conversations among the general people.

Table 26.2 Water availability in 1995 and 2025

Countries of the world	Per Capita water availability 1995 (m³/person/year)	Per Capita water availability 2025 (m³/person/year)
Africa		
Algeria	527	313▼
Burundi	594	292▼
Egypt	936	607▼
Ethiopia	1950	807▼
Rwanda	1215	485▼
Somalia	1422	570▼
Asia/Middle East		
Bahrain	162	104▼
Iran	1719	910▼
Jordan	318	144▼
Qatar	91	64▼
Singapore	180	142▼

Source: Zakar et al. (2020)

Fig. 26.2 Generating climate change initiatives: self-concept

On the other hand, experts on climate change have a history of seeing international society's awareness address global concerns. The pandemic's impact on the economy and global health was profoundly devastated. Leaders worldwide took advantage of the opportunity to raise concerns that climate change concerns are being disregarded. Despite this, climate change is still not receiving the attention it deserves, and few are speaking out strongly against its severity. We believe that leaders should also deploy their social identity and exchange to motivate their countries to take climate change seriously. Social media platforms, digital media, and print media should join forces in a comprehensive awareness campaign to foster social cognition against climate change challenges. Environmental catastrophes will not only affect the health and well-being of individuals. However, they will also substantially impact the global food supply and migration concerns due to climate change. We have proposed a conceptual framework for further research on improving social cognition through public education campaigns and communication awareness appeal as a change-agent social domain to address these issues (Fig. 26.2).

Limitations of the Study

In this chapter, we examined how Al Gore, one of the world's most recognized advocates of climate change education, has appeared notably in leading roles in developing social psychological awareness. However, the theoretical limitations of this study's focus on climate change's impact on his social identity have profound implications for researchers. This chapter covers many topics related to change agents that can be dangerous to global environments and call for action.

Recommendations

Because of the gravity of climate change, the authors of this study propose that COVID-19 awareness could be used in a social awareness campaign (Qazi et al. 2020). Research on social cognition (**Abbas et al. 2022b) could benefit from this study concept in the future, particularly in raising public awareness about climate change.

Conclusions

A wide range of literature explains how climate change is a multi-faceted problem intertwined with a wide range of social, political, economic, and global challenges. When the goal is to communicate the same vision to others, the task becomes more difficult. History of leadership theories is developed around the idea that only a few people could think of and lead others through such difficult times in human history. Since the problem of climate change is complicated and has regional and global ramifications, as a result, a global strategy has been clearly and convincingly required. A snowball effect will have a long-term impact on society and individuals due to this solution's use of identity. Educating others is the first step in this counseling sequence, and leaders can use their social identity to do so. When people realize that this is a severe problem, they will need to combat it.

Leaders like Al Gore may significantly impact public awareness campaigns, especially those focused on the global threat of climate change. Leaders with international reputations and ties to well-known for their values are the best role models for integrating the concept of social identity because of their position in social networks (Turner and Oakes 1986). Leaders of global identity can inspire any religious or minorities, sports, and organizational leaders to make inclusive judgments for nations and the greater well-being of the people. Suppose these leaders can devise strategies to mobilize central social identity bearers. In that case, they can help construct social cognition against climate change aversions that are taking place all across the globe. People need to understand that the dangers of climate change extend far beyond human health and well-being. They will have far-reaching consequences in the economy, water scarcity, and other environmental problems that will be dreadful for generations to come. As a result of these disputes, there could be massive rifts across regions. Because health awareness is not confined to a particular culture or country, a global concern should be raised worldwide. Expertise in coping with climate change's unique effects and dispositions requires a vast array of abilities.

Moreover, the decline in water resources, the rise in death tolls, and natural catastrophes have significantly impacted human lives since 2000. In order to increase public awareness, much attention in research should be paid to all these topics. This study is optimistic about how Al Gore's leadership formed his policies for public presence using his charisma to inspire others to attract the global climate morpheme. His unwavering commitment to this worldwide cause should be recognized as a key to the motivation of countless world leaders. Future researchers studying climate change may use leadership styles from all levels of government, including the regional, national, and perhaps international levels.

Acknowledgments We declare no conflict of interest and the non-commercial nature of this research project.

References

Abbas A, Saud M, Ekowati D, Usman I, Setia S (2020) Technology and stress: a proposed framework for coping with stress in Indonesian higher education. Int J Innov Creativity Change 13(4):373–390

Abbas A, Ekowati D, Suhariadi F (2021a) Managing individuals, organizations through leadership: diversity consciousness roadmap. In B Christiansen, H Chandan (eds.), Handbook of research on applied social psychology in multiculturalism (vol. 1, pp. 47–71). Hershey, Pennsylvania: IGI Global

Abbas A, Ekowati D, Suhariadi F (2021b) Individual psychological distance: a leadership task to assess and cope with invisible change. J Manag Dev 40(3):168–189. https://doi.org/10.1108/jmd-09-2020-0304

Abbas A, Saud M, Ekowati D, Suhariadi F (2021c) Social psychology and fabrication: a synthesis of individuals, society, and organization. In B Christiansen, H Chandan (eds.), Handbook of research on applied social psychology in multiculturalism (vol. 1, pp. 89–109). Hershey, Pennsylvania IGI Global

Abbas A, Saud M, Ekowati D, Usman I, Suhariadi F (2021d) Servant leadership: a strategic choice for organisational performance. An empirical discussion from Pakistan. Int J Product Qual Manag 34(4):468–485. https://doi.org/10.1504/IJPQM.2021.120599

Abbas A, Ekowati D, Suhariadi F, Fenitra RM, Fahlevi M (2022a) Human capital development in youth inspires us with a valuable lesson: self-care and wellbeing. In KL Clarke (ed.), Self-care and stress management for academic well-being (pp. 80–101). Hershey, Pennsylvania: IGI Global

**Abbas A, Ekowati D, Suhariadi F (2022b) Social perspective: leadership in changing society. In MI Hassan, S Sen Roy, U Chatterjee, S Chakraborty, U Singh (eds.), Social morphology, human welfare, and sustainability (pp. 89–107). Springer International Publishing, Cham

Adams WM, Sandbrook C (2013) Conservation, evidence and policy. Oryx 47(3):329–335. https://doi.org/10.1017/S0030605312001470

Ahmed M (2020) Introduction to modern climate change. Andrew E. Dessler: Cambridge University Press, 2011, 252 pp, ISBN-10: 0521173159. Sci Total Environ 734:139397. https://doi.org/10.1016/j.scitotenv.2020.139397

Anderson R, Bayer PE, Edwards D (2020) Climate change and the need for agricultural adaptation. Curr Opin Plant Biol 56:197–202. https://doi.org/10.1016/j.pbi.2019.12.006

Aswani R, Kar AK, Ilavarasan PV, Dwivedi YK (2018) Search engine marketing is not all gold: Insights from Twitter and SEO Clerks. Int J Inf Manag 38(1):107–116. https://doi.org/10.1016/j.ijinfomgt.2017.07.005

Ayanlade A, Sergi CM, Di Carlo P, Ayanlade OS, Agbalajobi DT (2020) When climate turns nasty, what are recent and future implications? Ecological and human health review of climate change impacts. Curr Clim Chang Rep 6:55–65. https://doi.org/10.1007/s40641-020-00158-8

Bardsley DK, Sweeney SM (2010) Guiding climate change adaptation within vulnerable natural resource management systems. Environ Manag 45(5):1127–1141. https://doi.org/10.1007/s00267-010-9487-1

Barnes ML, Wang P, Cinner JE, Graham NA, Guerrero AM, Jasny L et al (2020) Social determinants of adaptive and transformative responses to climate change. Nat Clim Chang 10(9):823–828. https://doi.org/10.1038/s41558-020-0871-4

Baron JS, Specht A, Garnier E, Bishop P, Campbell CA, Davis FW et al (2017) Synthesis centers as critical research infrastructure. Bioscience 67(8):750–759. https://doi.org/10.1093/biosci/bix053

Barron D, Smarrito-Menozzi C, Fumeaux R, Viton F (2012) Synthesis of dietary phenolic metabolites and isotopically labeled dietary phenolics. In: Jeremy P, Spencer E, Crozier A (eds) Flavonoids and related compounds: bioavailability and function, vol 30. CRC Press, Boca Raton, pp 233–293

Bennett NJ (2016) Using perceptions as evidence to improve conservation and environmental management. Conserv Biol 30(3):582–592. https://doi.org/10.1111/cobi.12681

Brem A, Voigt K-I (2009) Integration of market pull and technology push in the corporate front end and innovation management – insights from the German software industry. Technovation 29(5):351–367. https://doi.org/10.1016/j.technovation.2008.06.003

Brynjolfsson E, McAfee A (2014) The second machine age: work, progress, and prosperity in a time of brilliant technologies. Norton & Company, Manhattan

Campbell J, Barnett J (2010) Climate change and small island states: power, knowledge and the South Pacific. Routledge Publishing Inc., Oxfordshire

Chen H, Liu H, Chen X, Qiao Y (2020) Analysis on impacts of hydro-climatic changes and human activities on available water changes in Central Asia. Sci Total Environ 737:139779. https://doi.org/10.1016/j.scitotenv.2020.139779

Cinner J, Lau J, Bauman A, Feary D, Januchowski-Hartley F, Rojas C et al (2019) Sixteen years of social and ecological dynamics reveal challenges and opportunities for adaptive management in sustaining the commons. Proc Natl Acad Sci 116(52):26474–26483. https://doi.org/10.1073/pnas.1914812116

Cushman GT (2011) Humboldtian science, creole meteorology, and the discovery of human-caused climate change in South America. Osiris 26(1):19–44

Das U, Ghosh S (2020) Factors driving farmers' knowledge on climate change in a climatically vulnerable state of India. Nat Hazards 102(3):1419–1434. https://doi.org/10.1007/s11069-020-03973-2

Dawson L, Elbakidze M, Schellens M, Shkaruba A, Angelstam P (2021) Bogs, birds, and berries in Belarus: The governance and management dynamics of wetland restoration in a state-centric, top-down context. Ecol Soc 26(1):1–18. https://doi.org/10.5751/ES-12139-260108

De La Fuente A, Rojas M, Mac Lean C (2017) A human-scale perspective on global warming: Zero emission year and personal quotas. PLoS One 12(6):e0179705. https://doi.org/10.1371/journal.pone.0179705

Douville H, Chauvin F, Broqua H (2001) Influence of soil moisture on the Asian and African monsoons. Part I: Mean monsoon and daily precipitation. J Clim 14(11):2381–2403. https://doi.org/10.1175/1520-0442(2001)014<2381:IOSMOT>2.0.CO;2

Dunn S (2017) Praxes of "the human" and "the digital": spatial humanities and the digitization of place. GeoHumanities 3(1):88–107. https://doi.org/10.1080/2373566X.2016.1245107

Fauville G, Lantz-Andersson A, Mäkitalo Å, Dupont S, Säljö R (2016) The carbon footprint as a mediating tool in students' online reasoning about climate change. In: Erstad O, Kumpulainen K, Mäkitalo Å, Schrøder KC, Pruulmann-Vengerfeldt P, Jóhannsdóttir T (eds) Learning across contexts in the knowledge society. Sense Publishers, Rotterdam, pp 179–201

Fenitra RM, Tanti H, Gancar CP, Indrianawati U (2021a) Understanding younger tourist' intention toward environmentally responsible behavior. GeoJ Tour Geosites 36(2):646–653. https://doi.org/10.30892/gtg.362spl12-694

Fenitra RM, Tanti H, Gancar CP, Indrianawati U, Hartini S (2021b) Extended theory of planned behvaior to explain environmentally responsible behavior in context of nature-based rourism. GeoJ Tour Geosites 39:1507–1516. https://doi.org/10.30892/gtg.394spl22-795

Fielding KS, Hornsey MJ, Swim JK (2014) Developing a social psychology of climate change. Eur J Soc Psychol 44(5):413–420. https://doi.org/10.1002/ejsp.2058

Folke C, Hahn T, Olsson P, Norberg J (2005) Adaptive governance of social-ecological systems. Annu Rev Environ Resour 30:441–473. https://doi.org/10.1146/annurev.energy.30.050504.144511

Gore A (2017) An inconvenient sequel: truth to power: your action handbook to learn the science, find your voice, and help solve the climate crisis. Rodale Publishers, Emmaus

Gough I (2010) Economic crisis, climate change and the future of welfare states. Twenty First Cent Soc 5(1):51–64. https://doi.org/10.1080/17450140903484049

Griggs DJ, Noguer M (2002) Climate change 2001: the scientific basis. Contribution of working group I to the third assessment report of the intergovernmental panel on climate change. Weather 57(8):267–269. https://doi.org/10.1256/004316502320517344

Gross L (2018) Confronting climate change in the age of denial. PLoS Biol 16(10):e3000033

Grover P, Kar AK, Dwivedi YK, Janssen M (2019) Polarization and acculturation in US Election 2016 outcomes – can Twitter analytics predict changes in voting preferences. Technol Forecast Soc Chang 145:438–460. https://doi.org/10.1016/j.techfore.2018.09.009

Grover P, Kar AK, Gupta S, Modgil S (2021) Influence of political leaders on sustainable development goals–insights from Twitter. J Enterp Inf Manag 34:1893–1916. https://doi.org/10.1108/JEIM-07-2020-0304

Guile B, Pandya R (2018) Adapting to global warming: four national priorities. Issues Sci Technol 34(4):19–22

Hallegatte S, Rozenberg J (2017) Climate change through a poverty lens. Nat Clim Chang 7(4):250–256. https://doi.org/10.1038/nclimate3253

Hanlon HM, Bernie D, Carigi G, Lowe JA (2021) Future changes to high impact weather in the UK. Clim Chang 166(3):1–23. https://doi.org/10.1007/s10584-021-03100-5

Harvey S (2014) Creative synthesis: exploring the process of extraordinary group creativity. Acad Manag Rev 39(3):324–343. https://doi.org/10.5465/amr.2012.0224

Harvey M, Novicevic MM (2004) The development of political skill and political capital by global leaders through global assignments. Int J Hum Resour Manag 15(7):1173–1188. https://doi.org/10.1080/0958519042000238392

Harvey B, Huang Y-S, Araujo J, Vincent K, Roux J-P, Rouhaud E, Visman E (2021) Mobilizing climate information for decision-making in Africa: contrasting user-centered and knowledge-centered approaches. Front Clim 2:1–14. https://doi.org/10.3389/fclim.2020.589282

Heifetz RA, Laurie DL (1997) The work of leadership. Harv Bus Rev 75:124–134

Holling CS (1978) Adaptive environmental assessment and management. Wiley, Chichester

Hossain MS, Arshad M, Qian L, Kächele H, Khan I, Islam MDI, Mahboob MG (2020) Climate change impacts on farmland value in Bangladesh. Ecol Indic 112:106181. https://doi.org/10.1016/j.ecolind.2020.106181

Hrabok M, Delorme A, Agyapong VI (2020) Threats to mental health and well-being associated with climate change. J Anxiety Disord 76:102295. https://doi.org/10.1016/j.janxdis.2020.102295

Hugo G (2013) Migration and climate change. Canadian studies in population Edward Elgar, Cheltenham

Hussain M, Butt AR, Uzma F, Ahmed R, Irshad S, Rehman A, Yousaf B (2020) A comprehensive review of climate change impacts, adaptation, and mitigation on environmental and natural calamities in Pakistan. Environ Monit Assess 192(1):1–20. https://doi.org/10.1007/s10661-019-7956-4

Hwong AR, Kuhl EA, Compton WM, Benton T, Grzenda A, Doty B et al (2021) Climate change and mental health: implications for the psychiatric workforce. Psychiatr Serv 202100227. https://doi.org/10.1176/appi.ps.202100227

Iyengar S, Westwood SJ (2015) Fear and loathing across party lines: new evidence on group polarization. Am J Polit Sci 59(3):690–707. https://doi.org/10.1111/ajps.12152

Jiang L, Yu S, Zhao Y (2021) Incorporating digital multimodal composing through collaborative action research: challenges and coping strategies. Technol Pedagog Educ, pp 1–17. https://doi.org/10.1080/1475939X.2021.1978534

Johnson DG, Wetmore JM (2021) Technology and society: building our sociotechnical future. MIT Press, Cambridge, MA

Keith DA, Martin TG, McDonald-Madden E, Walters C (2011) Uncertainty and adaptive management for biodiversity conservation. Biol Conserv 144(4):1175–1178. https://doi.org/10.1016/j.biocon.2010.11.022

Khaksar SMS, Khosla R, Singaraju S, Slade B (2021) Carer's perception on social assistive technology acceptance and adoption: moderating effects of perceived risks. Behav Info Technol 40(4):337–360. https://doi.org/10.1080/0144929X.2019.1690046

Khare N, Singh D, Kant R, Khare P (2020) Global warming and biodiversity. In: Current state and future impacts of climate change on biodiversity. IGI Global, Hershey, pp 1–10

Kim Y (2015) Does disagreement mitigate polarization? How selective exposure and disagreement affect political polarization. J Mass Commun Q 92(4):915–937. https://doi.org/10.1177/1077699015596328

Kjeldsen JE (2013) Strategies of visual argumentation in slideshow presentations: the role of the visuals in an Al Gore presentation on climate change. Argumentation 27(4):425–443. https://doi.org/10.1007/s10503-013-9296-9

Koubi V (2019) Climate change and conflict. Annu Rev Polit Sci 22:343–360. https://doi.org/10.1146/annurev-polisci-050317-070830

Lata S, Nunn P (2012) Misperceptions of climate-change risk as barriers to climate-change adaptation: a case study from the Rewa Delta, Fiji. Clim Chang 110(1):169–186. https://doi.org/10.1007/s10584-011-0062-4

Lawrence E, Sides J, Farrell H (2010) Self-segregation or deliberation? Blog readership, participation, and polarization in American politics. Perspect Polit 8(1):141–157. https://doi.org/10.1017/S1537592709992714

Lecuyer L, White RM, Schmook B, Lemay V, Calmé S (2018) The construction of feelings of justice in environmental management: an empirical study of multiple biodiversity conflicts in Calakmul, Mexico. J Environ Manag 213:363–373. https://doi.org/10.1016/j.jenvman.2018.02.050

Lee KN (1999) Appraising adaptive management. Conserv Ecol 3(2)

Lee E-J (2007) Deindividuation effects on group polarization in computer-mediated communication: the role of group identification, public-self-awareness, and perceived argument quality. J Commun 57(2):385–403. https://doi.org/10.1111/j.1460-2466.2007.00348.x

Lee JK, Choi J, Kim C, Kim Y (2014) Social media, network heterogeneity, and opinion polarization. J Commun 64(4):702–722. https://doi.org/10.1111/jcom.12077

Lee TM, Markowitz EM, Howe PD, Ko C-Y, Leiserowitz AA (2015) Predictors of public climate change awareness and risk perception around the world. Nat Clim Chang 5(11):1014–1020. https://doi.org/10.1038/nclimate2728

Levitus S, Antonov JI, Boyer TP, Baranova OK, Garcia HE, Locarnini RA et al (2012) World ocean heat content and thermosteric sea level change (0–2000 m), 1955–2010. Geophys Res Lett 39(10). https://doi.org/10.1029/2012GL051106

Lewandowsky S (2021) Climate change disinformation and how to combat it. Annu Rev Public Health 42:1–21. https://doi.org/10.1146/annurev-publhealth-090419-102409

Liu X, Hao F, Portney K, Liu Y (2020) Examining public concern about global warming and climate change in China. China Q 242:460–486. https://doi.org/10.1017/S0305741019000845

Markowitsch J, Kollinger I, Warmerdam J, Moerel H, Konrad J, Burell C, Guile D (2002) Competence and human resource development in multinational companies in three European Union Member States: a comparative analysis between Austria, the Netherlands and the United Kingdom, CEDEFOP panorama series. CEDEFOP Panorama Series published by ERIC, Thessaloniki

Mazhin SA, Khankeh H, Farrokhi M, Aminizadeh M, Poursadeqiyan M (2020) Migration health crisis associated with climate change: a systematic review. J Educ Heal Promot 9:97–107. https://doi.org/10.4103/jehp.jehp_4_20

Mikhaylov A, Moiseev N, Aleshin K, Burkhardt T (2020) Global climate change and greenhouse effect. Entrepreneur Sustain Issue 7(4):2897–2913. https://doi.org/10.9770/jesi.2020.7.4(21)

Mukherjee S, Mishra A, Trenberth KE (2018) Climate change and drought: a perspective on drought indices. Curr Clim Chang Rep 4(2):145–163. https://doi.org/10.1007/s40641-018-0098-x

Nelson CH (2021) Climate change patterns. In: Witness to a changing Earth. Springer, Cham, pp 175–221

Ohring G, Wielicki B, Spencer R, Emery B, Datla R (2005) Satellite instrument calibration for measuring global climate change: report of a workshop. Bull Am Meteorol Soc 86(9):1303–1314. https://doi.org/10.1175/BAMS-86-9-1303

Ojala M (2021) Safe spaces or a pedagogy of discomfort? Senior high-school teachers' meta-emotion philosophies and climate change education. J Environ Educ 52(1):40–52. https://doi.org/10.1080/00958964.2020.1845589

Omer A, Elagib NA, Zhuguo M, Saleem F, Mohammed A (2020) Water scarcity in the Yellow River Basin under future climate change and human activities. Sci Total Environ 749:141446. https://doi.org/10.1016/j.scitotenv.2020.141446

Patz JA, Campbell-Lendrum D, Holloway T, Foley JA (2005) Impact of regional climate change on human health. Nature 438(7066):310–317. https://doi.org/10.1038/nature04188

Pielke R, Sarewitz D (2002) Wanted: scientific leadership on climate. Issues Sci Technol 19(2):27–30

Pierrehumbert RT (2005) Climate change: a catastrophe in slow motion. Chic J Int Law 6:573–596

Plantinga A, Scholtens B (2021) The financial impact of fossil fuel divestment. Clim Pol 21(1):107–119. https://doi.org/10.1080/14693062.2020.1806020

Qazi A, Qazi J, Naseer K, Zeeshan M, Hardaker G, Maitama JZ, Haruna K (2020) Analyzing situational awareness through public opinion to predict adoption of social distancing amid pandemic COVID-19. J Med Virol 92(7):849–855. https://doi.org/10.1002/jmv.25840

Qureshi ME, Hanjra MA, Ward J (2013) Impact of water scarcity in Australia on global food security in an era of climate change. Food Policy 38:136–145. https://doi.org/10.1016/j.foodpol.2012.11.003

Qureshi MI, Awan U, Arshad Z, Rasli AM, Zaman K, Khan F (2016) Dynamic linkages among energy consumption, air pollution, greenhouse gas emissions and agricultural production in Pakistan: sustainable agriculture key to policy success. Nat Hazards 84(1):367–381. https://doi.org/10.1007/s11069-016-2423-9

Reisch LA, Sunstein CR, Andor MA, Doebbe FC, Meier J, Haddaway NR (2021) Mitigating climate change via food consumption and food waste: a systematic map of behavioral interventions. J Clean Prod 279:123717. https://doi.org/10.1016/j.jclepro.2020.123717

Rooney-Varga JN, Sterman JD, Fracassi E, Franck T, Kapmeier F, Kurker V et al (2018) Combining role-play with interactive simulation to motivate informed climate action: evidence from the World Climate simulation. PLoS One 13(8):e0202877. https://doi.org/10.1371/journal.pone.0202877

Rubbani A, Awan A, Shamsi SMB (2021) Critical discourse analysis of the political speech of Prime Minister of Pakistan (PMOP) Imran Khan delivered on 27th of September 2019. Element Educ Online 20(5):6179–6185. https://doi.org/10.17051/ilkonline.2021.05.694

Saud M, Ida R, Abbas A, Ashfaq A, Ahmad Ramazan A (2020) The social media and digitalization of political participation in youths: an Indonesian perspective. *Society* 8(1):83–89. https://doi.org/10.33019/society.v8i1.160

Smith MB, Koppes Bryan L, Vodanovich SJ (2012) The counter-intuitive effects of flow on positive leadership and employee attitudes: incorporating positive psychology into the management of organizations. Psychol Manag J 15(3):174–198. https://doi.org/10.1080/10887156.2012.701129

Snyder H (2019) Literature review as a research methodology: an overview and guidelines. J Bus Res 104:333–339. https://doi.org/10.1016/j.jbusres.2019.07.039

Sousa MJ (2017) Human resources management skills needed by organizations. In: Benlamri R, Sparer M (eds) Leadership, innovation and entrepreneurship as driving forces of the global economy. Springer, Cham, pp 395–402

Sousa MJ, Rocha Á (2019) Digital learning: developing skills for digital transformation of organizations. Futur Gener Comput Syst 91:327–334. https://doi.org/10.1016/j.future.2018.08.048

Sousa MJ, Carmo M, Gonçalves AC, Cruz R, Martins JM (2019) Creating knowledge and entrepreneurial capacity for HE students with digital education methodologies: differences in the perceptions of students and entrepreneurs. J Bus Res 94:227–240. https://doi.org/10.1016/j.jbusres.2018.02.005

Swim JK, Stern PC, Doherty TJ, Clayton S, Reser JP, Weber EU et al (2011) Psychology's contributions to understanding and addressing global climate change. Am Psychol 66(4):241–250. https://doi.org/10.1037/a0023220

Tabari H (2020) Climate change impact on flood and extreme precipitation increases with water availability. Sci Rep 10(1):1–10. https://doi.org/10.1038/s41598-020-70816-2

Tan J, Andriessen K (2021) The experiences of grief and personal growth in university students: a qualitative study. Int J Environ Res Public Health 18(4):1899. https://doi.org/10.3390/ijerph18041899

Torres C, Jordà G, de Vílchez P, Vaquer-Sunyer R, Rita J, Canals V et al (2021) Climate change and their impacts in the Balearic Islands: a guide for policy design in Mediterranean regions. Reg Environ Chang 21(4):1–19. https://doi.org/10.1007/s10113-021-01810-1

Trenerry B, Chng S, Wang Y, Suhaila ZS, Lim SS, Lu HY, Oh PH (2021) Preparing workplaces for digital transformation: an integrative review and framework of multi-level factors. Front Psychol 12:822. https://doi.org/10.3389/fpsyg.2021.620766

Turner JC, Oakes PJ (1986) The significance of the social identity concept for social psychology with reference to individualism, interactionism and social influence. Br J Soc Psychol 25(3):237–252. https://doi.org/10.1111/j.2044-8309.1986.tb00732.x

Verma A (2021) Influence of climate change on balanced ecosystem, biodiversity and sustainable development: an overview. Int J Biol Innov 3(2):331–337

Walters CJ (1986) Adaptive management of renewable resources. Macmillan Publishers Ltd., Basingstoke

Woolway RI, Kraemer BM, Lenters JD, Merchant CJ, O'Reilly CM, Sharma S (2020) Global lake responses to climate change. Nat Rev Earth Environ 1(8):388–403. https://doi.org/10.1038/s43017-020-0067-5

Wyborn C, Louder E, Harrison J, Montambault J, Montana J, Ryan M et al (2018) Understanding the impacts of research synthesis. Environ Sci Policy 86:72–84. https://doi.org/10.1016/j.envsci.2018.04.013

Yardi S, Boyd D (2010) Dynamic debates: An analysis of group polarization over time on Twitter. Bull Sci Technol Soc 30(5):316–327. https://doi.org/10.1177/0270467610380011

Zakar MZ, Zakar DR, Fischer F (2020) Climate change-induced water scarcity: a threat to human health. South Asian Stud 27(2):293–312

Zandalinas SI, Fritschi FB, Mittler R (2021) Global warming, climate change, and environmental pollution: recipe for a multifactorial stress combination disaster. Trends Plant Sci 26(6):588–599. https://doi.org/10.1016/j.tplants.2021.02.011

Chapter 27
A Retrospective Cohort Study on Ambient Air Quality and Respiratory Morbidities

Shruti S. Tikhe and **Kanchan Khare**

Abstract Outdoor air pollution is one of the greatest environmental health issues and a major factor compromising the quality of life of urban population. Severe health effects are observed when the residents are subjected to acute and long-term exposure. Studies describing a relationship between changing air quality and human health status have been done by many researchers. The outcome of studies in the form of advanced information allows policymakers, health caregivers, and residents to plan strategies and take preventive steps.

The authors propose application of retrospective cohort study to investigate the potential association between air quality and respiratory diseases with changing seasonal patterns. Modeling options have been presented to assess the risk of respiratory diseases with variations in air quality.

A case study of Ahmedabad, a metropolitan city in India, is presented which demonstrates the strength of retrospective temporal and seasonal cohort study. Hazard ratio calculated for this study for all the models is greater than 1 which signifies increased risk of occurrence of respiratory diseases in different seasons with increase or decrease in the value of air quality index which indicates a positive association between air quality index and respiratory diseases.

Keywords Air quality · Respiratory diseases · Cohort study · Health status · Hazard ratio

---

S. S. Tikhe
Department of R & D (Research), DTK Hydronet Solutions, Pune, Maharashtra, India

K. Khare (✉)
Department of Civil Engineering, Symbiosis Institute of Technology, Pune, Maharashtra, India
e-mail: kanchan.khare@sitpune.edu.in

Introduction

Air is an essential requirement for survival and growth of all lives on the earth. It has a major impact on human health and is a major indicator responsible for economic development of the country (Yi Lin et al. 2018). The chemical state of the atmosphere at a particular time and place is described by real-time air quality. It continuously changes and accordingly fluctuates the human health status, especially in sensitive groups. Natural phenomenon and anthropogenic sources are the leading causes of air pollution. Air pollution is a mixture of gases and particles in harmful amounts that are released into the atmosphere as a result of either natural or human activities. It affects human health and the environment seriously. The global impact of air pollution can be assessed through climate change studies.

Majority of the metro cities of the world experiences rapid industrialization and urbanization which ultimately results in increasing air pollution levels and degraded air quality (Khare and Nagendra 2007). It has been observed that short-term exposure to high levels of pollution may pose severe health impacts such as eye irritation, difficulty in breathing, and cardiovascular and pulmonary health effects, whereas the health impacts such as the threat of cancer, risk of premature death and also damage to the body's immunity, respiratory, neurological, and reproductive system are closely associated with long-term exposure to polluted air (Manisalidis et al. 2020). Senior citizens, kids, and individuals with preexisting heart and lung maladies along with diabetics are called as sensitive groups who are at a greater risk for air pollution-related health effects (Zhang et al. 2012). As per the 2019 report of the National Health Portal of India, 41,996,260 cases of respiratory diseases and 3740 deaths from respiratory infections were recorded in 2018. Similarly, India contributes to 18% of the global population, with severe acute respiratory infection (SARI) as one of the prominent causes of mortality in children > 5 years of age (Rishabh Waghmode et al. 2021). The World Health Organization (WHO) has issued guidelines related to permissible levels of air pollutants so as to protect the human health and the environment. In spite of significant progress in understanding the process of emission and fates of the air pollutants as well as adopting measures in order to decrease their ambient levels in urban localities for the last five decades, it was observed that the lives of about four million people every year worldwide has been destroyed by air pollution (World Health Organisation 2020).

Rationale

Air quality and the level of air pollution in the environment are described by the criteria air pollutants (SO_2, NOx & RSPM) (Clean Air Act 1990). The trends and effects of air pollutants on human lives can be studied using air quality assessment programs. Air pollution control strategies can be well planned if source contribution is identified (Mishra et al. 2011). During the last five decades, various attempts have been made by the scientists all over the world in order to model and forecast the air pollution so as to establish a relationship between air quality and human health (Toyib Olaniyan et al. 2017; Zhe Mo et al. 2018; Spears et al. 2019). Questionnaire-based survey was carried out by Nakao et al. (2019) at two healthcare centers in Japan so as to assess the impact of degraded air quality on respiratory diseases. It was found that exposure to larger particles such as SPM has more impact on respiratory health as compared to smaller particles such as $PM_{2.5}$. Similarly, Yacong Bo et al. (2021) studied the effect of dynamic changes in air quality on the incidence of chronic kidney diseases using longitudinal cohort study in Taiwan during 2001–2016. The study revealed that the improvement of $PM_{2.5}$ is associated with the lower risk of chronic kidney diseases. Lai et al. (2016) have conducted a community-based cohort study in Taiwan for the period of 2005–2012 and estimated individual exposure to air pollution and the possible risk of tuberculosis. They concluded that exposure to fine particulate matter along with nitrogen oxides and carbon monoxide was associated with tuberculosis risk. Yu et al. (2018) examined the effect of changes in the air quality on sedentary behavior among college students in Beijing, China, during 2013–2017. Questionnaire was developed to assess the sedentary behavior and the pollution data were collected from Environmental Protection Department of China. It was observed that air pollution has resulted in sedentary behavior, especially more in female students than in male students. Various studies have been carried out by researchers to establish an association between changes in air quality and related health impacts at different metro cities of India such as Delhi, Mumbai, Chennai, and Ahmedabad. Mathew et al. (2015) tested the prevalence of respiratory symptoms with variations in air pollution of Delhi. Study revealed that there exists a positive relationship of PM_{10} with wheezing in the participants. Maji et al. (2017) carried out human health risk assessment due to air pollution in Mumbai using Air Q software and studied the health impacts of criteria air pollutants and found that particulate matter is more responsible for increased rate of mortality and morbidity than gaseous pollutants. Rahul Rajak and Aparajita Chattopadhyay (2020) evaluated the effect of short-term and long-term exposure of ambient air pollution recorded at 59 major locations in India on the respiratory morbidities. Research revealed that short-term exposure has strong association between chronic obstructive pulmonary diseases (COPD), respiratory illnesses, and higher rate of hospital visits whereas long-term exposure increases the risk of asthma, heart attack, cardiovascular mortality and premature mortality. The abovementioned studies identified the particulate matter as a major cause responsible for human health impacts in India and abroad. According to the study carried out by Vibha Gajjar et al. (2021), Ahmedabad City ranks third after Dhanbad and Ghaziabad in India as far as air pollution due to dust and pollen is concerned. It causes increased levels of asthma, bronchitis, cough and cold, especially during winter. In 2003, medical study was carried out by National Institute of Occupational health (NIOH) where they covered 679 families from residential commercial and industrial areas of Ahmedabad so as to assess the morbidity rate due to air pollution. The study concluded that the symptoms like cough and breathlessness were observed more in the commercial area whereas asthma and bronchitis were observed in industrial area. The limitation of the above study was that they could collect the medical data from hospitals, but the correct database regarding disease pattern with respect to variations in air pollution was not available. Studies by Vibha Gajjar et al. (2021) suggest that there is a need to examine a possible relationship between levels of daily air pollution and respiratory hospitalizations and also, it is necessary to compare the data from air quality monitoring sites at various localities. Considering this fact, a study has been presented in this chapter which aims at establishing a relationship between air quality and respiratory morbidities for Ahmedabad which could be used to plan the directed efforts to reduce the outdoor air pollution and its impacts.

Cohort Study

Cohort signifies a group of people who are studied over a period of time. It is an age-old powerful tool to carry out research on human population. It is a type of nonexperimental or observational study design. In this study, the participants are selected based on the exposure status of the individual. Then, they are followed to evaluate the occurrence of the outcome of the interest. Researchers also gather the data pertaining to social, biological, psychological, medical, environmental and genetic factors from the group over a period of time which may range from weeks, months, or years. The data are compared from the follow-up points to the baseline to see how different factors have affected the health of the participants (Aaron Candola 2021).

In cohort study, some of the study participants have exposure (defined as exposed) and others do not have the exposure (defined as unexposed). Over the period of time, some of the exposed individuals will develop the outcome and some of the unexposed individuals will also develop the outcome of interest which is compared in the two groups as shown in Fig. 27.1.

There are two types of cohort studies as follows:

1. Prospective Cohort Study: In this type of cohort study, all the data are collected prospectively. At the first instance, the investigator defines the population that will be included in the cohort. After that, the potential exposure of interest is measured. The participants are then classified as exposed or unexposed by the investigator. The investigator then follows these participants. At baseline and during follow-up, the investigator also collects information on other variables that are important for the study (such as confounding variables) which helps in the assessment of the outcome of interest in these individuals.
2. Retrospective Cohort Study: In this type of cohort study, the data are collected from records. Thus, the outcomes have occurred in the past. Even with the past outcome, the basic study design is essentially the same as prospective cohort study. Hence, the investigator starts with the exposure and other variables at baseline as well as at follow-up and then measures the outcome during the follow-up period.

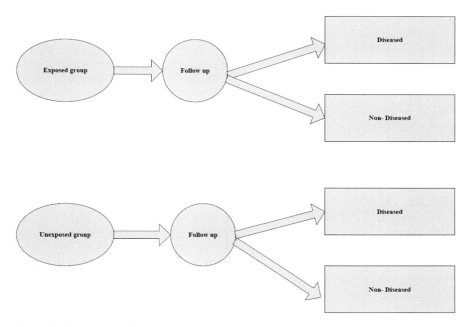

Fig. 27.1 Cohort study design

Prospective and retrospective cohort studies have been widely used to study incidence, causes, and prognosis in the medical field such as epidemiology, psychiatry, and oncology. Similarly, it has also been used to study the association of diseases with contamination/pollution. Niijkeuter et al. (2007) carried out prospective cohort study to understand the natural course of hemodynamically stable pulmonary embolism. Skull et al. (2009) studied community-acquired pneumonia in elderly Australian population. Dai et al. (2016) applied prospective cohort approach to study cancer prevention and control for Chinese subjects. Lamiece Hassan et al. (2011) investigated the prevalence and predictors of psychiatric symptoms among prisoners during early custody. Pasetto et al. (2013) conducted cohort study in Italy to assess the health profile of the residents living in the area of soil and ground water contamination. Yang et al. (2018) studied the effect of ambient air pollution on the risk of still birth using prospective birth cohort study in Wuhan, China.

Material and Methods

In the present case study, temporal and seasonal models have been developed and cohort analysis has been carried out to establish a relationship between air quality and respiratory diseases for Ahmedabad City (23° 2′ 1.9068″ N, 72° 35′ 6.0792″ E) which is one of the extremely polluted and fastest growing smart cities located in the Gujarat State of India. As per the 2011 census report, the city has more than 5 million population which is highest among the other cities of the state and seventh largest urban agglomeration in India (Vibha Gajjar et al. 2021). Exixtence of two thermal coal fired power plants in Ahmedabad city, namely Sabarmati Thermal Fired Station and Vatva Combined Cycle Power Plant adds to the air pollution. Rapid industrialization and urbanization has given rise to the exponential vehicular population growth which has further resulted in increased pollutant concentrations and worsening of the air quality of the study area for the last few years (NRDC 2017). A major study by Urban Emission, a research organization in 2012, found that the major sources for PM_{10} in Ahmedabad are as follows: 30% road dust; 25% power plants; 20% vehicle exhaust; 15% industry; 5% domestic cooking and heating; 2% diesel generator sets; 2% waste burning; and 1% construction activities (NRDC 2017). Central Pollution Control Board (CPCB) of India has also declared SO_2, NO_x, and RSPM (PM_{10}, $PM_{2.5}$) as the criteria air pollutants which decide the air quality (NAQI, CPCB 2020). The maximum permissible limit for daily average concentrations of SO_2, NO_x, and RSPM has been set as 50,40, and 60 μg/cum respectively, for residential urban areas as per the national ambient air quality standards. For the last five years, it has been observed that the upper limits have been crossed for NOx and RSPM with a maximum value recorded as high as 192 and 351 μg/cum respectively, for the study area even though SO_2 is found to be within the limits (NAQI, CPCB 2020).

Cumulative effect of criteria air pollutants is expressed as an air quality index (AQI). It is the yardstick that runs from 0 to 500. Higher the value of AQI, greater the level of air pollution and greater the health concern (World Health Organisation 2020). It is a tool that communicates information on air quality in qualitative terms (e.g., good, satisfactory, and poor) which makes citizens aware of associated health impacts and facilitates greater public participation in air quality improvement efforts. It is employed by cities, states, and countries all around the world to communicate present and future health risks of air pollution to residents. Additionally, the AQI also provides detailed data on how to protect health from air pollution and to guide pollution-reducing policies and regulations. Daily values of AQI as recorded from the pollution control board website indicate that for the last few years (2016–2020) the AQI value has continuously been above 200, indicating the poor air quality and associated health impacts on residents of the study area (Vibha Gajjar et al. 2021). Medical practitioners in the study area have confirmed that the deteriorating air quality had led to an increase in cases of upper respiratory tract infections and have called for initiating surveillance and regular monitoring of cases of respiratory illnesses (Ahmedabad Municipal Corporation 2017). Various initiatives are being taken by Ahmedabad Municipal Corporation to reduce air pollution and its effect on human health such as electric buses and wall to wall carpeting to eliminate dust (Ahmedabad Municipal Corporation 2017). Directed efforts can be concentrated more if they are planned on a scientific basis by studying the relationship between air quality and health. The present study aims at establishing a relationship between air quality and respiratory diseases using cohort analysis.

Study Population

The study area includes two locations in Ahmedabad (Gujarat State India), namely Chandkheda (23.11842 °N 72.586777 °E) and Maninagar (22° 59′ 46.2120″ N 72° 35′ 58.5096″ E). Chandkheda is a residential cum commercial locality of Ahmedabad, whereas Maninagar is emerging as a prime residential area in the southern part of the city which has an immediate vicinity to industrial areas. According to the 2011 census data, the area of Chandkheda is 11 km^2, whereas the area of Maninagar is 3.89 km^2. Population of Chandkheda is 100,000 and that of Maninagar is 126,002. Males constitute to 53% of the population and females 47% for both the study locations.

Air quality index of both the locations often falls in the category of poor and very poor since last few years (Vibha Gajjar et al. 2021). The present retrospective cohort studies were conducted at two hospitals, namely Shraddha Hospital and Suman Hospital in Chandkheda and Maninagar, respectively, in order to investigate the association between ambient air pollution and respiratory morbidities. It was observed from the hospital records of general information/history of the patients that there exists a community of similar socioeconomic status which comprises of information about education, social class, or income (Darin Mattson et al. 2017).

Data Collected

Air Monitoring Data
Air monitoring data, namely SO_2, NOx, $PM_{2.5}$, and also AQI for the period from Jan 1, 2016, to Dec 31, 2017, were collected. The data as available from the pollution control board of Ahmedabad were processed and used for the study (Data obtained at the nearest monitoring station of the current residential addresses of each participant were considered as the subject's environmental exposure (Nakao et al. 2019). For Shraddha Hospital, the nearest air quality monitoring station of Gujarat Pollution Board is LD Engineering College which is located at 13.2 km away from the hospital, whereas for Suman Hospital in Maninagar the nearest station is at Shardaben Hospital 8.2 km away from the Suman Hospital (Figs. 27.2 and 27.3).

Health Data
These data refer to the out-patient and in-patient data from Shraddha and Suman hospitals located in Chandkheda and Maninagar, respectively. The data consist of the information about the number of patients visiting the hospital for problems related to their respiratory system and their time of stay in case of inpatients along with their brief history (Skull et al. 2009; Xiaofeng Wang and Kattan 2020; Aaron Candola 2021).

Fig. 27.2 Shraddha Hospital and nearby air quality monitoring station at L D College of Engineering

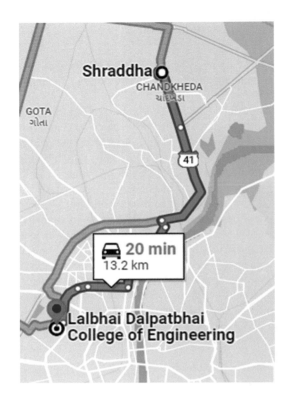

Fig. 27.3 Suman Hospital and the nearby air quality monitoring station at Shardaben Hospital

Cohort Modeling Strategy

For establishing a relationship between any incidence and a disease, cohort analysis is a common practice (FengChao et al. 2020).

In the cohort analysis, the model evaluates the effect of several factors on survival. It also examines how specified factors influence the rate of a particular event happening (e.g., infections and death) at a particular point of time (Mdzinarishvili et al. 2011). The rate is called as the hazard rate. Predictor variables are referred as the covariates, and the Cox model is expressed by the hazard function denoted by $h(t)$ and can be expressed as

$$h(t) = h0(t) \exp(b1x1 + b2x2 + \ldots + bpxp)$$

where t = survival time

$h(t)$ = hazard function determined by the set of p covariates $(x1, x2 \ldots \ldots xp)$

(27.1)

The coefficients $b1, b2 \ldots \ldots$ bn measure the impact of covariates

$$h0 = \text{baseline hazard}$$

the quantities exp (bi) are called as hazard ratios (HR) and interpreted as

HR = 1: No effect or lack of association
HR<1: Reduction in the hazard or smaller risk
HR>1: Increase in the hazard or increased risk

In the present study, the temporal analysis calculates the hazard ratio with covariates as the values of AQI corresponding to the previous week along with the corresponding number of admissions due to respiratory diseases. The seasonal analysis segregates the data pertaining to AQI and number of admission as per different seasons observed in a year so as to arrive at the hazard ratio.

Air monitoring data and hospital data were combined, and retrospective cohort analysis is performed in order to establish a relationship between air quality and respiratory diseases. The air quality changes with the season, and it also varies daily. Considering this fact, the following two types of cohorts were studied as under:

Temporal Cohort In temporal cohorts, the data pertaining to air quality are aggregated as the air quality index was combined with the number of admissions in the hospital due to respiratory diseases. The Cox proportional hazard analysis has been carried out for the aggregated data using SPSS and XLSTAT) so as to determine the value of hazard ratio which is the indicator of an association between air quality and respiratory diseases.

$$\text{HR(temporal models 2018)} = f(\text{Average AQI of the week,}$$
$$\text{No of admissions for that week due to respiratory diseases}) \quad (27.2)$$

$$\text{HR(temporal models 2019)} = f(\text{Average AQI of the week,}$$
$$\text{No of admissions for that week due to respiratory diseases}) \quad (27.3)$$

These temporal models helped us giving the complete picture about the relationship between air quality and respiratory diseases throughout the study period under consideration.

Seasonal Cohort Similarly, the air quality also changes with the season. As per India meteorological department guidelines (Murthy 2009), four seasons are observed in India, namely monsoon [Jun–Aug], post-monsoon [Sept–Nov], winter [Dec–Feb], and summer [Mar–May]. Every season has its own characteristics with respect to wind speed, humidity, and temperature which affect the drift, diffusion, and finally the dispersion of the pollutants (Weber 1982). As a result, the data were divided into four seasons for both the years 2018 and 2019 and hazard ratio was calculated.

$$\text{HR(Winter)} = f(\text{Average AQI of the winter,}$$
$$\text{No of admissions for winter due to respiratory diseases}) \quad (27.4)$$

$$\text{HR(Summer)} = f(\text{Average AQI of the winter,}$$
$$\text{No of admissions for winter due to respiratory diseases}) \quad (27.5)$$

$$\text{HR(Monsoon)} = f(\text{Average AQI of the winter,}$$
$$\text{No of admissions for winter due to respiratory diseases}) \quad (27.6)$$

$$\text{HR(Post monsoon)} = f(\text{Average AQI of the winter,}$$
$$\text{No of admissions for winter due to respiratory diseases}) \quad (27.7)$$

Experimentation

Cohort models were developed based on following assumptions:

- The population of Chandkheda and Maninagar has remained the same throughout the study period (i.e., from 2016 to 2019).
- The ultimate reason for the respiratory problems of the patients is the degraded air quality.
- Patient mortality rate is zero due to respiratory diseases alone during the study period.

- The pollutant exposure period is taken from Jan 1, 2016, to Dec 31, 2017, whereas the exposure assessment is carried out during Jan 1, 2018, to Dec 31, 2019.

A Cox proportional hazard analysis was performed with time-varying exposures on 2-year [annual] time scale accounting for temporal and seasonal variations of criteria air and related health effects in terms of respiratory diseases. The proportional hazard assumption in each Cox model was verified by evaluating the weighted Schoenfeld residuals, and we detected no violations with all p values >0.05.

Two-year follow-up period was considered for determining the association of respiratory diseases with the criteria air pollutants. Age, gender, education and profession and seasonal variations were used as the covariates within the multivariate-adjusted models. Cohort source was used as a stratum which addresses the effect of these parameters on incidence of respiratory diseases among different cohorts.

Hazard ratio with 95% confidence interval associated with increase in annual mean AQI was investigated, and the effects of temporal and seasonal exposures were also estimated.

To estimate the concentration response [CR] functions of criteria such as air pollutant exposure and incidence of respiratory diseases, AQI was fitted as a smooth term (Fengchao et al. 2020). The analysis was performed using IBM SPSS and XLSTAT.

Results and Discussion

We have demonstrated the application of retrospective cohort analysis, which is one of the most sought approaches to establish a relationship between any incidence and the disease in the medical field. Annual and seasonal models are developed and evaluated using hazard ratio as determined by the Cox proportional hazard method. Owing to the numerous advantages of cohort study such as clarity of temporal sequence, calculation of incidence, facilitating the study of rare exposure, examining multiple effects of a single exposure, etc., it has been commonly used to establish a relationship between air quality and health (Aaron Candola 2021). We have used retrospective cohort study with temporal and seasonal approaches to address the potential effect of air quality on respiratory health of the residents of Chandkheda and Maninagar location of Ahmedabad. Modeling options have been presented with this relatively new approach and the models have been assessed using hazard ratio (HR).

Modeling Options

Data quality and consistency decide the accuracy of the cohort studies. In developing countries, availability of continuous and accurate meteorological, pollutant concentration and health data is a challenging task. Some of the reasons can be listed as adverse meteorological conditions, instrumental errors, etc. (Tikhe et al. 2016). At the same time, confidentiality of the human data, hospital policies, etc., are some of the constraints which must be taken into account while modeling the air quality parameters and respiratory morbidities. Similarly, processing time for data collection and dissemination is quite high as a result; it becomes difficult to include meteorology as well as baseline characteristics of the study population as a model input in case of present work. Considering these facts, the models developed should be robust and quick enough to evaluate the effect of exposure to air pollution on respiratory morbidities. An attempt has been made to suggest two cohort options as temporal and seasonal cohort.

Analysis

During the study period (2018–2019), a total 108 respiratory incidents (60 from Shraddha hospital and 48 from Suman Hospital) were observed for the study location. The respiratory diseases were categorized as conservative diseases (patients were treated using conservative methods, i.e., using nonsurgical methods) and serious diseases (patients were given the support of ventilators and BiPAP treatments). Among them, 100 patients were suffering from mild form of respiratory disease and rest 8 were suffering from severe form of a disease due to severe exposure to ambient air pollutants. The gender diversity included 66% males and 34% females. Annual mean value of AQI at the study location was measured as 214 for the year 2018 and 170 for 2019. Similarly, max value of AQI was reported as 351 during post-monsoon season in 2018 and 265 during post-monsoon season in 2019. Summary statistics of the study population for each temporal and seasonal subcohort is presented in Tables 27.1 and 27.2. Similarly, hazard ratios associated with criteria air pollutants [aggregated as AQI] are presented in Table 27.3.

The figures in abovementioned table indicate % distribution of total respiratory incidences in four seasons of the year among males and females. From the data, it is clear that maximum patients were affected by respiratory problems during winter as

Table 27.1 Summary statistics of the study population for 2018

Parameter	Total	Temporal cohort 2018	Seasonal cohorts 2018			
			Summer	Monsoon	Post-monsoon	Winter
1. AQI	57–302	61–265	75–172	57–72	91–302	139–351
2. Population with incidence of respiratory diseases	1.26 Lakhs	1.26 Lakhs	1.26 Lakhs	1.26 Lakhs	1.26 Lakhs	1.26 Lakhs
3. Male %		70%	82%	65%	73%	71%
4. Female %		30%	18%	35%	27%	29%

Table 27.2 Summary statistics of the study population for 2019

Parameter	Total	Temporal cohort 2019	Seasonal cohorts 2019			
			Summer	Monsoon	Post-monsoon	Winter
1. AQI	61–265	61–265	65–128	61–141	130–265	76–166
2. Population with incidence of respiratory diseases	1.262 Lakhs	1.262 Lakhs	1.262 Lakhs	1.262 Lakhs	1.262 Lakhs	1.262 Lakhs
3. Male %		60%	74%	42%	75%	83%
4. Female %		40%	36%	58%	25%	17%

Table 27.3 Multivariate-adjusted hazard ratios (HRs) at 95% confidence interval (CI) of respiratory diseases associated with exposure to criteria air pollutants.

Cohorts	No of events	HR [SPSS]	HR [XLSTAT]
1. Temporal cohort			
2018	69	1.950	2.400
2019	37	1.900	1.946
2. Seasonal cohort 2018			
Summer	20	1.800	2.230
Monsoon	11	2.750	2.334
Post-monsoon	11	1.650	1.771
Winter	27	2.450	3.644
3. Seasonal cohort 2019			
Summer	11	1.350	2.112
Monsoon	12	0.950	1.232
Post-monsoon	8	1.200	1.435
Winter	6	0.900	1.336

compared to other seasons. Similarly, it is also consistently observed that male patients seem to suffer more commonly from most types of respiratory morbidities as compared to females which also matches with the findings of Matthew et al. (2007).

HRs and incidence of respiratory diseases in association with exposure to criteria air pollutants are presented in Table 27.3. The HR estimated for temporal 2018 models is 1.950 and 2.400, whereas for 2019 temporal models it is 1.900 and 1.946 using SPSS and XLSTAT, respectively. Hazard ratio > 1 indicates increase in the hazard or increased risk. The value of hazard ratios 1.9 and 1.946 for temporal model 2019 using both the tools indicates that the group of patients considered in the temporal model has 1.9/1.946 times the chance of getting affected by respiratory morbidities as compared to the healthy population of the area.

Similarly, for the seasonal models, it was observed that HR values for summer, monsoon, post-monsoon, and winter were 1.800/2.230, 2.750/2.334, 1.650/1.435, and 2.45/3.644 for 2018 seasons using SPSS and XLSTAT, respectively. Similarly, the HR values of summer, monsoon, post-monsoon, and winter were 1.350/2.112, 0.950/1.232, 1.200/1.771, and 0.900/1.336 for 2019 estimated using SPSS and XLSTAT, respectively.

The results for both the types of models are presented graphically as follows:

Temporal Models

The plot represents monthly variation of AQI and corresponding values of average hazard ratio during 2018 and 2019. It can be observed that HR values are recorded as > 3 for winter months when higher values of AQIs are recorded for both the years. Hazard ratio is less than 1 during early months of the year, i.e., January, February, and March, for 2019 and January and February in 2018. In contrast to this, higher AQIs are recorded in winter months because the dry and cold weather conditions increase respiratory tract infections and increase virus transmission by weakening the immune system (Audi et al. 2020). It can be further observed from the graph that during monsoon months even if the AQI is low the chances of respiratory infections are more as the damp and the humid weather is perfect for environmental and bacterial growth which is evident from the results (Fig. 27.4).

Seasonal Models

In Ahmedabad City, four major seasons are observed, namely winter, summer, monsoon, and post-monsoon. Winter is observed during January and February. With the prevalence of the inversion condition peculiar to winter air quality, it gets worsened during this time. When a layer of warmer air traps cold air near the ground, temperature inversion takes place. During inversion, the air becomes stagnant and pollutants are caught close to the land, which may result in the formation of smog (Murthy 2009). Smog can irritate eyes, nose, and throat and is also responsible for worsening heart and lung problems and may cause lung cancer with sustained long-term exposure (Kurt et al. 2017). This fact is evident from the results. As observed from the Table 27.1, hazard ratio is found to be maximum in winter with both the

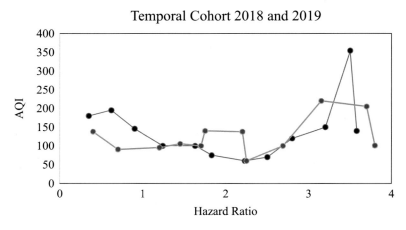

Fig. 27.4 AQI Vs hazard ratio for temporal annual models (2018 and 2019)

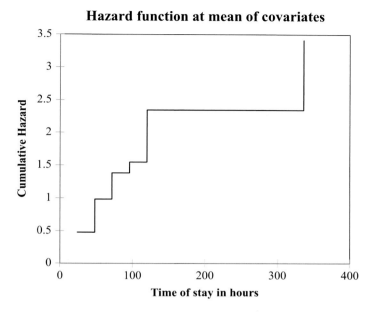

Fig. 27.5 Graph of cumulative hazard vs. time of stay for winter 2018

tools for 2018 when the maximum value of AQI (351) was recorded which has resulted in the increase in number of inpatients as well as their long stretch of stay in both the hospitals. In 2019, the hazard ratio (0.900 & 1.336) has lowered significantly in winter as compared to 2018 (2.450 & 3.500) indicating the stronger impact on nearer timeline.

From the plot (Figs. 27.5 and 27.6) for winter 2018, it can be seen that the maximum time of stay of the patients due to respiratory diseases is 336 h, whereas during winter 2019 the time of stay has been recorded as 72 h. The recorded pattern follows the fact that the degraded air quality has profound effect on immediate exposure.

As observed for summer 2018 (Fig. 27.7), the time of stay of patients due to respiratory illnesses increases throughout the season and the increasing pattern is recorded till 240th hour, whereas for 2019 (Fig. 27.8) the time of stay is more or less constant except spike in cumulative hazard recorded at 96th and 120th hour. The reason for summer respiratory infections lies in the fact that the long-term exposure to substances such as dust and fumes, vapors, and air pollution can irritate lungs and may result in bronchitis in summer resulting in admissions (Kurt et al. 2017).

Monsoon is observed during June to September in Ahmedabad. Rainfall results in washing away of the pollutants, and it acts as a cleansing mechanism (Gujarat Pollution Control Board 2018). Rainfall results in wet ground and makes it difficult for the dust to get mixed into the air again resulting in the lesser pollutant concentration. Ozone formation rate is also lower in monsoon as the rain clouds obstruct the

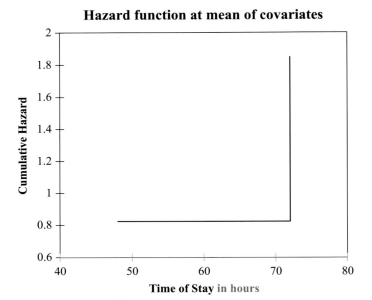

Fig. 27.6 Graph of cumulative hazard vs. time of stay for winter 2019

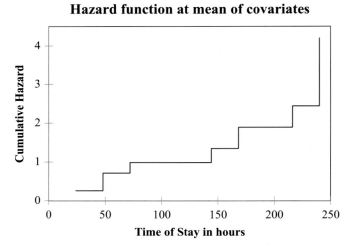

Fig. 27.7 Graph of cumulative hazard vs. time of stay for summer 2018

correct amount of temperature and sunlight required for ozone formation. During monsoon 2018, the respiratory illnesses follow an increasing pattern with a maximum time of stay recorded as high as 240, whereas for 2019 models, the hazard ratio is more or less constant except sudden increase on 48th and 72nd hour.

Fig. 27.8 Graph of cumulative hazard vs. time of stay for summer 2019

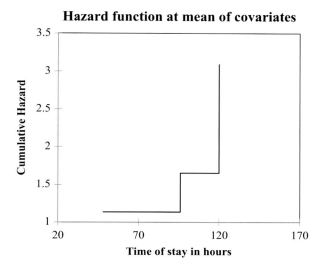

Similarly for post-monsoon models, the maximum time of stay is recorded as 120 and 336 h for 2018 and 2019, respectively. It indicates that the post-monsoon infections are observed more in 2019 as compared to 2018. The reason behind this lies in the fact that the drastic fluctuations in temperature during rainy season make the body susceptible to bacterial and viral attack resulting in cold and flu as observed immediately during post-monsoon season (Audi et al. 2020). October and November are the two months of post-monsoon or retreating monsoon in which the wind becomes weaker. Skies are clear, but humidity in the atmosphere increases. Diwali, which is one of the important festivals in India, is celebrated in this season. Poor air quality is observed in this season due to firecrackers. This condition increases the chance of sudden increase in certain pollutant concentrations resulting in pollution episodes and subsequent increase in the time of stay in the hospitals for both the years.

The above graphs (Figs. 27.5, 27.6, 27.7, 27.8, 27.9, 27.10, 27.11, and 27.12) represent the cumulative hazard as the incidence of respiratory diseases in both the forms, i.e., minor and severe for the period under consideration. The time of stay indicates number of hours the patient has stayed in the hospital due to the incidence of respiratory disease.

Both the tools reported hazard ratio [HR] greater than 1 which indicates close association between respiratory diseases and air quality as well as increased risk of respiratory diseases with degraded air quality. Temporal models indicate that the HR was higher in 2018 as compared to 2019 which further shows that the risk of respiratory diseases increases immediately after exposure to pollutants in 2016 and 2017.

All four seasons have different characteristics, and from the observation table, it can be seen that the seasonal models indicate max HR which is recorded in winter 2018 as the wind movements are lesser. In 2019, maximum HR is recorded for

Fig. 27.9 Graph of cumulative hazard vs. time of stay for monsoon 2018

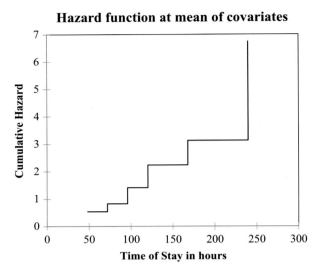

Fig. 27.10 Graph of cumulative hazard vs. time of stay for monsoon 2019

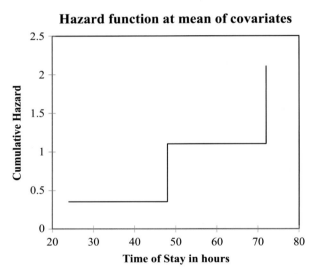

summer which matches with the fact that the rate of bacterial infections is more in summer due to highest spreading rate recorded for both the locations (Audi et al. 2020). Similarly, all the hazard ratio values for the year 2018 are more as compared to 2019, indicating close association between air quality and respiratory diseases especially immediately after exposure.

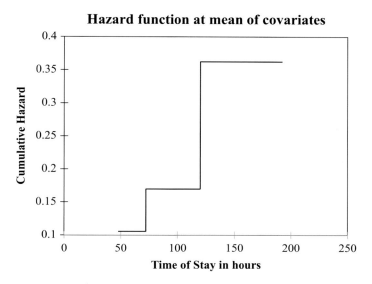

Fig. 27.11 Graph of cumulative hazard vs. time of stay for post-monsoon 2018

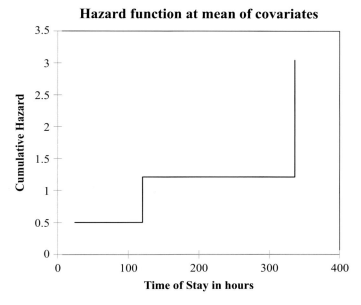

Fig. 27.12 Graph of cumulative hazard vs. time of stay for post-monsoon 2019

Conclusion and Future Directions

The application of retrospective cohort study with follow-up of two years in determining an association between ambient air quality and respiratory morbidities for Chandkheda and Maninagar locations of Ahmedabad that has been carried out through Cox proportional hazard method reported hazard ratio [HR] greater than 1 which indicates association between respiratory diseases and air quality as well as increased risk of respiratory diseases with degraded air quality. The results are also matching with the hospital visits/stay of the study population for the treatment of respiratory morbidities. Temporal and seasonal models analyzed using both the tools indicate that the risk of respiratory diseases increases immediately after exposure to pollutants in 2016 and 2017. Similarly, seasonal model results also match with the physics of the air quality and epidemiology characteristics of respiratory morbidities.

The study demonstrates the positive association between air quality index and respiratory diseases in the population of Chandkheda and Maninagar of Ahmedabad City (Gujarat State of India) and also supports the importance and urgency of improving ambient air quality. This would contribute toward highlighting the need for developing the strategies for the continued reduction of air pollutants from likely sources. Similar to the findings of the researchers namely Zhe Mo et al. 2018 for China, Planiyan et al. 2017 for South Africa, Nakao et al. 2019 for Japan there exists a close association between air pollutants and respiratory diseases. The study finding can be helpful for public authorities to take measures to improve air quality.

The absence of meteorological parameters and detailed demographic characteristics is the major limitation of this study. The present work is carried out only with the help of two-year data and is limited to only two hospitals located in Ahmedabad. Further, the study considers only air pollution as the cause of respiratory diseases and other causes are ignored. Certain assumptions were made while developing forecasting models such as uniform seasonal patterns, simplified meteorological processes, and linearity between the inputs, etc., which is seldom true. This study is an attempt to establish a relationship between respiratory diseases and air quality which can be considered as the benchmarking study, and further, in-depth studies can be carried out. Study can be extended for the other locations in Ahmedabad as well as for greater timeline so as to give a complete idea about the incidence of respiratory diseases with changes in air quality index. Follow-up and recording of the health status of selected cohort population are necessary which can be carried out with other important information such as preexisting ailments, age group, occupation, smoking habits, etc. Some mitigation measures such as taking promisong steps to conserve and protect the ecosystem and taking more green initiatives for Ahmedabad city is required to reduce pollution and to improve health status of the residents. Similarly, mass public transport system can be promoted with adequate efficiency and frequency. At the same time, solid waste management with proper segregation and disposal techniques should be practiced.

References

Aaron C (2021) Cohort studies: what they are, examples and types. Medical news today, article number 281703

Ahmedabad Municipal Corporation (2017) Protecting health from increasing air pollution in Ahmedabad. Ahmedabad

Audi A, Allbrahim KM, Hijazi G, Yassine HM, Zaraket H (2020) Seasonality of respiratory viral infections: will COVID-19 follow suit? Front Public Health:1–8

CPCB (2018–2019) Air quality report

Dai M, Bai Y, Pu H, Cheng N, Li H, He J (2016) Application of cohort study in cancer prevention and control. Zhonghua Liu Xing Bing Xue Za Zhi 37(3):303–3055. Chinese. https://doi.org/10.3760/cma.j.issn.0254-6450.2016.03.001. PMID: 27005524

Darin Mattson A, Fors S, Kareholt I (2017) Different indicators of socioeconomic status and their relative importance as determinants of health in old age. Int J Equity Health 16(173):1–11

Fengchao Liang, Fangchao Liu, Keyong Huang, Xueli Yang, Jianxin Li, Qingyang Xiao, Jichun Chen, Xiaoqing Liu, Jie Cao, Chong Shen, Ling Yu, Fanghong Lu, Xianping Wu, Xigui Wu, Ying Li, Dongsheng Hu, Jianfeng Huang, Yang Liu, Xiangfeng Lu, Dongfeng Gu (2020) Long – term exposure to fine particulate matter and cardiovascular disease in China. J Am Coll Cardiol 75(7):707–717

Gujarat pollution control board (2018) Action plan for control of air pollution in Nonattainment City of Gujarat (Ahmedabad). Gujarat pollution control board, Gandhinagar

Khare M, Nagendra SMS (2007) Artificial neural networks in vehicular pollution modelling. Springer, New York

Kurt OK, Zhang J, Pinkerton KE (2017) Pulmonary health effects of air pollution. Curr Opin Pulm Med 22(2):138–143

Lai TC, Chiang CY, Wu CF, Yang SL, Liu DP, Chan CC, Lin HH (2016) Ambient air pollution and risk of tuberculosis: a cohort study. Occup Environ Med 73(1):56–61. https://doi.org/10.1136/oemed-2015-102995. Epub 2015 Oct 29. PMID: 26514394

Lamiece H, Brimingham L, Harty MA, Jarrett M, Jones P, King C, Lathlean J, Lowthian C, Mills A, Senior J, Thornicroft G, Webb R, Shaw J (2011) Prospective cohort study of mental health during imprisonment. J Psychiatry 198(1):37–42

Maji KJ, Ak D, Chaudhary R (2017) Human health risk assessment due to air pollution in the megacity Mumbai in India. Asian J Atmos Environ 11(2):61–70

Manisalidis I, Stavropoulou E, Stavropoulos A, Bezirtzoglou E (2020) Environmental and health impacts of air pollution: a review. Front Public Health 8:14. https://doi.org/10.3389/fpubh.2020.00014

Mathew J, Goyal R, Taneja KK (2015) Air pollution and respiratory health of school children in industrial, commercial and residential areas of Delhi. Air Qual Atmos Health 8(4):421–427

Matthew EF, Mourtzoukou EG, Vardakas KZ (2007) Sex differences in the incidence and severity of respiratory tract infections. Respir Med 101(9):1845–1863

Mdzinarishvili T, Gleason MX, Kinarsky L, Sherman S (2011) Extension of Cox proportional Hazard Model for estimation of interrelated age period – Cohort effects on cancer survival. Libertas Acad Cancer Inf 10:31–44.

Mishra AS, Tijare DS, Asutkar GM (2011) Design of energy aware air pollution monitoring system using WSN. Int J Adv Eng Technol 1(2):107–116

Murthy PB (2009) Environmental meteorology. I. K. International Publishing House Pvt. Ltd., New Delhi

Nakao M, Yamauchi K, Mitsuma S, Omori H, Ishihara Y (2019) Relationship between perceived health status and ambient air quality parameters in healthy Japanese: a panel study. BMC Public Health 19(620):1–12

National Air Quality Index (2020) National Air Quality Index. Retrieved from Central Pollution Control Board. https://app.cpcbccr.com/AQI_India/

Niijkeuter M, Sohne M, Tick LW (2007) The natural course of hemodynamically stable pulmonary embolism: clinical outcome and risk factors in a large prospective cohort study. Chest 131(2): 517–523

NRDC (2017) Supporting research and analysis for the Ahmedabad air information & response. Protecting Health from increasing air pollution in Ahmedabad, pp 1–44

Pasetto R, Ranzi A, De Togni A, Ferretti S, Pasetti P, Angelini P, Comba P (2013) Cohort study of residents of a district with soil and groundwater industrial waste contamination. Ann Ist Super Sanita 49(4):354–357. https://doi.org/10.4415/ANN_13_04_06. PMID: 24334779

Rajak R, Chattopadhyay A (2020) Short and long term exposure to ambient air pollution and impact on health in India: a systematic review. Int J Environ Health Res 30(6):593–617. https://doi.org/10.1080/09603123.2019.1612042

Rushabh W, Jadhav S, Nema V (2021) The burden of respiratory viruses and their prevalence in different geographical regions of India: 1970–2020. Front Microbiol:1–12

Skull SA, Andrews RM, Byrnes GB (2009) Hospitalized community acquired pneumonia in the elderly: an Australian case – cohort study. Epidemiol Infect 137(2):194–202

Spears D, Dey S, Chowdhury S, Scovrnick N, Vyas S, Apte J (2019) The association of early – life exposure to ambient $PM_{2.5}$ and later – childhood height – for – age in India: an observational study. Environ Health 18(62):1–10

Tikhe S, Khare KC, Londhe SN(2016) Forecasting of air quality parameters using soft computing techniques. PhD thesis. Savitribai Phule Pune University, Pune

Toyib O, Jeebhay M, Röösl M, Naidoo R, Baatjies R, Künzil N, Tsai M, Davey M, de Hoogh K, Berman D, Parker B, Leaner J, Dalviel MA (2017) A prospective cohort study on ambient air pollution and respiratory morbidities including childhood asthma in adolescents from the western Cape Province: study protocol. BMC Public Health 17(712):1–13

Vibha Gajjar, Utpal Sharma M, Shah H (2021) Deliverable to improve air quality using NDVI analysis for Ahmedabad City: addressing the agendas of SDG. Curr Res Environ Sustain 3: 100036. https://doi.org/10.1016/j.crsust.2021.100036. ISSN 2666-0490

Weber E (1982) Air pollution, assessment methodology and modeling. Plenum, New York

World Health Organization (2020) Air pollution. Retrieved from World Health Organization. https://www.who.int/health-topics/air-pollution#tab=tab_1

Xiaofeng Wang, Kattan MW (2020) Cohort studies: design, analysis, and reporting. Chest 158- (1 Supplement):S72–S78. https://doi.org/10.1016/j.chest.2020.03.014. ISSN 0012-3692

Yacong Bo, Jeffery Robert Brook, Changqing Lin, Ly-yun Chang, Cui Guo, Yiqian Zeng, Zengli Yu, Tony Tam, Alexis K, Lau H, Xiang Qian Lao (2021) Reduced Ambinet PM 2.5 was associated with a decreased risk of chronic kidney disease: a longitudinal Cohort study. Environ Sci Technol 55(10):6876–6883

Yang S, Tan Y, Mei H, Wang F, Li NA, Zhao J, Zhang Y, Qian Z, Chang JJ, Syberg KM, Peng A, Mei H, Zhang D, Zhang Y, Xu S, Li Y, Zheng T, Zhang B (2018) Ambient air pollution the risk of stillbirth: a prospective birth cohort study in Wuhan, China. Int J Hyg Environ Health 221: 502–509

Yi Lin, Long Zhao, Haiyan Li, Yu Sun (2018) Air quality forecasting based on cloud model granulation. EURASIP J Wirel Commun Netw 106:1–10

Yu H, Cheng J, Gordon SP, An R, Yu M, Chen X, Yue Q, Qiu J (2018) Impact of air pollution on sedentary behavior: a cohort study of freshmen at a university in Beijing, China. Int J Environ Res Public Health 15:2811. https://doi.org/10.3390/ijerph15122811. PMID: 30544739; PMCID: PMC6313684

Zhang Y, Bocquet M, Mallet V, Seigneur C, Baklanov A (2012) Real-time air quality forecasting, part I: history, techniques, and current status. Atmos Environ 60:632–655

Zhe Mo, Qiuli Fu, Lifang Zhang, Danni Lyu, Guangming Mao, Lizhi Wu, Peiwei Xu, Zhifang Wang, Xuejiao Pan, Zhijian Chen, Xiaofeng Wang, Xiaoming Lou (2018) Acute effects of air pollution on respiratory disease mortalities and outpatients in Southeastern China. Sci Rep 8:3461

Web References

URL: http://safar.tropmet.res.in, Aug 2021
URL: https://cpcb.nic.in, Nov 2021
URL: https://en.wikipedia.org/wiki/Ahmedabad, Oct 2021
URL: https://gpcb.gujarat.gov.in, Nov 2021
URL: https://sphweb.bumc.bu.edu, May 2021
URL: https://timesofindia.indiatimes.com, Oct 2021
URL: https://www.ahmedabad.nic.in, Sept 2021
URL:https://www.epa.gov, Sept 2021
URL: https://www.ibm.com/in-en/products/spss-statistics, Apr 2021
URL: https://www.ncbi.nlm.nih.gov/pmc/articles, June 2021
URL: https://www.nhp.gov.in, July 2021
URL: https://www.who.int, Aug 2021
URL: https://www.xlstat.com/en/download, Apr 2021

Chapter 28
Coping Practices of Women Fisherfolk in Responses to Climate Change at UNESCO Declared World Heritage Site of Sundarbans

Anisa Mitra and **Prabal Barua**

Abstract The sustainable development of any socio-ecological system depends on the complex, mutual interaction of societal, economical, and ecological subsystems. In the ramification of climate change, the livelihood and well-being of individuals become endangered. Sundarbans, spanning across the border of India and Bangladesh, is one of the most vulnerable ecoregions, with diverse habitats of rich aquatic resources. Thus, the book chapter was conducted to assess the livelihood and ecological context of the climate change crisis for the women fisherfolk communities and their coping practices for their future uplifting in terms of the sustainable small-scale fishery. To achieve the goal of the study, the authors conducted a systematic literature review using a 6-step, systematic literature review protocol. Different journals, research papers, and books were reviewed for assessing the climate change impacts and coping practices of the women fisherfolk communities in the UNESCO heritage Sundarbans. A large rural population is dependent on climate-sensitive sectors of fishing activity and capture fishery in Sundarbans. As climate change directs ecosystem transformation, over the last two decades this region's small-scale fishery is confronting challenges of extreme natural events, anthropological, and socio-political conflicts. The influence of climate change is disproportionately high on women fisherfolk as they are more dependent on their livelihood on natural resources.

Keywords Sundarbans · Climate change · Women fisherfolk · Small-scale fisheries · Sustainability

---

A. Mitra
Department of Zoology, Sundarban Hazi Desarat College, South 24 Parganas, West Bengal, India

P. Barua (✉)
Department of Environmental Sciences, Jahangirnagar University, Dhaka, Bangladesh

Introduction

In the world, climate change becomes the most dreaded problem in the new millennium. The current trajectories of global climate change are imposing a serious threat worldwide, influencing the natural and social systems (IPCC 2014). The impacts of climate unpredictability manifested in floods, droughts, inexpedient rainfall, and extreme events. As the societal (human) and ecological (biophysical) subsystems are in mutual interaction, the terms vulnerability, resilience, and adaptive capacity are pertinent within the biophysical realm as well as within the social realm for analyzing the sustainable development of any socio-ecological system or SES (Gallopin 1991, 2006). Vulnerability and adaptation to the adverse impacts of climate change are the crucial environmental concerns of many developing countries. Climate change being a social challenge, jeopardizes social relationships, well-being, livelihoods, and survival of individuals over time in magnitude and intensity. Hence, the spatiotemporal pattern of vulnerability in the exposure of environmental distress should be discussed along social, cultural, economical, and political axes (Masika 2002; Seager 2015; Mason and Rigg 2019).

The developing countries experience colossal challenges because of their dependence on climate-sensitive economic sectors and their limited financial, technological, and human capacities (IPCC 2001). Especially the coastal areas of the developing countries are on the frontline of climate change. The most vulnerable people of this area are the poor living in the low-elevation coastal zone (LECZ) (Barbier 2015). Over 600 million people (around 10% of the world's population) live in this LECZ. Studies reported that nearly 2.4 billion people (about 40% of the world's population) live within 100 km (60 miles) of the coast. Due to the geographical locations, these ecosystems are more vulnerable to short-lived natural disasters with immediate and long-term impacts of climatic changes (Temmerman et al. 2013; Barbier 2015; Spalding et al. 2014). Along with limited assets, rapid and unregulated population growth, the likelihood of their various adaptation options to protect themselves against coastal hazards (sea level rise, storm surges, coastal erosion, and saltwater intrusion), lead to disproportionately high levels of death, social disruption, and economic damage (Atta-ur-Rahman and Khan 2013; Ullah et al. 2017) and eventually, become a significant socio-political context of national and global concern (Kais and Islam 2019). In many developing countries, the rural coastal populations and their livelihoods are supported and protected by coastal and near-shore ecosystems (mangroves, marsh, coral reefs, seagrass beds, and barrier islands). Climate change has threatened livelihoods due to the rapid disappearance of the ecosystem (Barbier 2015). Hence, there is an urgent need to understand the climatic impacts on natural and socioeconomic systems at regional/local levels, especially for the heterogeneous conditions of individual societies where rights and resources are unevenly distributed.

The Sundarbans landscape is also one of the LECZ with highly vulnerable socio-economic and ecological profile. It is the largest, tidally active lower deltaic plain in the world, formed by the rivers- Ganga, Meghna, and Brahmaputra (GBD), spanning over 10,247 km² across the two countries-Bangladesh and India (Danda 2019). This low elevated coastal landform characterized by the interplay between rivers, lands, and oceans that associate natural systems of diverse habitats, shapes an ideal mangrove ecosystem, harbors rich, aquatic resources of fish, shrimp, and edible crab. Therefore, a large rural population in the Sundarbans is dependent on climate-sensitive sectors of fishing activity and capture fishery. A total of 478,770 people are estimated to fish in the Sundarbans including the adjacent Bay of Bengal. Of these, 144,171 are active fisher. During the year of 2018–2019, Bangladesh earned BDT 2.46 crore revenue from the fishing activities of 7325 fishermen who completely depend on fishing in Sundarban, while the economic revenue increased BDT 3.00 crore from 8000 fishermen around the Bangladesh Sundarban area (Barua 2021). Fishing activities in the Sundarbans are primarily undertaken by small-scale fisheries (Islam and Chuenpagdee 2013). Small-scale fisheries contribute at least 25% of the world's fisheries catch, and most of them reside in developing countries, particularly Asia. Fish's contribution to food systems is much higher in developing coastal nations. During the year of 2018, there were 70 million tons of fisheries-related products that were exported worldwide and earned USD 164 billion. In this year, developing nations earned net income-generating revenues of USD 40 billion from the exporting fisheries-based products (FAO 2020). After the 2 years later, during 2020, globally there were 100 million tons of fishing products that were imported and obtained USD 200 billion revenue income and developing nations produced 50% of total exportable fisheries products (FAO 2021). At a global level, increased access to small-scale fisheries could be an essential way to increase people's food security, as a large portion of income in developing nations goes to obtaining food needs. But most of the small-scale fisheries are data-poor – especially in developing countries like India and Bangladesh.

This is found that Small-scale fisheries (SSF) have long been outshining through the apprehensions and professed significance of the industrial segment in the fisheries sectors and policy. At present, concentration to SSF is on the increase, patented through an explosion of different research-based scientific publications, the appearance of innovative worldwide strategies and policy outfits dedicated to the small-scale segment, and concentrated attempts to compute the effects and size of SSF on a worldwide extent. According to the different definitions, this is common that the idiom "small-scale fishery" (SSF) suggests a psychological picture of small, conventional fishing ability that is capable of harvesting fishes with the small mechanized fishing gear that necessitates the labor-intensive approach of fishing techniques. Fishermen characteristically the vital matter of this spiritual prospect, operating crafts independently or in few fishermen colleagues in the harvesting and detection of fisheries products (Chowdhury et al. 2008). The universal depiction of small-scale fishers (SSF) budding from this hard work suggests that SSF not means of word as "small." On the divergent, SSF are greatly outsized compared to the beforehand attention and come into view to have a great effect on individual life, health, poverty reduction, nutrition, occupations, as well as the seafood market structures (Barua and Chakraborty 2011). Rising accounts assert that SSF expected to land almost half the earth's seafood, contributing a significant function in the food safety and nutrition, particularly for the communities who are staying in the poverty circumstances (Bennett et al. 2018).

Being dependent on fishing as the main rural livelihood, the small-scale fisheries became "the occupation of last resort" for "the poorest of the poor" (Béné et al. 2007. The small-scale fisheries (SSFs) of Sundarbans are entrenched within coastal communities that support local traditions and social cohesion (Ruiz-Díaz et al. 2020). It contributes around 60% of the coastal fishery of the whole of eastern India and 80–90% to the total marine fisheries production in Bangladesh (Chatterjee et al. 2009). But the Sundarbans ecoregion is extremely vulnerable to climate change, owing to its typical geomorphology, highly vulnerable to the fluctuations of the sea level, river flow pattern along with the anthropogenic activities (Mitra et al. 2011 leading to acute multidimensional impacts on eco-socio-economic conditions (Neogi et al. 2016). It is projected that the absence of head-on discharge of Bidyadhari River channel, siltation on the river-bed, land subsidence, or the synergistic effects of all the factors could increase the SLR at an average rate of 3.14–3.5 mm per year (maximum centennial-scale Relative Sea Level Rise is estimated to be 0.9 ± 3.3 cm/year) over the next few decades (Chaudhuri and Choudhury 1994; Hazra et al. 2002; Danda 2019). reported that in the Sundarbans region, the sea surface temperature has been rising at the rate of 0.5 °C per decade over the past 30 years (eight times the rate of global warming rate of 0.06 °C per decade) which makes the Sundarbans one of the worst climate change hotspots on the globe.

Climate change-related vulnerability narrates direct paraphernalia such as frequent storms, abrupt rainfall or sea level rises that lead to displacement, and to indirect effects such as lower productivity from fluctuating ecosystems or disturbance to economic systems (Najam et al. 2003) which eventually intensifies engrained social inequality (Rosenzweig and Binswanger 1992; Dercon 2002). For the fishing communities, as the sea and the river are considered as their collective mother, they are integrally connected with seasonal problems and opportunity (Roy et al. 2017), hence more vulnerable to the effect of climate change. As climate change directs ecosystem transformation, over the last two decades the small-scale fisheries of this region are also confronting challenges, particularly vulnerable to extreme natural events, anthropological and socio-political interruptions (Chandra and Sagar 2003; Chavez et al. 2003; Defeo et al. 2013). The severe, recurrent cyclonic storms over the Bay of Bengal have increased by 26% over the last 20 years with most recent cyclones like Amphan, Yaas. Cyclonic storm surges in the coastal area cause annual floods and severely disturb estuarine systems. Embankments (*bund*s) in the Sundarbans act as a defense against cyclonic storms, sea level rise, and annual flooding. But the embankments built to protect the human habitation have been counterproductive as the structural integrity of embankments is very poor as they were built in the nineteenth century (Laha 2019) and the process of flooding promotes the siltation at the river mouths. During the last decade, the rate of coastal erosion in the Indian Sundarbans was about 6 km^2 year^{-1} (Hazra et al. 2002). The SLR in coastal areas, changes in sea surface temperatures (SST), and variance in salinity alter the salinity of estuarine habitats, which stances a retarding effect on mangrove growth (Mitra 2004) resulting in the loss of breeding ground and recruitment of the inhabitant fish species (Hlohowskyjl et al. 1996; Marshall and Elliot 1998; Abowei 2010; Mohammed and Uraguchi 2013).

The Sundarbans is considered as the principal defend that guards the assets and lives in responses to natural disasters frequently. Yet, few thought is waged to avert the obliteration of the mangrove forest, as this time cyclone *Amphan* has spoiled 85% of the forest scenery. This dramatic shifting includes reduce in vegetation of mangroves, agriculture croplands, coastal erosion, rising salinity intrusion in the sediments, and aquatic environment. Approximately 65% of the coastal shorelines of Bangladesh Sundarbans were submerged into the sea and on average, the coastal shoreline enthused 31.70 mm toward the land. The finding illustrates that maximum fractions of the Bangladesh Sundarbans have familiar of disintegration and degradation because of severe category different cyclonic storm since 2000–2020. There are 10,000 fishermen died in the Bangladesh Sundarban due to cyclones and coastal flood over the 20 years (Mishra et al. 2021).

Limitation of the Study

In this book chapter, the authors try to illustrate the overall scenario of climate change vulnerability and existing problem of women fisherfolk communities of Bangladesh and Indian Sundarbans. The authors reviewed the numbers of publication published in different sources and incorporate the major finding in the book chapter. Due to the COVID situation, although authors attempted to field visit and collect information directly from Sundarban communities, but due to lockdown situation all the process become stopped. So, with time limitation, the authors prepared this book chapter based on secondary literature review study.

Methodology

The authors conducted the review-based study on Sundarban areas (Fig. 28.1) to assess the vulnerability of climate change and coping practices by the women fisherfolk communities in responses to the effects of climate change.

The authors have reviewed the different research findings of the climate change impact and coping practices by the coastal communities around the world including the Indian Sundarban region that were published in different journals, book chapters, books, thesis, and newspapers. Then the authors prepared the review-based study in the context of climate change impact and coping practices for the women fisherfolk communities in the study areas and recommended some sustainable measures to resolve the crisis for the UNESCO declared world heritage site of Indian Sundarbans.

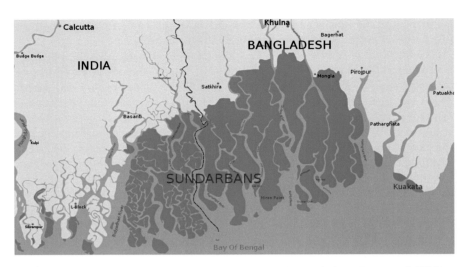

Fig. 28.1 Geographical location of the Bangladesh and Indian Sundarbans (Barua et al. 2020)

Results and Discussion

Changing of Climate Parameters and Sundarban Resources

The changes in ocean dynamics (salinity, pH and dissolved oxygen, primary production, secondary production, nutrient upwelling) lead to changes in fish distribution (through permanent movement or changes in migration patterns), fish abundance along with their compositional variation, phenology (timing of life-cycle events such as growth, reproduction, spawning, recruitment), and ultimately to extinction (Brander 2010; Kizhakudan 2014). The fluctuations of wind pattern and the intensity influence the water currents and upwelling processes, affect the fish movements. The staggered and irregular monsoon (the southwest monsoon remaining active into November) in consort with the abrupt and intense rainstorms, discharges a huge quantity of land pollutants (near industrial areas) into the sea, leading to mass mortality of nearshore fish (Barua 2021; Mukherjee et al. 2022; Salagrama 2012). Climate change in Sundarbans thus leads to reductions in fishing effort, with a decreasing trend in the catch per unit effort (CPUE) (Raha et al. 2012; Pitchaikani et al. 2017) and severely affects postharvesting processes (especially fish drying and processing). Increasing the numbers of mechanized boats, trawlers, tourist ships, seaports, large numbers of fish jetties, and storage centers built on the coastal fringes are prone to recurrent tidal and storm inundation in Sundarbans (Mitra et al. 2012; Barua et al. 2011).

The prawn/shrimp and crab capture fisheries of Sundarbans are also facing a serious problem due to climate change in recent years. The salt-water intrusion in the agricultural land increased the opportunity for prawn and shrimp farming. This alternate livelihood of coastal landless people (around 86%) has resulted in increased fishing pressure on postlarvae. The time series analysis reported that the engagement of the families in this occupation has increased in the last decades (Mondal and Bhaduri 2010). But the sea level rise impacted the distribution of the postlarvae along the riverbanks. With the increased weather temperatures and heatwaves, seed collectors have reduced fishing duration from an average of 6.25 h in 2006 to 6.07 h in 2011 (Ahmed and Troell 2010). The availability of postlarvae after cyclones was found to be considerably reduced for a period lasting from a week to a few weeks. During the rainy season, postlarvae marketing faces problem due to remote access and poor transport facilities. Hence, the average income from postlarvae fishing has declined by 22% over the last 5 years because of reducing catch rate and market price (Jackson et al. 2020). The average price of postlarvae was reported to be US$80 for 1000 in 2018, while it was US$38 in 2006 (Hasan et al. 2018). The catch of prawn postlarvae has gradually declined by an estimated 15% over the last few years. As most of the people in this region rely on fish and fisheries as a source of livelihood and protein, climate change intensifies their socio-economic and food-security problems (Dalton 2001; Allison et al. 2009; Salagrama 2012). Hence, understanding the dimensions of vulnerability, resilience in exposure to eco-socio-economic shocks of this natural resource-dependent community is crucial for planning relevant adaptations (Ruiz-Díaz et al. 2020).

Gender serves as a crucial dimension of both vulnerability and adaptation in response to climate change. Gender inequalities are persistent in the developing world. Systematically, it has different experiences based on the inequalities associated with social, economic, political, and cultural implications for individuals (Gender and Alliance 2016). In the larger nations, men dictate the term on which they participate and contribute to the reduction of increasing environmental problems, while women and smaller and poorer countries have little authority to change or influence. But in the context of changing environment and global economy, building resilience requires gender mainstreaming in access to policymaking (Gender and Alliance 2016).

As uncertainty is inherent to fisheries owing to their dependence on natural resources, the fisherfolk have had to adapt to the whims of weather and climate (Badjeck et al. 2010). But the small-scale fishers of Sundarbans are least adaptive to climate change and least "developed" (Salagrama 2012). One of the important dimensions which strongly influences adaptive capacity following any perturbation is the division of labor between men and women. Scholars across the globe described women as more vulnerable than men because they are poorer, less mobile, and they play a central role in securing household resources. They also have a higher reliance on climate-sensitive natural resources, but have restricted access to resources such as credit, and have a limited role in decision-making. As a result, the influence of climate change is disproportionately high in women and they are more challenged by the loss of lives, food insecurity, and livelihoods.

Women in the coastal region of developing countries belonging to rural or marginalized communities constitute one of the poorest and most disadvantaged groups in society. They are more dependent on natural resources for their livelihood and well-being that are threatened by climate change (Mishra and Bajaj 2020; Dankelman 2002; Khan et al. 2018). Previous researchers mainly focused on the climate change perspectives of small-scale fishing communities in the Indian part of Sundarbans (Salagrama 2012), discussed the coping strategies and risk negotiations of fishers in exposure to different shocks, and enlightened the poverty alleviation policy (Islam and Chuenpagdee 2013; Chacraverti 2014), regionally mapped the livelihoods of Sundarbans fishers (Hasan and Naser 2016) or described the socio-economic dynamics of small-scale fisheries of Sundarbans (Mozumder et al. 2018).

According to the 30% population of total employment in fisheries sector of the worldwide was women and that indicated 5.4 million women are involved in fishing in the world. While statistics from FAO (2020), 40% of the fisherfolk communities are women who are involved in fishing activities in the world at present that represented 8 million women are engaged in fishing in recent years. Dasgupta et al. (2022) mentioned that 1000,000 poor women are engaged in Indian Sundarban for collecting of wild prawn postlarvae and fishing in small-scale from rivers and creeks. They assess that health problems found noticeably greater for women in the Indian Sundarban due to always staying with high saline exposure from wild prawn postlarvae, crab larvae, and small-scale fishing and their drinking water is from the salinity-prone water sources. Sultana et al. (2011) found that only 5% of the total female communities of Bangladesh are engaged in fishing activities as a primary profession and their contribution in fisheries includes economic and social errands on around the family. Different research findings of Bangladesh Sundarbans indicated that 45% of women are playing role to fisheries-related activities as the alternative livelihood professions (Shelly and Costa 2020). Roy and Bhowmick found that 85% of the women were completely engaged in prawn collection, 49% in drying of fish, and 25% are involving fish-vending activities.

But there is no report on the status quo of the women fisherfolk of Sundarbans small-scale fisheries and how they are coping in the changing climate scenario. This is the first consolidated report to address the role of small-scale women fishers in Sundarbans and their resilience or adaptation in the shadow of climate change. The current review would be useful to understand the present situation of women fisherfolk in the eco-socio-ecological context of the climate change crisis and to implement adequate guidelines for their future upliftment in terms of the sustainable small-scale fishery.

Women of Sundarbans Small-Scale Fisheries- Valuing the Invisible

The gender differences are prominent in Sundarbans small-scale fisheries like other developing countries because of the patriarchal, patrilineal society (Roy and Bhaumik 2013; Kleiber et al. 2014; Mozumder 2018) (Fig. 28.3). Though the men and the women are engaged in overlapping activities like fishing, net mending, fish processing, and trading, the men are often considered as key providers as they target "cash crops" while women are "hunters gatherers", providing stability in the diet. Men dominate offshore and long-distance fishing activities, which are more hazardous but allied with higher-value fish species. But the women of Sundarbans have less control over profitable fishing activities. Fishing by women is only restricted to near the shore or from the small water channels mainly during the rainy season using simple fishing gears and complements the men's fishing activities, mainly for consumption purposes and as well as for sale (Fig. 28.2) like in the other parts of the world (Chapman 1987; Béné et al. 2009; Gereva and Vuki 2010; Roy and

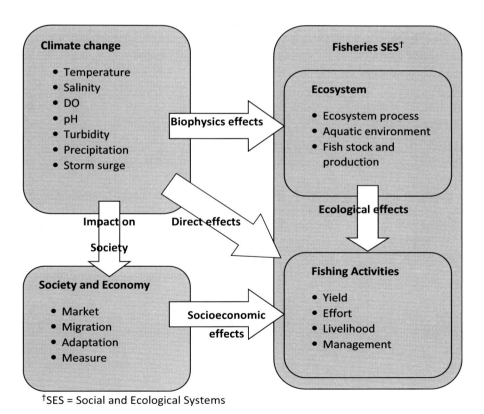

†SES = Social and Ecological Systems

Fig. 28.2 Impact of climate change on fishing communities (Barua et al. 2022)

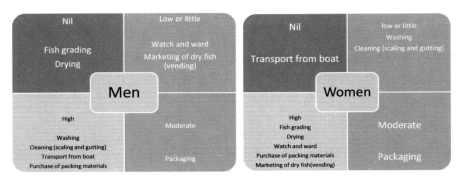

Fig. 28.3 Gender participation matrix of Sundarbans SSFs men and women fisher. (Adapted from Roy et al. 2017)

Bhaumik 2013). In the Indian part of Sundarbans, in the districts of North & South 24-Parganas, around 50,000 people are involved in fish seed collection, among them, around 75% are women (Roy et al. 2017) The analysis of the age structure of fisherwomen revealed that fish seed collection is done by the age group between 10 and 20 years, which is more than 30% of the total sample studied (Chatterjee et al. 2009). Fishing is a key source of livelihood for the communities especially those living in the fringe area of the Sundarbans Reserve Forests (Rajagopalan 2009). Around 2069 km^2 inside the Reserve Forest is considered ideal for riverine fishing using traditional methods (Mukherjee 2007). Despite rich resources prevailing in this area, the socio-economic condition of the rural women is very poor. A regional study by Sarkar in Satkhira district adjacent to Sundarbans West Reserve Forest, Bangladesh, reported that women fishers have a higher level of illiteracy, semi literacy (47%) than their male counter-part (7%). The women fishers often face sexual abuse, and the marriage prospects are stigmatized, and they become a family obligation due to the practice of dowry (BCAS 200). For the access of gear, boat, and money (even daily expenses), most of the artisans are dependent on the fish dealers and traders (Arotdars), middlemen, local private credit providers, and money lenders (*Dadondars*) at overpriced charges, who control the fish markets by fixing the prices. The market-based surveys have revealed that the fishers are verbally and physically abused when failing to supply enough and they hardly receive an appropriate price for their collections (Hoq 2007; Chacraverti 2014) (Fig. 28.3).

Invertebrate gleaning is a popular fishing method to collect invertebrates in many intertidal coastal areas of the mangrove ecosystem. But it is mostly the overlooked category of small-scale fisheries or considered as "forgotten fisheries" (Nessa and Rappe 2019) due to its data poor nature. Though the invertebrate gleaning has high social value, for women, this is often considered a group activity (Rabbitt et al. 2019).

The Invertebrate Gleaners

The Fry Collectors or Meendhar

They are considered as "the lowest of the low amongst the most penurious and marginalized people in coastal communities of Sundarbans. The women and the girls use handi, small fixed magnet, meen jal or even their hands to collect the *bagda* (tiger prawn or shrimp) seeds from the local creek or river (BOBP 1993). However, with the establishment of Sundarbans tiger reserve in 1973, many women lost their sources of livelihood, as they are barred from collecting the prawn seeds in the creeks near critical tiger habitat, reserve forests, and near the shores of inhabited islands (Patel and Rajagopalan 2009; Das et al. 2016). Estuarine and riverside shrimp fry collection engage an estimated 423,000 workforce including 40% of women and children in the last decade. Yearly earnings from shrimp fry collection are estimated as USD 375, which is 16% higher than the yearly wage of day labor (Islam et al. 2011) of this region. The capture of 20–40 fry or postlarvae helps to earn between US$0.45 and US$1.10 per day. However, female collectors earn half that of the male (around US$60 compared to around US$110 per year) and sometimes younger than their male counterparts (10.6 and 12.3 years, respectively) (Gammage et al. 2006).

The Crab Collectors and Fatteners

The mud crab (*Scylla serrata*) collection, fattening, and trading emerged as a profitable alternative for the poor coastal inhabitants due to natural disasters, disease outbreak, and a ban on shrimp export in the international market. Though, males dominate this sector, but gender affinity is slowly shifting toward gender equality. Nowadays, women are involved in planning, decision-making process in managing and controlling the sector in Bangladesh part of Sundarbans (Mozumder et al. 2018). But the lack of seed availability, exploitation by the middleman, nonexistence of proper transport facility, proper market information, insurance, and training facilities limit them to provide a better return on investment. As the women have a better capacity to endeavor long hours of work, the gender in mud crab feeding and fattening is also undergoing significant transformation (Hamid and Alauddin 1998; Sahu 2015. The most marginalized segment of the coastal population especially landless, widow, orphan, and girl child is reported to be involved in mud crab collection. The highest percentage (40%) of mud crab collectors belonged to the year class of 31–40 with a second major (24%) age group of 54–65 years (Molla et al. 2009). Though, in general the living standard of crab collectors is inferior to the fatteners and traders, depending on the season, availability, and mode of business, their income level fluctuates (Molla et al. 2009). A study conducted in the villages of Gosaba block of Indian Sundarbans revealed that still most the women are involved in feed application, rather than in marketing, due to social and religious norms and for the pressure of household works. They are either directly involved in mud crab fattening or they assist their husband or other family members in feeding, stocking, and other pre- and poststocking management (Sahu 2015). In the midst, the pre-existing village hierarchies, recent commotions in the global crab supply chain, and as a part of a cycle of intertwined moral indictments and environmental policymaking, the mud crab collectors have the lowest share of the economic gain but have received the most indictments for harvesting unsustainably.

The Fish Dryer

The processing of fish and invertebrates into marketable, tradable export products is labor-intensive, and most of the cheap labor in postharvest activities is provided by women (Harper et al. 2020b). During the lean period (mid-October to end of January or mid-February), women's secret savings from the fish processing/drying industry provide a form of insurance for their families (Bhaumik et al. 1996). Although technologically advanced fish processing using ice plants and transport systems threatens the women's realm of small-scale fish processing and trade (FAO and MRC 2003). Hence, sometimes women along with their children migrate to other parts to work as domestic help especially during the lean period (Das et al. 2016). The age groups of 22–54 years are mainly engaged in fish drying-related activities (65%). Below 21 years age group is regarded as the occasional working age group of fish drying activities (Molla et al. 2009). A case study based on the fish dry workers in Frezarganj areas of Sundarbans, India revealed that the dry-fish women workers belong to backward poor communities (44% schedule caste and 27% scheduled tribes) with a monthly average income of Rs. 3000/– per month during the fish drying season. About 87% of them are illiterate. They also have very low social participation (around 68%) (Roy et al. 2017). It has been observed that women are engaged in invertebrate gleaning and fish drying/processing because they have no other viable alternative.

Coping in the Shadow of Changing Climate
The impacts of the climate change or disaster crisis on women are manifested in their further marginalization and underestimation from fisheries exercises. Women and girls are described in the literature as more vulnerable to any kind of eco-sociological shocks, but they act as "shock absorbers" (Quisumbing et al. 2008; Islam and Chuenpagdee 2013). Owing to these shocks, they supplement household income through multiple professions, limit spending, and experience a multitude of disadvantages including physical hard work to secure food and livelihood; less control over income and assets; reducing their food consumption to leave more for other family members; exposed to violence and intimidation. Subordinate social status is less represented in politics and decision-making (Mendoza 2009; Roy et al. 2017). And ultimately malnutrition, adverse health, and lack of education lead to the intergenerational transmission of poverty, sustained health effects, stunting, and cognitive decline (Dercon 2006). With the limited opportunity in terms of income

generation and wage discrimination, female heads of households have the lowest income level in the local community. Their problems are exacerbated by extreme poverty, poor housing, health and sanitation, and limited access to safe drinking water, support services, resource opportunities, low literacy rates, and social exclusion. The aftermath of natural disasters results in further social ostracism including food and water insecurity, limited access to relief material, unemployment, and asset depletion. Even they suffer from several health issues including respiratory and other physical diseases, mental distress, and emotional health problems (FAO and ILO 2017).

As a result of climate change, the availability of resources along the riverbanks and water channels has been decreased, whereas there is increased fishing pressure. Consequently, this leads to the lower average income in the women fisherfolk. Hence, to meet the demand and increase the profit, the working time of the women and girls has been extended in recent years which reduces their engagement in the household activity. As a result, sometimes they are subjected to domestic violence and extortion. To collect the fish or prawn seeds, women and girls need to spend five to 6 h per day in the saline water which makes them vulnerable to skin diseases and gynecological problems including various reproductive tract infections. While catching spawns, they become easy prey to crocodiles and sharks (Kanjilal et al. 2010; Islam and Chuenpagdee 2013). However, it has been reported by Ahmed and Troell (2010) that, with the increased weather temperatures and heatwave, seed collectors have reduced fishing duration from an average of 6.25 h in 2006 to 6.07 h in 2011.

Climate change distresses the women fishers due to decline access to (particularly higher-valued) fish, higher competition at centralized locations from economically powerful traders and exporters to fish. The centralization of fish landings requires boats and traders. But for the women, it takes extra costs and time to travel long distances for harboring (Salagrama 2012) and this problem is aggravated after the natural disaster. Hence, women artisans suffer from unemployment and underemployment, leading to a lack of nourishment for their families, force migration, exploitative employment practices, and ultimately the trafficking (Salagrama 2012). Since women are stewards of household food, water, and fuel security, they have substantial potential to increase household adaptive capacity, through diversified livelihood strategies. The rapid physical changes of islands, loss of land due to land reforms, and land distribution by the government increase the pressure on land, distressing the small-scale fishers of Sundarbans necessitating migration. Sometimes the fishers are forced to retire from fishing prematurely due to climate change. Additionally, men being the main decision-maker of the family upon their migration, women in the community take over diversified livelihood profiles to cope with the challenges. They become involved in nontraditional activities like net mending, horticulture, poultry, husbandry, petty trade, and as industrial workers, domestic help or run minor businesses in the neighboring towns (Béné and Friend 2011; Salagrama 2012). Even in areas dominated by male migration, women have greater responsibility for fishing.

Fishing, in the tide-ruled water of Sundarbans, is a perilous profession than agriculture or any other land-based occupation due to the eco-vulnerability of this island. To get the most lucrative fish catches in the forest fringes or the region of nonpermissible reserve forest, women fishers are often attacked by the tigers, sharks (locally called *kamote*), and crocodiles which are not even reported (Das 2009; Mistri and Das 2015; Datta 2018). In the ramification of climate change, due to habitat loss and dwindling prey population, this human-animal conflict has become evident in recent years. Many of the survivors of tiger and crocodile attacks in addition to physical disability and disfigurement are leading to post-traumatic stress disorder (PTSD) (Chowdhury and Jadhav 2012; Chowdhury et al. 2016).

Future Perspectives

SSFs are complex systems where social and ecological systems are mutually reliant (Fidel et al. 2014). Ensuring the diversity and resilience of fisheries through adaptive policies while increasing the resilience of fisherfolk by supporting existing livelihood strategies is a top priority in a changing climate. This will bridge the gap between social vulnerability and ecological health (Brugère et al. 2008; Quiros et al. 2018; Ruiz-Díaz et al. 2020). However, the major challenge in managing the small-scale fisheries (SSFs) is inadequate knowledge regarding the role of women. This ultimately underestimates the values of small-scale fisheries as less attention is received for government support or management as industrial sectors (Jacquet and Pauly 2008; Schuhbauer et al. 2017). Thus, for successful planning and policy implementation of small-scale fisheries, it is imperative to understand the gender differences in access and control of resources, and potential benefits derived from them (Musinguzi et al. 2017).

Climate change adaptations for sustainable livelihood of the fisherfolk can be categorized as ecosystem-based approaches, built-environment solutions, and institutional-based or policy-based solutions. To provide optimal ecosystem services, an ecosystem-based approach supports the management, conservation, and restoration of fish habitats, diversifying fishing methods, and fish species. Whereas the built-environment adaptations include the design of coastal infrastructure to cope with rising sea levels, frequent cyclone, and flood. Where institutional or policy-based, practices and policies should include community-based climate-driven responses to maintain the sustainability of fisheries within established social governance and economic systems (Moosavi 2017; Mutombo and Ölçer 2017; Lam et al. 2020). In this section, probable recommendations have been discussed based on these approaches to uplift the condition of the women fishers in the shadow of climate change (Fig. 28.4).

28 Coping Practices of Women Fisherfolk in Responses to Climate Change... 715

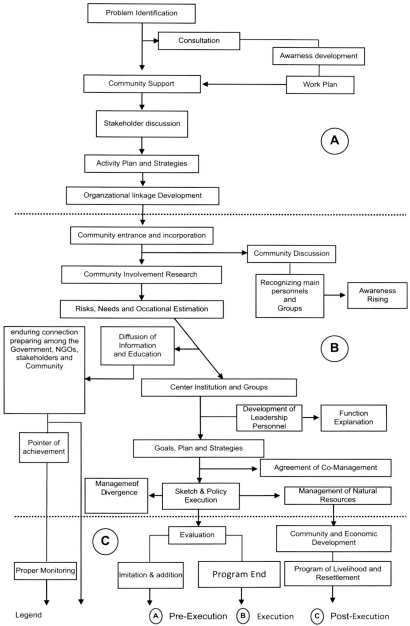

Fig. 28.4 Framework for women fisherfolk communities co-management for Sundarbans

Ecosystem-Based Approach

In this framework, a more holistic approach is considered that impacts fishing at the community and ecosystem level (Herrón et al. 2019). As the sustainability of SSF-based livelihoods is contested by different factors and processes, the policies that reduce fishing effort, rebuild the overexploited or depleted fish stocks along with the coastal development, reduce destructive fishing practices and pollution, in the context of the data, it can help maintain the reproductive potential of fish stocks and mitigate climate-induced declines in potential fishery production, while making livelihoods more resilient. Empirical or model-derived studies correlating several ecological factors with the dynamics of fish population and how they impact the fishing practices and the livelihood (Troadec 2000; Cochrane et al. 2011; Lam et al. 2020; Stacey et al. 2021) of the women fisherfolk are important for successful management in this rapidly changing climate condition. Site-specific mangrove conservation can be encouraged in the coastal areas to create natural barriers against sea level rise and extreme events. Strategic Integrated Coastal Zone Management (ICZM) plan to adapt to natural hazards is important for the protection of coastal communities and the environment (Pitchaikani 2020) by encouraging the construction or relocation of housing areas that are at risk of flood. This will reduce the susceptibility of the fisherfolk (Mitra 2019; Lam et al. 2020). Carbon credits for mangrove conservation could be advantageous for diversifying fisherfolk, livelihoods, and promoting restoration and conservation of ecosystems (Troadec 2000). Spatial and temporal periodic-harvested closures build ecological resilience and reduce social and ecological vulnerability to climate change. But often it fails to consider the impact of closures on women's access to resources and income. As fishing has traditionally been considered a male activity, there is a lack of voluntary reporting of regional or national data of sex-disaggregated official catch statistics for nearshore fisheries acknowledging the women fisherfolk (Bennett 2005; UN 2016; Harper et al. 2020a).

But these invisible catch statistics via household and community fishing surveys are very important for a comprehensive understanding of resource use patterns and ecosystem-based small-scale fisheries management methods (De la Torre-Castro et al. 2017; Gee and Bacher 2017; Frangoudes et al. 2019; Lam et al. 2020), especially in the context of changing climate.

Community-based management of experimentation and learning prioritizes local and indigenous knowledge to maintain fish stocks, fisheries yield, and catch efficiency (Carvalho et al. 2019) as adaptative solutions in the context of uncertainty. For conservation initiatives and promotions with a higher level of environmental awareness, the women fisherfolk community can play a big role in ecosystem conservation as they allocate more time and support (Gupta et al. 2016; Koralagama et al. 2017). Regional or local studies emphasizing the indigenous knowledge of women fisherfolk throughout the entire value chain, highlighting the stochasticity of the environment are thus necessary for sustainable management of Sundarbans SSFs.

Built Environment Approach

The built environment is considered as a set of five capital assets (natural, physical, human, financial, and social capital) with its access and utilization through different adaptation strategies to build resilience (Bebbington 1999; Daw et al. 2009) in the emergent threat. The quality of a built environment's capability is often designated as the capability of built-in resilience of a community (Hassler and Kohler 2014). Hence, the mutuality of the natural and the built environments should be acknowledged while policy drafting to conserve ecosystem services and uplift the women fisher's livelihood (Folke 2002). The livelihood framework approaches can help to understand the differential resilience of women fisherfolk to cope with crises (Allison and Ellis 2001). The sustainable fishery-dependent livelihoods include the occupational multiplicity of the women fisherfolk (several income-generating activities), the occupational mobility and diversification outside fisheries (entering or exiting the fishery sector), geographical mobility (migration), diversification with multispecies, multigears approach, and new harvesting tools and techniques (Allison et al. 2007; Brugère et al. 2008; McClanahan et al. 2008). Apart from the traditional and indigenous knowledge for developing adaptation strategies (Berkes and Jolly 2002; Vasquez-Leon 2002), technological break-through and predictive modeling for forecasting disaster or setting up Automatic Weather Stations (AWSs) can be a vital approach to predict climate variability and its impacts on the community (Pitchaikani 2020). Expanding access to climate information and forecasting through early warning systems will reduce vulnerability (Wooster and Zhang 2004). It has been observed that social cohesion and networks increase cooperation and can boost the adaptive capacity of fisherfolks' households (Perry et al. 2009). To sustain catches on mobile or fluctuating fish stocks in the changing climate scenario, geographical mobility is important to increase the span of economic benefits (Allison and Ellis 2001). But in the case of the women fisherfolk, as they play a significant role in their household, it becomes difficult or nearly impossible for them. In this context, an alternate livelihood can help them.

There is an urgent need to replace and repair breached earthen embankments of the eighteenth century to protect the fisher's village from sea level rise and flood. The proper road for transportation and cyclone-resistant building architecture is also necessary. Education and technology may enable women fishers to widen their choices, from participating in safe construction practices to potential risk assessments to reduce fatalities (Toya and Skidmore 2007). The additional investment women fishers need to adapt to climate change can bring direct and indirect benefits in the short- and long-term, providing a return on investment and a "win-win" situation (Srinivasan et al. 2010).

Institutional or Policy-Based Approach

To strike a balance between the sustainability of fisheries resources and the resilience of fishing communities, fisheries policy and governance must be more human-centric to recognize the interdependent nature of social and ecological systems (Berkes 2015; Bennett et al. 2019). Globally, there is a growing recognition of women's role in the feminization of fisheries (Kusakabe 2002). But in the case of Sundarbans SSFs, like in the other developing countries, the increased recognition of the multiple roles played by women and their contributions exists in stark contrast with the limited participation throughout the value chain (Alonso-Población and Siar 2018). In Sundarbans, where men and women target different species, use different gears, and fish in different habitats (Kleiber et al. 2015), a gender lens is necessary to assess the power relations and trade-offs of various management strategies. It has been evident from the literature that a comprehensive assessment of the women's contribution to fisheries-related economies must cover the entire length of the catch-to-consumption pathway (FAO 2013, 2017; Harper et al. 2020a). The regional or local level of case studies should be conducted to access information and knowledge that women have regarding the entire value chain (Kusakabe 2002). The public policies and civil society should support women's right to earn a fair profit from their work and recognize their contribution along the supply chain to promote more educated participation in planning and decision-making that impact their livelihood (Huysentruyt 2014; FAO 2017, 2020; Mishra and Bajaj 2020; Solano et al. 2021).

So far, the gender norms, beliefs, and practices are intertwined into the political framework of industrial environments and associated with the chief actors (Mangubhai and Lawless 2021). But the inclusion of women in decision-making can increase awareness of community wellbeing and economic development; increase access to fisheries management that is more equal and transparent; and promote for sustainability (Lentisco and Lee 2015; FAO 2017).

Low levels of technical development, combined with social, economic, and gender disparities, increase the vulnerability of primarily illiterate, unskilled, and resource-poor rural populations to current climatic concerns (Kesavan and Swaminathan 2006). Though the technological breakthroughs may increase catch and their independence that may increase the workload of women's income. This must be comprehended that women being a heterogeneous category, have different experiences due to different ages, ethnicity, and class, engaged in different activities about markets, tools, and techniques of catching fish. Even in families, both complementary roles and conflicts of interest are present between different members. Traditional divisions of labor and the household duties including the reproductive duties are still largely considered as the responsibility of women (FAO 2020) Therefore, expanding responsibility in fisheries, put new demands on their roles, effort, and time which make them less adaptable (Elson 1992; Kusakabe 2002).

With the attrition or declining access to common-pool resources of mangroves and grazing lands, it is becoming increasingly difficult to generate additional income sources (Badjeck et al. 2010). To reduce risk and cope with uncertainty, diversification through occupational multiplicity ensures a buffer for the marginalized communities and target-specific aid to women. The government and fishery managers should encourage small-scale women fishers as "professionals" to prohibit part-time fishing (Allison and Ellis 2001; FAO 2017). The Governmental and nongovernmental organizations by providing technical and financial help have started organizing women's Self-Help Groups (SHGs) or Mahila Mandal focusing on credit and savings to increase women's individual and collective self-respect and their visibility in the community by increasing their skill and knowledge (FAO 2020). In Bangladesh, Department of Fishing, Social welfare department, and district women welfare office provided different social security schemes for women of coastal areas like Sundarban fisherfolk communities like widow and deserted women allowance, vulnerable group feeding, old age allowance, maternity allowance, work for cash and work for food allowance, fisherfolk allowance during fishing band period, loan for income-generating activities for women and others, different facilities for economic empowerment, and livelihood for vulnerable women (Barua et al. 2020). The promotion of ornamental fish farming by establishing women Fisher cooperatives, women SHGs, and various ancillary activities can support women fishers (Mukherjee 2018). Vocational training can be offered for the other developing sectors like dairy, beekeeping, mushroom cultivation, poultry, livestock management, bio-diversity maintenance, wasteland development, pond management, nursery management, integrating farming systems, rural crafts, and entrepreneurial development for small trading enterprises (Roy et al. 2017; Chowdhury et al. 2011).

Social security protection should also be provided to elderly working women, widows, and abandoned of this sector to support their livelihood and to empower them in household activities with alternate livelihood options (FAO 2017). To rescue and rehabilitate girl children as labor, close community participation and implementation of child labor laws in fisheries should be fully aligned with international instruments. Rehabilitation and school education should be provided to the girls and women rescued from abusive work and trafficking in fisheries and establish systems to help reunite with their families and communities (FAO et al. 2010). As the facilitator of community cohesion, women can better communicate, resolve conflicts, collaborate, and unite with intercommunity and intracommunity communities/institutions. The initiative can be taken through local administration to support women in claiming their social and economic rights involving awareness and motivation, counseling, networking, and income generation (Chatterjee et al. 2009; Kilpatrick et al. 2015).

As Sundarbans is spanning across the border of two countries, overcoming the transboundary political and geographical delimitations for resource management is of utmost importance in the context of increased climate variability (Herbert 1995; Badjeck et al. 2010). The conception of a cross-country network of women's fishing organizations will aid in the exchange of information and support measures to ensure the safety of migrant or detained fishermen in other countries (FAO 2017).

Studies have revealed that under this constant threat of environmental stochasticity and overexploitation of the aquatic resources transforming the subsistence economy of the Sundarbans region into a remittance economy and this is important in rural areas. This remittance, financial investment in the fisheries sector can regulate the capitalization of fisheries. Supporting fiscal transfer mechanisms and flexible lending based on existing regional financial plans can provide adequate capital investment for fishery development and livelihoods (Allison and Ellis 2001; Hoq et al. 2003; Chatterjee et al. 2009; Dutta et al. 2017; Laha 2019). The women fisher (especially from indigenous groups or vulnerable and marginalized groups) should be provided with basic services (health, education, literacy, digital inclusion, and access to microcredit), especially after the disaster. In this context, private or public insurance systems can rebuild their asset base after climate disruption (Lam et al. 2020). The authors proposed three types of initiative like ecosystem-based approach, environmental approach, and institutional approach for the upliftment of Sundarbans small-scale women fisherfolk's livelihood in the shadow of climate change (Table 28.1).

The authors proposed community-based fisheries co-management initiative (Fig. 28.4) for durable solution of crisis for women fisherfolk in Bangladesh and Indian Sundarbans where people will be key actors, local community dependent, local and natural resource dependent, and based on partnership approach. Further, community-based fisheries co-management approach in Sundarbans ecosystem areas has obtained raising acknowledgment among the governments, local government institutions, nongovernment organizations, academicians, scientists, and respective researchers as a significant matter of the future integrated sustainable management system. But effective women-dependent fisheries co-management and effective process of partnerships could only happen during time of women empowerment. The execution of Sundarbans small-scale fisheries co-management could be observed as combination of three parts: pre-execution, execution, and postexecution moments (Fig. 28.4).

Conclusion

To date, limited studies have been done, focusing only on the regional socio-economic status of women and their fishing in the India and Bangladesh Sundarbans area. However, there are gaps in studying gender relations or how gender relations in families and communities affect small-scale fisheries-related activities in climate change scenarios. This chapter discusses the gender perspective of Sundarbans SSFs, emphasizing the status quo of women in the fishing community, including their ability to adapt in the shadow of climate change. However, more in-depth scientific research on climate change is needed, considering the vulnerabilities of different temporal and spatial situations related to the small-scale fishermen and women communities of Sundarbans to reveal the holistic view. There are few studies on women's role in community fisheries management, their access to resources, and

Table 28.1 Recommendation for the upliftment of Sundarbans small-scale women fisherfolk's livelihood in the shadow of climate change

Ecosystem-based approach	Built environment approach	Institutional or policy-based approach
Improving fisheries management and governance in accordance with climate change	Access and utilization of five capital assets (natural, physical, human, financial, and social capital) through different adaptation strategies and policies to built-in resilience	Comprehensive assessment of the contribution by women to fisheries-related economies including the entire length of the catch-to-consumption pathway
Policies to reduce fishing effort, rebuild the overexploited or depleted fish stocks along with the coastal development, reducing destructive fishing practices and pollution, maintaining the reproductive potential of fish stocks	Livelihood framework approaches to understand the differential resilience of fisherwomen folk to cope with crises	Increase women's participation in decision-making in the community and other organizations
Empirical or model-derived studies correlating ecological factors with the dynamics of fish population and how they impact the fishing practices and the livelihood	Diverse fisheries livelihood systems including occupational multiplicity, occupational mobility and diversification, geographical mobility (migration), diversification with multispecies, multigears approach, and with new harvesting tools and techniques	Monitor the changes in women's control over resources and their position in the household
Site-specific mangrove conservation and strategic integrated coastal zone management (ICZM) Using carbon credits for mangrove conservation	Technological breakthrough and predictive modeling for forecasting disaster or setting up automatic weather stations (AWSs)	Access to fishing areas and control over decision-making in the use and management of the fishing area, over-fishing gear, technology and extension services, microcredit, household income from fishing, and household expenditure
Spatial and temporal periodic harvested closures	Social cohesion and networks increase cooperation and capacity building	Analyze resource (credit, labor, markets, supply of raw material, tools and equipment, knowledge and information, extension services, and time) requirements for each activity relationships between different institutions
Fisher women community engagement in higher level of environmental awareness to prioritize local and indigenous knowledge	Replace and repair breached earthen embankments of eighteenth century	Support the fiscal transfer mechanism and flexible loans based on existing local financing schemes for capital investment

(continued)

Table 28.1 (continued)

Ecosystem-based approach	Built environment approach	Institutional or policy-based approach
Regional or local studies emphasizing the indigenous knowledge of fisherwomen folk throughout the entire value chain, highlighting impact of the environmental stochasticity	Proper road for transportation and cyclone-resistant architecture	Special grant and investment for women fisherfolk
Voluntary reporting of regional and national sex-disaggregated official catch statistics and their importance		Policies to pay women's right to earn a fair profit from their work
		Encouraging alternative livelihoods by self-help groups (SHGs) or mahila Mandal through vocational training
		Rescue and rehabilitate children engaged as girl child labor, close community participation, and implementation of child labor laws in fisheries
		Rehabilitation and school education, cross-country networks of women fishers' organizations, institutional collaboration, and private-public partnership

decision-making. The obstacles and restrictions that women face when participating in fisheries organizations as members and leaders should also be addressed to understand the complexities through the catch to consumption pathway. With a more integrative approach, in the coordination of fishery governance and climate change-related policies, it is possible to achieve inclusive development by improving the accuracy of future climate prediction and weather forecasts, as well as improving the catches, and to benefit household income.

References

Abowei JFN (2010) Salinity, dissolved oxygen, pH and surface water temperature conditions in Nkoro River, Niger Delta, Nigeria. Adv J Food Sci Technol 2(1):36–40

Ahmed N, Troell M (2010) Fishing for prawn larvae in Bangladesh: an important coastal livelihood causing negative effects on the environment. Ambio 39(1):20–29

Allison EH, Ellis F (2001) The livelihoods approach and management of small-scale fisheries. Mar Policy 25(5):377–388

Allison EH, Andrew NL, Oliver J (2007) Enhancing the resilience of inland fisheries and aquaculture systems to climate change. Fish Chim 35(5):80–95

Allison EH, Perry AL, Badjeck MC, Neil Adger W, Brown K, Conway D, Dulvy NK (2009) Vulnerability of national economies to the impacts of climate change on fisheries. Fish Fish 10(2):173–196

Alonso-Población E, Siar SV (2018) Women's participation and leadership in fisherfolk organizations and collective action in fisheries: a review of evidence on enablers, drivers, and barriers. FAO fisheries and aquaculture circular, (C1159), pp I–48

Atta-ur-Rahman, Khan AN (2013) Analysis of 2010-flood causes, nature and magnitude in the Khyber Pakhtunkhwa, Pakistan. Nat Hazards 66(5):887–904

Badjeck MC, Allison EH, Halls AS, Dulvy NK (2010) Impacts of climate variability and change on fishery-based livelihoods. Mar Policy 34(3):375–383

Barbier EB (2015) Climate change impacts on rural poverty in low-elevation coastal zones. Estuar Coast Shelf Sci 165(3):A1–A13

Barua P (2021) Coping practices of coastal fishermen in response to climate change for Southern Coastal Belt of Bangladesh. Soc Value Soc 3(2):74–40

Barua P, Chakraborty S (2011) Assessment of aquatic health index for coastal aquaculture activity in and around south-east coast of Bangladesh. Curr Biotica 5(2):180–195

Barua P, Mitra A, Banerjee K, Chowdhury MSN (2011) Seasonal variation of heavy metal concentration in water and oyster (*Saccostrea cucullata*) inhabiting Central and Western Sector of Indian Sundarbans. Environ Res J 5(3):121–130

Barua P, Rahman SH, Barua S, Rahman IMM (2020) Climate change vulnerability and responses of fisherfolk communities in the South-Eastern Coast of Bangladesh. J Water Conserv Manag 4(1): 45–65

Barua P, Rahman SH, Barua M (2022) Building resilience towards climate change vulnerability: a case study of fishing communities in southern Bangladesh. In: Islam MM (ed) Small in scale big in contributions: advancing knowledge of small-scale fisheries in Bangladesh. TBTI Global Publication, St. John's

Bebbington A (1999) Capitals and capabilities: a framework for analyzing peasant viability, rural livelihoods and poverty. World Dev 27(12):2021–2044

Béné C, Friend R (2011) Poverty in small-scale fisheries: old issue, new analysis. Prog Dev Stud 11(2):119–144

Béné C, Macfadyen G, Allison EH (2007) Increasing the contribution of small-scale fisheries to poverty alleviation and food security. Food Manage 40(3):80–95

Béné, C, Steel E, Kambala LB, Gordon A (2009) Fish as the "bank in the water": evidence from chronic-poor communities in Congo'. Food Policy 34(2):104–118

Bennett E (2005) Gender, fisheries and development. Mar Policy 29(5):451–459

Bennett A, Patil K, Rader D, Virdin J, Basruto X (2018) Contribution of fisheries to food and nutrition security: current knowledge, policy, and research. Fish Fish 40(5):50–65

Bennett NJ, Blythe J, Cisneros-Montemayor AM, Singh GG, Sumaila UR (2019) Just transformations to sustainability. Sustainability 11(14):3881–3890

Berkes F (2015) Coasts for people: interdisciplinary approaches to coastal and marine resource management. Mar Policy 39(4):920–935

Berkes F, Jolly D (2002) Adapting to climate change: social-ecological resilience in a Canadian western Arctic community. Conserv Ecol 5(2):80–98

Bhaumik U, Sen M, Chatterjee JG (1996) Participation of rural women of Sundarbans in decision-making process related to fishery. Nat Environ 12(11):45–47

BOBP (1993) A manual for operating a small-scale recirculation freshwater prawn hatchery. Bay of Bangal Programme, Chennai, p 22

Brander K (2010) Impacts of climate change on fisheries. J Mar Syst 79(3–4):389–402

Brugère C, Holvoet K, Allison EH (2008) Livelihood diversification in coastal and inland fishing communities: misconceptions, evidence and implications for fisheries management. Coast Manag 40(4):55–75

Carvalho P, Jupiter S, Januchowski-hartley F, Goetze J, Claudet J, Weeks R, White C (2019) Optimized fishing through periodically harvested closures author affiliation. J Appl Ecol 56(8): 1927–1936

Chacraverti S (2014) The Sundarbans fishers coping in an overly stressed mangrove estuary. International Collective in Support of Fishworkers, p 60

Chandra G, Sagar R (2003) Fisheries in Sundarbans: problems and prospects. Fish Chim 40(3): 50–60

Chatterjee P, Bhuinya N, Mondal S (2009) Traditional fishers in the Sundarban Tiger Reserve: A study on livelihood practice under protected area. Wetland Manag 60(3):80–95

Chaudhuri AB, Choudhury A (1994) Mangroves of the Sundarbans. Volume 1: India. International Union for Conservation of Nature and Natural Resources, p 90

Chapman MD (1987) Women's fishing in oceania. Hum Ecol 15(3):267–288

Chavez FP, Ryan J, Lluch-Cota E, Ñiquen M (2003) From anchovies to sardines and back: multidecadal change in the Pacific Ocean. Fish Manag 29(9):217–221

Chowdhury AN, Jadhav S (2012) Eco-psychiatry: culture, mental health and ecology with special reference to India. Indian J Ecol 40(3):50–65

Chowdhury MSN, Hossain MS, Barua P (2008) Artisanal fisheries status and sustainable management options in Teknaf coast, Bangladesh. Soc Chang 1(1):20–35

Chowdhury, M.S. N., Hossain, M. S., Das, N. G., and B. Prabal. 2011. Small-scale fishermen along the Naaf River, Bangladesh in crisis: a framework for management. Mesopotamian J Mar Sci, 26 (2): 146–169

Chowdhury AN, Brahma A, Mondal R, Biswas MK (2016) Stigma of tiger attack: study of tiger-widows from Sundarban Delta, India. Indian J Psycol 58(1):12–20

Cochrane KL, Andrew NL, Parma AM (2011) Primary fisheries management: a minimum requirement for provision of sustainable human benefits in small-scale fisheries. Fish Fish 12(3): 275–288

Dalton S (2001) Gender and the shifting ground of revolutionary politics: the case of Madame Roland. Can J Hist 36(2):259–282

Danda AA (2019) Environmental security in the Sundarban in the current climate change era: strengthening India-Bangladesh cooperation. Fish Manag 35(3):80–95

Dankelman I (2002) Climate change: learning from gender analysis and women's experiences of organising for sustainable development. Gend Dev 10(2):21–29

Das CS (2009) Spatio-temporal study of the hazards induced by tiger attack in Sundarban, West Bengal. Indian J Landsc Syst Ecol Stud 32(1):330–338

Das P, Das A, Roy S (2016) Shrimp fry (meen) farmers of Sundarban Mangrove Forest (India): a tale of ecological damage and economic hardship. Int J Agri Food Res 5(2):45–65

Dasgupta S, Wheeler D, Santadas G (2022) Fishing in salty waters: poverty, occupational saline exposure, and women's health in the Indian Sundarban. J Manag Sustain 12(1):50–65

Datta D (2018) Assessment of mangrove management alternatives in village-fringe forests of Indian Sundarbans: resilient initiatives or short-term nature exploitations? Wetland Ecol Manag 26(3): 399–413

Daw T, Adger WN, Brown K, Badjeck MC (2009) Climate change and capture fisheries: potential impacts, adaptation and mitigation. Curr Biotica 10(4):80–90

De la Torre-Castro M, Fröcklin S, Börjesson S, Okupnik J, Jiddawi NS (2017) Gender analysis for better coastal management–increasing our understanding of social-ecological seascapes. Mar Policy 83(4):62–74

Defeo O, Castrejón M, Ortega L, Kuhn AM, Gutiérrez NL, Castilla JC (2013) Impacts of climate variability on Latin American small-scale fisheries. Ecol Soc 18(4):65–85

Dercon S (2002) Income risk, coping strategies, and safety nets. World Bank Res Observ 17(2): 141–166

Dercon S (2006) Vulnerability: a micro perspective. Securing development in an unstable world. Soc Values 30(3):117–146

Dutta S, Chakraborty K, Hazra S (2017) Ecosystem structure and trophic dynamics of an exploited ecosystem of Bay of Bengal, Sundarban Estuary, India. Fish Sci 83(2):145–159

Elson D (1992) Male bias in structural adjustment. In: Afshar H, Dennis C (eds) Women and adjustment policies in the third world. Women's studies at York/Macmillan series. Palgrave Macmillan, London

FAO (2013) FAO statistical yearbook: world food and agriculture. Food and Agriculture Organization of the United Nations, Rome

FAO (2017) Nutrition-sensitive agriculture and food systems in practice. Food and Agriculture Organization, p 95

FAO (2020) The state of world fisheries and aquaculture 2020, Sustainability in action. Food and Agriculture Organization, Rome, p 80

FAO (2021) The state of world fisheries and aquaculture 2021, Sustainability in action. Food and Agriculture Organization, Rome, p 90

FAO, IFAD, ILO (2010) Gender dimensions of agriculture and rural employment: differentiated pathways out of poverty: status, trends and gaps, pp 120

FAO, ILO (2017) World social protection report 2017–19: universal social protection to achieve the sustainable development goals. FAO and ILO approach. Geneva

FAO, MRC (2003) New approaches for the improvement of inland capture fishery statistics in the Mekong Basin. Report of the Ad Hoc Expert Consultation held in Udon Thani, Thailand, 2–5 September 2002. FAO/RAP publication 2003/01. FAO/RAP, Bangkok, 145 pp

Folke C, Carpenter S, Elmqvist T, Gunderson L, Holling CS, Walker B (2002) Resilience and sustainable development: building adaptive capacity in a world of transformations. Ambio 31 (5):437–40

Fidel M, Kliskey A, Alessa L, Sutton OP (2014) Walrus harvest locations reflect adaptation: a contribution from a community-based observing network in the Bering Sea. Polar Geogr 37(1): 48–68

Frangoudes K, Gerrard S, Kleiber D (2019) Situated transformations of women and gender relations in small-scale fisheries and communities in a globalized world. Coast Res 25(5):80–90

Gallopin GC (1991) Human dimensions of global change: linking the global and the local processes. Int J Soc Sci 130(3):707–718

Gallopin GC (2006) Linkages between vulnerability, resilience, and adaptive capacity. Glob Environ Chang 16(3):293–303

Gammage S, Swanburg K, Khandkar M, Islam MZ, Zobair M, Muzareba AMA (2006) Gendered analysis of the shrimp sector in Bangladesh: greater access to trade and expansion. Bangladesh J Environ Sci 10(5):40–50

Gee J, Bacher K (2017) Engendering statistics for fisheries and aquaculture. Gender in aquaculture and fisheries: engendering security in fisheries and aquaculture. Fish Chim 40(3):277–290

Gender G, Alliance C (2016) Gender and climate change: a closer look at existing evidence. Global gender and climate alliance. Routledge, New York

Gereva S, Vuki V (2010) Women's fishing activities on Aniwa Island, Tafea Province, South Vanuatu. SPC Women Fish Inf Bull 21(2):17–21

Gupta N, Kanagavel A, Dandekar P, Dahanukar N, Sivakumar K, Mathur VB, Raghavan R (2016) God's fishes: religion, culture and freshwater fish conservation in India. Coast Manag 50(2): 244–249

Hamid A, Alauddin M (1998) Coming out of their homesteads? employment for rural women in shrimp aquaculture in coastal Bangladesh. Int J Soc Econ 25(2):314–337

Harper S, Adshade M, Lam VW, Pauly D, Sumaila UR (2020a) Valuing invisible catches: estimating the global contribution by women to small-scale marine capture fisheries production. PLoS One 15(3):150–175

Harper CA, Satchell LP, Fido D, Latzman RD (2020b) Functional fear predicts public health compliance in the COVID-19 pandemic. Int J Ment Heal Addict 10(4):1–14

Hasan R, Naser MN (2016) Fishermen livelihood and fishery resources of the Sundarbans reserved forest along the Mongla port area Bangladesh. Int J Fish Aquat Stud 4(3):468–475

Hasan M, Monjur BM, Sajjad HM, Jobaer A, Azam KC, Karim AA, Nuruddin KH (2018) The prospects of blue economy to promote Bangladesh into a middle-income country. Open J Mar Sci 8(3):355–369

Hassler U, Kohler N (2014) Resilience in the built environment. Build Res Inf 42(2):119–129

Hazra S, Ghosh T, DasGupta R, Sen G (2002) Sea level and associated changes in the Sundarbans. Sci Cult 68(3):309–321

Herbert GJ (1995) Fisheries relations in the Gulf of Maine implications of an arbitrated maritime boundary. Mar Policy 19(4):301–316

Herrón P, Castellanos-Galindo GA, Stäbler M, Díaz JM, Wolff M (2019) Toward ecosystem-based assessment and management of small-scale and multi-gear fisheries: insights from the tropical Eastern Pacific. Front Mar Sci 6(2):127–140

Hlohowskyjl I, Robert SM, Lackey T (1996) Methods for assessing the vulnerability of African fisheries resources to climate change. Clim Res 6(2):97–106

Hoq ME (2007) An analysis of fisheries exploitation and management practices in Sundarbans mangrove ecosystem, Bangladesh. Ocean Coast Manag 50(5–6):411–427

Hoq ME, Islam MN, Kamal M, Wahab MA (2003) Fisheries structure and management implications in Sundarbans mangrove reserve forest, Bangladesh. Indian J Fish 50(2):243–249

Huysentruyt M (2014) Women's social entrepreneurship and innovation. Society 30(3):5060

IPCC (2001) Climate change 2001. Impacts, adaptation and vulnerability. In: McCarthy JJ, Canziani OF, Leary NA, Dokken DJ, White KS (eds) Contribution of working group II to the third assessment report of the intergovernmental panel on climate change. Cambridge University Press, Cambridge, UK/New York

IPCC (2014) Climate change 2014. Impacts, adaptation, and vulnerability. Contribution of working group II to the fifth assessment report of the intergovernmental panel on climate change

Islam MM, Chuenpagdee R (2013) Negotiating risk and poverty in mangrove fishing communities of the Bangladesh Sundarbans. Wetland Manag 12(1):1–20

Islam MS, Rahman MS, Haque MM, Sharmin S (2011) Economic study on production and marketing of shrimp and prawn seed in Bangladesh. J Bangladesh Agri Univ 9(2):247–255

Jackson B, Doreen S, Boyd S, Christopher D, Ives K, Jessica L, Decker S, Giles M, Foody S, Stuart M, Marsh K (2020) Remote sensing of fish-processing in the Sundarbans reserve Forest, Bangladesh: an insight into the modern slavery-environment nexus in the coastal fringe. Mar Stud 19(3):429–444

Jacquet J, Pauly D (2008) Funding priorities: big barriers to small-scale fisheries. Conserv Biol 22(4):832–835

Kais SM, Islam MS (2019) Perception of climate change in shrimp-farming communities in Bangladesh: A critical assessment. Int J Environ Res Publ Heal 16(4):672

Kanjilal B, Mazumdar PG, Mukherjee M, Rahman MH (2010) Nutritional status of children in India: household socio-economic condition as the contextual determinant. Int J Equit Heal 9(1):1–13

Kesavan PC, Swaminathan MS (2006) Managing extreme natural disasters in coastal areas. Environ Manag 80(4):90–101

Khan A, Kemp GC, Currie-Alder B, Leone M (2018) Responding to uneven vulnerabilities: a synthesis of emerging insights from climate change hotspots. Ocean Coast Manag 85(5):80–110

Kilpatrick K, Reid K, Carter N, Donald F, Bryant-Lukosius D, Martin-Misener R, Kaasalainen S, Harbman P, Marshall DA, Charbonneau-Smith R, DiCenso, A (2015) A systematic review of the cost-effectiveness of clinical nurse specialists and nurse practitioners in inpatient roles. Can J Nurs Leadersh 28(3):56–76

Kizhakudan SJ (2014) Correlation between changes in sea surface temperature and fish catch along Tamil Nadu coast of India-an indication of impact of climate change on fisheries? Fish Chim 40(4):30–45

Kleiber D, Harris LM, Vincent AC (2015) Gender and small-scale fisheries: a case for counting women and beyond. Fish Fish 16(4):547–562

Kleiber D, Harris LM, Vincent ACJ (2014) Improving fisheries estimates by including women's catch in the central Philippines. Can J Fish Aquat Sci 71(2):1–9

Koralagama D, Gupta J, Pouw N (2017) Inclusive development from a gender perspective in small-scale fisheries. Curr Opin Environ Sustain 24(3):1–6

Kusakabe K (2002) Gender issues in small-scale inland fisheries in Asia: women as an important source of information. New approaches for the improvement of inland capture fishery statistics in the Mekong Basin. Asian J Fish 20(3):60–80

Laha A (2019) Mitigating Climate Change in Sundarbans. Fish Chim 30(3):80–95

Lam VW, Allison EH, Bell JD, Blythe J, Cheung WW, Frölicher TL, Sumaila UR (2020) Climate change, tropical fisheries and prospects for sustainable development. Natl Rev Earth Environ Sci 1(9):440–454

Lentisco A, Lee RU (2015) A review of women's access to fish in small-scale fisheries. FAO fisheries and aquaculture circular, (C1098), I

Mangubhai S, Lawless S (2021) Exploring gender inclusion in small-scale fisheries management and development in Melanesia. Mar Policy 123(3):104–120

Marshall S, Elliot M (1998) Environmental influences on the fish assemblage of the Humber estuary, U.K. Estuar Coast Shelf Sci 46(3):175–184

Masika R (ed) (2002) Gender, development, and climate change. Oxfam, Oxford

Mason LR, Rigg J (eds) (2019) People and climate change: vulnerability, adaptation, and social justice. Oxford University Press, Oxford

McClanahan TR, Cinner JE, Maina J, Graham NAJ, Daw TM, Stead SM, Polunin NVC (2008) Conservation action in a changing climate. Conservation 1(2):53–59

Mehta L, Adam HN, Srivastava S (2021) Unpacking uncertainty and climate change from 'above' and 'below'. Reg Environ Chang 19(2):1529–1532

Mendoza RU (2009) Aggregate shocks, poor households and children. Glob Soc Policy 9(1):55–78

Mishra S, Bajaj P (2020) On climate change and fisherwomen in India. Indian J Ecol 10(3):80–95

Mishra M, Acharyya T, Augusto C, Santos G, Richarde M, Dipika K, Kamal AHM, Susmita R (2021) Geo-ecological impact assessment of severe cyclonic storm Amphan on Sundarban mangrove forest using geospatial technology. Estuar Coast Shelf Sci 260(3):120–140

Mistri A, Das B (2015) Environmental legislations and livelihood conflicts of fishermen in Sundarban, India. Asian Prof 44(3):389–400

Mitra A (2004) Ecological profile of Indian Sundarbans: pelagic primary producer community, vol 1. World Wide Fund for Nature-India, West Bengal State Office

Mitra A (2019) Mangrove forests in India: exploring ecosystem services. Springer

Mitra A, Sengupta K, Banerjee K (2011) Standing biomass and carbon storage of above-ground structures in dominant mangrove trees in the Sundarbans. For Ecol Manag 26(7):1325–1335

Mitra A, Barua P, Zaman S, Banerjee K (2012) Analysis of trace metals in commercially important crustaceans collected from UNESCO protected world heritage site of Indian Sundarbans. Turk J Fish Aquat Sci 12(3):53–66

Mohammed EY, Uraguchi ZB (2013) Impacts of climate change on fisheries: implications for food security in Sub-Saharan Africa. In: Mohammed EY (ed) Global food security. Nova Science Publishers, Inc, Hauppauge, pp 114–135

Molla MAG, Islam MR, Islam S, Salam MA (2009) Socio-economic status of crab collectors and fatteners in the southwest region of Bangladesh. J Bangladesh Agri Univ 7(2):411–419

Möllmann C, Lindegren M, Blenckner T, Bergström L, Casini M, Diekmann R, Gårdmark A (2014) Implementing ecosystem-based fisheries management: from single-species to integrated ecosystem assessment and advice for Baltic Sea fish stocks. ICES J Mar Sci 71(5):1187–1197

Mondal B, Bhaduri S (2010) Effect of Tiger prawn seeds collection on the ecosystem of Indian Sundarban. Asian Stud 29:74–85. (ISSN-09707301)

Moosavi S (2017) Ecological coastal protection: pathways to living shorelines. Proc Eng 19(6):930–938

Mozumder MM (2018) Coastal ecosystems services in the Bay of Bengal and efforts to improve their management. Indian J Mar Sci 47(11):60–75

Mozumder MMH, Uddin MM, Schneider P, Islam MM, Shamsuzzaman M (2018) Fisheries-based ecotourism in Bangladesh: potentials and challenges. Resources 7(2):61–71

Mukherjee M (2018) Empowerment of fisherwomen of Sundarbans through ornamental fish farming. Asian Fish Sci 40(3):125–150

Mukherjee P, Mitra A, Zaman S, Mitra A (2022) Impact of climate-change-induced salinity alteration on Ichthyoplankton diversity of Indian Sundarbans. In: Filho WL et al (eds) Handbook of climate change management. Springer Nature, Cham

Mukherjee MN (2007) Sustainable development and global warming. Soc Work J 20(4):300–310

Musinguzi L, Natugonza V, Ogutu-Ohwayo R (2017) Paradigm shifts required to promote ecosystem modeling for ecosystem-based fishery management for African inland lakes. J Great Lakes Res 43(1):1–8

Mutombo K, Ölçer A (2017) Towards port infrastructure adaptation: A global port climate risk analysis. WMU J Marit Aff 16(2):161–173

Najam A, Rahman AA, Huq S, Sokona Y (2003) Integrating sustainable development into the fourth assessment report of the intergovernmental panel on climate change. Clim Pol 3(3):9–17

Neogi SB, Dey M, Lutful Kabir SM, Masum SJH, Kopprio GA, Yamasaki S, Lara RJ (2016) Sundarban mangroves: diversity, ecosystem services and climate change impacts. Indian J Mar Sci 30(3):30–50

Nessa FM, Rappe AM (2019) Invertebrate gleaning: forgotten fisheries. IOP Conf. Series: Earth Environ Sci 253(2):200–210

Patel V, Rajagopalan R (2009) Fishing community issues in the Sundarban Tiger Reserve: a case study. Sci Cult 50(4):40–50

Perry JL, Enaber T, Jun SY (2009) Back to the future? performance-related pay, empirical research, and the perils of persistence. Public Adm Rev 69(1):39–51

Pitchaikani JS (2020) Integrated coastal zone management practices for Sundarbans, India. Indian J Geo Mar Sci 49(3):352–356

Pitchaikani JS, Sarma KS, Bhattacharyya S (2017) First time report on the weather patterns over the Sundarban mangrove forest, East Coast of India. Springer, New York

Quiros TAL, Beck MW, Araw A, Croll DA, Tershy B (2018) Small-scale seagrass fisheries can reduce social vulnerability: a comparative case study. Ocean Coast Manag 85(4):56–67

Quisumbing AR, Meinzen-Dick RS, Bassett L, Usnick M, Pandolfelli L, Morden C, Alderman H (2008) Helping women respond to the global food price crisis. Society 40(3):50–65

Rabbitt S, Lilley I, Albert S, Tibbetts IR (2019) What's the catch in who fishes? Fisherwomen's contributions to fisheries and food security in Marovo lagoon, Solomon Islands. Mar Policy 108(3):103–120

Raha A, Das S, Banerjee K, Mitra A (2012) Climate change impacts on Indian Sundarbans: a time series analysis (1924–2008). Biodivers Conserv 21(5):1289–1307

Rajagopalan M (2009) Impact of rise in seawater temperature on the spawning of threadfin Breams. Coast Res 20(5):80–95

Ray A, Bhowmick S (2020) Green energy sources selection for sustainable planning: a case study. IEEE Trans Eng 69(4):110–120

Rosenzweig MR, Binswanger HP (1992) Wealth, weather risk, and the composition and profitability of agricultural investments, vol 1055. World Bank Publications

Roy A, Bhaumik U (2013) Participation of rural women of Sundarban in decision making related to fishery activities. Fish Chim 30(4):85–95

Roy A, Sharma AP, Bhaumik U, Pandit A, Singh SRK, Saha S, Mitra A (2017) Socio-economic features of womenfolk of Indian Sundarbans involved in fish drying. Indian J Ext Edu 53(2):142–146

Ruiz-Díaz R, Liu X, Aguión A, Macho G, de Castro M, Gómez-Gesteira M, Ojea E (2020) Social-ecological vulnerability to climate change in small-scale fisheries managed under spatial property rights systems. Mar Policy 121(4):104192

Sahu Sk K (2015) Rising critical emission of air pollutants from renewable biomass. Environ Res Lett 10(2):95–115

Salagrama V (2012) Climate change and fisheries: perspectives from small-scale fishing communities in India on measures to protect life and livelihood. Asian Fish Sci 40(4):130–150

Sarker AR (2011) Exploring the relationship between climate change and rice yield in Bangladesh: An analysis of time series data. Agr Syst 112(1):11–16

Schuhbauer A, Chuenpagdee R, Cheung WW, Greer K, Sumaila UR (2017) How subsidies affect the economic viability of small-scale fisheries. Mar Policy 82(3):114–121

Seager J (2015) Sex-disaggregated indicators for water assessment, monitoring and reporting. Water Manag 50(4):20–50

Shelly AB, Costa MD (2020) Women in aquaculture: initiatives of Caritas Bangladesh, Monographs. The WorldFish Center, number 36249

Solano N, Lopez-Ercilla I, Fernandez-Rivera Melo FJ, Torre J (2021) Unveiling women's roles and inclusion in Mexican small-scale fisheries (SSF). Front Mar Sci 7(4):120–135

Spalding MD, Ruffo S, Lacambra C, Meliane I, Hale LZ, Shepard CC, Beck M (2014) The role of ecosystems in coastal protection: adapting to climate change and coastal hazards. Ocean Coast Manag 90(3):50–57

Srinivasan UT, Cheung WW, Watson R, Sumaila UR (2010) Food security implications of global marine catch losses due to overfishing. J Bioecon 12(3):183–200

Stacey N, Gibson E, Loneragan NR, Warren C, Wiryawan B, Adhuri DS, Fitriana R (2021) Developing sustainable small-scale fisheries livelihoods in Indonesia: trends, enabling and constraining factors, and future opportunities. Mar Policy 132(3):104–120

Sultana MA, Nor EM, Zulkefli M (2011) Discrimination against women in the developing countries: a comparative study. Int J Soc Sci Humanit 2(3):256–259

Temmerman S, Meire P, Bouma TJ, Herman PM, Ysebaert T, De Vriend HJ (2013) Ecosystem-based coastal defence in the face of global change. Nature 504(7):79–83

Toya H, Skidmore M (2007) Economic development and the impacts of natural disasters. Econ Lett 94(1):20–25

Troadec JP (2000) Adaptation opportunities to climate variability and change in the exploitation and utilisation of marine living resources. Environ Monit Assess 61(1):101–112

Ullah R, Shivakoti GP, Zulfiqar F, Iqbal MN, Shah AA (2017) Disaster risk management in agriculture: tragedies of the smallholders. Nat Hazards 87(3):1361–1375

United Nations (2016) Integrating a Gender perspective into statistics United Nations. The Ocean Conference, New York

Vasquez-Leon M (2002) Assessing vulnerability to climate risk: the case of small-scale fishing in the Gulf of California, Mexico. Invest Mar Res 30(1):204–205

Wooster WS, Zhang CI (2004) Regime shifts in the North Pacific: early indications of the 1976–1977 event. Prog Oceanogr 60(2–4):183–200

Chapter 29
Climate Change and Health Impacts in the South Pacific: A Systematic Review

Mumtaz Alam , Mohammed Feroz Ali , Sakul Kundra ,
Unaisi Nabobo-Baba , and Mohammad Afsar Alam

Abstract The World Health Organization (WHO) estimates that climate change causes at least 150,000 fatalities each year, with that figure predicted to double by 2030. A common worldwide objective is to enhance everyone's health. Climate change is affecting human health globally. Not a risk creator or risk factor, rather a risk multiplier. Increase in the incidence of existing public health concerns, especially in communities with a high prevalence of climate-sensitive illnesses and disorders. Hotter and more frequent heatwaves raise the risk of myocardial infarction (heart attack), especially in individuals with coronary artery disease. Access to natural resources, services, and enrichment (such as culture) is required for civilizations and their inhabitants. In the South Pacific, negative health effects, especially non-communicable diseases (NCDs), are likely. Environments may cause illness and disease, as well as numerous harms, such as emotions of uneasiness and discontent. Changes in the environment affect the reproduction and spread of infectious organisms, vectors, human hosts, and disease reservoirs. Many Pacific Island nations confront significant difficulties today, including NCDs, infectious diseases, and climate change effects. According to the World Health Organization (WHO), NCD death rates in several Pacific Island countries (PICs) are already amongst the highest globally. There have been about 40 significant infectious disease outbreaks in the previous four years. Dengue, chikungunya, leptospirosis, and Zika virus illnesses are all connected to climate change. If climate change continues its current path, new patterns (and possibly new types) of health issues will develop when different thresholds are crossed later this century. This introspective study based on secondary data from books and publications as well as websites, web journals,

---

M. Alam (✉) · S. Kundra · M. A. Alam
Department of Social Sciences, Fiji National University, Suva, Fiji
e-mail: mumtaz.alam@fnu.ac.fj

M. F. Ali
Department of Sports Science and Education, Fiji National University, Suva, Fiji
e-mail: ali.feroz@fnu.ac.fj

U. Nabobo-Baba
Department of Education, Fiji National University, Suva, Fiji
e-mail: unaisi.baba@fnu.ac.fj

© The Author(s), under exclusive license to Springer Nature Switzerland AG 2022
U. Chatterjee et al. (eds.), *Ecological Footprints of Climate Change*, Springer Climate, https://doi.org/10.1007/978-3-031-15501-7_29

reports, and government agencies. The study's goal is to identify and characterize current and prospective health hazards in a community, population, or area. Assessing, planning, and evaluating adaptation (risk management) activities and assessing the extra-institutional, resource, and governance demands and needs on healthcare systems and larger population-based public health practice. This research could help Pacific islanders design and implement future climate change initiatives that improve structural and social determinants of health.

Keywords Environmental change · South Pacific · Health · NCDs · Diseases

Introduction

The greatest danger to human health is climate change, and health care providers throughout the world are already preparing to deal with the health consequences of this rapidly emerging problem. The WHO estimates that climate change causes at least 150,000 fatalities each year, with that figure predicted to double by 2030 (Fadda 2020; Thompson 2021; Hossain 2019). Malnutrition, malaria, diarrhoea, and heat stress are predicted to cause an additional 250,000 deaths per year by 2050 because of climate change. Responding to health threats throughout the world, healthcare professionals are adapting their services posed through this rapidly developing epidemic. To avert catastrophic health repercussions and prevent millions of deaths from climate change, the Intergovernmental Panel on Climate Change (IPCC) recommends limiting temperature rise to 1.5 °C. A certain degree of global temperature rises, and other climatic impacts are already a certainty because of previous emissions. Every extra tenth of a degree of global warming will have a devastating effect on human health and life, even if the temperature rises by just 1.5 °C. Global warming, as predicted by the United Nations Framework Convention on Climate Change (UNFCCC), will put the lives of most people, particularly those in the developing nations, in grave danger (Alam et al. 2021). In the late 1980s and early 1990s, academicians and other social scientists began examining globalization and climate change, and health was quickly recognized as a major issue (Alam 2021). People's health is among the many dangers posed by anthropogenic climate change because industrialization, economic progress, increasing numbers of people, and globalization all go hand in hand. It is impossible to overstate the danger presented by climate change to human health (McMichael 2014). The climate change consequences on Pacific Island countries (PICs) and regions are among the most severe in the world, including the likely detrimental impacts on health. Due to a convergence of geographical, demographic, and socio-economic variables, such as low elevation, small populations, and scarcity of resources, PICs and territories are particularly sensitive to the effects of climate change and health. People in the developing world suffer a triple burden of diseases: communicable diseases, NCDs, and the climate change effects. These consequences disproportionately affect specific parts of the population, including the poor and people with pre-existing diseases. They also disproportionately affect individuals who live in high-risk locations for occurrences triggered by changes in the climate, to give an example flooding, drought, and rise of sea level. This means that climate change can worsen socio-economic and health disparities. To safeguard Pacific Island people from climate change-related health concerns, a context-based evaluation of vulnerability is needed as part of an integrated set of adaptations across sectors. To combat climate change in the Pacific, the activities of several sectors must be coordinated (WHO 2015; McIver and Hanna 2015). There is no way for countries to regulate the physical variables that determine their exposure to climate change (Chen et al. 2015).

Materials and Methods

For this article, we outline PICs as the 15 low- and middle-income Pacific Islands Forum members: "Cook Islands, Federated States of Micronesia, Fiji, Kiribati, Nauru, Niue, Palau, Papua New Guinea, Republic of the Marshall Islands, Samoa, Solomon Islands, Tokelau, Tonga, Tuvalu, and Vanuatu." This study conducted introspective research using secondary data from the world factbook, books, and publications, as well as websites, web journals, reports, and government organizations. The research project's objective is to identify and characterize current and potential health hazards in a community, demographic, or geographic area. Evaluate and plan adaptation (risk management) actions while also considering the extra-institutional, resource, and governance demands and requirements on healthcare systems and broader population-based public health practice.

Conceptual Frame Work

For a national health response to climate change, the WHO proposes a framework that includes four critical components.

1. Governance policy and engagement for health protection from climate change.
2. Climate resilient health systems—robust surveillance, early warning, and response for climate-sensitive health impacts.
3. Enhanced management of environmental health interventions, services, and determinants.
4. Mobilizing support for the public health response to climate change.

Accordingly, this framework is meant mainly for public health sector authorities as well as managers and experts. Technical and policy assistance is the goal of this initiative, which aims to increase the capability of the public health system in the face of a rapidly changing environment. Decision makers in other health-related fields, such as food, water, and sanitation, municipalities, energy, transportation, and

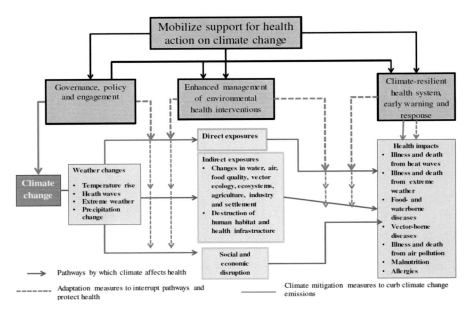

Fig. 29.1 Pathways for action on climate change and health. (Source: WHO (2017), climate change and health: a framework for action 2017–2021)

disaster management may all benefit from the framework's inclusion in their discussions and plans. The suggested framework's action paths are shown in Fig. 29.1.

Results and Discussion

Increased morbidity and mortality have already been reported in many parts of the world because of climate change and the warming trend that has occurred in recent decades. These include factors such as availability to clean air and water, healthy food, and a safe environment in which to live, as well as other social and environmental factors that will continue to be impacted by the climate health crisis. Since the 1960s, the number of weather-related natural catastrophes worldwide has more than quadrupled (Lomborg 2020). In the 1990s and early 2000s, pioneering work by McMichael and colleagues developed templates for conducting vulnerability assessments and developing strategies for the purpose of adaptation in order to protect human health from the repercussions of climate change. The human health element must be included in the study of the effects of climate change and adaptation strategies. Climate change and health risks and possibilities in the Pacific area are assessed using introspective methods. The results of these evaluations are presented; salient themes are highlighted, and they map the projected path of health adaptation to climate change in the South Pacific (McIver and Hanna 2015). People's health is already being affected by climate change in a variety of ways: extreme weather events like heatwaves, storms and floods, food systems being disrupted, a rise in zoonoses (zoonoses caused by food, water, and vectors), and mental health difficulties. Consequently, climate change is having an adverse effect on many of the socioeconomic determinants of health, such as income equality, access to healthcare, and social support networks, among other things. Females, children, ethnic minorities, low-income groups, migrants, or internally displaced persons (IDPs), older populations, and those with pre-existing medical conditions are particularly vulnerable to these climate-sensitive health challenges. Temperature extremes, increased precipitation, particularly from hurricanes, and rising sea level will all affect future human health. Concerns of climate-sensitive health in PICs include the impacts on people's mental well-being and health care systems as well as extreme weather occurrences and heat-related sickness (Altizer et al. 2013). Changing vector geographic ranges alter transmission patterns, which is another effect of climate change. Dengue fever and malaria, two illnesses spread by mosquitoes, are endemic in many of the PICs. Floods caused by cyclones, for example, can result in epidemics of both dengue fever and diarrhoeal illness (Taylor 2021). There are two main causes for climate and health issues:

1. **Direct:** Temperature-related deaths and disease, as well as flood and storm-related injuries.
2. **Indirect:** An increase in the number of illnesses carried by vectors, air pollution's negative effects on the lungs, a result of a lack of agricultural production because of migration and social dislocation.

Tropical Cyclones or Hurricanes

We are experiencing an increase in air and water evaporation due to increased heat in our atmosphere and seas. This precipitation causes rain, and the rising air generates pressure differences that lead to the wind. With more heat, our weather becomes even more dynamic and intense. Rapid evaporation from warm waters is what gives rise to tropical cyclones. Few places lying on the planet are more susceptible to the effects of climate change than the Pacific's low-lying atoll island countries. Tropical cyclones and hurricanes are the main threat to the health of Pacific Island nations. Mean sea levels, tropical storms, and ENSO (El Niño Southern Oscillation) patterns can all be affected by human-caused climate change, allowing for a wide range of future extreme sea levels. Sea levels will rise because of the warming of the oceans and the melting of ice caps and major ice sheets in Antarctica and Greenland. There is a possibility that climate change will alter tropical cyclone behaviour, affecting their frequency, intensity, and preferred locations of occurrence (Walsh et al. 2012; Salvat and Wilkinson 2011). There are no effective engineering solutions to protect vulnerable communities from the destructive power of these powerful tropical cyclones. It has been demonstrated that sea walls are ineffective against storm

Table 29.1 Tropical cyclones, categories, deaths, and damages

Season	Category of storm	Tropical cyclones or hurricanes	Deaths	Damages (USD)
2010–2011	4	Wilma	4	$25 million
2011–2012	4	Jasmine	13	$17.2 million
2012–2013	4	Sandra	17	$161 million
2013–2014	5	Ian	12	$48 million
2014–2015	5	Pam	16	> $250 million
2015–2016	5	Winston	50	≥ $1.41 billion
2016–2017	5	Donna	3	≥ $five million
2017–2018	5	Gita	11	$285 million
2018–2019	4	Pola	None	≥ $50 million
2019–2020	5	Harold	5	≥ $132 million
2020–2021	5	Yasa	7	>$246.7 million

Source: Regional Specialized Meteorological Centre Nadi—Tropical Cyclone Centre (2010–2021)

surges, that low-cost structures are ineffective against 200 km/h winds, and that drainage systems are incapable of removing floodwaters caused by such heavy rain. This anticipated increase in storm severity is beyond our ability to handle. As a result, drastic and rapid reductions in greenhouse gas emissions are critical. As these terrible weather disasters demonstrate, any further warming must be avoided. Governments all over the world must go above and beyond the Paris Agreement's minimum standards (Hughes et al. 2017).

After forming as Tropical Cyclone Harold over the tropical South Pacific in April 2020, it made landfall in the Solomon Islands before moving on to Vanuatu, Fiji, and finally Tonga. Overlooking Vanuatu, Harold demolished 80% of the dwellings on Espiritu Santo and Pentecost, moving more than 80,000 people and decimating the island's agricultural and animal industries in the process. Before the arrival of Hurricane Harold, rigorous border controls, travel restrictions, and port closures had kept the COVID-19 outbreak in Fiji to 32 people, with no more cases reported in the other three island nations. As a result, relief activities were slowed down significantly in terms of timing, scope, and efficiency. They were ordered out just weeks before Hurricane Irene and must now be quarantined before returning. Donations for disaster assistance fell because of the worldwide COVID-19 crisis's economic consequences (Shultz et al. 2020). Scheday et al. (2019) reveal that extreme heat events could spread vector- and water-borne diseases, as well as mental health problems linked to direct injuries and private losses (e.g., property damage). They could also cause non-communicable diseases (NCDs) like circulatory diseases, which can be caused by heat and air pollution.

According to the data in the preceding Table 29.1, there has been widespread destruction in PISC caused by cyclones and hurricanes between the years 2010 and 2021 in terms of both the human capital and financial resources.

Climate and Diseases in the Pacific

We are concerned about climate change and NCDs since climate change may increase the already significant and quickly growing burden of NCDs in low- and middle-income countries. Climate variability's impact on human health and the spread of illness has been brought to light because of the El Nino event. Malaria, cholera, dengue fever, and other new infectious illnesses have been connected to El Nino (Petrova et al. 2021; Patz et al. 1996; Bouma and van der Kaay 1996). A worldwide emergency has been declared for NCDs. We likened the growth of NCDs to climate change, an unprecedented global disaster. This is not an exaggeration; there is a good deal of evidence to back it up. NCDs are responsible for more than 70% of all global health-related fatalities and pose a grave danger to the worldwide population. They remain a foremost public health concern in all nations, especially low- and middle-income countries, which account for more than three-quarters of NCD mortality (Countdown NCD 2018; World Health Organization 2020). Around the world, more than 1 billion people suffer from high-blood pressure; about 800,000 dies by suicide; 425 million people have diabetes; and around 40% of persons in the adult population are fat or overweight (Zhou et al. 2017; 2–5). Most deaths in the Pacific Islands are due to non-communicable illnesses, including diabetes and obesity, which are among the world's highest (Colagiuri et al. 2008). WHO estimated that the 2030 Sustainable Development Goal (SDG) of reducing premature NCD mortality by a third would be a failure without significant new intervention (WHO 2018). NCDs and climate change share more than just numbers; there are causative aspects to consider as well. Both may be avoided. Human actions have a major role in both phenomena. A multi-sectoral approach is needed for both issues. However, neither is prioritized by responsible global or national policymakers (Climate Action Tracker 2018). As shown by the WHO's NCD Progress Monitor, which tracks the implementation of recommended national policies for the prevention of NCDs, progress has been inadequate (World Health Organization 2020).

Research shows that of the various vector-borne diseases, malaria is one of them. Dengue fever, chikungunya fever, Lymphatic Filariasis, Zika, Japanese Encephalitis, Murray Valley Encephalitis, Barmah Forest viral illness, Ross River fever, and malaria are only a few of the mosquito-borne diseases prevalent in the Pacific (Musso et al. 2018). Tropical locations are the hotbeds of disease transmission because of the perfect breeding and survival circumstances provided by a warm, humid environment for mosquito species. Probable vector breeding areas and environments for mosquito larvae may increase because of increased rainfall and floods. The larvae of mosquitoes could be washed away or destroyed by flash floods that occur when there is too much rain. Among the numerous concerns caused by climate change in PICs is the development of mosquito-borne illnesses, including dengue fever and Zika. Insight into how climate change involves the transmission of vectors is vital; however, it is also necessary to better comprehend the many health effects of climate change (Filho et al. 2019).

Health and Happiness: Tangible and Mental Welfare

Public awareness of climate change's mental health consequences is growing. People who have mental health problems do not just have mental diseases, illnesses, and disorders. They also have mental wellness, emotional resiliency, and psychosocial well-being, which are all parts of mental health (Friedli and WHO 2009; Butler et al. 2014). Extreme weather events have long been studied by psychologists, even before climate change was generally recognized as a factor in these occurrences, and a large body of data has amassed throughout that time. Depression and post-traumatic stress disorder are the most often reported mental health issues after severe weather events, according to several systematic studies of the literature (Clayton 2021; Lowe et al. 2019). People's mental and physical health can have more of an effect on the health of a community than climate (McMichael and Lindgren 2011; Bourque and Cunsolo Willox 2014; Shammin et al. 2022). People's opportunities for social interaction, their relationships with one another, and their relationships with nature are all changing owing to climate change (Clayton et al. 2017). The PICs are most susceptible to mental health. The problem is that it is thought that climate change will have a big impact on mental health, especially among people who are vulnerable both environmentally and economically. There have been several hypotheses on how climate change can affect mental health, but there has not been much research on the effects themselves (Gibson et al. 2020). A disaster-affected community's population is usually between 30 and 40% who suffers from some sort of poor mental health repercussions in a year of the occurrence of the catastrophe, which declines after the event but remains chronic in the population (Kelman et al. 2021; Rataj et al. 2016). While "Oceania" and other small island developing states (SIDS) included inside their respective continents were included in the global evaluation, particular nations were seldom referenced. Subsistence and non-subsistence activities including farming, fishing, forest management, and hunting can have a negative impact on Pacific Islanders' mental health and well-being when they are unable to adapt to climate change (Lund et al. 2010; McIver et al. 2017). Colonialism's systemic violence has exacerbated pre-existing vulnerabilities, including weak infrastructure and cash crops. Gender and age-specific impacts of food and water insecurity have been linked to mental health and well-being issues such as sadness and anxiety (Steel et al. 2009; Kelman et al. 2021; Weaver and Hadley 2009). Most people agree on the need for better mental health and well-being systems, services, training, and infrastructure, and Kelman et al. offer data from various Pacific Islands countries in support of that belief. It is necessary to find the right mix between (A) embracing local and indigenous ideas of health, and (B) affecting ingrained disgraces of mental health and well-being challenges to achieve this. As it is viewed as an external danger by SIDS peoples, climate change may provide an opportunity. It is, therefore, possible to raise awareness of the negative mental health and well-being concerns of climate change by bringing up the topic of climate change.

Pacific People and Their Public Health

Human Life Expectancy

The life expectancy at birth of a population is one of the most important indicators of how healthy the population is. Healthier people are more likely to live longer because mortality rates have decreased. For men, the Pacific life expectancy is 78.8 years, compared to New Zealand's 82.7 years for men and women. Both men and women in France live to the age of 78.1, while in Australia it is 79.3 and 83.9 years for men and women, respectively. A person in New Zealand, France, or Australia lives an average of 10 years longer than someone in Vanuatu, and 25 years longer than someone in Nauru, Kiribati, or Papua New Guinea (UNFPA).

Women across the world live longer than men. In the Pacific, this is likewise the case. Male newborn mortality rates are greater than female infant mortality rates at birth, which explains why male infants are more likely to die than female infants. To add insult to injury, the probability of a 15-year-old guy dying before attaining the age of 60 is substantially higher than that of a female, resulting in shorter male life expectancies (Aghai et al. 2020).

However, even though the populations of PICs are still relatively young, there is significant evidence that they are ageing, and the share of old people is growing. However, the proportion of old people in each PIC varies greatly, with Niue's elderly population being five times larger than that of Nauru (Anderson and Irava 2017).

Extended family is the primary caregiver and source of social security for the elderly in the Pacific, and this is likely to continue. There is still a strong sense of family, but it is eroding in the cities. As the population of the elderly rises, governments will have to devise strategies for supplementing family care with more formal institutional care (Andruske and O'Connor 2020).

A wide range of death trends may be seen in Pacific Island nations. Adults in low-mortality nations suffer from non-communicable illnesses and accidents, while infectious diseases and malnutrition are the primary causes of death in high-mortality populations. Urban–rural health disparities persist even in the world's poorest nations, where the prevalence of non-communicable illnesses is on the rise (Taylor et al. 1992). Table 29.2 reveals that in all PICs, women had a longer average lifespan than men. The longer life expectancy of people throughout the globe is due in large part to the advancement of medical technology during the past decade. However, despite the increase in life expectancy, the health impacts of climate change are more serious. NCDs and accidents are key reasons for death and disability among adults in Nauru, especially among men. Furthermore, the substantial sex disparities in death rates are explained by this cause structure of mortality, with men dying more often than females (Taylor and Thoma 1985; Carter et al. 2011) (Figs. 29.2 and 29.3).

With 19.1% of GDP spent on health, Tuvalu was the most expensive and most susceptible of the PICs. In contrast, PNG spent only 2.4% of its GDP on health (Table 29.3). As the expense of inpatient care continues to rise, so does the demand for hospital care, which necessitates careful planning of hospital capacity.

Table 29.2 Life expectancy 2021 (current available data)

Country	Population	Male	Female	Total population
Cook Islands	8327	74.05	79.88	76.89
Federated States of Micronesia	101,675	72.06	76.4	74.17
Fiji	939,535	71.32	76.82	74
Kiribati	113,001	65	70.3	67.59
Marshall Islands	78,831	72.12	76.76	74.38
Nauru	9770	64.06	71.3	67.62
Niue	2000	NA	NA	NA
Palau	21,613	71.19	77.75	74.38
Papua New Guinea	7,399,757	67.37	72.48	69.86
Samoa	204,898	72.01	77.98	74.92
Solomon Islands	690,598	73.78	79.25	76.45
Tokelau	1647	NA	NA	NA
Tonga	105,780	75.63	78.99	77.29
Tuvalu	11,448	65.67	70.59	68.07
Vanuatu	303,009	73.18	76.66	74.87

Source: The World Fact Book, https://www.cia.gov/the-world-factbook

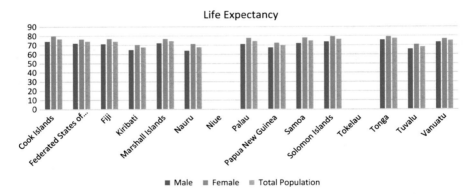

Fig. 29.2 Pacific Island small states—life expectancy in 2021. (Source: The World Fact Book, https://www.cia.gov/the-world-factbook)

Availability of inpatient treatment in many countries is primarily determined by bed capacity and its accompanying indicators such as bed occupancy and the ratio of beds to population. A hospital's bed capacity may be viewed as a sort of capital stock that is affected by the efficiency with which medical personnel and equipment fulfil their duties (Ravaghi et al. 2020). Gauld et al. (2018) mention that as part of the local health system, hospitals should be incorporated into the fabric of the system; they may take the lead in this by forming collaborative efforts with other healthcare institutions, such as primary care and private hospitals. Policymakers have a role to play in enabling this since it adds to the enhancement of the general health of the people. In the Pacific Island State Countries, we found that there is a shortage of

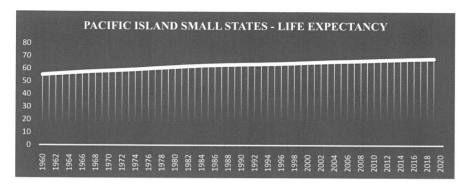

Fig. 29.3 Pacific Island small states—life expectancy (1960–2020). (Source: The World Bank—World development indicators; http://www.worldbank.org/)

Table 29.3 Health expenditures (% of GDP), hospital bed density and physicians

Country	Health expenditures 2018	Beds/1000	Physicians/1000
Cook Islands	2.9	NA	1.41
Federated States of Micronesia	12.6	3.2	NA
Fiji	3.4	2	0.86
Kiribati	12.1	1.9	0.2
Marshall Islands	NA	2.7	0.42
Nauru	9.6	5	1.35
Niue	8.3	NA	NA
Palau	10.9	4.8	1.42
Papua New Guinea	2.4	NA	0.07
Samoa	5.2	NA	0.35
Solomon Islands	4.5	1.4	0.19
Tokelau	NA	NA	2.72
Tonga	5.1	NA	0.54
Tuvalu	19.1	NA	0.91
Vanuatu	3.4	NA	0.17

Source: The World Fact Book, https://www.cia.gov/the-world-factbook

hospitals and beds/persons in pacific Island State Countries (Table 29.4). Therefore, administrators and legislators must engage in strategic planning to obtain a better grasp of the future of hospitals and the ideal number of beds available (Fig. 29.4).

COP 26: Health and Climate Resolution for Future

Climate change and health are inextricably linked, and the 10 proposals in the COP26 Special Report on Climate Change and Health urge governments and policymakers to address both issues as soon as possible. Over 150 groups, as well

Table 29.4 Population and hospitals

Country	Population	Hospitals
Fiji	935	180
Solomon Islands	685,097	303
Vanuatu	298,333	7
Federated States of Micronesia (FSM)	102,436	7
Kiribati	111,796	4
Marshall Islands	77,917	2
Nauru	11,000	2
Palau	21,685	1
Cook Islands	8574	2
Niue	2000	1
Samoa	203,774	12
Tokelau	1647	3
Tonga	106,095	3
Tuvalu	11,342	2

Source: The World Factbook, https://www.cia.gov/the-world-factbook/countries/

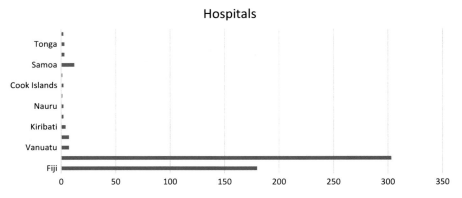

Fig. 29.4 Population and hospitals in PISC. (Source: The World Factbook, https://www.cia.gov/the-world-factbook/countries/)

as over 400 specialists and health professionals, were involved in the development of the proposals. Before the UNFCCC's 26th Conference of Parties (COP26), governments and other stakeholders will be informed, and several opportunities to prioritize health and fairness in the global climate development and sustainable development plan will be highlighted. Every proposal includes funds and case studies to assist policymakers and practitioners implement the proposed solutions. In approximately the course of 2 weeks, from the 31st of October to the 13th of November 2021, the COP26 UN climate summit in Glasgow attracted over 40,000 participants. The 2015 Paris Agreement on Climate Change and an improved aid package for vulnerable countries were agreed upon by representatives from 197 countries (What has COP26 achieved for health? (who.int)). Health and equity must be given top priority in the worldwide "climate movement and the sustainable development agenda," as evidenced by these 10 proposals:

1. Commit to a healthy recovery and a healthy, environmentally friendly, and just recovery from COVID-19.
2. There is no negotiating with our health: the UN climate negotiations must place health and social justice at the centre of their discussions.
3. To maximize the health, economic, and environmental advantages of climate action, prioritize those measures that have the greatest health, social, and environmental impact.
4. To better cope with the effects of climate change, it is important to strengthen the health of the population. Build health systems and infrastructure that are climate-resilient and environmentally sustainable and aid in the adaptation and resilience of the health sector.
5. Create energy systems that safeguard and promote the environment and human well-being: save lives by guiding an equitable and inclusive transition to renewable energy from coal combustion. Eliminate the use of inadequate amounts of energy in residences and medical institutions.
6. Transform the way people move around cities by focusing on active modes of transportation including walking, bicycling, and public transportation rather than fossil fuel-based modes.
7. As the cornerstone for a healthy life and a sustainable economy, we must protect and restore the environment.
8. Healthy, sustainable, and resilient food systems: ensure that food is produced in a way that is both environmentally and socially responsible.
9. It is time to move toward a wellness economy to save people's lives.
10. Pay attention to the medical community and suggest immediate climate action: support and mobilize the medical community to act on climate change.

Humanity's greatest health hazard is climate change. Climate change's health effects affect everyone, but the most vulnerable and disadvantaged people bear the brunt of the burden. To safeguard people's health from the consequences of climate change, we must make changes in every area, including energy, transportation, nature, food systems, and financial institutions. The advantages to public health that would result from enacting these bold climate measures overshadow the expenditures by a wide margin. As climate change worsens, health care officials globally are acting to safeguard their communities by cutting emissions and raising awareness of the threat. The 10 recommendations in this study outline the most important steps governments can take to combat climate change, restore biodiversity, and preserve public health. Our children, grandkids, and future generations will be better off because of these necessary reforms.

Conclusions

This research could help Pacific islanders plan and implement future climate change initiatives that improve the structural and social determinants of health, which are important for health. A lot of people who live on the Pacific Islands are very concerned about how changes in the weather will affect their health. Climate change and health processes, routes, and hazards have been explained in a new way by new information and ideas; changes in important characteristics of Pacific cultures such as institutional structures, economic growth, technology, and demography will necessitate periodic modifications of adaptation strategies. Doctors may play an important role in educating policymakers and the public by leveraging their well-established reputations as reputable members of the scientific community and society at large and the data mentioned in the paper showing shortage of doctors in PISC. One of the greatest opportunities of our time is for the medical community to encourage governments to take rapid action on climate change as a serious threat to global public health. Administrators and legislators must engage in strategic planning to obtain a better grasp of the future of hospitals and the ideal number of beds available. To further increase the climate change resilience of pregnant women, new mothers, and children and adolescents, Pacific health ministers signed up to the "Climate Change and Health Strategic Action Plan 2016–2020," which was endorsed by the "*KAILA*" "Pacific Voice for Action on Agenda 2030." (Vaka 2019). Many administrators do not believe in climate change and governments often lack the resources to implement long-term initiatives, so the most recent administrations have placed more emphasis on crisis management than on long-term planning. More significantly, it is critical to prioritize catastrophe-related mental health by increasing awareness and incorporating it into disaster management strategies at the local, national, and international levels.

Together we can fight and change the world in the era of the Anthropocene.

References

Aghai ZH, Goudar SS, Patel A et al (2020) Gender variations in neonatal and early infant mortality in India and Pakistan: a secondary analysis from the Global Network Maternal Newborn Health Registry. Reprod Heal 17:178. https://doi.org/10.1186/s12978-020-01028-0

Alam M (2021) History, historians and anthropocene. Scholar J Psychol Behav Sci 5(2):559–561.2

Alam MA, Alam M, Kundra S (2021) Knowing climate: Knowledge, Perceptions and Awareness (KPA) among higher education students in eritrea. Disast Adv 14(3):30–39

Altizer S, Ostfeld RS, Johnson PT, Kutz S, Harvell CD (2013) Climate change and infectious diseases: from evidence to a predictive framework. Science 341(6145):514–519

Anderson I, Irava W (2017) The implications of aging on the health systems of the Pacific Islands: challenges and opportunities. Heal Syst Refor 3(3):191–202. https://doi.org/10.1080/23288604.2017.1342179

Andruske CL, O'Connor D (2020) Family care across diverse cultures: Re-envisioning using a transnational lens. J Aging Stud 55:100892. https://doi.org/10.1016/j.jaging.2020.100892

Bouma MJ, van der Kaay HJ (1996) The EI Niño Southern Oscillation and the historic malaria epidemics on the Indian subcontinent and Sri Lanka: an early warning system for future epidemics? Tropical Med Int Health 1(1):86–96

Bourque F, Cunsolo Willox A (2014) Climate change: the next challenge for public mental health? Int Rev Psychiatry 26(4):415–422

Butler CD, Bowles DC, McIver L, Page L (2014) 26 mental health, cognition and the challenge of climate change. Clim Change Glob Health 26:251

Carter K, Soakai TS, Taylor R, Gadabu I, Rao C, Thoma K, Lopez AD (2011) Mortality trends and the epidemiological transition in Nauru. Asia Pacif J Publ Heal 23(1):10–23

Chen C, Noble I, Hellmann J, Coffee J, Murillo M, Chawla N (2015) University of Notre Dame global adaptation index country index technical report. South Bend, ND-GAIN

Clayton S (2021) Climate change and mental health. Curr Environ Heal Rep 8(1):1–6. https://doi.org/10.1007/s40572-020-00303-3

Clayton S, Manning C, Krygsman K, Speiser M (2017) Mental health and our changing climate: Impacts, implications, and guidance. American Psychological Association and ecoAmerica, Washington, DC. mental-health-climate.pdf (apa.org)

Climate Action Tracker (2018) Some progress since Paris, but not enough, as governments amble towards 3 C of warming. Climate Action Tracker. CAT Warming Projections Global Update - Dec 2018 (climateactiontracker.org)

Colagiuri S Palu T Viali S Hussain Z, Colagiuri R (2008) The epidemiology of diabetes in Pacific Island populations. The epidemiology of diabetes mellitus

Countdown NCD (2018) NCD Countdown 2030: worldwide trends in non-communicable disease mortality and progress towards Sustainable Development Goal target 3.4. Lancet (London, England) 392(10152):1072–1088

Fadda J (2020) Climate change: an overview of potential health impacts associated with climate change environmental driving forces. Renewable Energy and Sustainable Buildings (pp 77–119)

Filho WL, Scheday S, Boenecke J, Gogoi A, Maharaj A, & Korovou S (2019) Climate change, health and mosquito-borne diseases: Trends and implications to the pacific region. Int J Environ Res Public Health 16(24): 5114 https://doi.org/10.3390/ijerph16245114

Friedli L, World Health Organization (2009) Mental health, resilience and inequalities (No. EU/08/5087203). WHO Regional Office for Europe, Copenhagen. http://www.euro.who.int/__data/assets/pdf_file/0012/100821/E92227.pdf

Gauld R, Asgari-Jirhandeh N, Patcharanarumol W, Tangcharoensathien V (2018) Reshaping public hospitals: an agenda for reform in Asia and the Pacific. BMJ Glob Health 3(6):e001168

Gibson KE, Barnett J, Haslam N, Kaplan I (2020) The mental health impacts of climate change: findings from a Pacific Island atoll nation. J Anxiety Disord 73:102237

Hossain KA (2019) Global warming and impact to third world countries, Proceedings of the 2nd international conference on Industrial and Mechanical Engineering and Operations Management (IMEOM), Dhaka, Bangladesh, December 12–13, https://www.cia.gov/the-world-factbook/countries/fiji/

Hughes F, Hodkinson J, Montgomery H (2017) Tropical cyclones and public health. Br Med J 359

Kelman I, Ayeb-Karlsson S, Rose-Clarke K, Prost A, Ronneberg E, Wheeler N, Watts N (2021) A review of mental health and wellbeing under climate change in small island developing states (SIDS). Environ Res Lett

Lomborg B (2020) Welfare in the 21st century: Increasing development, reducing inequality, the impact of climate change, and the cost of climate policies. Technol Forecast Soc Chang 156:119981

Lowe SR, Bonumwezi JL, Valdespino-Hayden Z, Galea S (2019) Posttraumatic stress and depression in the aftermath of environmental disasters: a review of quantitative studies published in 2018. Curr Environ Heal Rep 6(4):344–360

Lund C, Breen A, Flisher AJ, Kakuma R, Corrigall J, Joska JA et al (2010) Poverty and common mental disorders in low- and middle-income countries: a systematic review. Soc Sci Med 71(3):517–528

McIver L, Hanna E (2015) Fragile paradise: health and climate change in the South Pacific. In: Butler CD, Dixon J, Capon AG (eds) Health of people, places and planet: reflections based on Tony McMichael's four decades of contribution to epidemiological understanding. Australia National University Press, Acton, pp 337–350

McIver L, Bowen K, Hanna E, Iddings S (2017) A 'Healthy Islands' framework for climate change in the Pacific. Health Promot Int 32(3):549–557

McMichael A (2014) Population health in the Anthropocene: gains, losses, and emerging trends. Anthropo Rev 1:44–56. https://doi.org/10.1177/2053019613514035

McMichael AJ, Lindgren E (2011) Climate change: present and future risks to health, and necessary responses. J Intern Med 270(5):401–413

Musso D, Rodriguez-Morales AJ, Levi JE, Cao-Lormeau VM, Gubler DJ (2018) Unexpected outbreaks of arbovirus infections: lessons learned from the Pacific and tropical America. Lancet Infect Dis 18(11):e355–e361

Patz JA, Epstein PR, Burke TA, Balbus JM (1996) Global climate change and emerging infectious diseases. JAMA 275(3):217–223

Petrova D, Rodó X, Sippy R, Ballester J, Mejía R, Beltrán-Ayala E et al (2021) The 2018–2019 weak El Niño: predicting the risk of a dengue outbreak in Machala, Ecuador. Int J Climatol 41(7):3813–3823

Rataj E, Kunzweiler K, Garthus-Niegel S (2016) Extreme weather events in developing countries and related injuries and mental health disorders-a systematic review. BMC Public Health 16(1): 1–12

Ravaghi H, Alidoost S, Mannion R et al (2020) Models and methods for determining the optimal number of beds in hospitals and regions: a systematic scoping review. BMC Health Serv Res 20: 186. https://doi.org/10.1186/s12913-020-5023-z

Salvat B, Wilkinson C (2011) Cyclones and climate change in the South Pacific. Revue d'écologie

Shammin MR, Enamul Haque AK, Faisal IM (2022) A framework for climate-resilient community-based adaptation. In: Climate change and community resilience. Springer, Singapore, pp 11–30

Shultz JM, Kossin JP, Ali A, Borowy V, Fugate C, Espinel Z, Galea S (2020) Superimposed threats to population health from tropical cyclones in the prevaccine era of COVID-19. Lancet Planet Heal 4(11):e506–e508

Steel Z, Chey T, Silove D, Marnane C, Bryant RA, Van Ommeren M (2009) Association of torture and other potentially traumatic events with mental health outcomes among populations exposed to mass conflict and displacement: a systematic review and meta-analysis. JAMA 302(5): 537–549

Taylor S (2021) The vulnerability of health infrastructure to the impacts of climate change and sea level rise in small Island countries in the South Pacific. Heal Serv Insight 14: 11786329211020857

Taylor R, Thoma K (1985) Mortality patterns in the modernized Pacific Island nation of Nauru. Am J Public Health 75(2):149–155

Taylor R, Badcock J, King H, Pargeter K, Zimmet P, Fred T et al (1992) Dietary intake, exercise, obesity and non-communicable disease in rural and urban populations of three Pacific Island countries. J Am Coll Nutr 11(3):283–293

The World Bank, World development indicators. http://www.worldbank.org/

Thompson HE (2021) Climate "Psychopathology". European Psychologist

UNFPA, Population and Development Profiles – Pacific Sub-Region, https://pacific.unfpa.org/sites/default/files/pub-pdf/web__140414_UNFPAPopulationandDevelopmentProfiles-PacificSub-RegionExtendedv1LRv2_0.pdf

VAKA, N (2019) IPPF PACIFIC. https://www.ippfeseaor.org/sites/ippfeseaor/files/2019-11/IPPF%20Pacific%20Strategy%20-%20Web%20(New).pdf

Walsh KJ, McInnes KL, McBride JL (2012) Climate change impacts on tropical cyclones and extreme sea levels in the South Pacific – a regional assessment. Glob Planet Chang 80:149–164

Weaver LJ, Hadley C (2009) Moving beyond hunger and nutrition: a systematic review of the evidence linking food insecurity and mental health in developing countries. Ecol Food Nutr 48(4):263–284

What has COP26 achieved for health? (who.int). https://www.who.int/news-room/feature-stories/detail/what-has-cop26-achieved-for-health

World Health Organization (2015) Human health and climate change in Pacific Island Countries

World Health Organization (2017) Climate change and health: framework for action 2017–2021 (No. EM/RC64/4). World Health Organization. Regional Office for the Eastern Mediterranean

World Health Organization (2018) Time to deliver a report of the WHO independent high-level commission on noncommunicable diseases

World Health Organization (2020) Noncommunicable diseases: progress monitor 2020. https://www.who.int/publications/i/item/9789240000490

Zhou B, Bentham J, Di Cesare M, Bixby H, Danaei G, Cowan MJ, Paciorek CJ, Singh G, Hajifathalian K, Bennett JE, Taddei C (2017) Worldwide trends in blood pressure from 1975 to 2015: a pooled analysis of 1479 population-based measurement studies with 19 1 million participants. Lancet 389(10064):37–55

Chapter 30
Changing Climate, Flood Footprints, and Climate-Related Actions: Effects on Ecosocial and Health Risks Along Ugbowo-Benin Road, Edo State, Nigeria

Angela Oyilieze Akanwa, Ngozi Joe-Ikechebelu, Angela Chinelo Enweruzor, Kenebechukwu Jane Okafor, Fredrick Aideniosa Omoruyi, Chinenye Blessing Oranu, and Uche Marian Umeh

Abstract Natural and human factors have had unprecedented impact on the earth climate system causing changes on a global scale and necessitating the need for climate-related actions. Nigeria has experienced the dire consequences of climate change resulting in large-scale shifts in weather patterns that have triggered increased rainfall intensity and flooding in almost every major city. Benin City, Edo State Capital for the past half a century, have experienced repeated flood events along transportation routes and its adjoining residential areas. These floods have occurred in diverse capacities causing devastating damages to lands, properties, infrastructure and roads. The transportation routes are usually in terrible conditions

---

A. O. Akanwa (✉)
Department of Environmental Management, Chukwuemeka Odumegwu Ojukwu University (COOU), Awka, Anambra, Nigeria

Spirit Filled Women International (SFWI). A non-governmental Organization (NGO) for Women Development, Anambra State, Nigeria

Nigeria Coalition on EcoSocial Health Research (NCEHR), Anambra, Nigeria
e-mail: angela.akanwa1@gmail.com; https://ncehr.org

N. Joe-Ikechebelu · U. M. Umeh
Department of Community Medicine and Primary Healthcare, Chukwuemeka Odumegwu, Ojukwu University (COOU) Teaching Hospital (COOUTH), Anambra, Nigeria

Nigeria Coalition on EcoSocial Health Research (NCEHR), Anambra, Nigeria
e-mail: nn.ikechebelu@coou.edu.ng; ngozijoeikechebelu@uvic.ca; https://ncehr.org; um.umeh@coou.edu.ng; https://ncehr.org

A. C. Enweruzor
Department of Sociology and Anthropology, University of Benin, Benin City, Edo State, Nigeria

Spirit Filled Women International (SFWI). A non-governmental Organization (NGO) for Women Development, Anambra State, Nigeria
e-mail: spokencreated@gmail.com

© The Author(s), under exclusive license to Springer Nature Switzerland AG 2022
U. Chatterjee et al. (eds.), *Ecological Footprints of Climate Change*, Springer Climate, https://doi.org/10.1007/978-3-031-15501-7_30

after floods disrupting movement and socio-economic activities. The concept of Climate-related action was employed as a remedy in this qualitative case study to investigate the socio-economic and health risks of flooding footprints on Ugbowo-Benin road failure in Eg or LGA, Benin City, Edo State, Nigeria. It includes the use of primary and secondary data sources. Primary data covered transects walks, observations, photographs and in-depth interviews of key informants. From our findings, the flooding has persisted for over five decades in the area, which may be (un)related to poor-road design, river intrusion, poor-road maintenance and rehabilitation, and worsened by changing climate, as well as destruction of lives, household goods and properties goods, infrastructures, disrupting human and vehicular movements and services. More so, the flooding has continued to worsen scenic views of the highway, degrade roads resulting to huge economic loss from the disrupted commercial services and submerged properties. Concerning health risks, the flooding heightens formation of water pockets along the highway aiding the breeding of mosquitoes worsening the occurrence of malarial diseases and upper-respiratory tract infections. Based on our study, we recommend that community is fully involved in the highway road intervention plan, as well as the integrated flood management plan by the national government.

Keywords Road failure · Flood footprints · Eco-social · Health · Highway · Climate action · Nigeria

Introduction

Globally, climate change has become the most serious environmental threat to the existence of mankind and societies, necessitating the need for *climate-related actions*. According to UNDP, climate-related actions are defined as the collective efforts made to reduce greenhouse gas (GHG) emissions and strengthen resilience and adaptive capacity to climate-induced impacts, including: climate-related hazards

---

K. J. Okafor
Management Science, Accountancy Department, Nnamdi Azikiwe University, Awka, Anambra, Nigeria

Spirit Filled Women International (SFWI). A non-governmental Organization (NGO) for Women Development, Anambra State, Nigeria
e-mail: kayokafor1@gmail.com

F. A. Omoruyi
Department of Statistics, Nnamdi Azikwe University, Awka, Anambra, Nigeria
e-mail: fa.omoruyi@unizik.edu.ng

C. B. Oranu
Department of Geology, Faculty of Physical Sciences, University of Benin, Benin City, Nigeria

Spirit Filled Women International (SFWI). A non-governmental Organization (NGO) for Women Development, Anambra State, Nigeria
e-mail: oranuchinny123@gmail.com

in all countries; integrating climate change measures into national polices, strategies, planning and improving education, awareness raising, human and institutional capacity with respect to climate change mitigation, adaptation, impact reduction and early warning. Scientists have recorded temperature increase across the globe. Nigeria's climate has experienced significant changes, and the projections for the coming decades reveal a significant increase in temperature over all the ecological zones in Nigeria (Akande et al. 2017).

Notably, the temperature of the nation has risen significantly since the 1980s (Enete 2014; Federal Ministry of Environment 2014). For the Nigeria's tropical climate, it has two precipitation regimes: the low precipitation in the North and high precipitation in parts of the Southwest and Southeast (See Fig. 30.1). Though climate variability is not the same across the country, the variation has influenced the northern and southern parts of Nigeria escalating extreme weather events across the nation. These weather variations are evident as increasing temperature, varying

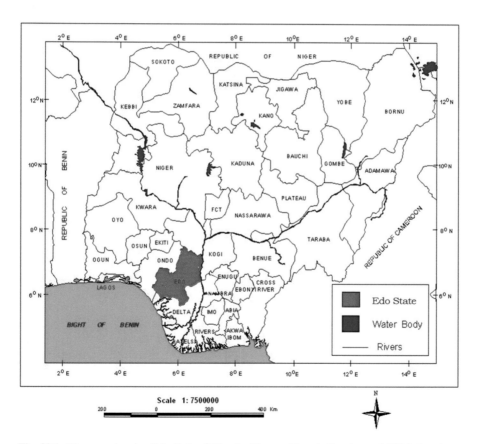

Fig. 30.1 The map showing Edo State of Nigeria. (*Source*: Remote Sensing and GIS Laboratory, Department of Environment Management, Chukwuemeka Odumegwu Ojukwu University, (2021))

rainfall, rising sea level and flooding resulting to drought and desertification, land degradation, degraded wetlands and loss of biodiversity (Elisha et al. 2017; Ebele and Emodi 2016; Olaniyi et al. 2013). However, there are distinctive climate problems in the nation whereby intense aridity, drought and desertification are particular in the north, while flooding and erosion are experienced in the South (Nkechi et al. 2016; Akande et al. 2017). A Climate change is increasing the intensity of rainfall producing large runoffs and floods in many places in Nigeria (Enete 2014). Flooding is a peculiar event in the southern Nigeria with high expectations of rising sea levels that can exacerbate flooding and submerge coastal lands (Akande et al. 2017; Ebele and Emodi 2016). Again, high sensitivity and porosity of the soils in the south-south and southeast zones have escalated the levels of flooding that even a short period of intense rainfalls can lead to flash floods and cause extreme havoc on the environment (Federal Ministry of Environment 2014).

Additionally, Nigeria is a mostly flat, low lying, swampy basin resulting in severe regular flooding which has led to a limited land area available for human habitation. Approximately, 7000 km^2 of nation's land cannot be used owing to flooding. Further, flooding has caused tremendous negative effects especially in urban areas. In 2012, flooding displaced more than one million Nigerians, damaged significant properties, caused traffic obstruction/delays, nuisance and health hazards, destruction of crops and farmlands, disruption of socio-economic activities heightening the cost of provision of goods and services. Within Southern Nigeria, the south-south (Niger Delta region) is the most vulnerable zone due to sea level rise, increased precipitation, coastal erosion and flooding, which has resulted in the displacement of many settlements, damage of infrastructure and properties (Matemilola 2019; Federal Ministry of Environment 2014; Sayne 2011).

Benin city, the Edo State capital for the past half a century, has experienced repeated flooding particularly along its major transportation routes and adjoining streets. Considering that, Edo state is one of the Nigeria's densely populated and increasingly urbanized cities within the south. The rapid growth of Benin City after its declaration as the Edo State capital led to increased human and vehicular traffic along Ugbowo Benin-Lagos Express Road. This multiplied commercial and business activities, government, state and private enterprises and administrative services. However, climate changes and increased activities coupled with human influx were not adequately supported by improved existing roads, drainage systems and infrastructure. These have escalated flood occurrence annually in diverse capacities causing devastating damages to lands, properties, infrastructure, roads and disruption of socio-economic activities, health risks and loss of human lives. Regrettably, according to Odemerho, floods have affected the built areas from buildings, access streets and the major roads since 1965. Again, with an initial count of two flood prone areas in 1965 according to the Master Plan for drainage and sewage, 38 flood prone areas were identified in 1981 and 45 in 1985.

More recently, a study carried out by Ogbonna et al. (2011) identified 35 streets and roads as areas prone to regular flooding. Eleven of these streets, including New Lagos-Uselu Road, Sakpoba-Oka Road, Plymouth-Oliha Road, Uwasota-Ogida road, Ogida-textile-Urubi Road, Television Road, Erhu Road, Upper Mission

Road, Wire Road, Forestry Road and Uselu-Ugbowo road, are identified as the most critical zones and currently under study. Following climate increased rainfall triggering annual flood crisis in the area, the conditions of the roads worsen. Notably, Ugbowo Road is 200 miles by road, East of Lagos, and plays an indispensable role in Benin City being within few kilometres away from the prestigious University of Benin. This makes it a carrier roadway for huge flow of people, materials, food, goods and services. It plays multi-dimensional functions from social and economic interactions to intra- and inter-regional movement covering major commercial, administrative, cultural and industrial cities in Nigeria such as Delta, Edo, Lagos, Anambra and Rivers States. To provide substantial solutions to climate-driven floods in Nigeria, particularly in Edo state, positive climate-related actions are indispensable in the wake of sustainable management of climate resources; otherwise, a vicious cycle of interconnected risks abounds (Anukwonke et al. 2021).

According to UNDP, the concentration of GHG emissions in the atmosphere is wreaking havoc across the world and threatening lives, economies, health and food. The world has been unable to secure a global temperature rise to below 2 °C as declared by the Paris Agreement. Moreover, the Nigerian government believes in the efficacy of political, institutional, public participation and legal frameworks in addressing climate change challenges. For instance, in addressing the risks of climate change and its interconnectedness to sustainable development, Anukwonke et al. (2021) opined that institutional platform and governance with core citizen participation are essential for arriving at a sustainable development. The climate-related actions will be appropriate, because it will aid the community to prepare for and respond to climatic stress by facilitating inclusive, sustainable and community-driven adaptation strategies. The principles for Nigeria's plan of action for climate change adaptation options illustrate a sustainable route of action. The plan incorporates all stakeholders' involvement, available adaptation patterns for communities and ecosystems to offset climate-induced disasters and the outright involvement of local experience (Daisy 2020). The Nigerian policy on climate change targets on halting the GHG emissions and their adverse impacts on vulnerable communities and sectors, increasing awareness of its adaptation strategies and programmes, strengthening socio-cultural, economic and environmental resilience.

The Nigerian climate policy targets to encourage research, build and strengthen capacity and heighten technological experience to optimize critical interventions at all government levels, incorporating their outcomes to the tenets of development planning across critical sectors. Ultimately, climate-related actions should be driven towards a sustainable and an effective flood-free transportation system with sufficient connectivity void of traffic congestion and discomfort during wet seasons, when floods are inevitable. Moreover, this will save travel time and allow for viable social and economic activities in a clean and safe environment. From the foregoing, repeated flood occurrence in 35 or more flood prone areas experienced in Benin for the past 54 years have remained persistent. Obviously, the consequent socio-economic and health risks of flooding are yet to be adequately addressed. This has necessitated this chapter on changing climate and flood footprints on eco-social and health risks along Ugbowo-Benin road failure in Edo State, Nigeria. This case study examined the challenges of climate change and flooding along Ugbowo transportation route, its ecosocial and health consequences as well as the adoption of the Nigerian climate change policy as a sustainable means of action. Finally, we made recommendations based on our study findings.

Climate-Related Actions, Community Involvement and Participation in Nigeria

Climate change has become a global issue, though it is felt on a local scale, yet requires public participation. In the sustainable development goals (SDG), goal 13 calls for urgent action to combat climate change and its impacts. This goal is closely connected to the 16 goals in the 2030 agenda for sustainable development. Central to climate-related action is the Paris Agreement (2015) that is built on the United Nations framework of Convention on Climate Change (UNFCCC) in 1994. For the first time, this convention brought all the UN member nations together to undertake ambitious efforts to combat climate change and adapt to its effects. The goal of the Paris Agreement was aimed at the member nations to plan and state their nationally determined contributions (NDCs) embodying each country to reduce national emissions, adapt and mitigate to the impacts of climate change. The importance of public participation in responding to climate change has featured in key international statements since the Rio Declaration developed in 1992 at the United Nations Conference on Environment and Development (UNCED). The statements included explicit goals for citizen participation and engagement in climate actions (Principle 10). Its significance was reiterated more than 28 years later in the Intergovernmental Panel on Climate Change (IPCC) Special Report on the impacts of global warming of 1.5 °C above pre-industrial levels, which specifically identified public participation in adaptation planning to enhance capacity to cope with climate change risks (IPCC 2022).

Similarly, issues related to public participation in climate action have long been a concern for researchers, although a paucity of empirical work in relation to public participation in adaptation planning is often lamented. Practitioners such as Francis and Hester argued that community participation can be involved in proactive practices governed by local people to emphasize their understanding of natural processes and social relational practices, which are brought in to solve climate-related problems within the locality and the larger environmental context. The critical climate-related actions in Nigeria borrow from the global antecedents and framework.

Through the Paris pledge, Nigeria has contributed immensely as a member actively participating in three deciding units at the Global Climatic discussions. These units are the G77 and China, the African group and the Coalition for Rainforest Nations (Daisy 2020). Since the ratification of the Paris Agreement in 2017, Nigeria, at the international level, has pledged allegiance to cut down her greenhouse pollutants by 20% by the year 2030 (Daisy 2020). Key climate actions are discussed by the United Nations through the 13th SDG. The climate action outlines target measures aimed at advocacy and readiness for the impacts of climate change; capable of mobilizing concerned communities for the required climatic shift (s) and adaptation choices. Additionally, the actions sought to halt the degree of risks on the most vulnerable communities; and to mainstream such actions into practices. For these related climate actions, there are potentials to offset the levels of GHG with commensurate climate actions capable of averting severe environmental, social and economic losses. Currently, in Nigeria, numerous climate-oriented actions have been in practice ranging from media advocacy, value reorientation and mitigation options to other practical means that address climate change impact across all country regions. Although the significant effects of climate change have been felt disproportionately between Northern and Southern Nigeria, climate change actions have been multiplied by climate activists, climate change leaders, volunteers, researchers, faith-based and youth organizations. Other actions are sourced and rooted through bodies such as Federal Ministry of Environment, State environmental protection agencies, National Environmental Standards and Regulations Enforcement Agency (NESREA), Nigerian Meteorological Agency (NIMET), National Emergency Management Agency (NEMA), Young African Leaders Initiative (YALI), Sustainable Development Groups and other international agencies like World Bank and United Nations. These sources of support have continued to strengthen capacities for climate-related actions in Nigeria. The actions, per se, resonate awareness campaigns and programs rooted in advocacy, such as tree planting campaigns and incentives to discourage deforestation and climate-related symposia hosted on annual environmental-based celebrations days, such as World Environmental Day on fifth June, Earth Day on 22nd April. With these exposures, climate-related actions have been domesticated at the grassroot community levels. Amongst all the climate-related actions promoted by the international climate agreements, the Nigerian government got a climate fund of $136 million USD dollars in 2016 from European countries to intensify climate actions in vulnerable communities (Daisy 2020). These funds have been maximized in harnessing the energy policy and the development of a credit profile for renewable energy projects, fuelwood management and hydropower infrastructure development. Though Nigeria is enlisted as leading in youth-friendly climate movement, intensive efforts are paramount to harness solar-powered infrastructure at all levels of governance. Through the Paris Agreement and pledge, practical options abound to support climate-related actions by increasing solar energy utilities, maximizing energy uses and efficiency, and halting gas flaring from the Niger Delta areas.

Study Area

Ugbowo is in Egor, LGA in Benin City, Edo State, the South-south geopolitical zone in Nigeria (See Figs. 30.1 and 30.2). The study covered Tomline and Adolo junctions both located along Ugbowo-Benin Express Road. The geographical

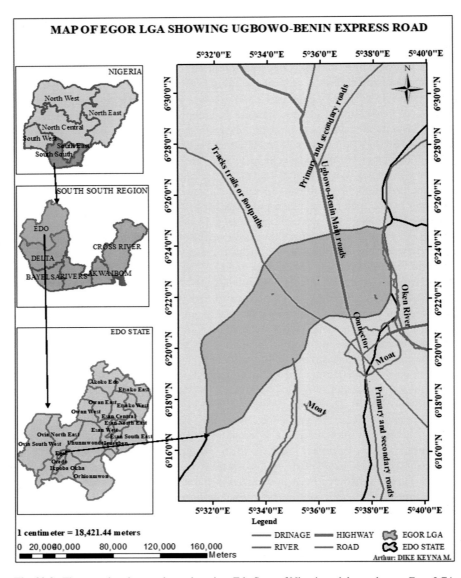

Fig. 30.2 The map showing south-south region, Edo State of Nigeria and the study area *Egor LGA*. (*Source*: Department of Environment Management, COOU, (2021))

co-ordinates for Tomline junction are latitudes 06°21′ 57 N and longitudes 05°37′ 02 E with an elevation of 197 m, while the co-ordinates for Adolo junction are latitudes 06°23′ 21 N and longitudes 05°36′ 39 E with an elevation of 225 m. The city has a population of 1,086,882 and a projected population of 1,481,310 in 2016 at 3.5% growth rate. Apart from its thriving population, it serves as an administrative, commercial, industrial, institutional and a cultural city. Ugbowo flourished during the thirteenth to nineteenth century to a multi-functional city, but the town has not been adequately provided with the attendant infrastructures necessary for a principal, colonial/historical city of the Edo kingdom.

However, Ugbowo town is made up of sedimentary rock with the Benin formation of the Miocene-Pleistocene age. Its soil is characterized as majorly clay (52%), sand (28%) and silt (19%) with low permeability, coupled with a flat and lowing terrain that aids flooding (Ogbonna et al. 2011). The city is drained by two major rivers namely: the Ikpoba River draining the eastern part and the Ogba River draining the western part of the city. It is in the tropical rain forest zone of Nigeria and experiences both wet and dry seasons (Ikhile and Olorode 2012). The wet season is from late February to October with a break in August giving rise to double maximum rainfall. Rain falls for about 9 months in the year with its peak in the month of July which explains its vulnerability to flood. The dry season lasts from November to March with a cold Harmattan period between December and January. There are two temperature ranges throughout the year, the higher average temperature in the dry season and lower in the raining season. Mean annual temperature ranges from 21 to 30 °C. The mean rainfall is above 2000 mm (Akujieze 2004). The high-rainfall intensity and spread are exacerbated by climate change and other natural factors such as the flat terrain, clay soil, low permeability and sedimentary formation. Other human actions, such as waste dumping on drainages, building of infrastructures and physical properties on floodplains with the poor road networks, maintenance and draining capacities have contributed to persistent flooding on roads and built areas for 54 years.

Materials and Methods

This study employed a qualitative approach covering sampling strategies, data collection analysis, synthesis, integration and reporting (Creswell 1995). However, our data collection covered socio-demographic and environmental data. Qualitative data collected through semi-structured in-depth interviews from key informants, observations, photographs and informal interactive dialogues with participants were recorded. Using purposive sampling technique (convenient and snowball methods), we collaborated with the Ugbowo Street Market Union Leader to recruit participants for this study. We used the following inclusion criteria for the 20 participants: 18 years of age and above, confirmed residents of 20 years and above, participants operating daily socio-economic activities along Ugbowo Express Road and their ability to understand and speak English.

Members of the research team interviewed study participants using a semi-structured interview guide and each interview lasted about 15 min. The participants include married/single people, traders, cosmetic and clothes dealers, fruit dealers, and local dealers involved in ATM and phone recharge card, mechanics, tailors, residents operating and living along Ugbowo Express Road. Further, for our key informant interview, we interviewed four persons; Union Leader for Market Traders, Head of Ministry of Works Edo State, Religious Leader and the Chairman of Egor local Government Area (LGA).

With our sample size for this study ($N = 20$), our socio-demographic data will be descriptively analyzed. The qualitative component of this study yielded rich data and will be the focus of the findings reported here. The interview guide included questions on repeated flooding menace, effects of flooding, socioeconomic/health risks, income levels and environmental conditions of roads among others. Interviews were audio recorded and transcribed verbatim. Secondary data were obtained from relevant papers, journals or publications related to climate-induced flooding and social dimensions of flooding among others.

Results

Working with the local Ugbowo Street Market Union as collaborators, we were able to recruit 20 participants who completed the survey. Five participants consented to in-depth individual follow-up interviews with our researcher. The demographic variables, such as sex, age, income and status, were asked, and the summary of socio-economic, environment and health data gathered from our research methods with participants was provided. Key informant interviews, discussions, observations and photographs were manually transcribed and analyzed from the detailed transcripts from qualitative data generated by the researchers. The analyses were presented and discussed using quotes from the interviewed four key informants, which were adopted to illustrate their unique experiences over the persistent flood issue.

Table 30.1 summarizes the key characteristics of participants. Quantitative findings were descriptively analyzed to provide contextual information alongside the qualitative findings. For the highest age distribution, 30% of the participants were between 18 and 30 years, while the lowest age group was 10% and within the age bracket of 60 years and above. Majority of the participants were young and active. For the sex of our parrticipants, 60% were males, while the female participants were 40%. Again, 40% of the participants had their highest educational level to be the West African School Certificate (WASC), which may mean that they finished their secondary school education. For the next educational level, 35% had the First School Leaving Certificate (FSLC), implying they may have stopped at their primary education level. While 10% of them had no educational background.

Married participants represented 40% of the sample, while the single population represented 30%. Only low fractions of the sampled population were widowed, divorced and separated represented by 15%, 10% and 5%, respectively. For the number of children within families, the highest households had children between 1 and 3 representing 60%, followed by families with children between 3 and 5 indicating 25%, and the lowest household size was 15% representing families with 5 children and above. This showed that majority of the married participants had

Table 30.1 Socio-economic/environmental data of participants along Uselu-Ugbowo Road

Variables	Frequency (20)	Percentage (100%)
Age		
18–30 years	06	30
30–40 years	05	25
40–50 years	04	20
50–60 years	03	15
60 and above	02	10
Gender		
Male	12	60
Female	08	40
Education level		
FSLC	07	35
WAEC/WASC	08	40
NCE/OND	03	15
No education	02	10
Marital status		
Single	06	30
Married	08	40
Widow/ed	03	15
Divorced	02	10
Separated	01	05
Land use activities		
Residential area	07	35
Commercial area	10	50
Institutional area	03	15
Household size		
1–3	12	60
3–5	05	25
5 and above	03	15
Years of living in the community		
1–2	01	05
2–5	02	10
5–10	05	25
10 years & above	12	60
Income level (monthly)		
₦1000–10,000	01	05
₦11,000–₦20,000	02	10
₦21,000–₦30,000	02	10
₦31,000–₦40,000	04	20
₦41,000–50,000	05	25
₦50,000 and above	06	30
No of hours for flood to recede		

(continued)

Table 30.1 (continued)

Variables	Frequency (20)	Percentage (100%)
Time		
1–12 h	04	20
12–24 h	06	30
24 h and above	10	50
Environmental risks		
Poor aesthetics	07	35
Flooded roads	06	30
Blocked drainage system	04	20
Offensive odour	03	15
Health risks		
Cold/fever	08	40
Catarrh/cough	05	25
Malaria	04	20
Typhoid	03	15
Local adaptation strategies during floods		
Shut down shops/buildings	06	30
Raise pavements	03	15
Temporary relocation	07	35
Residents stay indoors	04	20

Source: Researchers' Analysis (2021)

wards and children to cater for. Further, the highest land use activities along Ugbowo Express Road were commercial activities representing 50%, next was residential 35% and lowest was institutional uses which was 15%. This revealed that the study area is mostly a commercial area characterized by shops, business centers and retail/wholesale dealers. Majority of the sample population have domiciled in the study area for 10 years and above. This was recorded as the highest with 60% and the lowest was 5% of people with 2 years of residency. Also, using the http://wise.comcurrency-converter, the highest monthly income for 30% of the participants was 50,000 naira (120.25 USD), while the lowest income for the 5% was within 10,000 naira (24.05 USD) representing 30% and 5%, respectively.

For the environmental date, 50% of the sampled population reported that when flooding is intense, the floods last a day or more before it recedes during wet seasons, while the lowest interval of flooding can last between 1 and 12 h. Again, the floods are responsible for poor aesthetics in the study area as indicated by 35%, while flooded roads and streets indicate 30%, blocked drainages revealed 20% and offensive odors was 15%. In the last 6 months before our study, for their health challenges, 40% reported that they have suffered from cold and catarrh, while 25% and 15% had suffered from malarial and typhoid fever, respectively, as evidenced by laboratory results. The remaining 25% report that they face risks from cold, catarrh, cough and fever. Some of the local adaptation strategies adopted in the study area by 35% of the participants were temporary relocation of businesses during wet seasons,

30% shutdown their business activities, 20% stay indoors more often and 15% raise their pavements higher to avoid the flood submerging their properties and buildings.

Discussion

A Young Population

There is a high possibility of sustainable climate-related actions by the large population of youths in our study area in Egor LGA, Edo State. According to the National Bureau of Statistics (2012), the population of youths in Nigeria ranges between 15 and 35 years and is estimated to be at 64 million youths out of 200 million Nigerians. The United Nation defines "youth" as one between the ages of 15 and 24 (UN 2020). Further, the United Nations projections on the population of the young people in the world stands at roughly 1.21 billion youths, which may imply that majority of the world's youth are suffering from the detrimental impacts of climate change particularly intense weather events like flooding. Young people make up the largest population in many countries like Nigeria. They are becoming a driving force in pursuing a low carbon and climate resilient future (UN 2013).

The concept of sustainable development recognizes the valuable contributions (positive and negative) of the present generation that affect the future. As the rising levels of climate change impact intensify over time, it is unavoidably certain that the younger generation which represents the future have a role to play. The world has a population of 1.8 billion youths (UN 2020), and this is the largest generation of young people in history. Added to their rising awareness and participation in climate change solutions and action, this is a positive step in achieving sustainable actions towards climate crisis. The increased mobilization of the efforts of youths globally on climate issues is an indication of their power as decision makers and stakeholders. Young people in the Edo state can become positive contributors to climate action and agents of change, innovators, entrepreneurs through science, technology, research, sensitization on recurring flood problems.

Ecosocial Risks of Flooding

Our study found that 40% of our sampled population are men, and with 60% of them married with an minimum of three children from the result. Further, more than half of our respondents have lived in the study area for over a decade. Ideally, they can give a fair representation of their lived experiences in a flooded city. With over 50% of respondents dominating in commercial activities, we can deduce the diverse representation of commerce activities. Of note, the minimum wage pay earned in Nigerian Civil Service is 30,000 naira (72.13 USD), we can infer that our study area is lucrative, because 30% of the participants earn a monthly income of about 50,000 naira (120.25 USD).

Notably, the sampled population are about 40% male and 60% of them are married with a minimum of three children. This is an indication that they have families to feed and 60% of them have lived in the area for over 10 years. Added, to the long years of domcile, they have also attained the basic education, unaguably, they can provide relevant information concerning repeated occurrence of floodwaters in the study area. Also, the study area is dominated by diverse commercial activities representing 50%, thus, it is a highly economic active zone.

For the inhabitants of Ugbowo-Benin highway, the wet seasons make the study area to be full of difficulties relative to movement of people goods and transport. Economic activities almost come to a halt having been disrupted by the floods. The Union Leader, one of the key informants mentioned that:

- Our rainy seasons is the worse season of the year for us here. Our businesses are greatly affected especially the high reductions in our monthly income to over 50%. Most times, the reduction levels in our income are determined by the intensity of the flood. We sell only when the floods recede or in worst scenarios, we remove some of our fragile goods such as clothing, food products and stationary among others. Most of us relocate to other street markets until the rains are gone with their pain.

Moreover, 50% of the sampled population reported that the floods last a day or more before it recedes during wet seasons when it is most intense. While the lowest of the floods can last between 1 and 12 h representing 20%. This is an indication that the floodwaters disrupt the flow of business activities in the area. Another key informant participant, a religious leader complained that social activities are also negatively affected.

During rainy season, the floods are so much that people don't even attend weekly and Sunday (church) services because the streets are blocked with floodwater. People can neither walk nor drive their cars through the streets. In fact, all social, religious and recreational activities located along the study area are brought to a halt.

Majority of the participants agreed that the roads are flooded, and this increases transportation costs while increasing transit time. Notably, transportation and socio-economic development are closely related, and the society depends largely on the nature, structure and functionality of the roads (Omofonwan 1992). Accessible transportation network aids in socio-economic exchange of people, raw materials and finished goods among others. However, flooding on the Ugbowo-Benin Express Road is responsible for traffic congestion, poor environmental conditions and degradation, degradation of the road and disruption of socio-economic activities (See Figs. 30.3, 30.5, 30.6, and 30.7).

Majority of the shops and markets along the road networks are shut down for hours; and in most cases, it takes a whole day for the submerged water to subside. Majority of the people involved in commercial activities in the study area are self-employed and depend on the daily business activities for their family upkeep. The

Fig. 30.3 Showing flooded streets and deserted road inhibiting social activities and movement of people along Tomline, Ugbowo Lagos-Benin Express Road

business types include shopping plazas, supermarkets, restaurants, markets, transportation business, phone recharge-card vendors and food vendors, who operate under umbrellas and make-shift buildings (Fig. 30.4).

Findings from the study also reveal that houses, household properties, roads, electricity poles and quality of life are greatly affected. This is consistent with Okechi, whose study identified the negative effect of flooding, such as loss of livelihood from physical properties to daily businesses, disruptions to infrastructures (e.g. electrical utilities and transportation systems). An assessment by Oruonye further revealed some socio-economic effects such as loss of lives and internal displacements.

Environmental Risks

Findings reveal that the environmental conditions of the study area are negatively affected by floods along Uselu-Ugbowo road, which is an important interstate road within the Ugbowo Lagos-Benin Road network, as well as the frequent flooding from other adjoining roads that also worsen the heavy vehicular traffic problems along this major road. About 35% of the participants reported that floods are responsible for poor aesthetics, 30% indicated that the roads are flooded, 20% mentioned that the drainages are blocked with floodwaters and 15% identified that there is the release of offensive odour by the floods running off with garbage, breaking up sewage with spill offs of contents from sceptic tanks into the streets (See Fig. 30.7).

Fig. 30.4 Showing interactive sessions during interview with participants who are involved in small-scale businesses along Ugbowo, Lagos-Benin Express Road

Fig. 30.5 Showing flooded houses and traffic problems hindering human and vehicular movement along Adolo junction, Ugbowo Lagos-Benin Express Road

Many of the flood ponds gradually turn to stagnant pool of waters that produces foul odour, breed mosquitoes and sometimes obstruct the movement of people and goods. These pools of stagnant water can lead to surface and ground water contamination. Participants expressed their concerns over the situation, but they have

Fig. 30.6 Showing flooded Ugbowo Lagos-Benin Express Road and disruption of commercial activities and street market along Adolo junction

Fig. 30.7 Showing poorly designed and garbage-blocked drainage system along Adolo junction, Ugbowo Lagos-Benin Express Road

climate-related actions to ease their adaptation. Some common adaptation strategies reported that 35% of the participants relocated their social-economic activities temporarily until the end of the flood season. 30% shut down their buildings/shops, while 15% raised their pavements as a means of protecting their building from being submerged. It was also noted that the drainage systems were not properly designed and, hence, unable to control the flood rates. Also, solid waste management was reported as another environmental challenge that worsen the aesthetic road values, besides the health risks they pose. The solid wastes are indiscriminately disposed along street corners and in gutters. Easily, the floods carry them along, depositing these wastes on the streets or in gutters (See Fig. 30.7). According to Ogbonna, solid waste materials, like plastic bags and bottles, broken pieces of wood, saw dust, are deposited by floodwaters in streets and in poorly developed gutters and they inevitably block the gutters and drainage systems.

Health Risks

Generally, floods are the most common naturally occurring hazard and are responsible for a greater number of fatalities globally (EM-DAT 2011). Floods have more destructive effects when compared to other natural events. Findings from the study showed that majority of the participants experienced health risks during floods in the study area. 40% of the participants indicated that they had experienced cold and fever, 25% had catarrh and cough, 20% were affected by malaria and 15% participants suffered from typhoid. Mostly water-borne diseases such as typhoid and vector-borne diseases such as malaria were experienced in the study area. However, three participants reported that floodwaters serve as immediate dangers to their health particularly injuries, mental discomfort, malaria and extreme cold conditions. One of the participants spoke of the trauma of almost drowning during one of the episodes of floods:

> *I was so scared on how fast the water level rose and while I was trying to escape, I fell into one of the open gutters that was covered with water. I sustained injuries from the fall that almost drowned me*

Notably, floodwater can harbor bacteria and disease-carrying organisms that pose risk to people. Climate change has brought about disorder in the sequence of precipitation, and sea level rise is expected to increase the frequency and intensity of floods in many regions of the world (IPCC 2001). After flood crisis, there are unsanitary conditions that increase the risk of disease outbreaks. Again, flooding affects drinking water facilities resulting into diseases such as typhoid and cholera. According to Paterson et al. (2018), exposures to floodwaters pose direct dangers to human health including the risk of malaria infection. There is relationship between flooding and spread of disease. Many important infections are transmitted by mosquitoes, which breed in, or close to, stagnant or slow-moving water. In the study area, most of the floodwater is collected in open drainages and could last for a long time (See Figs. 30.5 and 30.7) aiding the rapid breeding of mosquitoes' parasites. Another participant noted that most of the floodwaters were increased due to many surrounding rivers overflowing their banks:

> *After I had direct contact with the polluted water during the flood, I found out that I contracted cold and started having feverish condition, added to the stuffiness of my ears and nose and I started having throat infections.*

Concerning mental health, the WHO (2012) in their definition of mental health defined it as a state of well-being in which the individual realizes his or her own abilities, can cope with normal stress of life, can work productively and fruitfully, and is able to make a contribution to his or her community. In their definition, mental health is described relative to symptoms such as anxiety, depression, aggression, stress, intrusive memories, irritability and sleeplessness. These symptoms can be worsened during the flooding events. In this study, one of the participants described their despair and mental discomfort during the flood crisis:

> *I felt so stressed and full of anxiety because of the flood, again, I had to face the discomfort of transferring my goods and properties. I was also worried about their safety and that of my children. It is usually a very turbulent period for me, their outcomes always make me stressed.*

Drawing from in-depth interviews with the Head of Ministry of Environment and other staffs in Edo State, they reported that Benin City, the Capital of Edo State is located in the Niger Delta region which experiences peak rainfalls during the wet seasons coupled with the changing climate that has induced increased storm surges. He stated that . . .

> *Edo State has a plain relief area, that is low lying and has a swampy terrain accelerating the high intensity of storm run-off and hence, triggers flood disasters rising into built up areas and transportation routes. It has been grouped as a major flood prone areas with historical records of severe flood occurrences traceable to changing climate as well.*

The south-south (Niger Delta region) is regarded as the most vulnerable in the south (Federal Ministry of Environment 2014). This was confirmed by Sayne (2011) that the Niger Delta region is within the southern coast having a vulnerable flooded network of estuaries, rivers, creeks, and streams, sits particularly low less than 20 feet above sea level. The rising sea levels, attributable to climate change, and precipitation are expected to be higher in coastal areas. Coastal areas have already experienced sea level increment of almost one foot in the past five decades, with forecasts indicating that the increment could be as high as 3 feet within the next nine decades. Also, Matemilola (2019) identified the Niger Delta region as highly vulnerable to impacts from climate change, stemming from sea level rise, increased precipitation and intensive industrial activities from oil exploration. According to the Head of Environment/Infrastructure, Ministry of works Edo State, he stated that the government have been several efforts to address the flooding disasters. He mentioned that. . .

> *a storm water project started but was not approved in 2009, the budget estimated for the project was about 30 billion naira. In 2015 the project was taken up by NDDC, a*

non-governmental organisation and was concluded around 2016/2017. The work done by NDDC was poor, the method adopted by them was the channelling of the flood water into the aquifer which led to more complications.

The project was finally approved by the government of Edo State in 2018/2019 and was taken up by the Ministry of Works. They plan to channel the storm water from Ugbowo and Ekehun into the existing Ogba River. The project has started with over 27 billion naira spent though without accomplishing much and the problem of flood persists. The population and spatial growth in Benin City have grown beyond the rate of infrastructural provision and this has exacerbated flood crisis in the area. More areas are being exposed to flooding as the city extends beyond its traditional city walls (the moat). According to Odemerho, this is despite the huge capital outlay, where over 80% of the yearly allocation for flood control for the entire State is spent on Benin City alone in other to curtail the flood menace. Generally, it can be deduced that flooding is a huge threat to sustainable development and urban transportation movement in the country. It affects social, economic, environment and health aspects of the society. It is unfortunate that its effect cuts across most of SDG. There is need for the country to develop and provide a flood management policy that can address the present reoccurring flood disaster in majority of its cities. It is expedient that a legal and policy framework firmly related to urban planning and development should be provided, adopted and enforced. This would also be useful when it is combined with the Nigerian national policy on climate change.

Nigerian National Policy on Climate Change

According to the National Policy on Climate Change, the guiding modalities to the attainment of the climate change targets are numerous. These principles include local intervention packages and responses, impressive public participation from grassroots, international synergy and technical networks, judicious management of environmental resources, gender equality roles and approved interventions. Other factors of concern are transparency, equity, integrity among stakeholders and coordination within the stakeholders from the government, individuals and corporate organizations (FME 2021). The enablers from the Nigerian Government provide policies, plans and strategies that handle all adaptation and mitigation measures to pursue a climate-resilient economy. The legal framework provides impetus for strict compliance with all climate actions enshrined in her laws. Other frameworks are to outline and clarify all roles and duties of levels of governance to climate actions, improve coalition of all sectors having an interest in climate mitigation, provide extant laws for compliance, harmonize the synergy of climate change mitigation with taxes, energy, transportation and other related matters among others. Nigeria has enabled a holistic legal framework on climate action to support Nigeria's future climate targets bordering on zero tolerance to carbon emissions, national climate resilience, adequate finance for climate actions. Finally, it incorporates mitigation options into

National policy and implementation (Okereke and Onuigbo 2021). The dictates of the law enacted seek to inculcate a multi-dimensional approach in coordinating different sectors, businesses and corporate bodies required to meet the future climate objectives in Nigeria. Furthermore, the law has a comprehensive systematic mechanism to handle the critical climatic challenges, risks and threats like vulnerable environments within agro-ecological zones in Nigeria and support the individual capacities of communities to adapt to the conditions. According to the National Adaptation Strategy and Plan of Action on Climate Change for Nigeria (NASPA-CCN) implementation and policy, the Federal government must lead and carter for all climate change adaptation in Nigeria through different means. These key roles include:

- Reviewing existing policies on climate change, programmes and projects.
- Implementing already existing climate change programmes and projects.
- Research output, knowledge sharing and data handling.
- Strategic programmes.
- Building of infrastructures.
- Funding

Over time, strategic programs have been developed in Nigeria targeted at minimizing the effects of climate change in Nigeria. Some examples include: community-based climate change adaptation, integrated water resources management program, community-based natural resources management program, early warning projects, among other numerous linkages of all policies related to climate action abound and are in full practice. Typical examples are the National Renewable Energy and Energy Efficiency Policy (NREEEP) 2015; National Gas Policy (2017); National Forest Policy (NFP) 2010; National Climate Change Policy and Respond Strategy (NCCPRS) 2012; and REDD+ Strategy, 2019 amongst others FME (2021).

Conclusion

Findings from the study showed that floodwaters can destroy homes and businesses, disrupt social and economic activities, and create health risks such as malaria, typhoid fever, physical injuries and mental risks. Also, the floods damaged drainage and sewage systems, thereby releasing pollutants into the boreholes and wells that can spread water-borne diseases. In our study area, the income, employment and livelihoods of the residents are seasonally affected particularly during the wet season consequent of the floods. Many businesses and properties were affected by the floods resulting in poor sales, damage of goods, little or no movement of people and exchange of goods and services. Also, the floods affected the aesthetics value and beauty of the environment. Climate-related action has become expedient to tackle the flood situation. There is need for collaborative efforts geared toward climate change by all especially youths, men and women, stakeholders and the community to a sustainably developed environment. Nigerian government have made significant efforts toward climate action by ratifying the 2015 Paris Agreement in the year 2017, as well as a gradual effort made toward to a low-carbon economy. However, the Nigerian policy and legal frameworks must consolidate responsive efforts towards to the challenges of climate change particularly flooding suffered by the Niger Delta region-Edo State. So that these vulnerable areas and the people can toward achieve a flood resilient environment.

References

Akande A (2017) Geospatial analysis of extreme weather events in Nigeria (1985–2015) using self-organizing maps. Adv Meteorol 2017. https://doi.org/10.1155/2017/8576150

Akujieze CN (2004) Effects of anthropogenic activities (sand quarrying and waste disposal) on urban groundwater system and aquifer vulnerability assessment in Benin City, Edo State, Nigeria. PhD thesis, University of Benin

Anukwonke CC, Tambe EB, Nwafor DC, Malik KT (2021) Climate change and interconnected risks to sustainable development in 'climate change, the social and scientific construct'. Springer Nature Switzerland AG. ISBN: 978-3-030-86290-9

Creswell JW (1995) Research design: qualitative and quantitative approaches. Sage, Thousand Oaks

Daisy D (2020) The carbon brief profile: Nigeria. https://www.carbonbrief.org/the-carbon-brief-profile-nigeria

Ebele NE, Emodi NV (2016) Climate change and its impact in Nigerian economy. J Sci Res Rep 10(6):1–13. http://www.journaljsrr.com/index.php/JSRR/article/view/21917/40737

Elisha I et al (2017) Evidence of climate change and adaptation strategies among grain farmers in Sokoto State, Nigeria. IOSR J Environ Sci Toxicol Food Technol (IOSR-JESTFT) 11(3):1–7. http://www.iosrjournals.org/iosr-jestft/papers/vol11-issue%203/Version-2/A1103020107.pdf

EM-DAT (2011) The OFDA/CRED international disaster database, viewed 07 Aug 2017, from www.emdat.be/database

Enete IC (2014) Impacts of climate change on agricultural production in Enugu State, Nigeria. J Earth Sci Clim Chang 5(9):234. https://www.omicsonline.org/open-access/impacts-of-climate-change-on-agricultural-production-in-enugu-state-nigeria-2157-7617.1000234.php?aid=32633

Federal Ministry of Environment (2014) United Nations Climate Change Nigeria. National Communication (NC). NC 2. 2014. https://unfccc.int/sites/default/files/resource/nganc2.pdf

FME, (Federal Ministry of Environment) (2021) National Climate Change Policy for Nigeria, (2021–2030) Department of Climate Change

Ikhile CI, Olorode DO (2012) Climate change and water balance in the Osse-Ossiomo sub-basin of S. W. Nigeria. Port Harcourt J Soc Sci 3:98–109

Intergovernmental Panel on Climate Change (2001) Climate change impacts, adaptation, and vulnerability. Cambridge University Press, Cambridge, p 2001

IPCC, Intergovernmental Panel on Climate Change (2022) The IPCC, 2022 report, IPPC-54 BIS. Working group

Matemilola S (2019) Mainstreaming climate change into the EIA process in Nigeria: perspectives from projects in the Niger Delta region. Climate 7(2):29. https://doi.org/10.3390/cli7020029

National Bureau of Statistics (2012) Federal Ministry of Youth Development. 2012 National Baseline Youth Survey. Final report. The Federal Republic of Nigeria

Nkechi O et al (2016) Mitigating climate change in Nigeria: African traditional religious values in focus. Mediterr J Soc Sci 7(6):299–308. https://www.mcser.org/journal/index.php/mjss/article/view/9612

Ogbonna DN, Amangabara GT, Itulua PA (2011) Study of the nature of urban flood in Benin City, Edo, State, Nigeria. Global J Pure Appl Sci 17(1):7–21

Okereke C, Onuigbo S (2021) Opinion: significance of Nigeria's new climate change law. https://www.vanguardngr.com/2021/11/opinion-significance-of-nigerias-new-climate-change-law/

Olaniyi OA et al (2013) Review of climate change and its effect on Nigeria ecosystem. Int J Afr Asian Stud 1:57. https://pdfs.semanticscholar.org/f9bd/9c18dfb45724a2a946a3854c756e62ad9f6b.pdf

Omofonwan SI (1992) Indicators for measuring the quality of life in rural areas: a case study of Orhionwon Local Government of Bendel State. Nigerian J Soc Sci 1(1):5–17

Paterson DL, Wright H, Harris PNA (2018) Health risks of flood disasters. Clin Infect Dis 67:1450–1454

Sayne A (2011) Climate change adaptation and conflict in Nigeria. USIP, Washington, DC. https://www.usip.org/sites/default/files/Climate_Change_Nigeria.pdf

United Nations (2013) Joint framework initiative on children, youth and climate change. "Youth and Climate"; Fact sheet (Dec 2013), p 1. Available at http://www.un.org/esa/socdev/documents/youth.fact-sheet/youth-climatechange.pdf

United Nations (2020) Framework convention on climate change. "Young People are boosting global action" 12 Aug 2020. Available at http://unfcc.int/news/young-people-are-boosting-globalclimate-action

WHO (2012) Public health risk assessment and interventions. Flooding disaster: Nigeria, viewed 20 Aug 2018, from http://www.who.int/hac/crises/nga/RA_Nigeria_1Nov2012a.pdf

Index

A

Acidification, 11, 32, 40, 124, 326
Adaptive capacity, 51–53, 56, 273, 279–281, 284, 301, 302, 374, 382–384, 392, 449, 702, 707, 713, 717, 750
Agricultural vulnerability assessment (AVA), 375, 376, 378–380, 382–385, 388, 389, 392, 393
Agriculture, forestry, and other land uses (AFOLU), 34, 35, 474, 476, 482
Agro-ecology, 74
Agroforestry, 74, 143, 334, 338, 342, 452, 453, 456
Air quality index (AQI), 682, 683, 685, 687–691, 696
Annual normal rainy days (ANRD), 327
Anthropocene, 66, 159, 744
Aquatic fauna preservation, 16
Atlantic Meridional Overturning Circulation (AMOC), 40
Atlantic Warm Pool (AWP), 7

B

Bangladesh Delta Plan 2100, 232
Bangladesh Sundarbans, 703, 705, 708, 720
Bhuvan Geo-Platform, 251
Bias correction constructed analogs (BCCA), 98
Biocapacity, ix, 4, 5, 9, 10, 19, 20, 23, 24
Biodiversity index, 19, 20
Biodiversity repositories, 400
Biogenic, 73
Biome shifts, 79
Boolean-Fuzzy Logic Model, 561
Brahmaputra floodplains, 259

C

Carbon credits, 466, 468, 716, 721
Carbon emissions, 4, 5, 9, 18, 19, 24, 71, 136, 321, 468, 474, 481, 485, 486, 768
Carbon footprint (CF), 17–19, 24, 98, 105, 107–110, 351, 352, 453, 470, 487
Carbon neutrality, 66, 466–469, 472, 488, 489
Carbon pools, 11, 135, 331
Carbon sequestration, 9, 11, 12, 15, 55, 66, 69, 79, 272, 286, 302, 330, 331, 334, 350, 449, 451, 453, 466–468, 474, 477, 485
Carbon smart technologies, 452–453, 457
Chronic obstructive pulmonary diseases (COPD), 679
Climate health crisis, 734
Climate sensitivity, 71, 80, 283
Climate-smart agriculture (CSA), 338, 375, 393, 446–459
Climate smart villages (CSV), 338
Climatic anomalies, 447
Climatic stressors, 180
Cluster analysis (CA), 354
Coastal ecosystems, 5, 104, 159, 163, 167, 170, 521
Coastal hazard wheel (CHW) decision support system, 163–164, 166–171, 174, 176
CO_2 capture and storage (CCS), 49, 50
Cohort studies, xi, 678–696
Concentration response (CR), 432, 567, 687
Conditioner application, 142
Conservation agriculture (CA), 73, 74, 330, 355, 456, 503
Coupled Model Intercomparison Project (CMIP), 70, 97, 132, 611
Cox model, 685, 687
Cox proportional hazard method, 687

Crop simulation models, 333, 341
Cryosols, 12

D
Decarbonization, 34, 467
Delphi method, 189, 194
Desertification, 11, 12, 22, 23, 76, 77, 85, 120, 124, 141, 422, 752

E
Ecocrop model, 449
Ecological footprint (EF), ix, x, 5, 9–20, 24
Ecological resilience, 159, 716
Eco-socio-economic shocks, 707
Ecosystem disruption, 159, 163, 164, 167, 171–172, 174–176
Ecosystem services, xi, 11–13, 15, 19, 21, 24, 51, 74, 76–78, 120, 139–142, 159, 172, 272, 278, 284, 287, 295, 296, 334, 450, 455, 714, 717
Edaphic, 4, 317
Energy Efficiency Index, 468
Energy smart technologies, 457
Environmental water requirements (EWR), 104–105, 110
Equilibrium climate sensitivity (ECS), 41
Erosion Productivity Impact Calculator (EPIC) model, 130
European Centre for Medium-Range Weather Forecasts (ECMWF), 6, 9
Extrinsic vulnerability, 189, 190, 197

F
Farmers' fertilizer practices (FFP), 451
Financial pillow, 361
Five-point Likert scale, 632
Food insecurity, x, 7, 8, 324, 446, 447, 455, 628, 666, 708
Forest-dependent people, 416
Forest fire risk index, 422, 423, 434
Forest fragmentation, 400, 401, 403, 407, 410–418
Forgotten fisheries, 710
Furrow irrigated raised bed system (FIRBS), 450
Fuzzy Analytical Hierarchy Process (Fuzzy AHP), 561

G
General circulation model (GCM), 37, 81, 96–98, 100, 105, 111, 112, 132–134, 136, 144, 272, 284, 290–292, 323, 611, 612, 615
Glacial lake outburst floods (GLOFs), 75
Global agro-ecological zoning (GAEZ), 317, 318, 340
Global warming, ix, xi, 5, 6, 12, 14, 15, 39, 40, 44, 52, 56, 64–86, 132, 149, 159, 162, 222, 282, 288, 289, 301, 321, 324–326, 328, 330, 350, 366, 400, 439, 466, 467, 654, 656–658, 663, 704, 732, 754
Global warming potential (GWP), 34, 35, 65, 82, 132, 134, 331, 451, 452
Green Road Policy, 468
Groundwater vulnerable index, 182

H
Hazard assessment, 164, 173–176, 439, 578
Hazard ratios, 685–687, 689–694, 696
Household Diversity Dietary Score (HDDS), 456
Household Food Consumption Score (HFCS), 456
Human-animal conflict, 233, 417, 714
Hydrologic extremes, 75

I
Injection wells, 196, 197
Insensitive tourism, 162
Integrated nutrient management (INM), 331
Integrated Valuation of Ecosystem Services and Trade-Offs (InVEST), 138, 295
Intrinsic vulnerability, 189, 297

K
Kernel density estimation (KDE), 434, 437–439

L
Land degradation, xi, 5, 6, 9, 11, 12, 22, 23, 65, 75–77, 85, 120–149, 288, 294, 302, 314, 316, 319, 320, 334, 752
Landscape shape index (LSI), 405
Land surface albedo, 519
Limited area models (LAM), 97
Lung maladies, 678

Index

M
Mann Kendall test, 530
MarkSim weather generator, 82, 83, 147
Mass Rapid Transport Systems (MRTS), 468
MODIS fire spots, 423, 434
Morgan-Morgan-Finney Model, 130
Multifaceted Poverty Index, 67

N
Nash-Sutcliff coefficient (ENS), 103, 106, 107, 111
Natural capital, 5, 19, 22, 23, 68
Net zero emissions, 39, 66, 466, 474
Non-communicable diseases (NCDs), 732, 736, 737
Nutrient exhaustion, 124

O
Ocean acidification, 33, 34, 42
Ocean acidifications, 5, 68
Olive Ridley turtles, 172, 521

P
Passive behavior factor, 357
Perturbation regimes, 21, 79
Post-traumatic stress disorder (PTSD), 626, 628, 640, 646, 714, 738
Precipitation anomalies, 9
Precipitation intensity, 32, 34, 291
Principal component analysis (PCA), 283, 323, 354, 355, 357, 359, 380, 381, 499, 586, 587, 633, 640, 645

Q
Quantitative microbial risk assessment (QMRA), 609–614, 620, 622

R
Radiative forcing, 65, 66, 81, 101, 102, 132, 133, 136, 321, 322
Radioactive tracers, 128
Rainfall erosivity, 125, 138, 144, 145, 292, 336
Rainfall kinetic energy, 125
Rainfall-runoff models, 111
Ramsar Convention, 23
Random consistency index, 432
Ravenous lands, 127
Reducing Emissions from Deforestation And Forest Degradation (REDD+), 23, 85
Regenerative capacity, 10, 24
Representative concentration pathway (RCP), 37–42, 46, 47, 50, 83–85, 96–98, 101, 102, 105, 107–111, 133, 137, 138, 147, 148, 321, 322, 324, 608, 609, 615
Respiratory morbidities, 679, 680, 683, 688, 689, 696
Revised Universal Soil Loss Equation (RUSLE model), 129, 144, 145, 286
Rotating fishery closure system, 659

S
Sacred forest, 477
Salinization and compaction, 11
Saltwater intrusion, 71, 162, 164, 171, 175, 176, 702
Sedentary behaviour, 679
Sediment balance, 163, 164, 166, 167, 170–172
Sediment River Network Model (SedNet), 130
Semi-automated Tasselled Cap technique, 160
Self-help groups (SHGs), 478, 483, 719, 722
Shared socioeconomic pathways (SSPs), 37, 38, 321
Simpson Index, 379, 387, 390
Site-specific nutrient management (SSNM), 451
Small-scale fisheries (SSFs), 703, 704, 708–710, 714, 716, 718, 720
Social benefit-cost ratio (BCR), 450
Social cohesion, 659, 704, 717, 721
Social security, 719, 739
Societal cohesion, 663
Socio-ecological system (SES), 366, 702
Socio-political interruptions, 704
Soil inorganic carbon (SIC), 135
Soil organic matter (SOM), 11, 12, 71, 73, 76, 124, 136, 143, 277, 325, 326, 330, 383, 451
Soil resilience, 139
Soil water assessment tool (SWAT), 98, 130, 138, 286, 290, 294, 295
Special purpose vehicle (SPV), 487
Special Report on Emissions Scenarios (SRES), 37, 38, 83–85, 101, 102, 132, 300, 321, 324
Statistical downscaling model (SDSM), 82, 83, 98, 100–101, 105, 107–112, 134, 138, 144, 323, 324

Sustainable development goal (SDG), 9, 19, 22, 24, 67, 121, 123, 140–143, 149, 216, 229, 232, 288, 302, 317, 662, 737, 754, 768
System of Rice Intensification (SRI), 330

T
Terrain characterization index, 96
Transboundary, 221, 224, 226, 229, 236, 376, 719
Tree banking project, 476, 482
Tsunami risk index, 580
Tsunami vulnerability index, 580, 587

U
Unceasing decarbonization, 34
Universal Soil Loss Equation (USLE), 129, 137

Urban heat island (UHI), xi, 81, 519, 543–557
Urban hydrology, 519

V
Variable rate technologies (VRT), 335
Vicissitudes, 78, 273, 286

W
Water Quality Monitoring protocol, 210
Watershed sustainability index, 287
Water smart technologies, 450, 457

Z
Zero net land degradation (ZNLD), 12
Zoonoses, 735